INTERCORRELATED SATELLITE OBSERVATIONS
RELATED TO SOLAR EVENTS

ASTROPHYSICS AND
SPACE SCIENCE LIBRARY

A SERIES OF BOOKS ON THE RECENT DEVELOPMENTS

OF SPACE SCIENCE AND OF GENERAL GEOPHYSICS AND ASTROPHYSICS

PUBLISHED IN CONNECTION WITH THE JOURNAL

SPACE SCIENCE REVIEWS

VOLUME 19

INTERCORRELATED SATELLITE OBSERVATIONS RELATED TO SOLAR EVENTS

PROCEEDINGS OF THE THIRD ESLAB/ESRIN SYMPOSIUM
HELD IN NOORDWIJK, THE NETHERLANDS,
SEPTEMBER 16–19, 1969

Edited by

V. MANNO

and

D. E. PAGE

Space Science Department (ESLAB),
European Space Research and Technology Centre, Noordwijk, The Netherlands

D. REIDEL PUBLISHING COMPANY

DORDRECHT-HOLLAND

The Symposium was jointly sponsored by the Space Science Department (ESLAB), Noordwijk, The Netherlands, and the European Space Research Institute (ESRIN), Frascati, Italy, of the European Space Research Organisation (ESRO)

Library of Congress Catalog Card Number 77-118134

ISBN-13: 978-94-010-3280-3   e-ISBN-13: 978-94-010-3278-0
DOI: 10.1007/ 978-94-010-3278-0

FOREWORD

The European Space Research Organisation put its first satellite into orbit in March 1968 and was successful with two more before the end of that year. It was not entirely surprising therefore that the third annual ESLAB/ESRIN Symposium should in some way deal with the results obtained. This book is the Proceedings of that Symposium which, for reasons which Dr. Trendelenburg explains in the introductory talk, concentrated on intercorrelating phenomena occurring during solar events and in particular during the event of 25 February 1969. It is generally acknowledged that space data could yield a much more fruitful harvest if measurements taken simultaneously in different regions of space could be compared and the detectors intercalibrated. ESRO therefore sought right from the start to encourage this comparison of data.

The first two days of the Symposium (16 and 17 September 1969) were devoted to review lectures on inter-related phenomena occurring on the sun, through interplanetary space and the magnetosphere right down to the earth's ionosphere. The last two days were used to hear papers presenting results obtained from the ESRO and certain U.S.S.R. and U.S.A. satellites during the solar events around 25 February 1969. The Proceedings published here follow the same sequence as the Symposium presentations.

Because of the short time interval between the solar event and the Symposium, some of the data presented in the second part must be regarded as preliminary. However, the editors thought it worthwhile to publish in this form in the hope that the collection of data in one place would stimulate further and deeper study of the 25 February 1969 and other events.

The editors are pleased to note that the Symposium appears to have achieved its aim of prompting experimenters to get together to correlate their data and they wish to thank all who contributed to this success. They would especially acknowledge the help of their ESLAB colleagues in the preparation of this book.

D. E. PAGE
V. MANNO

Noordwijk, January 1970

TABLE OF CONTENTS

8. LECTURE IN CONCLUSION

Professor J. ROEDERER gives his summary talk. In the foreground a model of the HEOS I satellite.

LIST OF PARTICIPANTS

Anderegg, M., Space Science Department, ESTEC, Domeinweg, Noordwijk, Holland

Anscombe, J., ESOC, Robert-Bosch-Strasse 5, 61 – Darmstadt, Germany

Axford, W. I., Dept. of Applied Electrophysics, Univ. of Calif., San Diego, La Jolla, Calif. 92037, U.S.A.

Bahnsen, A., Danish Space Research Institute, Lundtoftevej 7, 2800 Lyngby, Denmark

Balogh, A., Imperial College of Science and Technology, Prince Consort Road, London, S.W.7, England

Beek, H. F. van, Laboratorium voor Ruimte-onderzoek, Huizingalaan 121, Utrecht, Holland

Boeckel, J. J. van, Space Science Department, ESTEC, Domeinweg, Noordwijk, Holland

Bondi, H., ESRO, 114 Avenue de Neuilly, 92 – Neuilly-sur-Seine, France

Bonetti, A., Officine Galileo, Via Carlo Bini 44, Florence, Italy

Borg, H., Kiruna Geophysical Observatory, Kiruna-C, Sweden

Bostrom, C. O., Johns Hopkins Univ. Applied Physics Lab., 8621 Georgia Avenue, Silver Spring, Md. 20910, U.S.A.

Brinkman, A. C., Laboratorium voor Ruimte-onderzoek, Huizingalaan 121, Utrecht, Holland

Cambou, F., Faculté des Sciences, Université de Toulouse, B.P. 4057, Toulouse, France

Clark, D. D., Space Science Department, ESTEC, Domeinweg, Noordwijk, Holland

Coleman, P. J. Jr., Inst. of Geophysics and Planetary Physics, Univ. of California, Los Angeles, Calif. 90024, U.S.A.

Czechowsky, P., Max-Planck-Institut für Aeronomie, Postfach 20, 3411 Lindau/Harz, Germany

Dalziel, R., Radio and Space Research Station, Ditton Park, Slough, Bucks., England

Davis, L. R., NASA, Goddard Space Flight Center, Code 601, Greenbelt, Md. 20771, U.S.A.

Davis, T. N., Geophysical Institute, University of Alaska, College, Alaska 99701, U.S.A.

Durney, A. C., Space Science Department, ESTEC, Domeinweg, Noordwijk, Holland

Egeland, A., Norwegian Institute of Cosmic Physics, P.O. Box 1048, Blindern, Oslo 3, Norway

Egidi, A., Laboratory for Space Research – CNR, Piazzale delle Scienze 5, Rome, Italy

Engel, A. R., Imperial College of Science and Technology, Prince Consort Road, London, S.W.7, England

Engelmann, J., Centre d'Etudes Nucléaires de Saclay, B.P. Nr. 2, 91 – Gif-sur-Yvette, France

Feiter, L. de, Laboratorium voor Ruimte-onderzoek, Huizingalaan 121, Utrecht, Holland

Feuerbacher, B. P., Space Science Department, ESTEC, Domeinweg, Noordwijk, Holland

Fitton, B., Space Science Department, ESTEC, Domeinweg, Noordwijk, Holland

Fooks, G., ESOC, Robert-Bosch-Strasse 5, 61 – Darmstadt, Germany

Formisano, V.,Istituto di Fisica 'Guglielmo Marconi', Laboratory for Space Research – CNR, Piazzale delle Scienze 5, Rome, Italy

Gendrin, R., Groupe de Recherches Ionosphériques, 2 Avenue de la République, 92–Issy-les-Moulineaux, France

Graaf, W. de, Laboratorium voor Ruimte-onderzoek, Huizingalaan 121, Utrecht-Holland

Grard, R., Space Science Department, ESTEC, Domeinweg, Noordwijk, Holland

Green, T. S., ESRIN, Casella Postale N. 64, I 00044 Frascati/Roma, Italy

Gringauz, K. I., Radiotechnical Inst. of the Academy of Sciences of the U.S.S.R., Moscow A-83, U.S.S.R.

Grünwaldt, H., Max-Planck-Institut für Extraterrestrische Physik, 8046 Garching bei München, Germany

Gustaffson, G., Kiruna Geophysical Observatory, Kiruna-C, Sweden

Hedgecock, P., Imperial College of Science and Technology, Prince Consort Road, London, S.W.7, England

Hones, E. W. Jr., Los Alamos Scientific Laboratory, P.O. Box 1663, Los Alamos, N. M. 87544, U.S.A.

Hultqvist, B., Kiruna Geophysical Observatory, Kiruna-C, Sweden

Hundhausen, A. J., Los Alamos Scientific Laboratory, P.O. Box 1663, Los Alamos, N. M. 87544, U.S.A.

Israel, G., ESRO, 114 Avenue de Neuilly, 92 – Neuilly-sur-Seine, France

Jager, C. de, Laboratorium voor Ruimte-onderzoek, Huizingalaan 121, Utrecht, Holland

Jakeways, R., Physics Department, The University, Leeds 2, Yorks., England

Jordan, H. L., ESRIN, Casella Postale N. 64, I 00044, Frascati/Roma, Italy

Knott, K., Space Science Department, ESTEC, Domeinweg, Noordwijk, Holland

Koch, L., Centre d'Etudes Nucléaires de Saclay, P.B. Nr. 2, 91 – Gif-sur-Yvette, France

Köhn, D., Space Science Department, ESTEC, Domeinweg, Noordwijk, Holland

Landmark, B., Norwegian Defence Research Estab., P.O. Box 25, 2007 Kjeller, Norway

Lange-Hesse, G., Max-Planck-Institut für Aeronomie, 3411 Lindau/Harz, Postfach 20, Germany

Lanzerotti, L. J., Bell Telephone Laboratories, Murray Hill, New Jersey 07901, U.S.A.

Lenhart, K., ESOC, Robert-Bosch-Strasse 5, 61 – Darmstadt, Germany

Maccagni, D., Istituto di Scienze Fisiche 'Aldo Pontremoli'. Via Celoria 16, 20133 Milan, Italy

Manno, V., Space Science Department, ESTEC, Domeinweg, Noordwijk, Holland

Mariani, F., Laboratory for Space Research, Piazzale delle Scienze 5, Roma, Italy

Marsden, P. L., Physics Department, The University, Leeds LS2-9 JT, Yorks., England

Meiner, R. C., Space Science Department, ESTEC, Domeinweg, Noordwijk, Holland

Meredith, L. H., Code 610, NASA, Goddard Space Flight Center, Greenbelt, Md. 20771, U.S.A.

Michard, R., Observatoire de Paris, 92 – Meudon, France

Montalenti, U., ESOC, Robert-Bosch-Strasse 5, 61 – Darmstadt, Germany

Moreno, G., Istituto di Fisica 'Guglielmo Marconi', Laboratory for Space Research – CNR, Piazzale delle Scienze 5, Rome, Italy

Münch, J., Max-Planck-Institut für Aeronomie, Postfach 20, 3411 Lindau/Harz, Germany

Murdin, J., Mullard Space Science Laboratory, Holmbury St. Mary, Dorking, Surrey, England

McDonald, F. B., High Energy Branch, NASA, Goddard Space Flight Center, Greenbelt, Md. 20771, U.S.A.

Nishida, A., Inst. of Space and Aeronautical Science, University of Tokyo, Komaba, Tokyo, Japan

Norman, K., Mullard Space Science Laboratory, Holmbury St. Mary, Dorking, Surrey, England

Occhialini-Dilworth, C., Istituto di Scienze Fisiche, Via Celoria 16, Milan, Italy

Ortner, J., ESRO, 114 Avenue de Neuilly, 92 – Neuilly-sur-Seine, France

Oster, A., Space Science Department, ESTEC, Domeinweg, Noordwijk, Holland

Page, D. E., Space Science Department, ESTEC, Domeinweg, Noordwijk, Holland

Parks, G. K., C.E.S.R., 118 Route de Narbonne, 31 – Toulouse, France

Paulikas, G. A., Aerospace Corporation, P.O. Box 95085, Los Angeles, Calif. 90045, U.S.A.

Pedersen, A., Space Science Department, ESTEC, Domeinweg, Noordwijk, Holland

Peters, B., Danish Space Research Institute, Lundtoftevej 7, 2800 Lyngby, Denmark

Pieper, G. F., NASA, Goddard Space Flight Center, Greenbelt, Md. 20771, U.S.A.

Raitt, J., Mullard Space Science Laboratory, Holmbury St. Mary, Dorking, Surrey, England

Raviart, A., Centre d'Etudes Nucléaires de Saclay, B.P. Nr. 2, 91 – Gif-sur-Yvette, France

Reid, G. C., ESSA Research Laboratories, Boulder, Colo. 80302, U.S.A.

Riedler, W., Technische Hochschule Graz, Institut für Nachrichtentechnik und Wellenausbreitung, Krenngasse 37, A-8010 Graz, Austria

Roederer, J. G., Department of Physics, University of Denver, Denver, Colo. 80210, U.S.A.

Rosenbauer, H., Max-Planck-Institut für Extraterrestrische Physik, 8046 Garching bei München, Germany

Rossberg, L., Max-Planck-Institut für Aeronomie, Postfach 20, 3411 Lindau/Harz, Germany

Rothwell, P., Physics Department, University of Southampton, Southampton, England

Schindler, K., ESRIN, Casella Postale N. 64, I 00044 Frascati/Roma, Italy

Shaw, M. L., Space Science Department, ESTEC, Domeinweg, Noordwijk, Holland

Simon, P., Observatoire de Paris, 92 – Meudon, France

Singer, S., Los Alamos Scientific Laboratory, P.O. Box 1663, Los Alamos, N.M. 87544, U.S.A.

Smart, D. F., Space Physics Laboratory, Laurence G. Hanscom Field, Bedford, Mass. 01730, U.S.A.

Søraas, F., Department of Physics, University of Bergen, Bergen, Norway

Stauning, P., Ionosphere Lab., Technical Univ., Lundtoftevej 100, Bldg. 348, 2800 Lyngby, Denmark

Švestka, Z., Astronomical Institute, Ondřejov, Czechoslovakia

Swanenburg, B., Kamerlingh Onnes Laboratorium, Nieuwsteeg 18, Leiden, Holland

Taylor, B. G., Space Science Department, ESTEC, Domeinweg, Noordwijk, Holland

Tanzi, E., Istituto di Scienze Fisiche, 'Aldo Pontremoti', Via Celoria 16, 20133 Milan Italy

Thomas, G., Radio and Space Research Station, Ditton Park, Slough, Bucks., England

Trainor, J. H., NASA, Goddard Space Flight Center, Code 611, Greenbelt, Md. 20771, U.S.A.

Trendelenburg, E. A., Space Science Department, ESTEC, Domeinweg, Noordwijk, Holland

Tunaley, J., Space Science Department, ESTEC, Domeinweg, Noordwijk, Holland

Velut, P. M., Centre National d'Etudes Spatiales, B. P. Nr. 4, 91 – Brétigny-sur-Orge, France

Vernov, S. N., Moscow State University, Institute of Nuclear Physics, Moscow V-234, U.S.S.R.

Vette, J. I., NASA, Goddard Space Flight Center, Code 601, Greenbelt, Md. 20771, U.S.A.

Völk, H., Max-Planck-Institut für Physik und Astrophysik, Institut für Extraterrestrische Physik, 8046 Garching bei München, Germany

Wenzel, K.-P., Space Science Department, ESTEC, Domeinweg, Noordwijk, Holland

Wibberenz, G., Institut für Kernphysik der Universität Kiel, Olshausenstrasse 40/60, Gebäude 32, 23 – Kiel, Germany

Wilcox, J. M., Space Sciences Laboratory, University of California, Berkeley, Calif. 94720, U.S.A.

Witte, M., Institut für Kernphysik der Universität Kiel, Olshausenstrasse 40/60, Gebäude 32, 23 – Kiel, Germany

Zhulin, I. A., IZMIRAN, p/o Academgorodok, Moscow Region, U.S.S.R.

INTRODUCTION

Early this year when we discussed possible subjects for the 1969 ESLAB/ESRIN Symposium, ESRO was in a position quite different from that during previous years. In 1968 we successfully launched three European satellites, and even if this still looks like a rather moderate achievement compared with the large number of missions performed by the U.S. and the Soviet Union and even if some national satellites had been flown earlier in Europe, it was for us a substantial step forward. It was therefore natural that our Symposium this year should take this into account, in particular since we had the three satellites, ESRO I, ESRO II and HEOS A, operating at the same time. All three satellites basically deal with solar-terrestrial relations and therefore the Symposium will also deal with this subject.

Since the first Sputnik went into orbit in 1957 several hundred scientific satellites have been launched, providing us with a wealth of new and quite frequently unexpected information. This information is scattered over many different journals and reports and deals usually with results obtained either from single experiments or from different experiments flown in a single spacecraft. What is to a large extent lacking, however, is the intercorrelation of data obtained by different spacecraft either at a particular time or in a particular location. I think it is obvious to everyone sitting in this room that this correlation of data will in many cases be the key to the gradual solution of many of the complex problems not so far understood in solar-terrestrial relations. As far as I know, this Symposium is the first attempt to discuss this problem from a broader point of view, and the calibre of our participants is sufficient proof that this subject is not only interesting but also urgent.

It is not my intention at this moment to presuppose the possible conclusions of the Symposium since we will discuss them on Friday afternoon. However, I would like to offer for consideration a few of the ideas which we have discussed internally during the past year and which could be developed further.

(1) As was already said, a correlated analysis of data obtained by different spacecraft will without question provide a far better understanding of solar-terrestrial phenomena than the analysis of any individual data. Consideration should therefore be given to simplifying such correlation studies by providing concise data in a standardised format. This would enable in particular those scientists who are not building and flying hardware but are interested in space research in general to work on such correlation studies.

(2) The next step would be to plan such co-ordinated missions 'a priori'. In other words, machinery should be set up so that at an early stage of planning the different

V. Manno and D. E. Page (eds.), Intercorrelated Satellite Observations Related to Solar Events, 1–3. All Rights Reserved
Copyright © 1970 by D. Reidel Publishing Company, Dordrecht-Holland

space agencies could co-ordinate missions on different orbits as far as the type of instrumentation and the time schedule are concerned.

(3) In future consideration might also be given to flying the second or third flight models of a particular spacecraft in different orbits but with identical instrumentation. There are two reasons for this. First of all, a second and third identical flight model is much cheaper than a modified version of the same spacecraft even when the modifications are only slight. Secondly, everyone is aware of the fact that comparison of data obtained from different experiments is quite frequently rather difficult since ground calibrations of different instruments are in many cases not very accurate. By flying identical experiments at the same time in different orbits one would therefore increase the relative accuracy of the data being compared.

Apart from the scientific value of such a type of co-ordination, there are also po-litical reasons which make this problem so urgent. Firstly, there is an increasing tendency to transfer money from purely scientific missions towards applications problems, since it is felt that in the latter field a much quicker return can be expected. It is up to the scientists to prove that they obtain the maximum amount of results for the large sums of money which are invested in their projects. A second point is that the money available for scientific research in general is not unlimited. Therefore, different branches of science are fighting for their proper share, and the share that goes to space research will certainly be based more and more on the results obtained. This will apply increasingly as the glamour of sending hardware into space gradually decreases in the future. It would be regrettable if we, the space scientists, neglected the opportunities available to us because we had not found the means necessary to co-ordinate our work. Space science is almost the only field of scientific activity where scientists of completely different backgrounds with completely different interests are working and have to work together in order to achieve their objectives. Apart from the direct scientific return, I feel this is of great value, since we are all aware of the increasing specialisation in this modern world which is, to some extent, unavoidable. Our field still gives scientists an opportunity to come together and forces them to express their views and accept other people's opinions in an undogmatic way. It therefore provides an opportunity for cross-fertilisation between many branches such as astrophysics, plasma physics, high energy nuclear physics, ionospheric physics and, last but not least, the fields of modern engineering and management, in a way which cannot be compared with any other field of scientific activity.

Let me finally go back to our Symposium Programme. In order to correlate satellite data on solar-terrestrial relations one obviously needs solar events around which the observations are centred. For the purpose of our Symposium, we have concentrated on a single solar event – namely that of February 25, 1969. We chose this event not because it was particularly spectacular but because it was the first solar event to be well covered by our three satellites in orbit. As can be seen from the Programme, however, we felt that apart from dealing with this particular solar event we should have a series of general lectures covering the whole space between the sun and the ionosphere. Although many of the things which will be presented here in these survey lectures may be very well known to many of you, we felt it would be useful to have this type of 'tour d'horizon' since even those who are engaged in solar-terrestrial research tend quite frequently to specialise in their own subjects and do not find time to keep abreast of developments in other areas.

E. A. TRENDELENBURG

September 16, 1969

1. SUMMARY LECTURE

A SURVEY OF INTERPLANETARY AND TERRESTRIAL PHENOMENA ASSOCIATED WITH SOLAR FLARES

W. I. AXFORD

Dept. of Physics, Dept. of Applied Physics and Information Science, Institute for Pure and Applied Physical Sciences, University of California, San Diego, La Jolla, Calif., U.S.A.

1. Introduction

In this review the various phenomena that are associated with solar flares are described briefly, with particular attention being paid to effects that can be detected from satellites, space probes and rockets. Since most space physicists are quite well aware of the wide scope of their subject, it is not necessary to describe in detail everything that is likely to happen during and following a solar event. Instead, certain problem areas are discussed at some length, and other topics which are reasonably well understood are mentioned only in passing, if at all. A complete set of references for every topic is not given, since that would require a much too extensive bibliography; instead, review articles are cited where they are available, together with some key references which should make it easy for a reader interested in a particular topic to discover the literature for himself if he wishes.

2. Solar Flares

The phenomenon which is of most importance in the present context is the solar flare. Flares have been studied both directly and indirectly for many years [1, 2, 3], but although we have a great deal of observational information available, our understanding of the flare mechanism remains quite rudimentary. This is rather unfortunate since the flare phenomenon seems to occur frequently in nature, and as far as solar flares in particular are concerned, there are a number of significant effects which are of practical as well as scientific importance.

The closest equivalent to a solar flare in terrestrial terms is the magnetospheric substorm [4]. This analogy has been pointed out in the past as a means of illustrating some of the basic features of substorms [5, 6], but since recent observations of the latter have greatly enhanced our understanding of the phenomenon [4, 7], it is now more appropriately turned around the other way. That is, we can regard a flare as being in many respects the equivalent of a 'substorm' which occurs in the 'magnetosphere' associated with a sunspot configuration. Accordingly, the energy involved in the flare can be considered as having been stored in the form of magnetic energy, and released suddenly when the configuration of magnetic field and plasma in the upper chromosphere and lower corona discovers that it can proceed to a state of lower total energy without further alteration of the field at photospheric levels. The change in the configuration involves the reconnection of magnetic field lines at neutral points

V. Manno and D. E. Page (eds.), Intercorrelated Satellite Observations Related to Solar Events, 7–22. All Rights Reserved
Copyright © 1970 by D. Reidel Publishing Company, Dordrecht-Holland

and lines, the production of regions of very hot plasma, of high energy particles and possibly an 'aurora'. At the earth we observe thermal and non-thermal electro-magnetic radiation (especially radio, UV, and X-ray emissions), energetic particles (electrons, protons and other nuclei), and a disturbance in the interplanetary medium (possibly associated with a geomagnetic storm).

The chief differences between solar flares and magnetospheric substorms are that (i) magnetic field lines in the earth's magnetosphere are not firmly tied to the solid earth (i.e., they are free to convect [8–10]) whereas the solar magnetic field lines are tied to the photosphere, and (ii) the energy involved in a flare is transmitted from the photosphere, while the energy involved in a magnetospheric substorm is presumably extracted from the solar wind. In both cases however, we are concerned with the stability of a magneto hydrostatic configuration satisfying $\nabla p = \mathbf{j} \times \mathbf{B}$ and subject to certain boundary conditions. It should be noted that the plasma pressure (p) plays an important role in determining whether or not a particular magnetic field configuration is stable even though the field might be almost force-free ($\mathbf{j} \times \mathbf{B} = 0$) throughout much of the region concerned. In the case of the magnetosphere, the instability appears to be controlled by the pressure of the plasma in the plasma-sheet while in the case of solar flares, there are presumably similar regions containing hot, trapped plasma which are at least temporarily capable of preventing rapid reconnection of magnetic field lines and thus permit a meta-stable state of equilibrium to exist.

If we pursue this analogy between flares and substorms further, we see that it is reasonable to expect that in many cases the plasma/magnetic field configuration in the flare region should 'tear' into two distinctly different pieces. One piece must remain tied to the sun, and contain trapped plasma heated to say $\sim 50 \times 10^6$ K as a result of simple adiabatic compression associated with the contraction of newly reconnected magnetic field lines. The second piece is magnetically detached from the sun and is free to be ejected upwards as a result of its buoyancy with respect to its surroundings and (perhaps) of the contraction of magnetic field lines which stretch out into interplanetary space. In the region where magnetic field line reconnection takes place, the electric current is very large (i.e., there is a 'discharge'), and one might expect some special effects to occur, such as the acceleration of particles to very high energies [11]. Note however, that as in the case of the magnetosphere, there is no reason to believe that there is only one process which leads to the production of very energetic particles.

A schematic and undoubtedly over-simplified sketch of the situation described is shown in Figure 1. It is evident that this sort of picture at least provides reasonable possibilities for the explanation of various phenomena associated with flares, notably the electromagnetic radiation and the ejected plasma. Whether or not the quantitative aspects of such a model are satisfactory is still unclear however. A good review of the requirements of a solar flare theory has been given by Sturrock and Coppi [12]; note though that the theory proposed by these authors is rather dubious and some of their arguments against reconnection theories are not completely sound. Our understanding of the reconnection process is still limited although some progress has been made

recently [13, 14]. The real problem is, however, not concerned simply with the reconnection of magnetic field lines, but with the stability of equilibria when reconnection
is permitted to occur.

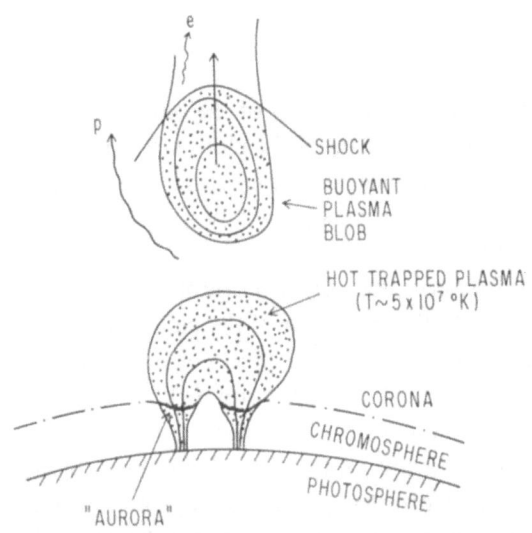

Fig. 1.   Schematic diagram of a solar flare immediately following the tearing phase.

3. Disturbances in the Solar Wind Following a Flare

There are two distinct sources of disturbance associated with a solar flare as far as the
flow of the solar wind is concerned. The first corresponds to a simple dumping of
energy into a localized region of the lower corona. This should lead to an impulsively
driven blast wave [15, 16], which should be felt everywhere in the hemisphere above
the flare to a degree that depends upon the azimuth and distance of the observer.

The second source of disturbance is associated with the flare ejecta itself more or less
in the manner described by Akasofu [21] and by Hirshberg [17]. Although the plasma
blob which is detached from the flare region probably does not have any significant
initial momentum, it is likely to be very hot and contain a strong magnetic field. The
total pressure within the blob should be approximately the same as that of the corona
on its exterior, and hence it must be extremely buoyant with respect to its surroundings. It is to be expected therefore, that the blob should be ejected directly away from
the sun, and in the absence of any drag could attain a radial velocity of the order of
$v_c/\sqrt{x}$ (where v_c is the escape speed in the lower corona and x is the ratio of the density
in the plasma blob to that outside initially). Since $v_c \approx 600$ km sec^{-1}, it is necessary
only that $x \lesssim 0.1$ if the plasma blob is to achieve a speed of 1000–2000 km sec^{-1} as
appears to be necessary. Ultimately of course, the plasma blob must be slowed as it
passes through the ambient solar wind plasma since the drag is not negligible, and
indeed can be very large if the relative motion is supersonic.

These latter ideas are relatively new and they have so far not been given much

careful thought. Furthermore, a detailed analysis of the behaviour of such a plasma
blob is not easy to carry out, although some very rough treatments are available [18].
Nevertheless, it seems evident that the blob should be ejected at first in the form of a
very narrowly confined, high velocity jet, which, as the effects of drag become more
pronounced suddenly expands and slows to approximately the ambient solar wind
speed. The drag on the blob is approximately $\varrho A v^2$, where ϱ is the density of the
ambient plasma, v is the relative velocity of the blob and A is essentially determined
by pressure balance at the boundary of the blob; since the internal pressure decreases
rapidly as the blob expands whereas the external pressure is of order ϱv^2, one sees
that the drag is likely to be an inverse power of ϱv^2 and hence becomes progressively
larger as ϱ and v decrease.

The reader should observe that this description is in some respects similar to that
which might be applied to the behaviour of the ball of hot gas produced by a nuclear
explosion in the atmosphere. There the effects of buoyancy in producing an initally
well-collimated, vertically-moving jet are quite striking. Other examples of apparently
similar phenomena are the jets seen in certain strong radio sources (notably M87);
we have discussed the behaviour of the ejecta from such objects elsewhere [19].

The observational evidence that plasma blobs are ejected by solar flares in the
manner described is limited, but fairly convincing. An early observation of the scintillation of small diameter radio sources following a solar flare suggests that the
disturbance might be very highly directed (see Figure 2). It has also been shown that
certain characteristics of geomagnetic storms are strongly enhanced if the associated
solar flare occurs close to central meridian (Figure 3). Finally it has been observed
that disturbance velocities estimated from the time delay between solar flares and the
resulting sudden commencement of a geomagnetic storm are not easy to make consistent with directly observed plasma characteristics [22–25]. Radio source scintilla-

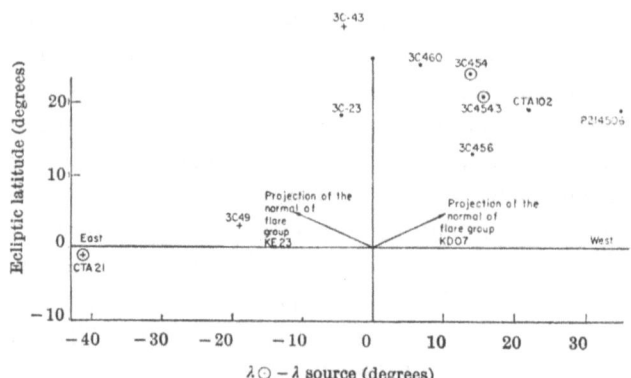

Fig. 2.   The relative position of radio sources on 27 March 1966 (dots) and on 31 March 1966 (crosses),
with respect to the sun and the ecliptic as seen from the earth. Sources which exhibited a large increase in their scintillation index, apparently as a result of flare activity, are indicated by circles. It is
interesting to note that on 27 March 1969 for instance, only two of the eight sources observed showed
an increase in scintillation index, suggesting that the disturbance in the interplanetary medium was
quite well collimated. (Taken from a paper by Sharp and Harris [20].)

tions offer by far the best means of detecting the motion of the blobs and we therefore expect some very interesting results to be obtained from the observations now being carried out by the University of California at San Diego. Direct detection of the blobs by spacecraft is possible if composition measurements are carried out, since there is no reason to expect them to have the same composition (in terms of $[H^+]:[He^+]$) as the ambient solar wind. Such abrupt changes of composition have been, observed on several occasions in association with geomagnetic storms [26–28].

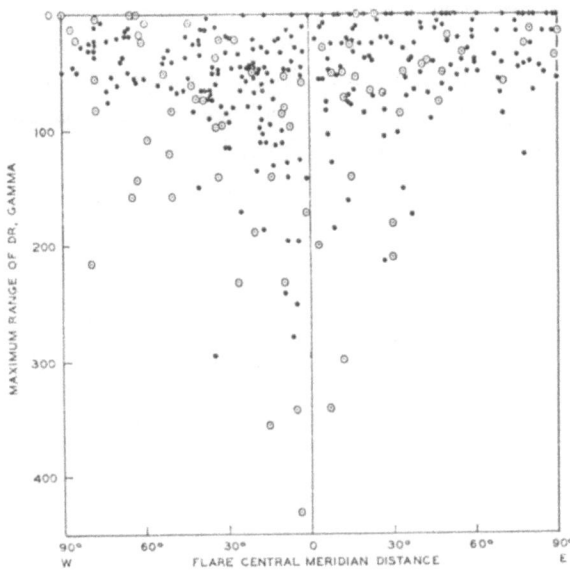

Fig. 3.   The magnitude of main phase decreases during geomagnetic storms plotted against the longitude of the responsible flares. (Taken from Akasofu [21].)

4. Solar X-Ray Bursts

X-ray bursts commonly occur in association with solar flares, especially (but not necessarily always) in cases where microwave bursts, sudden ionospheric disturbances, and energetic electrons are also observed [29–36]. Presumably most of the hard X-ray flux arises from free-free transitions occurring in the hot plasma which remains tied to the flare region as indicated in Figure 1, although free-bound and line emission must also contribute to the softer component of the flux. Since type III radio bursts and solar electron events are likely to be associated with the escaping plasma blob, the poorer correlation between the occurrence of these and of X-ray bursts is perhaps to be expected.

A very nice example of a solar X-ray burst is shown in Figure 4. The spike of hard emission is nearly coincident with the peak in the microwave burst, and has a non-thermal spectrum (roughly a power law in energy). The electrons involved in this spike must also have had a non-thermal spectrum, and perhaps were produced directly in the region of magnetic field line reconnection (i.e. the 'discharge'). In many cases the softer, more slowly-varying component has an approximately thermal spectrum in the decay phase, although it is also probably non-thermal in the increasing phase [30, 31, 36]. The low energy trace in Figure 4 contains variations which could be interpreted as damped oscillations of the flare region, following the sudden change in its topology.

The electron temperature in the emitting regions can be estimated if the spectrum is reasonably well known, and is commonly found to exceed 5×10^7 K. The emission measure ($n_i n_e \times$ volume) is roughly constant ($\sim 2.5 \times 10^{47}$ cm^{-3}) during a given burst, and (surprisingly) also from burst-to-burst [30]; constancy of the emission measure implies that the variations in flux are chiefly due to variations of temperature. Furthermore, there is a very high correlation between the occurrence of Hα flares and solar X-rays [30] which suggests that the 'auroral' phenomenon indicated in Figure 1 may be involved, although an interpretation involving filamentary structures with neighbouring hot and cold filaments is also possible.

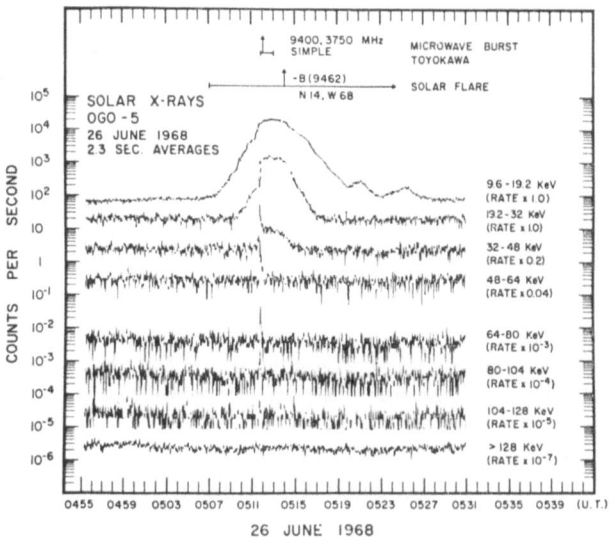

Fig. 4.   An example of a solar X-ray burst observed on 26 June 1968. (Taken from Kane [31].)

5. Energetic Solar Particles

Until recently it seemed that the propagation of energetic particles released during solar flares (i.e., solar 'cosmic rays') could be adequately explained in the basis of relatively simple anisotropic diffusion models [37, 38]. It is a simple matter to amend such models (i) by allowing an absorbing (free-escape) boundary, or alternatively a partially-reflecting boundary at some arbitrary heliocentric distance, and (ii) by replacing the thin isotropically-diffusing layer surrounding the sun by an equivalent anisotropically-diffusing layer. In this manner it is possible to account for most of the observed characteristics of solar cosmic ray events, viz., (i) the east-west effect, (ii) the behaviour of the intensity as a function of time, (iii) the direction of the anisotropy, and (iv) the decay of the anisotropy. The effects of convection and energy loss do not appear to be very important as far as these features are concerned [39].

While this type of model works rather well on the whole, it has become evident from observations of the behaviour of the anisotropy of relatively low energy (~ 10 MeV) particles, that some difficulties remain [40, 41, 25]. It is extremely important in comparing the predictions of a particular model with observations, to take into account the behaviour of the anisotropy as well as that of the mean intensity [42]. Indeed, McCracken and his colleagues [41] have shown that in one case it is possible to derive two values for the diffusion coefficient near the earth which differ by a factor $\sim 10^2$, using two distinct methods of analysis (one for the intensity and one for the anisotropy). It should be noted that the diffusion models usually involve enough free parameters to permit a fair fit to the intensity-time profile of almost any event.

The main problems at present are that during the highly anisotropic phase of some events, an apparent reflection of the particles occurs at some distance beyond the earth, and also the anisotropy often remains large for much longer periods than one might have expected. These effects can be explained by assuming that the interplanetary magnetic field is relatively smooth near the earth, with most of the diffusion taking place near the sun and well beyond the orbit of earth [43]. On the other hand, the observed power spectra of interplanetary magnetic field fluctuations suggest that the diffusion coefficient is usually quite small and consistent with estimates based on the temporal behaviour of mean intensity [44–47]. It is possible, of course, that the interplanetary magnetic field is quite variable in its characteristics [45], so that all types of behaviour occur; however, there is a need for more detailed observations to establish that this could be the case. Jokipii [48] has suggested that the difficulty with the anisotropy might be due to the fact that the diffusion coefficient is pitch-angle dependent and becomes large for pitch-angles near $90°$; if this is so, additional theoretical problems arise which require investigation.

If the diffusion coefficient is small for 1–10^2 MeV protons, the radial gradient of galactic cosmic rays must be large. However, there is no consensus in this point at present. A possibility which would permit relatively small radial gradients to exist together with small diffusion coefficients is that most of the particles with energies less than ~ 30 MeV observed in the interplanetary medium are of solar origin [25, 49, 50]. There is evidence that very low energy ($\gtrsim 0.3$ MeV) protons are solar [51], but the situation at higher energies is only just becoming clear (see McDonald, this volume, p. 34).

There are fewer problems as far as the electrons emitted by solar flares are concerned, possibly because the observations are less detailed than those for protons. There is evidence that 30–100 keV electrons undergo strong scattering in the interplanetary medium [52–54] and indeed they appear to behave as if undergoing diffusive propagation. At energies of a few MeV, there is again evidence for diffusive propagation, but observations of the anisotropy have yet to be made [55]. It should be noted that since these electrons have very low rigidities, they tend to follow magnetic field lines rather closely. Thus they must somehow escape from the flare region along 'open' field lines (see Figure 1), or else those observed near the earth might be accelerated by the shock front rather than in the flare itself.

Nuclei, other than protons, have not yet exhibited any anomalous behaviour, although there have been relatively few observations to date [56]. Observations of α-particles (for example), are extremely useful as a means of determining the form of the diffusion coefficient, and to some extent, of distinguishing between source and propagation effects (see Wibberenz, this volume, p. 499). Neutrons have given us no problems whatsoever, since all attempts to detect them have so far failed.

Low energy solar cosmic rays are often strongly affected by the passage of disturbances through the interplanetary medium. This was first noticed independently from Explorer XII proton observations, and riometer data [57–59]. In contrast to the case of galactic cosmic rays where the most obvious effect is the Forbush decrease, low energy solar particles often show a substantial increase of intensity, beginning up to an hour before the shock wave arrives at its point of observation, and lasting for perhaps a few hours afterwards (Figure 5). The simplest explanation for this phenomenon is that the particles energy and density are enhanced by simple compression at the shock front, with the pre-SC effect being due to leakage upstream and possibly to Fermi acceleration between the shock and approaching magnetic field irregularities. On some occasions the intensity of solar cosmic rays decreases very rapidly following an energetic storm particle event; this probably represents the passage of the expanded flare ejecta past the point of observation.

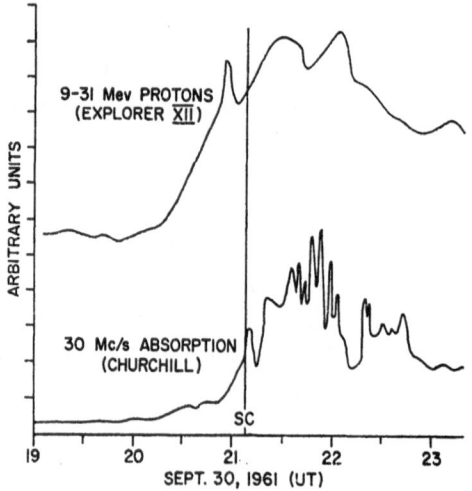

Fig. 5. An example of an energetic storm particle event observed from Explorer 12 [57] and using a riometer [58, 59].

6. Galactic Cosmic Rays

The most obvious effect of solar events on galactic cosmic rays is the Forbush decrease. This is probably the result of a combination of reflection of particles by the shock wave which precedes the disturbance in the interplanetary medium, and of enhanced convection and adiabatic deceleration of particles in the disturbed region behind the shock wave [15, 60, 61]. If one attempts to construct a reasonably detailed model for Forbush decreases on this basis, one finds immediately that there are additional effects, namely a small pre-increase which should be evident for many hours prior to the passage of the shock wave, and a corresponding anisotropy [62]. In fact such effects are observed (see Figure 6), as well as pre-decreases and accordingly it should be possible to use them as a means of probing a large volume of interplanetary space [63–65].

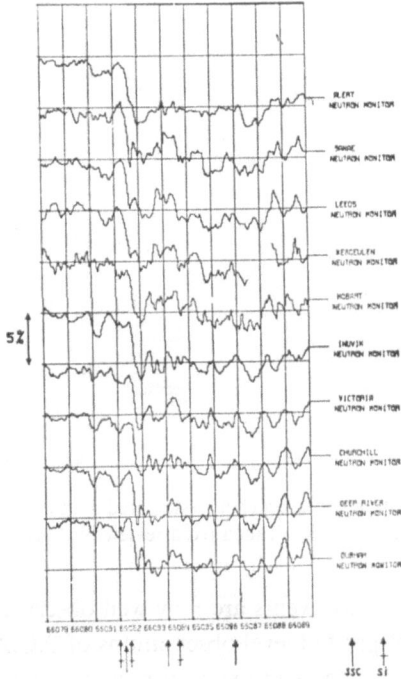

Fig. 6.   An example of a Forbush decrease, showing the enhanced anisotropies usually observed at such times. (Taken from a paper by Mercer and Wilson [101].)

There is some evidence that the low energy ($\lesssim 50$ MeV) component of the cosmic ray spectrum might in fact be largely of solar origin, with the particles being produced in flares and other active regions on the sun almost continuously (see McDonald, this volume, p. 34). It is important to establish whether or not this is the case; this might require measurements of the radial gradient, anisotropy and composition of the particles, as well as observations of their temporal behaviour.

7. Ionospheric Effects

The most important ionospheric effects associated with solar events are (i) sudden ionospheric disturbances (S.I.D.) due to the enhanced UV and X-ray emission from the solar flare, (ii) polar cap absorption (P.C.A.) due to 1–10^2 MeV solar protons, and (iii) magnetic storm effects (see also Section 8).

The S.I.D. is a direct result of the ionizing radiation emitted by a solar flare, and its most obvious effects apart from enhancements of the electron density in the E- and D-regions of the sunlit ionosphere are (a) increased radio-wave absorption, (b) changes in propagation characteristics of radio-waves, especially at low frequencies, and (c) a geomagnetic 'crochet' due to the temporarily enhanced conductivity of the 'dynamo' region of the ionosphere (see Figure 7).

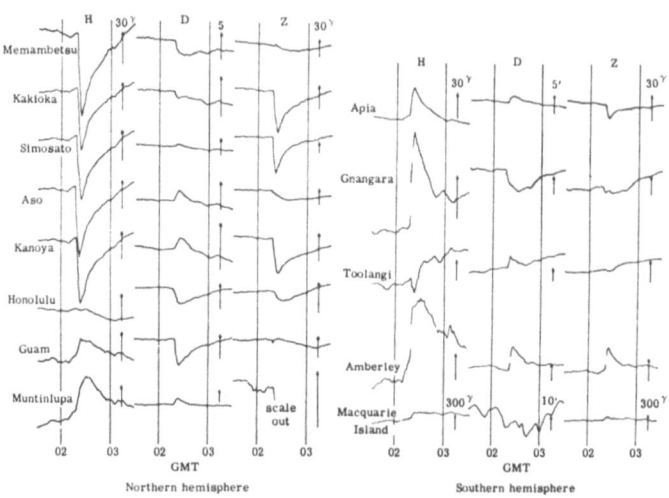

Fig. 7.   An example of a geomagnetic 'crochet', associated with the 3 + flare beginning at 0207 UT on 15 November 1960. (Taken from review by Matsushita [102].)

Since most solar cosmic ray events are now well observed by satellites and space-probes, the main value of ground level observations of P.C.A. events, as far as extra-terrestrial effects are concerned, is in the information they yield on the configuration of the geomagnetic field and in particular on the degree to which geomagnetic and interplanetary magnetic field lines are linked [66]. The wide and continuous coverage afforded by the world riometer network provides very useful supplementary results to those obtained from satellite observations in this respect.

The ionosphere is affected in a number of ways during magnetic storms. Particle precipitation, especially that associated with aurora, produces drastic changes in the electron density profiles [67]. Convective electric fields cause the F-region trough (associated with the plasma-pause and possibly with middle latitude red arcs) to move to lower latitudes [68], and also tend to inhibit some low-latitude phenomena which appear to be produced by the quiet-time tidally-induced electric fields (i.e. equatorial sporadic E, spread F, and the equatorial electrojet). Auroral zone heating together with the $\mathbf{j} \times \mathbf{B}$ force associated with the electrojets, produce large scale disturbances in the neutral atmosphere (gravity waves) which in turn can be detected in the F-region as 'travelling ionospheric disturbances' or T.I.D.'s [69–71]. Heating of the neutral atmosphere produces changes in its scale height and hence increases of satellite drag [72].

8. Geomagnetic Storms

The essential features of a geomagnetic storm are indicated in Figure 8. The first well-defined effect is the sudden commencement (SC) due to the impinging of the interplanetary shock wave on the magnetosphere [73]. The SC is most evident on geomagnetic records from low-latitude stations, but there are other effects, notably sudden cosmic noise absorption (SCNA) observed in the auroral zone on the dayside of the magnetosphere [74]. The magnetosphere remains compressed for some hours as a result of the additional pressure exerted by the enhanced flow of solar plasma [75]. During this initial phase it is believed that the tail of the magnetosphere grows at the expense of the doughnut-shaped region which surrounds the earth; and there is some evidence that the high latitude day-side aurora is unusually active in this period [76].

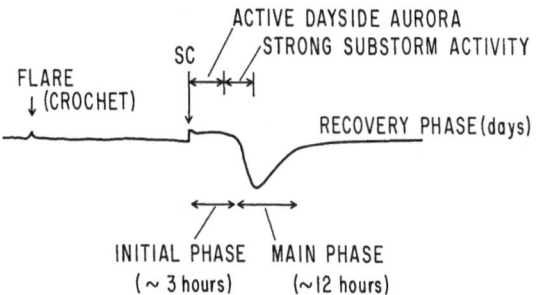

Fig. 8. The characteristic features of a geomagnetic storm as it might be observed at a temperate latitude magnetic observatory.

The main phase of a geomagnetic storm begins with a flurry of substorms [77] associated with the formation of an asymmetric ring current. Additional particles are injected into the radiation belts following each substorm, with energetic protons appearing in the afternoon side of the magnetosphere and electrons on the morning side [84–87]. Large-scale electric fields appear to play an important role in energizing and injecting these particles into the magnetosphere [10]. The electric fields can be detected directly by means of probes, or barium ion clouds, and also indirectly through their influence in the shape and size of the plasma-pause [78–82].

It is believed that substorms represent a gross instability of the entire magnetosphere which is associated with the increase of magnetic flux in the tail when field lines are eroded from the front of the magnetosphere and dragged downstream by the solar wind. The magnetosphere is apparently unable to recover from such a change of configuration in a continuous fashion due to the resistance afforded by the pressure of plasma in the plasma sheet and perhaps by 'line-tying' in the ionosphere [10]. Equilibrium is restored spasmodically when the support provided by the plasma sheet is removed by precipitation of particles into the atmosphere, and magnetic flux is returned to the front of the magnetosphere from the tail. It is interesting to note that there appear to be precursor effects prior to the explosive phase of substorms, which are rather similar to certain features of solar flares (see Figures 4 and 9). The inner edge of the plasma sheet is observed to drift inwards at the beginning of substorms in much the same manner as the plasma-pause [80, 83]. During the explosive phase, energetic particles appear throughout large regions of the magnetosphere (especially in the midnight, morning and day-side regions), essentially in coincidence with the occurrence of auroral-zone absorption events [84–87].

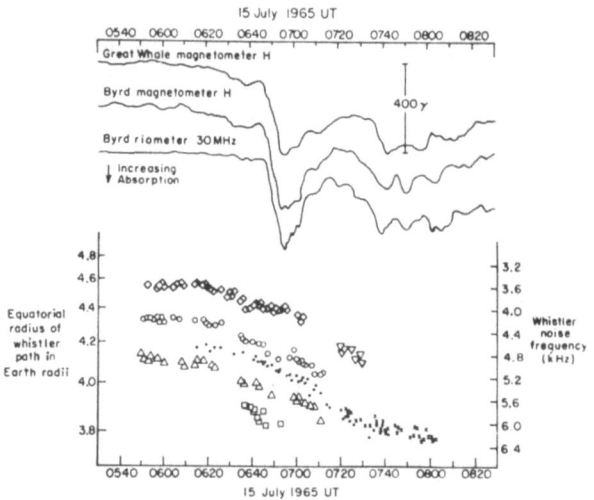

Fig. 9.   Inward movement of whistler ducts associated with a polar substorm. Note that the electric field appears at about 0620 UT, which is some 25 min prior to the abrupt decrease of the horizontal component of the geomagnetic field at Byrd and Great Whale. (Taken from a paper by Carpenter and Stone [80].)

The characteristic feature of the main phase of a geomagnetic storm (i.e., the 'ring' current) is caused by the inflation of the doughnut-shaped portion of the magneto-sphere due to the presence of newly-trapped particles (mostly protons with energies of a few tens of keV). The main phase persists as long as the solar wind continues to try and enlarge the tail, thus inducing substorms and continued injection of particles into the inner magnetosphere. Ring current particles can be lost by a variety of processes, notably convection out of the magnetosphere, charge-exchange with geocoronal neutral hydrogen atoms, and precipitation into the atmosphere. These processes dominate during the recovery phase of the storm which may last for several days.

Most of the energy dissipated during geomagnetic storms is involved in auroral precipitation, atmospheric heating by various means, and the formation of the ring current. The power is typically 3×10^{18} ergs sec^{-1}, which is about one percent of the incident solar wind energy flux. A small fraction of this energy goes into enhancing the fluxes of the more energetic particles in the radiation belt. Particles which are not stably trapped tend to be lost during the early phases of a storm, and others 'disappear' temporarily as a result of betatron deceleration during the main phase. However, the fresh particles that are introduced during substorms, together with the effects of changing convective electric fields (which cause the more energetic particles to diffuse into the magnetosphere as a result of third invariant breakdown), are more than sufficient to replenish the radiation belts [88, 89].

One of the interesting aspects of geomagnetic storms is their influence on cosmic ray cut-offs [90–92]. One would expect the cut-offs to be raised slightly during the initial phase of a storm and lowered during the main phase. However, the depression of the cut-off is far greater than one would expect on the basis of Stoermer type calculations, even if allowance is made for the asymmetry of the real magnetosphere [93, 94]. It has been suggested in the past that diffusive entry might be required to explain this effect [95, 96], and there is now some evidence that this might be the case (see Lanzerotti, this volume, p. 205, and Bostrom this volume, p. 229). Cut-off increases have been observed during substorms, which is perhaps reasonable in view of satellite observations which show that the magnetic field strength at the equator on auroral zone field lines, increases at such times [97].

Solar cosmic rays are especially useful probes of the distant parts of the tail of the magnetosphere [66, 98]. In particular it is possible in principle to make deductions concerning the extent to which geomagnetic and interplanetary magnetic field lines are connected (see Paulikas *et al.*, this volume, p. 193, Bostrom, this volume, p. 229, and Singer, this volume, p. 571). It seems to have been adequately demonstrated that there is such a connection [98–100], although there are still some who argue for purely diffusive entry of even solar electrons into the tail. It should be noted that much of the tail is comprised of closed field lines (notably the region containing the plasma sheet and associated with the auroral zones); however, it is reasonable to suppose that the field lines which intersect the earth at geomagnetic latitudes greater than about 75° are normally open.

Acknowledgements

This work was supported in part by the Advanced Research Projects Agency of the Department of Defense and was monitored by the U.S. Army Research Office-Durham under Contract DA-31-124-ARO-D-257, and in part by the National Aeronautics and Space Administration under Contract NGR-05-009-081.

References

[1] 1964, AAS-NASA Symposium on the Physics of Solar Flares, Washington, D.C., NASA SP-50.
[2] COSPAR: 1969, Symposium on Solar Flares and Space Research, Tokyo, *Space Research IX*, North-Holland Pub. Co., Amsterdam.
[3] Smith, H. J. and Smith, E. v. P.: 1963, *Solar Flares*, Macmillan, New York.
[4] Akasofu, S.-I.: 1969, *Polar and Magnetospheric Substorms*, Reidel, Dordrecht, Holland.
[5] Axford, W. I.: 1967, *Space Sci. Rev.* 7, 149.
[6] Axford, W. I.: 1967, in *Aurora and Airglow* (ed. by B. M. McCormac), Reinhold Pub. Corp., New York, p. 499.

[7] 1969, International Symposium on the Physics of the Magnetosphere, Washington, D.C., *Rev. Geophys.* **7**.

[8] Axford, W. I. and Hines, C. O.: 1961, *Canadian J. Phys.* **39**, 1433.

[9] Axford, W. I.: 1967, in *Physics of Geomagnetic Phenomena* (ed. by S. Matsushita and W. H. Campbell), Academic Press, New York, p. 1243.

[10] Axford, W. I.: 1969, *Rev. Geophys.* **7**, 421.

[11] Dungey, J. W.: 1958, *Cosmical Electrodynamics*, Cambridge University Press, Cambridge.

[12] Sturrock, P. A. and Coppi, B.: 1966, *Astrophys. J.* **143**, 3.

[13] Yeh, T. and Axford, W. I.: 1969, *J. Plasma Phys.* (in press).

[14] Sonnerup, B. U. O.: 1969, *J. Plasma Phys.* (in press).

[15] Parker, E. N.: 1963, *Interplanetary Dynamical Processes*, Interscience, New York.

[16] Hundhausen, A. J. and Gentry, R. A.: 1962, *J. Geophys. Res.* **74**, 2908.

[17] Hirshberg, J.: 1968, *Planetary Space Sci.* **16**, 309.

[18] Obayashi, T.: Private communication.

[19] De Young, D. S. and Axford, W. I.: 1967, *Nature* **216**, 129.

[20] Sharp, L. E. and Harris, D. E.: 1967, *Nature* **213**, 377.

[21] Akasofu, S.-I.: 1966, *Space Sci. Rev.* **6**, 21.

[22] Gosling, J. T., Asbridge, J. R., Bame, S. J., Hundhausen, A. J., and Strong, I. B.: 1968, *J. Geophys. Res.* **73**, 43.

[23] Ogilvie, K. W. and Burlaga, L. F.: 1969, *Solar Phys.* **8**, 422.

[24] Lubimov, G. P.: 1968, *Astron. Tsirkular* **4**, No. 488.

[25] Vernov, S. N., Chudakov, A. E., Vakulov, P. V., Gorchakov, E. V., Kontor, N. N., Logachev, Yu. I., Lyubimov, G. P., Pereslegina, N. V., and Timofeev, G. A.: 1969, Paper presented at the XIth International Conference on Cosmic Rays, Budapest, August, 1969.

[26] Bame, S. J., Asbridge, J. R., Hundhausen, A. J., and Strong, I. B.: 1968, *J. Geophys. Res.* **73**, 5761.

[27] Hundhausen, A. J.: 1968, *Space Sci. Rev.* **8**, 690.

[28] Axford, W. I.: 1968, *Space Sci. Rev.* **8**, 331.

[29] Chubb, T. A., Kreplin, R. W., and Friedman, H.: 1966, *J. Geophys. Res.* **71**, 3611.

[30] Hudson, H. S., Peterson, L. E., and Schwartz, D. A.: 1969, *Astrophys. J.* **157**, 389.

[31] Kane, S. R.: 1969, *Astrophys. J.* **157**, L139.

[32] Arnoldy, R. L., Kane, S. R., and Winckler, J. R.: 1968, *Astrophys. J.* **151**, 711.

[33] Mandel'stam, S. L.: 1965, *Space Sci. Rev.* **4**, 587.

[34] De Jager, C.: 1965, *Ann. Astrophys.* **28**, 125.

[35] Kundu, M. R.: 1963, *Space Sci. Rev.* **2**, 438.

[36] Cline, T. L., Holt, S. S., and Hones, E. W., Jr.: 1968, *J. Geophys. Res.* **73**, 434.

[37] Reid, G. C.: 1969, *J. Geophys. Res.* **69**, 2659.

[38] Axford, W. I.: 1965, *Planetary Space Sci.* **13**, 1301.

[39] Fisk, L. A., and Axford, W. I.: 1968, *J. Geophys. Res.* **73**, 4396.

[40] Fan, C. Y., Pick, M., Pyle, R., Simpson, J. A., and Smith, D. R.: 1968, *J. Geophys. Res.* **73**, 1555.

[41] McCracken, K. G., Rao, U. R., and Bukata, R. P.: 1967, *J. Geophys. Res.* **72**, 4293.

[42] Fisk, L. A. and Axford, W. I.: 1969, *Solar Phys.* **7**, 486.

[43] Burlaga, L. F.: 1969, Paper presented at XIth International Conference on Cosmic Rays, Budapest, August, 1969.

[44] Sari, J. W. and Ness, N. F.: 1969, *Solar Phys.* **8**, 155.

[45] Sari, J. W. and Ness, N. F.: 1969, Paper given at XIth International Conference on Cosmic Rays, Budapest, August, 1969.

[46] Wibberenz, G., Hasselmann, K., and Hasselmann, D.: 1969, Paper given at the XIth International Conference on Cosmic Rays, Budapest, August, 1969.

[47] Quenby, J. J., Balogh, A., Engel, A. R., Hedgecock, P. C., Hynds, R. J., Morfill, G., and Elliot, H.: 1969, Paper given at XIth International Conference on Cosmic Rays, Budapest, August, 1969.

[48] Jokipii, J. R.: 1968, *Astrophys. J.* **152**, 997.

[49] Meyer, P. and Vogt, R.: 1963, *Phys. Rev.* **129**, 2275.

[50] Krimigis, S. M. and Venkatesan, D.: 1969, *J. Geophys. Res.* **74**, 4129.

[51] Krimigis, S. M.: 1969, Paper given at XIth International Conference on Cosmic Rays, Budapest, August, 1969.

[52] Lin, R. P.: 1968, *J. Geophys. Res.* **73**, 3066.
[53] Van Allen, J. A. and Ness, N. F.: 1969, *J. Geophys. Res.* **79**, 71.
[54] Allum, F. R., Palmeira, R. A. R., Rao, U. R., and McCracken, K. G.: 1969, Paper given at XIth International Conference on Cosmic Rays, Budapest, August, 1969.
[55] Cline, T. L. and McDonald, F. B.: 1968, *Solar Phys.* **5**, 507.
[56] Lanzerotti, L. J.: 1969, *J. Geophys. Res.* **74**, 2815.
[57] Bryant, D. A., Cline, T. L., Desai, U. D., and McDonald, F. B.: 1962, *J. Geophys. Res.* **67**, 4983.
[58] Axford, W. I. and Reid, G. C.: 1962, *J. Geophys. Res.* **67**, 1692.
[59] Axford, W. I. and Reid, G. C.: 1963, *J. Geophys. Res.* **68**, 1793.
[60] Morrison, P.: 1956, *Phys. Rev.* **101**, 1397.
[61] Laster, H., Lenchek, A. M., and Singer, S. F.: 1962, *J. Geophys. Res.* **67**, 2639.
[62] Fisk, L. A. and Axford, W. I.: 1969, *Trans. Am. Geophys. Union* **50**, 306.
[63] Lindgren, S.: 1969, Paper given at XIth International Conference on Cosmic Rays, Budapest, August, 1969.
[64] Mathews, T., Mercer, T. B., and Venkatesan, D.: 1968, *Canad. J. Phys.* **46**, S854.
[65] Castagnoli, G. Cini and Dodero, M. A.: 1969, *J. Geophys. Res.* **74**, 2414.
[66] Reid, G. C. and Sauer, H. H.: 1967, *J. Geophys. Res.* **72**, 197.
[67] Hartz, T. R. and Brice, N. M.: 1967, *Planetary Space Sci.* **15**, 301.
[68] Rycroft, M.: 1969, Paper given at International Conference on the Physics of the Magnetosphere, Washington, D.C., 1969.
[69] Newton, G. P., Pelz, D. T., and Volland, H.: 1969, *J. Geophys. Res.* **74**, 183.
[70] Harris, K. K., Sharp, G. W., and Knudsen, W. C.: 1969, *J. Geophys. Res.* **74**, 197.
[71] Thome, G. D.: 1964, *J. Geophys. Res.* **69**, 4047.
[72] Newton, G. P., Horowitz, R., and Priester, W.: 1965, *Planetary Space Sci.* **13**, 599.
[73] Gold, T.: 1955, in *Gas Dynamics of Cosmic Clouds* (ed. by H. C. van de Hulst and J. M. Burgus), North-Holland Pub. Co., Amsterdam, p. 193.
[74] Brown, R. R., Hartz, T. R., Landmark, B., Leinbach, H., and Ortner, J.: 1961, *J. Geophys. Res.* **66**, 1035.
[75] Chapman, S. and Ferraro, V. C. A.: 1931, *Terrest. Mag. Atmos. Elec.* **36**, 77, 171; 1932, **37**, 147, 421; 1933, **38**, 79.
[76] Lassen, K.: 1967, in *Aurora and Airglow* (ed. by B. M. McCormac), Reinhold Pub. Corp., New York, p. 453.
[77] Davis, T. N. and Parthasarathy, R.: 1967, *J. Geophys. Res.* **72**, 5825.
[78] Föppl, Haerendel, G., Haser, L., Lüst, R., Melzner, F., Meyer, B., Neuss, H., Rabben, H.-H., Rieger, E., Stöcker, J., and Stoffregen, W.: 1968, *J. Geophys. Res.* **73**, 21.
[79] Aggson, T. L.: 1969, in *Atmospheric Emissions* (ed. by B. M. McCormac and A. Omholt), Reinhold Pub. Co., New York, p. 305.
[80] Carpenter, D. L. and Stone, K.: 1968, Paper given at International Symposium on the Physics of the Magnetosphere, Washington, D.C., 1968.
[81] Wescott, E. M., Stolarik, J. D., and Heppner, J. P.: 1969, *J. Geophys. Res.* **74**, 3469.
[82] Mozer, F. S. and Serlin, R.: 1969, *J. Geophys. Res.* **74**, 4739.
[83] Vasyliunas, V. M.: 1969, to appear in *Production and Maintenance of the Polar Ionosphere* (ed. by G. Skovli).
[84] Jelley, D. and Brice, N. M.: 1967, *J. Geophys. Res.* **72**, 5919.
[85] McDiarmid, I. B., Burrows, J. R., and Wilson, M. D.: 1969, *J. Geophys. Res.* **74**, 1749.
[86] Lin, W. C., McDiarmid, I. B., and Burrows, J. R.: 1968, *Canadian J. Phys.* **46**, 80.
[87] Lezniak, T. W., Arnoldy, R. L., Parks, G. K., and Winckler, J. R.: 1968, *Radio Sci.* **3**, 710.
[88] Van Allen, J. A.: 1969, *Rev. Geophys.* **7**, 233.
[89] Vernov, S. N., Gorchakov, E. V., Kuznetsov, S. N., Logacher, Yu. I., Sosnovets, E. N., and Stolpovsky, V. G.: 1969, *Rev. Geophys.* **7**, 257.
[90] Akasofu, S.-I., Lin, W. C., and Van Allen, J. A.: 1963, *J. Geophys. Res.* **68**, 5327.
[91] Lanzerotti, L.: 1968, *Phys. Rev. Letters* **21**, 929.
[92] Paulikas, G. A. and Blake, J. B.: 1969, *J. Geophys. Res.* **74**, 2161.
[93] Ray, E. C.: 1956, *Phys. Rev.* **101**, 1142.
[94] Smart, D. F., Shea, M. F., and Gall, R.: 1969, *J. Geophys. Res.* **74**, 4731.
[95] Rothwell, P.: 1959, *J. Geophys. Res.* **64**, 2026.

[96] Ray, E. C.: 1964, *J. Geophys. Res.* **69**, 1737.

[97] Barcus, J. R.: 1969, *J. Geophys. Res.* **74**, 4699.

[98] Lin, R. P. and Anderson, K. A.: 1966, *J. Geophys. Res.* **71**, 4213.

[99] Montgomery, M. D. and Singer, S.: 1969, *J. Geophys. Res.* **74**, 2869.

[100] Van Allen, J. A.: 1970, *J. Geophys. Res.* **75**, 29.

[101] Mercer, J. B. and Wilson, B. G.: 1968, *Canadian J. Phys.* **46**, S849.

[102] Matsushita, S.: 1967, in *Physics of Geomagnetic Phenomena* (ed. by S. Matsushita and W. H. Campbell), Academic Press, New York, p. 793.

2. SOLAR PARTICLES AND ELECTROMAGNETIC RADIATION

HIGH ENERGY FLARE RADIATION

C. DE JAGER

The Astronomical Institute, Utrecht, The Netherlands

Abstract. Flare-associated photon and particle fluxes are characterized by short rise times and longer lasting decays. High energy X-ray photon emission decays faster than low energy X-radiation, which, together with the observed shapes of the energy spectra suggests that the emission is due to a high energy flare associated plasma, whose projected position is, certainly in the first phase, closely associated with the first appearing bright elements of the optical flare. It is suggested that this plasma emits both the non-thermal (hard) X-burst as well as the quasi-thermal softer X-bursts, but the non-thermal component is mainly confined to the first phase of the flare. During the further thermalization of the plasma the average particle energy decreases while the high energy plasma cloud seems to expand and to exchange energy with the ambient gas, so that the number of involved particles increases.

The phase of energetic particle acceleration may be explained with Alfvén and Carlqvist's circuit-interruption model.

1. The Flare-Associated X-Radiation

A schematical review of intensities in the solar X-ray spectrum is given in Figure 1. The spectrum contains a quasi thermal component in the region of soft X-bursts and has also a short lived emission component in the region of harder X-bursts. In particular in that region the flux spectrum decreases rapidly towards higher energies, as was shown for the first time by Chubb *et al.* (1960) for the 2^+ flare of 31 August 1959. This was confirmed later for many other bursts, in particular by Arnoldy *et al.* (1968) and Hudson *et al.* (1969).

2. Time Development of the Flare-Associated X-Radiation

A characteristic example of the time development of a solar X-burst in the range 7.7–210 keV is given in Figure 2 after Hudson *et al.* (1969). The most typical aspect of these observations is the rapid decay of the intensity at large values of the energy and the much slower decay at lower energies. There is even an indication of a broad secondary maximum of the intensity in the lowest energy band.

This can be explained. In the last few years various investigations have been made of the expected X-radiation of a hot plasma confined by a magnetic field and losing energy by collisions, synchrotron and X-radiation and by run-away processes. See in particular Takakura and Kai (1966), Snijders (1968), Holt and Ramathy (1969), Lingenfelter (1969) and others. Snijders gives a fine example of computed time profiles that compare fairly well with the observations by Hudson *et al.* (see Figure 3). Since synchrotron losses are dominant for the time history of the high energy flare plasma it is the magnetic field that greatly defines the shape of the curves. Some of Snijders' figures compare very well with those of Hudson *et al.*, but it should be remarked that the time scale is greatly different. This could perhaps only be explained

V. Manno and D. E. Page (eds.), Intercorrelated Satellite Observations Related to Solar Events, 25–33. All Rights Reserved
Copyright © 1970 by D. Reidel Publishing Company, Dordrecht-Holland

by assuming smaller values of the magnetic field, than those assumed by Snijders.

An important question is that of the thermalization of the high energy flare plasma. All computations show that even a plasma with an initially completely non-thermal energy distribution quite rapidly thermalizes; Snijders could show that even an initially mono-energetic plasma will have a near-Maxwellian energy distribution after a few times the average collision time of the plasma.

The computations described here deal mainly with the question what happens to a high energy flare plasma initially excited to a given energy spectrum. The question how the plasma is accelerated is not touched upon. The final answer to that fundamental question may perhaps only be obtained after a thorough discussion of the structure and physical conditions in the high energy flare plasma.

3. Physical Parameters of the High Energy Flare Plasma

Since theoretical computations show the plasma to thermalize quite rapidly it is allowed to define a temperature. From various investigations this temperature appears to be of the order of 10–50×10^6 K (Hudson *et al.*, 1969); in one particularly

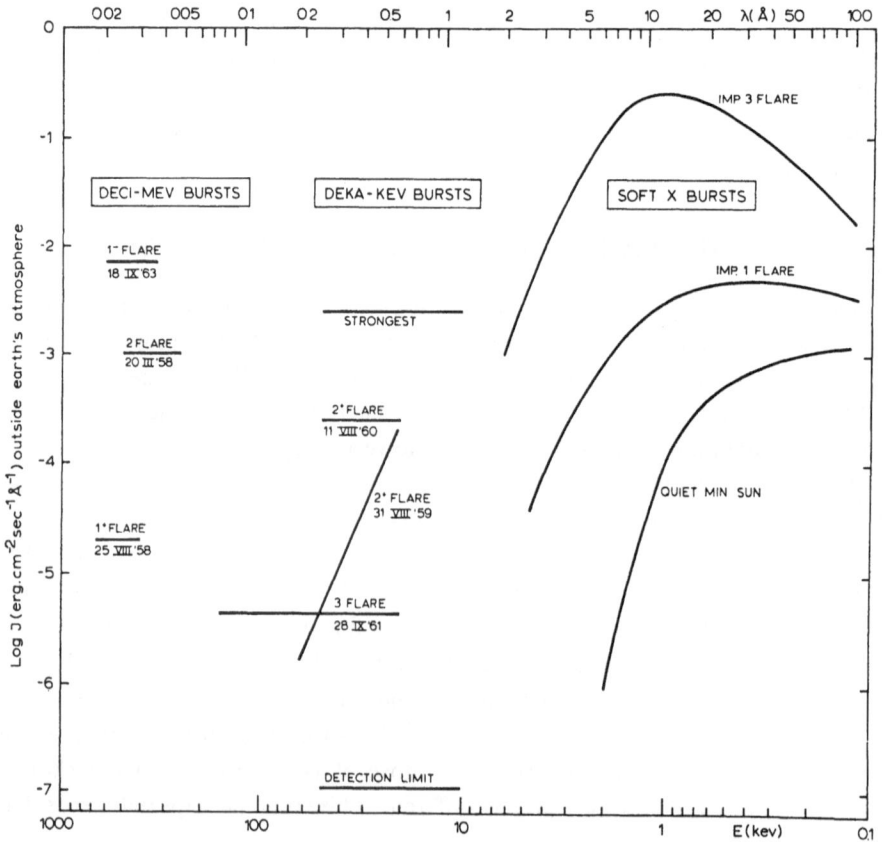

Fig. 1.   The solar spectrum between 0.1 and 1000 keV (De Jager, 1967).

well studied case Zirin and Ingham (1969) found 27×10^6 K. The electron density in the excited plasma cloud can be determined from the decay times, while the integrated flux yields the emission measure

$$\int N_e N_i \, dV \,.$$

A comparison of data obtained by various authors in different energy ranges yields the result shown in Table I (see De Jager (1967) for older observations; further Hudson *et al.* (1969) and Holt and Ramathy (1969) for more recent data).

This table shows that the total energy of the high energy plasma cloud is of the order of 10^{38} keV, virtually independent of the average energy of the accelerated electrons. This may be interpreted in two ways. First it may seem as if nature always puts the same amount of energy into the high energy plasma cloud, independent of the average energy of the particles. However, the result may also be interpreted by the assumption that the various plasma clouds listed in Table I are all aspects of one

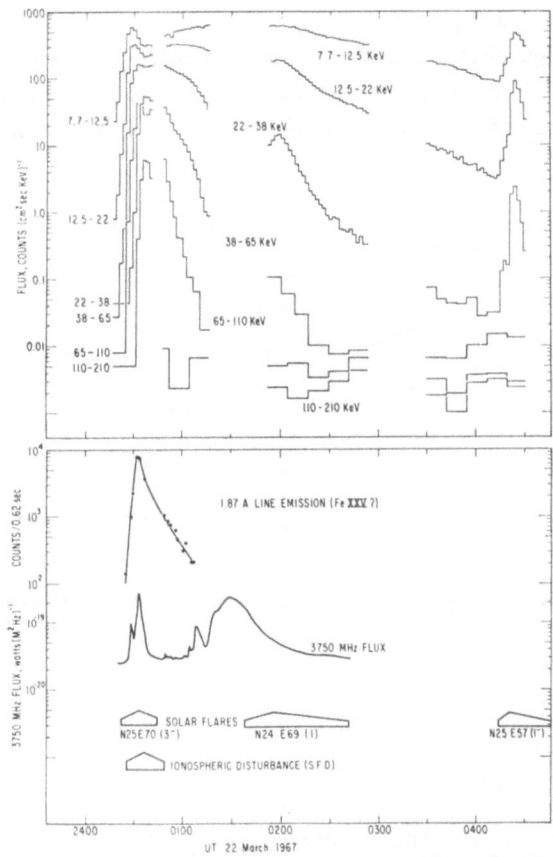

Fig. 2. Observed time profiles of the flare of 22 March 1967 in many channels between 7 and 210 keV (Hudson *et al.*, 1969).

TABLE I

Main physical parameters of the high energy
flare plasma

Average energy (keV)	$\log \int N_e^2 \, dV$ (cm^{-3})	$\log(N_e \bar{E} V)$ (keV)
1	48–49	38–39
20	46.5	38
100	47	39
300	45–45.5	37.5–38

and the same cloud in the course of its evolution and observed in various energy ranges. In that case one obtains the picture of high energy electrons injected in small numbers into a restricted volume. In the course of the decay and thermalization the number of involved particles and the volume increases while the total energy of the cloud, of course, starts to decrease or remains approximately constant. Although a decision is as yet difficult to take, the second hypothesis seems more probable to us than the first, in particular with a view to the recently observed time profiles of flares like the one shown in Figure 2.

Since the average electron density in the high energy flare plasma is mostly of the order of 10^{10} cm^{-3}, the volume of the high energy flare plasma also seems to increase with decreasing particle energy. For deci-MeV bursts this volume is of the order of

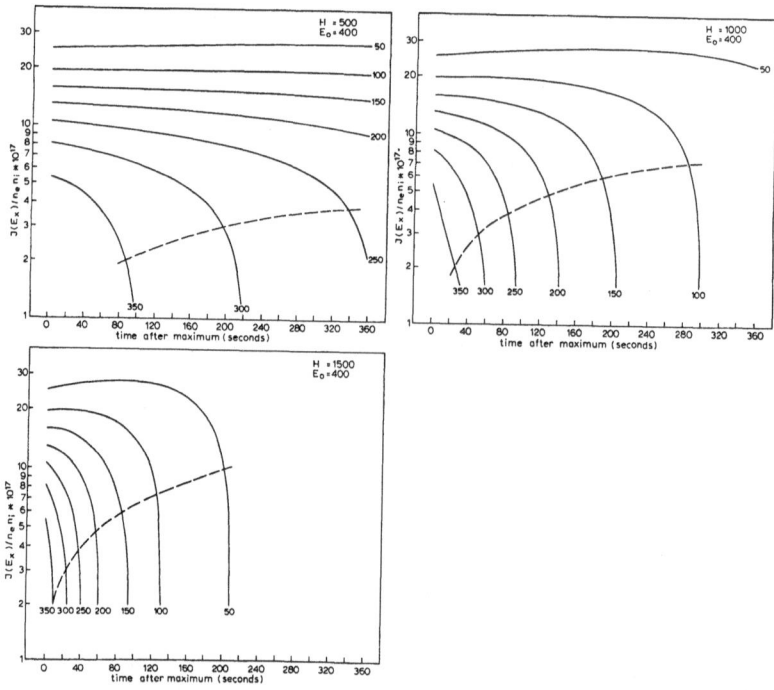

Fig. 3.   Theoretical time profiles of the differential intensity of bremsstrahlung emission generated
by 400 keV electrons for different assumed magnetic fields (Snijders, 1968).

10^{25} cm, which would make a spherical cloud with a radius of 2000 km. This radius is of the same order as the sizes of initial flare brightenings. The flare plasma with softer X-ray energies (≈ 1 keV) has a volume, about 1000 times larger, and a radius about 10 times larger. This is comparable to the size of a flare in its full development.

We therefore think that the most probable hypothesis is that the life history of the high energy flare plasma may show three consecutive phases:

(a) Primordial acceleration yields a relatively small number of highly energetic non-thermal excited electrons. The excited flare plasma occupies a relatively small volume.

(b) In subsequent thermalization the cloud expands and exchanges energy with the ambient gas. Hence the average number of particles involved in the flare plasma becomes greater, but the average number decreases.

(c) In the end phase the volume of the high energy flare plasma is 2 or 3 orders of magnitude larger than the initial plasma cloud. Also the particle energy is about 100 times smaller than in the first acceleration phase.

4. The Structure of the Flare and Its Environment

The two parts of Figure 4 give a possible schematic description of the structure of a flare. The upper part gives the flaring region seen from above. The dashed line indicates the line of zero vertical field components. Dotted are parts of lines of equal vertical field components. Observations by various authors show the flaring elements to originate fairly close to local maxima of the lines of equal vertical field components in small areas, called 'evolving magnetic regions' by Martres *et al.* (1968). Observations also show that these initial flaring areas may be connected by a bright bridge seen in Hα. We assume this bridge to be identical with the often observed flux tube which connects the two flaring areas. Observations by Severny and coworkers show the flaring elements to be the seats of an electric current system with currents of at least a few times 10^{11} A. It seems probable that the flux tube connecting the two flare elements is twisted and force-free. Otherwise, the flux tube could not be stable and no current could flow between the two flare elements.

A vertical cross section through the flare elements is given in Figure 4b. The densely dotted part is assumed to be the seat of the Hα emission. It does not extend higher than about 5000 km. Observations of limb flares show an extension of the emission to above 10^4 km, but with a much smaller electron density than in the bright flare elements (10^{12} vs. 3×10^{13} cm^{-3}). The emission of decimeter radio waves should occur in still higher regions and is certainly related to the high energy flare plasma of which the density is of the order of 10^{10} cm^{-3}. It is the same cloud which should emit X-radiation, but this radiation should be emitted in denser, lower layers where the density and hence the collision frequency is larger than in the radio wave emitting regions.

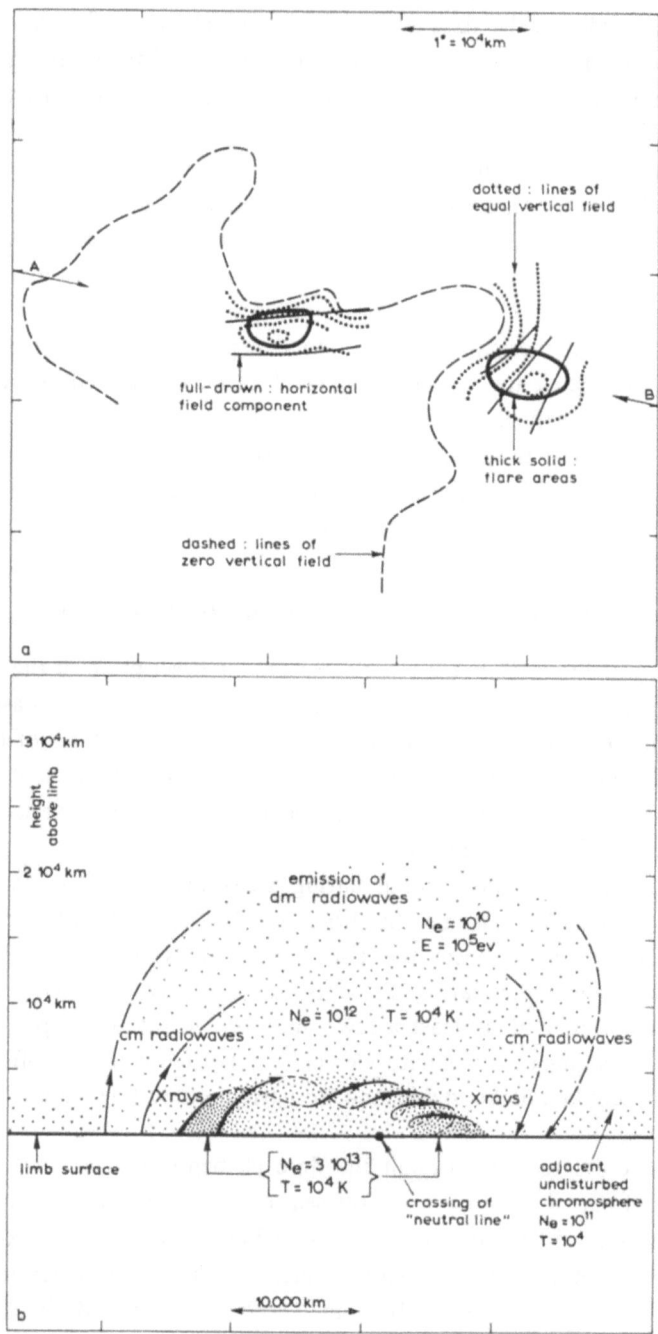

Fig. 4.   Model of a solar flare (De Jager, 1969).

5. Particle Acceleration

The structure as shown in Figure 4 is the natural environment for an acceleration as described by Alfvén and Carlqvist (1967), and Carlqvist (1968) where the origin of the high energy flare plasma is thought to be due to an instability in the current system, leading to a current interruption which may yield large electric voltages and consequent acceleration.

A quantitative calculation of this mechanism by Carlqvist (1968) shows a relation between the e-folding time of the acceleration, the intensity of the current and the total energy of the current system (see Table II).

TABLE II

Carlqvist's relation between the e-folding time t of the acceleration, the electric current I, and the current's energy W.

t (sec)	I (A)	W (erg)
30	10^{11}	10^{30}
100	10^{12}	10^{32}

The observed rise time of hard X-bursts is as a rule very short. Hudson *et al.* (1969) give an average e-folding time of the rising part of 65 sec. Interpolation between the data in Table II yields agreement with the observed fact that the current in regions where flares occur is at least a few times 10^{11} A; also the total kinetic energy of large flares is observed to be of the order of 10^{31}–10^{32} ergs. The greater part of this kinetic energy seems to go into the interplanetary plasma cloud that may leave the sun after large flares.

As stated above the total energy contained in the high energy flare plasma is of the order of 10^{38} keV, hence 10^{29} ergs. This brings us to the problem of particle escape from the sun; it can evidently not be due to the high energy flare plasma, but the two processes may have a similar origin.

6. Particle Emission by the Sun

A review of the various kinds of particles leaving the sun after a flare is excellently given in a fairly old diagram due to Obayashi (1964) (see Figures 5 and 6). Apart from quiet particle emission (the solar wind) the flare-associated particle spectrum consists of the solar plasma cloud with interplanetary velocities of 500 km/sec for singly emitted plasma clouds to 1000 km/sec for clouds emitted soon after another. These clouds may contain energetic storm protons. Furthermore, occasionally sub-relativistic particles are emitted. Their rigidity spectra may be described by the expression

$$J = J_0 \exp(-P/P_0).$$

The schematic time variation of the characteristic rigidity P_0 and the flux parameter J_0 is given for three kinds of parameters in Figure 6. The initial acceleration of the plasma cloud in the corona may yield a shock wave, intimately related to the initial flare-producing instability, and followed by a buoyancy of the accelerated gas, now detached from the flaring region, as suggested by Axford (this volume, p. 7). The

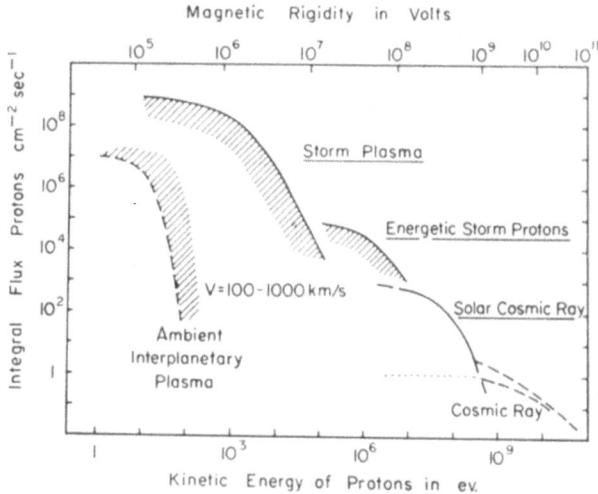

Fig. 5.   The energy spectrum of solar particles near the earth's orbit, for a typical moderate subrelativistic solar particle event (Obayashi, 1964).

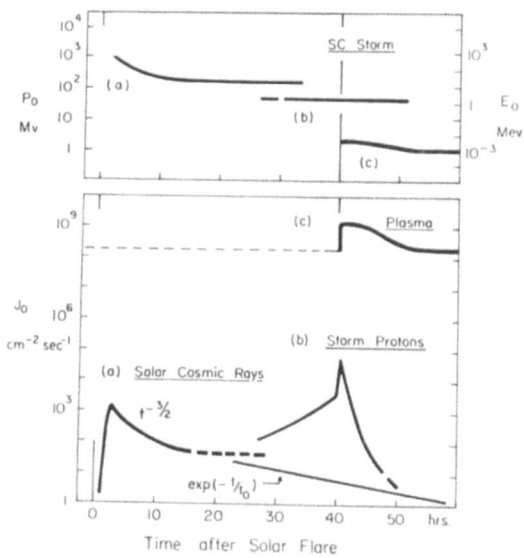

Fig. 6.   Schematic time variation of the characteristic rigidity P_0, and the characteristic flux J_0 for (a) solar subrelativistic particles; (b) energetic storm protons and (c) the solar plasma cloud (Obayashi, 1964).

particles of higher energy may be due to the high energy flare plasma and should be magnetically trapped to be contained in the plasma cloud.

A particular case are the electron clouds emitted by solar flares. Apparently they are related to type III radio bursts. These too are due to jets of accelerated electrons with energies of the order of 100 keV, shooting upward through the corona. In the range of energies exceeding 40 keV the electron fluxes range between 10 and 10^3 cm^{-2} sec^{-1} $ster^{-1}$, whereas in the range 3–13 MeV the flux ranges between 0.1 and 10 in the same units. A simple estimate, taking into account the observed lifetimes of the plasma clouds and the extent of their possible cone of propagation shows that the number of these electrons is at least by two or three orders of magnitude smaller than the actual number of excited electrons in the high energy flare plasma. Hence only a very small fraction of the initially excited electrons leaves the sun.

The problem of the escape of excited electrons and protons from the sun, briefly touched here, is as yet virtually unsolved and deserves careful attention of theorists and observers.

References

Alfvén, H. and Carlqvist, P.: 1967, *Solar Phys.* **1**, 220.
Arnoldy, R. L., Kang, S. R., and Winckler, J. R.: 1968, *Astrophys. J.* **151**, 711.
Carlqvist, P.: 1968, in *Mass Motions in Solar Flares and Related Phenomena*, Nobel Symposium 9 (ed. by Y. Öhman), p. 163.
Chubb, T. A., Kreplin, R. W., and Friedman, H.: 1960, *J. Geophys. Res.* **65**, 3611.
De Jager, C.: 1967a, in *Electromagnetic Radiation in Space* (ed. by J. G. Emming), Reidel, Dordrecht, p. 101.
De Jager, C.: 1967b, *Solar Phys.* **2**, 347.
De Jager, C.: 1969, in *Solar Flares and Space Research* (ed. by C. de Jager and Z. Švestka), North-Holland Publishing Co., Amsterdam, p. 1.
Holt, S. S. and Ramathy, R.: 1969, *Solar Phys.* **8**, 119.
Hudson, H. S., Peterson, L. E., and Schwarz, D. A.: 1969, *Astrophys. J.* **111**.
Lingenfelter, R. E.: 1969, *Solar Phys.* **8**, 341.
Martres, M. J., Michard, R., Soru-Iscovici, I., and Tsap, T. T.: 1968, *Solar Phys* **5**, 187.
Moreton, G. E. and Severny, A. B.: 1968, *Solar Phys.* **3**, 282.
Obayashi, T.: 1964, *Space Sci. Rev.* **3**, 79.
Snijders, R.: 1968, *Solar Phys.* **4**, 432.
Takakura, T. and Kai, K.: 1966, *Publ. Astron. Soc. Japan* **18**, 57.
Zirin, H. and Ingham, W.: 1969, *Astrophys. J.* (in the press).

Discussion

Parks: Some results presented by Ken Frost of Goddard at the recent OSO workshop at Boulder Colorado on solar X-rays show that X-rays of very high energy have the characteristics of the type III radio burst, and also show a semi-quasi-periodic behaviour.

Zhülin: Last year in Tokyo, Prof. Elliot from Imperial College presented a sort of model of the quasi-solar magnetosphere, and the precipitation of particles from this magnetosphere into the denser layers of the solar atmosphere. Could you say some words about the difference between your approach, confirming Alfvén's model, and Elliot's model?

De Jager: The general picture might be the same. Everyone tends to assume there is a magnetic field which confines the particles, but I think the difference between the two models is that Alfvén's picture assumes that suddenly a large electric field originates due to an instability, which is able to accelerate the particles which then fill this magnetic configuration. Meanwhile, Elliot thinks in terms of a process which gradually builds up in the course of hours or days. I have a slight preference, as I have already said, for Alfvén's picture because I see a way to understand it quantitatively.

SATELLITE OBSERVATIONS OF SOLAR COSMIC RAYS

FRANK B. McDONALD

NASA/Goddard Space Flight Center, Greenbelt, Md., U.S.A.

1. Introduction

The first observations of solar cosmic rays were made by Lange and Forbush (1942) in February and March 1942 using sea level, shielded ion chambers with a mean threshold of some 6 GeV. There has been a steady improvement in detector techniques and an energy threshold of 600 keV is typical of many present day satellite experiments. With the higher sensitivity, and lower energy thresholds combined with the long term monitoring capability of satellite-borne particle detectors, it has become increasingly obvious that the solar production of energetic particles is not a rare, isolated occurrence but that the sun is essentially a continuous source of MeV particles. These appear to manifest themselves in four distinct types of phenomena.

(a) *Flare-associated events:* Display a close association with solar flares, X-ray and micro wave radio emission and a diffusive particle propagation for both electrons and nucleons.

(b) *Recurrent events:* Particle increases which sometimes appear on the next rotation after a flare associated event and contain electrons and protons displaying essentially the same time history. These particles apparently originate in the same active region as that producing the original flare.

(c) *Energetic particles associated with an active center:* These display no velocity dispersion and appear to be co-rotating with the active center. They often persist from 3–14 days and have steep energy spectra. There appears to be a striking anticorrelation between the MeV electron and proton increases which separates this type from the recurrence events.

(d) *Energetic storm particle events:* Large increases of low energy electrons and protons which appear immediately behind strong interplanetary shock waves on certain occasions.

It would be surprising if there were not a strong interrelationship between these four classes of events. For example, Schatzmann (1967) has suggested that the protons stored in an active center might serve as a high energy injection mechanism for flare associated events, i.e., that the particle acceleration is a two step process. The data also suggests that the energetic storm particle increase may represent the superposition of an interplanetary shock wave and a co-rotating particle stream.

2. Evidence that the Sun is a Continuous Source of Solar Cosmic Rays

The first precise determination of the low energy cosmic ray proton and alpha spectra at energies below 20 MeV nucleon were made by the University of Chicago on the

V. Manno and D. E. Page (eds.), Intercorrelated Satellite Observations Related to Solar Events, 34–52. All Rights Reserved
Copyright © 1970 by D. Reidel Publishing Company, Dordrecht-Holland

OGO-I satellite (Fan *et al.*, 1965). These spectra (Figure 1) were obtained during periods of relatively low solar activity. The unexpected feature is a minimum in the 15–20 MeV region with a steep negative slope at lower energies. The number of counts in the low energy region is small and counts must be integrated over long periods to obtain meaningful measurements. It was not possible from this data to determine whether the low energy component was galactic or solar in origin. This is vital since if it is galactic in nature, significant changes would be required in modulation concepts. The alternative is that the sun is a quasi-continuous source of MeV particles.

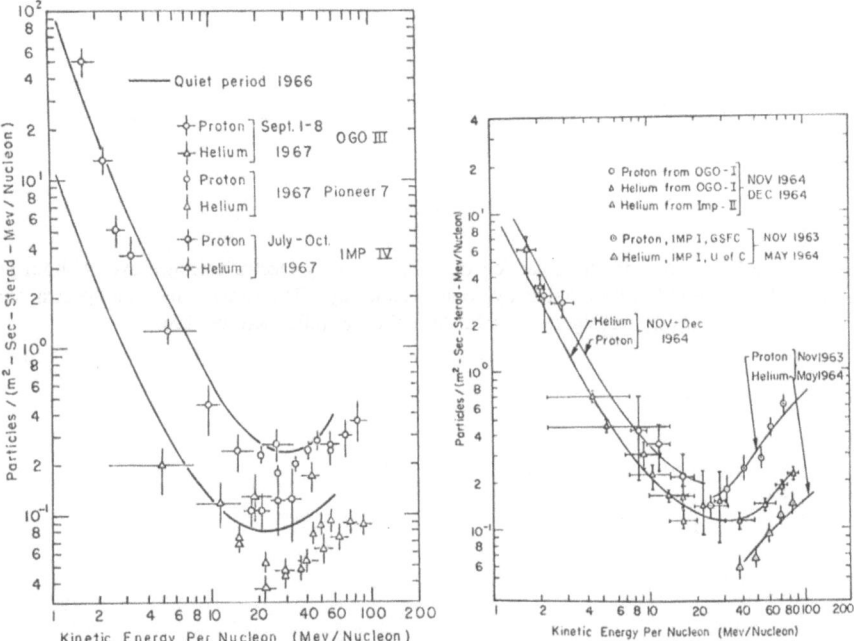

Fig. 1.   Differential energy spectra of protons and helium nuclei measured in 1967 by University of Chicago Cosmic Ray Group on OGO Pioneer and IMP. The spectra measured in 1966 have been plotted as solid lines for comparison (after Fan *et al.*, 1969).

Recently Kinsey (1969) has used the data from the Goddard cosmic ray experiment on IMP IV to examine this question. The energy interval covered was 4–80 MeV. It was found that 96-hour periods were necessary to obtain statistically significant spectra during quiet times. Four typical spectra are shown in Figure 2 for protons and alphas. The 13–17 June is a period of high solar activity while the other three represent varying degrees of solar quiet periods. All four day averages from May 1967 – May 1969 display a negative slope in the low energy proton region. Kinsey has proposed that the observed fluxes can be represented as the sum of two power laws in kinetic energy and the differential flux can be expressed in the form

$$dJ/dE = F_s E^{-s} + F_g E^g.$$

FRANK B. MCDONALD

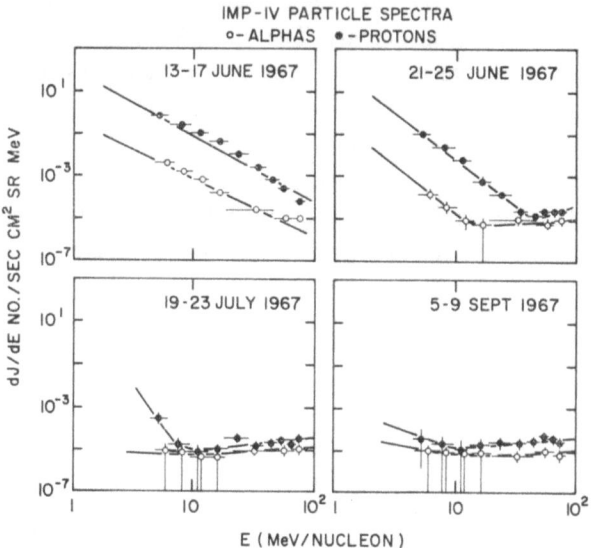

Fig. 2.   IMP IV proton and alpha particle differential energy spectra averaged over 96 hour periods. The June 13–17 period is dominated by solar cosmic rays. The three remaining spectra represent varying degrees of solar 'quiet' times (after Kinsey, 1969).

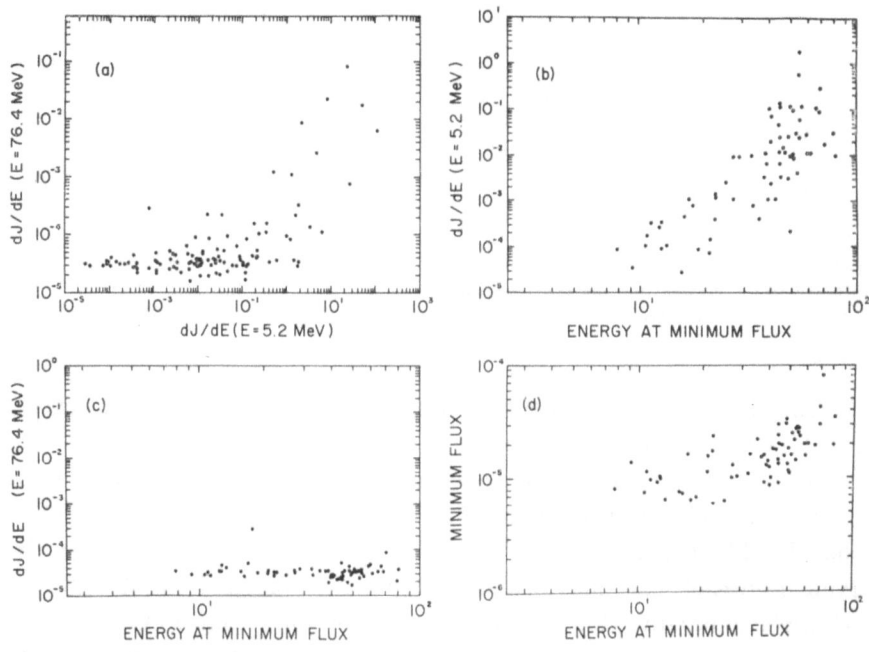

Fig. 3.   IMP IV proton flux correlation results for 96 hour averages covering the period from 24 May 1967 – 20 August 1968. Flux units are proton/cm²-sr-MeV-sec. Kinetic energy units are MeV (after Kinsey, 1969).

Further analysis suggests that the first term represents a solar contribution and the second is the galactic component. Using this model, Kinsey concludes that the following behavior would be expected for the particle spectra observed in the energy interval under discussion:

(1) Independence of the lowest and highest energy fluxes for all cases except the occurrence of a purely solar spectrum,

(2) direct dependence of the lowest energy flux upon the energy at minimum flux,

(3) independence of the highest energy flux with respect to the energy at the minimum flux, and

(4) the minimum flux as a function of the energy at minimum flux should have the same power law index as the galactic component.

In Figure 3 it is obvious that Kinsey's predictions are strikingly verified. While these represent the proton data, the alpha data is identical. It is apparent that the 5 MeV and

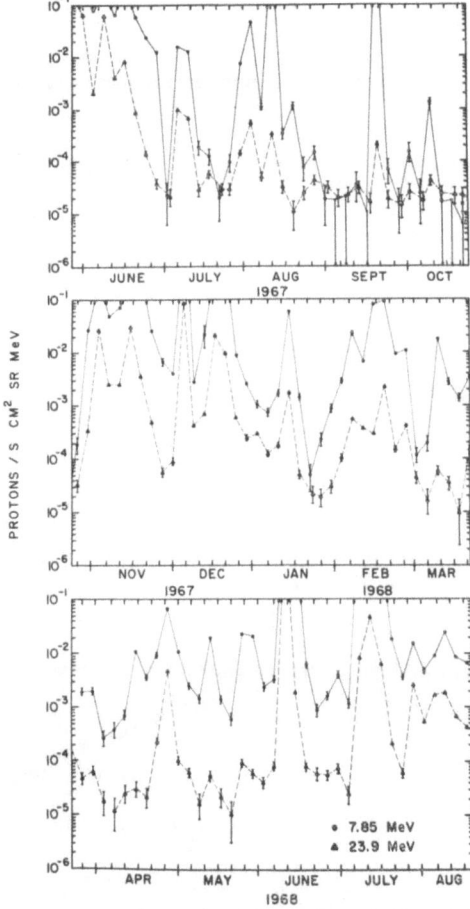

Fig. 4.   IMP IV proton fluxes (96-hour averages) for 8 and 24 MeV components for period June 1967 to September 1969 (after Kinsey, 1969).

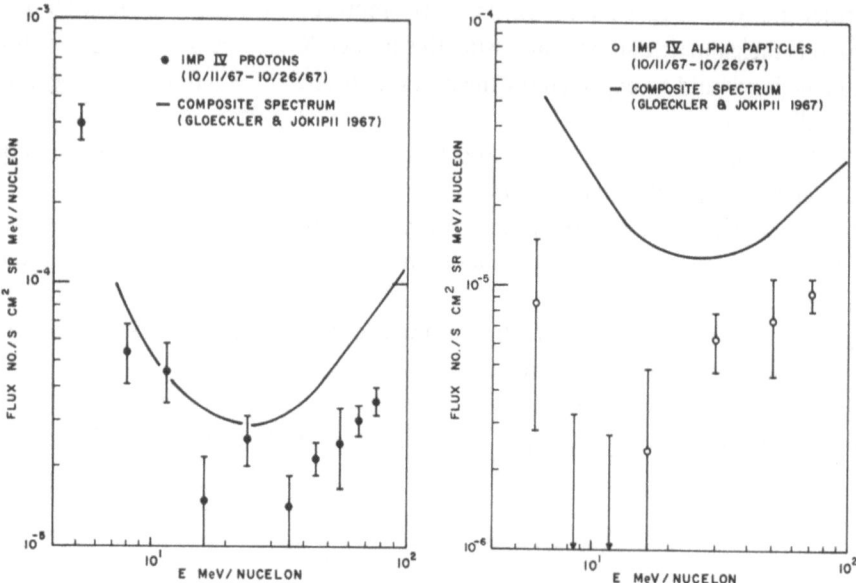

Fig. 5.   Quiet time proton and alpha energy spectra (after Kinsey, 1969).

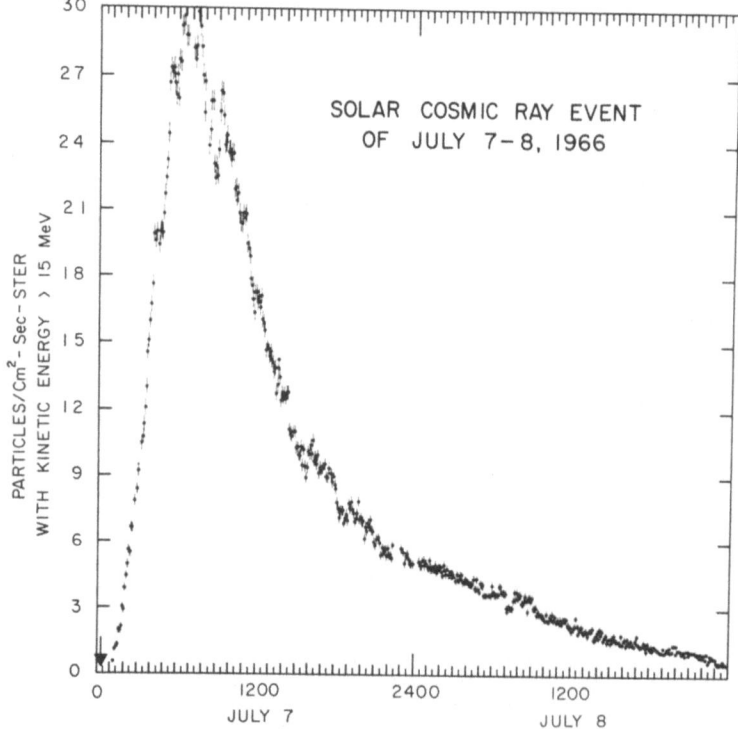

Fig. 6.   The event of 7 July 1966 as observed by the Goddard cosmic ray experiment on IMP III.
The curve represents the integral intensity of protons> 18 MeV and the flux of electrons> 3 MeV
(after Fichtel and McDonald, 1967).

76 MeV data are uncorrelated over nearly 2 orders of magnitude variation in the 5 MeV component. The low energy component shows a good correlation with the energy at which the minimum flux occurs while the 76 MeV component shows no correlation. From this statistical study it is clear that one is dealing with two distinct components. The most plausible conclusion is that the lowest energy component is solar in origin.

In Figure 4 are plotted the 96 hour averages from June 1967 – August 1968 for 8 and 24 MeV protons. There are large increases due to solar events, small increases and several quiet periods in September and October, 1967. There is also a significant long term increase in the intensity of the 8 MeV component.

One of these quiet time periods (extending from October 11–26, 1967) was used for the determination of the quiet time proton and alpha spectra (Figure 5). Even these quietest periods show an increase in dJ/dE at the lowest energies. These conclusions are not changed if the data is extended through June 1969. When viewed in four day increments over a two year period, the sun must be regarded as a continuous source of MeV particles.

3. Classification of Solar Particle Events

As discussed in the introduction, the present observations of solar particle phenomena can be grouped into four categories. We shall discuss these in terms of typical events. The proposed classification is a tentative one and will certainly evolve as more comprehensive measurements are made.

4. Flare-Associated Events

The larger events which produce energetic particles with energies greater than 20 MeV have displayed an excellent correlation with solar flare and solar radio-burst activity. For the purpose of classification, a threshold for these events has been defined such that the peak flux of protons with energies above 20 MeV must exceed 0.5 proton/cm²-sec-sr (Fichtel and McDonald, 1967). Although these energy and intensity thresholds are arbitrary, they give a reasonable separation of these events from the normal recurrence events and other events which cannot be directly associated with flares. A few events of moderate size which have exceeded these thresholds did not have an obvious parent flare. In general, however, the term 'solar cosmic-ray events' will imply flare-associated events unless specifically noted otherwise.

The character of these events as defined by the energy spectrum of the particles and its time variation differs markedly from one occurrence to another. The 7 July 1966 event illustrates the properties of a medium size event. Both the microwave radio emission and the hard (~ 100 keV) X-ray emission, which coincided in time profile and peaked in intensity at 0037 UT (Cline et al., 1968), were unusually intense. The Alert neutron monitor showed a very small increase in high-energy particle intensity. Even at proton energies down to a few MeV, this appeared to be a modest-sized event;

however, it turned out to be the largest particle event between September 1963 and September 1966. The solar longitude of the flare was between 45° and 48° West, near the probable origin of the earth-intercepting field line. Figure 6 shows the integral counting rate of protons above 18 MeV. There is a well defined rise to maximum with some structure in the peak. The time history of the electrons indicates (in Figure 7) that although the rise and decay times are relatively short (matters of minutes and hours respectively) the onset of the event is actually not a very prompt one. Considering the 48° West location of the flare, the half-hour delay in onset for relativistic electrons and the total two hour time to maximum seen here indicate that considerable trapping and diffusion of the particles is taking place. This behavior is similar to that

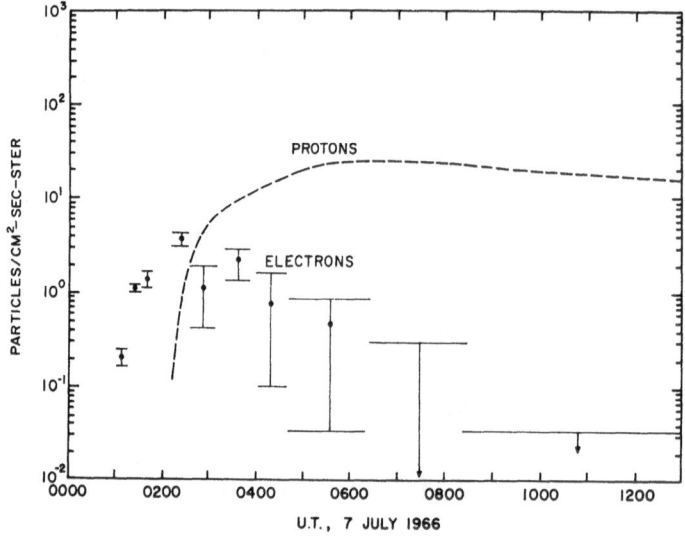

Fig. 7. The time history of > 3 MeV solar electrons contrasted with the 18–80 MeV solar proton time history for the 7 July 1966 event. There electrons are seen to be well towards maximum intensity before the lower-velocity protons begin to arrive (after Cline and McDonald, 1968).

observed for relativistic protons (Bryant *et al.*, 1965) in that for both cases the intensity maxima occur at a time delay equivalent to 12 to 15 AU in travel. This appears to confirm the fact that for certain events the containment, or diffusion, of the particles is on a velocity basis, rather than with respect to kinetic energy, total energy, or rigidity. Because of this time delay, we can then conclude that considerable storage of the particles must have taken place near the sun if they were produced at the time of the microwave and X-ray burst. This conclusion is consistent with that reached by Lin *et al.* (1967) in their study using three spacecraft at different locations during the same flare event, wherein they conclude that the spatial interplanetary intensity geometry reflects a near-solar profile translated towards 1 AU along the spiral field lines. Although one cannot make a distinction between the interplanetary diffusion picture and a near-solar diffusion picture on the basis of time histories alone, it is instructive to use the standard diffusion plot. Figure 8 shows that a straight line fit of

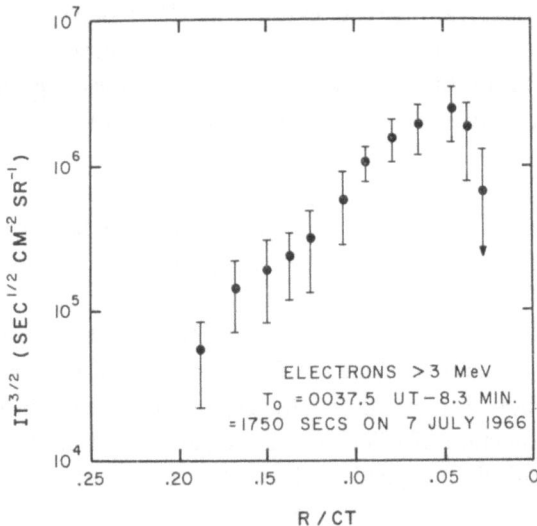

Fig. 8. A plot of $\ln(It^{1.5})$ vs. t^{-1} for the 7 July event, showing the straight-line fit to a standard diffusion equation. Here $t = T - T_0$, in which the zero of time was chosen to be 8.3 min (the transit time across 1 AU) before the observed 0037.5 UT maximum of energetic X-rays. The slope yields a mean free path of 0.027 R and the intercept is 5×10^{31} particles in the energy range observed (after Cline and McDonald, 1968).

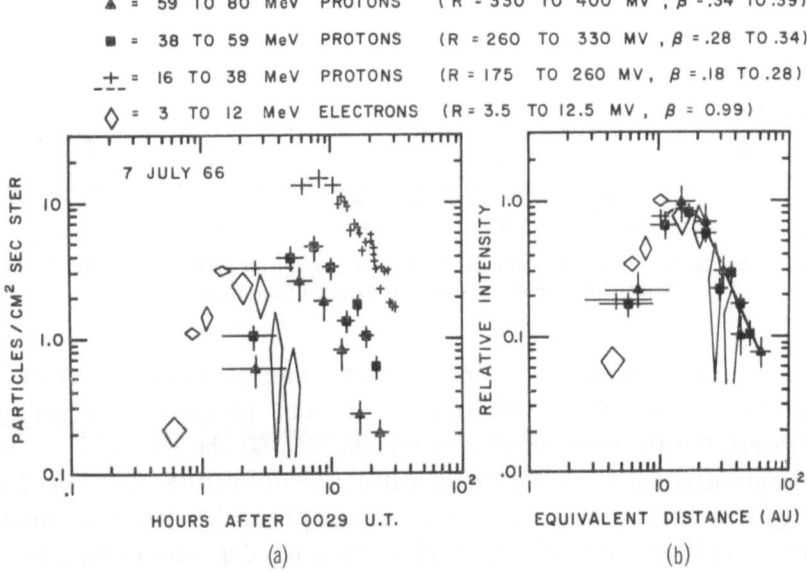

Fig. 9. (a) Profiles of the observed intensity of 16–38 MeV, 38–59 MeV, 59–80 MeV protons, and 3–12 MeV electrons plotted against time; (b) profiles of each relative intensity, $I/I_{\max}$, plotted against distance traveled, $x = vt$, where v is the mean velocity for each energy group. The fit to a common curve is clearly seen. The velocities of the protons are between 0.18 and 0.4 that of the electrons ($\approx c$), but the rigidities of the protons (175–400 MV) are up to two orders of magnitude higher than those of the electrons (> 3.5 MV) (after Cline and McDonald, 1968).

$\ln[I(T-T_0)^{1.5}]$ against $(T-T_0)^{-1}$ does result, and has a slope within 30% of agreement with the former relativistic proton fits. We do not believe this supports a classical interplanetary diffusion but claim rather that the relativistic protons and relativistic electrons appear to travel in a similar manner, wherever the trapping and propagation take place. This can be best illustrated by comparing the time histories of relativistic electrons and low energy protons (Figure 9). These proton components in the energy range 16–18 MeV (rigidity interval 175–400 MV and v/c between 0.18 and 0.40) are plotted along with the intensity of low energy relativistic electrons ($E > 3$ MeV; rigidity ≥ 3.5 MV, $\beta = 0.99$). Figure 9a shows the observed time history for these four components. In Figure 9b the abscissa has been transformed to represent the distance traveled of a given component and the four components here have been normalized to their peak values. For this event, the solar particle time history for these four components is a function of velocity, and the path length distribution and the mean free paths are relatively independent. It must be emphasized that this analysis does not apply to all events and will also break down for very low electron and proton energies.

The frequency of occurrence of flare-associated events is shown in Figure 10 which

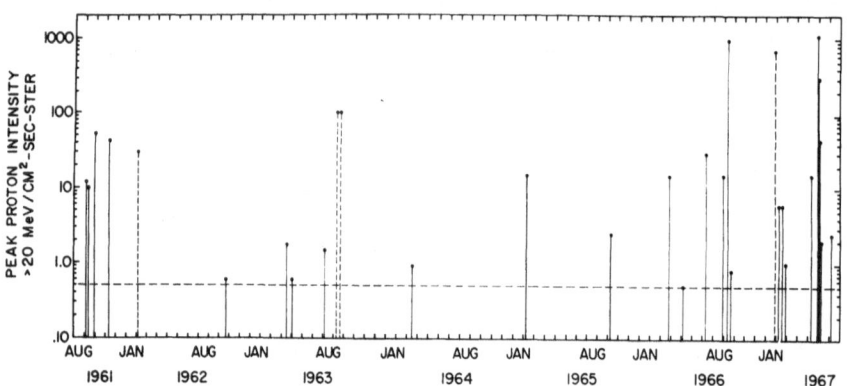

Fig. 10.   Peak intensity > 20 MeV for solar cosmic ray events in period August 1961 – May 1967
Dashed lines represent data extrapolated from riometer data.

displays the peak intensity > 20 MeV versus the time of occurrence for the period August 1961 – May 1967. The 1964–65 solar minimum period is represented by three small events. For the previous solar maximum, 1956–60, the integrated flux of solar particles at energies above 20 MeV is estimated to be 10^{10} particles/cm^2-year. For 1964 the integrated intensity above 20 MeV was some 10^5 particles/cm^2-year and this increased to $> 10^7$ particles/cm^2-year in 1966. There is a difference of some five orders of magnitude in the production of solar cosmic rays above 20 MeV going from solar maximum to solar minimum. Not only are the events smaller in size at solar minimum, but also the frequency of occurrence is much less.

Of the 31 'flare-associated type events' included in Fig. 10, all but four can actually be associated with a flare on the visible disk and appropriate type-IV solar emission.

There is a reasonable correlation between the intensity of the radio emission and the integrated particle flux. The four events are most probably produced by flares on the non-visible side of the disk. While no study has been made of this group, it is expected these will exhibit flatter energy spectra.

If the observations are extended from 0.5 to 0.001 particles/cm²-sec-ster (Figure 11), there is an increase in the number of events and the solar correlation starts to break

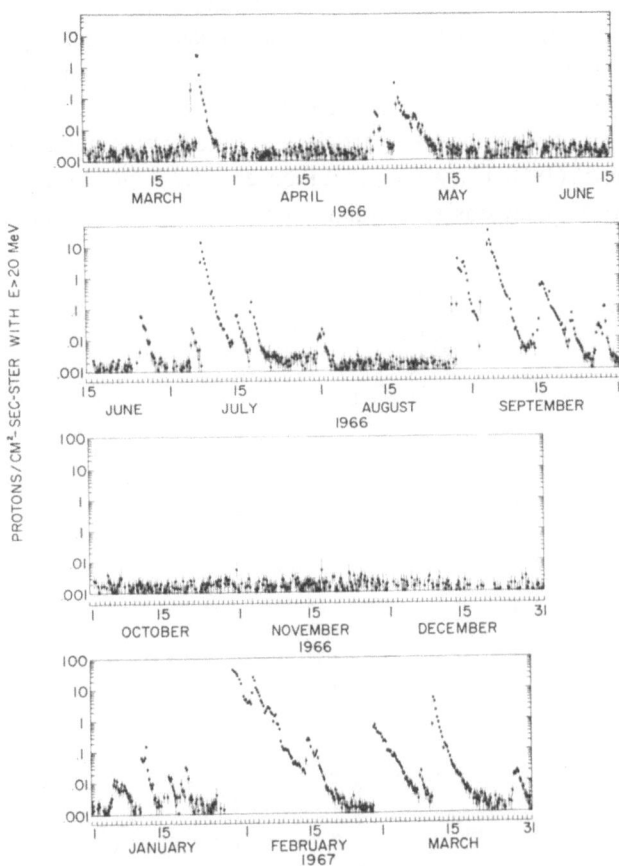

Fig. 11. 6 hour integral fluxes for energies > 20 MeV from the Goddard cosmic ray experiment on IMP III (Kinsey, 1969).

down. Figure 13 is a plot of 6 hour averages of the integral flux above 20 MeV for the period March 1966 – March 1967. These averages will distort the rise time data but generally represent the longer decay phase quite accurately. No definitive study has yet been made of this 'micro-event' group. Most of these time histories closely resemble the larger events – i.e., rapid rise time and exponential decays. There is a small class of events on 1 August and 27 September 1966 and 6 January 1967 that resemble recurrence events. These will be discussed in a later section.

5. Solar Particles Associated with Active Centers

If the energy threshold is lowered from 20 MeV to 5 MeV, there is a dramatic change in the general character of the observed increases. In Figure 12 the results of Kinsey (1969a) for 24-hour average proton fluxes at 5 MeV are plotted for the period June 1967 – August 1968. The increases represent the superposition of several effects (a) the injection of flare particles (b) protons associated with a center of activity. A rough representation of the two effects is made by indicating the time of large flares with an *F* and also indicating the time of central meridian passage of large plage regions. The data of Figure 12 suggest that a significant number of the low energy proton increases can be associated with a given active center.

The first report of energetic solar particles associated with centers of solar activity was made by Fan *et al.* (1968) using data from Pioneer 6 and 7 and covering an energy range from 0.6–13 MeV. They observed enhanced proton fluxes from the sun continuously

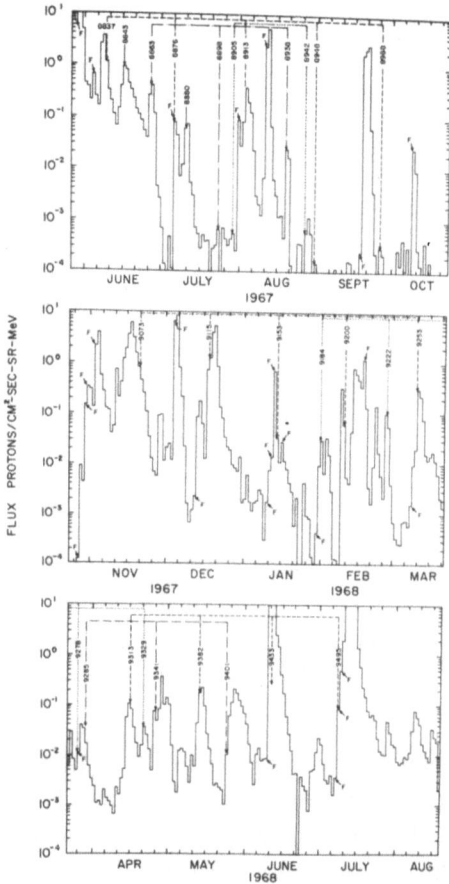

Fig. 12.   The 24 hour average proton fluxes at 5.2 MeV for IMP IV. The times when large confirmed solar flares occurred are marked *F*. Central meridian passage of calcium plage regions are marked by the plage region number. Recurring regions are connected by horizontal lines (Kinsey, 1969).

over heliocentric longitude ranges as great as 180° in association with specific active regions. In general the increases began when the active region was 60–70° east of central meridian and were observed until the region was 100–130° west of central meridian. Figure 13 shows the 13 March – 11 April 1966 period. There are two low energy increases attributed to solar region 8207 and 8223. When region 8207 reaches 37° west, there is a 3B flare on 24 March which produces a moderate size flare associated event. The critical question now becomes, were the MeV particles in the active center accelerated to energies of several hundred MeV? One could postulate a two stage acceleration mechanism with the first stage providing the quasi-steady state MeV

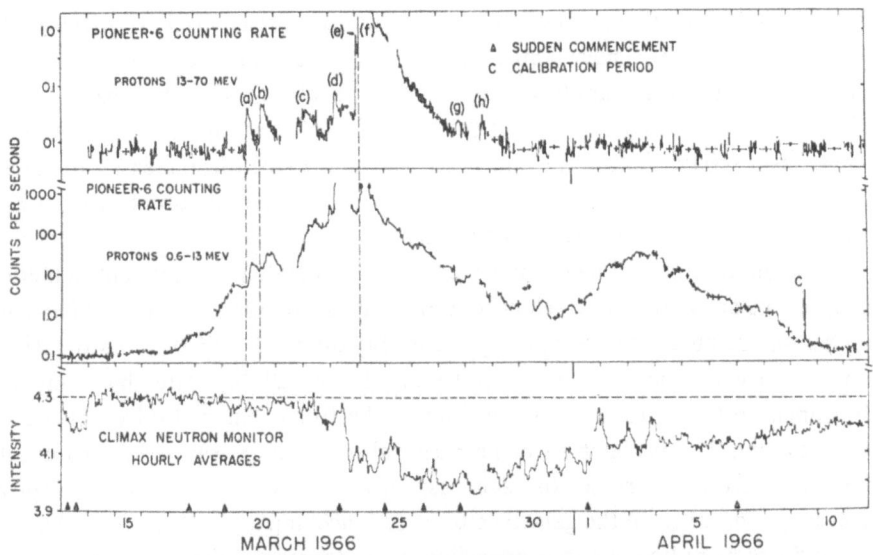

Fig. 13. Thirty-min. averages of the counting rates of protons 13–70 and 0.6–13 MeV. The enhanced flux of 0.6–13 MeV protons during March 15–31 is attributed to solar region 8207. The first evidence of enhanced flux from the following region (8223) as well as a magnetic sector boundary appears on March 31. This coincides with an abrupt change in the level of modulation at the climax neutron monitor. (a)–(h) denotes discrete flare and shock wave events seen at 13–70 MeV. Region 8207 is also the source of the large flare associated event on 24 March which occurs almost at the mid point of the low energy increase (after Fan et al., 1968).

protons stored above the active center and the second, or flare stage, increasing the energy by a factor of 10–30 and also providing an ejection mechanism. This was first suggested by Schatzmann (1967) and by Fichtel and McDonald (1967) based on the observations of Bryant et al. (1965b) of long lived streams of MeV protons. Since that time, a number of satellite experiments have established the fact that energetic protons constitute a characteristic feature of many centers of solar activity. A necessary condition that the two step acceleration process must satisfy is that the charge composition of the active center emission should be the same as that observed in flare-associated events. As Bertsch et al. (1969) and Biswas and Fichtel (1965) have reported, the measured charge spectrum in the helium-iron range appears to be similar from event to event and shows remarkable agreement with the photospheric abundances. Helium has already been detected in the low energy non-flare events (Fan *et al.*, 1965; Kinsey, 1969). If a two-step process is occurring, then low energy carbon and heavier elements up to iron will be detected in the co-rotating increases. However, this also requires that either energy losses in the storage region are negligible even for Fe nuclei with energies of several MeV/nucleon or that the acceleration process is a continuous one and losses are negligible. The extension of the charge composition studies to low energies and small increases is one of the important problems in solar cosmic ray studies.

6. Recurrence Events

In a later section it will be shown that the co-rotating streams of low energy protons associated with active centers anti-correlate with quiet time increases of 3–12 MeV electrons. This does not apply to lower energy electrons. For this reason, the recurrence events observed on the next solar rotation after a flare-associated event are classified as a separate phenomenon. If the quiet time electron increases are galactic in nature, then the recurrence events would become a special subset of co-rotating proton increases associated with active centers.

The first observations of recurrence events were associated with the central meridian passage of the active region which produced the 28 September 1961 and 10 November 1961 flare-associated solar cosmic ray event (Bryant *et al.*, 1963). Thus the 28 September event which occurred at 30° east was followed 29 days later by a small but well defined proton increase. In a like manner, the 10 November 1961 event occurred at 90° west and 21 days later at the next central meridian passage of the parent flare there was a proton increase. In each case there was a large magnetic storm and Forbush type decrease of the galactic cosmic rays accompanying the events, but there were no flares or radio emissions preceding them. In general the recurrence has been confined to the next rotation following the flare associated event and does not appear at a detectable level following all events.

The July 7, 1966 event (Figure 6) occurred at 46° west and was followed some 25 days later by a small particle event (Figure 11, 14) which is markedly different from the other micro events. It is also obvious (Figure 16) that there is a significant increase of electrons in the 3–10 MeV region and Anderson (1968) has reported a large increase in the flux of low energy electrons.

In Figure 14 the vertical line represents the passage of the active region producing the 7 July 1966 event. The recurrent event occurs when the original source is still one day east of central meridian. A similar pattern has been established for the two 1961 events (Bryant *et al.*, 1963). It is assumed that the active center producing the flare associated event is also the source of the recurrent event in the following solar rotation. This relation between the energetic particle stream and the parent active region is an unexpected one. It is well established that transverse diffusion in the interplanetary region is limited. Normally it would be expected that the peak of the particle event would occur when the active region is some three days west of central meridian. The most plausible explanation is that the space above the active area is closed due to the high magnetic fields and that particles preferentially escape from the preceding part of the active region.

There is a difference between the time histories observed for proton increases associated with active centers and recurrence events. In general the 'active center' observations have been at lower energies. Whether this accounts for the larger time scale (corresponding to some 180° of solar longitude) is not clear. The behavior for the 3–12 MeV electron component is also different.

It is important to note that the intensity level for recurrence events is very low. Nevertheless, the fact that particles are stored above the flare site would suggest that events where the parent flare occurs east of central meridian should differ from those at solar longitude 30° west. In general these differences appear in two ways.

(a) The more eastern events display a longer decay time.

(b) The existence of energetic storm particles. Both of these effects will be illustrated by the event of July 6–14, 1968 in the following section.

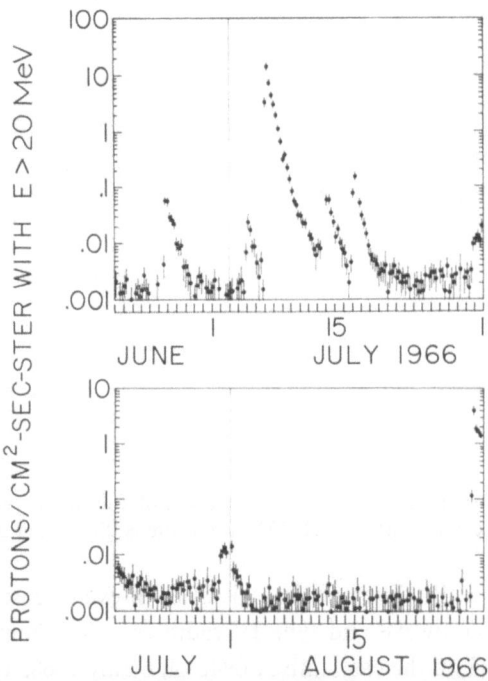

Fig. 14. IMP III integral proton flux > 20 MeV plotted for 2 solar rotations. The vertical line represents central meridian passage of the active region producing the 7 July 1966 event. Note that the recurrence event is quite different from the other micro events.

7. Energetic Storm Particle Events

This type of event represents the superposition of a co-rotating particle stream and a blast wave. It was first observed by Bryant *et al.* (1963) from Explorer XII particle data and from riometer data of Axford and Reid (1963) for the 28 September 1961 event. The parent flare for this event occurred at 28° east on 28 September 1961. There was a marked increase in the flux of low energy particles some 40 hours after the flare and immediately following the blast wave from the flare. Rao *et al.* (1967) and Anderson (1968) have more than seven additional examples of intense flux of ~ 10 MeV protons associated with disturbances in the interplanetary medium. The time scale for these events is on the order of 6–12 hours. They always follow strong shocks which also produce magnetic storms and Forbush decreases.

These particles cannot be stored behind the shock front since adiabatic cooling would dominate and this trapping could not produce the observed effect. However, most of the storm particle events are produced by events which are close to or east of central meridian. This suggests that one has a co-rotating stream of MeV protons which also represents a solar source. If one superimposes a shock wave, then this source is continually injecting particles into the turbulent region behind the shock. It is not clear yet whether some additional interplanetary acceleration is necessary.

The event of July 6–14 (Figure 15) illustrates both the energetic storm particle

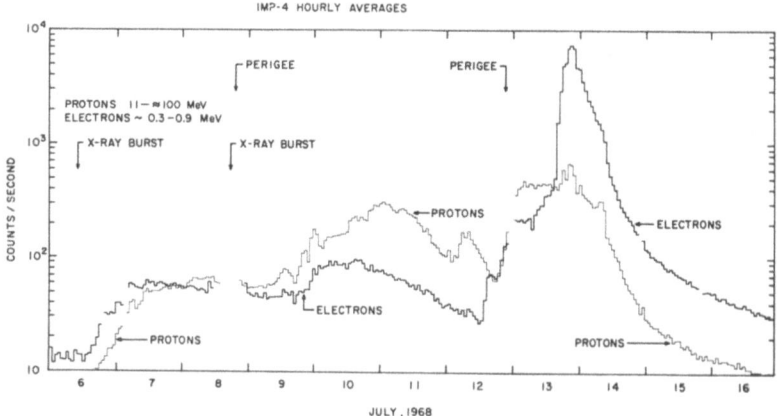

Fig. 15. Time histories from IMP IV of the solar event of July 6–14, 1968 for the 0.3–0.9 MeV electron components and 11–100 MeV protons (Simnett *et al.*, 1969).

increase and the generally different time histories of events east of CMP. There was a flare and strong microwave and type IV radio emission from a flare in McMath plage region MP 9503 on the east limb at 0946 on 6 July 1968. There were additional flares from this region on the 8th, 9th and 12th of July (Simnett *et al.*, 1969). From Figure 15 it appears there is a fresh injection of energetic protons following the event on the 9th and a small increase after the flare of the 12th. The impression is one of a broad plateau as the region approaches CMP. The general levels start to increase late on the 12th and this can be attributed to the satellite entering the co-rotating stream as the active region reaches central meridian. There is a magnetic storm at 1610 on the 13th of July which produces a strong energetic storm particle event. This event illustrates that storm particle increases require the coincidence of a co-rotating particle stream and an interplanetary shock wave.

8. Electron Increases

The interplanetary electron intensity between 3 and 12 MeV has been monitored with the IMP I, III and IV spacecraft from November 1963 – April 1964, June 1965 – April 1967 and May 1967 – April 1969 respectively. There are a variety of types of time variations in the intensity which include flare-associated and recurrent solar electron events and what is termed 'quiet time' increases. The flare associated increases have already been observed in conjunction with the 7 July 1966 event.

A plot of the electron counting rate (24 hour average) over the period covered by IMP's I, III and IV is shown in Figure 16. It is evident from this figure that there are many occasions when the intensity increases by an order of magnitude or greater above the general background level. Almost without exception these are correlated with 'flare-associated' proton increases. However, there are many other events in Figure 15 where the intensity of electrons increases significantly above the background

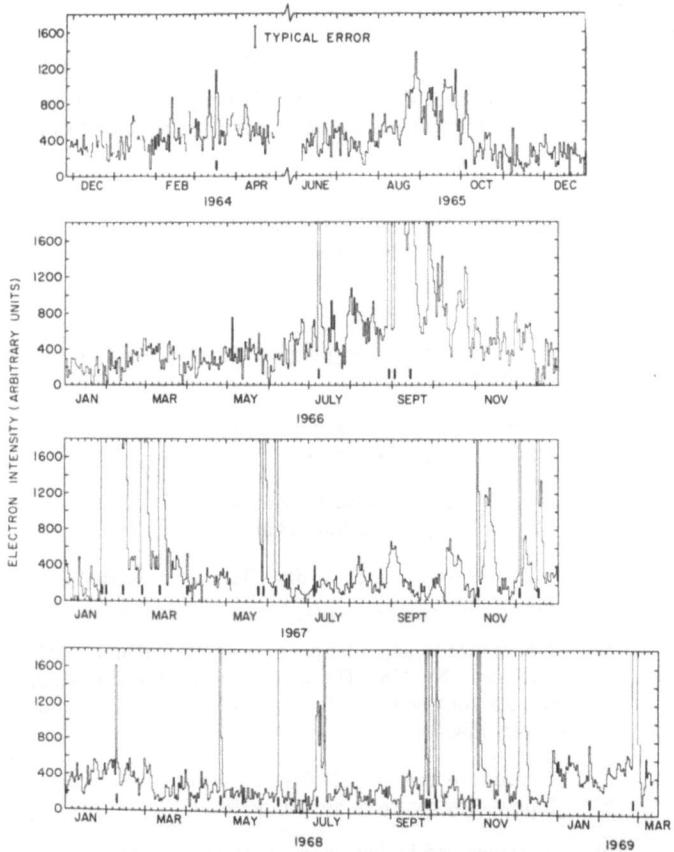

Fig. 16. The intensity of 4–12 MeV interplanetary electrons observed between November 1963 and December 1968. The hatched lines represent 'quiet time' electron events (Simnett *et al.*, 1969).

level ranging from a few days to around two weeks. The identification of these events is subjective, but the basic criterion used was to select times when the intensity was a factor of two above the running mean for either the preceding or succeeding weekly period (excluding flare-associated events). A summary of events is given in Table I.

The intensity increases which are not directly related to flare events can, as mentioned in the introduction, be grouped into two classes: recurrent events which sometimes appear on the next solar rotation after a flare increase, and quiet time increases. In general, the first type contains an energetic nucleon component while the latter type is anticorrelated with low energy proton events. The time histories of the two types tend to be similar, but are completely different from that of the typical flare-associated increase. In general, a flare event shows a very rapid rise to maximum (under 2 hours) and a power law decay; the quiet-time and recurrent electron events are more symmetrical and can last for periods as long as about 14 days. Series of quiet time increases were observed from IMP I in early 1964, IMP III in August to October 1965 and IMP IV in the latter half of 1967 (Table I). Some increases in these series did reappear on

TABLE I

Date of Maximum		Comments
Feb. 12,	'64	Series of short-lived increases (2–4 days) separated by 27 days. Not coinci-
Mar. 11,	'64	dent with recurrent proton increases observed by Fan *et al.* (1968)
Apr. 7,	'64	
May 3,	'64	
Aug. 28,	'65	Periodic set of increases separated by 27 days and not coincident with proton
Sep. 6–9,	'65	events observed by O'Gallagher and Simpson (5), maximum intensity 5 ×
Sep. 17–26,	'65	background.
Oct. 4,	'65	
Aug. 1,	'66	Recurrent event in plage region MP 8413, which was the return of MP 8362 (region for July 7 event).
Oct. 4–7,	'66	Recurrent event in plage region MP 8527, which was the return of MP 8484 (region for Sep. 14 event).
Oct. 19–25,	'66	The increase is coincident with MP 8551, which was the third rotation of MP 8461 (region for Aug. 28 and Sep. 2 events).
Aug. 7,	'67	Increases separated by 27 days. The latter increase is at a time of extremely
Aug. 31–Sep. 3	'67	low proton activity.
Oct. 13,	'67	Not related to any solar phenomena other than it is 27 days before the next increase in Nov. '67. The intensity is high for 13 days, during which time the 0.3–0.9 MeV electron intensity is high for 2 days and the proton intensity is very low.
Nov. 8–11,	'67	Increases separated by 27 days, occurring at a time of high solar activity and
Dec. 6–7,	'67	followed by recurrent proton increases.
Dec. 28	'68	Accompanied by low-energy electrons and protons.

(After Simnett *et al.*, 1969)

a 27-day basis. The electron recurrent event on 1 August 1966 is similar to that shown in Figure 14 for protons. However, the remaining increases in Table I do not have obvious solar correlation. Neither do they show any correlation with Forbush decreases or magnetic activity. Because of this, they have been termed quiet time increases (Simnett *et al.*, 1969). The energy spectra for these events is almost as flat as that observed for the 4–12 MeV galactic component (Simnett and McDonald, 1969). They also display a strong anticorrelation with the active center proton increases. In Figure 17 the 5 electron 'quiet time' increases occurring in 1967 have been marked

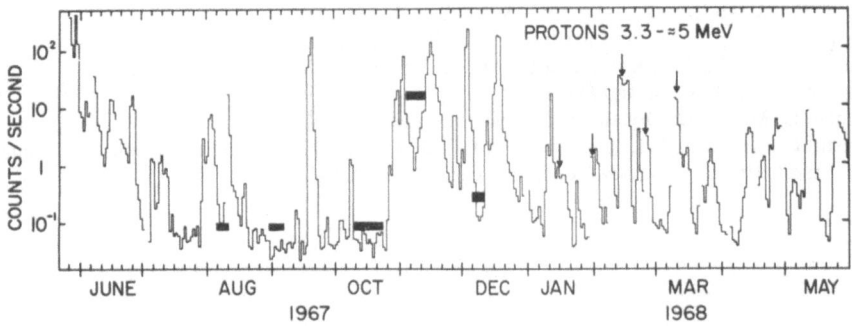

Fig. 17. Intensity of 4.5 MeV proton component observed on IMP IV for the second half of 1967 and early 1968. Also shown from Figure 16 is the occurrence of quiet time maximum and minimum points. The maximum points occur between proton events while the electron minimum points fall on proton peaks.

with a rectangular box. In addition, 5 electron minima are shown by arrows in early 1968 when the electron increases are small and not as easily identified. These 5 'quiet time' and minimum periods are superimposed on a graph of the 4.5 MeV proton data from IMP IV. The anticorrelation between the quiet time electron increases and the 'active center' proton events is very striking. This will break down as one goes to lower electron energies as Lin and Anderson (1967) have reported low energy ($E > 40$ keV) electron increases associated with large centers of activity. So far there is no obvious solar feature that can be associated with the 4–12 MeV electron increases. This might indicate they are galatic in origin. This would require a rather drastic change in the concepts on cosmic ray modulation. Another interpretation is in terms of energy loss of particles in the interplanetary medium (Gleeson and Axford, 1968). That in the increase in K, the diffusion coefficient, means a smaller loss in the interplanetary medium and thus a significant increase in intensity for the electron component. This is highly speculative and further study of these events is required.

References

Anderson, K. A.: 1968, *Solar Phys.*
Axford, W. I. and Reid, G. C.: 1963, *J. Geophys. Res.* **68**, 1793.
Bertsch, D. L., Fichtel, C. E., and Reames, D. V.: 1969, *Astrophys. J.* **157**, L53.
Biswas, S. and Fichtel, C. E.: 1969, *Space Sci. Rev.* **4**, 709.

Bryant, D. A., Cline, T. L., Desai, U. D., and McDonald, F. B.: 1965a, *Astrophys. J.* **141**, 478.
Bryant, D. A., Cline, T. L., Desai, U. D., and McDonald, F. B.: 1965b, *Phys. Rev. Letters* **14**, 481.
Cline, T. L. and McDonald, F. B.: 1968, *Solar Phys.* **5**, 507.
Fan, C. Y., Gloeckler, G., and Simpson, J. A.: 1966, Proc. Ninth International Conference on Cosmic
    Rays, London Institute of Physics and Physical Society **1**, 109.
Fan, C. Y., Pick, M., Pyle, R., Simpson, J. A., and Smith, D. R.: 1968, *J. Geophys. Res.* **73**, 1555.
Fan, C. Y., Gloeckler, G., McKibben, B., and Simpson, J. A.: 1969, 11th International Cosmic Ray
    Conference, Budapest, Hungary.
Fichtel, C. E. and McDonald, F. B.: 1967, *Ann. Rev. Astron. Astrophys.* **5**, 351.
Gleeson, L. J. and Axford, W. I.: 1968, *Astrophys. J.* **154**, 1011.
Kinsey, J. H.: 1969, Ph.D. Thesis, University of Maryland.
Lange, I. and Forbush, S. E.: 1942, *Terrest. Magnetism Atmos. Elec.* **47**, 185.
Lin, R. P. and Anderson, K. A.: 1967, *Solar Phys.* **1**, 446.
Lin, R. P., Kahler, S. W., and Roelof, E. C.: 1967, *Solar Phys.* **4**, 338.
Rao, U. R., McCracken, K. G., and Bukata, R. P.: 1967, *J. Geophys. Res.* **72**, 4325.
Schatzmann, E.: 1967, *Solar Phys.* **1**, 411.
Simnett, G. M. and McDonald, F. B.: 1969, *Astrophys. J.* **157**, 1435.
Simnett, G. M., Cline, T. L., and McDonald, F. B.: 1969, 11th International Cosmic Ray Conference,
    Budapest, Hungary: NASA Goddard Preprint X-611-69-413, p. 51.

PROPAGATION OF SOLAR AND GALACTIC COSMIC RAYS OF LOW ENERGIES IN INTERPLANETARY MEDIUM

S. N. VERNOV, A. E. CHUDAKOV, P. V. VAKULOV, E. V. GORCHAKOV,
N. N. KONTOR, Yu. I. LOGACHEV, G. P. LYUBIMOV,
N. V. PERESLEGINA, and G. A. TIMOFEEV

Moscow State University, Moscow, U.S.S.R.

Abstract. The report summarizes the results of measurements of low-energy cosmic rays from Zond-3, Venus-2, Venus-3, and Venus-4 space probes for the period from 1965 to 1967 as well as preliminary results from Venus-5 and Venus-6 space probes. It has been shown on the basis of solar cosmic ray bursts and Forbush-decreases of galactic cosmic rays that the solar cosmic rays propagate rapidly from the ejection region near the sun along the beam of magnetic lines of force emerging from the active region on the sun and slowly when they are retarded in the region of the field knee created by the shock wave from the burst and move together with the knee at the velocity of the wave. It has been shown that the observed retardation of shock waves from the bursts results in disappearance of Forbush-decreases and accumulation of solar cosmic rays at distances of ~ 2 AU from the sun. It has been shown on the basis of the observations of variations in galactic cosmic rays that the variation parameters (amplitude, steepness of the fall) are determined by the dynamics, latitude, and development phase of active regions.

1. Introduction

The studies of solar and galactic cosmic rays as well as trapped radiation near planets are one of the ways of getting to know the structure and physical conditions of the solar system. This method of investigation is based on the interaction of cosmic rays with quasistationary and moving magnetic fields of interplanetary medium.

The flights of automatic space probes present a unique possibility for this since these flights make it possible to detect not only energetic cosmic and solar particles but also their soft component. It has been found as a result of the flights of Soviet and U.S. space probes that the solar system is filled with protons of ~ 1 MeV energies and the increases in the proton fluxes of such energies occur more frequently and are of higher intensity than those for high-energy protons detected on the earth.

The magnetic field of the interplanetary medium is mainly of solar origin. The lines of force of the field are on the average of the form of Archimedes spirals twisted in the direction of the sun's rotation. These lines are pulled out from the sun by the solar plasma stream which is continuous in time and space and moves almost radially away from the sun at a velocity (in quiet periods) of about 300 km/sec.

The solar plasma streams, the so called 'corpuscular streams', are ejected at high velocity from the active regions on the sun. These streams carry more strong magnetic fields.

Even stronger magnetic field disturbances and shock waves moving almost radially from the sun at velocities up to 3000 km/sec are caused by solar flares.

The solar interplanetary magnetic field forms some region around the sun into which the galactic cosmic rays find it difficult to enter. This region is on the average in the state of dynamic equilibrium, i.e. the galactic cosmic rays diffuse from without to the solar system and the moving inhomogeneous magnetic fields throw them in the direction outward from the sun. The size of the cosmic ray flux modulation region is expanded during the increase and contracts during the decrease of solar activity.

These processes give rise to the galactic cosmic ray gradient inside the modulation region. The gradient of soft protons of solar origin which diffuse back to the sun after flares is also explained in such a way.

The shock waves from chromospheric flares, when propagating from the sun, cause an effect of Forbush-decrease type in the cosmic ray flux characterized by large amplitude and steepness of the galactic cosmic ray intensity fall. The influence of the corpuscular flux is less obvious.

2. The Results of Cosmic Ray Studies in 1965–67

Presented below are the results obtained from Zond-3, Venus-2, Venus-3, Venus-4 space probes which flew in 1965–67.

Propagation of the low energy solar cosmic ray characterizes the state of the interplanetary medium within the propagation region. Since the lines of force of magnetic field are on the average of the form of an Archimedes spiral then, in the absence of magnetic inhomogeneities with sizes comparable to the Larmour radius, the particles will move along these lines at velocities determined by the energies and pitch angle of these particles. In this case the region of propagation of such particles will be determined by their distribution near the sun and by the specific forms of Archimedes spirals. In such a case of charged particle propagation in the interplanetary space we would fail to observe particles produced in the flares which are not connected with observation point by a line of force. However, increases in the intensities of protons with $E_p = 1–5$ MeV relating to the flares occurring on the eastern half of the solar disc have been experimentally found.

It is likely that every flare produces protons which propagate in two ways: (1) a portion of particles with high velocity and strong anisotropy of the flux move along lines of force still undisturbed by shock wave and having the Archimedes spiral form; (2) quasi-trapped particles, the flux of which is almost isotropic, propagate following the shock wave within a large region of the space. It will be noted that in the case of a 'corpuscular stream' from active regions the protons are likely to propagate in the way close to the first one but along an Archimedes spiral straightened more than usual, which is explained by a higher velocity of 'solar wind'.

Observation of one or another way of particle propagation from the sun depends on the location of the flare on the sun with respect to detector, i.e. on the latitude of the flare with respect to the visible central meridian of the sun [1, 2].

Consider at first the second way of particle propagation which is more complex for explanation, i.e. transfer at a shock wave velocity. In this case the solar cosmic rays are observed 1–3 days after the eruption on the sun and the anisotropy of proton flux with $E_p = 1–5$ MeV is usually not higher than $\sim 20\%$.

A Forbush-decrease of galactic cosmic rays, an increase in the strength and change in the direction of magnetic field, and an increase in the solar plasma stream are observed simultaneously with the appearance of solar cosmic rays.

These events correspond to the arrival of the shock wave and plasma from the flare.

One of the authors (Lyubimov) proposes the following explanation.

Shown conditionally in the diagram (Figure 1a) are the Forbush-decrease and three regions of the space corresponding: (1) to undisturbed medium opposing the shock wave front; (2) to the phase of the galactic cosmic ray intensity fall, i.e. to the magnetic field of increased strength squeezed between the shock wave front and plasma front; (3) to the phase of the intensity recovery, i.e. to the region between back edge of the shock wave and the sun.

The diagram of Figure 1b shows the same regions for the magnetic field configuration in the ecliptic plane (view from the North pole of the ecliptic). The sections of spiral lines represent the lines of force of the magnetic field and the dotted arcs represent the fronts of the shock wave and plasma. The diagram has been plotted in accordance with Parker's model for the propagation of a shock wave from the solar flare.

The particles of galactic or solar cosmic rays which remain from previous flares or quiet stationary fluxes and accelerated at the front edge of the shock wave or the particle passing through the front edge may be observed in region (1). These particles may form a peak before the onset of Forbush-decrease.

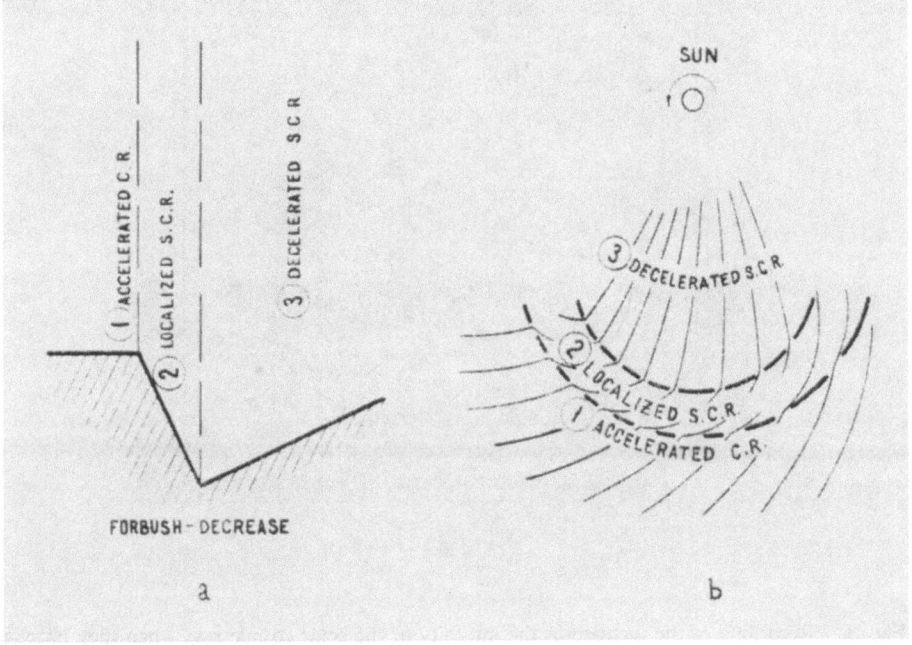

Fig. 1. Idealized picture of the change in the galactic cosmic ray intensity during the passage of shock wave (a) and the character of the change in the interplanetary magnetic field in the region of the shock wave (b).

Localized solar cosmic rays generated by the flare but retarded between the shock wave and hot plasma front are observed in region (2). These particles propagate together with layer (2) at a velocity of the shock wave and plasma.

When propagating and moving along the knee of the magnetic lines of force the solar cosmic rays may partly pass through the knee, i.e. flow from the localization layer and enrich region (1).

Thus, the solar proton anisotropy may be fairly high along the field direction and

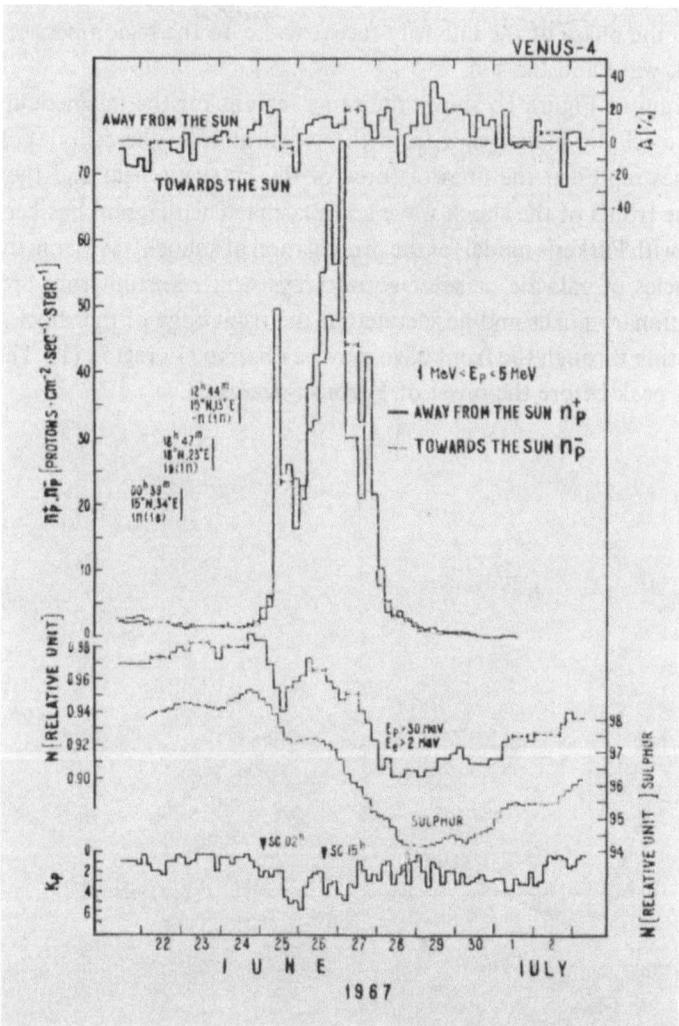

Fig. 2a. Recording of the increase in the intensity of the solar cosmic rays when they propagate behind the shock front. Noted with vertical lines are the moments of the flares on the sun. The upper curve represents the flux anisotropy. Shown below is the solar cosmic ray intensity in the direction from the sun (thick line) and to the sun (thin line). Shown further are the galactic cosmic ray intensity ($E_p > 30$ MeV), neutron monitor data (Sulphur) and K_p-index. Shown with the triangle are the moments of the sudden commencements of magnetic storms (SC).

low perpendicular to this direction, i.e. along the direction of the wave front propagation.

The particles which escaped from layer (2), injected by the next flare, or leaking quietly from active regions reside in region (3).

The first two sources, i.e. the particles which escaped from the layer and those injected by the next flare should be characterized by a double direction anisotropy due to particle motion between the layer and the sun. The solar cosmic rays continuously leaking from the sun should be characterized by the presence of a positive anisotropy due to the action of the source.

The solar cosmic ray burst observed from Venus-4 in June, 1967 will be considered as an illustration of this way of particle propagation.

Figure 2a is a plot of proton fluxes with $E_p = 1$–5 MeV streaming from the sun (thick lines) and to the sun (thin lines) vs. time; shown above is the anisotropy in percent (above the zero line the positive one, below the negative one); intensity of protons with $E_p > 30$ MeV, i.e. practically only galactic cosmic rays; the vertical shadowing represents solar flares, the time of commencement, coordinates and class being indicated.

Beside these main characteristics the figure shows: intensity of galactic cosmic rays detected on the earth, i.e. that of protons with $E_p > 1$ GeV and K_p – the index of geomagnetic field disturbance. Three consecutive flares on the sun (the first, second, and third) occurred in the same active region located eastwards from the central meridian.

About 2.5 days after the first flare the Venus-4 detectors observed a Forbush-decrease of galactic cosmic rays and a burst of solar protons.

The second and third Forbush-decreases and, respectively, the second and third solar cosmic ray maxima were observed after the recovery phase of the galactic cosmic ray intensity. During the whole increase phase the anisotropy did not exceed 20% and the duration of the increase corresponded to the duration of the decrease phase of the galactic cosmic rays.

The results of magnetic field measurements on Venus-4 carried out in the same period [3] showed an increase in the magnetic field and an augmentation in the force line inclination; according to the data of [4] also obtained from Venus-4 an increase in the plasma stream was observed. Arrivals of shock waves and magnetic storms were recorded on the earth.

The diagram in Figure 2b shows, as a projection on the ecliptic plane from its Northern pole, positions of the earth, Venus-4, sun, and the shock wave fronts from the first and second flares at the moment of occurrence of the third flare.

Turn now to the description of the direct particle motion along magnetic lines of force. In this case we observe solar cosmic rays several hours after a flare on the sun. The particles arrive mainly from the direction toward the sun, the anisotropy is close to 100%. The level of galactic cosmic rays does not show considerable decrease.

The cases of observation of four flares appearing in the western part of solar disk

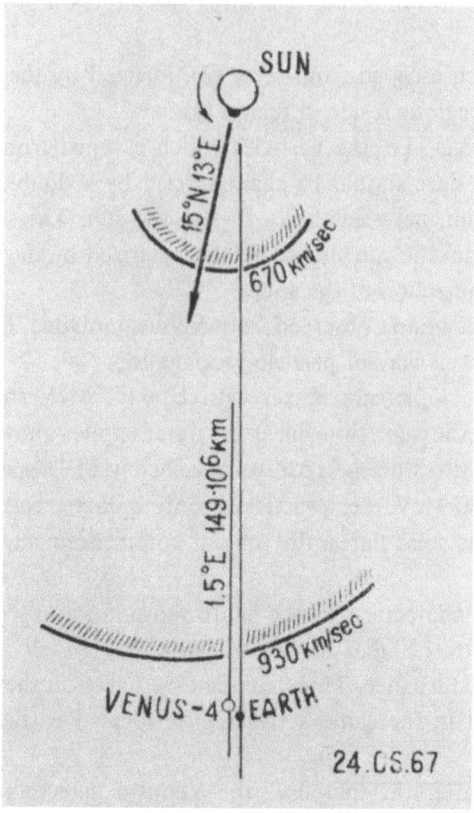

Fig. 2b.   Interpretation of the increase in the solar cosmic ray intensity. Shown is the picture of the position of the shock front from two solar flares in the ecliptic plane at the moment of occurrence of the third flare.

are a very striking illustration of this mode of propagation (see Figure 3). As in the previous case this figure shows the solar proton anisotropy, particle flux from and to the sun, and galactic cosmic ray intensity.

Four flares occurred in the same active region. In about 3 hours after each flare the Venus-4 detectors observed four increases in solar protons with $E_p = 1$–5 MeV. The increases were not accompanied by Forbush-decreases of the galactic cosmic rays. On the contrary, a small flare of solar protons with $E_p > 30$ MeV was observed. These particles came earlier in accordance with their higher velocity.

The diagram in Figure 3b shows the positions of the earth, Venus-4, sun, and a magnetic line of force for a 350 km/sec solar wind velocity emerging from the region of the flare at the moment of its appearance (for the first flare). The proton propagation along the Archimedes spiral corresponds to observation of the maximum, equalling to almost 100%, anisotropy. A delay of particle flux returning to the sun with respect to the direct flux was simultaneously observed. This is indicative of the presence of a certain kind of 'mirror' located at about 1.5 AU from the sun. It should be noted

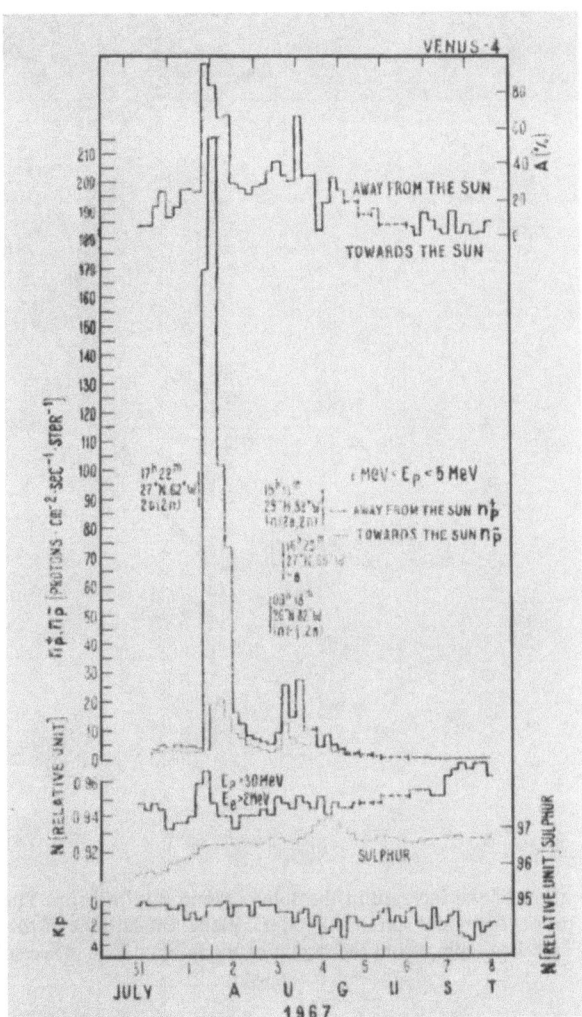

Fig. 3a.   Recording of the increase in the intensity of solar cosmic rays when they propagate along
the magnetic lines of force. Noted with vertical lines are the moments of flares on the sun.
The upper curve represents the flux anisotropy. Shown below are the intensity,
neutron monitor data, and K_p-index.

that in other cases the backward flux was observed simultaneously with the direct
flux which is indicative of the reflection and scattering of particles from a number of
distributed and semi-transparent mirrors, i.e. the presence of a number of inhomogeneities on the magnetic lines of force near the detector.

The peak value of the anisotropy for the cases of the direct arrival of solar protons
corresponds to an intensity increase.

The anisotropy, however, remained high (higher than 60%; see Figure 3a) during

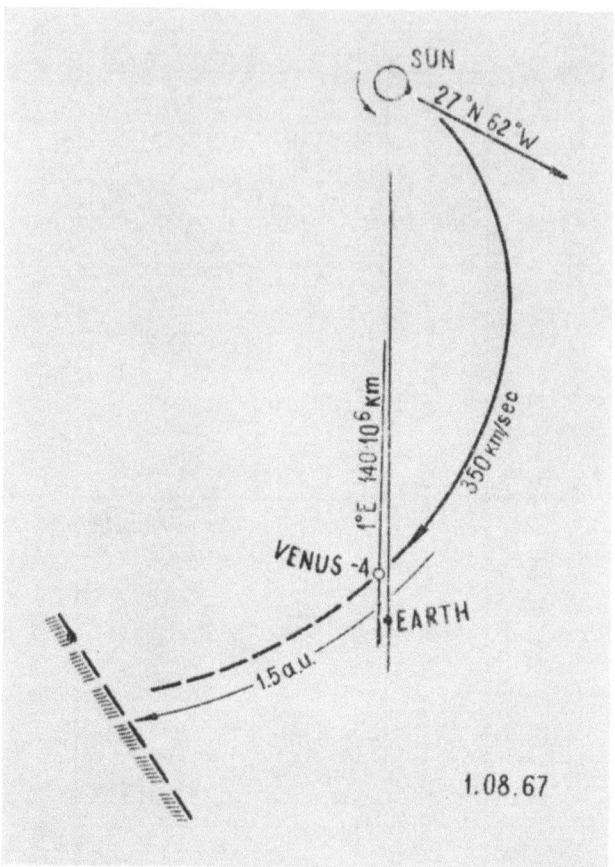

Fig. 3b.   Interpretation of the increase in the solar cosmic ray intensity. The mutual position of the Venus-4 space probe, earth and sun in the ecliptic plane, the latitude of the flare on the sun and the Archimedes spiral emerging from the region of the flare at 350 km/sec solar wind velocity.

16 hours. According to the Monte Carlo calculations set forth below such a high value of the anisotropy is possible only in the beginning of a flare. Therefore the authors of work [4a] suggest that the injection of protons with $E_p = 1$–5 MeV is not instantaneous but lasts for about 10 hours.

Besides that, the prolonged observation of the particles may be ensured by the replenishment of the flux with a secondary source of particles retarded behind the shock wave front. It is difficult, however, to exclude the proton generation by the shock wave from the flare [4b].

The diagram in Figure 3b shows a single line of force. But actually quite a beam of the magnetic lines of force is populated by the particles during a flare. Probably they are irregularly populated due to directed ejection of particles during a flare. The corotation of the force line beam and the sun with respect to the observation point corresponds to the passage of the force lines with different particle populations through the detector. Thus, if the leakage of particles with time is neglected the intensity profile of particle distribution over the cross-section of the spiral flux will be determined.

Figure 4a shows a case of observation of solar protons after a single flare occurring to the Western hemisphere of the sun but at a longitude of 38°, i.e. lower than the optimum longitude of about 60° at which the maximum particle flux is directed to the observation point. The particles came early and then an intensity increase connected with the approach of flux core began.

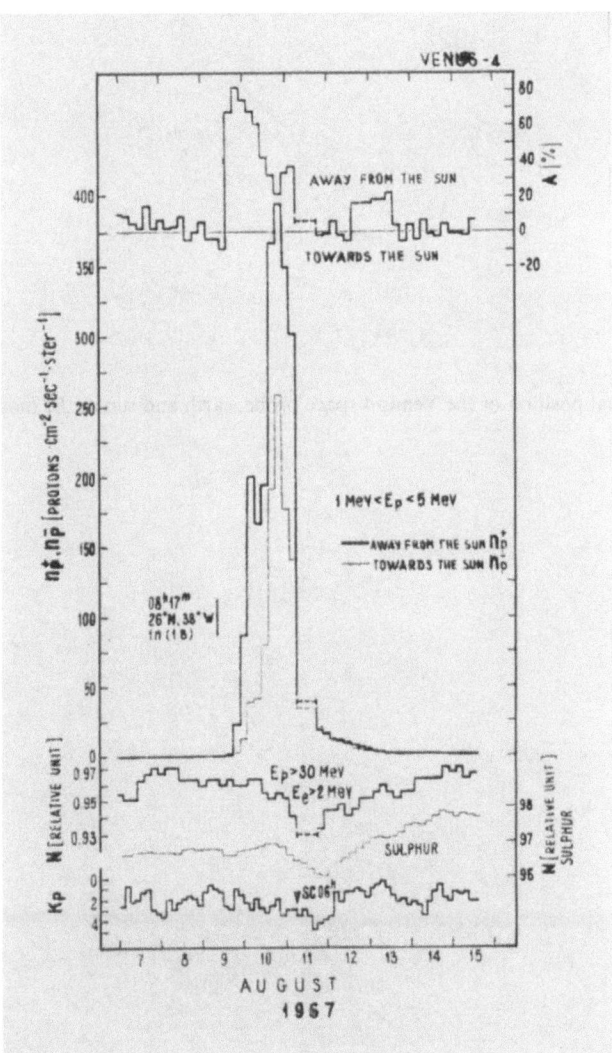

Fig. 4a.   Increase in the intensity of the solar cosmic rays from the flare on the sun on 9 August 1967. The upper curve represents the flux anisotropy. Noted with the vertical line is the moment of the flare on the sun. The upper curve represents the flux anisotropy. Shown below are the intensity, the neutron monitor data, and K_p-index. Shown with the triangle is the moment of the sudden commencement of the magnetic storm (SC).

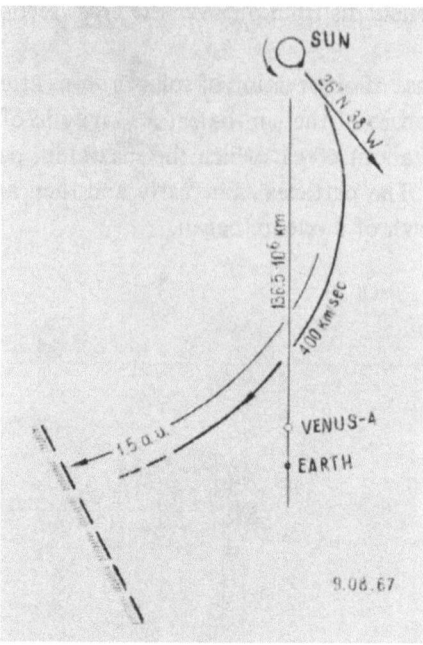

Fig. 4b.   Mutual position of the Venus-4 space probe, earth and sun at the moment of the flare.

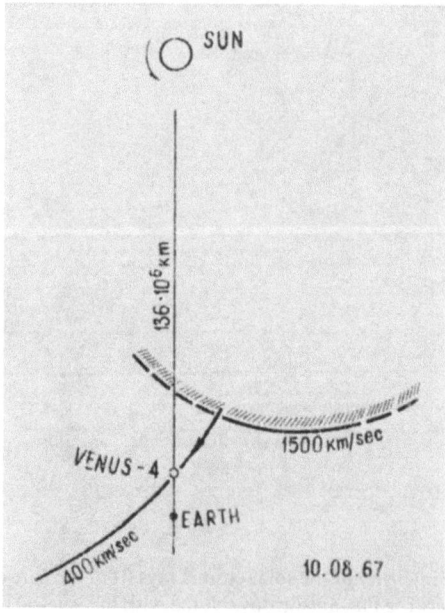

Fig. 4c.   Position of the shock front from the solar flare on 9 August 1967 not long before this front
crossed the Venus-4 and the earth.

The presence of the first peak corresponds to the arrival of the central line of force with maximum population seen by the detector and after this the intensity decreases. The further decrease of the intensity connected with the going away of the main channel is stopped with the arrival of the shock wave and we can observe a Forbush-decrease of the galactic cosmic rays with a new increase in the flux of solar protons arriving behind the front of this wave. A pronounced anisotropy decrease for the particles transferred behind the shock wave can be clearly seen.

The following diagrams (Figures 4b and 4c) illustrate the above statements; one of the diagrams shows the moment of the appearance of a flare on the sun. It can be seen that the main line of force is to the east from the detector. The next diagram represents the moment of matching of this line with Venus-4 in the process of its corotation with the sun. The same diagram shows also the position of the shock wave front.

Some solar cosmic ray bursts were accompanied by an increase in the protons with $E_p > 30$ MeV. In these cases the energy spectra in the 1–30 MeV range were determined. Figure 5 shows the dependence of spectrum indices γ, for various solar cosmic ray bursts, on time elapsed from the moment of the commencement of the flare on the sun. The large circles correspond to the peaks in the intensity of protons with $E_p > 30$ MeV. The small black circles correspond to the peaks in protons with energies $1 < E_p < 5$ MeV.

For the cases of direct arrival, i.e. for the flares on 4 October 1965, 1 and 3 August 1967 in the region when the intensity is increased with time, γ is also increased with time, i.e. the spectrum becomes softer. This is connected with the fact that the protons of lower energies arrive at the observation point later, i.e. the particles are separated

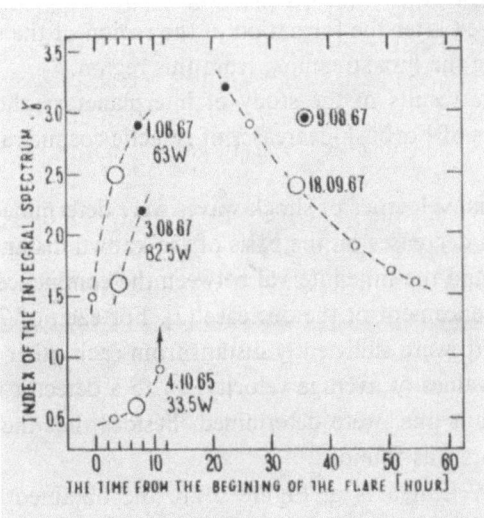

Fig. 5.   The change in the index of the energy spectrum (γ) of the solar cosmic rays for various flares as a function of time elapsed from the beginning of the flare. The large circles correspond to the spectrum indices obtained in the intensity peak, $E_p > 30$ MeV. The small black circles denote those in the proton intensity at 1 MeV $< E_p < 5$ MeV.

in space in accordance with the difference in their velocities determined by kinetic energy. For the flare on 18 September 1967, i.e. for the arrival of solar cosmic rays behind the shock wave front the spectrum becomes more rigid. Such a change in the spectrum may be due to the more rapid fall of the intensity of the lower energy particles [5]. But this fall may be considered as spatial due to more effective penetration of higher energy particles from layer (2) into region (3) (Figure 1).

The rate of changes of the spectral indices and their values in the peaks of flares are different for the three Western flares. This difference may be connected with the longitudinal position of the flares and with different ways of propagation of particles with different energies (assuming that the injection spectrum for all flares was the same). The flare on 1 August 1967 was at optimum longitudinal position for the particle motion along the lines of force and the flare on 4 October 1965 was located in a position more preferable for the diffusion wave propagation. Thus, the protons with $E_p = 1$–5 MeV, when propagating along the lines of force, arrived sufficiently early from the optimum direction in the flare on 1 August 1967 whereas in the flare on 4 October 1965 they markedly fell behind the protons with $E_p > 30$ MeV propagating in the diffusion mode with sufficient effectiveness which resulted in a more rigid spectrum for this flare. Different rates of the change in γ may be explained also by the longitude effect, assuming a small decrease of the diffusion rate for $E_p = > 30$ MeV the flare move Westwards and a comparatively greater increase in the intensity of protons with $E_p = 1$–5 MeV when the flare approaches the optimum direction.

It follows from Figure 5 that the spectra for the localized solar cosmic rays are softer. The same result has been obtained in other work [6]. After a flare on the sun the spectrum becomes softer due to the impulsive character of injection and the rapid going away of the diffusion wave of the rigid solar cosmic rays. This spectrum is likely to be stabilized after the formation of the region of the solar cosmic ray localization replenishing the flux streaming from this region.

Turn now to the results of the study of interplanetary shock wave propagation and to observations of Forbush-decreases of galactic cosmic rays at various distances from the sun.

The average radial velocities of shock waves were determined for the flares accompanied by Forbush-decreases on the basis of the known distance from the sun to the observation point and the time interval between the commencement of a flare on the sun and the commencement of the decrease [7]. For each of 7 cases when the space probes and the earth were sufficiently distant from each other in the radial direction from the sun two values of average velocity, $\bar{V}_1$ to a detector nearest to the sun and $\bar{V}_2$ to a more distant one, were determined. Besides that the average velocity, $\bar{V}_3$, between these points was found.

Shown on the vertical axis in Figure 6 are the obtained values of the average velocities as a function of averaged times of their observations (a) and of averaged velocities at which the observations were carried out (b).

Two values of the velocity, $\bar{V}_1$ and $\bar{V}_2$ correspond to each flare denoted by the given number.

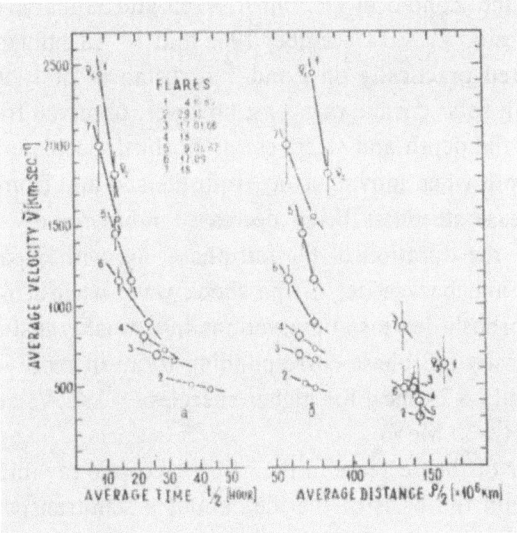

Fig. 6. The average velocities of the shock wave movement from the sun as a function of the average time between the moment of the flare on the sun and the beginning of Forbush-decrease (a) and of the average distance between the detector and the sun (b). $\bar{V}_1$ is the average velocity of the Forbush-decrease arrival to the detector located nearer to the sun; $\bar{V}_2$ is the average velocity of the Forbush-decrease arrival to the detector located farther from the sun; $\bar{V}_3$ is the average velocity of the Forbush-decrease movement between the detectors.

The values of $\bar{V}_3$ are given in the right part of this figure.

It follows from the presented plot that:

(1) The values of velocities $\bar{V}_2$ for each flare are always less than the values of $\bar{V}_1$ which is indicative of a decrease in the velocity with time and distance;

(2) Velocities $\bar{V}_3$, i.e. the velocities near the earth's orbit, are considerably lower than the average velocities $\bar{V}_1$, and $\bar{V}_2$; they are close to the solar wind velocities which indicates that the shock waves damp at a distance of 1 AU from the sun.

(3) If one considers that the average velocity is linearly decreased with time, i.e. in the way shown in Figure 6 with straight lines connecting $\bar{V}_1$, and $\bar{V}_2$ for each flare it is possible to find initial velocities the decelerations α and distances S_0 at which the shock waves decay.

It appeared that the shock waves decayed approximately between the earth and Mars orbits, the decelerations being changed approximately from 140 to 4000 cm sec^{-2} and the value of deceleration depended on the square of the initial velocity, i.e. on the ejection energy.

The shock wave deceleration results in a gradual decrease in the effect of the Forbush-decrease of galactic cosmic rays with distance, due to the decrease in the propagation velocity and the decay of the waves at some distance from the sun.

As the shock waves are being damped the solar cosmic rays localized beyond the shock wave fronts should be gradually released and accumulated near the modulation region boundary. These effects were experimentally confirmed in the region of joint flight of Venus-2 and Zond-3 in December, 1965 and January, 1966. The Forbush-decrease from flares on 29, 30 December 1965 and 17, 18 January 1966 observed on Venus-2 disappeared practically on Zond-3 at distances of 1.29–1.31 AU from the sun; the increase in solar cosmic rays was, however, observed from Zond-3 [2].

The analysis of the depth and steepness of Forbush-decreases reveal a trend to a decrease in the depth when moving away from the sun and from the flare direction; the Forbush-decrease steepness being decreased more rapidly which is connected with a decrease in the duration of the fall phase, i.e. with expansion of the region between the front and back edges of the shock wave when it moves from the sun.

The cases of Forbush-decrease observations in a weakly disturbed medium show a knee in the intensity fall phase corresponding to an increase in the steepness; the increase being small (~ 2 times) for higher energies (> 1 GeV) and great (6–8 times) for lower energies (> 30 MeV).

Such a character of the decrease in the galactic cosmic ray intensity may be qualitatively explained on the basis of the idea about a semitransparent mirror moving from the sun to the detector. Such a mirror will shield the solar cosmic ray flux incident on the detector. After the mirror arrived at the detector the major decrease will take place inside the mirror. It is obvious that for the low energy cosmic rays the transparency coefficient will be less than for high energies. Thus, for the lower energy cosmic rays the intensity fall in front of and behind the mirror will be gentler and steeper respectively than for higher energies.

Turn now to consideration of the results of analysis of the galactic cosmic ray variations, i.e. galactic cosmic ray decreases of small depth caused by magnetic fields of quasi-stationary corpuscular streams from active regions corotating with the Sun.

During the period of Zond-3 and Venus-2 flights, i.e. for about 6 months, about 30 intensity decreases were distinguished, except for Forbush-decreases associated with flares. One of the authors (Pereslegina) connected each decrease with a certain active region on the sun [8]. The confrontation of the moments of successive passages of active regions through the sun's central meridian with the moments of the commencements of the decreases in the galactic cosmic ray intensity (taking account of a 5-day average time shift for the corpuscular stream propagation) indicates a strong connection of these events.

The average radial velocities of the corpuscular stream movement were determined on the basis of the known heliocentric coordinates of the space probes and the delays of the moments of observations of the commencement of the galactic cosmic ray decreases with respect to the moments of passages of active regions through the sun's central meridian. It appeared that a dependence of these velocities on the phase of the region development was observed. Higher plasma velocities correspond to the maximum development of active regions.

Also studied were the 'power' of active region, i.e. the product of its area by brightness and the depth of the galactic cosmic ray decrease. This study showed a sufficiently strong connection between these parameters.

It will be noted that a number of cases fall out of this dependence. These cases, however, are explainable.

For example, the equatorial regions on the sun cause more considerable decreases at a lower power due to the straight line propagation of the plasma.

Thus, the comparison of the galactic cosmic ray variations with active regions on the sun and with dynamical processes of their development reveals broad possibilities for the study of the sun and interplanetary space.

The study of the galactic cosmic rays at two and more points in space supplements this method and increases its possibilities, permitting the large scale structure of the field to be determined.

The method for the determination of the field structure supposes it to be in the form of Archimedes spiral. The characteristic details, for example peaks, minima, steps, etc., were visually marked in the curves of the galactic cosmic ray intensities on Venus-2 and Zond-3 [9]. The moments of recording of these details were determined and time shifts between the details from Venus-2 and Zond-3 were found. A certain fictitious velocity V_c determining a step of the Archimedes spiral was determined for each moment of time. It will be noted that V_c, when characterizing the position of front of the corpuscular stream magnetic field may have values within the $0-\infty$ range in various sections of magnetic tubes bent in the form of a 'zig-zag' or loop. The value of V_c will coincide with the value of the plasma stream only in the undisturbed sections in the case of regular plasma streaming from active regions [8].

Non-flare solar cosmic ray fluxes were observed in the same region of the joint flight of Venus-2 and Zond-3. Individual peaks of solar cosmic rays in Zond-3 are located differently with respect to the galactic cosmic ray variations as compared to Venus-2; the difference being a lesser time shift between the solar cosmic ray peaks as compared to the shifts between the details of galactic cosmic rays at the same points. In accordance with time shifts between the solar cosmic ray fluxes one may calculate velocities V_c. In all cases V_c for the solar cosmic ray fluxes are higher than the corpuscular stream velocities corresponding to them, i.e. the solar cosmic rays propagate along spirals with greater speed. Thus, in the case of non-flare solar cosmic ray flux propagation a Westward 'drift' of solar cosmic rays with respect to the spiral magnetic fields is observed. The same effect was also observed in the region of the joint flight of IMP-III, Zond-3, and Mariner-IV [9].

Thus, the sounding of interplanetary space by low-energy cosmic rays reveals a spiral structure of the interplanetary magnetic fields outcoming from solar active regions corotating with the sun. These fields are distorted by the flare shock wave running on them. The shock waves damp rapidly and decay between the orbits of the earth and Mars. When propagating along the spiral fields the solar cosmic rays undergo a Westward 'drift'. The low-energy solar cosmic rays can be effectively transferred from the flares beyond the shock wave fronts.

The regions of the joint flight of the space probes carrying the same equipment enabled us to determine the radial gradient of cosmic rays.

The treatment of the data has shown that in the end of 1965 the radial gradient for the protons with energies higher than 30 MeV was 5% per 1 AU [10].

In 1967 the intensity of protons with $E_p > 30$ MeV was measured during the flights of the Venus-4 space probe and Explorer-34 artificial earth satellite [11]. These measurements permitted the radial gradient of cosmic rays in the interplanetary space to be measured. These space vehicles flew in the beginning of a new solar activity cycle. During the Venus-4 flight the distance of the space probe from the sun decreased gradually from 150 million to 108 million km. During this period the intensity at the earth's orbit was decreased (according to Explorer-34 data) by about 7.4% whereas the Venus-4 detected only a 3.3% decrease (see Figure 7). It can be seen from the figure that the time variation in the intensity of protons with $E_p > 30$ MeV are of similar character on Venus-4 and Explorer-34 which enables one to assume that the different value of the intensity change is connected with the change in the distance from the sun. If follows from this that in 1967 the cosmic ray intensity was increased when approaching the sun, i.e. the radial gradient of protons with $E_p > 30$ MeV as compared to that in the years of solar activity decrease [10, 18] had another sign and was -13.5% per 1 AU. This result may mean that in the beginning of a new activity cycle the sun is an additional source of high energy protons.

In this case the solar proton flux with $E_p > 30$ MeV which is decreased when moving away from the sun proportionally to the square of the distance from the sun will be

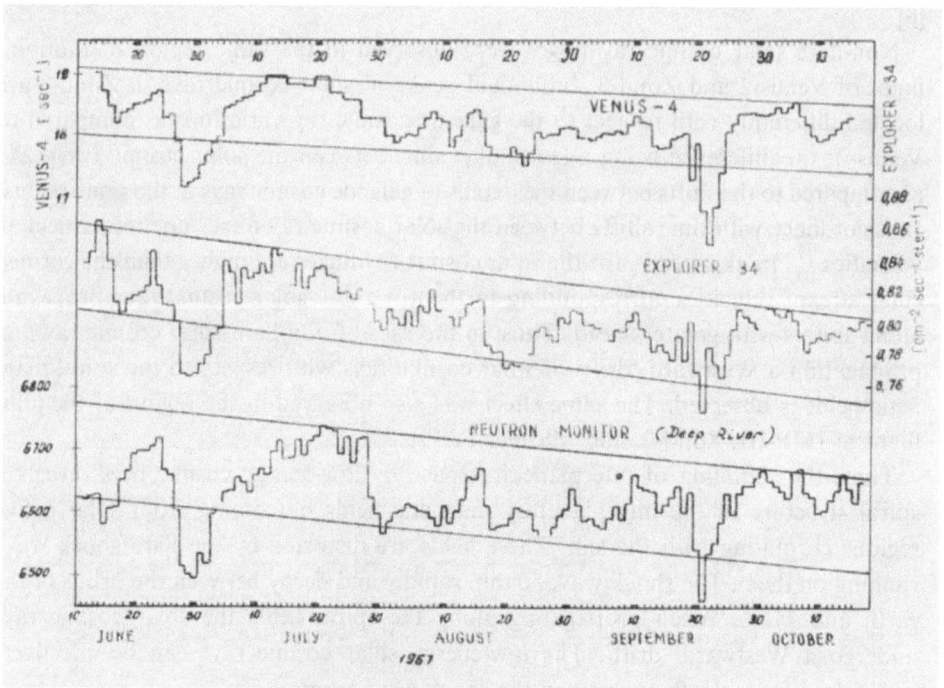

Fig. 7.   The change in the galactic cosmic ray intensity in the period from July to October, 1967 according to the data from Venus-4, Explorer-34 and neutron monitor.

4% of the galactic cosmic ray flux at the earth's orbit. It is natural to suppose that the protons are ejected by the sun not uniformly from the whole surface but only from some of the 'hottest' regions. In this case the 'galactic' cosmic ray flux at the earth's orbit would undergo variations with an amplitude up to 4% and, in general, the variations in the counting rates of a neutron monitor and counter of protons with $E_p > 30$ MeV would be of different character. The readings from the detectors of protons with $E_p > 30$ MeV on Venus-4 and Explorer-34 and the Deep River neutron monitor (see Figure 7) undergo almost the same long-period variations which are due to the same, likely modulation, mechanism. Meantime, the difference in the variations of the intensity protons with $E_p > 30$ MeV and that of protons recorded with the neutron monitor ($\geqslant 5$ GeV) reaches 1–2% and more for short periods in the order of 2–5 days which supports the assumption about the ejection of high energy protons by individual regions on the sun.

Another possibility to explain the negative cosmic ray gradient consists in the assumption that a large cosmic ray intensity gradient exists in the direction perpendicular to the solar equatorial plane*. The prolonged measurements of the galactic cosmic ray fluxes at the earth's orbit are, at the same time, the measurements of the transverse gradient of these fluxes with respect to the solar equatorial plane within $\pm 7°$ angular interval (the angle of inclination of sun's rotation axis to the ecliptic plane). This method permits the conclusion to be drawn that the transverse gradient near the solar equatorial plane is practically absent [19]. The Venus-4 flight is characterized by the fact that the distance of the space probe from the ecliptic plane was added to the distance of the earth from the solar equatorial plane so that the maximum distance of Venus-4 from the solar equatorial plane was about 10°. Thus, if a transverse gradient having a strong non-linearity (in the $\pm 7°$ interval the gradient is equal to zero and when moving further away from the solar equatorial plane is 100% per 1 AU) exists the results from Venus-4 and Explorer-34 may be coordinated. The absence of a transverse gradient in the $\pm 7°$ angular interval may be explained by the fact that the solar magnetic field focussed in the equatorial plane [20] hampers completely the penetration of galactic particles into this region in the directions which are at a considerable angle to the ecliptic plane.

Both explanations set forth above for the increase in the cosmic ray flux detected on Venus-4 when approaching the sun require new events: a stable solar cosmic ray flux or non-linear transverse gradient of galactic cosmic rays. The question as to which of these or which other possibilities are realized in nature is still unsolved.

3. Preliminary Results Obtained in 1969

During the flights of Venus-5 and Venus-6 space probes in the beginning of 1969 the solar and galactic cosmic rays were measured using equipment similar to that installed on Zond-3, Venus-2 and Venus-4.

---

* This explanation was proposed by G. P. Luybimov and N. V. Pereslegina.

This period corresponds approximately to the maximum of the 20th solar activity cycle. Therefore the galactic cosmic ray intensity ($E_p > 30$ MeV) was probably close to the minimum.

This intensity is 15% lower than that measured during the Venus-4 flight in June–October, 1967 and 40% lower than that during the Zond-3 flight which took place at a lower solar activity. During January–May, 1969 the Venus-6 detectors recorded a great number of increases in protons with $E_p = 1$–4 MeV and 6 bursts of solar cosmic rays with energies > 30 MeV (Figure 8).

The increases (9–10 January, 14–16 January, 20–21 January, 9 February, 23–24 February, 2–12 May 1969) have intensities (in the peak) from 0.7 to 10 cm^{-2} sec^{-1}

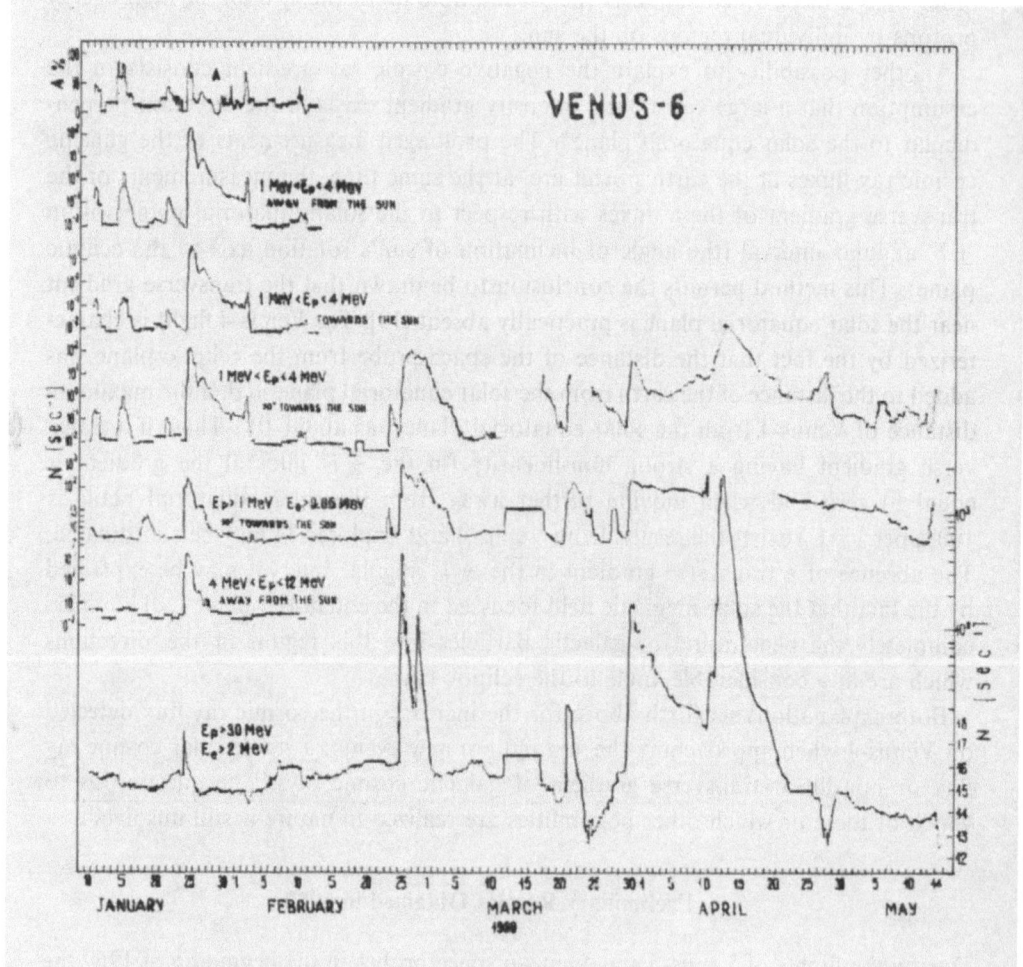

Fig. 8. Intensities of the solar and galactic cosmic rays recorded with detectors of the Venus-6 space probes. The ranges of the recorded energies are shown in the figure. Double scale is used for the curve for $E_p > 30$ MeV and $E_e > 0.05$ MeV: linear scale up to the intensities of 18cts/sec and 13cts/sec respectively and logarithmic scale for higher intensities.

ster^{-1} and are accompanied by small but prolonged Forbush-decreases ($E_p > 30$ MeV) in the galactic cosmic rays.

The analysis of solar activity [11, 12] in this period has shown that the noted galactic cosmic ray decreases are likely to correspond to the passage through the detector of corpuscular streams from 2 groups of powerful active regions located in the Northern hemisphere of the sun at longitude from 60° to 120° and from 230° to 320°.

The observed small proton fluxes are also likely to be connected with their steady streaming from some active regions which is confirmed by gradual dependence of anisotropy at low value of the anisotropy and the absence of the corresponding chromospheric flares.

Beginning from the middle of March the increases in the protons with $E_p = 1$–4 MeV become very frequent, reach a considerable intensity and are due mainly to considerable proton flares.

The first considerable increase began on 24 January 1969 and was recorded with all detectors simultaneously within ± 2 hours. Hence, this increase contained the solar cosmic ray protons with $E_p = 1$–30 MeV and higher. Besides that the electrons with $E_e = 0.05$–2 MeV were also present in the increase. The maximum intensity in this increase was: for 1–4 MeV protons 940 cm^{-2} sec^{-1} ster^{-1}, for 4–12 MeV protons 66 cm^{-2} sec^{-1} ster^{-1}, for > 30 MeV protons 0.13 cm^{-2} sec^{-1} ster^{-1}, for 0.05–2 MeV electrons 2200 cm^{-2} sec^{-1} ster^{-1}.

The estimate of the proton spectrum is indicative of the fact that in the beginning of the increase the energy spectrum was rigid and then began to become softer rapidly and was stabilized. The index of the integral spectrum for the first 16 hours of the flare increased from 1.7 to 3.0 and was stabilized at this level. The whole increase lasted for about 9 days.

Two flares occurred almost simultaneously in the examined active longitude zone less than an hour before the beginning of the increase. The flares had the following characteristics:

(1) Beginning 07^{h}06^m (20°N, 08°W), class 2B.
(2) Beginning 07^{h}10^m (15°N, 50°W), class 1n.

An X-ray burst in the 8–20 Å range was detected at the moment of the occurrence of the flares. In accordance with the model of the solar cosmic ray propagation from a flare on the sun set forth in Section 2 we interpret the form of the examined increase as follows. The protons and electrons injected by flare '2' reach the Venus-5 and Venus-6 detectors in (2 ± 2) hours after the beginning of the flare moving along the lines of force of the interplanetary magnetic field. A great anisotropy of the protons streaming from and to the sun (about 90% for the 1–4 MeV protons) is observed during this period. Since the longitude of the flare is 50°W the central magnetic channel, the intensity in which is maximum, is about 10° to the east from the detectors of the space probes. Therefore the further intensity increase is due to the approach of the central channel corotating with the sun at a velocity of ~ 400 km/sec to the detectors. The intensity peak was observed in (20 ± 2) hours after the beginning of the flare. Then the intensity begins to fall since the central channel has crossed the detectors. The peak of protons with $E_p > 30$ MeV is delayed by (3 ± 2) hours with respect to the solar flare which corresponds to a diffusion coefficient of 3.5×10^{21} cm^2 sec^{-1} at a diffusion velocity of 14000 km/sec. The shock wave formed by flare '1', when propagating in space, reaches the detectors in about 30 hours and is detected as a Forbush-decrease. A pronounced increase in the anisotropy corresponds to this moment. A 1400 km/sec mean velocity of the shock wave corresponds to such delay. The shock wave from flare '2' cannot be recorded since it passes to the West from the detectors. The protons and electrons captured in the trap between the shock wave front and plasma front and then the particles confined between the shock wave front and the sun arrive together with the shock wave from flare '1'. Anisotropy in this flux does not exceed 30%. The intensity decreases not gradually but spasmodically (two leaps at the back edge of the second subincrease) which is indicative of the presence of several magnetic regions inside the zone which serve as 'magnetic semi-transparent partitions'. The whole increase ends in a prolonged part with isotropic distribution of particles and a constant rather high intensity and is finished in an abrupt burst (on 2 February 1969) with an anisotropy reaching 60%. This burst comes abruptly to an end which corresponds in time to the beginning of relative small (4.4%) but rather abrupt (0.36% per hour) Forbush-decrease in the intensity of protons with $E_p > 30$ MeV. The whole decrease lasts for about 1 day (fall and recovery of 12-hour duration each). The Deep River neutron monitor [13] also records a Forbush-decrease in the galactic cosmic rays with a 2.8% depth and abruptness of 0.23% per hour. Analysis of solar data does not reveal a suitable solar flare in the environment of these events. Preliminary study of the active region dynamics permits the conclusion to be drawn that the observed Forbush-decrease is similar to the cases studied earlier [8] which were caused by 'quasi-flares', i.e. the active regions developed rapidly near the central meridian. In fact, such an active region No. 9890 (13°N) passed through the sun's central meridian on 27 January 1969. The activity of this region increased pronouncedly near that moment. The front of the plasma ejected by this active region caused a Forbush-decrease in both solar and galactic cosmic rays, i.e. in the protons with $E_p = 1$–4 MeV from previous flares. The intensity burst of protons with $E_p = 1$–4 MeV observed at the Forbush-decrease front is due to acceleration of these protons [14, 15].

Before turning to description of the next great flare we shall consider an interesting small increase in electrons with $E_e > 0.05$ MeV occurring earlier, on 17 January 1969.

Intensity increase of electrons with $E_e > 0.05$ MeV was observed from 17 to 19 January 1969. The electron flux in the peak was ~ 35 particles cm^{-2} sec^{-1} ster^{-1}. This increase was not accompanied by a marked increase in the proton fluxes with $E_p = 1$–4 MeV and $E_p > 30$ MeV. The peak of the electron increase is delayed by ~ 25 hours with respect to the beginning of the increase. The beginning of the increase coincides with the moment of the solar flare occurring on 17 January 1969 at 17^h03^m (17°N, 45°E) within the measurement accuracy (± 2 hours). The flare had class 1B and was accompanied by powerful radioemission bursts in the centimeter range of wavelengths. It is possible to assume that the above mentioned flare and radioemission burst caused the generation of a very great flux of electrons which diffused, owing to their high mobility, from the region of the flare to Venus-6 whereas the protons, if generated, passed along spiral lines of force of the magnetic field considerably to the east of the space probe location. Proceeding from the delay time of the electron flux peak with respect to the flare on the sun we obtain the diffusion coefficient equal to $\sim 4.10^{20}$ cm^2 sec^{-1} and diffusion velocity of ~ 1700 km/sec. The diffusion velocity is very low and close to the shock wave velocity. The beginning of a small decrease in intensity of the protons with $E_p > 30$ MeV having $\sim 1\%$ depth and one day duration may be noted in the region of the electron increase peak. The increase in the proton flux with $E_p = 1$–4 MeV arriving from the sun with the peak of 0.17 cm^{-2} sec^{-1} ster^{-1} corresponds to this decrease. If all these events are assumed to be connected the observed electron flux increase, they may be explained mainly by the transfer of the electrons behind the shock wave and the diffusion through the wave front during the transfer. Then the diffusion coefficient will be considerably lower.

From 23 February to 6 March 1969 the Venus-5 and Venus-6 detectors were within the corpuscular stream of the same active region group which was responsible for the increase on 24 January–5 February 1969. From 24 to 28 February all detectors recorded a great increase in protons and electrons. This increase is of complex structure with four peaks for protons with $E_p > 30$ MeV and electrons with $E_e > 0.05$ MeV.

The protons with $E_p = 1$–4 MeV show an increase with a single broad peak.

The intensity peaks have the following values:

for the 1–4 MeV protons 800 cm^{-2} sec^{-1} ster^{-1},

the 1st peak for the 30 MeV protons 16 cm^{-2} sec^{-1} ster^{-1},

the 2nd peak for the 30 MeV protons 8 cm^{-2} sec^{-1} ster^{-1},

the 3rd peak for the 30 MeV protons 5 cm^{-2} sec^{-1} ster^{-1},

the 4th peak for the 30 MeV protons 9 cm^{-2} sec^{-1} ster^{-1}.

From 1 to 6 March, 1969 the detectors recorded only the protons with $E_p = 1$–4 MeV the intensity of which was decreased in the same manner as during the second part of the increase of 24 January–2 February 1969.

In the examined passage through the solar disk one of the active regions (No. 9946) of the group was in the state of increased activity. During the period from 24 to 28 February 1969 six considerable flares accompanied by radioemission bursts in a broad wavelength range (from centimeters to meters) occurred in this active region.

The first powerful flare in this region occurred on 24 February at 23^h00^m (12°N 31°W) and had class 2b. The flare was accompanied by II and IV types of radioemission and X-rays. The intensity increase for protons with $E_p > 30$ MeV and electrons with $E_e > 0.05$ MeV began on 24 Feburary 1969 at 24 ± 2^h and that for protons with $E_p = 1$–4 MeV began on 25 February 1969 at 4 ± 2^h. Hence, the particles came immediately after the flare which injected them within a broad angle.

Further flares followed:

on 25 February 1969 at 09^h00^m, 13°N 31°W, 2b;

on 25 February 1969 at 16^h47^m, 12 °N 41 °W, 1;

on 25 February 1969 at 19^h36^m, 12 °N 43 °W, 1b;

on 26 February 1969 at 04^h18^m, 13 °N 45 °W, 1b,

which were accompanied by type IV of radioemission; on 27 February 1969 at 13^h48^m, 13 °N 65 °W, 2b which was accompanied by powerful bursts of II and IV types of radioemission and X-rays.

These flares injected the particles which caused the subsequent increases. We associate the second main peak of protons with $E_p > 30$ MeV (on 27 February 1969) with the flare on 27 February 1969 at 13^h48^m. The comparison of maximum intensities of protons in the 1–4 MeV and >30 MeV ranges gives the index of the integral proton spectrum in the beginning of the increase on 25 February 1969 of about 0.3, i.e. the spectrum is very rigid.

During the second main peak of protons with $E_p > 30$ MeV the index of the integral spectrum was changed from 1.0 to 2.3.

The first peak of protons with $E_p > 30$ MeV (on 25 February 1969) is delayed by (10 ± 2) hours with respect to the solar flare on 24 February 1969 and the second main peak on 27 February 1969 is delayed by (3.5 ± 2) hours with respect to the solar flare occurring on the same day. Respectively, the coefficients and velocities of the diffusion for the first and second cases will be: 0.9×10^{21} cm^2 sec^{-1} at 3900 km/sec and 2.7×10^{21} cm^2 sec^{-1} at 11 000 km/sec. Besides that, the anisotropy during the increase in the protons with $E_p = 1–4$ MeV is small, not exceeding 20%. These data indicate the following mechanism for propagation of the solar cosmic rays from these groups of flares. The protons with $E_p > 30$ MeV travel rapidly to the detectors in a diffusive way and along the lines of force of the spiral magnetic field passing ahead of the shock wave front. The small value of the index of the spectrum and anisotropy is indicative of the fact that the protons with $E_p = 1–4$ MeV are effectively accumulated in the magnetic trap formed by many flares of the examined active region group and only a small portion of them can penetrate through the wall of this trap and penetrate along the spiral lines of force of the magnetic field. The shock wave arrived between 25 and 26 February 1969 and the increase in the protons trapped behind its front is probably noted with the peak of protons with $E_p = 1–4$ MeV and, possibly, with the second peak of protons with $E_p > 30$ MeV. In this case the index of the integral spectrum is increased up to 2.3.

On 25 February 1969 at 09^h42^m the Deep River neutron monitor recorded a 15% intensity increase and the beginning of a Forbush-decrease on 26 February 1969 at 03^h; SC was observed on 26 February 1969 at 01^h58^m. Thus, the average velocity of the shock wave from the flare of 24 February 1969 which began at 09^h00^m corresponds to 1600 km/sec.

The next, less considerable, increase in the protons with $E_p > 30$ MeV occurred on 21 March, 1969 and lasted for about 2 days. On 21 March, 1969 at 13^h02^m a proton flare occurred on the sun (20 °N, 10 °E; class 2b) accompanied by power radioemission burst of II and IV types and an X-ray burst. Apart from this burst two less considerable bursts were observed somewhat earlier: at 09^h30^m (20 °N, 20 °E; class 1n) and at 09^h59^m ($30°S$, $20°E$; class 1b). The increases in the protons with $E_p = 1$–4 MeV and > 30 MeV began simultaneously within ± 2 hours. The peak of the proton flux with $E_p > 30$ MeV is delayed by only (1 ± 2) hours with respect to the peak of the first power proton flare of class 2b. An attempt may be made to explain such early arrival of particles from the eastern flare on the assumption that the lines of the interplanetary magnetic field during this flare were extended as a loop by the plasma ejected by a flare occurring earlier (for example on 18 March 1969). On 23 March 1969 at 18^h27^m the sudden commencement (SC) of a considerable (maximum $K_p = 8$) and prolonged (about a day) magnetic storm was observed on the earth. At the same moment the Deep River neutron monitor recorded a Forbush-decrease. On 23 March 1969 Venus-6 also observed a Forbush-decrease in the intensity of protons with $E_p > 30$ MeV of $\sim 18\%$ depth and with about 1 day duration of the decrease phase at an abruptness of $\geqslant 0.75\%$ per hour. The average velocity of the shock wave calculated from a ~ 53 hour delay of SC with respect to the flare of class 2b on 21 March 1969 is 790 km/sec. The long duration of the decrease phase of the Forbush-decrease is probably connected with the group of three flares and with the power burst of II type.

The next flare of protons with $E_p > 30$ MeV on 30 March 1969 whose flux was ~ 12.6 cm^{-2} sec^{-1} ster^{-1} is close to that of the flare of 25 February 1969 but differs from this flare in a very long duration. The intensity decreased from the peak during about 12 days. During the first ~ 40 hours after the peak the intensity falls according to an exponent with a time constant of ~ 45 hours, during the next ~ 100 hours the constant is ~ 90 hours and then the decrease follows the law $t^{-1.5}$. The intensity increase lasts for (16 ± 2) hours. On 30 March 1969 at 15^h30^m the Deep River neutron monitor recorded a 5% peak at 10 hour duration of the increase.

Two chromospheric eruptions were observed before this solar cosmic ray burst. The first one occurred on 29 March 1969 at 19^h23^m ($10°N$, $54°E$), was of class 1b and was accompanied by a burst in the cm wavelength range and by X-rays. The second flare occurred on 30 March 1969 at 03^h30^m ($26°N$, $25°E$) had class 1n and was accompanied by a II type burst and X-rays. Probably, both solar flares were a source of generation of the observed solar cosmic ray burst. The particles were effectively captured in the trap behind the front of the shock waves from these flares. A comparatively great delay of the peak and slow prolonged intensity fall are explained by inconsiderable transparency of this trap even for the ~ 1 GeV particles.

In the beginning of the increase the index of the integral spectrum is close to unity.

Determination of the velocity and diffusion coefficient for this flare on the basis of the peak delay is meaningless since the major flux of particles arrive by means of the transfer behind the shock front. To determine the above said values one should include the velocity of the shock front which results in an underestimation of the velocity and diffusion coefficient.

The most considerable burst of protons with $E_p > 30$ MeV began on Venus-6 on 11 April 1969. The intensity increase lasted for two days. The flux in the peak reached 177 cm^{-2} sec^{-1} ster^{-1}. The proton flux with $E_p = 1$–4 MeV was $\sim 1.4 \times 10^4$ cm^{-2}

sec^{-1} $ster^{-1}$ (average over 1 day). These values of the fluxes are in agreement with a ~ 1.4 index of the integral spectrum. The total duration of the flare was ~ 12 days.

A powerful burst in the meter wavelength range and a less powerful one in the centimeter range were observed on 10 April 1969 at 10^h56^m. Work [11] does not contain information about considerable solar flares during this period.

A SC and the beginning of Forbush-decrease was observed on the earth on 12 April 1969 at 21^h (Deep River). According to the neutron monitor data the decrease in the galactic cosmic ray intensity was $\sim 6\%$ and lasted for ~ 13 hours which correspond to a decrease abruptness of $\sim 0.45\%$ per hour. The intensity was slowly recovered.

The characteristics of the solar cosmic ray burst considered above and ground events enable one to assume that a power chromospheric eruption in the eastern part of the solar disk was a source of the observed solar cosmic ray burst and that the protons with $E_p > 30$ MeV were effectively retarded behind the front of the shock wave formed by this flare and were transferred in the trap in the region close to the wave front.

The intensity fall is exponential with ~ 13 hour time constant during ~ 20 hours after the peak, further a knee is observed and during ~ 70 hours the time constant is 26 hours and after this the intensity fall becomes slower. Such a character of the intensity decrease is connected with the structure of the space after the solar flare, i.e. the regions with different inclinations of the intensity decrease front are in a good agreement with the regions of different steepnesses of the recovery of the galactic cosmic ray intensity after the Forbush-decrease according to the Deep River monitor data.

This solar cosmic ray flare can be compared with its observations from Pioneer-VI and -VII published in [11].

Figure 9a shows as a function of time the intensity of the protons of various energies measured on Venus-6 (stepped thick curve), Pioneer-VII (thin fractured line) and Pioneer-VI (dash line). According to [11] during this flare the Pioneer-VI was at a distance of 1 AU from the (sun) at $\sim 155°E$ with respect to the earth and Pioneer-VII was, respectively at ~ 1.1 AU and $\sim 95°E$. At that time the Venus-6 was at a distance of $\sim 120 \times 10^6$ km from the sun and at angle of $\sim 9°W$ to the earth (see Figure 9b). Thus, the assumed solar flare was to the east with respect to Venus-6, central with respect to Pioneer-VII and to the west with respect to Pioneer-VI.

Figure 9 illustrates well the model of the solar cosmic ray propagation from the flare even for the particles of higher (than 1–5 MeV) energies as set forth in Section 2. The first peak in the end of 10 and beginning of 11 April, 1969 observed from Pioneer-VI and -VII corresponds to the early arrival of the diffusion wave along the lines of force of the magnetic field. Only the beginning of the increase corresponds to this moment on Venus-6 since the velocity of the diffusion across the lines of force is considerably lower than that along them. In the beginning of 13 April 1969 only the peak of the particles transferred behind the shock front was observed on Venus-6 (the first leap after the beginning of the increase may contain a small diffusion peak but this is not seen because of insufficient time resolution in this region). In the beginning of 14 April 1969 the second peak was distinctly seen on Pioneer-VII in the range of 0.6–13 and especially 13–70 MeV. This second peak is delayed on Pioneer-VII with respect to the peak on Venus-6 due to a great radial distance of Pioneer-VII

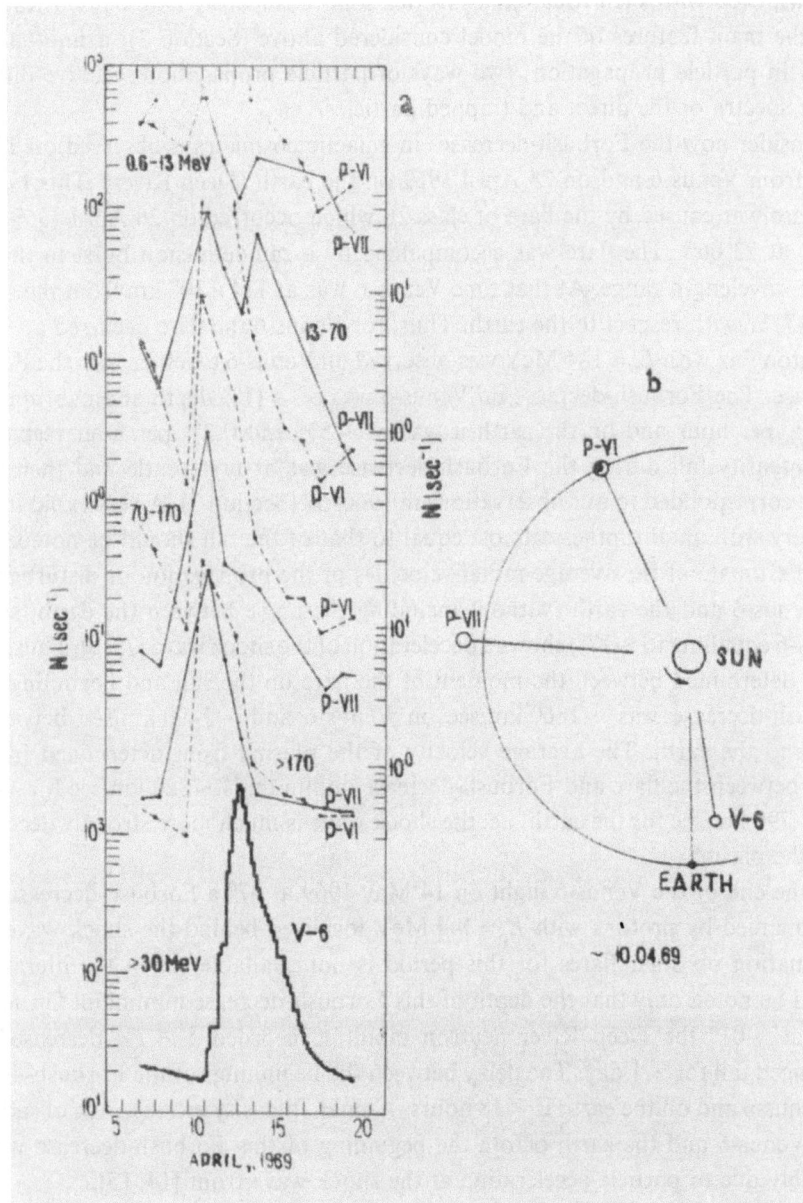

Fig. 9. (a) Intensities of the solar cosmic rays recorded with the detectors of the Venus-6 space probe (lower curve), Pioneer-VI (dotted line) and Pioneer-VII (solid line). – (b) Mutual position of the space probes and the earth in the ecliptic plane at the moment of the recording of the examined flare.

from the sun. A knee on the fall of the intensity of the first peak for the 70–170 MeV protons can be seen on Pioneer-VI in the end of 12 April 1969. This knee is probably indicative of the presence of the second peak for the particles retarded behind the shock front. As can be seen from the curves presented for Pioneer-VII the particle spectrum in the second peak was softer than that in the first peak.

Thus, even with such rough analysis this solar cosmic ray flare demonstrates very well the main features of the model considered above (Section 2): azimuthal asymmetry in particle propagation; two ways of particle propagation; relative difference in the spectra of the direct and trapped particles.

Consider now the Forbush-decreases in galactic cosmic rays observed on 27 April 1969 from Venus-6 and on 28 April 1969 on the earth (Deep River). This Forbush-decrease was caused by the flare of class 2b which occurred on 26 April 1969 (07°N, 40°E) at 23^h06^m. The flare was accompanied by a radioemission burst in the centimeter wavelength range. At that time Venus-6 was at 114×10^6 km from the sun and at $\sim 17°W$ with respect to the earth. Thus, for Venus-6 the flare occurred at $\sim 57°E$. A proton flux with $E_p = 1$–4 MeV was observed on Venus-6 together with the Forbush-decrease. The Forbush-decrease on Venus-6 was of $\geqslant 11\%$ depth at an abruptness of $\sim 0.6\%$ per hour and on the earth it was of $\sim 5\%$ and 0.2% per hour respectively. The intensity fall during the Forbush-decrease was at first gentle and then abrupt which corresponded to our observations in 1966–67 (Section 2). A very rapid intensity recovery with an abruptness almost equal to that of the fall should be noted.

The estimate of the average radial velocities of the propagation of disturbances to the Venus-6 and the earth (without including the angle between the earth, sun and Venus-6 equalling to $\sim 17°$) shows a deceleration of the shock wave [7, 16]. Thus, the velocity determined between the moment of the flare on the sun and beginning of the Forbush-decrease was ~ 1860 km/sec on Venus-6 and ~ 1490 km/sec between the flare and the earth. The average velocity of the plasma front determined from the delay between the flare and Forbush-decrease minimum is ~ 820 km/sec for Venus-6 and ~ 790 km/sec for the earth, i.e. the shock wave is much more strongly decelerated than the plasma.

In the end of the Venus-6 flight on 14 May 1969 at 07^h a Forbush-decrease began accompanied by protons with $E_p = 1$–4 MeV localized behind the shock wave front. Information on solar flares for this period is not available yet in the literature. It should be noted only that the depth of this Forbush-decrease minimum. On 15 May, 1969 at $\sim 01^h$ the Deep River neutron monitor recorded a $\sim 7\%$ decrease and a prolonged fall for ~ 1 day. The delay between the beginnings of the Forbush-decrease on Venus-6 and on the earth is ~ 18 hours. A small intensity increase was observed on both Venus-6 and the earth before the beginning of this Forbush-decrease which is probably due to particle acceleration at the shock wave front [14, 15].

It should be noted in concluding the description of preliminary results obtained from Venus-5 and -6 that during this period of the solar cycle: (1) A more rigid spectrum of solar cosmic rays due to more effective generation of high energy particles is observed. (2) The solar cosmic ray particles are effectively (for a long period) retarded at radial distance of 1 AU from the sun. (3) Considerable proton fluxes with $E_p > 30$ MeV are retarded behind the shock wave fronts. (4) Particle acceleration in the interplanetary space is observed. It should be also noted that the character of the intensity change in the considerable flares (on 30 March and 11 April 1969) may be indicative of the realization of Gold's model [17] for these cases. To elucidate this, however, it is necessary to measure the magnetic field and particle anisotropy.

4. On Propagation of Protons with Energy of 1.5 MeV Generated During Solar Flares

The problem of the injection and propagation of solar protons in the interplanetary space is of considerable interest in connection with the recent studies of cosmic ray bursts detected at distances in the order of 1 AU [9, 21, 22].

It was assumed when solving the problem about the motion of protons injected by a solar flare that a regular magnetic field which is changed depending on distance to the sun as $1/R^2$ existed in the interplanetary space.

Thus, the azimuthal component of the regular solar field was not included by us. Further we assumed that in this regular magnetic field of the sun inhomogeneities existed the mean distance between which corresponded to the length of the free path l.

When a particle moves between the inhomogeneities its magnetic momentum is conserved which determines the trajectory and pitch angle of the particle. When a particle interacts with magnetic field inhomogeneities it is scattered and its magnetic momentum is immediately changed without changing the energy. We characterized an elementary scattering act by the average scattering angle σ. The transport path, i.e. the distance sufficient to scatter of a particle within angle $\pi/2$ will be determined in our case by the formula

$$\Lambda = l(\pi/2\sigma)^2 .$$

We assumed also that the particles were injected into a sufficiently considerable group of lines of force the transverse dimension of which was much more than the Larmour radius, of the examined particles. Because of this we did not consider the transverse diffusion but included only motion and scattering of particles along the lines of force. The problem was solved using Monte-Carlo method. The proton paths between magnetic field inhomogeneities and the angles of scattering over inhomogeneities were successively drawn. Such an approach to the solution of the problem permits various conditions of particle propagation in the interplanetary medium to be easily simulated including the case when the value of the transport path depends on distance.

The initial and boundary conditions were present as follows. It was assumed that at the moment of injection a proton was at a distance of 10^{-2} AU from the center but if it moved away to a distance greater than 2 AU it was removed from consideration. Having traced a sufficient number of histories we found the average proton number at a distance of 1 AU from the sun which crossed a line of force at angle θ in the angular range for the time Δt. In all versions of calculations the angular range was assumed to be $\Delta\theta = \pi/3$. The value was varied depending on the transport path value. Some results of calculations carried out with such stating of the problem are presented in [4a, 23, 24].

The values of the transport path $\Lambda = 0.1$, 0.3, 1.0 AU were considered by us at the greatest length. At first we considered instantaneous proton injection.

Figure 10 presents histograms of the angular distribution of protons at various time moments for the proton transport path of 1 AU (Figure 10 has been plotted according to the results of 5000 histories). It can be seen from Figure 10 that when the flare intensity is high the major portion of protons stream within the 30–60° interval of angles to the line of force. When the proton intensity is decreased the particles at first fill more uniformly the front hemisphere and then the distribution becomes isotropic. (It will be noted that the distribution peak is at 45° angles due to the fact that the particle number is related to the angular range and not to the solid angle.)

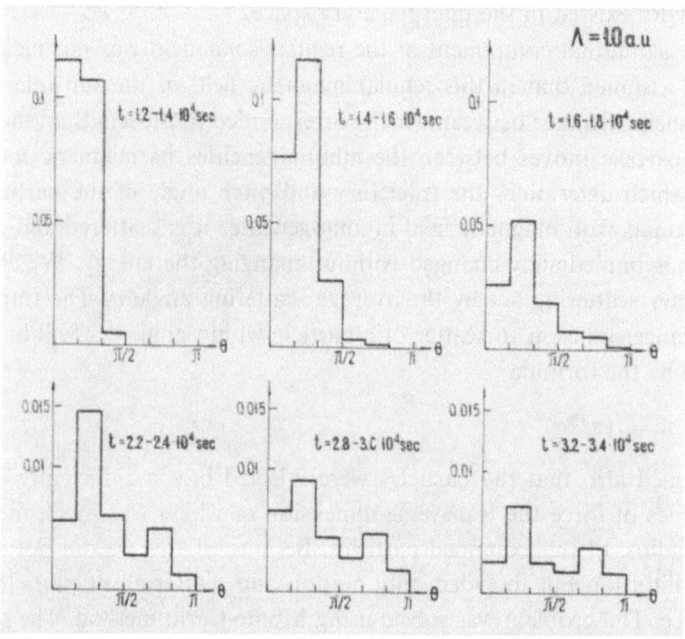

Fig. 10. Histogram of the angular distributions of protons with energy of 1 MeV at a distance of 1 AU for various moments at the transport path $\Lambda = 1$ AU. The histogram has been calculated using Monte-Carlo method, over 5000 histories. The proton injection was assumed to be at a distance of 10^{-2} AU from the sun at an angle of 30° to the magnetic line of force.

Figure 11 presents the time dependence of the proton intensity for three different values of the transport path length $\Lambda = 1.0$, 0.3, 0.1 AU (for $\Lambda = 1.0$ Figure 11 has been plotted according to the results of 5000 histories; for $\Lambda = 0.3$ and 0.1 AU it has been plotted according to the results of 250 histories). As it can be seen from Figure 11 the time when the intensity peak sets in respectively 1.3×10^4 sec, 3×10^4 sec, 7×10^4 sec. (The time of the direct arrival of protons with energies of ~ 1 MeV is

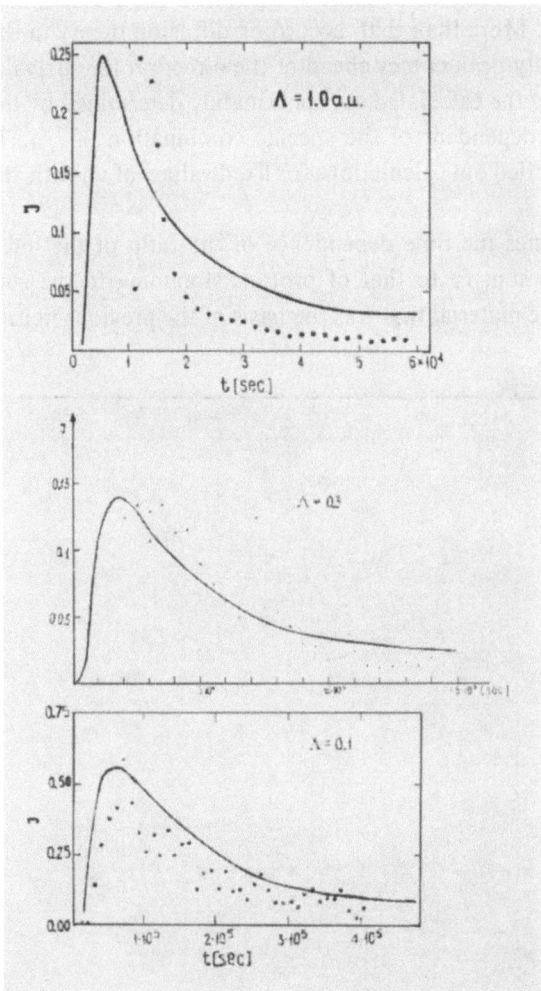

Fig. 11.   Time dependence of the proton intensity for three values of the transport paths $\Lambda = 1.0$ AU, $\Lambda = 0.3$ AU, $\Lambda = 0.1$ AU. The points represent the data obtained with Monte-Carlo method. The solid curve shows the data according to the diffusion theory.

$\sim 10^4$ sec.) The flare intensity is decreased by a factor of 3 as compared to the intensity in the peak during the periods of 3×10^3 sec, 5×10^4 sec, 1×10^5 sec for the paths $\Lambda = 1.0, 0.3, 0.1$ AU respectively. The results of our work make it possible to elucidate the truthfulness of the use of the diffusion theory to solve the problem of the proton propagation in the solar magnetic field. Figure 11 presents the curves plotted on the basis of the diffusion theory. It can be seen from the comparison of these curves with the results of our calculations that the diffusion theory may be used for the transport path $\Lambda = 0.1$ AU. The agreement is violated when $\Lambda = 0.3$ AU whereas for $\Lambda = 1.0$ AU the time dependences have nothing in common. It is expedient to note here that according to the results of the diffusion theory in the cases of $\Lambda = 1.0$, 0.3 AU the intensity may increase considerably before the protons streaming from the sun reach the examined point. More than that, according diffusion theory in the case of $\Lambda = 1.0$ AU even the intensity peak comes ahead of the direct proton arrival. It will be noted that the character of the calculated curves is mainly determined by the value of transport path and is independent of the specific combination of l, σ. To elucidate this circumstance we carried out calculations for fixed values of Λ with different values of l and σ.

Figure 12 illustrates the time dependence of the ratio of the intensity of protons streaming from the sun, I_+ to that of protons streaming to the sun I_-, plotted on the basis of the same material that was the basis of the previous figures. The character

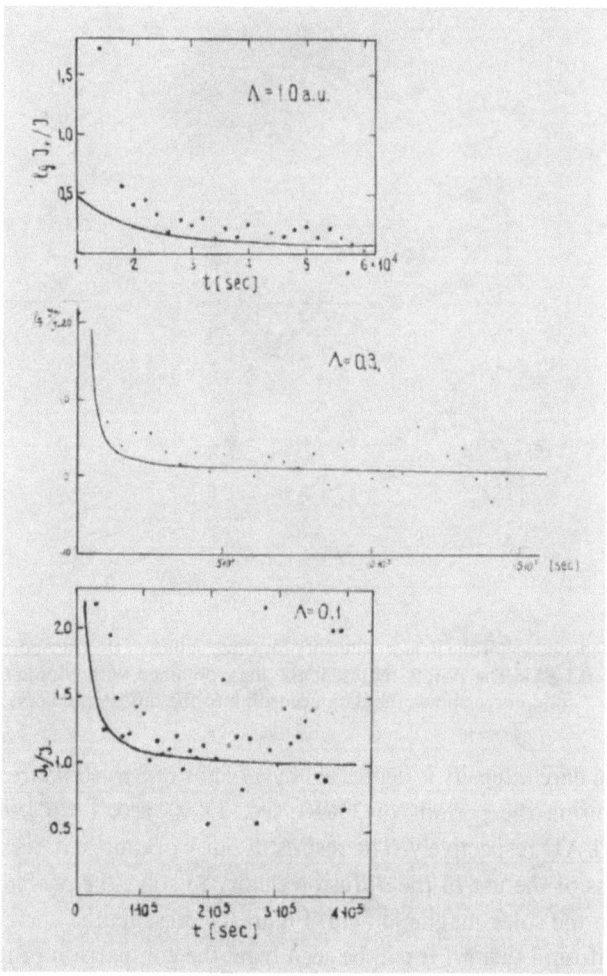

Fig. 12.   Time dependence of the flux anisotropy for the values of transport paths $\Lambda = 1.0$ AU, $\Lambda = 0.3$ AU, $\Lambda = 0.1$ AU. The anisotropy was determined as the ratio I_+/I_- i.e. that of the flux streaming from the sun to the flux streaming to the sun. The points represent the calculation with Monte-Carlo method, solid line shows the data according to the diffusion theory.

of time dependence of the anisotropy is also determined by the value of transport path. For the transport path $\Lambda = 1$ AU the ratio of the number of protons streaming from the sun to the number of protons streaming to the sun remains considerable during the whole flare. For the transport paths $\Lambda = 0.1$, 0.3 AU this ratio is close to unity during almost the whole flare except for the very initial phase. Figure 12 presents the plots of such ratios of proton flux streaming from the sun to the flux of the returning protons as a function of time on the basis of the diffusion theory (solid lines). For the transport path $\Lambda = 0.1$ AU the agreement is satisfactory which confirms once more the truthfulness of the use of the diffusion theory in this case.

The data on the detection of proton flares for $E_p \gtrsim 1$ MeV near the earth's orbit presented above are indicative of the excessive variety of flares. Of most interest are the most intensive proton flares with considerable anisotropy $A \approx 50$–80%.

$$A = (I_+ - I_-)/(I_+ + I_-).$$

Such flares may last for several days. If one proceeds from the assumption about the constancy of the transport path then, according to our calculations, the path $\Lambda = 1.0$ AU corresponds to the flares with larger anisotropy. But at such a value of the path in the case of instantaneous injection the flare lasts for only $\sim 10^4$ sec.

This provokes the idea that either the proton injection by the Sun is not instantaneous or the transport path is not constant and depends on the distance to the sun. In fact, if the transport path is dependent of the distance and is so dependent that the transport path length increases with increasing distance (there are physical grounds for such assumption) then even at instantaneous injection the particles will be retarded near the sun for a long time which is in practice equivalent to the smearing of the injection function in time. When reaching the observation point the particle distribution will have a considerable anisotropy due to an increase in the occurrence frequency of collisions with inhomogeneities. It is natural that in order to distinguish these two possibilities it is necessary to carry out simultaneous measurements of particle fluxes and their anisotropy at several points of interplanetary space.

The problem about propagation of the protons injected by the sun for a period of 5×10^4 sec uniformly during the whole injection period (rectangle impulse) has been considered on the basis of the calculation that had been already carried out. The dependence obtained as a result of examination of instantaneous flare is a function of the source of the problem. Figure 13 presents the time dependence of the intensity for the same values of transport path $\Lambda = 1.0$, 0.3, 0.1 AU. (I_+ is denoted as $\times$; I_- is denoted as $\bullet$; $I_+ + I_-$ is denoted as $+$). In the examined case of the rectangle impulse of injection one may distinctly see in what manner the shape of the source function determines the character of the time dependence of the intensity.

Thus for example, in the case of the injection with rectangle impulse at $\Lambda = 1$ AU the intensity is rapidly increased in the beginning of the flare and is even more rapidly decreased in the end of the flare. Such picture can be explained by the character of proton flare from an instantaneous source: an abrupt increase in the beginning of the process and a slower decrease. The plateau on the curves corresponding to the trans-

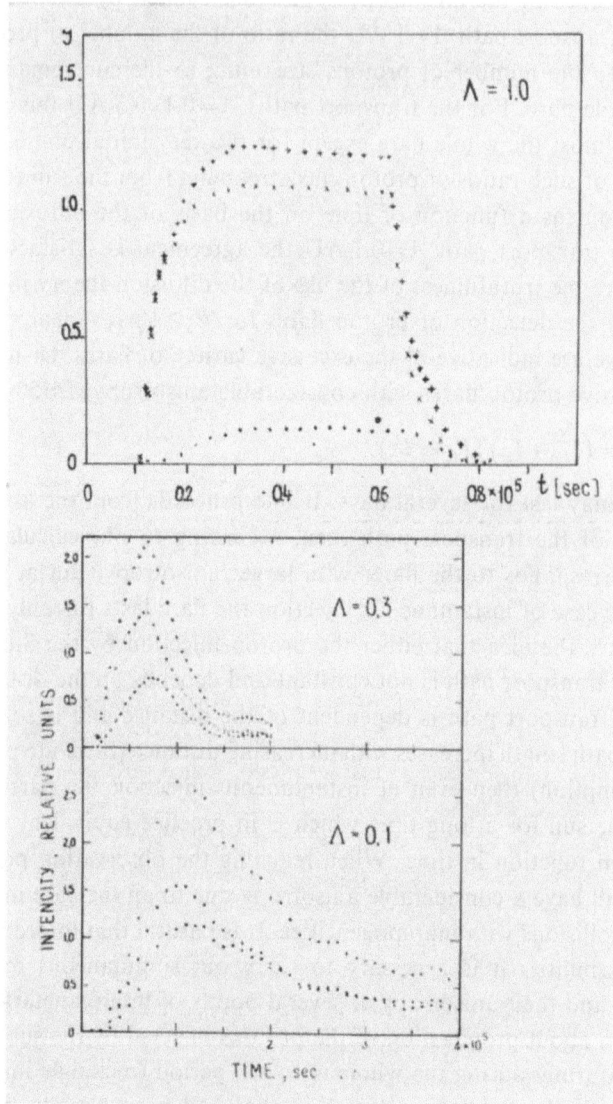

Fig. 13. Relative changes in the intensity with time for constant injection during 5×10^5 sec for the values of the transport paths $\varLambda = 1$ AU, $\varLambda = 0.3$ AU, $\varLambda = 0.1$ AU.

port path $\varLambda = 1.0$ AU is connected with the fact that the injection occurs during a period longer than the period during which the flare from an instantaneous source damps.

As it should be expected in the case of prolonged injection the flare is the more prolonged the longer is the period of a flare from instantaneous injection (i.e. the shorter is the transport path length). The dependence of anisotropy on the value of transport path is illustrated in Figure 14. For the transport path $\varLambda = 1.0$ AU the aniso- tropy is large during the phase of the flare when the proton intensity is increased or

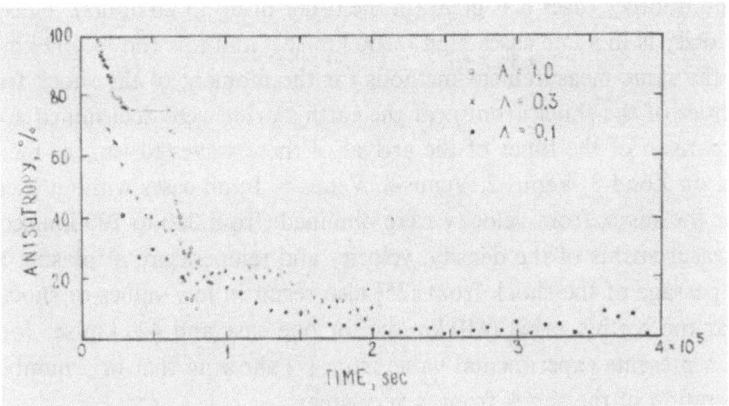

Fig. 14.   Time dependence of the anisotropy for the values of the transport paths $\varLambda = 1$ AU, $\varLambda = 0.3$ AU, $\varLambda = 0.1$ AU.

remains constant. It will be noted that the time dependence of the intensity of protons streaming to the sun is similar to the time dependence of the total intensity.

One may suggest on the basis of the calculation results that the observed flares of protons with $E_p \approx 1$–5 MeV of several days duration and with larger anisotropy during the whole flare correspond to prolonged injection of protons and their further propagation with a transport path $\varLambda = 1.0$ AU. At the same time the form of the experimentally detected flares is not probably in agreement with the form of the flares obtained in calculations. For example, for some detected flares the proton intensity increases more rapidly and decreases more slowly than it is shown in Figure 13. In our opinion an agreement between the experiment and the results of our calculations may be obtained if the injection by the rectangular impulse is rejected and a time dependence is found for each flare at a constant transport path or a suitable dependence of the transport path on the distance is selected.

5. Necessity for Particle Acceleration to Explain the Shock Wave Deceleration and Charakhchyan's Effect

As it was indicated in Section 2, it was shown in [7, 25] that the shock waves appearing as a result of flares on the sun are considerably decelerated when passing through the interplanetary medium.

This conclusion was drawn by measuring the time of the arrival of the shock front according to the measurements of:

(1) beginning of a Forbush-decrease in cosmic rays [7];

(2) sudden commencement of magnetic storms [7, 25];

(3) spasmodic change of the properties of interplanetary plasma [25].

The analysis of the events observed on the sun permitted the observed shock waves to be matched to certain flares on the sun. The time of the shock front arrival to a meter located at a distance of about 1 AU is in some cases small (in the order of 20 hours) and in other cases it is great (in the order of up to 80 hours). Thus, the mean shock velocity is in some cases high (2500 km/sec) and low (up to 600 km/sec).

Using the same measurement methods for the moment of the shock front arrival the velocities of the shock front near the earth's orbit were determined according to intercomparison of the times of the arrival of these waves to various meters on the earth and on Zond-3, Venus-2, Venus-4, Venus-6. In all cases without exception low values for the shock front velocity were obtained (from 300 to 700 km/sec).

The measurements of the density, velocity and temperature of plasma before and after the passage of the shock front [25] also result in low values of shock front velocity near the earth's orbit (410 km/sec for one case and 420 km/sec for another).

Figure 6 presents experimental value from [7] showing that in a number of cases the deceleration of the shock front is very great.

For theoretical reasons one should have expected much greater deceleration of the shock waves.

In fact, in *Mechanics of Compact Media*, p. 482, Landau and Lifshits considered the deceleration of the shock front obtained works by L. I. Sedov for the medium with constant density subject to practically instantaneous energy release within a region the size of which is small as compared to the size of the region within which the shock wave is examined. The shock wave properties including the shock front velocity may be a function of the density of medium (ϱ), energy of burst (E) time variables (t) and distance from the point of the burst (r_0). Only a single dimensionless parameter $\xi = r_0 (\varrho_1/Et^2)^{1/5}$ may be composed of these values. Thus, it follows from the most general considerations that the shock front velocity $V = dr/dt \sim r^{-1.5}$ at $\varrho = \text{const} \neq f(r_0)$. If we take into account that the density in the interplanetary medium is decreased inversely proportional to the square of the distance (i.e. $\varrho_1 \sim r_0^{-2}$) we obtain on the basis of the same considerations that $V \sim r_0^{-0.5}$.

A similar slow dependence of the shock front velocity was obtained by Parker [5]. Parker inserts the parameter λ including the change in the shock wave energy with time. At $\lambda = \frac{3}{2}$ the shock wave energy is constant which corresponds to the blast character of the shock wave formation. At $\lambda < \frac{3}{2}$ the shock wave energy is increased with time which corresponds to the pushing of the shock wave by plasma ejected by the sun at high velocities already after the moment of the shock wave formation. Parker obtained the shock front velocity $V \sim r^{-\alpha}$ where $\alpha = \lambda - 1$. At $\lambda = \frac{3}{2}$ we obtained $\alpha = 0.5$, as above. At $\lambda < \frac{3}{2}$, $\alpha < 0.5$ since the decrease in the shock front velocity due to interaction with interplanetary medium is compensated with pushing of the shock wave by plasma.

From the experimental data [7] shown in Figure 6 for the flare Nos. 1, 5, 6, 7 we obtain $\alpha = 3.5, 3.3, 2, 1.5$ respectively assuming that $V \sim r^{-\alpha}$ and the least possible value (200 km/sec) for the solar wind velocity at the moment preceding a flare.

To explain this disagreement between the experiment $(\alpha = 3.5 - 1.5)$ and theory $(\alpha = 0.5)$ far going assumptions should be made.

One of such assumptions is the hypothesis set forth by Tverskoy [4b] on the particle acceleration with respect to high energies $(\gtrsim 10^5 \text{ eV})$ in the interplanetary space due to the presence of considerable magnetic field variations with an inhomogeneity scale of the order of magnitude of the Larmour radius of the accelerated particles in the shock waves. In this case the disagreement between the theory and experiment is settled in the way that the particle acceleration takes so high an energy from the shock wave that this results in the shock wave deceleration. If this hypothesis is adopted Charakhchyan's effect can be first explained.

As far back as 1959 Charakhchyan *et al.* [6] observed first the coincidence between the time of charged particle arrival and the beginning of Forbush-decrease due to the shock wave arrival. This hypothesis explains the presence of shock waves with only low velocities near *t* the earth's orbit by the fact that at high velocities of shock wave the particles are so strongly accelerated that the shock front velocity is always decreased at 1 AU down to the values corresponding to the shock wave with a low Mach number. To analyze the experimental data obtained in [7] and to compare them

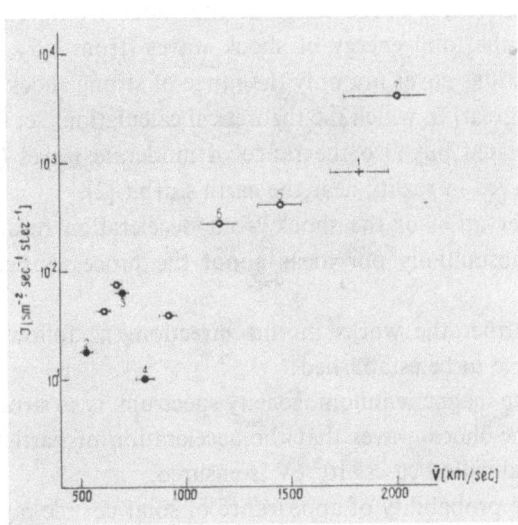

Fig. 15. Dependence of the solar cosmic ray fluxes with energies of 1–5 MeV on the shock front velocity for the cases presented in Figure 6. The fluxes for cases 2, 3, 4 have been obtained by recalculating the Pioneer-VI data for the energy spectrum $N (> E) \sim E^{-3}$. For case 1 the data have been normalized for $E_p > 30$ MeV in accordance with case 5. Three non-numbered cases have been obtained from the Venus-4 data.

with the hypotheses set forth above Figure 15 presents the dependence of the flux (j) of accelerated particles with energies from 1 to 5 MeV as a function of the shock front velocity. Only the lower limit of the particle flux could be evaluated for flares Nos. 1, 2, 3, 4 (Figure 6). Therefore, the value of the flux for flares Nos. 2, 3, 4 was obtained by rescaling on the basis of simultaneous measurements carried out on Pioneer-VI for the protons with $E_p = 7.5$–45 MeV and that for flares No. 1 was obtained from the data of Geiger counter on Zond-3. It was adopted for both rescalings that the proton spectrum could be represented in the form $N(>E) \sim E^{-3}$ where E

is the kinetic energy of protons. As it can be seen from Figure 15 an explicit dependence of N on V is present in the range of change in V from 500 to 2500 km/sec.

The data presented in Figure 15 support Tverskoy's hypothesis set forth above [4b]. In fact, if the particle acceleration is abruptly increased one may understand the strong deceleration of the shock waves and practical absence of shock waves with velocities higher than 700 km/sec near the earth's orbit.

It should be noted that the total energy of particles with energies of $\gtrsim 10^6$ eV is about 2 orders of magnitude lower than the shock wave energy. Therefore, to explain the shock wave deceleration by particle acceleration one should admit a very strong acceleration of particles with energies of $\sim 10^5$ eV (for example at the integral spectrum with $\gamma = 3$ as it takes place for higher energies) which, to the point, also follows from Tverskoy's theory.

It should be noted in conclusion that in work [26] appearing recently very important numerical calculations were carried out for the shock wave propagation in the interplanetary space in a broad range of the shock front velocities (from 3000 to 500 km/sec) and the total energy of shock waves (from 1.6×10^{33} to 5×10^{30} erg). Thus, the calculations cover not only the range of strong shock waves (for which the Mach number is great) to which the theoretical calculations set forth in the beginning of this report related but also the range of moderate waves (Mach numbers ~ 2) which were observed in reality near the earth's orbit [2].

Thus, the observation of the shock front deceleration results in the hypotheses which change substantially our ideas about the processes occurring in the interplanetary space.

To develop further the works in this direction the following points would be extremely desirable to be established:

(1) Whether the magnetic inhomogeneity spectrum, is so strongly increased during the passage of the shock waves that the acceleration of particles up to energies of 10^6 eV and, in individual cases, 10^8 eV is ensured.

(2) What is the probability of appearance of solar cosmic rays during observation of the shock wave using the plasma methods.

(3) What is the probability of observation of the shock wave using the plasma methods during appearance of solar cosmic rays.

References

[1] S. N. Vernov, N. N. Kontor, S. N. Kuznetsov, Yu. I. Logatchev, G. P. Lyubimov, and N. V. Pereslegina: 1968, 'Observations of the solar cosmic ray bursts and Forbush-decreases of the galactic cosmic rays in space from 1965 to 1967', Report at *International Symposium on Solar-Terrestrial Physics*, Crimea, U.S.S.R., 1968.
[2] S. N. Vernov, G. P. Lyubimov, and N. V. Pereslegina: 1969, 'The structure of interplanetary space according to low-energy cosmic ray observations in 1965–1967', Report at VIth summer school of cosmophysicists, Apatity.
[3] Sh. Sh. Dolginov, E. G. Eroshenko, and L. N. Zhuzgov: 1968, *Space Explorations*, No. 4.
[4] K. I. Gringaus and E. K. Solomatina: 1968, *Space Explorations*, No. 4.
[4a] S. N. Vernov, E. V. Gorchakov, and G. A. Timofeev: 1969, *Geomagnetism and Aeronomy*, No. 5.
[4b] B. A. Tverskoy: 1967, *Doklady Akad. Nauk S.S.S.R.* **53**, 1417.

[5] E. K. Parker: 1965, *Dynamical Processes in the Interplanetary Medium*, Mir, Moscow.
[6] A. N. Charakhchyan, V. F. Tulinov, and T. N. Charakhchyan: 1961, *Zh. Eksp. Teor. Fiz.* **41**, 735.
[7] G. P. Lyubimov: 1968, Astronomic Circular, Academy of Sciences of U.S.S.R., No. 488, p. 4.
[8] S. N. Vernov, Yu. I. Logachev, G. P. Lyubimov, and N. V. Pereslegina: 1968, 'Galactic cosmic ray variations and quasi-stationary streams of solar plasma in space', Report at *International Symposium on Solar-Terrestrial Physics*, Crimea, U.S.S.R., 1968.
[9] S. N. Vernov, A. E. Chudakov, P. V. Vakulov, E. V. Gorchakov, N. N. Kontor, S. N. Kuznetsov, Yu. I. Logachev, G. P. Lyubimov, A. G. Nikolaev, N. V. Pereslegina, and B. A. Tverskoy: 1968, *Proc. V All-Union summer school on cosmophysics*, Apatity, p. 5.
[10] S. N. Vernov, A. E. Chudakov, P. V. Vakulov, Yu. I. Logachev, G. P. Lyubimov, and N. V. Pereslegina: 1967, *Izv. Akad. Nauk S.S.S.R.* **31**, 1259.
[11] Compilation of *Solar-Geophys. Data*, Boulder, Colo.
[12] *Map of the Sun*, Fraunhofer Institute.
[13] The neutron monitor data, WDC-B2.
[14] L. I. Dorman: 1963, *Cosmic ray variations and space exploration*, Academy of Sciences of U.S.S.R.
[15] V. R. Rao, K. G. McCracken, and R. P. Bukata: 1967, *J. Geophys. Res.* **72**, 4325.
[16] A. I. Hundhausen and R. A. Gentry: 'Numerical Simulation of Flare-Generated disturbances in the solar wind', Preprint LA-DC-10064.
[17] T. J. Gold: 1959, *J. Geophys. Res.* **64**, 1665.
[18] S. N. Vernov, A. E. Chudakov, P. V. Vakulov, E. V. Gorchakov, Yu. I. Logachev, G. P. Lyubimov, and A. G. Nikolaev: 1964, *Space Exploration* **2**, 633.
[19] R. P. Kane: 1968, *Nuovo Cimento* **B57**, 36.
[20] L. S. Levitsky and B. M. Vladimirsky: 1969, *Izv. krymsk. Astrofiz. Observ.* **40**.
[21] S. N. Vernov *et al.*: 1967, *Izv. Akad. Nauk S.S.S.R.*, ser. fiz. **31**, 1296.
[22] A. N. Charakhchyan and T. N. Charakkchyan: 1966, *Geomagnetism and Aeronomy* **6**, 406.
[23] I. V. Getselev, E. V. Gorchakov, and G. A. Timofeev: 1968, *Proc. International Symposium of socialist countries on solar-terrestrial relations*, Crimea, U.S.S.R., 1968.
[24] S. N. Vernov, P. V. Vakulov, E. V. Gorchakov, N. N. Kontor, Yu I. Logatchev, G. P. Lyubimov, N. V. Pereslegina, G. A. Timogeev, and A. E. Chudakov: 1969, 'The study of low-energy cosmic rays in the interplanetary space', Report at the Seminar of Nuclear Physics Branch of Academy of Sciences of U.S.S.R., Leningrad.
[25] J. T. Gosling, J. R. Asbridge, S. J. Bame, A. J. Hundhausen, and I. B. Strong: 1968, 'Satellite Observation of Interplanetary Shock Waves', *J. Geophys. Res.* **72**, 43.
[26] A. J. Hundhausen and R. A. Gentry: 'Numerial Simulation of Flare-generated Disturbances in the Solar Wind', Preprint from L.S.L. of U.C.

PARTICLE EVENT FORECASTING

ZDENĚK ŠVESTKA

Astronomical Institute of the Czech. Academy of Sciences, Ondřejov, Czechoslovakia

Abstract. The paper reviews the present state of the forecasting problem. The solar particle events are first classified according to the classification scheme proposed by Smart and Shea. Main characteristics of proton-active regions are then summarized: magnetically complex type of the region, increase of the magnetic-field gradient, δ-configuration and A-configuration of the sunspot group, intensification and hardening of the microwave and X-ray spectrum, increased activity in the X-ray and microwave spectral regions, permanent emission of ~ 1 MeV proton flux, and loop-prominence-system occurrence. Characteristic properties of proton flares in the Hα-light, X-ray and radio spectrum also are mentioned, as well as the influence of the flare position on the solar disk on the particle flux recorded in the earth's surroundings. Basic principles of the long-term forecasts are explained in the last paragraph: active longitudes, complexes of activity, permanent proton flux, and travelling impulses of solar activity are mentioned.

This paper reviews the present state of the problem of forecasting of solar particle events. Before doing so, however, we must make clear what is to be understood under the 'particle events', for which the methods of forecasts have been developed.

1. Classification of Solar Particle Events

The very much increased number of solar particle events of greatly different size detected in the last years calls for some reasonable system of classification, since measurements by different methods and in widely different energy ranges usually do not give a clear picture about the actual size of the particle event in question. Such a classification has been proposed by Smart and Shea (1969), who distinguish five classes of proton events, schematically shown in Figure 1 (Švestka, 1969). Here the total flux of the recorded protons is plotted against the proton kinetic energy. Two orders of magnitude in the particle counts are adopted for a change in the classification. All the events of importance 1 to 4 produce PCA effects and the classes 3 and 4 also produce increases in the neutron monitor records. Thus these four classes all together correspond to the particle events, which often are called as *proton flares*. Events that do not produce any measurable polar cap absorption and generally are associated with a peak flux less than 3 protons above 10 MeV per cm² sec ster, are classified as of importance 0.

In order to give the reader some idea on the distribution of events of different size, I have plotted in this figure the peak flux of protons with energy exceeding 10 MeV for all particle events recorded during the year 1966, on which fairly complete data are available (McCracken *et al.*, 1967; Kinsey and McDonald, 1968; Fan *et al.*, 1968; Durgaprasad *et al.*, 1968; Masley and Goedeke, 1967). For two typical events of importance 1 and 2 respectively, the whole energy spectra are plotted in the graph. We observe that there was no event of importance 3 or 4 during that year, one or two events of importance 2, three or four events of importance 1, that is altogether 5 proton flares, and 25 events of importance 0. Of course, the actual number of the

V. Manno and D. E. Page (eds.), Intercorrelated Satellite Observations Related to Solar Events, 90–101. All Rights Reserved
Copyright © 1970 by D. Reidel Publishing Company, Dordrecht-Holland

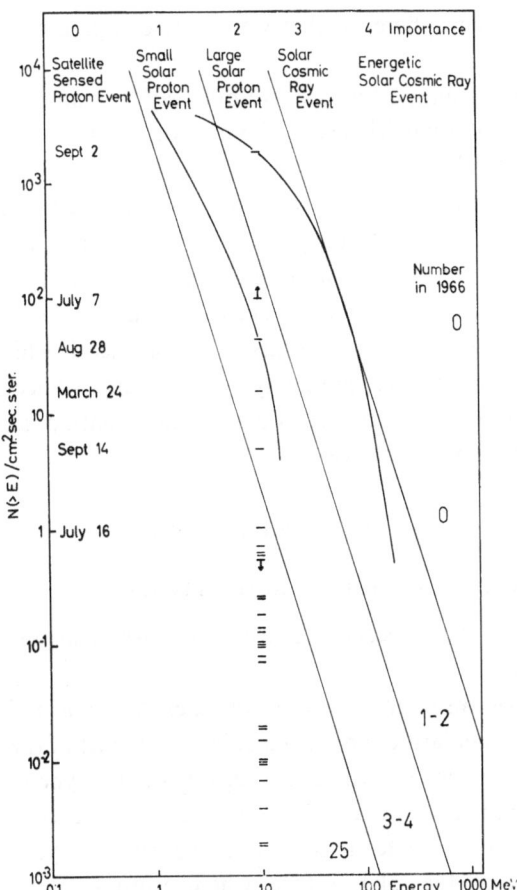

Fig. 1. Scheme of the Smart and Shea's (1969) classification of solar proton events. The peak flux of protons with energy exceeding 10 MeV is plotted here for all events recorded in 1966 and dates of the most important events are given on the left-hand side. The whole energy spectra are shown for the events of 1966 September 2 (importance 2) and August 28 (importance 1). (After Švestka, 1969.)

events of importance 0 certainly would be still higher if the sensitivity of the detectors were further improved.

Apart from these discrete proton events, 8 active regions were sources of permanent proton flux in 1966 (Fan *et al.*, 1968) and during the same period satellites and space probes detected twice relativistic electrons (Cline and McDonald, 1968), and 34 electron events were observed in the energy range above 40 keV (Lin and Anderson, 1967; Anderson, 1969). These electron fluxes were mostly associated with the proton events. In 7 cases, however, only electrons were recorded near the earth, without any measurable proton component (Švestka, 1969).

Out of all these various types of particle ejections from the sun, forecasting methods have been developed for the proton flares only. As the other, smaller particle events are concerned, our knowledge of the flares and active regions that produce them, is unsatisfactory. And some of these small events also come from sources situated on the invisible solar hemisphere, so that we hardly can predict them (cf. e.g., Dodson *et al.*, 1969).

Of course, when issuing a forecast, we only say that an active region is *suspected* from a production of a particle event. We cannot say, how strong a particle event is to be expected. It can be a particle event of importance 4, 3, 2, 1, and sometimes even only of importance zero. And sometimes the suspected active region dissolves again, without producing any recognizable particle event at all. But if we look at the problem from the opposite point of view, we find that there are only very exceptional events of importance 4, 3, and 2, which could occur without any forecast issued 1 or 2 days in advance, that there are very few events of importance 1, which could appear without any forecast, but that there are many events of importance 0 which could not be suspected in advance. This summarizes the general situation, and now I can describe the forecasting methods themselves.

2. Short-Term Forecasts

2.1. CHARACTERISTICS OF THE PROTON-ACTIVE REGION

We know that proton flares occur in active regions which are characterized by several specific properties:

(1) The proton-active region is always of *magnetically complex* type, i.e. sunspots of different magnetic polarities are mixed at least in a part of the active center (Figure 2). As Martres (1968) has shown, active regions of this type only occur, when they form within the remnants of an old active center. The complexity of the newly formed active region is greater if the old active center is younger. Thus, there are always some areas on the solar surface, where proton-active regions can – but, of course, need not – be born, and other areas, where the occurrence of such regions can be completely excluded.

(2) Of course, only some of the complex sunspot groups produce particle events of a recognizable size. A necessary condition of it is a substantial *increase of the gradient* of the magnetic field in the active region, as has been shown by Severny (1963) and his coworkers (Gopasyuk *et al.*, 1963). Without any magnetic measurements, an easily recognizable manifestation of this fact is a close approach of at least

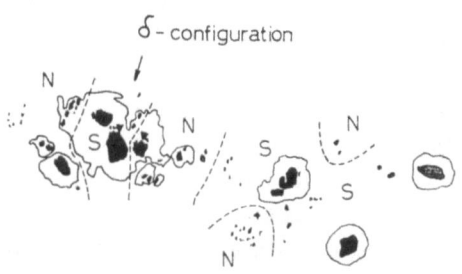

Fig. 2.   Example of a magnetically complex active region, in which the proton flare of May 23, 1967, occurred. N and S denote regions with northern and southern magnetic polarity, respectively. A δ-configuration is formed on the left-hand side of the figure. (After McIntosh, 1969).

two sunspots of opposite magnetic polarity in the sunspot group (Avignon *et al.*, 1963), and the formation of a common penumbra, embedding spots of different polarities, the so-called *δ-configuration* of a complex sunspot group (Warwick, 1966). In that case the probability of a particle event occurrence is very much increased. But the most important events occur when high gradient of the magnetic field exists within a large portion of the active region, which is manifested optically by a close approach of two rows of spots of opposite polarity. Such a situation has been called by Avignon *et al.* (1965) an *A-configuration* of the sunspot group (Figure 3).

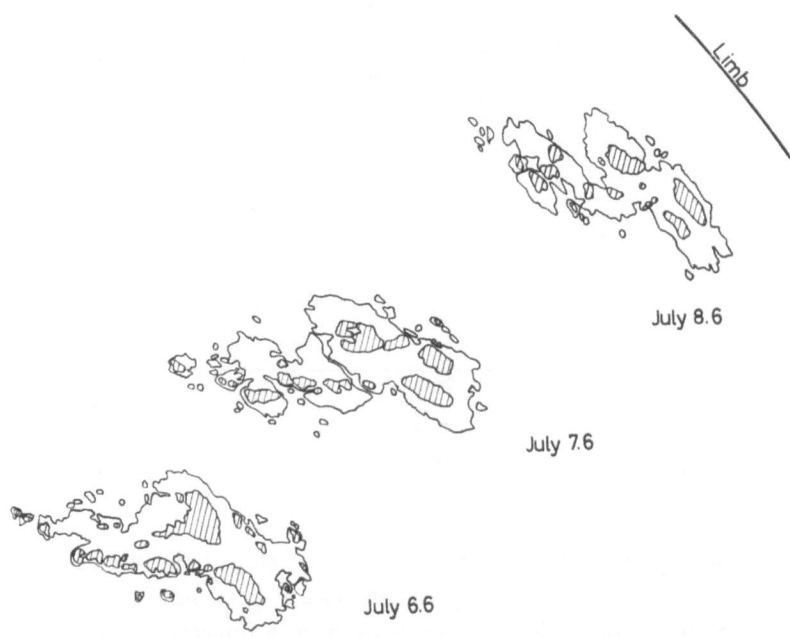

Fig. 3. Example of an A-configuration in a sunspot group. The shape of the group, in which the proton flare of July 7, 1966 appeared, is shown here for three successive days. The upper row of spots had a southern, and the lower row predominantly a northern magnetic polarity. (After Sawyer, 1968.)

(3) This development of the active region is associated with an intensification and a hardening both of the microwave and X-ray spectrum. In the microwave range the radio emission strongly increases (Pick, 1961), and the usual maximum close to 10 cm is suppressed by the very much increased flux at shorter wavelengths (Tanaka and Kakinuma, 1964) so that the microwave spectrum flattens, as one can see in Figure 4 (Tanaka *et al.*, 1969). Here, the change in the microwave spectrum is shown for 3 days preceding the proton flare of 7 July, 1966. This is probably due to an increased electron density in the coronal condensation above the active region (Krüger, 1969). At the same time, the X-ray flux also increases. This increase often exceeds one order of magnitude in the spectral range of a few Ångströms, decreases with the increasing wavelength and becomes insignificant for wavelengths above 40 Å (Friedman and Kreplin, 1969; Křivský and Nestorov, 1968; Švestka and Simon, 1969).

(4) During the same time, the association of flares with X-ray and microwave bursts is strongly increased, in some cases at a factor of about three (Eliseev and Moiseev, 1965; Křivský and Nestorov, 1968).

Apart from these basic characteristic features, several other properties of proton-active regions have been observed and described (cf. e.g., Banin *et al.*, 1969; Simon and Švestka, 1969; Severny and Steshenko, 1969), but – at least as far as we know at the present time – they do not seem to characterize the particle-activity in such a general way as the four items mentioned above.

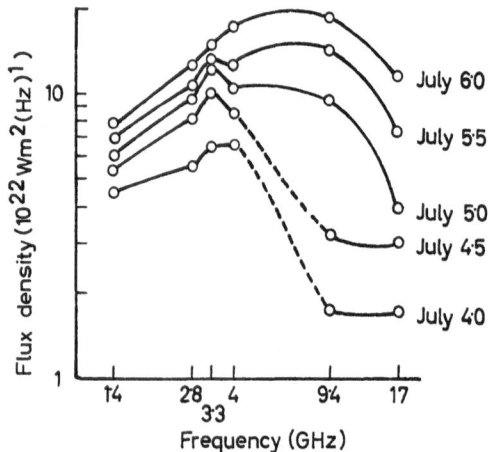

Fig. 4. Variation of spectrum of the slowly varying radio component preceding the proton flare of July 7, 1966. (After Tanaka *et al.*, 1969.)

All these basic characteristic features begin to develop about three days before the proton flare occurrence and usually become clearly recognizable one or two days before the proton flare, when in consequence of it, a forecast can be issued.

Quite often (and maybe always, but we do not yet have enough data on this subject) the active region also becomes a permanent source of the flux of protons with peak energy of the order of 1 MeV. It may be that this particle flux comes continuously from the active region in this phase of development. But it seems more probable that this particle stream is a combined product of several smaller flares preceding the proton flare which already become source of weak particle ejections (Švestka and Simon, 1969). In fact, it has been observed several times, by Bruzek (1969) and others, that the proton flare is often preceded by smaller flares which already possess some of the characteristic properties of the proton flare itself.

2.2. The proton flare

When a proton flare appears in the Hα light, it has the characteristic shape of *two bright ribbons* (Ellison *et al.*, 1961) shown in Figure 5, which separate and move one from the other with decreasing velocity until the ribbons cover sunspot umbrae (Malville and Moreton, 1963; Švestka, 1968b). This phase seems to be the period when the particles are accelerated in the flare region (Dodson and Hedeman, 1949;

Křivský, 1963; Švestka and Simon, 1969). In the radio range, the flare is characterized by a strong *Type IV burst* covering wide range of frequencies from millimetre to metre wavelengths. The burst has usually a typical U-type spectrum, with minimum flux near 2000 MHz and increasing in intensity both towards the lower and higher frequencies (Castelli *et al.*, 1967; Švestka and Simon, 1969). In the X-ray region these flares seem to be associated with very hard bursts above 50 keV. Thus, the proton flare usually can be easily recognized on the solar disk earlier than the particles arrive at the earth.

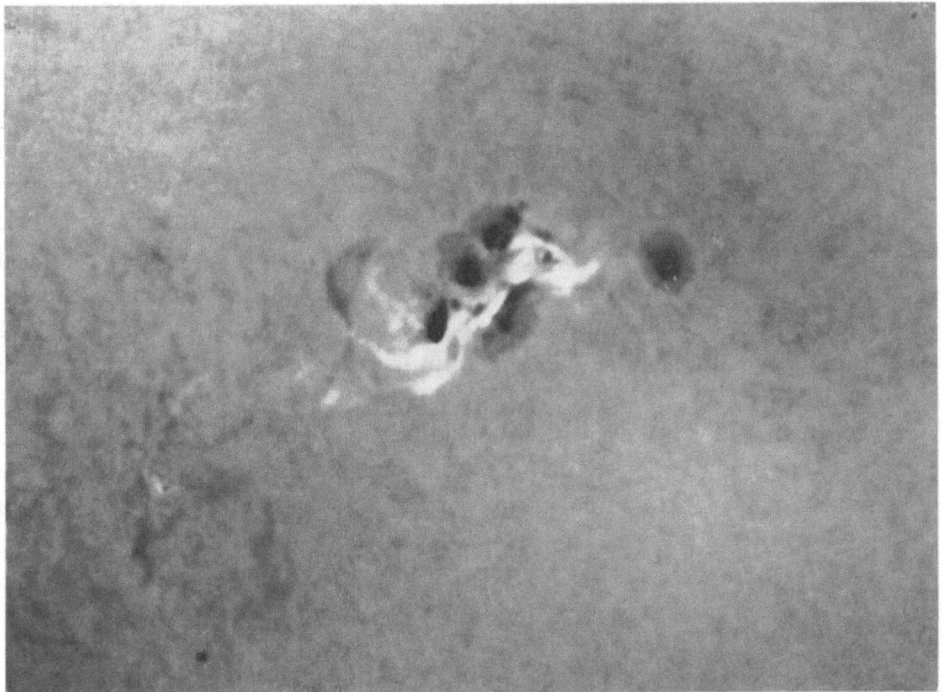

Fig. 5.   Example of a two-ribbon proton flare (of July 12, 1961). (Courtesy of K. Hermann-Otavský, Černošice, Czechoslovakia.)

2.3. Series of Proton Flares

This is the situation, when a new proton-active region develops on the sun and produces a proton flare event, and a description of the background on which a forecast of this first coming proton event can be made. Of course, quite often several proton flares appear in the same active region and then the situation becomes fairly complicated. The only conclusion one can make is that the active region is capable of another proton flare production as long as the complex magnetic structure and the increased X-ray and radio flux are maintained. But we usually cannot say *whether* and *when* another proton flare in the region appears.

Sometimes, of course, such a developed proton active region appears on the eastern solar limb, and the initial critical phase of its development could not be observed. In that case, apart from the complex magnetic configuration, which however hardly can be recognized close to the limb, the region becomes suspicious through its *permanent flux* of ~1 MeV protons, which probably are the remnants of earlier particle events in the region, and sometimes one also can observe an occurrence of *loop prominences*, which accompany many flares in the proton-active region, even flares that cannot be classified as proton flares themselves (Bruzek, 1964; Švestka, 1968a).

2.4. PARTICLE FLUX AT THE EARTH

So far, we have spoken only on the forecasts of the occurrence of proton flares on the solar disk. Our main interest, however, is in the forecasts of particle streams from the sun that hit the earth. And then, of course, the *flare position* on the solar disk also has to be taken into account. Figure 6 shows the distribution of flares causing strong PCA-events, according to their distance from the central solar meridian, for a period of 25 years (Švestka, 1966). One can see that there are very few flares situated more than 15° East from the central meridian, which cause such a strong particle event at the earth, and 84% of flares that produce a strong PCA event are located between 15° East and the western solar limb. That means that, statistically, a proton-active region becomes dangerous for the surroundings of the earth from about one day before its central-meridian passage up to its disappearance behind the western solar limb.

Of course, in individual events, there are exceptions from this rule. From time to time, a strong flux of energetic particles can come to the earth not only from the eastern hemisphere, but also from proton flares that occur fairly far *behind the western*

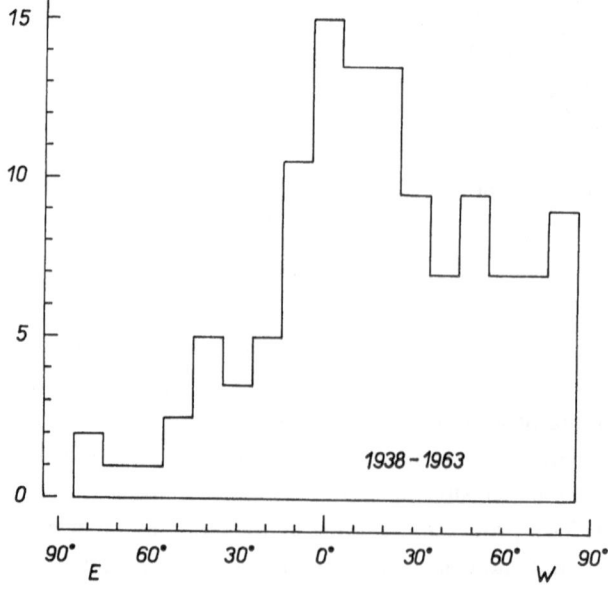

Fig. 6.   The distribution of flares that produced strong PCA effects at the earth, according to their
distance from the central solar meridian. (After Švestka, 1966.)

solar limb. The most conspicuous event of this type occurred on 28th January 1967, when a strong GLE was produced at the earth by a flare that was located – in the most favourable case – 60° behind the western solar limb (Dodson and Hedeman, 1969b; Lockwood, 1968), and one even cannot exclude that the flare position was still farther on the invisible solar hemisphere (Švestka, 1969).

This, of course, means a very serious trouble for the forecasts. We have seen that the whole preparation of an active region for the formation of a proton flare takes about three days. Therefore, a well-developed proton flare can occur 60° behind the solar limb without any slightest indication of such an activity when the region is disappearing on the western solar limb. Therefore, it is of great importance to check, how often the energetic particles can get to the earth from such positions on the solar surface, whether this 1967 event was an exceptional case, or whether such particle propagation can occur quite regularly, as it would follow, for example, from Burlaga's (1967) theory of anisotropic diffusion (cf. Švestka, 1969).

3. Long-Term Forecasts

Another very difficult problem is the long-term forecast of proton flares. We have seen that so far we are able to make fairly good forecasts of proton flares one or two days before the proton flare occurrence. But more than three days in advance there may be – and often is – completely nothing on the solar disk, which would indicate a development leading to a proton flare formation. Therefore, the long-term forecasts must be based on quite different principles than the short-term forecasts.

3.1. Active longitudes

The basic principle is the existence of *active longitudes* on the sun. It has been proved by several authors that the probability of formation of new active regions is not distributed homogeneously in the solar longitude. At any time there exist one or more sectors in the longitude, where this probability is increased, and this probability increase is most striking for the most important active phenomena (Dodson and Hedeman, 1969a; Vitinskij, 1969; Švestka, 1969). This can be easily understood if we take into account that the most important activity, and the particle events in particular, occur in complex active regions and that these regions only form, if a new group originates within the remnants of an old active center. Thus, in this case we have to consider the probability that two active regions roughly coincide in their positions. If one denotes as R the probability ratio of the formation of an active region within an active longitude and outside of it, then this ratio becomes approximately R^2 for the formation of complex active regions.

The existence of active longitudes can be best demonstrated for years close to the solar minimum. For example, since the beginning of the present solar cycle in 1964 the activity predominantly appeared on the northern solar hemisphere and essentially all the important active phenomena occurred with a restricted longitude sector of about 150° width centered at 230° of the heliographic longitude (Švestka, 1969).

Figure 7 illustrates this behaviour on the sun as evidenced in the 2800 MHz radio emission records in Ottawa (*Solar-Geophysical Data*, Boulder). Here, intensity of 2800 MHz radio flux is plotted in dependence of the 27-day period from 1964 to 1967. Each curve represents a sum of twelve 27-day rotations, and the date gives the first day of this period. Shadowed parts of the curves are those exceeding the average flux during the twelve rotations. At the bottom we plot the number of days on which Pioneers 6 and 7 recorded increased flux of >0.6 MeV protons, from December 1965 to August 1966 (Fan *et al.*, 1968). Shaded squares show the CMP days of the regions responsible for this permanent particle emission and circles show CMP days of active regions that produced proton flares on the northern solar hemisphere, from 1964.5 to 1967.5.

The figure shows that this striking restriction of solar activity to preferred active longitudes lasted for about three years and only dissolved in the first half of 1967, when other parts of the solar surface began to take part in producing important active phenomena.

The definition of this active sector of longitudes on the solar surface is the best

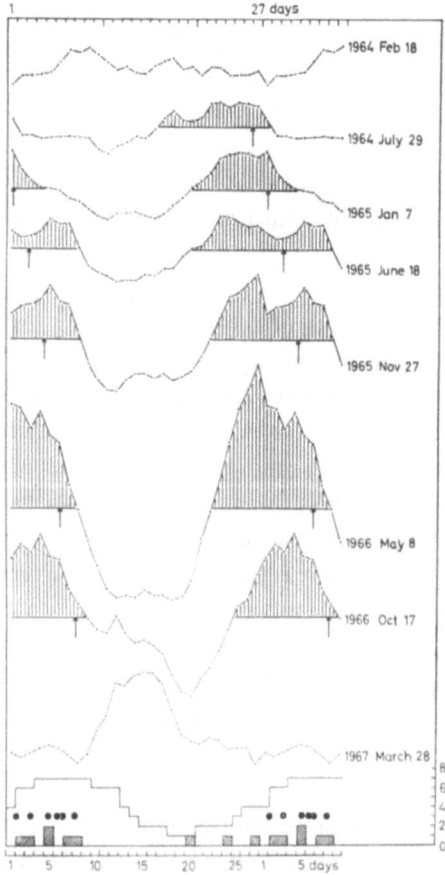

Fig. 7.   Intensity of 2800 MHz radio flux plotted in dependence on the 27-day period from 1964 to 1967. See the text for a detailed description. (After Švestka and Simon, 1969.)

for the regions in which proton flares occurred (Figure 8). Thus, for almost three years at the beginning of the present solar cycle, only a restricted part of the solar surface was dangerous, as the production of strong particle events was concerned. A similar situation also existed during the preceding solar minimum, for about four years (Švestka, 1969).

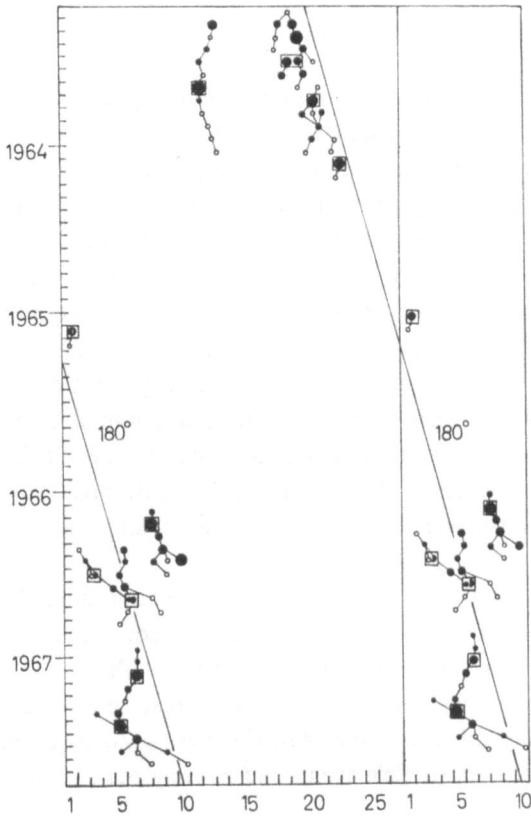

Fig. 8. Days of central-meridian passage of regions on the northern hemisphere that gave rise to proton flares, from 1963 to the half of 1967. (After Švestka, 1969.)

3.2. COMPLEXES OF ACTIVITY

When the activity is higher, the active longitudes usually cannot be distinguished so easily. Nevertheless, even at that time, the proton-active regions tend to occur in *complexes of activity*, which stay at a given location on the sun for many months and even years (Švestka, 1968c). A proton-active region develops and decays in such an activity complex, another region of this type is born close-by, and this repeats again and again, sometimes for a very long time.

This behaviour of solar activity certainly is favourable for the long-term forecasts, because we know, at least during some periods, the longitudes on the sun, where the proton flares most probably occur. But unfortunately, we do not know, *when* in such an activity complex a proton-active region will be formed. That means that such an active longitude on the sun is dangerous *permanently*. Particles from it can arrive at the earth in high number as soon as the longitude approaches about 15° east of the central meridian up to the western limb or even behind it, that is for at least 9 days every solar rotation. And since there may be two or even more such activity longitudes on the sun at the same time when the activity is high, the practical use of this our knowledge for the forecasts is very small.

3.3. OUTLOOK

Nevertheless, it seems that some progress can be expected in this respect, maybe in a near future. Space probes situated far from the sun-earth line can give us information on the permanent proton flux from regions approaching the eastern solar limb, and as far as we know, such a permanent flux always precedes a proton flare occurrence in the region (Švestka, 1969). Of course, sometimes the onset of the flux precedes the proton flare for few days only, so that neither this observation is helpful in all cases. What I have in mind, however, as the most hopeful perspective, is the suspected existence of *impulses* of activity which travel around the sun in semi-regular periods and lead to a formation of great increases in activity when crossing the active longi-tudes. Bumba has succeeded to show this effect at a number of proton-active regions, on the grounds of the synoptic charts of the solar magnetic fields published by him-self, Howard and Sara Smith (Howard *et al.*, 1967; Bumba, 1969). A confirmation of this effect and its better understanding, however, still waits for a more extensive study of this fairly difficult problem.

As short-term forecasts are concerned, great attention should be paid to the periods of about 12 hours that immediately precede the proton flare occurrence. There are some indications that important changes occur in the active region at that time (Simon and Švestka, 1969), which might be a great help for a more precise identi-fication of the time, when a strong particle event is to be expected, maybe also in cases when several such particle events successively occur in the same active region.

References

Anderson, K. A.: 1969, *Solar Phys.* **6**, 111.
Avignon, Y., Caroubalos, C., Martres, M. J., and Pick, M.: 1965, IAU Symposium No. **22**, 373.
Avignon, Y., Martres, M. J., and Pick, M.: 1963, *Comptes Rendus* **256**, 2112.
Banin, V., de Feiter, L. D., Fokker, A. D., Martres, M. J., and Pick, M.: 1969, *Ann. IQSY* **3**, 229.
Bruzek, A.: 1964, *Astrophys. J.* **140**, 746.
Bruzek, A.: 1969, *Ann. IQSY* **3**, 82.
Bumba, V.: 1969, 'Extension of the Atlas of Solar Magnetic Fields', unpublished.
Burlaga, L. F.: 1967, *J. Geophys. Res.* **72**, 4449.
Castelli, J. P., Aarons, J., and Michael, G. A.: 1967, *J. Geophys. Res.* **72**, 5491.
Cline, T. L. and McDonald, F. B.: 1968, *Solar Phys.* **5**, 507.
Dodson, H. W. and Hedeman, E. R.: 1949, *Astrophys. J.* **110**, 249.
Dodson, H. W. and Hedeman, E. R.: 1969a, *Ann. IQSY* **4**, 3.
Dodson, H. W. and Hedeman, E. R.: 1969b, *Solar Phys.* **9**, 278.
Dodson, H. W., Hedeman, E. R., Kahler, S. W., and Lin, R. P.: 1969, *Solar Phys.* **6**, 294.
Durgaprasad, N., Fichtel, C. E., Guss, D. E., and Reames, D. V.: 1968, *Astrophys. J.* **154**, 307.
Eliseev, G. F. and Moiseev, I. G.: 1965, *Izv. Krymsk. Astrofiz. Observ.* **34**, 3.

Ellison, M. A., McKenna, M. P., and Reid, J. H.: 1961, *Publ. Dunsink Observ.* **1**, 53.
Fan, C. Y., Pick, M., Pyle, R., Simpson, J. A., and Smith, D. R.: 1968, *J. Geophys. Res.* **73**, 1555.
Friedman, H. and Kreplin, R. W.: 1969, *Ann. IQSY* **3**, 78.
Gopasyuk, S., Ogir, M., Severny, A. B., and Shaposhnikova, E.: 1963, *Izv. Krymsk. Astrofiz. Observ.* **29**, 15.
Howard, R., Bumba, V., and Smith, S. F.: 1967, Atlas of Solar Magnetic Fields, Carnegie Inst. Washington Publ. 626.
Kinsey, J. H. and McDonald, F. B.: 1968, IAU Symposium No. **35**, 536.
Křivský, L.: 1963, *Bull. Astron. Inst. Czech.* **14**, 77.
Křivský, L. and Nestorov, G.: 1968, *Bull. Astron. Inst. Czech.* **19**, 197.
Krüger, A.: 1969, *Ann. IQSY* **3**, 70.
Lin, R. F. and Anderson, K. A.: 1967, *Solar Phys.* **1**, 446.
Lockwood, J. A.: 1968, *J. Geophys. Res.* **73**, 4247.
Malville, J. M. and Moreton, G. E.: 1963, *Publ. Astron. Soc. Pacific* **75**, 176.
Martres, M. J.: 1968, IAU Symposium No. **35**, 25.
Masley, A. J. and Goedeke, A. D.: 1967, 'The 1966–1967 Increase in Solar Cosmic Ray Activity', Tenth Intern. Conf. on Cosmic Rays, Calgary, June 1967.
McCracken, K. G., Rao, U. R., and Bukata, R. P.: 1967, *J. Geophys. Res.* **72**, 4293.
McIntosh, P. S.: 1969, WDC-A Report UAG-5, p. 14.
Pick, M.: 1961, *Ann. Astrophys.* **24**, 183.
Sawyer, C. S.: 1968, IAU Symposium No. **35**, 543.
Severny, A. B.: 1963, *AAS-NASA Symposium on the Physics of Solar Flares*, p. 95.
Severny, A. B. and Steshenko, N. V.: 1969, 'On the Short-Term Forecasting of Flares', *STP Notes*, in press.
Simon, P. and Švestka, Z.: 1969, *Ann. IQSY* **3**, 469.
Smart, D. F. and Shea, M. A.: 1969, 'Proposed Solar Proton Classification System', presented at IUCSTP Assembly, London, January 1969.
Švestka, Z.: 1966, *Bull. Astron. Inst. Czech.* **17**, 262.
Švestka, Z.: 1968a, IAU Symposium No. **35**, 287.
Švestka, Z.: 1968b, Nobel Symposium **9**, 17.
Švestka, Z.: 1968c, *Solar Phys.* **4**, 18.
Švestka, Z.: 1969, 'Solar Particle Events', Annual Review Lecture at the Twelfth Plenary Meeting of COSPAR, Prague, May 11–24, 1969.
Švestka, Z. and Simon, P.: 1969, *Solar Phys.* **10**, 3.
Tanaka, H. and Kakinuma, T.: 1964, *Rep. Ionosph. Space Res. Japan* **18**, 1964.
Tanaka, H., Kakinuma, T., and Enome, S.: 1969, *Ann. IQSY* **3**, 63.
Vitinskij, J. I.: 1969, *Solar Phys.* **7**, 210.
Warwick, C. S.: 1966, *Astrophys. J.* **145**, 215.

Discussion

Roederer: What can be said about correlation between flares and planetary positions, looking at the tidal effects on the sun.

Švestka: There is some ground in that. There are some effects which show a one year period lasting for several decades and one might assume this to be caused by the rotation of the earth around the sun.

One can, in fact, guess that by means of forces, which are difficult to describe, the planets in their orbits modulate the activity of the sun.

Wilcox: Figure 6 which shows the number of PCA events as a function of heliographic longitude, features a clear peak on the Western hemisphere. What happens beyond the Western limb? One might expect PCA events there also. Although we do not see the flares we might still observe the effects.

Švestka: You are referring to the function that shows the number of strong PCA events. When we look at all PCA events, there is almost no dependence on longitude. The strong PCA events are associated with visible flares. When considering all PCA events there are quite a number of small events which cannot be associated with anything on the visible hemisphere. In these cases you might want to consider events on the other side of the sun. These statistics do, however, show that very few strong PCA events originate on the far side of the sun.

PROPOSED SOLAR PROTON EVENT CLASSIFICATION SYSTEM

D. F. SMART and M. A. SHEA

Air Force Cambridge Research Laboratories, Bedford, Mass., U.S.A.

Considerable confusion exists on what is the 'size' of a proton event. With the greatly improved sensitivity of detectors on satellites, it appears that almost every solar 'hiccup' is being reported as a solar proton event with the descriptive adjective 'large' being applied to some events that are very much smaller in particle flux and energy than some of the classic solar cosmic-ray events of the 19th solar cycle. Additional confusion arises when discussing specific events with respect to their detection by various ground-based sensors. For example the May 23, 1967 event which was extremely large to the radio astronomers was not detected by ground-based neutron monitors. Conversely, the large neutron monitor increase associated with the May 4, 1960 event was followed by only a 'moderate' riometer absorption.

In view of this confusion it appears that some type of classification system for solar proton events would be beneficial to scientists utilizing solar proton information for various geophysical analyses. A preliminary proton event classification system based upon the proton flux in the immediate vicinity of the earth was presented at the Inter-Union Commission on Solar-Terrestrial Physics Plenary Meeting held in London in January 1969. Working Group 3 (Interplanetary Disturbances) recommended that such a system be adopted by IUCSTP as soon as it could be finalized, and we were appointed to refine the preliminary system and present it for formal consideration and recommendation for international adoption at the next IUCSTP meeting to be held in May 1970.

In the past few months the preliminary system has been considered by various researchers and its use by some scientists (Švestka, 1969) has resulted in several recommended changes. We have modified the original system incorporating as many suggestions as practical. The revision presented here is *not* to be construed as a final classification system, but as a revised preliminary system.

This proposed classification system (Table I) is based on the logarithm (base 10) of the solar proton flux in units of $cm^{-2} sec^{-1} ster^{-1}$. A reference level of 10 MeV was recommended by members of IUCSTP. Basically each solar proton event would be classified by a three digit number; the first digit representing neutron monitor response, the second representing riometer response and the third describing the proton intensity with $E > 10$ MeV. Since we are mainly concerned with these events as they affect the earth, this three digit number is confined to the response of earth-based sensors and satellites in an earth-moon system orbit.

The basic criteria for this classification system is represented by the third digit – the satellite measured proton intensity. In order to incorporate the suggestions of Švestka (1969) on the increasing sensitivity of satellite instrumentation, we have

V. Manno and D. E. Page (eds.), Intercorrelated Satellite Observations Related to Solar Events, 102–107. All Rights Reserved
Copyright © 1970 by D. Reidel Publishing Company, Dordrecht-Holland

TABLE I

Proposed solar proton event classification system

Index number	Sea level neutron monitor increase	Daylight Polar 30 MHz riometer absorption	$E > 10$ MeV Satellite measured proton intensity P/cm²-sec ster
-3	–	–	From 10^{-3} to 10^{-2}
-2	–	–	From 10^{-2} to 10^{-1}
-1	–	–	From 10^{-1} to 10^{0}
0	No measurable increase	No measurable increase	From 10^{0} to 10^{1}
1	Less than 3 %	Less than 1.5 db	From 10^{1} to 10^{2}
2	From 3 % to 10 %	From 1.5 db to 4.6 db	From 10^{2} to 10^{3}
3	From 10 % to 100 %	From 4.6 db to 15 db	From 10^{3} to 10^{4}
4	Greater than 100 %	Greater than 15 db	Greater than 10^{4}

included negative index numbers to allow for the classification of small events. Since satellite measurements at 10 MeV are not always available, it is suggested that a power law of -3 be used in extrapolating from other energies (if spectral data are not available) to estimate the proton intensity at 10 MeV and that the index number be selected on the basis of this extrapolated value.

For the second digit we have roughly translated integral flux levels into effects observed by riometers. It has been shown by various scientists (Bailey, 1964; Van Allen *et al.*, 1964; Fichtel *et al.*, 1963; Masley and Goedeke, 1968; Juday and Adams, 1969) that the riometer absorption is proportional to the square root of the integral particle flux above some threshold, so the conversion from particle flux with $E > 10$ MeV to an appropriate riometer absorption is not impractical. The equivalence between riometer absorption and particle flux given in Table I utilizes the conversion $J_0 = 47A^2$ (Juday and Adams, 1969). The data from riometers should be hourly averages of absorption from a daylight riometer located inside the auroral oval with the daytime cutoff less than about 5 MeV (equivalent to an L value of 12 or greater). The preferred frequency is 30 MHz; for locations where a different frequency is used the 30 MHz absorption can be estimated by the following relationship:

$$A_1/A_2 = (f_2/f_1)^{1.8}.$$

A partial list of riometer locations meeting the necessary criteria for this classification system is given in Table II.

TABLE II

Station	Geographic		L Value	Invariant Latitude
	Lat.	Long.		
Thule, Greenland	76 N	68 W	182	86
Godhavn, Greenland	69 N	54 W	21	77
Shepherd Bay, Canada	68 N	94 W	30	80
Frobisher Bay, Canada	63 N	67 W	15	75
Vostok, Antarctica	78 S	106 W	75	83
McMurdo, Antarctica	78 S	167 E	33	80
South Pole, Antarctica	90 S	0 E	13	74

The first digit, corresponding to neutron monitor response, is somewhat more difficult to relate to proton intensity. Consequently, the index number assigned has been selected primarily with respect to a percentage increase in neutron monitor intensity above quiescent background. The neutron monitor data should represent at least a 15-minute average from a sea level monitor with essentially an atmospheric cutoff ($\sim$1 GV or less). A partial list of neutron monitor locations acceptable for this classification system is given in Table III.

TABLE III

Station	Geographic		Vertical
	Lat.	Long. (E)	cutoff rigidity (GV)
Alert, Canada	82.50	297.67	0.00
Apatity, U.S.S.R.	67.55	33.33	0.65
Cape Hallet, Antarctica	−72.32	170.22	0.04
Churchill, Canada	58.75	265.91	0.21
Deep River, Canada	46.10	282.50	1.02
Dumont Durville, Antarctica	−66.67	140.02	0.01
Ellsworth, Antarctica	−77.72	318.88	0.79
Goose Bay	53.33	299.58	0.52
Heiss Island, U.S.S.R.	80.33	57.80	0.10
Inuvik, Canada	68.35	226.27	0.18
Mawson, Antarctica	−67.60	62.88	0.22
McMurdo, Antarctica	−77.85	166.72	0.01
Mirny, Antarctica	−66.55	93.00	0.04
Murmansk, U.S.S.R.	68.97	33.08	0.50
Oulu, Finland	65.00	25.42	0.81
Resolute, Canada	74.69	265.09	0.00
Syowa, Antarctica	−69.03	39.60	0.42
Thule, Greenland	76.55	291.16	0.00
Tixie Bay, U.S.S.R.	71.55	128.90	0.53
Vostok, Antarctica	−78.47	106.87	0.00

One of the tests of any classification system is its application to specific events. We have selected the list of Bailey (1964) for a preliminary application of this system; this list is given in Table IV. For events of the 19th solar cycle there are very few direct particle measurements and so available particle data are estimated from indirect measurements by riometers or balloon sensors and the interpolation of these data vary. For some events in this table we illustrate the difficulty in utilizing data from the 19th solar cycle by comparing the classification resulting from the particle fluxes estimated by Bailey (1964) to those given in the NASA *Solar Proton Manual* (1963).

An extension of this proton event classification system would be to apply it to events detected by instrumentation aboard space probes. It is suggested that these events be classified by a one digit number equivalent in value to the proton intensity index number with $E > 10$ MeV. This one digit number could be followed by the letter P indicating data were obtained by a space probe which is not within the earth-moon system.

TABLE IV

Proton event classification applied to list of Polar cap absorption events as compiled by Bailey (1964)

Date		Time of onset (UT)	Neutron monitor		Riometer		$E > 10$ MeV cm^{-2} sec^{-1} ster^{-1} estimated by Bailey		Index est fr NASA	NRP Index
			Increase %	Index	Absorption db	Index				
1956	23 Feb.	0400	2000	4	13.0	3	1570	3	3	433
	10 Mar.	0900		0	3.5	2	96	1		021
	31 Aug.	1430	2	1	4.9	3	192	2		132
	13 Nov.	2000		0	5.4	3	240	2		032
1957	20 Jan.	1500		0	4.1	2	135	2		022
	3 Apr.	1330		0	3.9	2	120	2		022
	6 Apr.	0800		0	3.2	2	80	1		021
	22 Jun.	0500		0	5.0	3	203	2		032
	3 Jul.	1000		0	9.2	3	740	2		032
	9 Aug.	1600		0	3.1	2	76	1		021
	29 Aug.	0000		0	3.2	2	80	1		021
	29 Aug.	1400		0	8.2	3	580	2	3	032
	31 Aug.	1500		0	4.9	3	192	2		032
	2 Sep.	1700		0	7.2	3	440	2		032
	21 Sep.	1700		0	5.1	3	214	2		032
	20 Oct.	2100		0	7.8	3	515	2		032
	5 Nov.	0200		0	2.6	2	52	1		021
1958	10 Feb.	0600		0	3.2	2	80	1		021
	23 Mar.	1500		0	3.2	2	80	2		032
	25 Mar.	1530		0	10.0	3	890	2		032
	10 Apr.	0900		0	4.4	2	153	2		022
	7 Jul.	0300		0	23.7	4	6700	3	4	043
	16 Aug.	0600		0	12.1	3	1350	3		033
	22 Aug.	1530		0	10.6	3	1010	3		033
	26 Aug.	0330		0	16.6	4	2800	3		043
	22 Sep.	1400		0	5.0	3	204	2		032

Table IV (Continued)

Date		Time of onset (UT)	Neutron monitor		Riometer		$E > 10$ MeV cm^{-2} sec^{-1} ster^{-1} estimated by Bailey		Index est fr NASA	NRP Index
			Increase %	Index	Absorption db	Index				
1959	13 Feb.	0800		0	2.6	2	52	1		021
	11 May	0030		0	22.0	4	5500	3	4	043
	10 Jul.	0700		0	20	4	4300	3	4	043
	14 Jul.	0730		0	23.7	4	6700	3	4	043
	17 Jul.	0000	7	2	21.2	4	5000	3		243
1960	29 Mar.	0800		0	2.6	2	52	1		021
	1 Apr.	1000		0	3.6	2	104	2		022
	5 Apr.	0700		0	3.1	2	76	1		021
	28 Apr.	0230		0	2.5	2	47	1		021
	29 Apr.	0500		0	11.2	3	1130	3	3	033
	4 May	1030	295	4	3.4	2	88	1	3	421
	6 May	1800		0	8.7	3	660	2	3	032
	13 May	0730		0	3.6	2	104	2		022
	3 Sep.	0500	2	1	2.7	2	58	1		121
	12 Nov.	1400	115	4	21.2	4	5000	3	4	443
	15 Nov.	0430	70	3	20.0	4	4300	3	4	343
	20 Nov.	2200	5	2	3.0	2	69	1		221
1961	12 Jul.	1300		0	17.0	4	2900	3	3	043
	18 Jul.	1130	11	3	8.7	3	660	2	3	332
	a20 Jul.		4	2	4	2			2	222
	10 Sep.	2100		0	2.9	2	65	1		021
	a28 Sept.			0	1.8	2			2	022

a Event not on Bailey's list. Data from NASA *Solar Proton Manual*.

References

Bailey, D. K.: 1964, 'Polar Cap Absorption', *Planetary Space Sci.* **12**, 495.

Fichtel, C. E., Guss, D. E., and Ogilvie, K. W.: 1963, 'Details of Individual Solar Particle Events', in *Solar Proton Manual* (ed. by F. B. McDonald), NASA TR R-169, Dec., Chapter 2.

Juday, R. C. and Adams, G. W.: 1969, 'Riometer Measurements, Solar Proton Intensities and Radiation Dose Rates', *Planetary Space Sci.* **17**, 1313.

Masley, A. J. and Goedeke, A. D.: 1968, 'The 1966–67 Increase in Solar Cosmic-Ray Activity', *Can. J. Phys.* **46**, S766.

Švestka, Z.: 1969, Solar Particle Events, preprint of Annual Rev. Lecture R7, COSPAR, Prague, May 1969.

Van Allen, J. A., Lin, W. C., and Leinbach, H.: 1964, 'On the Relationship Between Absolute Solar Cosmic Ray Intensity and Riometer Absorption', *J. Geophys. Res.* **69**, 4481.

3. SOLAR WIND VARIATIONS DURING AND AFTER SOLAR EVENTS

SOLAR WIND DISTURBANCES ASSOCIATED
WITH SOLAR ACTIVITY

A. J. HUNDHAUSEN

University of California, Los Alamos Scientific Laboratory, Los Alamos, N.M., U.S.A.

1. Introduction

Two different classes of geomagnetic activity have long been attributed to particle emission from active solar regions. The occurrence of geomagnetic storms subsequent to large solar flares has been explained by the arrival at earth of material ejected by the flare. Recurrent geomagnetic activity has been interpreted as the effect of persistent streams emitted from long-lived active regions on the rotating sun. Within the past decade, both flare-associated and recurrent disturbances have been directly observed on spacecraft probing interplanetary space. This paper will review and interpret these observations, with major emphasis on the flare-associated phenomena. The role of multiple satellite measurements in studying the propagation and spatial structure of interplanetary disturbances will be considered.

2. Flare-associated Disturbances

The existence of an ambient interplanetary plasma (the solar wind) with fluid-like properties implies that a shock wave will propagate ahead of material ejected from a flare (Gold, 1955). Attention here will first be given to the observed properties of these shock waves. The pattern of variations in the solar wind flow behind the shock will then be described. Finally, some implications of these observed characteristics of interplanetary disturbances will be examined by comparison with theoretical models. The relevance of these phenomena to solar and interplanetary physics will be emphasized, and their significance in the study of magnetospheric physics will be briefly mentioned.

A. INTERPLANETARY SHOCK WAVES

An interplanetary shock wave observed on January 20, 1965, by both the LASL electrostatic analyzer on Vela 3A (Gosling *et al.*, 1968) and the GSFC magnetometer on Imp 3 (Taylor, 1969) will serve as an initial and typical example. Figure 1 shows the positive ion properties measured on the former spacecraft; the abrupt jumps in proton temperature, density, and solar wind speed near 0210 UT indicate the passage of the shock. An abrupt rise in the horizontal component of the ground-level geomagnetic field was recorded shortly thereafter. The solar wind properties measured just before and after the shock are given in Table I, including the magnetic field intensities from Imp 3. The observed changes in these properties are consistent with

V. Manno and D. E. Page (eds.), Intercorrelated Satellite Observations Related to Solar Events, 111–129. *All Rights Reserved*
Copyright © 1970 by D. Reidel Publishing Company, Dordrecht-Holland

TABLE I

Solar wind properties measured before and after the interplanetary shock
of January 20, 1965 (Gosling *et al.*, 1968; Taylor, 1969)

	Pre-Shock	Post-Shock
Density	9 cm^{-3}	23 cm^{-3}
Flow Speed	340 km sec^{-1}	385 km sec^{-1}
Proton Temperature	3×10^4 K	8×10^4 K
Magnetic Field Intensity	5.9 γ	12.3 γ

the hydrodynamic Rankine-Hugoniot relations (neglecting the magnetic field) to a surprising degree of accuracy (Gosling *et al.*, 1968).

If the plasma density and flow velocity are known on both sides of a shock, the equation of mass continuity can be used to compute the propagation speed of the shock front. The density and speed values given in Table I yield a shock speed of 413 km sec^{-1} if the shock normal is assumed to be in the radial direction. The ambient solar wind plasma is then flowing into the shock at 73 km sec^{-1}, whereas the sound

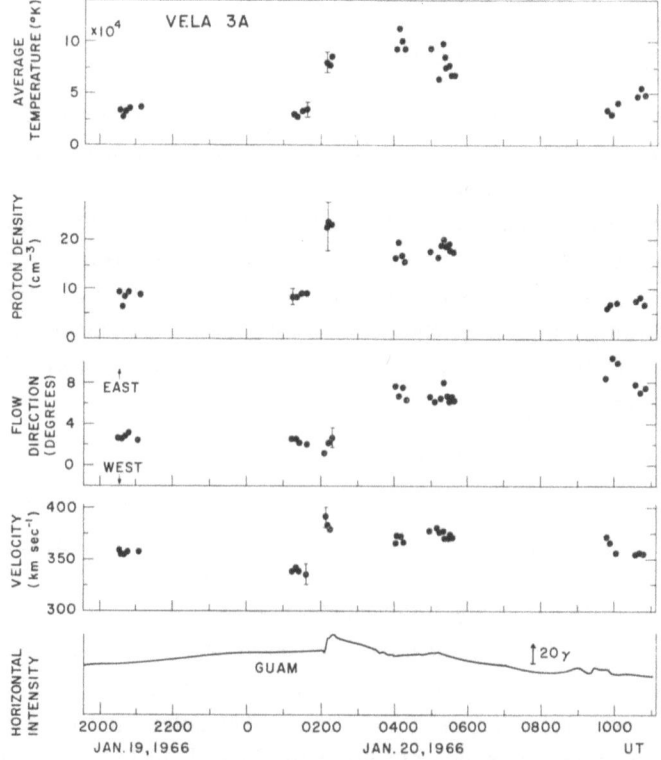

Fig. 1. Vela 3A measurements of the solar wind proton temperature, density, flow direction, and flow speed, and a trace of the horizontal component of the geomagnetic field measured at Guam, on January 19–20, 1966 (Gosling *et al.*, 1968).

and Alfvén speeds in the ambient gas are respectively 29 and 42 km sec^{-1}. The partially thermalized plasma flows out of the shock at 28 km sec^{-1}, whereas the sound and Alfvén speeds in the post-shock gas are 47 and 56 km sec^{-1}. The Mach number of the inflowing plasma is then approximately 2, that of the outflowing plasma approximately $\frac{1}{2}$.

Gosling *et al.* (1968) suggested that either of two solar flares, occurring at 27.3 and 63.5 hours before the observation of the shock at 1 AU, might be associated with the event. Taylor (1969) favored the earlier flare. The mean propagation speed during transit from the sun to 1 AU implied by the 63.5 hour flare association is 685 km sec^{-1}. This is significantly higher than the shock speed inferred from the Vela observation, suggesting that the shock has been decelerated in propagating through interplanetary space.

Published measurements of interplanetary shock properties are summarized in Table II (from Sonett *et al.*, 1966; Bame *et al.*, 1968; Gosling *et al.*, 1968; Lazarus and Binsack, 1968; Ogilvie *et al.*, 1968; Hirshberg *et al.*, 1969; Ogilvie and Burlaga, 1969; Taylor, 1969). The source and date of each observation is listed along with the shock speed (derived either from the mass continuity equation, as in the example above, or from the delay between two spatially separated observations), and the Mach number of the ambient plasma flow into the shock. Where flare associations have been made, the transit time from the sun to 1 AU and the implied mean shock propagation speed are listed. On the basis of this sample, the typical interplanetary shock wave is found to be moving at about 500 km sec^{-1} near 1 AU, and is of intermediate strength (i.e., the Mach number of the inflowing gas is not large). The shock arrives at 1 AU two to three days after its presumed flare origin. The mean propagation speed is higher than the speed measured at 1 AU, indicating that flare-associated disturbances are usually decelerated in interplanetary space.

Figure 2 presents an additional interplanetary shock observation which shows, for the first time, the properties of the electron component of the plasma and directly illustrates the intermediate strength of the shock. Vela 4B measurements of the solar wind speed, proton density, proton temperature, and electron temperature are shown for a seven-hour period on June 5, 1967. Two values of both the proton and electron temperature are given, as the distribution functions of both particle species are often anisotropic. The interplanetary shock passage occurred at 0915 UT. While the proton temperature jumped by a factor of three at the shock, the electron temperature increased by only about 15%. Earlier, at 0815 UT, the satellite had crossed the earth's bow shock into the magnetosheath, emerging again into the solar wind at 0830 UT. At the bow shock crossings the proton and electron temperatures changed by factors of 30 and 3 respectively. The contrast between the large temperature changes at the bow shock (whose stationary nature implies a Mach number of about 8) and the small temperature changes at the interplanetary shock gives direct evidence of the weaker nature of the latter.

Several direct determinations of the orientation of interplanetary shocks have been made, and are of interest both as a check on the assumption (frequently made in the

TABLE II

The properties of directly observed interplanetary shock waves

Source	Date	Shock speed (km sec⁻¹)	Sonic Mach No.	Alfvén Mach No.	Transit time (hours)	Mean propagation speed (km sec⁻¹)
Sonett et al.	Oct. 7, 1962	509	3.0	4.2		
Gosling et al.	Oct. 5, 1965	420	2.3		17.2	2500
	Jan. 20, 1966	410			27.3	1670
					or 63.5	or 685
Taylor	Oct. 7, 1965				49.5	836
	Jan. 20, 1966				63.5	652
	July 8, 1966				44.6	928
	July 15, 1966				102.0	406
	Aug. 29, 1966				67.2	616
	Sept. 3, 1966				39.5	1048
	Jan. 6, 1967				55.7	743
	Jan. 7, 1967				45.6	908
Lazarus and Binsack	July 8, 1966	750				
	July 9, 1966	500				
	July 10, 1966	830				
Bame et al.	Jan. 13, 1967	463			58	717
Hirshberg et al.	Feb. 15, 1967	505			53.5	661
Ogilvie et al.	May 30, 1967	506			56.2	735
	June 5, 1967	347			59.5	700
	June 25, 1967					
Ogilvie and Burlaga	June 26, 1967	482		3.1		
	Aug. 29, 1967	496		15		
	Sept. 13, 1967	416		1.0		
	Sept. 19, 1967	497		1.3		
	Nov. 29, 1967	394		1.7		
	Jan. 11, 1968	524		.11.3		

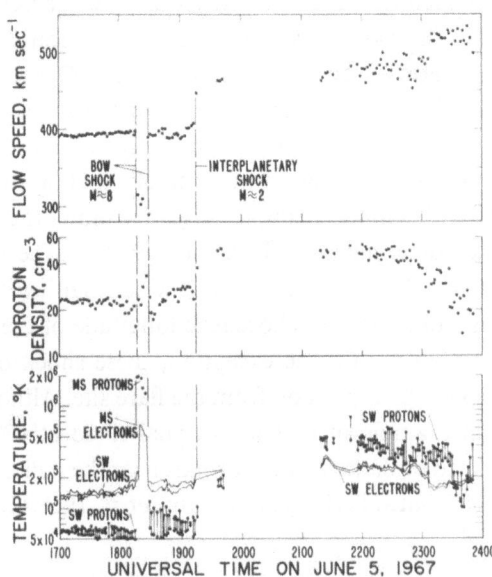

Fig. 2.   Vela 4B measurements of the flow speed, proton density, proton temperature, and electron temperature on June 5, 1967. The two values shown for both the proton and electron temperature are the maximum and minimum temperatures describing the anisotropic distribution function. After a brief excursion in the magnetosheath between 1815 and 1830 UT, an interplanetary shock is observed at 1915 UT.

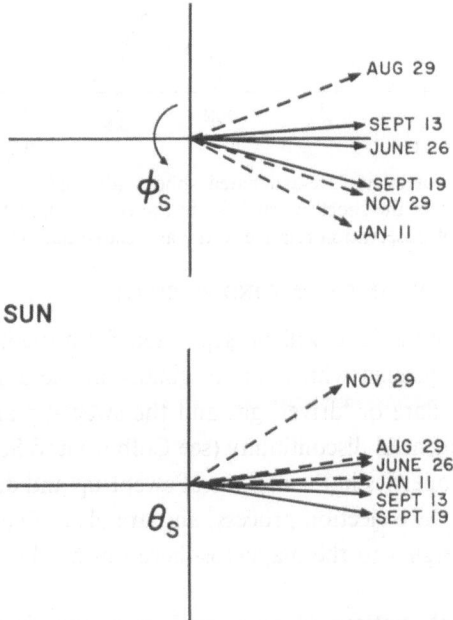

Fig. 3.   Six interplanetary shock normals derived by Ogilvie and Burlaga (1969) from the observed change in magnetic field at the shock (solid arrows) or from multiple satellite observations (dashed arrows). The angle ϕ_s is solar ecliptic longitude; θ_s is solar ecliptic latitude.

analysis of shock measurements and in the derivation of theoretical models) that the shock normal points in the radial direction and as an indication of the shape of the shock surface. Figure 3 shows 6 shock normals derived by Ogilvie and Burlaga (1969a) either from the observed change in magnetic field at the shock (the solid arrows) or from multiple satellite observations (the dashed arrows). All 6 are clustered about the radial direction, and the assumption of a radial shock normal would be reasonable for any member of this limited sample. Figure 4 shows the orientations of eight flare-associated shocks, determined by Taylor (1969) from the measured changes in the magnetic field, under the assumption that the normal lies in the ecliptic plane. Each determination is shown at the heliocentric longitude of the observation relative to the associated flare. With only one exception, these shock orientations are consistent with a nearly spherical expansion from the flare site. Although Hirshberg *et al.* (1969) have shown an example of a shock normal tilted at 60° out of the ecliptic plane, the present limited evidence favors normals near the radial direction, and hence a nearly spherical shock front at 1 AU (see also Bame *et al.*, 1968, Greenstadt *et al.*, 1969).

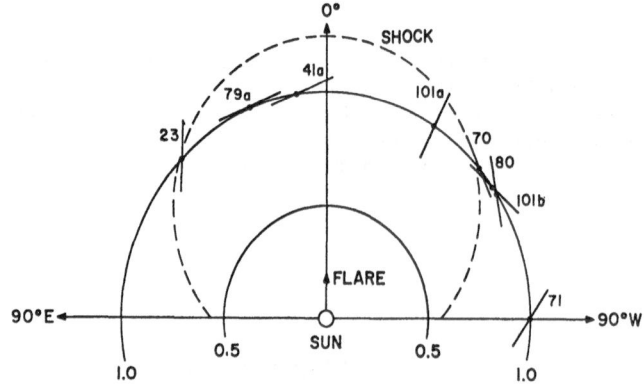

Fig. 4.   The orientations of eight flare-associated shocks (derived from the measured change in magnetic field, assuming that the shock normal lies in the ecliptic plane) plotted at the heliocentric longitude of observation relative to the associated flare (Taylor, 1969).

B. POST-SHOCK VARIATIONS IN SOLAR WIND PROPERTIES

The material ejected by a flare will be separated from the interplanetary shock by a region of swept-up plasma which was originally in the ambient solar wind. The interface between the flare or 'driver' gas and the swept-up ambient gas is expected to be a contact or tangential discontinuity (see Colburn and Sonett, 1966). The variations in solar wind properties in the regions of swept-up and driver gas contain information regarding the flare ejection process, and are also of considerable geophysical interest as the input signal to the magnetosphere during the development of a geomagnetic storm.

A particularly simple pattern of post-shock variations is displayed in Figure 1. The density, solar wind speed, and proton temperature all reached their maximum values immediately or soon after the shock and then decreased within eight hours to nearly the same values measured before the shock. This simple pattern is not that most often observed; a more typical case is shown in Figure 2. In this example the flow speed continued to rise throughout the five hours after shock passage, while the density and temperatures peaked about three hours after the shock, declining thereafter.

The solar wind variations associated with an interplanetary disturbance on February 15–16, 1967, have been particularly well documented and will be presented here as a detailed example. The positive-ion properties were measured by the LASL electrostatic analyzer on Vela 3A (see Hirshberg *et al.*, 1969) and the magnetic field properties by the ARC magnetometer on Explorer 33 (see Hirshberg and Colburn, 1969). Figure 5 shows the proton density N, the solar wind speed V_0, the proton temperature T, and the flow direction ϕ during the period of interest. The solar wind properties on February 15 showed only slow small changes until about 2350 UT, when abrupt rises in the density, flow speed, and temperature signaled the passage of a shock. All of the positive ion properties underwent large changes during the 14 hours of data transmission after the shock. Most striking of these changes were the large, steady rise in flow speed and the 20° shift in flow direction.

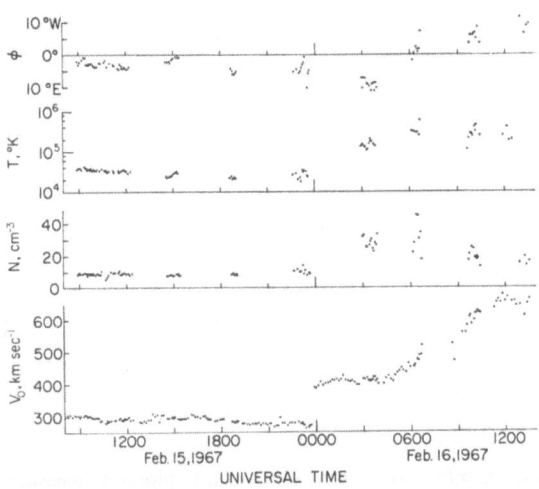

Fig. 5.   Vela 3A measurements of the solar wind flow direction ϕ, the proton temperature T, density N, and flow speed V_0 from February 15–16, 1967.

Figure 6 (from Hirshberg *et al.*, 1969) repeats the time history of the solar wind speed during the post-shock period and adds the magnetic field intensity measured on Explorer 33 and the helium content of the plasma (the ratio of the number densities of He^{++} and H^+) measured on Vela 3. The magnetic field increased abruptly at the shock (as expected) and reached very high values within the next 9 hours. The field pointed northward from the ecliptic plane until 0839 UT when a shift to the south occurred. At about 0910 UT the field magnitude dropped from 42 to 6γ in 5 min. Thereafter the field showed considerable fluctuation and remained southward pointing until 1112 UT. The helium content of the plasma underwent a series of changes best described with the aid of Figure 7, which shows a series of energy-per-charge spectra obtained during different stages of this disturbance. The flux peak due to He^{++} will occur at twice the energy per charge of the H^{+} peak. The helium content in the pre-shock and swept-up solar wind was near 1%, as seen from Figure 6 or the small secondary peak in the first three spectra of Figure 7. After 0841 UT Vela 3 entered the magnetosheath, but following the abrupt change in the magnetic field at 0910 UT, again observed the solar wind plasma. The first spectrum obtained after this change (the 4th on Figure 7) shows a large increase in the helium peak. The points defining the spectrum, obtained in a memory mode of spacecraft operation, are too widely spaced to give a good quantitative value of the helium content. This situation prevails until a spectrum is obtained in a direct data transmission mode 24 min. later. This spectrum (the last in Figure 7) shows the same large He^{++} peak, and gives a helium content of 22%. The helium-hydrogen ratio declines thereafter, but remains above 10% until 1008 UT.

The geomagnetic response to this complex pattern of post-shock variations is quite interesting (see discussion at the end of this paper); the interested reader is referred

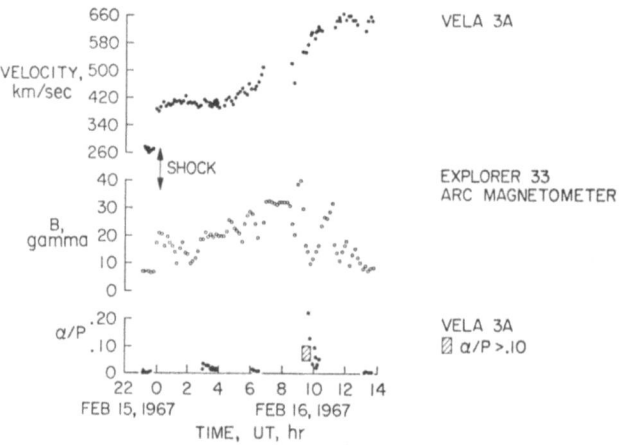

Fig. 6. Vela 3A measurements of the solar wind speed, Explorer 33 measurements of the magnetic field intensity, and Vela 3A measurements of the solar wind helium content (the ratio of number densities of He^{++} and H^{+}) from February 15–16, 1967. The shaded area on the helium content graph indicates a high but not quantitatively determined ratio (Hirshberg *et al.*, 1969).

to Hirshberg and Colburn (1969) and Hirshberg *et al.* (1969). Similar large enhancements of the solar wind helium content have been observed from 5 to 12 hours after an interplanetary shock on several other occasions (Gosling *et al.*, 1967; Bame *et al.*, 1968; Ogilvie *et al.*, 1968; Robbins *et al.*, 1969). Hirshberg *et al.* (1969) identify the plasma of high helium content observed on Feburary 16 with the driver gas ejected by the flare. The flare ejecta would be expected to be rich in helium because of its transient origin in the low corona, which is postulated to be enriched with helium by the process which lowers the normal solar wind helium content below the accepted solar value. The magnetic field changes observed just prior to the helium enrichment are ascribed to the tangential discontinuity separating the helium-rich driver gas from the helium-poor ambient gas.

C. COMPARISON WITH THEORETICAL MODELS

Theoretical models of interplanetary shock waves have been derived by Parker (1961), Simon and Axford (1966), Dryer and Jones (1968), and Lee and Chen (1968) by finding similarity solutions of the hydrodynamic equations. Implicit in all of these

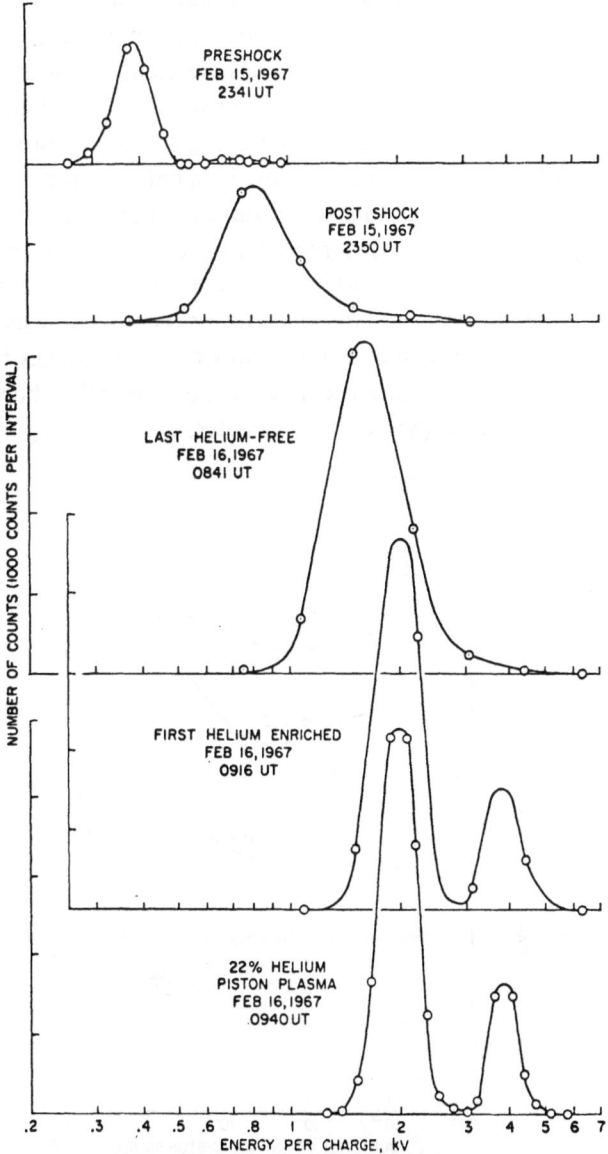

Fig. 7. Vela 3A energy-per-charge spectra obtained during the February 15–16 interplanetary disturbance. The large spectral peak is due to H⁺; the smaller peak at twice the energy per charge of the primary peak is due to He⁺⁺ (Hirshberg *et al.*, 1969).

models is the assumption that the flow speed of the ambient solar wind is negligible compared to the propagation speed of the shock wave. However, most of the observed interplanetary shocks listed in Table II have propagation speeds near 500 km sec^{-1}, whereas a typical ambient solar wind speed is about 400 km sec^{-1}. Thus the results of theories based on the neglect of the ambient motion are of limited applicability. To overcome this difficulty, Hundhausen and Gentry (1969) have used numerical integration of the hydrodynamic equations to derive models of shock propagation in a more realistic ambient solar wind. Numerous simplifying assumptions are retained (e.g., spherical symmetry about the sun, neglect of magnetic forces) which should not be forgotten when the models are compared with observations.

The ejection of material by a flare is simulated in the Hundhausen and Gentry models by the introduction of an outward-moving shock, followed by rapidly outflowing plasma, at 0.1 AU. The rapid outflow is maintained for a given duration of time, after which ambient conditions are resumed. The numerical calculation then follows the propagation of this disturbance in the ambient solar wind. Figure 8 shows the transit time to 1 AU for the shock at the front of a disturbance with given total energy and initial duration. An adiabatic ambient solar wind with a proton density of 12 cm^{-3} and a flow speed of 400 km sec^{-1} at 1 AU has been assumed. If the ejection of material by the flare occurs during a sufficiently short time ($\Delta \lesssim 10^{-2}$ on Figure 8), the transit time (and other parameters characterizing the shock) is only

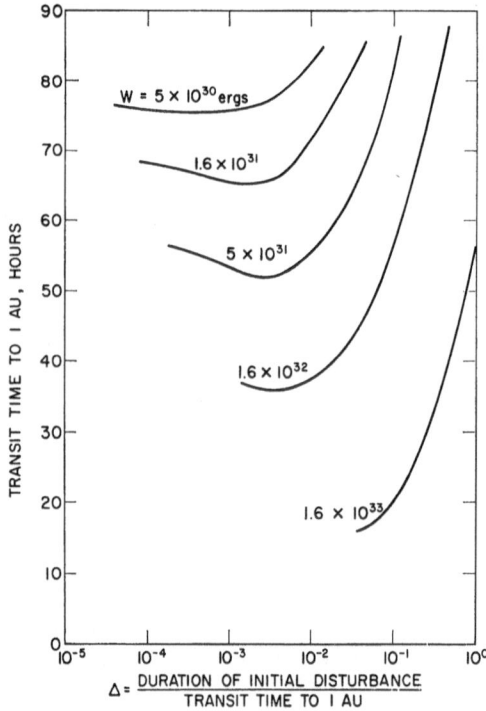

Fig. 8.   The transit time to 1 AU for the shock at the leading edge of an interplanetary disturbance with total energy W and initial duration Δ (normalized to the transit time). For $\Delta \lesssim 10^{-2}$, the transit time is only weakly dependent on Δ (Hundhausen and Gentry, 1969).

weakly dependent on the initial duration. In this limit of impulsive initiation of the flare-associated disturbance, the properties of the resulting interplanetary 'blast wave' depend only on the energy in the disturbance. Figure 9 shows the transit time to 1 AU as function of the shock propagation speed at 1 AU for this blast-wave limit and for the case of constant propagation speed. The blast waves can reach 1 AU earlier because they are moving more rapidly near the sun; momentum transfer to the swept-up ambient solar wind slows them to their 1 AU speed.

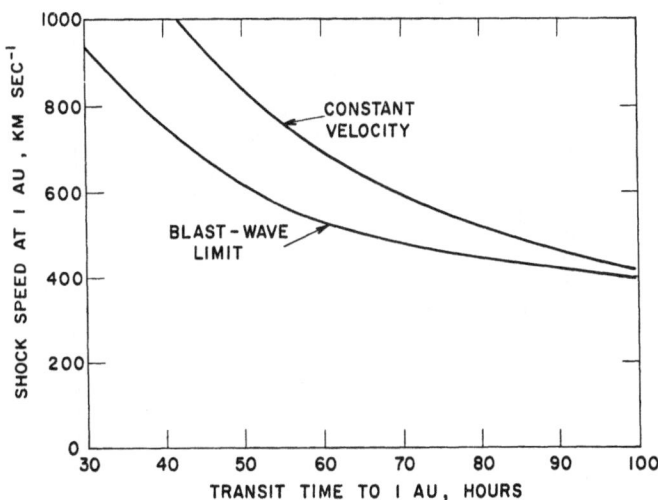

Fig. 9.   The transit time to 1 AU as a function of shock speed at 1 AU for (1) constant propagation speed and (2) the 'blast-wave limit' of Hundhausen and Gentry (1969).

A sh ock wave traveling at the typical observed speed of 500 km sec^{-1} at 1 AU is predicted to require 60 to 80 hours for transit from the sun. Examination of Table II reveals that the transit time inferred from flare associations are generally in the 50 to 70 hour range. Thus the Hundhausen-Gentry models and the observations are in reasonable agreement, and it is plausible to use the results of these calculations to infer from the interplanetary observations some properties of the flare ejection process.

Figure 10 gives the transit time to 1 AU as a function of energy in a blast wave. This figure applies directly to the 400 km sec^{-1}, 12 cm^{-3} ambient solar wind, and is easily adapted to another ambient density n through multiplication of the energy by $n/12$. For disturbances with finite initial durations, hereafter referred to as 'driven', a given transit time implies a somewhat higher energy. The typical transit time of 50 to 70 hours then corresponds to an energy of about 10^{31} (using an ambient solar wind density of 5 cm^{-3}) for blast waves, of a few times 10^{31} ergs for driven disturbances. Both the similarity and numerical calculations predict that the density, flow speed, and particle temperature decrease monotonically behind the shock in a blast wave, while one or more of these parameters will continue to rise behind the shock in a driven disturbance. The interplanetary disturbance shown in Figure 1 has the appearance of a blast wave, whereas those shown in Figures 2 and 5 follow the pattern predicted for driven shock waves. As the latter cases are more frequently observed, it follows that most interplanetary disturbances are of the driven type. Thus the energy released in material by the flare is a few times 10^{31} ergs and the material ejection must last for more than half an hour (about 10^{-2} of the transit time).

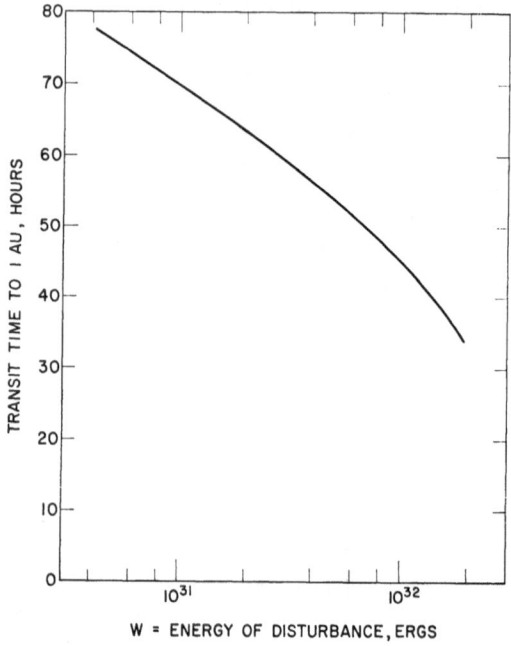

Fig. 10.   The transit time to 1 AU as a function of energy in the disturbance. This relation is directly applicable to an ambient solar wind density of 12 cm^{-3} at 1 AU. For an arbitrary density n, the energy must be multiplied by $n/12$ (Hundhausen and Gentry, 1969).

3. Recurrent Disturbances

The existence of a continuous outflow of coronal material to form the solar wind requires only a slight modification of the traditional explanation for recurrent geomagnetic storms. The stream of emitted particles is replaced by a solar wind stream, with some special property, emitted from a solar active region. Such solar wind streams were identified in the first long-term, continuous observations of the interplanetary plasma. Figure 11 (from Neugebauer and Snyder, 1966) shows 3-hour averages of the solar wind speed and proton density measured on the Mariner-2 Venus probe in 1962. The time coordinate has been broken into 27-day solar rotation periods. Large peaks in the flow speed recur on successive solar rotations, with corresponding density peaks at the rising portion of the velocity peak (Snyder *et al.*, 1963). Recurrent geomagnetic activity was associated with each of these 'high-velocity streams'. Attempts to relate these streams to specific regions of solar activity have not been successful.

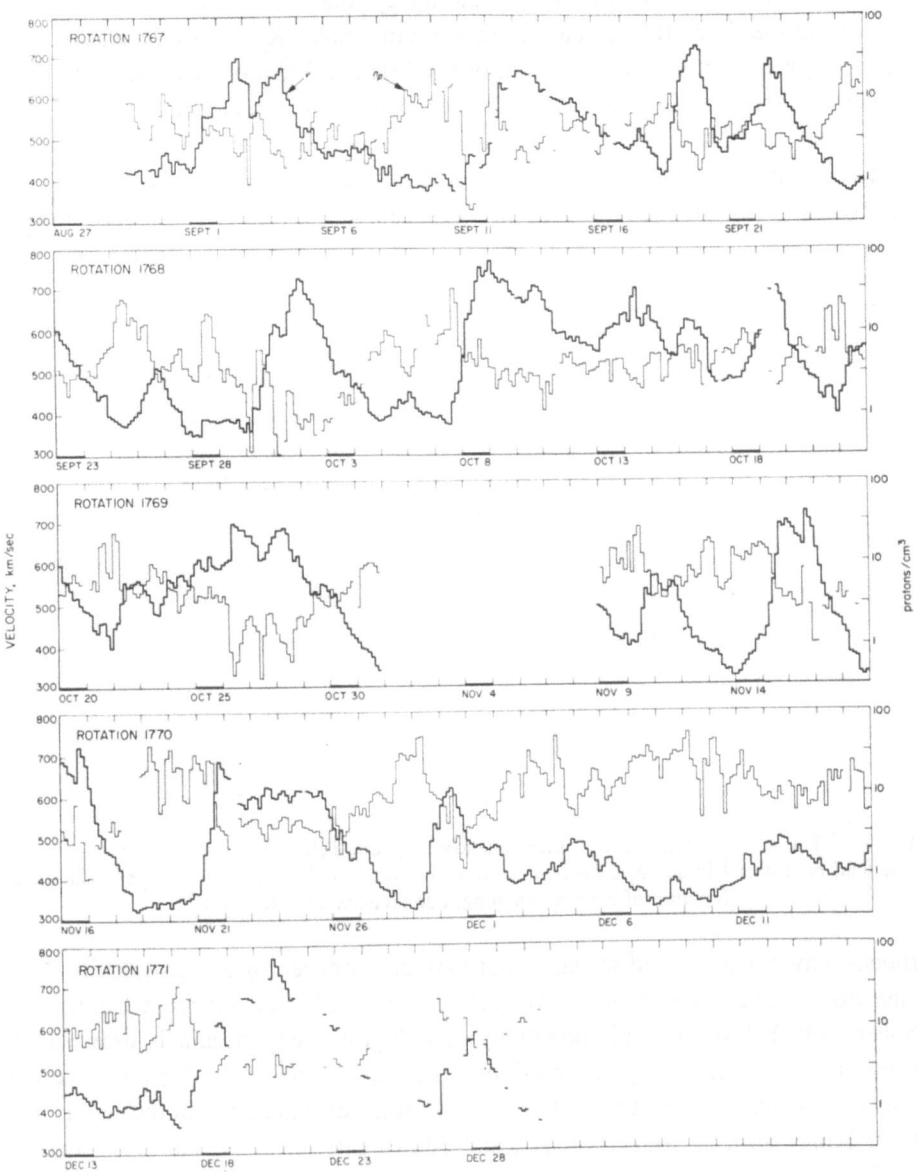

Fig. 11.   Three-hour averages of the solar wind speed and density observed on Mariner 2 in 1962
(Neugebauer and Snyder, 1966).

The interplanetary magnetic field measured within each Mariner-2 high velocity
stream was found to have a dominant polarity, pointing either toward or away from
the sun (Coleman *et al.*, 1966; Davis *et al.*, 1966). A similar pattern in the magnetic
field was observed on the Imp 1 satellite in 1963–64. Figure 12 (Wilcox and Ness,
1965) shows the Imp 1 magnetic field polarity (+ indicates a field away from the sun,
− a field toward the sun) as a function of heliocentric longitude. The recurrence of

the pattern with only minor changes during $2\frac{1}{2}$ solar rotations is clearly seen. The solar wind speed and the geomagnetic disturbance index K_p followed a pattern within each 'magnetic sector' similar to that of the Mariner-2 high-velocity streams.

This recurrent pattern has been discussed extensively in a recent review by Wilcox (1968). Of particular interest to the present discussion is the fact that the first ob- served interplanetary shock, on October 7, 1962 (see Table II) occurred at the leading edge of a recurrent high-velocity stream (see Figure 11). Thus all interplanetary

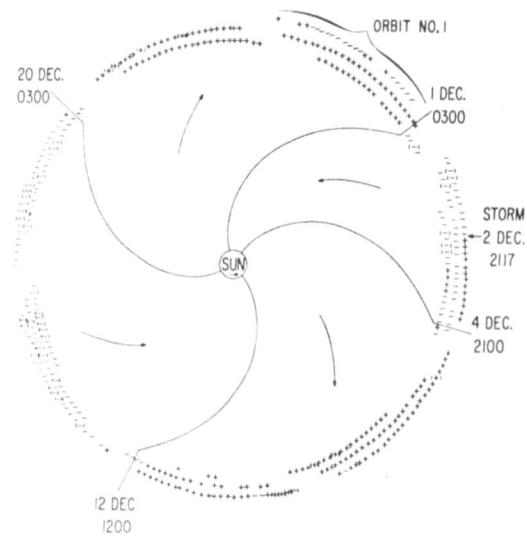

Fig. 12.   The sector pattern in the interplanetary magnetic field measured on Imp 1 in 1963–64. The field polarity ($+$ indicates a field away from the sun, $-$ a field toward the sun) is shown as a function of heliocentric longitude (Wilcox and Ness, 1965).

shocks may not be related to flares, but may be produced by a high-velocity stream emanating from a long-lived active region. Figure 13 (adapted from Colburn and Sonett, 1966) illustrates this possibility. The high-velocity stream is drawn into a spiral-like configuration by the rotation of the source region, and overtakes the slow- moving ambient plasma at the interface T. At some distance A from the sun, a shock front S may form in front of the stream, following its spiral-like configuration.

The shock waves related to solar flares and to long-lived source regions would be expected to have radically different spacial configurations. Due to the large energy density in the driver gas, the flare-associated interplanetary disturbance should ex- pand in a configuration roughly symmetric about the radial direction. In contrast, the shock preceding a recurrent high-velocity stream will have the spiral form shown in Figure 13. The normals of the flare-associated shocks should cluster about the radial direction while the normals of shocks associated with high-velocity streams should lie near 45° from the radial (as shown in Figure 13). The limited number of shock normal determinations described in Section 2A follow the first pattern, indi- cating a flare origin for most interplanetary shock waves. These results are not consistent with the recent hypothesis of Ballif and Jones (1969) that there is not a cause and effect relationship between flares and geomagnetic activity.

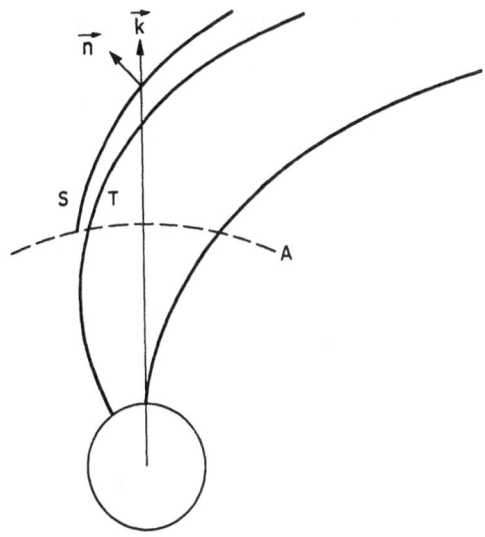

Fig. 13.   The geometry of a high velocity stream continuously flowing from a region on the rotating sun. The stream is drawn into a spiral-like configuration and overtakes the ambient solar wind along the front *T*. A shock *S* may form beyond some heliocentric distance *A*, following the general shape of the stream. The shock normal **n** will be tilted at about 45° from the radial direction **k** (adapted from Coburn and Sonett, 1966).

1
4. Multiple Satellite Observations

Observations of an interplanetary shock wave made at different satellites have occasionally been combined to give a more complete documentation of solar wind properties or used to determine shock properties (e.g., the propagation speed or the shock orientation). More interesting possibilities arise when the spacecraft involved are sufficiently separated so that spatial properties, such as shock decelerations or large-scale configurations, can be studied. These possibilities will be illustrated with two examples, one of which confirms some of the conclusions already drawn above, the other of which suggests new complications.

   The first example, combining observations made with the MIT plasma cup on Mariner 5 and the GSFC – U. of Maryland plasma experiment on Explorer 34, has been discussed by Ogilvie and Burlaga (1969). Figure 14 shows hourly averages of the solar wind speed, proton density, and proton mean thermal speed measured at both satellites on August 11–13, 1967. The two spacecraft were near the earth-sun line, with Mariner 5 closer to the sun by 1.6×10^7 km, or about 0.1 AU. Each Mariner 5 hourly average has been shifted to the time of arrival at the earth assuming propagation at the observed flow speed, and the densities have been corrected for an inverse-square dependence on heliocentric distance. An interplanetary shock was observed on Explorer 34 at 0555 UT on August 11 and at Mariner 5 during a gap in data transmission between 0000 UT and 0600 UT (shifted time) on August 12. The difference in time remaining after correction for motion at the solar wind speed is due to the propagation of the shock through the plasma. The material in the region behind the Explorer 34 shock but ahead of the Mariner 5 shock is that which has been swept up by the shock in moving 0.1 AU. The similarity of the variations in solar wind properties observed at satellites 0.1 AU apart indicates that the flow pattern retains its identity over this distance, i.e., the interplanetary plasma near 1 AU behaves like a fluid on a 0.1 AU scale. The flow pattern is also similar to that predicted by Hundhausen and Gentry (1969) for a driven shock.

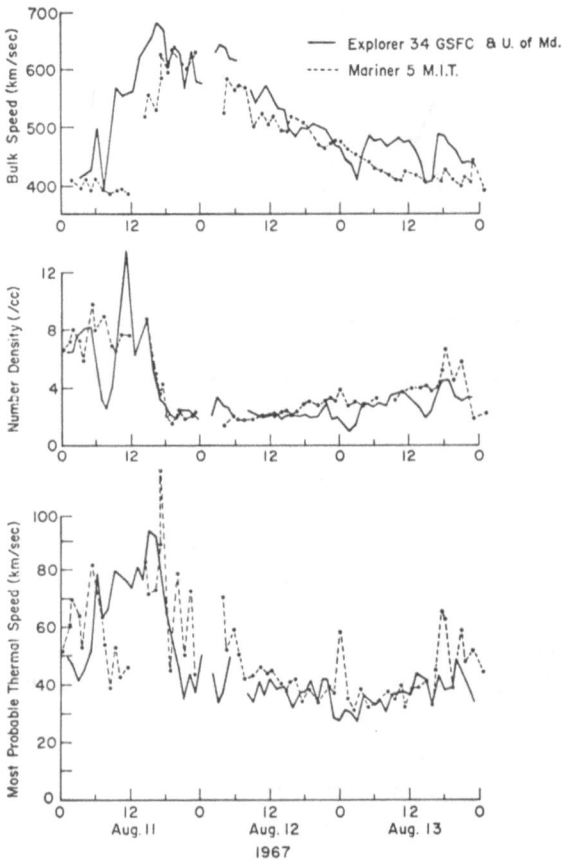

Fig. 14.   Hourly averages of the solar wind speed, proton density, and most probable proton thermal speed measured on Explorer 34 and Mariner 5, August 11–13, 1967.

The second example has been discussed by Lazarus and Binsack (1968). Figure 15 shows the positions of the two satellites involved, Explorer 33 near the earth, and Pioneer 6 at 0.84 AU heliocentric distance and 43.7° heliocentric longitude west of the earth. An interplanetary shock was observed on Explorer 33 at 2106 UT on July 8, 1966. Interpretation of the Pioneer 6 data is complicated by a gap in transmission, but it appears that no similar shock wave reached this spacecraft until 45 hours after the Explorer 33 observation, despite the fact that Pioneer 6 was 0.14 AU closer to the sun. Lazarus and Binsack point out that the boundary of an interplanetary magnetic sector was positioned as shown in Figure 15, and suggest that the spatial configuration of the shock wave was distorted by its presence. The plasma is generally dense near a sector boundary (as at the leading edge of the high-velocity streams of Figure 11), and the region of the shock front propagating toward Pioneer 6 would have reached this dense plasma earlier than that propagating toward the earth. The former region would travel most of the path to Pioneer 6 at the lower speed produced by passing through the dense shell, while the latter would travel most of the way to the earth at the faster original speed. This example may then illustrate the channeling of an interplanetary disturbance by the non-homogeneity of the ambient solar wind, here associated with the sector pattern.

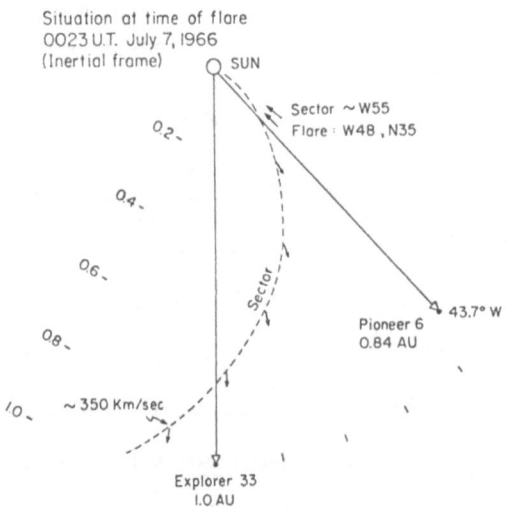

Fig. 15. The positions of Explorer 33, Pioneer 6, and a magnetic sector boundary on July 7, 1966 (Lazarus and Binsack, 1968).

5. Conclusions

A reasonably complete phenomenological description of a typical flare-associated interplanetary disturbance can be construed from satellite observations. An intermediate strength shock, propagating at about 500 km sec^{-1}, forms the leading edge of the disturbance. The observed shock orientations are generally consistent with a nearly spherical expansion from the flare site. Several solar wind properties, and in particular the flow speed, continue to rise after the shock passage. A region of high helium content often follows the shock by five to twelve hours, and has been tentatively identified as the 'driver gas' originally ejected by the flare. By comparison with theoretical models, these observations imply an energy of a few times 10^{31} ergs in the disturbance, with ejection from the flare lasting longer than $\frac{1}{2}$ hour.

Recurrent interplanetary disturbances, presumably associated with long-lived active regions, have also been observed. Shocks may also form at the leading edge of these structures, but the observed shock orientations can be interpreted as evidence that such shocks are not common.

A number of interesting questions concerning the propagation and spatial configuration of interplanetary disturbances remain open. Multiple satellite observations provide a means of answering these questions. The few multiple satellite studies which have been carried out both confirm the general description drawn above and indicate that inhomogeneities in the ambient solar wind may have important effects on the propagation and spatial structure of interplanetary shock waves.

Acknowledgements

This work was performed under the auspices of the United States Atomic Energy Commission. The author wishes to thank Dr. S. J. Bame of the Los Alamos Scientific Laboratory for permission to use the unpublished Vela data which appear in this paper.

References

Ballif, J. R. and Jones, D. E.: 1969, *J. Geophys. Res.* **74**, 3499.
Bame, S. J., Asbridge, J. R., Hundhausen, A. J., and Strong, I. B.: 1968, *J. Geophys. Res.* **73**, 5761.
Colburn, D. S. and Sonett, C. P.: 1966, *Space Sci. Rev.* **5**, 439.
Coleman, P. J., Jr., Davis, L. Jr., Smith, E. J., and Jones, D. E.: 1966, *J. Geophys. Res.* **71**, 2831.
Davis, L., Jr., Smith, E. J., Coleman, P. J., Jr., and Sonett, C. P.: 1966, in *The Solar Wind* (ed. by R. J. Mackin and M. Neugebauer), Pergamon Press, N.Y., p. 35.
Dryer, M. and Jones, D. L.: 1968, *J. Geophys. Res.* **73**, 4875.
Gold, T.: 1955, in *Gas Dynamics of Cosmic Clouds* (ed. by H. C. van de Hulst and J. M. Burgers), North-Holland Publishing Co., Amsterdam, p. 103.
Gosling, J. T., Asbridge, J. R., Bame, S. J., Hundhausen, A. J., and Strong, I. B.: 1967, *J. Geophys. Res.* **72**, 1813.
Gosling, J. T., Asbridge, J. R., Bame, S. J., Hundhausen, A. J., and Strong, I. B.: 1968, *J. Geophys. Res.* **73**, 43.
Greenstadt, E. W., Green, I. M., Inonye, G. T., and Sonett, C. P.: 1969, *Planet. Space Sci.* (in press).
Hirshberg, J. and Colburn, D. S.: 1969, *Planet. Space Sci.* **17**, 1183.
Hirshberg, J., Alksne, A., Colburn, D. S., Bame, S. J., and Hundhausen, A. J.: 1969, submitted to *J. Geophys. Res.*
Hundhausen, A. J. and Gentry, R. A.: 1969, *J. Geophys. Res.* **74**, 2908.
Lazarus, A. J. and Binsack, J. H.: 1968, to be published in *Space Res.*
Lee, T. S. and Chen, T.: 1968, *Planet. Space Sci.* **16**, 1483.
Neugebauer, M. and Snyder, C. W.: 1966, *J. Geophys. Res.* **71**, 4469.
Ogilvie, K. W. and Burlaga, L. F.: 1969a, Goddard Space Flight Center preprint X-616-69-29.
Ogilvie, K. W. and Burlaga, L. F.: 1969b, Goddard Space Flight Center preprint X-616-69-303.
Ogilvie, K. W., Burlaga, L. F., and Wilkerson, T. D.: 1968, *J. Geophys. Res.* **73**, 6809.
Parker, E. N.: 1961, *Astrophys. J.* **133**, 1014.
Robbins, D. E., Hundhausen, A. J., and Bame, S. J.: 1969, *Trans. Am. Geophys. Union* **50**, 302.
Simon, M. and Axford, W. I.: 1966, *Planet. Space Sci.* **14**, 901.
Snyder, C. W., Neugebauer, M., and Rao, U. R.: 1963, *J. Geophys. Res.* **68**, 6361.
Sonett, C. P., Colburn, D. S., and Briggs, B. R.: 1966 in *The Solar Wind* (ed. by R. J. Mackin and M. Neugebauer), Pergamon Press, New York, p. 165.
Taylor, H. E.: 1969, *Solar Phys.* **6**, 320.
Wilcox, J. M.: 1968, *Space Sci. Rev.* **8**, 258.
Wilcox, J. M. and Ness, N. F.: 1965, *J. Geophys. Res.* **69**, 1769.

Discussion

Lanzerotti: Recent work on solar alpha particles at somewhat higher energies, around 600 keV, at the time of shocks show the same type of behaviour as Dr. Hundhausen has shown for the solar wind, and we expect to use the α to proton ratio to demonstrate this short time scale modulation of the more energetic particles by the shock structure.

Roederer: Have you looked whether there is a correlation between the arrival of the He and the start of the main phase of a storm?

Hundhausen: The following Figure 1 shows that when the shock wave occurs, and the magnetosphere is compressed, during this early time between the shock wave and the arrival of the driver gas, the magnetic field rises and points north out of the ecliptic plane. Then one has the initial phase of the storm. But when the Helium arrives and the field points southwards, then very quickly things start happening, which involve internal currents in the magnetosphere and there is the main phase of the storm. We are tempted to associate the arrival of this driver gas with the development of the main phase of the storm. Those who believe in field line reconnection will recognise that there has been a sudden turn south of the field, which has done it. However, other things did happen. There were more fluctuations in the field, and one has to note that with a content of 22 % He, one has as much energy in He as one would have in Hydrogen and that this energy in particles with different mass might have a strange effect in the magnetosphere. But in fact this event is particularly interesting, because one sees clearly a correlation between the start of the main phase of the storm and something which happened in the solar wind. The magnetosphere is responding to changes in the solar wind.

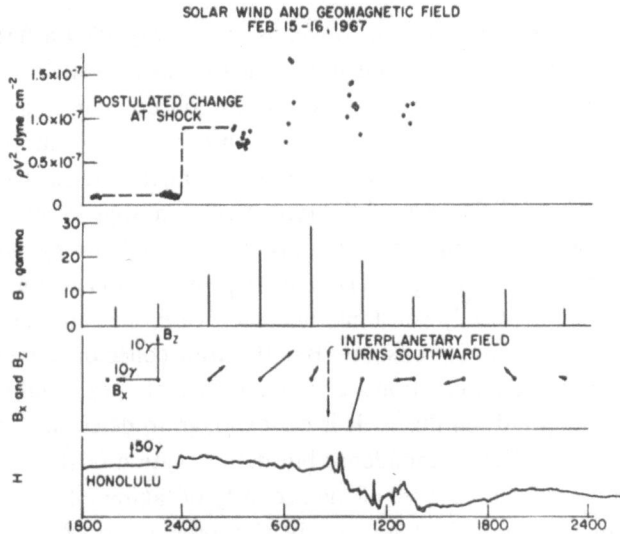

Fig. 1. – (a) Momentum flux of the solar wind. – (b) Three hours average magnitudes of interplanetary field. – (c) The Z and X components in solar magnetospheric coordinates (parallel to the dipole of earth and perpendicular to it towards the sun). – (d) Low latitude magnetometer trace.

Coleman: We on ATS1 in January 1967 have seen a case in which the picture you have outlined was not exactly reproduced. The inward motion of the magnetosphere boundary to the spacecraft was coincident with a large increase of the He flux but this occurred after the main phase of the storm had been in progress for some time.

VARIATIONS OF SOLAR WIND FLUXES OBSERVED ON BOARD
VENERA-5 AND VENERA-6 FROM JANUARY 21 TO MARCH 21 1969
AND PULSATIONS OF THE EARTH'S ELECTROMAGNETIC
FIELD CAUSED BY THESE VARIATIONS

K. I. GRINGAUZ, V. A. TROITSKAYA, E. K. SOLOMATINA and
R. V. SHCHEPETNOV

Radiotechnical Institute, Institute of the Physics of the Earth,
Academy of Sciences of the U.S.S.R., Moscow

Abstract. This paper describes the preliminary results of comparison of data obtained from recording of solar wind fluxes on board Venera-5 and Venera-6 from January 21 to March 21, 1969 and from recording of pulsations and short-period disturbances of the earth's electromagnetic field at the Soviet Observatory 'Borok' ($\Phi = 53°$, $\Lambda = 123°$). The considerable increases of solar wind fluxes as a rule were accompanied by considerable disturbances of the earth's electromagnetic field so that there is no doubt about their causality.

1. Some Information on Solar Wind Measurements
Carried out on Venera-5 and Venera-6

On each of the spacecraft launched to Venus in January 1969 a flat four-electrode integral trap was installed which made it possible to register positive ions with energies $E > 50$ eV. On the way from the earth to Venus the collector currents of the traps were measured; the results of measurements were conserved in a memorizing device, information from which was recorded on the earth during radiotelemetric transmissions.

The collector current of a trap of the type used, is always lower than the current created by ion fluxes affecting the trap. This is accounted for by photoemission of electrons from the suppressor grid [1]. The trap orientation to the sun during the measurements was practically constant; in such a case one can consider the above-mentioned photoemission as constant. But the trap collector currents varied to a great extent during the observations, giving evidence of occasional considerable increases of the solar wind ion fluxes. It is easy enough to determine variations of the ion fluxes ΔN_i with sufficient confidence but determination of the absolute values of ion fluxes is more difficult because of the necessity of taking into account the above-mentioned photoemission. That is why, in this preliminary paper, only values of ΔN_i above some level N_{i0} are given. The value N_{i0} (10^8 cm^{-2} sec^{-1} in order of magnitude) is not defined more precisely here.

2. On Earth's Electromagnetic Field Pulsations and
Their Registration at the Observatory 'Borok'

For comparison with the observations of solar wind fluxes from Venera-5 and Venera-6 the record of electromagnetic field pulsations of P_c-type and its variations over longer

V. Manno and D. E. Page (eds.), Intercorrelated Satellite Observations Related to Solar Events, 130–139. *All Rights Reserved*
Copyright © 1970 by D. Reidel Publishing Company, Dordrecht-Holland

periods [2] were used. According to recent ideas the P_c-type pulsations result from the interaction of solar wind fluxes with the outermost region of the subsolar part of the magnetosphere. The available experimental data give evidence that the values and periods of these pulsations depend both on the physical parameters of the magnetosphere in which magnetohydrodynamic waves propagate (radius of subsolar part of magnetosphere, space distribution of plasma density and magnetic field) and on the parameters of the interplanetary plasma interacting with it (bulk velocity and density of solar plasma, magnitude and direction of interplanetary magnetic field).

It is supposed that plasma parameters have a tangential discontinuity at the shock wave front, which may be unstable and responsible for the generation of surface waves; i.e. conditions may occur under which the amplitude of surface waves can increase considerably.

Waves generated at the magnetospheric boundary surface are transformed to magnetohydrodynamic waves by means of different processes. These waves propagate through the magnetosphere to the lower boundary of the ionosphere, and can create standing waves in 'magnetospheric resonators'. Magnetopause – plasmapause and plasmapause – lower boundary of the ionosphere serve as the walls of these resonators. It is worth noting that as the quality of these resonators is low ($Q \sim 5$–10) the life-time of these waves must completely depend on the time during which some solar wind inhomogeneity is interacting with the magnetosphere.

As the lower boundary of the ionosphere is separated from the earth by a non-conducting region, the magnetohydrodynamic waves in the ionosphere induce electromagnetic waves in this region. That is why magnetohydrodynamic waves in the magnetosphere can be detected on the surface of the earth by means of measurements of pulsations of electric and magnetic fields which are fluctuating synchronously.

It is convenient to observe electric field pulsations by recording the potential differences created by telluric currents induced in a well-conducting surface layer of the earth. At the observatory 'Borok' records of telluric currents are made by means of devices with much broader amplitude-frequency characteristics than those used to record the magnetic field (these devices enable one to record oscillations with periods from 10 sec to over 1h practically without distortion of amplitudes). For our preliminary comparison with solar wind data the most interesting feature was the general level of the electromagnetic field disturbance in a broad frequency band; that is why the telluric currents records were used for this purpose. The registration of telluric currents was carried out continuously on photopaper, moving with a speed of 90 mm per hour; the scale of the electric field horizontal components was the following:

the North–South direction

$$E_x = 0.036 \text{ mv/mm km}$$

the West–East direction

$$E_y = 0.131 \text{ mv/mm km.}$$

Such sensitivity corresponds to registration of similar magnetic field variations of magnitude ~ 0.08–0.1 γ/mm. So in the following comparison the records were used with a sensitivity exceeding the sensitivity of the standard magnetograms by more than an order of magnitude.

3. Results of the Observations and Brief Discussion

The results of the Venera-6 observations of solar wind ion fluxes for the period considered are presented in Figure 1 (the upper graph). Directly underneath the time intervals corresponding to increases of ion fluxes during which ΔN_i reached maximum value, equal to or more than 3×10^8 cm^{-2} sec^{-1} are marked. The lowest diagram shows time intervals during which pulsations and short-period disturbances of electromagnetic field were observed at the observatory 'Borok'. Only those with indices of disturbance level $E_d > 1$ are shown. The dimensionless index E_d was defined as the ratio of the maximum amplitude of the pulsations horizontal component of the electric field (equal to $\sqrt{(E_x^2 + E_y^2)}$) during one hour to its value averaged over a year from the 'Borok' data.

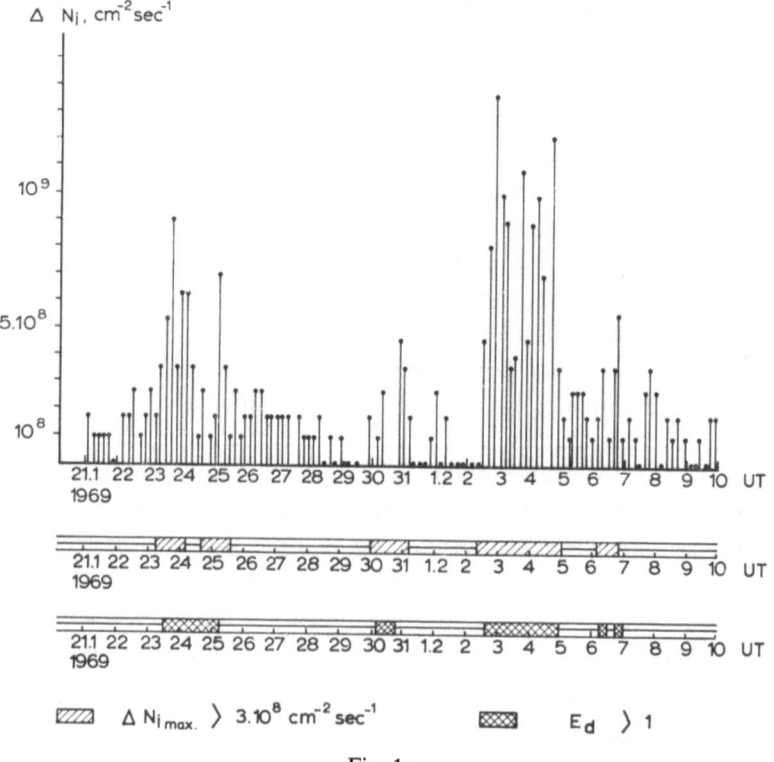

Fig. 1a.

Fig. 1.   Variations of solar wind ion fluxes measured on board Venera-6 for the period from January 21 to March 21, 1969 (upper graph). The time intervals corresponding to ion fluxes increases during which ΔN_i reached maximum value $\Delta N_{i_{max}} > 3 \times 10^8$ cm^{-2} sec^{-1} are marked lower; the time intervals during which pulsations and short-period disturbances with indices $E_d > 1$ were observed at the observatory 'Borok' are shown at the bottom of the graph.

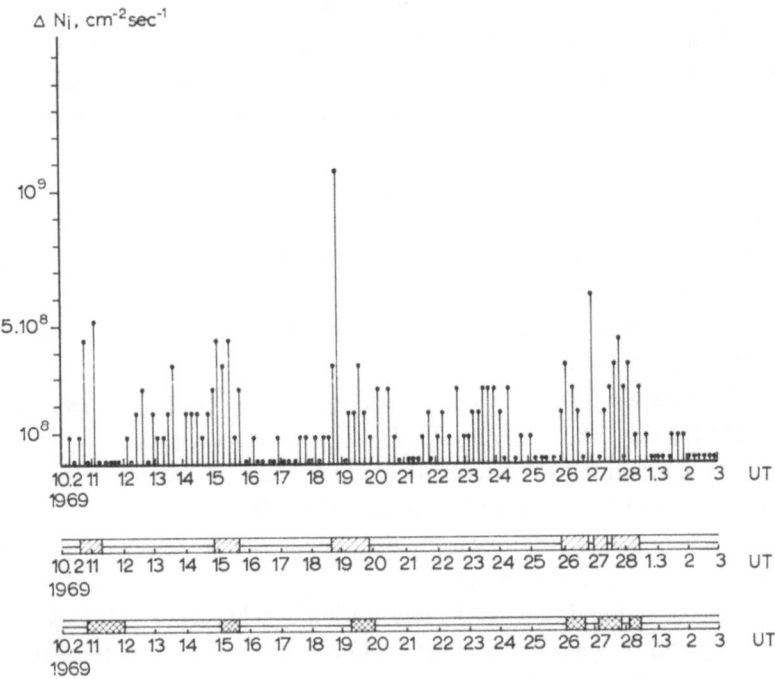

Fig. 1b.

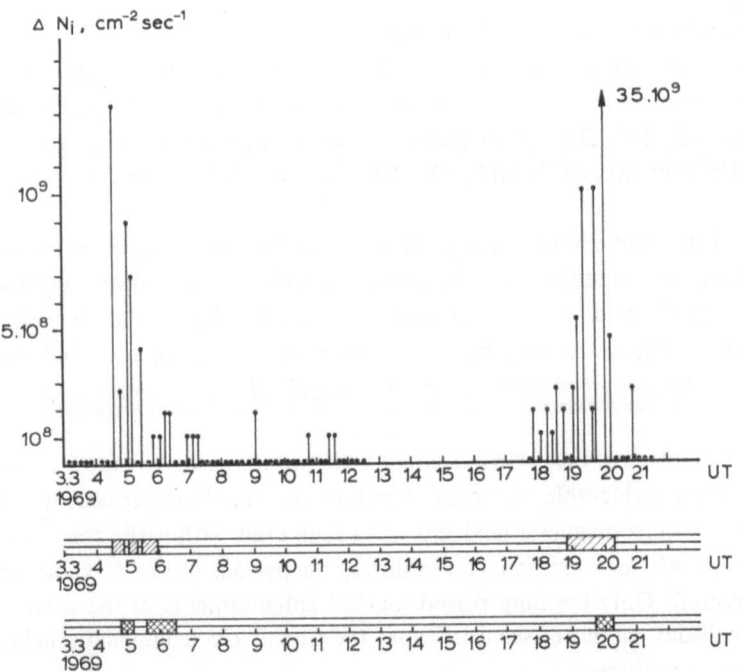

Fig. 1c.

In this paper universal time is given in all cases.

From 9 h March 12 to 17 h March 17 measurements were not carried out. A graph on which the data obtained from Venera-5 are plotted in a similar manner is very much like that given in Figure 1 and is not presented here. It differs from the graph given in Figure 1 in some details, because the spacecraft were at the distance of some millions of kilometers from one another and the times of ion flux measurements on both spacecraft were different.

In Figures 2, 3 and 4 samples of telluric currents data records used for this investigation are presented. The upper curve corresponds to the component of the horizontal electric field E_x (in the direction North–South), and the lower one to E_y (in the direction West–East).

Figure 2 corresponds to very small pulsations of the electromagnetic field of the earth during the day 3.8.1969 ($\Delta N_i \sim 0$, $E_d \sim 0$, see Figure 1).

Figure 3 corresponds to the sudden commencement of an electromagnetic storm (transition from quiet to disturbed state at $1^h 58^m$ 2.26.1969).

In Figures 4a and 4b transitions from disturbed to quiet states corresponding to the days 2.27 and 2.28.1969 are shown.

One must bear in mind that, as the registration of telluric currents was continuous, the beginning and the end of any event could be determined using these records with an error of some minutes. The variations of solar wind fluxes were measured only once every 4 h, that is why variations with shorter period could not be revealed and beginnings and ends of disturbances could be determined with correspondingly low accuracy.

Let us return to the graphs in Figure 1.

One can see that the periods when the solar wind ion fluxes were comparatively low as a rule correspond to periods with low level of the earth's electromagnetic field pulsations ($E_d \leqslant 1$). During all periods under consideration there were 10 cases of substantial increases of N_i ($\Delta N_i \gtrsim 3 \times 10^8$ cm^{-2} sec^{-1}) with duration more than 8 h each.

Each of these time intervals (including the interval following the event on February 25, 1969) is accompanied by a time interval with a disturbed and pulsating earth's electromagnetic field. In all cases mentioned, beginnings of the increases of N_i observed on spacecraft located between the sun and the earth preceded the commencements of the periods of the disturbances and pulsations of the earth's electromagnetic field.

A consideration of the earth current records within these two months showed that in daytime the considerable increases of pulsations and short-periodic disturbances of the earth's electromagnetic field did not occur even within the four-hour intervals between the ΔN_i measurements taken during the periods of low N_i values recorded on the spacecraft. Only the long period marked enhancements of the solar wind fluxes (about half-day or more) seem to cause the earth's electromagnetic field pulsations with large amplitudes.

It should be noted that comparison between N_i-values and K_p-indices (shown on

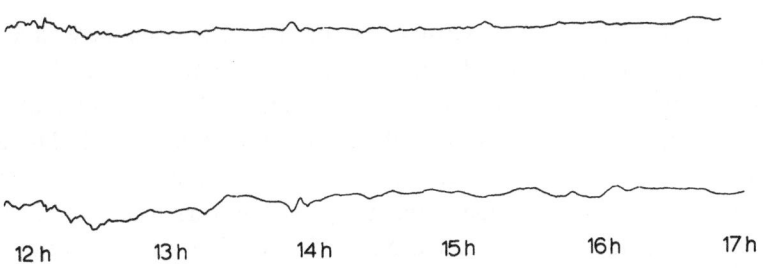

Fig. 2.   An example of telluric currents data records from the observatory 'Borok'. The upper curve corresponds to the component value of horizontal electric field E_x (in the direction North–South), and lower one to E_y (in the direction West–East). The state of the earth's electromagnetic field 3.8.1969 is very quiet ($\Delta N_i \sim 0$, $E_d \sim 0$).

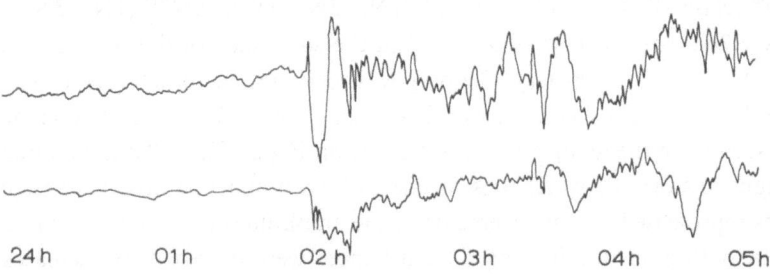

Fig. 3.   An example of telluric currents data records corresponding to sudden commencement of electromagnetic storm (the transition from quiet to disturbed state at 1^h58^m 2.26.1969).

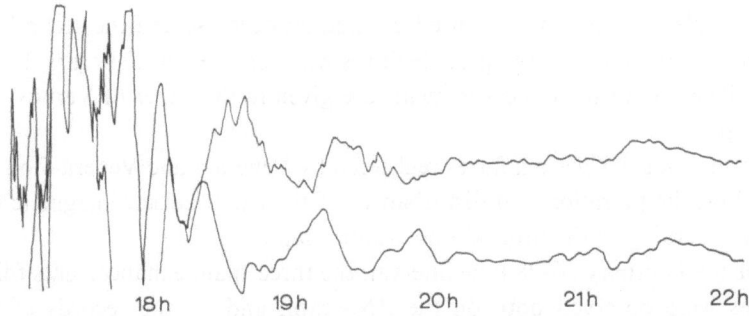

Fig. 4a.   An example of telluric currents data records corresponding to the transition from disturbed to quiet state 2.27.1969.

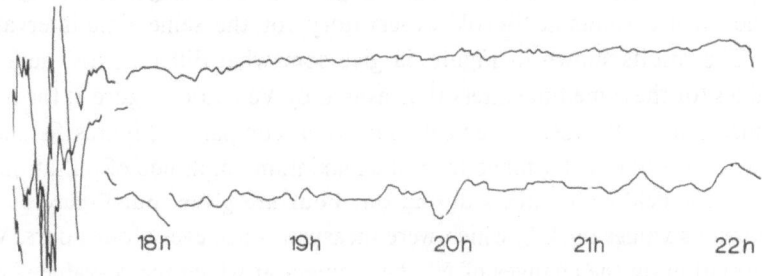

Fig. 4b.   The transition similar to the one shown in Figure 4a for 2.28.1969.

Figure 1) reveals that there is no satisfactory correlation between these data (although the values of $N_i \gtrsim 10^9$ cm^{-2} sec^{-1} generally corresponds to $K_p \sim 5$).

The good agreement of ΔN_i-values with the results of the registration of pulsations and disturbances with short periods of the earth's electromagnetic field together with the poor correlation with K_p-indices may be related to the fact that K_p-indices are defined from the data of a standard magnetograph network whereas the registration of pulsations is made, as was pointed out in Section 2, with instruments of higher sensitivity.

4. On the Events of February 25–27, 1969

As it is known a number of solar flares were observed on February 25 (including those at 9^{h}20^m (class 2B, 14N, 36W), 16^{h}59^m (1N, 13N, 41W), 19^{h}43^m (1N, 13N, 43W). On February 26 a solar flare was observed in the same part of the solar disk at 04^{h}46^m (1B, 13N, 46W). On February 27 a flare was registered at 14^{h}08^m (2B, 14N, 65W) [3].

At 01^{h}58^m on February 26 a world magnetic storm Sc began (its commencement can be seen from the telluric currents records on Figure 3). At 00^{h}38^m on February 27 the magnetic storm Sc which was registered by a number of observatories began [3] and was represented by enhancements of the amplitudes of the telluric current pulsations. In both cases the indicated disturbances were preceded by increases of ΔN_i-values observed on Venera-5 and Venera-6. In addition on February 25 the solar flares were observed aboard Venera-5 and Venera-6 with devices aimed to observe protons of energy $\leqslant 30$ MeV at ~ 10 h. These appeared in the form of a sharp enhancement of the particle counting rate followed by gradual decrease and the second sharp enhancement of the same energy particle fluxes was observed on February 27 at ~ 20 h. The detailed data of these measurements are given in the paper of Vernov *et al.*, this volume, p. 53.

The values of solar plasma fluxes registered by Venera-5 and Venera-6 on February 25 were low, the pulsation and disturbances of the earth's electromagnetic field were also low according to the 'Borok' observatory data.

As for the February 26–28 time interval, the three main enhancements followed by decreases were observed both on the ΔN_i-graph and on the records of the earth electromagnetic field pulsations (Figure 1).

In Figure 5a the results of measurements of ΔN_i aboard the Venera-5 are given for the period February 25–28, 1969, and in Figure 5b the changes of hourly E_d-indices from the earth currents at 'Borok' observatory for the same time interval are presented. The results shown in Figure 5a give somewhat different ΔN_i data from the ΔN_i-values for the same time interval measured by Venera-6 (Figure 1) for the reasons mentioned above. It must be remembered when comparing Figures 5a and 5b that in Figure 5b – values determined from the maximum amplitude of pulsations of horizontal electric field components during one hour are given, but Figure 5a gives the instantaneous values of ΔN_i, which were measured once every four hours. With such an information on the changes of N_i, the moment at which the N_i-value is maximum can be missed because it can occur between two measurements.

Nevertheless in Figures 5a and 5b one can clearly see the above mentioned increases and decreases of solar wind fluxes and earth's electromagnetic field pulsations during the February 26–28 period.

We think that it is difficult to establish which of the numerous solar flares that occurred over the period of February 24–27 (in part they are noted in the beginning of this section) caused each of the enhancements of the solar wind flux, geomagnetic field disturbances and increase of pulsations observed over the time interval considered. Some considerations on the identification of the solar flares causing the magnetic

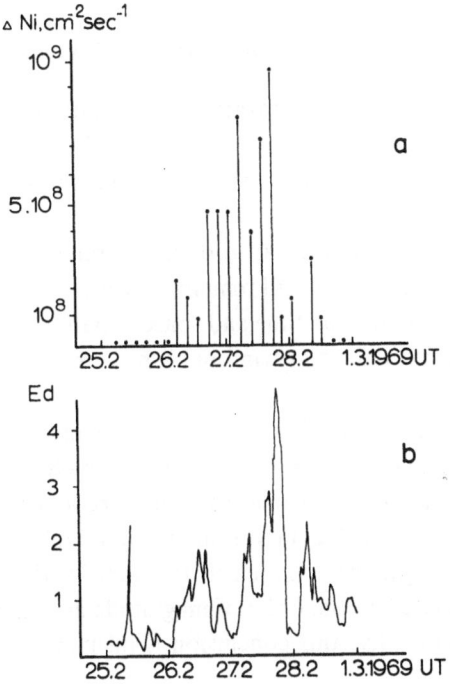

Fig. 5

Fig. 5.   Variations of solar wind ion fluxes measured on board Venera-5 (upper graph) and the pulsations and short-period disturbances of earth's electromagnetic field from the observatory 'Borok' for the period February 25–28, 1969.

storms Sc observed on February 25 and 27 are given in the paper of Vernov *et al.*, in the present volume, p. 53. We should like to note the following. If the above-mentioned magnetic storms Sc are caused by the arrival of shock waves fronts to the earth from interplanetary plasma (which in turn are created by solar plasma eruptions accompanying the flares) then it follows from the data shown in Figures 1 and 5a that the eruptions were continuous (lasting many hours). This is because in both cases the ΔN_i values reached the maximum only after many hours after the beginning of their increase which may be connected with the commencement of the storm.

According to numerical calculations of Hundhausen and Gentry [4] pertaining to the phenomena related to the interplanetary plasma disturbances caused by solar flares, the maximum solar wind ion flux occurs during long disturbances not at the shock wave front (where the ion velocity is greatest) but sufficiently far beyond the front. One can see it from the values of velocities and concentrations of ions in Figure 3 [4].

This conclusion is in qualitative agreement with the observational results of Figure 5.

The maximum ΔN_i value was observed about 12^h after magnetic storm Sc on February 27; the maximum electric field E_d disturbance occurred also at 21^h on the same day, i.e. a long time after the commencement of the geomagnetic disturbance.

5. Note on Solar Wind Fluxes Values

In the last few years the important characteristics of interplanetary plasma are under extensive and successful investigation. Those are the chemical composition, the variations of directions of the bulk velocity, the temperature anisotropy, the discontinuities of plasma parameters. We should like, however, to draw attention to the fact that some problems related to such an ordinary parameter as the value of integral flux of solar wind particles $N = nv$ still remain unclear.

There is no clear information on time and space variations of the solar wind particle flux in the excellent reviews on solar wind published recently by Axford [5], Ness [6] and Hundhausen [7].

It seems to be evident that the flux value should change inversely with the square of the distance from the sun and this was concluded by Snyder and Neugebauer [8] from the data of Mariner 2. But if one takes into account that, as a result of some misunderstanding in published data on solar wind ion flux values measured on Mariner-4 during the flight from the earth to Mars by the MIT group [10] these values were underestimated by the order of magnitude as related by Prof. A. Lazarus [9], then one may think that this conclusion is insufficiently supported by experiments.

It is not clear enough whether the quality of the plasma emitted by the sun (i.e. mean solar wind flux) changes with the phase of solar activity since the conclusion made in [11] is not supported by the preliminary Venera-5 and Venera-6 data.

It was pointed out previously that considerable magnetic storms sometimes occur during large variations of solar wind flux without substantial changes of the solar wind bulk velocity [12].

The comparison of results of different experiments, even when they are made simultaneously, presents a problem because of differences between methods and devices used for solar wind investigations and non-identity of data-processing procedures.

Considering that direct measurements of fluxes of solar wind particles N_i are interesting and important (this is supported by the data of this paper) one can conclude that every solar wind measurement must include determination of integral particle flux with the maximum obtainable precision.

Conclusion

There is a close relationship between the variations of solar wind ion fluxes ΔN_i and pulsations and short-periodic disturbances of the earth's electromagnetic field. The comparison between the ΔN_i measurements aboard Venera-5 and Venera-6 and the records of telluric currents and magnetic field pulsations of 'Borok' observatory revealed that increases of the amplitudes of earth's electromagnetic field pulsations ($E_d > 1$) corresponded to each case when $\Delta N_i \geqslant 3 \times 10^8$ cm^{-2} sec^{-1}. In future the simultaneous analysis of data on variations of solar wind fluxes and earth's electromagnetic field pulsations will be continued; in particular the behaviour of pulsations with different periods will be analysed.

The consideration of the results of ΔN_i registrations and the records of earth's electromagnetic field pulsations over the period of February 25–28, 1969, and taking into account the solar events data and results of energetic particles ($E \leqslant 30$ MeV) registration enable one to believe that during this period the earth's magnetosphere was affected by at least two shock wave fronts in interplanetary plasma (2.26.69 and 2.27.69) caused by solar flares. Each of these flares seemed to be accompanied by a long (with durations of many hours) cold plasma eruption and in each case the maximum flux values occurred rather far beyond the shock wave front (this is in qualitative agreement with [6]).

Together with the investigations of the fine details of the solar wind structure the vagueness related to space and time variations of solar wind integral flux values $N_i = n_i v$ and to their geophysical effects should be removed.

References

[1] Gringauz, K. I.: 1961, *Space Res.* vol. 2, North-Holland Publ. Co., Amsterdam, p. 539.
[2] Troitskaya, V. A.: 1967, *Solar-Terrestrial Physics* (ed. by T. W. King and W. S. Newman), p. 213.
[3] *Solar-Geophysical Data*, March, 1969.
[4] Hundhausen, A. J. and Gentry, R. A.: 1969, *J. Geophys. Res.* **74**, 2908.
[5] Axford, W. I.: 1968, *Space Sci. Rev.* **8**, 331.
[6] Ness, N. F.: 1968, *Ann. Rev. Astron. Astrophys.* **79**.
[7] Hundhausen, A. J.: 1968, *Space Sci. Rev.* **8**, 690.
[8] Snyder, C. W. and Neugebauer, M.: 1964, *Space Res.* vol. 4, 89.
[9] Private communication.
[10] Lazarus, A. J., Bridge, H. S., and Davis, J. M.: 1967, *Space Res.* vol. 7, p. 1296.
[11] Gringauz, K. I. and Solomatina, E. K.: 1968, *Kosmitcheskie Issledovaniya*, **6**, 586.
[12] Gringauz, K. I., Bezrukikh, V. V., and Musatov, L. S.: 1967, *Kosmitcheskie Issledovaniya* **5**, 251.

Discussion

Question: What was the distance from the earth of the spacecraft on February 26?

Gringauz: Venera 6 at 3 million km at the beginning of the period and 17 million km at the end. Venera 5 was farther from the earth because launched some days earlier.

Question: Was the time lag corrected for in relating to the solar wind parameters and geomagnetic response?

Answer: No, observations refer to real time.

MAGNETIC FIELD VARIATIONS IN INTERPLANETARY SPACE

JOHN M. WILCOX

Space Sciences Laboratory, University of California, Berkeley, Calif., U.S.A.

Abstract. Magnetic field variations and structures in interplanetary space are described and related to the sun. Discrete flare-related effects in the interplanetary medium occur within a large-scale interplanetary sector pattern related to a solar sector pattern that is ordered over a large extent in heliographic latitude and longitude. A random walk transport of photospheric field lines slightly modifies the over-all ordered pattern, but sector boundaries appear to resist the random walk process. There is a delay of approximately one solar rotation between appearance of an active region and its possible effect on the interplanetary sector pattern. A flare occurrence is most probable near a solar sector boundary.

In an other paper at this Symposium Hundhausen (1970) has already discussed the magnetic variations incident to interplanetary shock waves, so as to describe these structures completely in terms of both plasma and field variations. Solar wind disturbances associated with flares have also recently been reviewed by Wilcox (1969). The

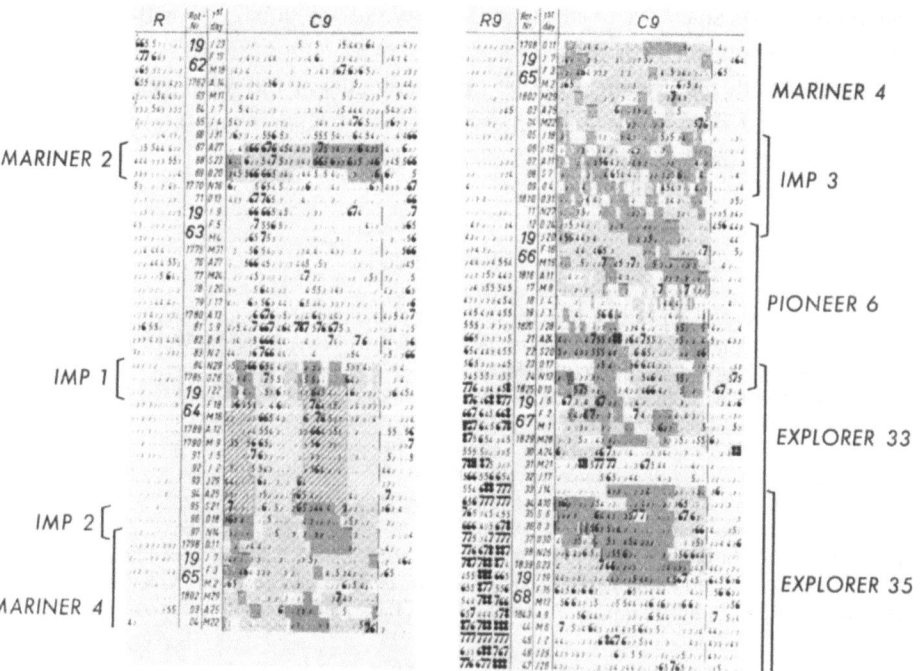

Fig. 1. Observed sector structure of the interplanetary magnetic field, overlayed on the daily geomagnetic character index C9, as prepared by the Geophysikalisches Institut in Göttingen. Light shading indicates sectors with field predominantly away from the sun, and dark shading indicates sectors with field predominantly toward the sun. Diagonal bars indicate an assumed quasi-stationary structure during 1964 (Wilcox and Colburn, 1969).

V. Manno and D. E. Page (eds.), Intercorrelated Satellite Observations Related to Solar Events, 140–151. *All Rights Reserved*
Copyright © 1970 by D. Reidel Publishing Company, Dordrecht-Holland

present discussion therefore gives a broader view of the configuration of the inter-planetary magnetic field and its relation to the sun.

We may first observe that the sector structure of the interplanetary magnetic field has persisted up to the most recent observations, as shown in Figure 1. For several consecutive days the interplanetary field will be varying about an average Archimedes spiral directed away from the sun. Then at a sector boundary there is an abrupt change of direction to an average spiral field directed toward the sun for the next several days. It is within this large-scale sector pattern that the specific solar flare-related effects occur.

A recent investigation of the solar origin of the interplanetary field has utilized observations at the Crimean Astrophysical Observatory by Severny (1969) of the mean photospheric magnetic field. In these observations Severny simply admits sun-light to the solar magnetograph and observes the resulting Zeeman splitting. In this observation each area of the visible solar disk has equal weight in proportion to its brightness. A comparison of this mean photospheric field with the direction of the interplanetary magnetic field is shown in Figure 2, after Wilcox *et al.* (1969). A very close correspondence between these two magnetic fields can be observed in Figure 2. The interplanetary observations have been adjusted to allow for an average 5-day transit time of solar wind plasma from sun to earth, as has been previously observed (Wilcox, 1968).

This agreement between solar and interplanetary fields could exist only if the solar source of the interplanetary magnetic field was part of a field pattern that was ordered over an appreciable portion of the solar disk. (If the solar pattern that was the source

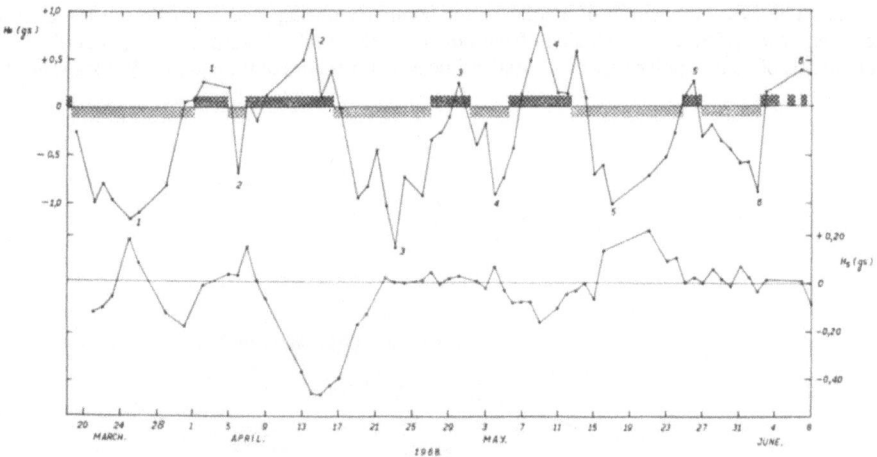

Fig. 2. (Top) Mean value of the solar magnetic field observed at the Crimean Astrophysical Observatory (dots) and polarity of the interplanetary magnetic field (bars), displaced to allow for a 5-day transit time for the solar wind plasma from the sun to the earth. – (Bottom) Contribution of sunspot magnetic fields to the mean solar field shown above. The polarity of the sunspot fields tends to be opposite to the polarity of the mean solar field and the interplanetary magnetic field. This suggests that sunspot fields are not *directly* related to the source of the interplanetary magnetic field (Wilcox *et al.*, 1969).

of the interplanetary field existed only over a small range of latitudes this pattern would contribute only a small part of Severny's observations, and would not control the time of polarity reversals.) Such a large-scale solar sector pattern has been proposed by Wilcox and Howard (1968) and by Schatten *et al.* (1969). Figure 3 shows the position of a solar sector boundary as a function of heliographic latitude observed by Schatten *et al.* (1969). The weak large-scale photospheric magnetic field has the same predominant polarity over a wide range of latitudes on both sides of the solar equator, and the predominant polarity changes at the sector boundary as shown in Figure 3.

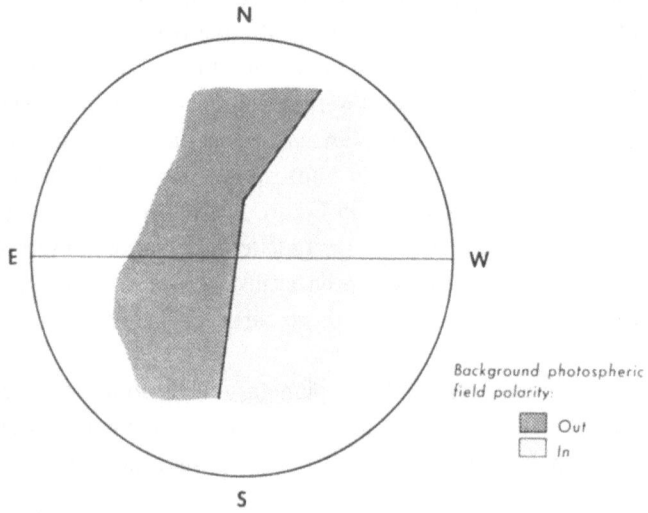

Fig. 3.   The average position of a solar sector boundary during 1965 according to the analysis of Schatten *et al.* (1969). On each side of the boundary the weak photospheric magnetic field is predominantly of a single polarity in equatorial latitudes on both sides of the equator (Wilcox *et al.*, 1969).

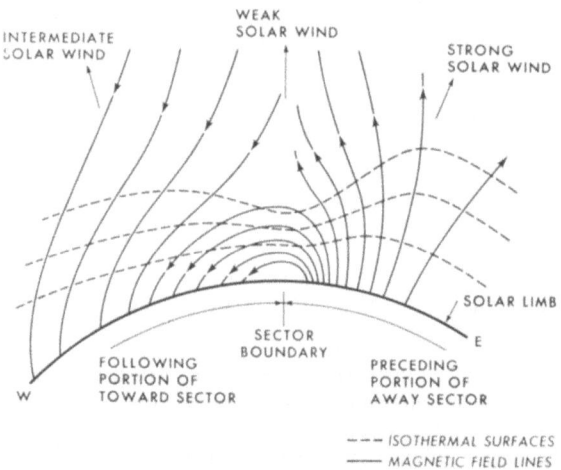

Fig. 4.   Plausible magnetic-field pattern and isothermal surfaces near a solar sector boundary (Wilcox, 1968).

The large-scale configuration in longitude of the solar sector magnetic field may be similar to that shown in Figure 4, after Wilcox (1968). The large-scale pattern of the solar field may be modified by a random walk process recently proposed by Jokipii and Parker (1969). Consider the bundle of magnetic field lines associated with a small element of solar wind plasma during its four or five day transit time from the sun to the earth. This bundle of field lines will not remain fixed on the sun in the small element of area from which this solar wind plasma left the sun, but rather the field lines will be dispersed in the photosphere by a random walk process associated with the supergranulation as first described by Leighton (1964). A schematic of this process is shown in Figure 5. During the transit time of an element of solar wind plasma from the sun to the earth the field lines that originally passed through this element in the photosphere will be dispersed in a Gaussian pattern having a $1/e$ width that corresponds to the amount of solar rotation in approximately two days.

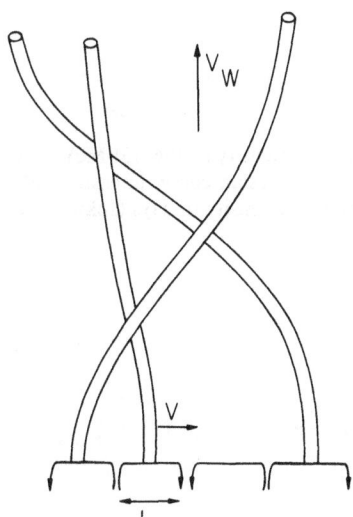

Photospheric Supergranulation Pattern

Fig. 5. Schematic illustration of the field-line random walk generated by the turbulent motions in the photosphere. A typical element of fluid moves a distance L at velocity V, and a cell lasts a time $t = L/V$. The field lines are convected out by the solar wind at velocity V_w, which is several times the Alfvén velocity (Jokipii and Parker, 1969).

Fan et al. (1968) have observed the intensity of 0.6–13 MeV protons in the interplanetary medium near the earth. Often such protons are present in a stream lasting for several days as the interplanetary magnetic field pattern corotates with the sun past the earth. Figure 6 shows the intensity of one of these streams observed during April 1966. If we assume that these protons were accelerated in a localized active region, the agreement of these observations with the Gaussian curve having a $1/e$ width of 2.5 days is consistent with the mechanism proposed by Jokipii and Parker.

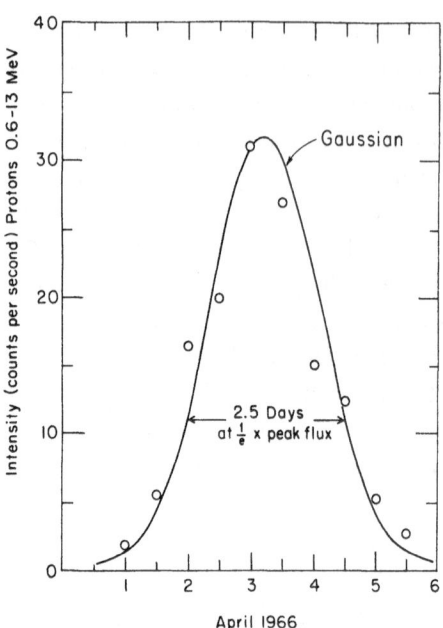

Fig. 6.   Experimental points are the intensity of (0.6–13)-MeV protons in 1966 observed with Pioneer 6 by Fan *et al.* (1968). *Solid line*, Gaussian curve, intensity $= 32 \times \exp[-(\Delta t/1.25)^2]$, where Δt is time for maximum intensity in days (Jokipii and Parker, 1969).

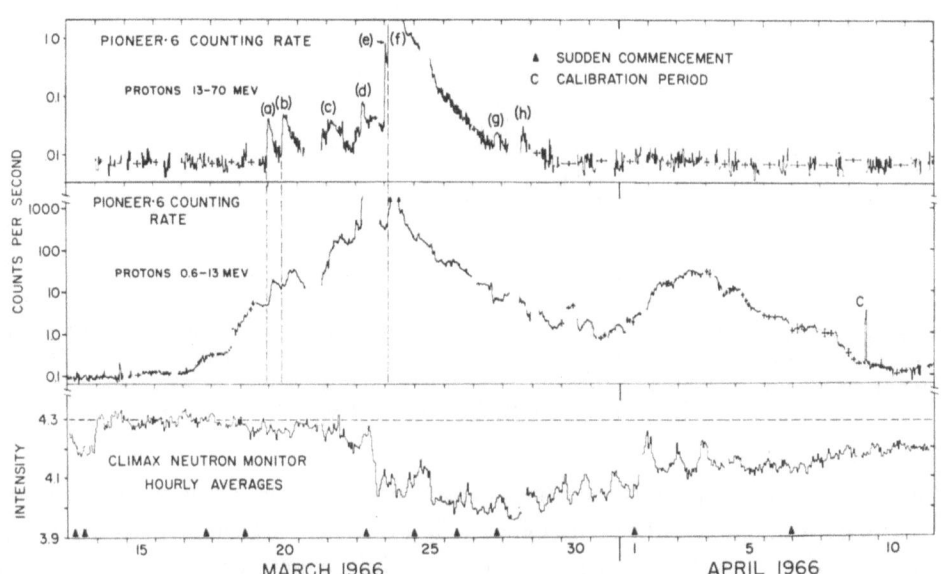

Fig. 7.   Thirty-min averages of the counting rates of protons 13–70 and 0.6–13 MeV. The enhanced flux of 0.6- to 13-MeV protons during March 15–31 is attributed to solar region 8207. The first evidence of enhanced flux from the following region (8223) appears on March 31. A magnetic sector boundary occurs on March 31. Note at this time the abrupt change in the level of modulation at the Climax neutron monitor. (a)–(h) denote discrete flare and shock-wave events seen at 13–70 MeV (Fan *et al.*, 1968).

Figure 7 shows another of the MeV interplanetary proton streams observed by Fan *et al.* (1968). Notice that the ordinate in this figure is logarithmic, as compared with the linear ordinate shown in Figure 6. The spikes in the intensity of protons 13–70 MeV labelled (a)–(h) denote discrete flare and shock-wave events. The physical interpretation of Figure 7 (see also Wilcox and Schatten, 1969) is that the envelope of the proton stream observed at 0.6–13 MeV is caused by protons more or less continually accelerated at an active region by sub-flare processes that have followed interplanetary field lines dispersed by the Jokipii-Parker process into the wide longitudinal section of the interplanetary medium. Protons accelerated in the flares (a)–(h) at the active region were able to almost immediately reach the interplanetary medium near the earth by travelling along the field lines dispersed by the Jokipii-Parker process.

A magnetic sector boundary was observed on March 31, 1966. Notice that this appears to mark the division between two proton streams shown in Figure 7. Fan *et al.* (1968) conclude that an interplanetary magnetic sector may occupy most but not necessarily all of the region associated with the proton fluxes. Further evidence of the relation between a sector boundary and energetic proton fluxes is shown in Figure 8, which shows observations by McCracken *et al.* (1968) of the temporal

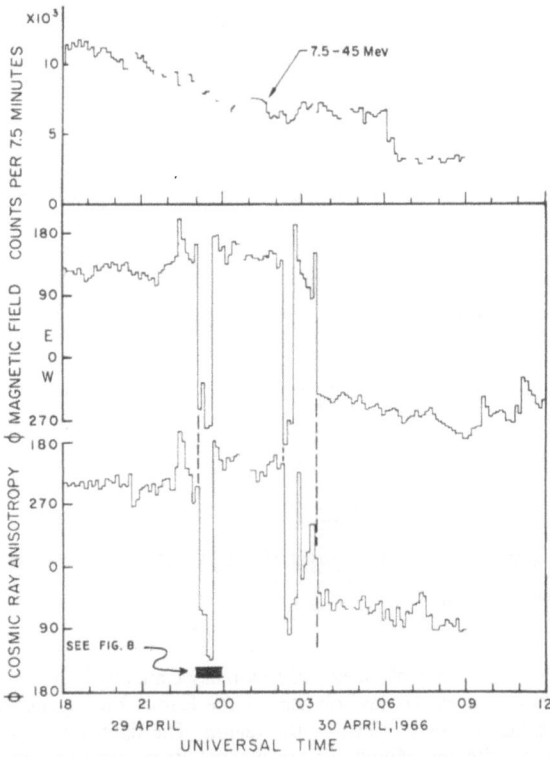

Fig. 8.   The temporal dependence of the flux of 7.5- to 45-MeV protons, of the azimuth of the magnetic field, and of the cosmic-ray anisotropy, during the passage of a magnetic sector boundary past the Pioneer 6 spacecraft. Note the high degree of correlation between the changes in the magnetic and cosmic-ray azimuths (McCracken *et al.*, 1968).

dependence of the flux of 7.5–45 MeV interplanetary protons, of the azimuth of the interplanetary magnetic field, and of the cosmic-ray anisotropy. The marked change of cosmic-ray anisotropy at the sector boundary is evident. Notice that the flux of energetic protons has an abrupt decrease approximately $2\frac{1}{2}$ hours after the sector boundary has passed. Lanzerotti (1969) has also observed changes in energetic proton fluxes related to a magnetic sector boundary.

It appears that a sector boundary may be an exception to the magnetic random walk process proposed by Jokipii and Parker. The persistence of a given sector boundary essentially unchanged for a year or more would not appear to be consistent with a diffusion of field lines caused by the random walk process. The energetic proton observations discussed above also indicate that the random walk process does not produce field lines crossing sector boundaries. The termination of energetic interplanetary proton streams near but not necessarily at sector boundaries remains an unexplained problem. Diffusion might cause some spilling over of protons across a sector boundary, but presumably would not yield a sharp termination near but not at a sector boundary.

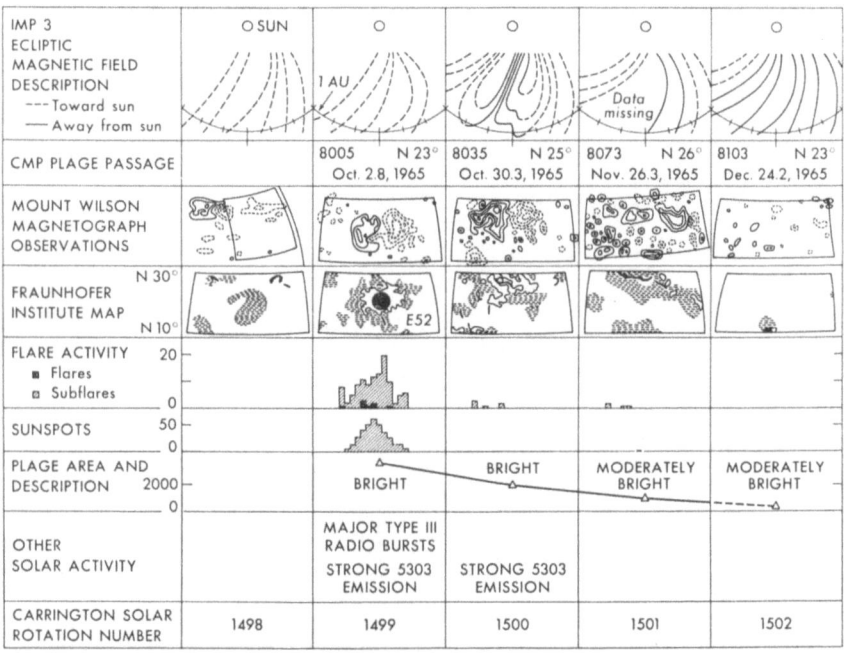

Fig. 9.   Chart showing the history of the active region associated with the interplanetary magnetic-loop event. Each column shows the development of the feature during successive solar rotations. Each row describes different observations of the region. The figures are centered on the central meridian plage passage, with the Mount Wilson magnetograph observations and the Fraunhofer Institute maps extending over a scale of 40° in longitude and 20° in latitude. The first contour level on the Mount Wilson magnetogram for solar rotation 1502 has been omitted due to an increase in noise during that time period. The plage area is graphed on a scale of millionths of the solar disk
(Schatten *et al.*, 1968).

It seems reasonably well established that there tends to be a delay of approximately one solar rotation between the appearance of a magnetic feature in the photosphere and a corresponding change in the interplanetary sector pattern. A summary of an investigation by Schatten *et al.* (1968) of the influence of a solar active region on the interplanetary magnetic field is shown in Figure 9. Each column represents a solar rotation and each row represents a particular observation. The top row represents a portion of a map in the ecliptic plane of the interplanetary magnetic field, showing in rotation 1500 the appearance of a loop pattern convected outward by the solar wind. In the following rotation a new interplanetary sector appears, which increases in size in subsequent rotations. Development of a bipolar photospheric magnetic region and of plage and sunspot activity is shown in the 3rd and 4th rows of Figure 9. The important point for the present discussion is shown in the 5th row, in which one can see that almost all of the flares associated with the active region occurred *before* the change in the interplanetary field that gave rise to a new sector. In other words the shock waves and accelerated particles associated with a flare tend to be sent out into the pre-existing large scale pattern of the interplanetary magnetic field. This effect has also been observed on a statistical basis during 9 solar rotations by Schatten *et al.* (1968).

Bumba and Obridko (1969) have shown that flare activity and especially proton-flare activity is concentrated in the neighborhood closest to the solar sector boundaries. Figure 10 is a histogram of frequency distribution of the time differences between

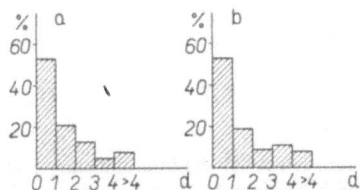

Fig. 10.   The histogram of frequency distribution of the time differences between the central meridian passage of spot groups and that of the interplanetary magnetic field sector boundary for the groups: (a) with flares of importance 1 + or greater; (b) with the number of flares equal or greater than ten (Bumba and Obridko, 1969).

the central meridian passage of spot groups and that of the interplanetary magnetic field sector boundaries for the case of groups with flares of importance of 1+ or greater and with the number of flares equal to or greater than 10. About one half of the proton-flare region developments were accompanied in the interplanetary magnetic structure by fast and short lived changes of polarity around the boundary of sectors.

In addition to the interplanetary shock wave events that have been discussed at this Symposium by Hundhausen (1970), interplanetary large-amplitude, aperiodic Alfvén waves propagating outward from the sun along the average magnetic field direction have been observed by Belcher *et al.* (1969). These waves appear to be present at least 30% of the time during 5 months of observations by Mariner 5 in

1967. Figure 11 shows plots of the radial component of the interplanetary magnetic field and of the solar wind velocity. A very detailed correlation of the nonsinusoidal fluctuations can be observed. Belcher *et al.* (1969) give a number of reasons for interpreting the observed fluctuations as Alfvén waves. In particular they observe a strong tendency for the fluctuations of the magnetic field to be normal to the average magnetic field. This is required for any superposition of Alfvén modes but not for the magneto-acoustic modes. Figure 12 shows the distribution of the angle between the field averaged over a 6-hour interval and the direction of minimum fluctuation of B during this interval, showing that the fluctuations are predominantly normal to the average field direction.

Power spectra of the interplanetary magnetic field have been obtained by Coleman (1966) with data from Mariner 2 in 1962, by Siscoe *et al.* (1968) with Mariner 4 data

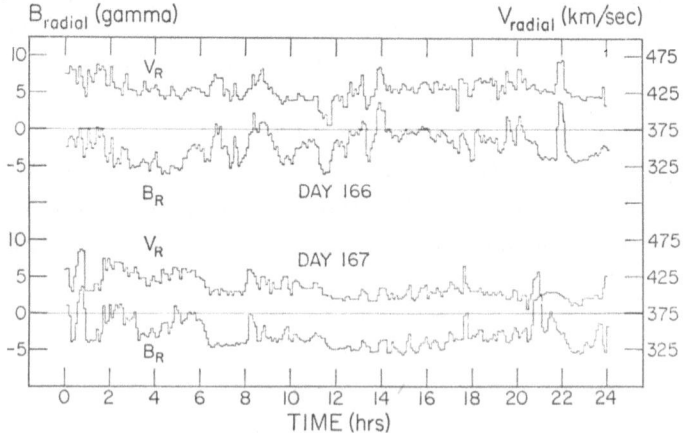

Fig. 11.   Plots of the radial components of the interplanetary magnetic field, B_R, and of the solar wind velocity, V_R, observed with Mariner 5. Note the very good detailed correlation of the non-sinusoidal fluctuations (Belcher *et al.*, 1969).

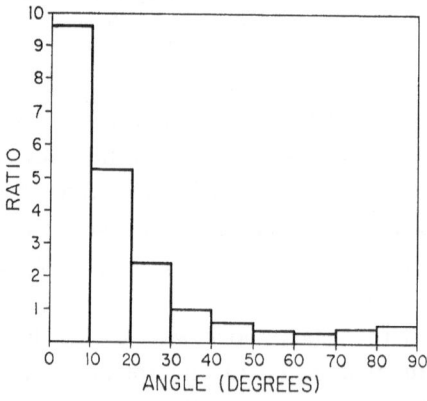

Fig. 12.   Distribution of the angle between the field averaged over a 6-hour interval, and the direction of minimum fluctuation of the field during this interval. The ordinate is the ratio of the number observed to that expected for an isotropic distribution (Belcher *et al.*, 1969).

in 1964, and by Sari and Ness (1969) with Pioneer 6 data in 1966. The results are summarized in Figure 13. In the higher frequency range the spectral slope and magnitude are dominated by the presence of microstructural discontinuities on a scale less than 0.01 AU. The discrepancy between the various observations shown in Figure 13 is not understood, but could be related to a change with time (and with the solar cycle) of the microstructure of the interplanetary medium. The observed power spectra have been related to the theoretical determination of the cosmic ray diffusion coefficient by Jokipii (1966) and by Roelof (1966).

We may conclude by mentioning an interesting new technique for observing fluctuations in the interplanetary medium. Levy *et al.* (1969) have observed the Faraday rotation of the telemetry carrier signal from Pioneer 6 when it was occulted by the sun in the last half of November 1968. Figure 14 shows observations of polarization versus time when the line-of-sight distance from the sun was 6.2 solar radii. An increase in the electron content of the ionosphere could only cause an increase in

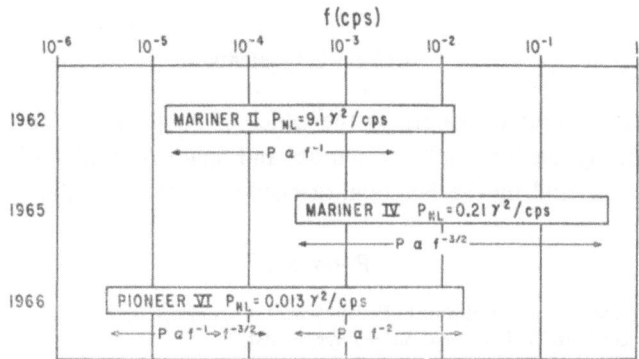

Fig. 13.   Comparison of power spectra of the interplanetary field obtained by Coleman (1966) with data from Mariner 2 in 1962, by Siscoe *et al.* (1968) with Mariner 4 data in 1964, and by Sari and Ness (1969) with data from Pioneer 6 in 1966. The noise levels $T_{\rm NL}$ are also indicated (Sari and Ness, 1969).

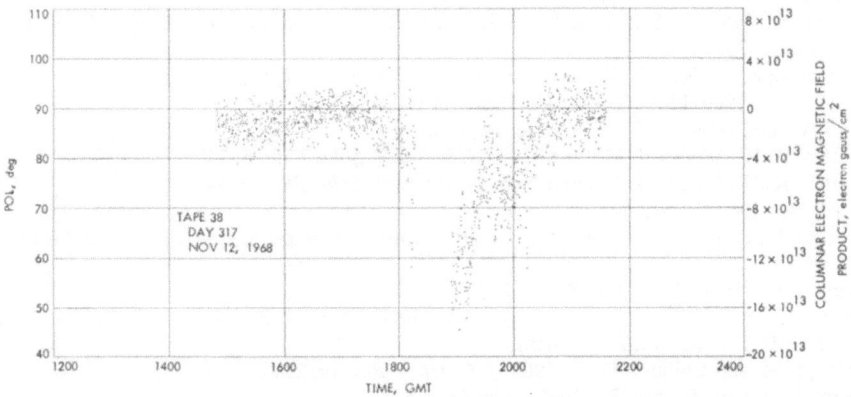

Fig. 14.   Faraday rotation of the telemetry carrier signal from Pioneer 6 near solar occultation when the line-of-sight distance from the sun was 6.2 solar radii (Levy *et al.*, 1969).

polarization angle above 90°. In Figure 13 the base line is approximately 87°. The reason for this appears to be an increase in the steady-state plasma density and magnetic field as the line-of-sight approached the sun. These steady-state phenomena are now being analysed and compared with the interplanetary magnetic sector pattern. The transient event is probably associated with a moving plasma concentration ejected by the sun. It appears to be associated with solar dekametric radio bursts. If this is correct the plasma velocity is of the order of several hundred km/sec.

In summary, it appears that discrete flare-related effects in the interplanetary medium occur within a large-scale interplanetary sector pattern related to a solar sector pattern that is ordered over a large extent in heliographic latitude and longitude. A random walk transport of photospheric field lines slightly modifies the overall ordered pattern, but sector boundaries appear to resist the random walk process. There is a delay of approximately one solar rotation between appearance of an active region and its possible effect on the interplanetary sector pattern. A flare occurrence is most probable near a solar sector boundary.

Acknowledgements

This work was supported in part by the Office of Naval Research under Contract Nonr 3656(26), by the National Aeronautics and Space Administration under Grant NGR 05-003-230, and by the National Science Foundation under Grant GA-1319.

References

Belcher, J. W., Davis, L., Jr., and Smith, E. J.: 1969, *J. Geophys. Res.* **74**, 2302.
Bumba, V. and Obridko, V. N.: 1969, *Solar Phys.* **6**, 104.
Coleman, P. J., Jr.: 1966, *J. Geophys. Res.* **71**, 5509.
Fan, C. Y., Pick, M., Pyle, R., Simpson, J.A., and Smith, D. R.: 1968, *J. Geophys. Res.* **73**, 1555.
Hundhausen, A. J.: 1970, this volume, p. 111.
Jokipii, J. R.: 1966, *Astrophys. J.* **146**, 480.
Jokipii, J. R. and Parker, E. N.: 1969, *Astrophys. J.* **155**, 777.
Lanzerotti, L. J.: 1969, *J. Geophys. Res.* **74**, 2851.
Leighton, R. B.: 1964, *Astrophys. J.* **140**, 1547.
Levy, G. S., Sato, T., Seidel, B. L., Stelzried, C. T., Ohlson, J. E., and Rusch, W. V. T.: 1969, *Science*, **166**, 596.
McCracken, K. C., Rao, U. R., and Ness, N. F.: 1968, *J. Geophys. Res.* **73**, 4159.
Roelof, E. C.: 1966, Ph.D. Thesis, University of California, Berkeley.
Sari, J. W. and Ness, N. F.: 1969, *Solar Phys.* **8**, 155.
Schatten, K. H., Wilcox, J. M., and Ness, N. F.: 1968, *Solar Phys.* **5**, 240.
Schatten, K. H., Wilcox, J. M., and Ness, N. F.: 1969, *Solar Phys.* **6**, 442.
Severny, A.: 1969, *Nature*, **224**, 53.
Siscoe, G. L., Davis, L., Jr., Coleman, P. J., Jr., Smith, E. J., and Jones, D. E.: 1968, *J. Geophys. Res.* **73**, 61.
Wilcox, J. M.: 1968, *Space Sci. Rev.* **8**, 258.
Wilcox, J. M.: 1969, *Solar Flares and Space Research* (ed. by Z. Švestka and C. De Jager), North-Holland Publishing Co., Amsterdam.
Wilcox, J. M. and Colburn, D. S.: 1969, *J. Geophys. Res.* **74**, 2388.
Wilcox, J. M. and Howard, R.: 1968, *Solar Phys.* **5**, 564.
Wilcox, J. M. and Schatten, K. H.: 1969, *J. Geophys. Res.* **74**, 2449.
Wilcox, J. M., Severny, A., and Colburn, D. S.: 1969, *Nature*, **224**, 353.

Discussion

Schindler: Some time ago you gave an upper limit to the width of the sector boundary from IMP-1 data as being 150000 km. Do you have any more recent information, i.e. can you now resolve the thickness of the neutral sheet?

Wilcox: The sector boundary is very thin and no detailed study has been made. Usually the crossing is made within a minute or so, consistent with thickness of 150000 km. However, it is not uncommon for the boundary to be very complex with a number of field reversals of a few minutes duration over a period of 2 to 3 hours, before the field finally changes over. Each discrete change is very narrow.

Axford: If the sector boundary on the sun is at constant longitude, how does this survive differential rotation?

Wilcox: The sector boundary resists differential rotation, in other words, it is fixed in a rigidly rotating system over the range from 40°N and 40°S. We have seen this constant longitude boundary in observations in 1964 and 1965, and the agreement is reasonable.

4. BOW SHOCK VARIATIONS FOLLOWING SOLAR EVENTS

PLASMA MEASUREMENTS ACROSS THE BOW SHOCK
AND IN THE MAGNETOSHEATH

A. J. HUNDHAUSEN

University of California, Los Alamos Scientific Laboratory, Los Alamos, N.M., U.S.A.

1. Introduction

The existence of a standing bow shock wave at which the incoming flow of supersonic solar wind is deflected around the geomagnetic cavity has become well established. Plasma and magnetic field properties near the bow shock and in the magnetosheath, the region of altered flow between the shock and the nearly impenetrable boundary of the geomagnetic field (the magnetopause), have been observed by instruments on numerous satellites and space probes (see reviews by Ness, 1967, and Wolfe and Intriligator, 1969). Both the spatial configuration and the plasma flow pattern of the magnetosheath are found to be in basic agreement with hydrodynamic models of hypersonic flow around an obstacle (see the review by Spreiter and Alksne, 1969).

This discussion will center on two areas of interest related to the bow shock and magnetosheath. The first will be the changes in plasma properties which occur at the bow shock; this topic is pertinent here as it is the post-shock or magnetosheath plasma which is in actual contact with the magnetopause and which transmits interplanetary disturbances to the magnetosphere. The second area of interest will be the changes in bow shock and magnetopause position which occur in response to solar wind variations, and which in turn induce changes in the geomagnetic field.

2. Plasma Changes at the Bow Shock

The bow shock is defined observationally by the abrupt changes which occur in the characteristics of the plasma and magnetic field. These changes will be discussed here with emphasis on the nature of the post-shock plasma flowing around the magnetosphere and on the distinct nature of the pre-shock and post-shock states. Both the detailed information contained in the particle distribution functions derived from observed fluxes and the simplified representation given by the conventional fluid parameters used to describe the distribution functions will be presented. The discussion will be based on examples which will be drawn from Vela satellite data, with which the author has a working familiarity. The reader is referred to the review papers already cited for related observations made on other spacecraft.

A. POSITIVE ION DISTRIBUTION FUNCTIONS

The positive ions are of primary interest in this discussion, as they carry most of the momentum and energy associated with the solar wind. Figure 1 shows a typical set

V. Manno and D. E. Page (eds.), Intercorrelated Satellite Observations Related to Solar Events, 155–169. *All Rights Reserved*
Copyright © 1970 by D. Reidel Publishing Company, Dordrecht-Holland

of positive ion flux measurements made in the solar wind by the LASL electrostatic analyzer on Vela 3A. On the upper half of the figure is an energy-per-charge spectrum, or the measured flux as a function of energy per charge in a given direction (in this case the direction of maximum flux). The channels in which the flux is collected are very narrow, having a full width at half-maximum transmission of about one-fourth the interchannel spacing. The large flux peak is due to H^+; the smaller peak at twice

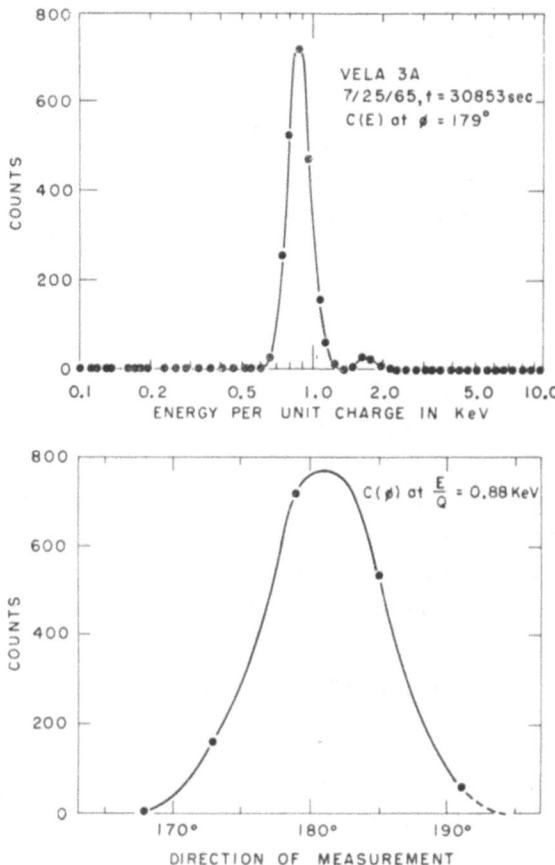

Fig. 1.   A typical positive ion energy-per-charge spectrum and directional distribution obtained in the solar wind by LASL electrostatic analyzer on Vela 3A (Hundhausen *et al.*, 1967).

the energy per charge of the primary peak is due to He^{++}. On the lower half of the figure is shown a directional distribution, or the measured flux as a function of the orientation of the 5° wide acceptance window of the detector, at a constant energy per charge (in this case at the peak of the proton flux in the spectrum above). The 180° direction is that of particles coming from the sun. These narrow flux peaks in spectra and directional distributions are characteristic of the solar wind, where the mean or flow speed of the ions is an order of magnitude greater than the mean random or thermal speed.

Figure 2 shows a typical energy-per-charge spectrum obtained in the magneto-sheath by the same Vela 3 detector system. The magnetosheath spectral peak is distinctly wider (full width at half-maximum ~ 1 kV) than the solar wind spectral peak in Figure 1 (full width at half maximum ~ 0.2 kV). Thus the chief characteristic which differentiates magnetosheath (post-shock) ions from solar wind (pre-shock) ions, i.e., the lower ratio of flow speed to mean random speed, is apparent even in a comparison of simple plots of the data. This difference is large in the above example despite the fact that the Vela orbits cross the magnetosheath well away from the subsolar point, observing an oblique bow shock and a supersonic magnetosheath flow.

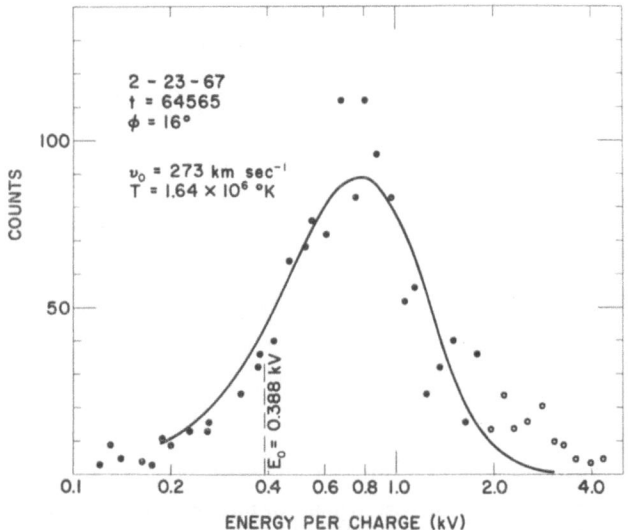

Fig. 2.   A typical positive ion energy-per-charge spectrum obtained in the magnetosheath by the LASL electrostatic analyzer on Vela 3A (Hundhausen *et al.*, 1969).

The solid curve in Figure 2 shows the proton spectrum expected from the bi-Max-wellian distribution function which best fits (in a least-squares sense) the data matrix (flux vs. direction *and* energy per charge) of which Figure 2 is a part. Two clearly visible deviations of the individual measured fluxes from the curve indicate two further characteristics of post-shock positive ions. The large fluctuations about the best-fit curve reveal that magnetosheath properties vary on the time scale comparable to that on which these data were collected (tens of sec). These variations are super-posed on the basic flow pattern. The elevation of the measured fluxes at high energy per charge ($\gtrsim 2$ kV here) above the best-fit curve is a common feature, and shows the presence of a 'high-energy tail' on the distribution. Note that the proton spectral peak is so spread that no distinct second peak due to He^{++} can be distinguished.

B.   ELECTRON DISTRIBUTION FUNCTIONS

The electrons are of lesser interest here because they carry only a small amount of momentum and energy. They are, however, of considerable interest in attempts to understand the mechanism by which directed energy is converted to random energy at the shock; for a discussion with emphasis on this phenomenon see Montgomery and Bame (1969) and Montgomery *et al.* (1969). Figure 3 shows a typical solar wind electron distribution function derived from Vela 4B flux measurements. This 'slice' through the function uses data obtained looking away from the sun (left branch) and toward the sun (right branch). The bell-shaped nature of the inner part (at small random speeds) of the distribution function is characteristic of solar wind electrons. In fact, a Maxwellian distribution gives an extremely good fit to the measured fluxes

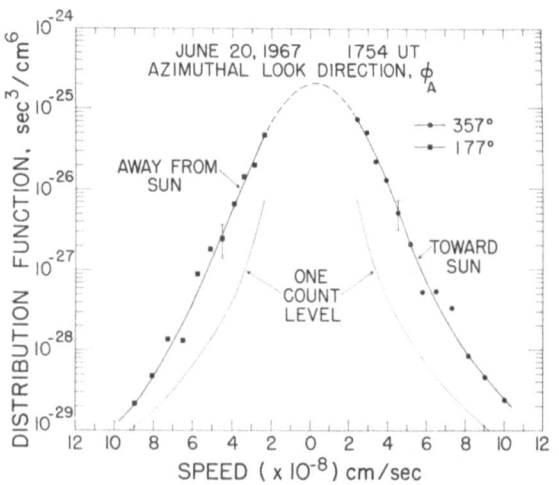

Fig. 3.   A typical solar wind electron distribution function derived from data obtained by the LASL electrostatic analyzer on Vela 4B (Montgomery *et al.*, 1969).

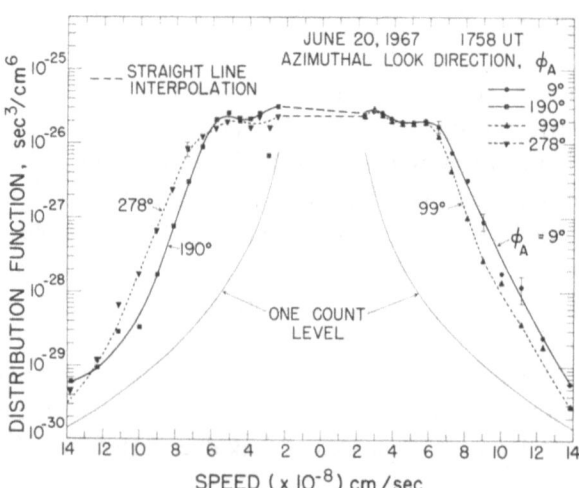

Fig. 4.   A typical magnetosheath electron distribution function derived from data obtained by the LASL electrostatic analyzer on Vela 4B (Montgomery *et al.*, 1969).

at small particle speeds (Montgomery *et al.*, 1968). A definite elevated tail is generally present at higher energies or random speeds.

Figure 4 shows a magnetosheath electron distribution function derived from Vela 4B flux measurements made only 4 min after the solar wind observations of Figure 3. The greater spread in the distribution function shows that the electrons have been heated in passing through the shock. The most striking change in the distribution function, however, is the flattening at low random speeds. This flattened type of distribution function is observed throughout much of the magnetosheath.

C. FLUID PARAMETERS

Particle distribution functions, such as those illustrated above, are conveniently described by a series of moments which correspond to the parameters used in fluid mechanics. For the purposes of this discussion five such parameters will suffice: the number density, the flow speed and direction (the requirement of charge neutrality implies that these three parameters are nearly equal for the electrons and the protons, the latter constituting about 95% of the positive ion component), the proton temperature, and the electron temperature. Table I gives typical values of these parameters obtained, under quiet conditions, in the solar wind and in that region of the magnetosheath (near 90° from the subsolar point) probed by the Vela satellites. The changes in fluid parameters observed in Vela satellite bow shock crossings (under quiet conditions) can be inferred from this table.

TABLE I

Solar wind and magnetosheath properties under quiet conditions

	Solar wind	Magnetosheath
Density	$5 \, \text{cm}^{-3}$	$15 \, \text{cm}^{-3}$
Flow speed	$320 \, \text{km sec}^{-1}$	$250 \, \text{km sec}^{-1}$
Flow direction	Nearly radial from sun	Deflected by $\sim 20°$ from sun-earth line
Proton temperature	$5 \times 10^4 \, \text{K}$	$10^6 \, \text{K}$
Electron temperature	$1.5 \times 10^5 \, \text{K}$	$5 \times 10^5 \, \text{K}$

Figure 5 shows some specific examples of Vela 4B shock crossings. Of particular interest is the series of ten crossings during a $1\frac{1}{2}$ hour period on May 15, 1967. Although the spacecraft was probably in close proximity to the shock throughout this period, the solar wind and magnetosheath plasmas are readily distinguishable, and the transition between these two distinct states is abrupt on the four minute time scale of the measurements. There is no evidence for the diffuse change in properties reported by Wolfe *et al.* (1966). The distinct nature of the pre-shock and post-shock states is further illustrated on Figure 6, which shows the flow speed and proton temperature measurements made in the solar wind and magnetosheath by both Vela 3 satellites on February 4, 1967. Although Vela 3A crosses the bow shock several times, the measured points fall into two distinct groups, separated by an order of magnitude difference in temperature. The Vela 3B magnetosheath measurements, made well away from the expected shock location on the opposite side of the earth from Vela 3A, fall into the same group as those made on the later spacecraft.

The changes in positive ion properties at thirteen Vela 3 shock crossings have been discussed by Argo *et al.* (1967) and Spreiter *et al.* (1968). Despite a wide range of upstream solar wind conditions, the changes produced at the bow shock were in close agreement with the predictions of hydrodynamic theory. This leads to a rather simple conclusion regarding the variations in magnetosheath plasma properties during the interplanetary disturbances associated with solar events. The variations in solar

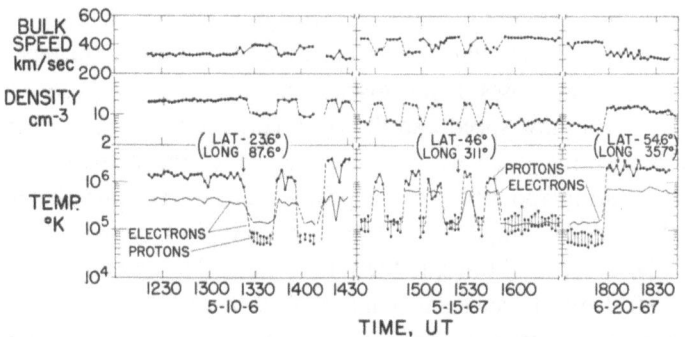

Fig. 5.   Plasma properties measured near the bow shock by the Vela 4B satellite. The abrupt changes in properties indicate bow shock crossings (Montgomery *et al.*, 1969).

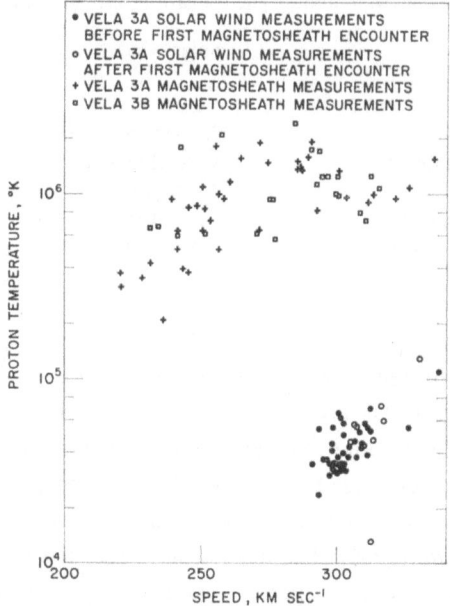

Fig. 6.   Flow speed and proton temperature measurements made near the bow shock by the Vela 3 satellites on Feb. 4, 1967. The measurements fall into two well-separated groups, showing that the pre-shock and post-shock states are distinct.

wind density, flow speed, and proton temperature will be transmitted, with modifications similar to those given by the models of Spreiter *et al.* (1966) and Dryer and Heckman (1967), through the bow shock and magnetosheath to the magnetosphere. This relationship between solar wind and magnetosheath properties has been qualitatively confirmed by observations made during geomagnetic storms (e.g., Bame *et al.*, 1968). Fluctuations are either generated or amplified at the bow shock and within the magnetosheath, producing a noisier signal than originally present in the interplanetary region.

D. THE MAGNETIC FIELD

No attempt will be made here to give a comprehensive review of the magnetic field changes at the bow shock or within the magnetosheath (see the reviews cited in the introduction for such a treatment). The observations pertinent to this discussion can be summarized as follows (see Fairfield, 1967):

(1) The magnitude of the magnetic field jumps at the bow shock in the expected manner.

(2) The magnetic field within the magnetosheath tends to be draped around the magnetosphere, with the field lines drawn out tangent to the magnetopause. However, the field orientation relative to the ecliptic plane (i.e., the sign of the solar ecliptic latitude of the field) is generally preserved in passage through the bow shock and magnetosheath.

(3) Interplanetary structures, such as discontinuities, retain their identities in passage through the bow shock and magnetosheath.

(4) Upon this orderly field pattern is superposed a level of fluctuations which is considerably higher than that in the solar wind.

We are thus led to conclusions similar to those of the previous section. Although the magnetic field properties in the magnetosheath differ from those in the solar wind, the former depend upon the latter, and temporal variations in the solar wind produce in response similar variations in the magnetosheath field. Of particular significance is the preservation of the northward or southward (relative to the ecliptic plane) character of the magnetic field. This property of the field has been shown to be generally correlated with geomagnetic activity (Fairfield and Cahill, 1966; Fairfield, 1968), and to be related to the development of bay activity and the main phase of a geomagnetic storm (Hirshberg and Colburn, 1969). Fluctuations are again produced or amplified above the solar wind level.

3. The Spatial Configuration of the Magnetosheath

The spatial locations of bow shock and magnetopause crossings (which together define the configuration of the magnetosheath) have been mapped by many satellites. Figure 7 shows the results of the first extensive mapping, performed on the Imp 1 satellite in 1963–64. The positions of the boundary crossings identified by the GSFC magnetometer (see Ness *et al.*, 1964) have been rotated about the sun-earth line into

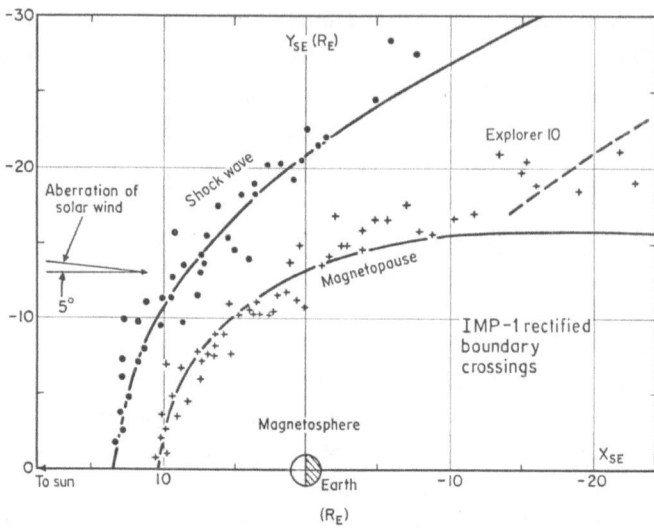

Fig. 7.   Bow shock and magnetopause locations, identified by the GSFC magnetometer on Imp 1, rotated into the ecliptic plane (Ness, 1967). The theoretical boundaries indicated by solid lines were based on the predictions of Spreiter and Jones (1963).

the ecliptic plane. The theoretical curve labeled 'magnetopause' was derived by Spreiter and Jones (1963) under the so-called 'Newtonian assumption'; i.e., it is the surface along which the normal component of the solar wind momentum flux is balanced by the pressure of the tangential component of the geomagnetic field compressed to twice its normal dipole value (see Beard, 1960; Spreiter and Briggs, 1962). The theoretical curve labeled 'shock wave' is based on a computation by Spreiter and Jones (1963) of the hydrodynamic flow around an obstacle whose shape is that of the magnetosphere, as derived above. The observed crossings define a magnetosheath configuration with an average shape close to the Spreiter and Jones prediction. This general agreement has been confirmed by subsequent observations.

The large amount of scatter present in the boundary locations shown in Figure 7 and in other similar mappings has been attributed to variations in the properties of the solar wind incident on the earth and to hypothetical waves on the surface of the bow shock and magnetopause. The effect of solar wind variations can be predicted on the basis of the same models which have successfully predicted the general shape of the magnetosphere. Several recent observational studies of this effect are of interest here both as tests of the models and as examples of the use of multiple satellite observations.

A. SLOW CHANGES IN SOLAR WIND CONDITIONS

The basic configuration of the magnetosheath results from a balance between the pressures in the magnetosheath and in the magnetosphere. Most theoretical models are based on the simplifying assumption that the former is entirely due to particles, and can be approximated by the normal component of the solar wind momentum flux, while the latter is entirely due to the geomagnetic field. For example, Williams and Mead (1965) have derived such a model in which the field is determined by a terrestrial dipole, the current at the magnetopause, and the current sheet implied by the existence of the geomagnetic tail.

Under conditions which permit the neglect of other currents (e.g., ring currents), the magnetosphere should respond to slow variations by assuming the steady state configuration appropriate to prevailing conditions. Roederer *et al.* (1968) have used the magnetic fields measured by the ATS-1 satellite, the Williams-Mead model, and Vela 3 boundary crossings to examine this behavior. The standoff distance R_s to the subsolar point of the magnetosphere is computed from the field measurements by using the model. The resulting magnetosheath configuration is then checked by comparison with Vela 3 bow shock and magnetopause crossings from the same time period. Figure 8 shows the positions of a series of crossings, rotated into the ecliptic

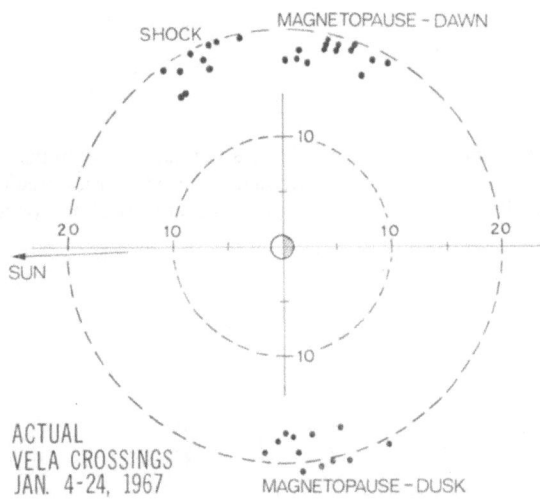

Fig. 8.   Vela 3 boundary crossings during January, 1967 (Roederer, 1969).

plane. A scatter similar to that in Figure 7 is present. Figure 9 shows the same series of crossings with each radius vector normalized to a standard set of conditions for which the subsolar point is at 10 earth radii (i.e., multiplied by the factor $10/R_s$, where R_s is determined near the time of the crossing). The scatter is considerably reduced, indicating that the R_s values do describe the magnetosheath configuration and that the field variations at ATS-1 are due to changes in this configuration. The boundary locations given by the Williams-Mead model, with the subsolar point at 10 earth radii, are also shown, but with the symmetry axis rotated 7° west of the sun-earth line. This rotation is based on the observed symmetry of the ATS-1 fields (Roederer *et al.*, 1968) and leads to an impressive agreement between the observed (normalized) crossings and the predicted boundaries.

These results are strong evidence that the gross configuration of the magnetosheath changes in a coherent manner. The changes are presumably in response to slow

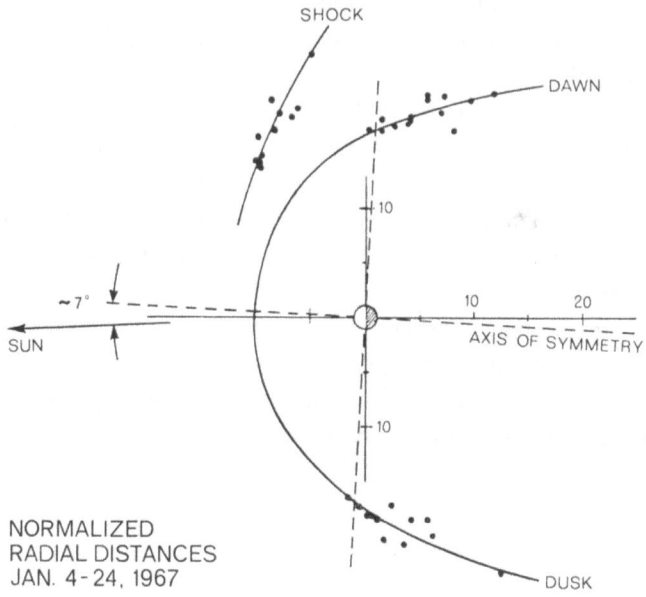

Fig. 9. The Vela 3 boundary crossings of Figure 8 normalized by means of the magnetospheric dimensions derived from ATS-1 magnetometer data. The boundaries predicted by Williams and Mead (1965), tilted by 7° from the sun-earth line, are also shown (Roederer, 1969).

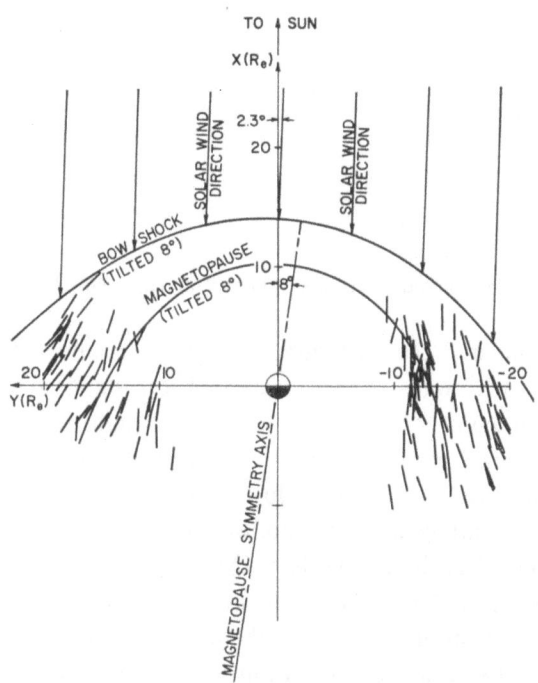

Fig. 10. The plasma flow pattern measured in the magnetosheath by the Vela 3 satellites. Each flow line shows a 3-hour average of the flow direction. The pattern is symmetric about a line tilted 8° from the sun-earth line, and 6° from the observed average direction of solar wind flow (Hundhausen *et al.*, 1969).

variations in the solar wind momentum flux, although this relationship has not been directly established. The 7° tilt of the basic configuration has also been proposed on the basis of Vela 3 observations of the flow pattern in the magnetosheath (Hundhausen *et al.*, 1969). Figure 10 indicates the measured flow directions (3-hour averages) by a short flow line at the point of observation (in solar ecliptic coordinates). The flow pattern is symmetric about a line tilted 8° from the sun-earth line and 6° from the observed average direction of the solar wind. This tilt may be the effect of the interplanetary magnetic field (see Walters, 1964).

B. ABRUPT CHANGES IN SOLAR WIND CONDITIONS

Abrupt changes in solar wind properties, including the momentum flux, occur at interplanetary shock waves and discontinuities. These changes would be expected to produce abrupt compressions or expansions of the magnetosphere and impulsive changes in the geomagnetic field, the latter recorded at ground level as sudden commencements or sudden impulses.

An example of this behavior involving measurements from three satellites has been discussed by Binsack and Vasyliunas (1968). Figure 11 shows the positions of Imp 2

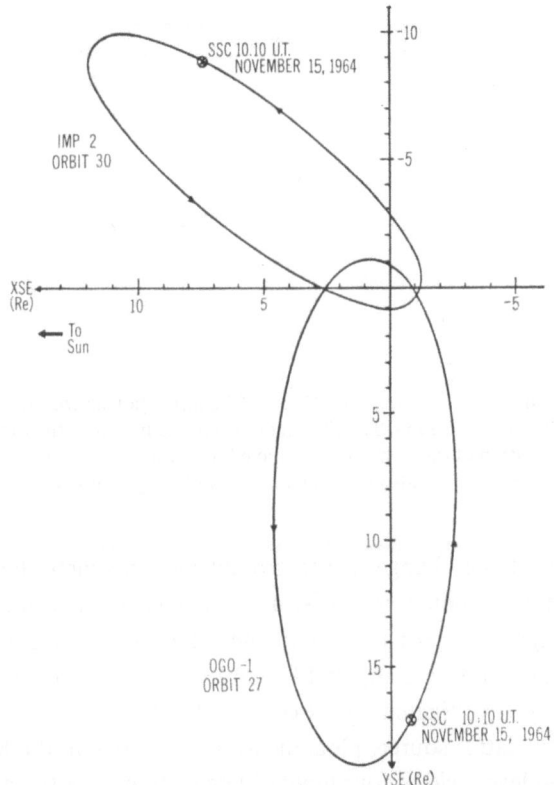

Fig. 11.   Orbits and positions of Imp 2 and Ogo 1, in solar ecliptic *X* and *Y* coordinates, on Nov. 15, 1964 (Binsack and Vasyliunas, 1968).

and Ogo 1, in solar ecliptic X and Y coordinates, on November 15, 1964. During a small geomagnetic storm on that date, a series of bow shock crossings occurred which are illustrated in Figure 12. The radial distance of Ogo 1 is shown as a function of time. The radial distance of Imp 2 has been normalized by the ratio of the quiet-time bow shock distances at the Imp 2 and Ogo 1 position angles from the subsolar point, and this normalized distance is shown on Figure 12 as a function of time. The shading of each radial distance curve indicates the plasma regime (interplanetary or magneto-sheath) observed at the spacecraft. The radial distance to the bow shock at the Ogo 1 position angle has been computed using the solar wind momentum flux measured at the Imp 1 satellite, which was outside the bow shock near the sun-earth line at the time of interest. This predicted bow shock position is also shown as a function of time on Figure 12. The observed Imp 2 and Ogo 1 crossings of the bow shock occurred close to the predicted locations at the magnetospheric compressions and expansions associated with the sudden changes in the solar wind momentum flux observed at Imp 1. The predicted compression of the magnetosphere between about 1010 and 1500 UT is verified by the Ogo 1 observation of the solar wind at positions from 4 to 7 earth radii inside of the normal position of the shock.

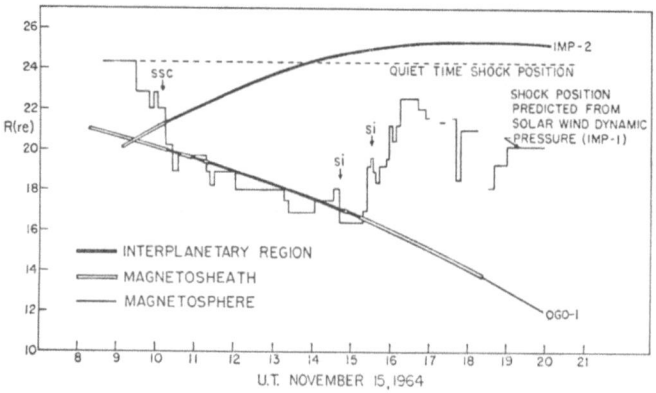

Fig. 12.   The radial distances of Imp 2 and Ogo 1 (the latter normalized to the Imp 2 angle from the subsolar point) on Nov. 15, 1964. The shading of each line indicates the plasma regime observed by the satellite. The position of the bow shock predicted from the solar wind momentum flux measured on Imp 1 is also shown (Binsack and Vasyliunas, 1968).

The magnitude of the change in the ground level magnetic field produced by a sudden magnetospheric compression is, under the Newtonian assumption, proportional to the change in the square root of the solar wind momentum flux. The proportionality constant can be computed for any specific model. This relationship has been quantitatively investigated by Siscoe *et al.* (1968) and Ogilvie *et al.* (1968). Figure 13, from the latter source, plots the average change in the horizontal component of the ground level field at four low-latitude stations as a function of the directly observed change in the solar wind momentum flux. These two quantities are found to be roughly proportional, but with a proportionality constant approximately one-half that expected on the basis of most theoretical models. Siscoe *et al.* (1968) attribute this difference to particle currents other than that at the magnetopause, which are not considered in the models.

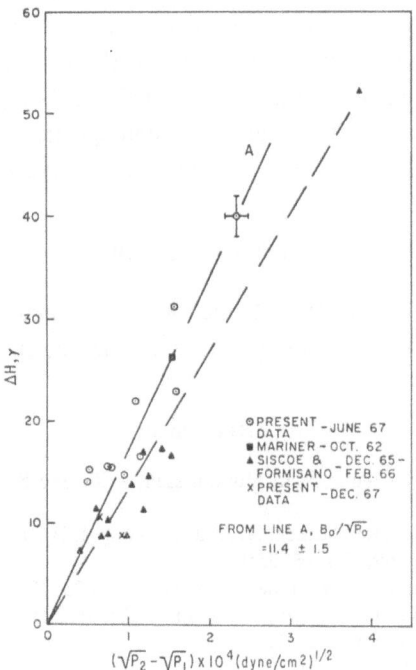

Fig. 13.   Observed values of the change in the square root of the solar wind momentum flux and the magnitude of the ground-level field change in the associated SC or SI (Ogilvie *et al.*, 1968).

4. Conclusions

Some observations of the plasma changes occurring at the earth's bow shock have been discussed. These observations show the existence of distinct pre-shock (solar wind) and post-shock (magnetosheath) states; these states are found to be related, over a wide range of solar wind parameters, in the manner predicted by theoretical models of hypersonic fluid flow around an obstacle. Changes in the solar wind plasma or magnetic field produce the expected changes in the properties of the magnetosheath plasma and magnetic field.

Observations of the spatial configurations of the bow shock and magnetopause under varying solar wind conditions have also been discussed. The locations of these boundaries are found to change with variations in the solar wind momentum flux in a manner again consistent with theoretical models. This conclusion can be drawn for both slow and abrupt solar wind changes. In the latter case the magnitude of the associated impulsive change in the ground-level geomagnetic field (the SC or SI) shows the proper relationship to the solar wind momentum flux change, but is smaller by a factor of two than predicted by the models.

These observations indicate a relatively straightforward response of the bow-shock and magnetosheath to solar wind variations associated with solar events. The basic features of an interplanetary disturbance will be transmitted through the bow shock and magnetosheath, producing an 'input signal' to the magnetosphere which is not much different from the original solar wind variations. The bow shock and magneto-pause configurations will change in response to the varying solar wind momentum flux, producing small changes in the geomagnetic field. These processes, which, for example, occur during the initial phase of a geomagnetic storm, have been described in a quantitative sense by the use of multiple satellite observations.

Acknowledgements

The author wishes to thank Dr. S. J. Bame for permission to include unpublished Vela 3 data in this paper. This work was performed under the auspices of the United States Atomic Energy Commission.

References

Argo, H. V., Asbridge, J. R., Bame, S. J., Hundhausen, A. J., and Strong, I. B.: 1967, *J. Geophys. Res.* **72**, 1989.

Bame, S. J., Asbridge, J. R., Hundhausen, A. J., and Strong, I. B.: 1968, *J. Geophys. Res.* **73**, 5761.

Beard, D. B.: 1960, *J. Geophys. Res.* **65**, 3559.

Binsack, J. H. and Vasyliunas, V. M.: 1968, *J. Geophys. Res.* **73**, 429.

Dryer, M. and Heckman, G. R.: 1967, *Planetary Space Sci.* **15**, 515.

Fairfield, D. H.: 1967, *J. Geophys. Res.* **72**, 5865.

Fairfield, D. H.: 1968, *Space Res.* **8**, 107.

Fairfield, D. H. and Cahill, L. J., Jr.: 1966, *J. Geophys. Res.* **71**, 155.

Hirshberg, J. and Colburn, D. S.: 1969, *Planet. Space Sci.* **17**, 1183.

Hundhausen, A. J., Asbridge, J. R., Bame, S. J., Gilbert, H. E., and Strong, I. B.: 1967, *J. Geophys. Res.* **72**, 87.

Hundhausen, A. J., Bame, S. J., and Asbridge, J. R.: 1969, *J. Geophys. Res.* **74**, 2799.

Montgomery, M. D. and Bame, S. J.: 1969, to be published in the *Proceedings of the Study Group on Collision-Free Shocks*, Frascati, June, 1969.

Montgomery, M. D., Bame, S. J., and Hundhausen, A. J.: 1968, *J. Geophys. Res.* **73**, 4999.

Montgomery, M. D., Asbridge, J. R., and Bame, S. J.: 1969, Los Alamos Scientific Laboratory pre-print LA-DC-10339, submitted to *J. Geophys. Res.*

Ness, N. F.: 1967, in *Solar-Terrestrial Physics* (ed. by J. W. King and W. S. Newmann), Academic Press, London, p. 57.

Ness, N. F., Scearce, C. S., and Seek, J. B.: 1964, *J. Geophys. Res.* **69**, 3531.

Ogilvie, K. W., Burlaga, L. F., and Wilkerson, T. D.: 1968, *J. Geophys. Res.* **73**, 6809.

Roederer, J. G.: 1969, *Rev. Geophys.* **7**, 77.

Roederer, J. G., Coleman, P. J., Cummings, W. D., and Robbins, M. F.: 1968, *Trans. Am. Geophys. Union* **49**, 227.

Siscoe, G. L., Formisano, V., and Lazarus, A. J.: 1968, *J. Geophys. Res.* **73**, 4869.

Spreiter, J. R. and Alksne, A. Y.: 1969, *Rev. Geophys.* **7**, 11.

Spreiter, J. R. and Briggs, B. R.: 1962, *J. Geophys. Res.* **67**, 37.

Spreiter, J. R. and Jones, W. P.: 1963, *J. Geophys. Res.* **68**, 3555.

Spreiter, J. R., Summers, A. L., and Alksne, A. Y.: 1966, *Planetary Space Sci.* **14**, 223.

Spreiter, J. R., Summers, A. L., and Alksne, A. Y.: 1968, *J. Geophys. Res.* **73**, 1851.

Walters, G. K.: 1964, *J. Geophys. Res.* **69**, 1769.

Williams, D. J. and Mead, G. D.: 1965, *J. Geophys. Res.* **70**, 3017.

Wolfe, J. H. and Intriligator, D. S.: 1969, to be published in *Space Sci. Rev.*

Wolfe, J. H., Silva, R. W., and Myers, M. A.: 1966, *J. Geophys. Res.* **71**, 1319.

Discussion

Lanzerotti: Does the direction of the solar wind flow across the bow shock change discontinuously or gradually?

Hundhausen: The flow direction changes between successive measurements of the positive ions, these measurements being spaced in time by a few minutes.

Wibberenz: Are there more superthermal particles in the magnetosheath, which means that a large part of the energy is transferred?

Hundhausen: The actual belief is that in the magnetosheath a large superthermal tail is always present. When one tries to do Maxwellian fits, this works pretty well in the solar wind, no matter where one cuts off the points. However, in the magnetosheath if one includes points down the tail, one gets a very obvious bad fit.

Gringauz: What is the maximum increase of solar ion fluxes across the shock?

Hundhausen: We have not looked at the flux values, but the density changes by a factor of 2.5. Vela crosses the bow shock about 60° from the subsolar point, and the flow behind the bow shock is actually supersonic. The flow velocity drops by about 15%, which indicates a change in flux of a factor between 2 and 2.5.

INTERACTION OF SOLAR ENERGETIC PARTICLES
WITH THE EARTH'S BOW SHOCK AND MAGNETOPAUSE

SIDNEY SINGER

University of California, Los Alamos Scientific Laboratory, Los Alamos, N.M., U.S.A.

1. Introduction

For many years, energetic particles emitted by the sun have been studied for the purposes of determining the acceleration process producing them and the properties of the interplanetary medium as deduced from the propagation features of the particles. As a result, it is now generally recognized that the sun accelerates protons, He nuclei, and electrons impulsively as well as continuously to energies of many MeV, that these particles initially populate a zone of limited solar longitude, and that propagation to the vicinity of the earth occurs by streaming along and diffusion across the interplanetary field lines.

Although the interaction of energetic particles with the geomagnetic field can be traced to the rigidity cut-off calculations of Stoermer (1955), only recently has the interaction with specific features of the field come under study. Most of these efforts (Lin and Anderson, 1966; Krimigis *et al.*, 1967; Lanzerotti, 1968; Montgomery and Singer, 1969) have been concerned with questions of the topology of the field, i.e., whether it is open or closed to the interplanetary field. Also, recent studies (Reid and Sauer, 1967; Taylor, 1967; Gall *et al.*, 1968) of the high-latitude rigidity cut-off have been employed to test the accuracy of existing models for the magnetic field generated by the magnetotail current sheet and the magnetotail surface currents.

In this paper, the temporal and directional properties and the energy spectrum of energetic protons in the interplanetary medium, the magnetosheath, and the magnetotail are intercompared in order to determine the interactions of energetic particles with the bow shock and the magnetopause. Most of the data to be evaluated are taken from the Vela 4 satellite observations of the May, 1967, and the February, 1969, solar particle events, and the results of the study may be summarized as follows:

(a) Detectable interactions with the earth's bow shock are usually observed only at proton energies below a few MeV.

(b) The bow shock does not act as a barrier that slows the penetration of energetic particles. The only delay that can be observed between fluctuations seen in the interplanetary medium and their subsequent detection in the magnetosheath can be attributed to differences in the convection (at solar wind speed) path of the associated disturbances to each of the Vela 4 spacecraft.

(c) The interaction most commonly observed is an enhancement of the lower energy fluxes in the magnetosheath, which sometimes is as large as a factor of 2.

(d) Certain changes in the directional characteristics of the protons show no consistent trends. When the protons are noted to be streaming in the interplanetary

V. Manno and D. E. Page (eds.), Intercorrelated Satellite Observations Related to Solar Events, 170–180. All Rights Reserved
Copyright © 1970 by D. Reidel Publishing Company, Dordrecht-Holland

medium, the crossing into the magnetosheath is usually accompanied by a change in streaming direction which is presumably related to the shift in magnetic field direction. At other times, the crossing is accompanied by the appearance of a nearly back-streaming group of protons.

(e) The magnetopause tends to act as a multiple-scattering or diffusing barrier for energetic particles. As the spacecraft approaches the magnetopause, the large aniso-tropies often seen outside the bow shock tend to diminish even though the aniso-tropy in the interplanetary medium remains unchanged. When the spacecraft crosses the magnetopause and enters the magnetotail, the residual anisotropy frequently shows an abrupt decrease and the particle angular distributions become nearly isotropic.

2. Instrumentation

Details of the Vela 4 satellite system and the Vela 4 instrumentation have been des-cribed in detail elsewhere (Montgomery *et al.*, 1968; Singer *et al.*, 1969); for com-pleteness, we mention here only those aspects of the instrumentation which are pertinent to the data being described.

Most of the solar energetic particle data are taken from the Vela Energetic Particle (ST) experiment. This consists primarily of a guarded 6-element semiconductor telescope that possesses a full angle of 30° and a geometric factor of 0.2 cm²-ster. The detector identifies protons from 0.5 to 100 MeV and counts them in 13 logarithmically spaced differential energy channels. During each 64 sec rotation of the satellite, eight uniformly spaced 4-sec data accumulations are obtained in an analysis plane that is perpendicular to the spacecraft spin axis. Additional information concerning the directional distribution of the protons is obtained from the 4 Geiger counters that are contained in the Electron-Proton Spectrometer (EPS) experiment. The Geiger coun-ters, which have a proton counting threshold of approximately 1 MeV, are placed in a single spacecraft meridian plane at spacecraft latitudes $\pm 22.5°$ and $\pm 67.5°$. The counting rates from the Geiger counters are sampled once each 2 sec, so that 64 samples of the distribution over the 4π solid angle are obtained every spacecraft rotation.

The location of the spacecraft (interplanetary medium, magnetosheath, or magneto-tail) can be definitively determined by employing the low energy ion and electron data from the plasma analyzer contained in the EPS (Bame, 1968; Coon, 1966; Mont-gomery *et al.*, 1968). The assignment is made by comparing the density and spectral characteristics of the two particle species with those expected in the various regions of space. For example, a crossing from the interplanetary medium through the bow shock into the magnetosheath would be identified by the detection of an abrupt increase in the temperature and density of the solar wind ions and electrons, and a reduction in their bulk velocity.

3. Observations

A. INTERACTION OF PROTONS WITH THE EARTH'S BOW SHOCK

In Figure 1, data from spacecraft 4A are plotted for an 18-hour period on May 25–26,

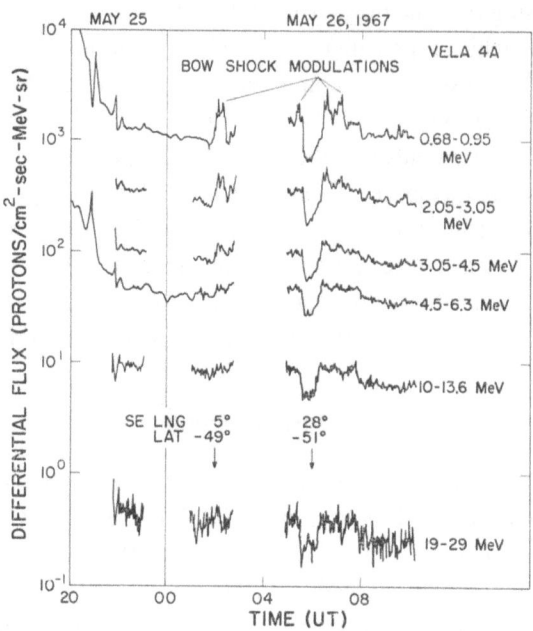

Fig. 1.   Interactions of solar energetic particles with the earth's bow shock. Six channels from
0.68–29 MeV, from the Vela 4 Energetic Particle experiment, are plotted during the decay phase of
the complex series of solar energetic particle events which began late on May 21, 1967. The effect
of the interaction is seen primarily at the lower energies, and it consists of amplitude modulations
of the flux such as the rectangular step near 0200 UT on May 26, and the many peaks and spikes
from 0500–1000 UT. The highest intensities occur when the spacecraft is in the magnetosheath. The
gap in some of the plots near 2400 UT on May 25 arises from the limited information storage
rate available when the spacecraft is operated in the memory mode. No data are available from
0300–0500 UT on May 26.

1967. During the last half of May 25, the energetic particle flux intensity decreased
from its peak value at noon to a nearly constant level by ∼2200 UT. In the period
0145–1300 UT on May 26, the spacecraft crossed the bow shock at least 17 times, and
most crossings were accompanied by flux intensity changes that were sometimes as
large as a factor of 2 or 3. Increases in flux intensity were observed upon entry of the
spacecraft into the magnetosheath, and decreases were seen when the spacecraft
re-entered the interplanetary medium. In this time period, the data from spacecraft
4B (which was in the magnetotail) showed no traces of similar intensity variations;
hence the fluctuations must have arisen from a local interaction and not a source
change. The magnitude of the interaction depends on the particle energy, as can be
clearly seen from the crossings near 0200 and 0700 UT. At energies ⩽1 MeV, the
spikes associated with the multiple crossings are evident; at ∼3 MeV, they are hardly
detectable; and at energies >10 MeV, they do not appear. Thus, the interaction
appears to be important only for particles whose cyclotron radius is less than the
magnetosheath thickness near the subsolar point. Since the observed flux is proportional to the particle density times velocity, it is not possible to determine unambigu-

ously whether the intensity changes are produced by density changes (brought about by the increased interplanetary field inside the magnetosheath) or velocity changes produced by a local acceleration that is most effective for particles with small cyclotron radii.

In Figure 2, we show that changes in the source strength propagate through the bow shock without important delay. Plotted are data from both spacecraft in the period 1930–2030 UT on May 24, 1967. From 1950–2020 UT, a series of strong fluctuations were detected in the interplanetary proton flux intensity. During that interval, 4B was in the magnetosheath whereas 4A was in the interplanetary medium for the first half of the period and in the magnetosheath for the second half. Both spacecraft saw almost exactly the same signal shape. It is difficult to imagine local interactions that could produce nearly identical flux variations at spacecraft separated by nearly 40 R_E; instead, it seems probable that the variations arose from upstream source changes. In Figure 2 it can be seen that the maximum time difference between corresponding features in the fluctuations was ∼50 sec, which is comparable to the time resolution of 64 sec set by the data averaging process. It is further apparent that

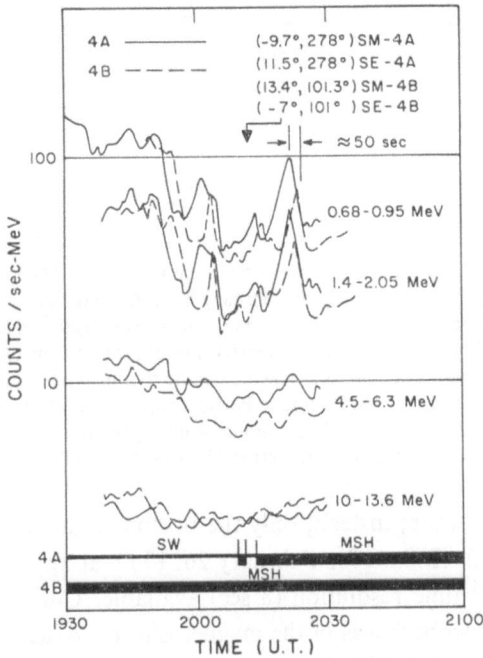

Fig. 2. Propagation of energetic protons through the bow shock. Data from four energetic proton channels are plotted from both Vela 4 spacecraft for a brief period on May 24, 1967. The legend at the bottom of the figure shows that spacecraft 4B was in the magnetosheath (MSH) throughout the entire period, whereas 4A crossed from the interplanetary medium (SW) through the bow shock into the magnetosheath at ∼2010 UT. The excellent agreement between the signals from the two spacecraft shows that the bow shock does not effectively exclude protons of energies < 10 MeV. The spacecraft locations in solar ecliptic (SE) and solar magnetospheric (SM) coordinates are given at the top of the figure.

delays between the two signals did not depend on whether spacecraft 4A was in the magnetosheath or the interplanetary medium; hence the bow shock did not act as a barrier even for <1 MeV protons. The absence of fluctuations at higher particle energies suggests that the source change was produced by perturbed solar wind plasma or a small scale filamentary structure that affected only particles with small cyclotron radii. The source change therefore would propagate at solar wind speed, and 1 min delays between widely separated spacecraft could easily occur.

Figure 3 shows an example of a bow shock crossing where a large amplitude

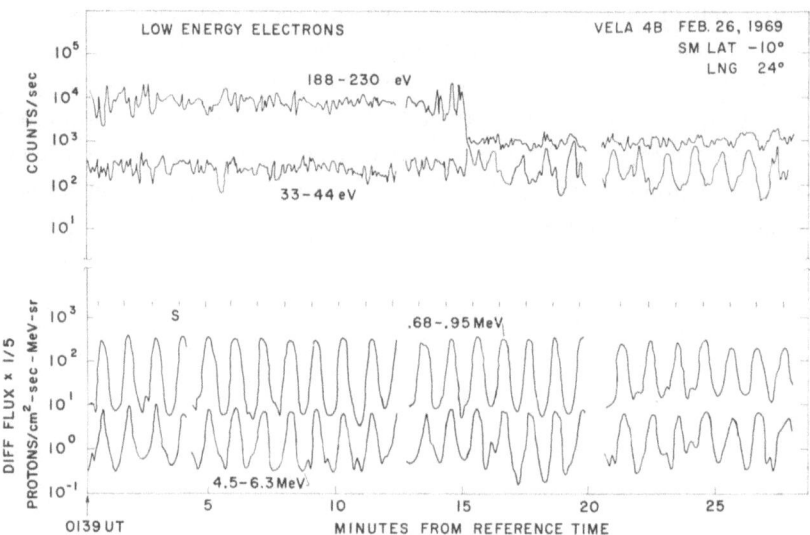

Fig. 3.   Interaction of solar energetic particles with the bow shock. The change in counting rate of the low energy electron flux at 0144 UT on February 26, 1969, indicates a crossing of the spacecraft from the magnetosheath into the interplanetary medium. The energetic protons, plotted at the bottom of the figure, showed almost no detectable intensity modulation, but instead underwent a change in streaming direction. The reference marks labelled S denote that instant during each spacecraft rotation at which the solar aspect angle of the Energetic Particle experiment is a minimum. The change in the timing between the intensity peaks and the minimum solar aspect angle location shows that the proton streaming direction changed by $\sim 45°$.

modulation did not occur; instead, only the proton streaming direction changed. In this figure, spacecraft 4B data for February 26, 1969, at approximately 0139 UT are plotted with the full time resolution (8 sec) available. Until 15 min from reference time $(R + 15)$, the spacecraft was in the magnetosheath. After $R + 15$, a change in the relative intensity of the 188–230 eV and 33–44 eV electron flux, and the cooling of the 33–44 eV electrons, indicated that the spacecraft had crossed the bow shock and was in the interplanetary medium. During this entire period, the anisotropy ratio for 0.68–0.95 MeV and 4.5–6.3 MeV protons remained approximately constant at about 30:1, and the change in averaged flux intensity at the crossing was just detectable. The most significant change occurred in the streaming direction of the protons. In the figure, the tics labelled S denote that instant during each spacecraft rotation when the solar aspect angle was a minimum. When the spacecraft was in the interplanetary medium (after $R+15$), the peak in the proton counting rate occurred when the detector axis was nearly aligned with the sunward direction. However, when the spacecraft was in the magnetosheath, the peak occurred at a solar aspect angle of approximately 45° from the sun. Thus, the effect of the bow shock interaction was to cause the proton streaming direction to change by ~45° as measured in the plane of analysis of the instrument. This direction change is presumably related to the shift in the interplanetary field direction across the shock, and the energetic protons were able to follow the direction change without much alteration of their pitch angle distribution.

Some crossings of the bow shock are more complex in that both directional and pitch angle changes occur. In Figure 4 we return to the May 26, 1967, crossings that occurred near 0516 UT. The spacecraft was in the magnetosheath until the crossing at $R+13$, after which it was in the interplanetary medium. When in interplanetary space and (presumably) far from the shock, the proton directional characteristics were typical of particles streaming in a direction nearly parallel to the satellite-sun line. However, in the magnetosheath, not only did the streaming direction change, but a second counting rate peak of different amplitude appeared that was almost 180° out of phase with the first peak.

In order to further study the changes in proton characteristics that occurred during the bow shock crossing of Figure 4, the GM counter data have been added to the ST

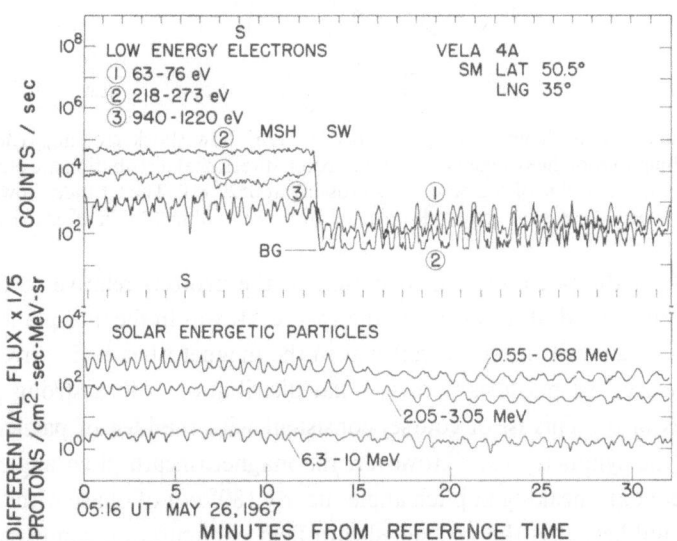

Fig. 4.   Backscattered protons in the magnetosheath. The low energy electron data show that Vela 4A crossed the bow shock and entered the interplanetary medium at 0529 UT on May 26, 1967. The solar energetic particles were streaming in interplanetary space, but had a bi-directional characteristic when the spacecraft was in the magnetosheath. The time between the minimum solar aspect angle marks S corresponds to one spacecraft rotation; in the interplanetary medium the proton directional distribution has one peak per rotation, whereas in the magnetosheath it has two peaks unequal in magnitude.
The changes in the particle properties are almost undetectable above 6 MeV.

energetic proton data to give a combined data set that better defines the proton directional distribution. The principal axis theorem was applied to the combined data in order to estimate the direction of the axis of symmetry (which is parallel to the magnetic field) of the particle population. In Figure 5, the solar ecliptic latitude and longitude of the symmetry axis are plotted for a 30-min period in the vicinity of the bow shock crossing. Prior to the crossing at $R+13$, the longitude of the symmetry direction was approximately 278°; within 10 min after the crossing, the direction of this axis changed to approximately 0°. The latitude of the axis shows little average change except possibly for some fluctuations when the spacecraft emerges into the interplanetary medium.

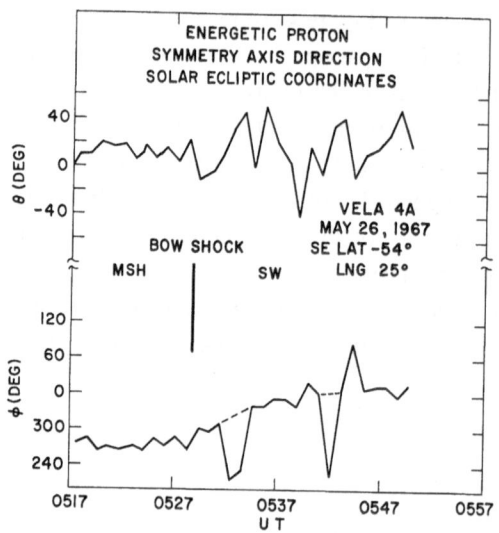

Fig. 5. Symmetry axis change during the May 26, 1967 bow shock crossing. The solar ecliptic latitude and longitude of the symmetry axis of the proton directional distribution are plotted for a short period of time on each side of the bow shock crossing at 0529 UT. The latitude shows no important net change, whereas the longitude changed by ~ 90° within 10 min after the crossing.

In Figure 6, the pitch angle distribution of the protons relative to the symmetry axis has been plotted at a time when the spacecraft was in the interplanetary medium (Figure 6a), and when the spacecraft was in the magnetosheath (Figure 6b). In interplanetary space, the particles possess a flat distribution with a strong peaking near pitch angles of 0°. This is, of course, consistent with the idea of particles streaming parallel to the symmetry axis. However, the magnetosheath pitch angle distribution shows a secondary peaking at pitch angles nearly 180° out of phase with the first peak.

It seems unlikely that the magnetosheath field configuration can even temporarily sustain particle trapping of the kind that occurs in the radiation belts. An interpretation of the backstreaming particles that uses concepts such as magnetic bottles or back-scattering from the magnetopause encounters certain difficulties that we discuss below.

(a) *Magnetic bottle*

A backstreaming group of particles could be expected to arise if a magnetic mirror

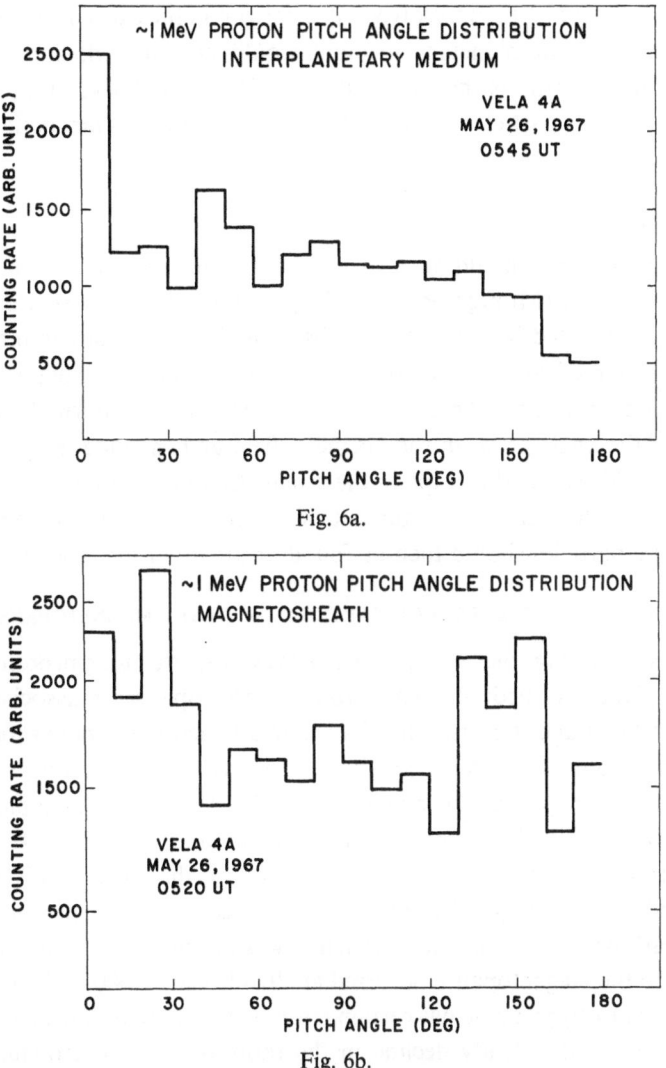

Fig. 6a.

Fig. 6b.

Fig. 6.   Proton pitch angle distributions for the May 26, 1967 bow shock crossing. The distribution of pitch angles relative to the symmetry axis of the particle population are plotted before and after the bow shock crossing at 0529 UT. – (a) Interplanetary Space, 0545 UT. The pitch angle distribution shows a nearly isotropic background and strong peak at 0°, which corresponds to protons streaming nearly parallel to the magnetic field. – (b) Magnetosheath, 0520 UT. The major change from (a) is the appearance of a second peak almost 180° out of phase with the first.

point existed in the magnetosheath downstream of the bow shock. Particles which mirrored there could propagate back up the field lines to the bow shock. Since interplanetary protons can penetrate the shock with little difficulty, backstreaming particles should then be able to emerge through the shock into interplanetary space. However, as backstreaming particles are not usually observed upstream of the bow shock, these particles are somehow not able to emerge through the shock. This difficulty could be overcome by assuming that another mirror point exists at the bow shock; yet the essential difficulty remains unresolved: if a mirror point existed at the bow shock, particles from the interplanetary medium would be reflected, and backstreaming particles should nevertheless be seen upstream of the bow shock. Thus, interpretation of the backstreaming particles in terms of an interaction with a magnetic bottle cannot easily be made consistent with observation.

(b) *Scattering from the magnetopause*

Backstreaming particles could arise as a result of backscattering from the magneto-pause. The evidence (Montgomery and Singer, 1969) for such scattering, some of which will be presented below, indicates that it is *diffuse*, i.e., streaming particles are scattered into all pitch angles. The sort of scattering required to explain backstreaming particles, however, is one where specular reflection from a 'normal' surface occurs regardless of the angle between the magnetic field and the magnetopause. Even if a mechanism could be found to explain the attenuation of just those particles scattered through large pitch angles, the argument of the previous paragraph appplies in that backstreaming particles should then be found upstream of the bow shock.

B. INTERACTION OF SOLAR ENERGETIC PARTICLES WITH THE MAGNETOPAUSE

At various times during solar energetic particle events, streaming protons are detected in the interplanetary medium. The proton anisotropy ratio associated with the streaming particles does not ususally change greatly across the bow shock. However, as the magnetopause is approached, the ratio tends to decrease and the anisotropy may disappear entirely after the magnetopause is crossed. An example is shown in Figure 7, where the EPS low energy electron flux and the 0.65–0.91 MeV proton anisotropy ratio are plotted for a 2.5-hour period on the morning of May 29, 1967. The shape of the low energy electron spectrum reveals that the spacecraft was in the magnetosheath until 1050 UT, after which it was in the magnetotail; however, near approaches to the magnetopause occurred at 0920 UT and 1000–1015 UT. The shape of the proton anisotropy curve has certain resemblances to that of the electrons, but its essential feature is the steady decline in the ratio as the magnetopause crossing is approached. Especially after 0950 UT, it is likely that the spacecraft was in the immediate vicinity of the boundary, and the proton pitch angle distribution was filled in by protons scattered from the magnetopause.

At the boundary crossing, the proton anisotropy ratio frequently decreases abruptly and the directional distribution inside the magnetotail is usually almost isotropic (Montgomery and Singer, 1969). In Figure 8, EPS electron and ST energetic proton data are plotted at maximum time resolution for a 13 min period near 1251 UT on May 25, 1967. Prior to $\sim R + 9.5$, the 43–57 eV electron data indicate that the space-craft was in the magnetosheath. The magnetopause was crossed at $R + 9.5$, after which time the spacecraft was in the magnetotail. Prior to $R + 9.5$, the spin modulation indicated that the protons were streaming along field lines at an angle of $\sim 60°$ to the sun (measured in the spacecraft equatorial plane). Beginning about 3–4 min from the time of crossing, the pitch angle distribution broadened and the streaming direc-

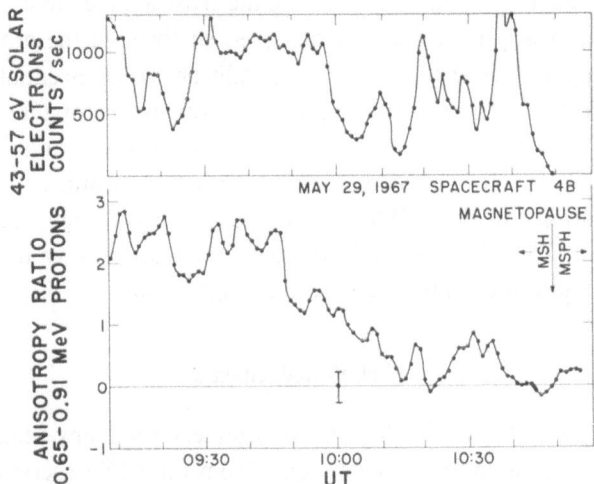

Fig. 7. Scattering of energetic protons by the magnetopause. The 43–57 eV electron flux shows that Vela 4B was in the magnetosheath at 0900 UT on May 29, 1967, and crossed into the magnetotail (MSPH) at 1050 UT. Close approaches to the magnetopause occurred at 0920 UT and 1000–1015 UT. The energetic proton anisotropy ratio (defined as R-1, where R is the ratio of the maximum to minimum spin modulated flux during one spin period) shows a steady decrease as the magnetopause crossing is approached. This phenomenon is interpreted as a filling-in of the streaming pitch angle distribution by protons scattered from the magnetopause.

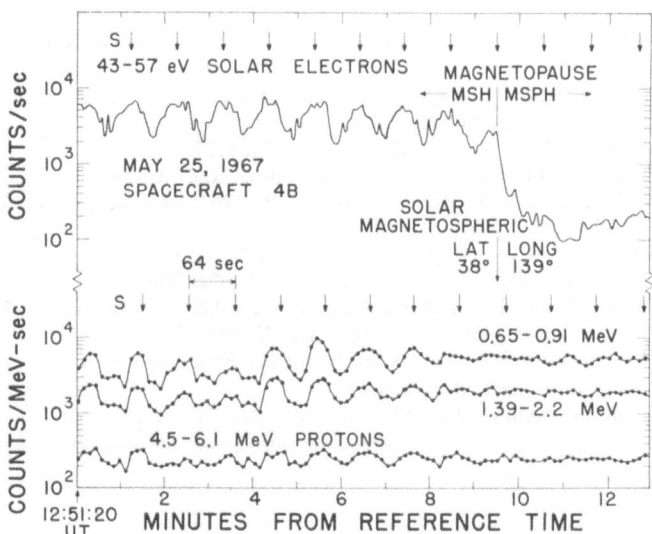

Fig. 8. Change in energetic proton directional distribution at the magnetopause. Plotted here are the low energy electron and energetic proton counting rates for a magnetopause crossing that occurred at 13:00:20 UT on May 25, 1967. The arrows labelled S show the instant of minimum solar aspect angle. Beginning about 8 min after reference time, the proton anisotropy ratio decreases abruptly; and by the time the magnetopause has been crossed and the spacecraft is in the magnetotail (MSPH), the directional distribution is almost isotropic. When such occurrences are observed, the proton anisotropy in the interplanetary medium (inferred from data from the other Vela 4 spacecraft) usually show no significant change.

tion shifted towards the sun; at ~ 2 min from the crossing, the directional distribution suddenly became nearly isotropic and remained so through the crossing. Since the proton anisotropy in interplanetary space generally shows no essential change at these times, it is the scattering and perhaps diffusion at the magnetopause that isolates the nearly-isotropic magnetotail protons from the streaming protons in the magneto-sheath. The boundary also impedes the penetration of protons into the magneto-sphere, since various workers (Montgomery and Singer, 1969; Lanzerotti, 1968; Evans and Stone, 1969; Williams and Bostrom, 1967) have shown that these particles do not have free and immediate access to the magnetotail.

Acknowledgements

The author wishes to thank Dr. M. D. Montgomery for many illuminating discussions, and Dr. S. J. Bame for providing data from the EPS experiment. Thanks are also due Earl Tech, Jean Dabney, Wynoka Miller and Rose Watts of the Vela Data Reduction Group, whose efforts made possible the orderly analysis of the data.

This research was done as a part of the Vela nuclear test detection satellite program, jointly administered by the Advanced Research Projects Agency of the Department of Defense and the U.S. Atomic Energy Commission, and managed by the U.S. Air Force.

References

Bame, S. J.: 1968, in *Earth's Particles and Fields* (ed. by B. M. McCormac), Reinhold Book Corporation, New York, p. 373.
Coon, J. H.: 1966, in *Radiation Trapped in the Earth's Magnetic Field* (ed. by B. M. McCormac), Reidel, Dordrecht, The Netherlands, p. 231.
Evans, L. C. and Stone, E. C.: 1969, *J. Geophys. Res.* **74**, 5127.
Gall, R., Jeménez, J., and Camacho, L.: 1968, *J. Geophys. Res.* **73**, 1593.
Krimigis, S. M., Van Allen, J. A., and Armstrong, T. P.: 1967, *Phys. Rev. Letters* **18**, 1204.
Lanzerotti, L. J.: 1968, *Phys. Rev. Letters* **13**, 929.
Lin, R. P. and Anderson, K. A.: 1966, *J. Geophys. Res.* **71**, 4213.
Montgomery, M. D. and Singer, S.: 1969, *J. Geophys. Res.* **74**, 2869.
Montgomery, M. D., Bame, S. J., and Hundhausen, A. J.: 1968, *J. Geophys. Res.* **73**, 4999.
Reid, G. C. and Sauer, H. H.: 1967, *J. Geophys. Res.* **72**, 197.
Stoermer, C.: 1955, *The Polar Aurora*, Oxford University Press, London.
Taylor, H. A.: 1967, *J. Geophys. Res.* **73**, 4467.
Williams, D. J. and Bostrom, C. O.: 1967, *J. Geophys. Res.* **72**, 4497.

MAGNETIC AND ELECTRIC FIELD CHANGES ACROSS
THE SHOCK AND IN THE MAGNETOSHEATH

F. L. SCARF*, P. J. COLEMAN†, R. W. FREDRICKS*,
C. F. KENNEL†, and C. T. RUSSELL†

1. Introduction

In recent months correlated measurements from the OGO-5 particle, field and wave experiments have been used to study in detail the microscopic structure of the earth's bow shock. As described in a series of preliminary reports [1, 2, 3, 4], instabilities driven by currents at steep magnetic field gradients generally produce large amplitude electrostatic waves. The electrostatic turbulence then interacts strongly with the plasma particles, and this provides a primary shock dissipation mechanism.

Here we summarize the bow shock observations briefly, and we then consider a number of closely related topics. Some intercorrelated spacecraft observations of solar events are discussed using Pioneer 8 and OGO-5 data from March 14 and April 5, 1968. It is also shown that the two-stream current instability can be triggered by strong magnetic field compressions that are not directly associated with the bow shock. Field configurations that yield intense VLF electrostatic turbulence include those near 'null regions' (probably associated with reconnection) and those associated with large amplitude oblique magnetosonic waves.

2. Bow Shock Structure

The most commonly observed form of bow shock is one in which large amplitude electrostatic waves provide a large share of the proton dissipation over short distances. Figure 1 shows the magnetic field profile for a typical shock crossing of this type. As described on the figure, a fairly well-defined sequence of changes in particle and wave characteristics goes along with the variation in $|\mathbf{B}|$. On the upstream side, the JPL plasma probe measurements reveal that the positive ions first begin to slow down (in this case near 2248:40), and shortly thereafter the Lockheed spectrometer generally shows an abrupt increase in the proton thermal energy. For Figure 1, the largest change in the proton 'temperature' is found to occur near 2248:43 to 2248:44, where the TRW dipoles detect the peak electrostatic noise ($f \simeq 500$ Hz to about 2 kHz). We note that in almost all cases the JPL-UCLA search coil does detect some moderate low frequency noise (the TRW loop at 560 Hz usually remains at its background value throughout), and that some additional proton heating occurs over much longer distances in the downstream region. However, for this type of 'electrostatic shock' it appears that the major positive ion dissipation is provided by the interaction of the

---

* Space Sciences Laboratory, TRW Systems, Redondo Beach, Calif., U.S.A.
† University of California, Los Angeles, Calif., U.S.A.

V. Manno and D. E. Page (eds.), Intercorrelated Satellite Observations Related to Solar Events, 181–189. All Rights Reserved
Copyright © 1970 by D. Reidel Publishing Company, Dordrecht-Holland

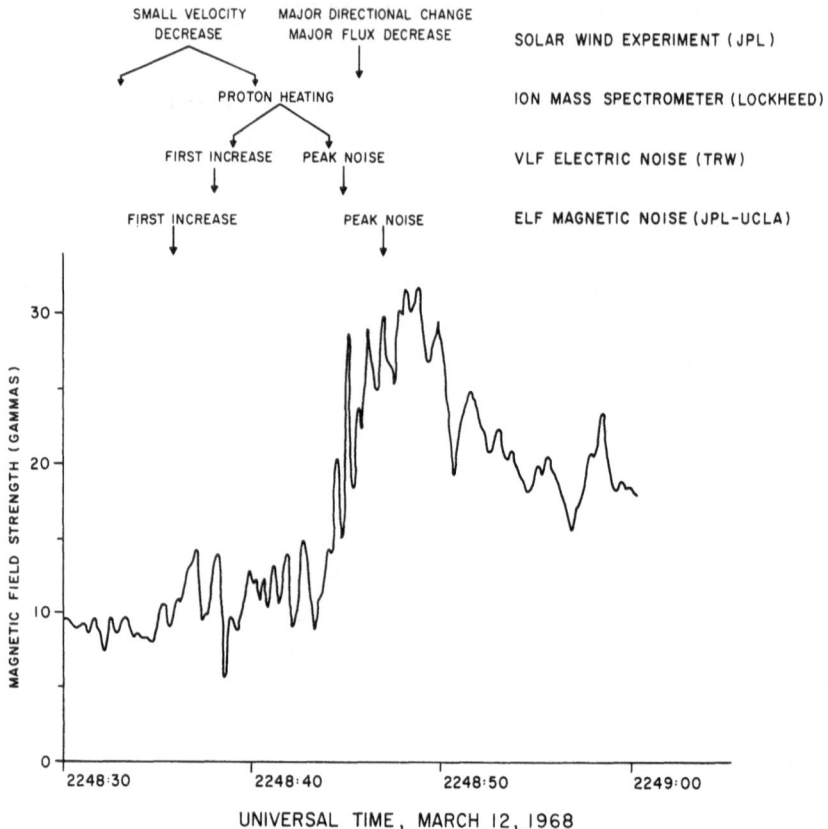

Fig. 1.   Magnetic profile of a typical 'electrostatic' type of bow shock crossing. The response of a few individual diagnostic instruments measuring related particle and wave properties is described above. The peak VLF electric noise occurs at 2248:42–45 with E(max) $\simeq$ 10 mV/m in 15 % bandpass channels centered at 1.3 kHz and at 3.0 kHz.

particles with electrostatic waves. It is also found that the peak E-field amplitudes are detected near the steepest gradients in $|\mathbf{B}|$, and it seems that the scale length of the strongest interaction may be as small as 2–5 km in the shock frame.

Some interpretations of these observations have been formulated [1, 2, 3, 4], and Figure 2 illustrates a probable sequence of events. A strong field gradient at the shock is produced by formation of a solitary pulse, by non-linear growth of a standing whistler mode wave to a large amplitude, or by some other, as yet unidentified, mechanism. At any rate, as the plasma flows into this region, we may consider that the initial sequence of events resembles the formation of an idealized Chapman-Ferraro sheath. The $\mathbf{V} \times \mathbf{B}$ forces produce a charge separation (dc) electric field that tends to slow down the positive ions. The electrons acquire additional energy, and a current system is naturally set up to maintain the gradient in $|\mathbf{B}|$. The thickness of this sheath region is about (1–2) c/ω_{pe}, where c is the speed of light and $\omega_{\mathrm{pe}}/2\pi = 9 \times 10^3 \, (N)^{1/2}$ Hz is the electron plasma frequency. For nominal solar wind densities, the collisionless electron inertial length $\delta (\simeq c/\omega_{\mathrm{pe}})$ is about 2–4 km, in agreement with many observations.

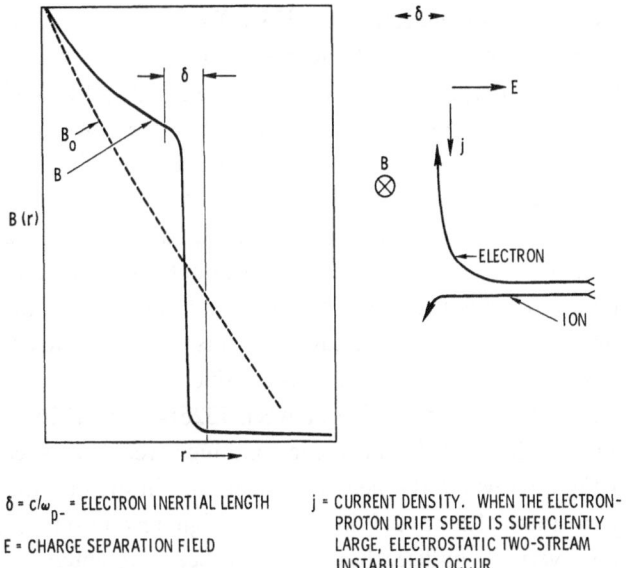

$\delta = c/\omega_{p-}$ = ELECTRON INERTIAL LENGTH

E = CHARGE SEPARATION FIELD

j = CURRENT DENSITY. WHEN THE ELECTRON-PROTON DRIFT SPEED IS SUFFICIENTLY LARGE, ELECTROSTATIC TWO-STREAM INSTABILITIES OCCUR

Fig. 2. Schematic representation of charge separation processes that can lead to a two-stream instability when a steep magnetic gradient is encountered.

Electrostatic waves develop in this region because the current is large enough to trigger a two-stream instability [1], and it appears that the oscillations are ion sound waves having **k** aligned with **j**. For the shock crossing of Figure 1, the E-field spatial distribution is, in fact, confined to the region where one would expect strong currents on the basis of ($\mathbf{V} \times \mathbf{B}$). However, it must be realized that the actual instability is only described in a very general sense by this model. Recent high speed and high resolution measurements of the electron distribution functions on Vela 4 and 5 strongly suggest that resonant wave-particle interactions develop within the shock [5]. The 'flat top' electron distribution is indeed unstable with respect to ion acoustic waves, but the instability is somewhat more complex than the one suggested in the early discussions [1, 2].

3. The Bow Shock during a Storm

The VLF electric field experiments on Pioneer 8 and 9 allow us to explore the development of electrostatic noise in deep space when low Mach number interplanetary shocks are encountered [6, 7], and in several cases we can examine the changes in the bow shock region as the disturbed solar wind arrives at the earth. The lower part of Figure 3 shows the Pioneer 8 response in the (qualitative) broadband E-field channel for the period 13 March 1968 through 8 April 1968. At the beginning of this solar rotation (1842) the Pioneer 8 solar ecliptic coordinates were $X_{SE} = -1145\ R_e$, $Y_{SE} = 1172\ R_e$, and on April 8 we had $X_{SE} = -1356\ R_e$, $Y_{SE} = 2155\ R_e$, respectively. The Pioneer data displayed here represent maximum and minimum potential amplitudes in hourly samples, and the solid and open triangles show where prominent SC or SI events were detected on the ground.

The ground sudden commencement at 1328 UT on April 5 is clearly related to the enhanced noise signals detected on Pioneer 8, as shown by the heavy vertical arrow. Moreover, when this interplanetary shock front reached 1 AU, OGO-5 was in the outer magnetosheath and the storm encounter pushed the bow shock past the OGO, so that the spacecraft suddenly found itself in the solar wind. The upper part of Figure 3 shows the magnetic field profile, the E-field amplitudes in the 15% bandpass channels, and the E-field dynamic spectrum (the latter extends from 1 kHz to 22 kHz, and the closely-spaced, nearly horizontal lines below about 1 kHz represent the Rubidium magnetometer lines for this period). The high resolution capabilities of the OGO-5 instrumentation (8 kilobits/sec digital data rate plus broadband telemetry) thus allow us to study an interplanetary shock in considerable detail.

The shock front passed over OGO-5 between 1326:40 and 1326:50 UT and in the spacecraft frame the electric field turbulence had high frequency spectral peaks. This can be seen in Figure 3 where the nearly vertical lines in the $f(t)$ diagram have little amplitude below about 1.5 to 2.5 kHz. Very high resolution spectrograms with sweep repetition every 50 millisec have recently been presented for this event (see Figure 3 of [4], and it has been shown that the E-field spectrum actually changed

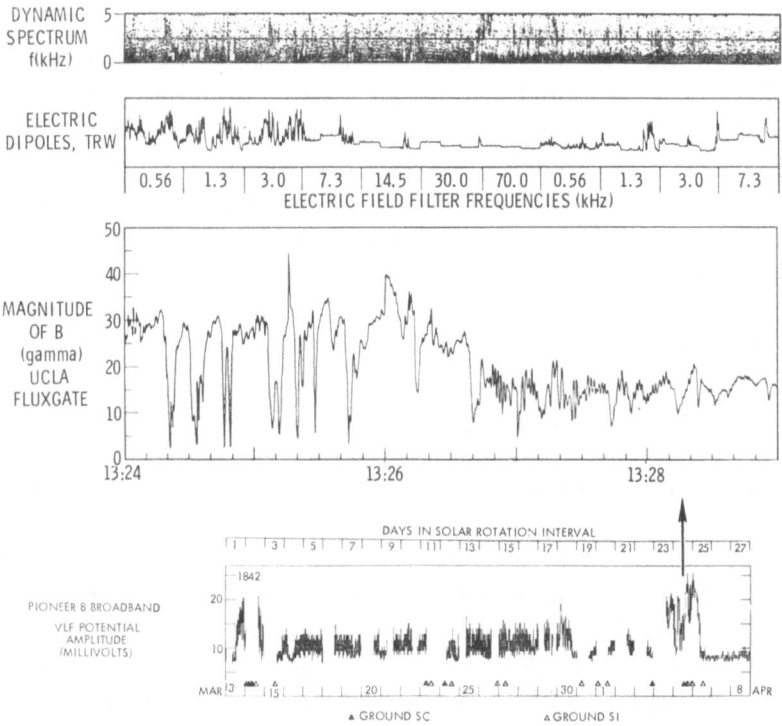

Fig. 3.   Pioneer 8 and OGO-5 measurements of the April 5, 1968 storm. The Pioneer broadband channels give a qualitative measure of the noise intensity for $f \geqslant 100$ Hz. The relative amplitudes in bandpass channels are shown for the OGO-5 electric field noise bursts and the f-t diagram gives the dynamic spectrum.

rapidly over these fine time-scales. This suggests that very short wavelength ($\lambda \simeq 2$–3 Debye lengths) oscillations with $\omega/k \ll 400$ km/sec have been detected, and that Doppler effects provide significant frequency shifts. At this time the upstream solar wind plasma density was about 20 cm^{-3} (M. Neugebauer, private communication), so that f_p^+ was nearly 1 kHz; reasonable Doppler shifts could then easily give an apparent frequency of (2–5) kHz, and we conclude that the April 5 observations are compatible with the interpretation that intense electrostatic ion sound waves develop in the shock region. Furthermore, the Pioneer 8 data show that more moderate enhanced noise levels do persist in the solar wind for many hours after the front has passed.

4. Whistler Mode Shocks

Figure 4 shows some Pioneer 8 and OGO-5 data taken during another sequence of solar disturbances early in rotation 1842. On March 14 a number of sudden commencements and sudden impulses were detected on the ground, and during this period the Pioneer 8 electric field experiment measured significantly enhanced VLF potentials. OGO-5 found that the magnetopause and shock were encountered at abnormally small radial distances, and the shock crossings were marked by an unusual and distinctive structure. The upper box in Figure 4 shows a magnetic field profile that

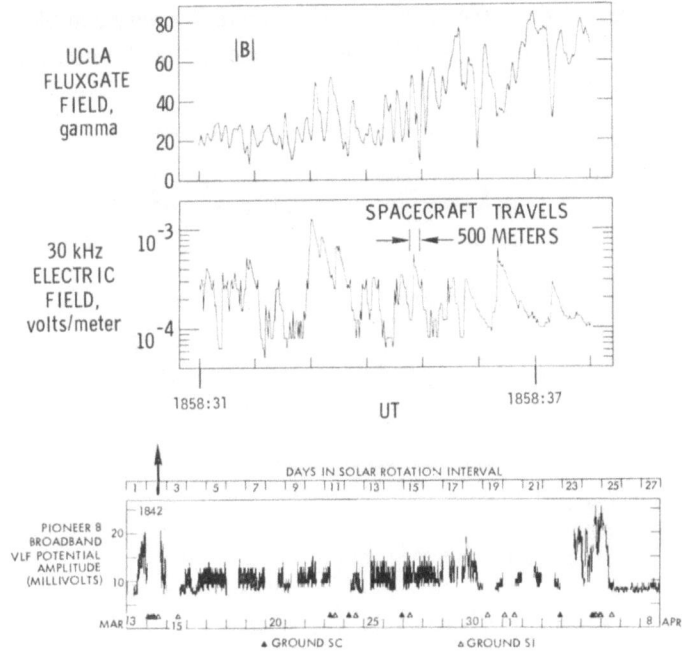

Fig. 4. Data similar to Figure 3 for the disturbed March 14, 1968 bow shock crossing. The only strong low frequency ($f < 22$ kHz) electric fields detected here were the electric components of whistler mode bursts. Special purpose telemetry is not available to us on OGO-5 for $f > 22$ kHz, and we cannot display the dynamic spectrum for these 30 kHz signals.

bears a superficial resemblance to the one contained in Figure 1. However, there are two significant differences: (a) the field magnitude is approximately 2 to 2.5 times greater on March 14 than on March 12, and; (b) the frequency content of the magnetic compressions is much higher on March 14 (note that Figure 1 displays 30 sec worth of data while Figure 4 shows only 7 sec worth).

The central box in Figure 4 illustrates that moderate amplitude ($E \leqslant 1$ mV/M) high frequency waves were generated in the shock region. These may represent electrostatic electron plasma oscillations (for N near $16/cm^3$ (M. Neugebauer, private communication), $\omega_{pe}/2\pi \simeq 36$ kHz) and it can be seen that many of the impulses are found in the steep B-field minima. Although these high frequency waves are generally found near the bow shock, the March 14 event (and a few other crossings) are marked by an unusual *absence* of strong low frequency electrostatic noise related to ion sound waves. The only significant low frequency E-field waves detected here were the electric components of whistler mode waves with $f \ll 560$ Hz.

We do not yet understand what conditions determine when the bow shock microstructure is to be governed by electrostatic or electromagnetic wave excitation. However, the measurements displayed in Figure 4 show that at least on some crossings, whistler mode growth predominates.

5. Null Regions

On the downstream side of the April 5 shock (Figure 3) a very complex magnetic field profile is seen to be associated with intense VLF electric field noise. Figure 5 contains

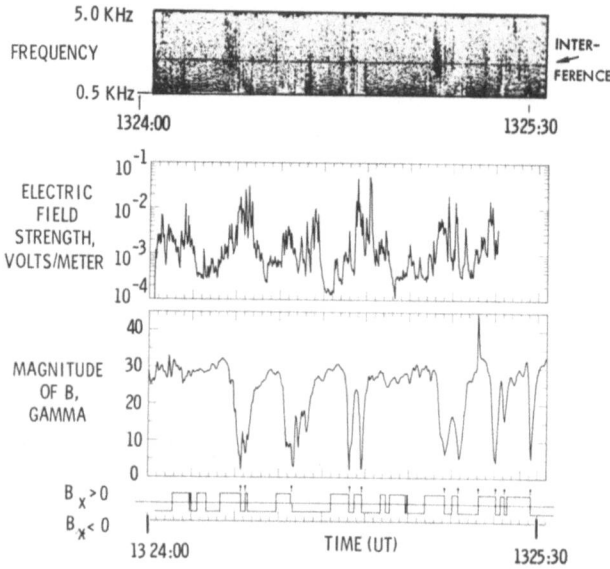

Fig. 5.   Enlarged view of a section of magnetosheath data for April 5, 1968. The field 'nulls' are found when the field rotates. The large amplitude electrostatic noise bursts are again detected where the steepest magnetic gradients are found.

an expanded plot of these data, and here the E-field calibrations are specifically inserted. The digital data rate for Figure 5 is 8 kilobits/sec, the E, B samples are obtained every 144 millisec, and it is clear that the full structure could not have been determined with a significantly lower sampling rate.

The vector magnetometer data on OGO-5 show that these configurations where $|\mathbf{B}|$ dips nearly to zero are commonly detected in the outer regions of the magnetosheath. We call these field 'nulls' (a sharp increase in $|\mathbf{B}|$, such as that observed at 1325:15 UT, is found very rarely), and we observe that the nulls always separate regions where $\mathbf{B}$ has different orientations. This is illustrated in Figure 5 by the lowest box. It can be seen that rotations in $\mathbf{B}$ do not always yield nulls, but that null regions require the rotations. Thus, it appears that these magnetosheath features are X-type nulls, as first proposed by Sonett [8]. (It is rather remarkable that in 1963 Sonett was already able to decide on the presence of X-type nulls in the magnetosheath using data from Pioneer 1. This probe made a single traversal of the sheath, and a spinning search coil was used so that $|\mathbf{B}|$ was not known. However, the Pioneer 1 experiment had a high effective data rate and an extremely comprehensive analysis was performed.)

During the Pioneer 9 outbound pass of November 8, 1968, similar magnetic features were observed and superthermal protons were detected in the low field region [9]. One might try to interpret such an event as a simple diamagnetic null, however the $|\mathbf{B}|$-field profiles generally argue against this evaluation. Increased field magnitudes are rarely found along the outer slopes, but for diamagnetic repulsion the 'frozen-out field' would have to appear on the shoulders.

The E-field amplitude distributions of Figure 5 indicate that very intense electrostatic noise develops on the flanks where the B-field compression produces a steep gradient. The top box shows that the E-field bursts have relatively high frequencies and they appear to exhibit little dispersion on this time-scale. (The horizontal line labeled 'interference' is the 2461 Hz OGO converter frequency. Note that the $f(t)$ diagram of Figure 5 has not been precisely aligned with the digital data; however, the field nulls show up as dropouts in the nearly horizontal Rubidium magnetometer lines, and extremely precise correlation can be achieved by comparing the dropouts with the $|\mathbf{B}|$ plot.) We interpret the intense electrostatic noise as ion acoustic waves or related Buneman modes produced by current-driven two-stream instabilities, as discussed previously.

It is reasonable to suggest that the data shown in Figure 5 can illustrate the phenomenon of magnetic field annihilation or reconnection. When one component of the $\mathbf{B}$ field reverses across a skewed current pinch of thickness l, field lines diffuse through the pinch to merge in time t, given by

$$t \sim \sigma l^2. \tag{1}$$

Here σ is presumably the anomalous conductivity associated with the drift instability and with scattering of particles from the electrostatic waves. Some summary discussions of general reconnection phenomena are given by Axford, Dungey and Petschek in *The Solar Wind* [10], and it now seems likely that ion acoustic waves provide a major dissipation mechanism that allows the reconnection to proceed.

6. Wave-Wave Interactions

The idea that large amplitude oblique (compressional) whistler mode waves or solitary magnetic pulses with appropriate phase speeds can produce a standing bow shock was advanced in Section 2. If proposals of this type have any validity, then we should expect to detect occasional moving shock-like structures associated with whistlers, or solutions having 'wrong' phase speeds. Large amplitude magnetic gradients could still trigger two-stream electrostatic instabilities, but these events would lead to propagating disturbances, rather than the standing bow shock.

It now seems that such interactions are indeed present in the upstream solar wind. Long period oscillations ($T \simeq 20$–60 sec in the spacecraft frame) with $|\varDelta\mathbf{B}|/|\mathbf{B}| \gtrsim 0.5$ have been detected on Explorers 33–35, Vela 3, and finally on OGO-5. The high data rate capability of OGO-5 also allows detection of high frequency ($T \simeq 1$–2 sec) damped electromagnetic tones that appear in the midst of the long period waves. The study of these bursts is far from complete at present, but it does appear that the magnetosonic waves in the upstream region can have field gradients steep enough to trigger electrostatic turbulence. An example of this upstream wave-wave interaction is given in Figure 2 of [7]. In that case (0925 on March 10, 1968), the OGO-5 LEPEDEA of Dr. L. A. Frank showed that significant fluxes of 4–7 keV protons were also present in the wind. Investigations of these wave-particle interactions are being pursued intensively at this time and no definitive conclusions are presently available. However, the results do suggest that the extremely low fluxes of energetic protons detectable with a LEPEDEA can have a profound effect on the upstream or interplanetary environment.

Acknowledgements

We gratefully acknowledge support of this work by the OGO Project Office at Goddard Space Flight Center under National Aeronautics and Space Administration Contracts NAS5-9278 (TRW) and NAS5-9098 (UCLA). The Pioneer 8 and 9 data analysis, leading to the lower boxes in Figures 3 and 4, was supported by the Pioneer Project at Ames Research Center under NAS2-4673.

References

[1] Fredricks, R. W., Kennel, C. F., Scarf, F. L., Crook, G. M., and Green, I. M.: 1968, *Phys. Rev Letters* **21**, 1761.
[2] Fredricks, R. W. and Coleman, Jr., P. J.: 1969, in *Proc. Int. Conference on Plasma Instabilities in Astrophysics* (in press).
[3] Fredricks, R. W., and Scarf, F. L.: 1969, in *Proc. of the ESRIN Conference on Collisionless Shocks* (Frascati) (in press).
[4] Scarf, F. L., Fredricks, R. W., and Kennel, C. F.: 1970, in *Earth's Particles and Fields,* 1969 (ed. by B. M. McCormac), Reidel, Dordrecht, The Netherlands.
[5] Montgomery, M.: 1970, in *Earth's Particles and Fields* (ed. by B. M. McCormac), Reidel, Dordrecht, The Netherlands.
[6] Scarf, F. L., Crook, G. M., Green, I. M., and Virobik, P. F.: 1968, *J. Geophys. Res.* **73**, 6665.
[7] Scarf, F. L.: 1970, in *Proceedings of the XVIth General Assembly of URSI* (in press).

[8] Sonett, C. P.: 1963, *J. Geophys. Res.* **68**, 1265.
[9] Scarf, F. L., Green, I. M., Colburn, D., Sonett, C. P., Wolfe, J. H., and Intriligator, D.: 1969, *Trans. AGU* **50**, 278.
[10] Axford, W. I., Dungey, J. W., and Petschek, H. E.: 1966, in *The Solar Wind* (ed. by R. Mackin and M. Neugebauer), Pergamon, London.

[8] ...
[9] ...
[10] ...

5. THE MAGNETOSPHERE DURING AND AFTER SOLAR EVENTS

SOLAR PARTICLE OBSERVATIONS OVER THE POLAR CAPS

G. A. PAULIKAS, J. B. BLAKE and A. L. VAMPOLA

Space Physics Laboratory, Aerospace Corporation, El Segundo, Calif., U.S.A.

1. Introduction

The increasing incidence of energetic particle emission from the sun has, since 1966, provided a number of opportunities to measure the spatial, angular and energy distribution of these particles at high latitudes with instruments flown aboard low altitude polar orbiting satellites. Such measurements provide insight into the structure of the distant magnetosphere and the mechanisms of solar particle access into the transmission through the magnetosphere. This work complements the measurements carried out by high altitude spacecraft.

The phenomenology exhibited by solar particles observed over the polar caps becomes increasingly complex as one studies lower energy particles. Of special interest, because of their small gyroradii, are solar protons and alpha particles with energies less than 10 MeV/nucleon. Solar electrons with energies down to 40 keV have also been observed over the polar caps and provide an additional means (a very low rigidity particle) for studying solar particle access to the polar caps. This paper will summarize some of the recent measurements of spatial (latitude) distributions and angular distributions. We will also compare solar proton and electron data and discuss the connection between the limits of trapping and the access and cutoff latitudes for solar particles. The bulk of the data to be discussed were obtained aboard the USAF-OAR satellites 1966-70A (OV3-3), 1967-72D (OV1–12) and 1968-59A (OV1–15).

2. Connection between Trapping Boundary and the Cutoff Latitude

The idea of the last closed drift shell (the last family of lines upon which a low rigidity particle may be permanently trapped) may be used as a definition of the limits of access of extraterrestrial particles. The concept can be experimentally tested by comparing a trapping boundary of low energy electrons with the cutoff latitude of low energy protons. We have performed this experiment and illustrate the results in Figure 1 for the midnight meridian and in Figure 2 for the noon meridian. In the case of near midnight, the disappearance of the trapped electrons is directly connected with the appearance of solar protons. Studies of trapped particle trajectories in a Mead-Williams model of the magnetosphere using Roederer's (1967) program SPLIT indicate that near local midnight the last closed field line also defines the last closed drift shell for particles mirroring at high latitude. Near local noon, however, solar protons arriving via longitudinal drift (quasi-trapped particles) appear deeper in the electron trapping region and, in general, exhibit a very ill-defined cutoff. Figure

V. Manno and D. E. Page (eds.), Intercorrelated Satellite Observations Related to Solar Events, 193–204. All Rights Reserved
Copyright © 1970 by D. Reidel Publishing Company, Dordrecht-Holland

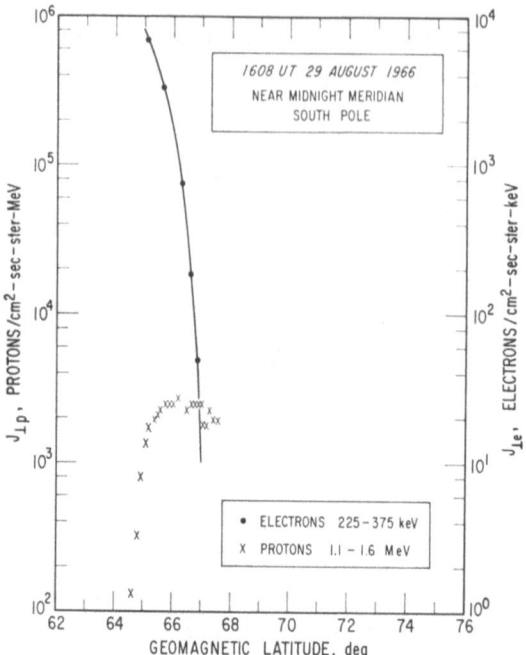

Fig. 1.   Boundary of trapped electrons and cutoff of 1.1–1.6 MeV solar protons. Data taken by
OV3-3 satellite near local midnight.

3 summarizes the comparison between the trapping termination and cutoff latitude
near local midnight for quiet as well as disturbed times. The tracking of the Λ_p with
Λ_e is evident; the small difference $\Delta\lambda$ (about 1.5°) can be quantitatively connected
to the non-zero gyroradius of the protons on these field lines near the magnetic
equator. Near local noon (Figure 4) definition of the cutoff becomes more difficult.
Solar protons penetrate 3° to 4° past the apparent limit of electron trapping. If the
electron boundary at local noon can truly be identified with the last *closed* drift shell,
then proton scattering (and hence radial diffusion because of shell splitting) need be
invoked in order to account for the experimental results.

During magnetically disturbed times, as the proton cutoff latitudes is depressed,
the solar proton flux merges with the stably trapped particle population (Figure 5).
However no permanent injection of energetic solar particles appears to occur. This
has been established by examining the trapped alpha particle population. Since the
solar energetic alpha flux is higher than the trapped alpha particle flux, a small
injection efficiency might be detectable. No such injection was found (Blake *et al.*,
1968).

A summary of data showing cutoffs (near local midnight) as a function of proton
energy is presented in Figure 6. During quiet or moderately disturbed condition the
cutoff energy tends to zero near $\approx 67°$. However there exists no evidence that appreciable ($>10^3$ cm^{-2} sec^{-1} sr^{-1} keV^{-1}) fluxes of heated (2–12 keV) solar wind protons
from the magnetosheath can reach the polar caps. Thus the interesting question of

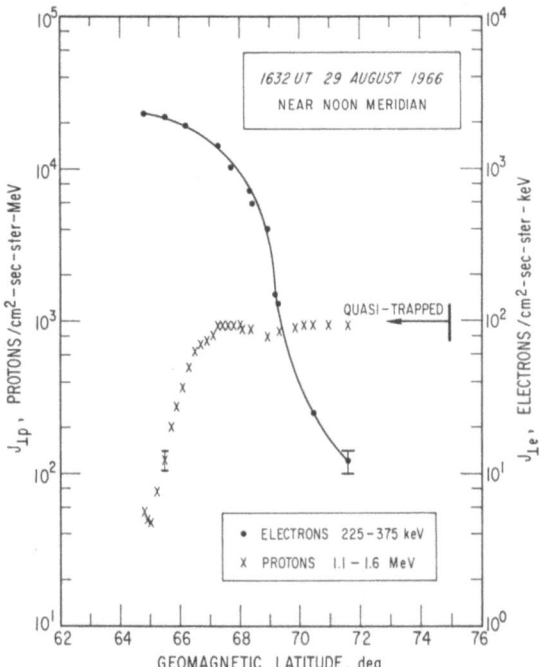

Fig. 2. Boundary of trapped electrons and cutoff of solar protons near local noon. Quasi-trapped region (pitch angle distributions peak perpendicular to magnetic field) is accessible to solar protons by longitudinal drift.

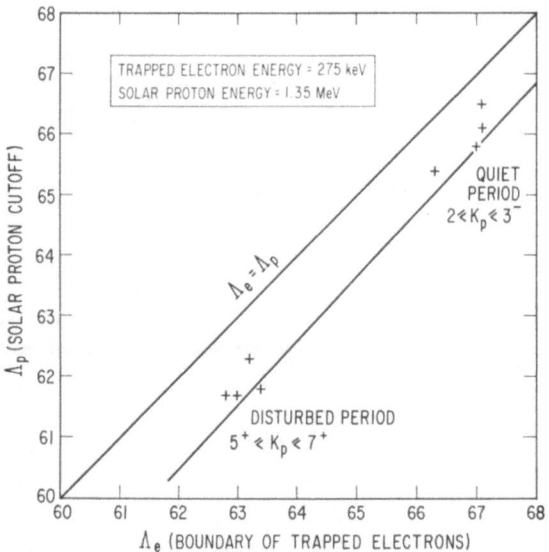

Fig. 3. Comparison of proton cutoff and outer zone boundary for magnetically quiet as well as disturbed periods. Data were obtained in the August 28 – September 3, 1966 interval near *local midnight*. The lower solid line indicates cutoff latitude shift assuming protons penetrate one cyclotron radius (in a 20 γ equatorial field) past the last closed field line.

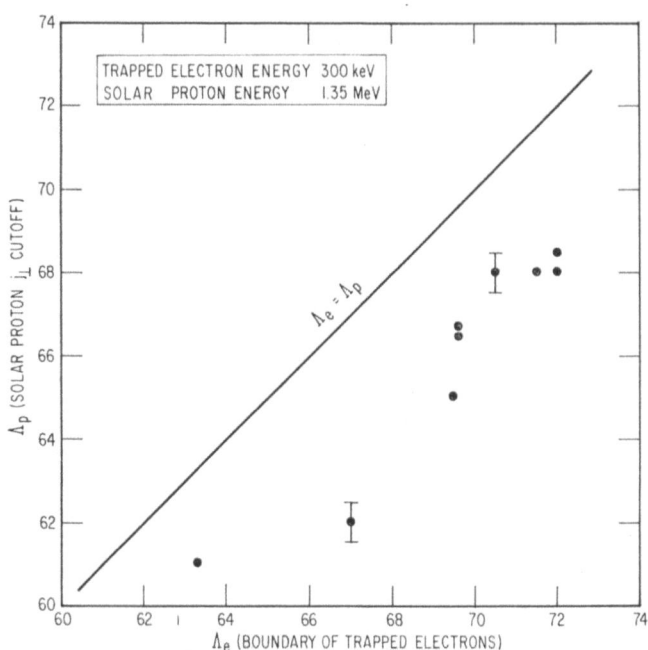

Fig. 4.   Comparison of proton cutoff ($j_\perp$) and outer zone boundary near *local noon*. Data were obtained in the August 28 – September 3, 1966, interval.

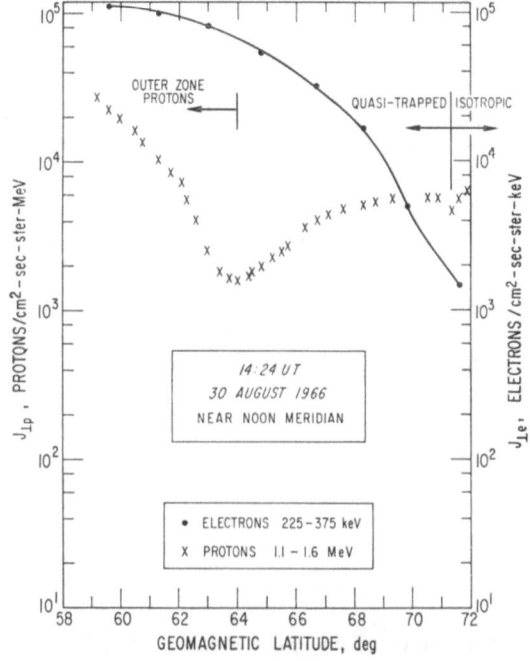

Fig. 5.   Merging of solar proton fluxes and radiation belt proton fluxes. Angular distributions for $\Lambda \gtrsim 71°$ show peaking of fluxes perpendicular to local field vector.

the solar proton cutoffs in the 1–100 keV interval remains: at what energy does a test particle (proton) placed in the solar wind cease to behave as a member of the solar wind plasma streaming supersonically past the magnetosphere and assume a single particle mode of motion which enables it to reach the polar cap?

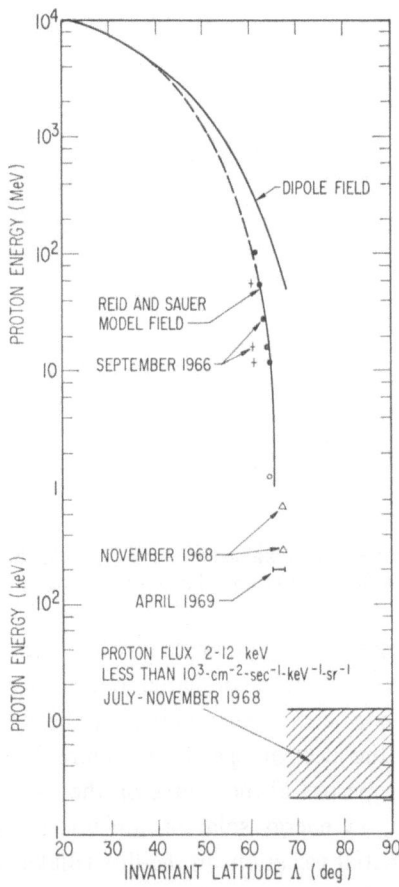

Fig. 6. Proton cutoff latitude near local midnight as function of particle energy. Nightside cutoff calculation of Reid and Sauer (1967) as well as cutoffs in a dipole field are indicated. The upper limits for 2–12 keV proton flux at high latitudes were determined in the July to November 1968 interval by Hilton (private communication).

3. Penetration of Solar Electrons to the Polar Caps

It has been found that solar electrons (Vampola, 1969) in addition to the well documented low energy protons and α's have ready access to the polar caps. The measurements have now been extended down to electron energies of 40 keV; a representative electron spectrum obtained on April 13, 1969, is shown in Figure 7. The salient features of the solar electron observations over the polar caps are summarized here:

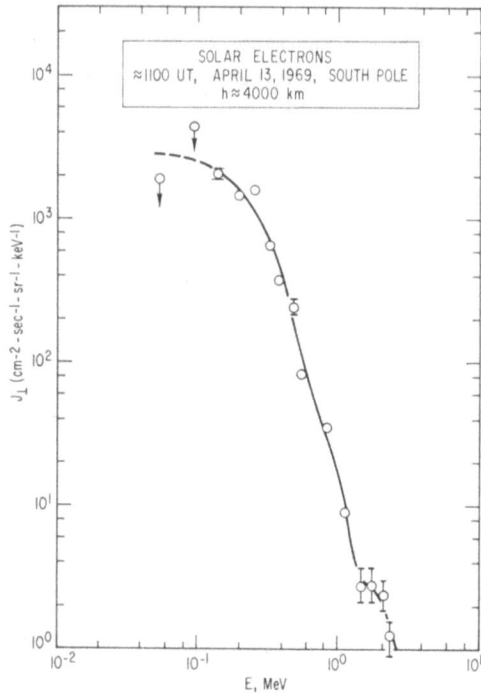

Fig. 7.   Preliminary electron spectrum obtained on April 13, 1969, aboard the polar orbiting OV1-19
(1969-27D). A flux of electrons illuminated the polar cap uniformly.

(1) Electrons have access to the polar caps with high efficiency. Measurements outside the magnetosphere have been compared to measurements over the polar caps, and show a transmission efficiency near unity. At the present time it is not clear whether the turnover in the electron spectrum (Figure 7) is a magnetospheric effect or associated with the properties of the source on the sun.

(2) In contrast to the low energy solar proton fluxes whose latitude profiles are often characterized by spatial variations in the flux (Blake *et al.*, 1968; Williams and Bostrom 1969; Bostrom, 1970) most solar electron observations show featureless flux vs latitude profiles; solar electrons can be distinguished from the trapped electron population on the fringes of the trapping region by examining the latitude-intensity profiles. Such nonuniformities as have been detected appear, in most cases, to be associated with motions past the satellite of the trapping boundary containing terrestrial electrons.

(3) Diurnal asymmetry in the polar electron flux is sometimes observed – usually in the early stages of the polar particle event. The polar electron fluxes merge abruptly with the trapped outer zone electron fluxes (as determined by a change in the pitch angle distribution) near the local midnight meridian; in contrast, near the noon meridian a more gradual connection between the solar flux and the outer zone electrons is found. A drop in the solar electron flux spatially coincident with the beginning of an anisotropic pitch angle distribution for solar protons on the dayside hemisphere

is often observed (Figure 8). Measurements of electron pitch angle distributions here show a peaking of the flux perpendicular to the magnetic field. There is no latitude gap between the domain of solar electrons and the domain of terrestrial electrons.

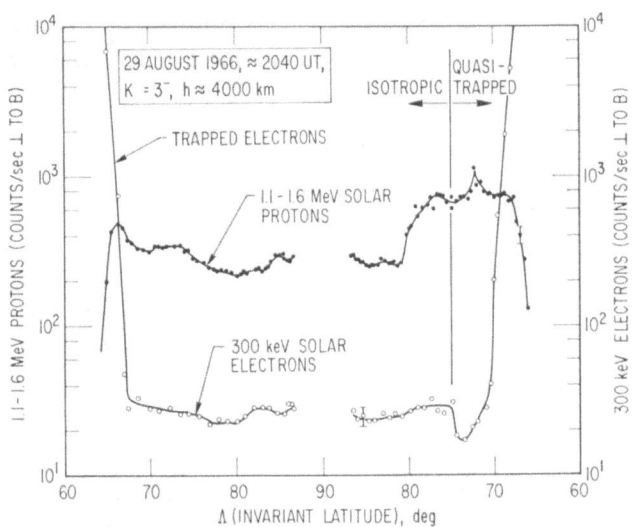

Fig. 8.   Comparison of solar electrons (225 keV $\leqslant E_e \leqslant$ 375 keV) and solar protons (1.1 MeV $\leqslant E_p$ $\leqslant$ 1.6 MeV) on August 29, 1966. Data were obtained aboard OV3-3 (1966-70A). Note the alignment of the electron flux depression and the beginning of the quasi-trapped region for electrons and protons. These data were obtained over the south polar cap. The dayside flux enhancement contains, in part, an anisotropic pitch angle distribution.

4. Latitude and Pitch Angle Distributions

Latitude profiles of low energy (< 10 MeV) solar proton fluxes observed over the polar cap exhibit an amazing range of behavior. An indication of the possibilities has already been presented in Figure 8. Compilation of all our data obtained since 1966

TABLE I

Solar particle flux profile summary*

	August 1966 June 1967		July 1967 December 1967	July 1968 November 1968
	Protons 1.1–1.6 MeV	Electrons 225–375 keV	Protons 3–10 MeV	Protons 0.75–1.7 MeV
Non-uniform profiles	49	4	21	15
Flat profiles	14	55	2	7

* These observations are weighted (because of satellite operational considerations and instrument counting statistics) toward the later phases of large solar particle events.

indicate that spatial non-uniformity is the rule for solar protons, but the exception for solar electrons (Table I).

Frequently one observes an enhancement in the solar proton flux in a latitude interval at higher latitudes than the cutoff. Examples of such enhancement are found at all local times. The enhancements may have isotropic or anisotropic pitch angle distributions. Figure 9 illustrates a flux enhancement which, near local noon, shows

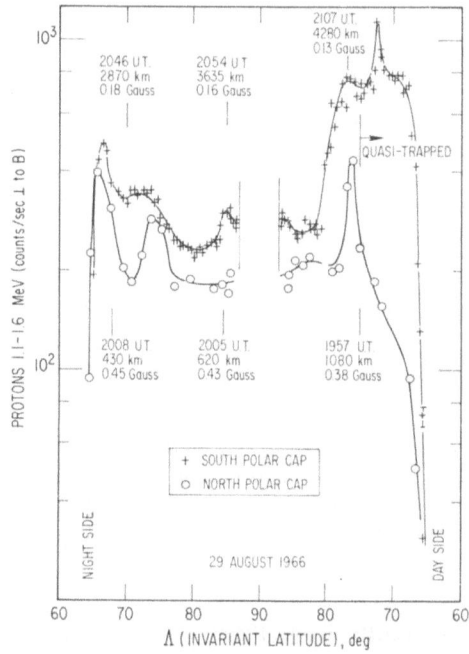

Fig. 9.   Proton latitude – flux profiles over opposite polar caps. Large difference in fluxes (altitude effect) on day side results because particles mirroring near 4000 km do not reach lower altitudes. Width of region containing anisotropic fluxes is indicated for high altitude pass. The region of pitch angle anisotropy is only a few degrees wide at low altitudes.

both species of angular distributions. Measurement for protons and electrons made at altitudes > 1000 km and away from local midnight show anisotropic pitch angle distributions with the particle flux peaking in a direction perpendicular to the local field. The width of this region of pitch angle anisotropy approaches 10° in latitude near local noon at high ($\approx$ 4000 km) altitudes. The latitude interval of anisotropic fluxes is only a few degrees wide at low altitudes. These distributions indicate temporary trapping of particles and have been discussed (for protons) by various authors (Taylor 1967; Gall *et al.*, 1968). A simple way to describe the generation of an anisotropic solar particle flux on the dayside is by inverting the arguments used by Roederer (1967) to describe the motion of trapped particles: drift shells of particles with low altitude mirror points near local noon connect to open field lines in the magnetic tail (Figure 10). The motion of electrons (in the range of these measurements) is

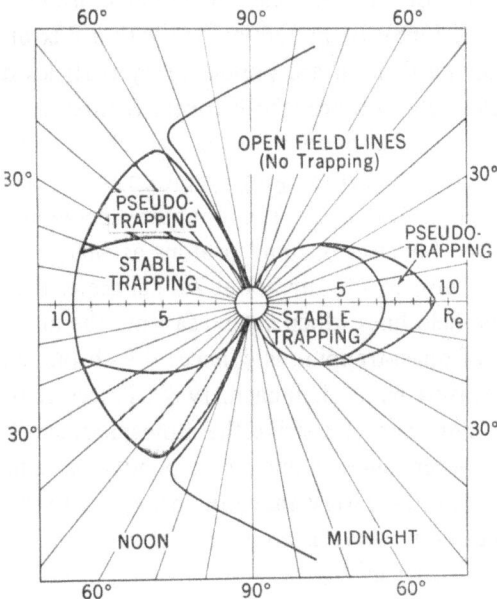

Fig. 10.   Domains of quasi-trapped particles (Roederer, 1967). Particles mirroring in high latitude pockets are connected by drift paths to open field lines on nightside of magnetosphere. Pseudo-trapping region at low latitudes near local midnight is accessible (by drift) to extraterrestrial particles entering through the flanks of the magnetosphere.

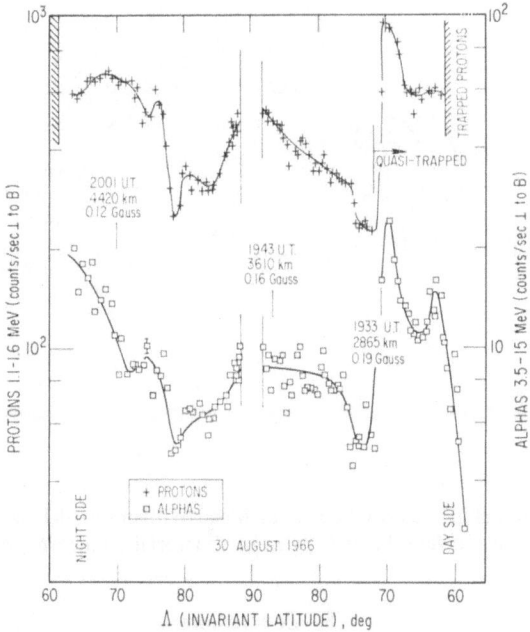

Fig. 11.   Proton and α latitude-flux profiles. α's and protons in the same energy/nucleon (same velocity) interval behave similarly with respect to polar cap spatial non-uniformities and transition to anisotropic pitch angle distributions.

adiabatic (in the absence of wave-particle interactions) throughout most of the magnetosphere (Gall *et al.*, 1968). If the tail is filled with a flux of solar particles, then field lines closed near local noon are populated by particles drifting in longitude. Note that no particle scattering need be invoked. Similar arguments also apply to ≈ 1 MeV protons, although their motion does not satisfy the adiabatic criterion exactly and protons may suffer from scattering by cyclotron radius sized field curvatures in the limit of almost open field lines. The boundary between isotropic and quasi-trapped fluxes observed for electrons and protons as well as α-particles (Figure 11) occurring at latitudes between 73° and about 78° delineates the separation of locally closed field lines from field lines which extend far into the magnetotail.

The phenomenon of quasi-trapping is connected with the mid-day recoveries observed in riometer measurements (Leinbach, 1967). Referring to Figure 9, it is clear that near the cutoff more particles strike the atmosphere near local midnight than near local noon. However, since riometers respond most efficiently to ionization caused by protons in the 10–20 MeV range the present observations of the behavior at ~ 1 MeV protons can only serve as an indirect guide to the PCA mid-day recovery phenomenology.

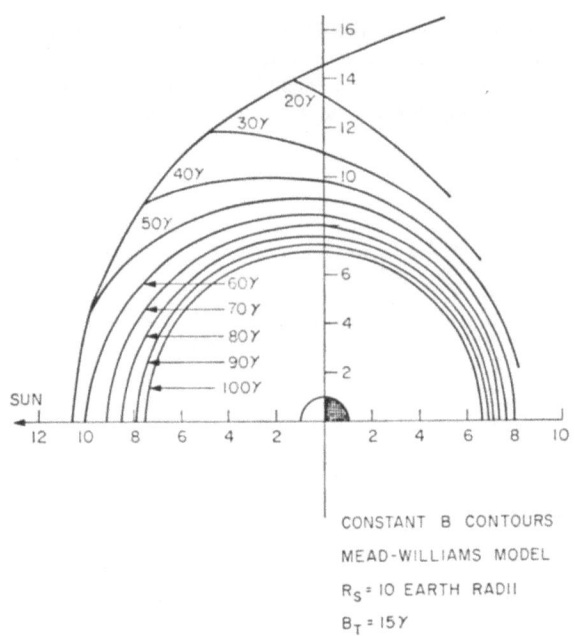

Fig. 12.   Isofield contours in equatorial plane for Mead-Williams model magnetosphere (Roederer, 1969). These contours also define the drift paths of equatorially mirroring quasi-trapped particles.

5. Conclusions

The results of this work, taken together with the work of others, serve to define some of the properties of the magnetosphere for the transmission of solar particles.

Montgomery and Singer (1969), as well as Kane *et al.* (1968), find that solar proton temporal variations and angular distribution anisotropies, although not eliminated, are smoothed in transit to the interior of the magnetosphere near 18–20 R_e. Assuming adiabatic motion for simplicity, remaining spatial gradients in the particle angular distribution function across the magnetotail will be translated into non-uniform latitude-intensity profiles across the polar cap which is the usual observation. Note however, that because of the high mirror ratio (≈ 1000) only a small portion of the angular distribution at the Vela orbit (those particles moving parallel to the field) reach the polar caps. The uniformity of electron latitude-intensity profiles in the domain of open field lines over the pole suggests that, in contrast, the magnetotail is filled very uniformly with solar electrons. Anderson and Lin (1969) have demonstrated that the magnetotail is filled with solar electrons travelling along field lines directly connected to the interplanetary field rather than strong diffusion.

The very sharp and temporally stable separation (as seen by solar electron fluxes, Figure 8) of the region of open field lines from locally closed field lines and the abrupt changes in pitch angle distributions suggest that diffusion by solar electrons in the region near the earth where open field lines are in contact with closed field lines is slow (less than ≈ 70 km per longitudinal electron drift period). Solar protons penetrate deeper into the region of closed field lines than pure trajectory tracing permits. We postulate that scattering of protons by (< 1 cps) hydromagnetic waves is the mechanism which transports protons radially.

One of the persistent observations, ours as well as that of others (Blake *et al.*, 1968; Williams and Bostrom, 1967; Bostrom, 1970) has been the enhancement of the solar proton flux at latitudes just above the cutoff, i.e., on open field lines and in the quasi-trapping region. This has been interpreted as indicating, at times, diffusive entry of solar protons into the magnetosphere (Williams and Bostrom, 1969). The time development of the latitude-intensity profile required for this explanation is one in which a flat latitude profile evolves with time. Observations exist where such evolution did not proceed; the flux enhancement near the cutoffs persisted throughout a proton event. There are also indications that the flux intensities near the cutoffs follow the interplanetary flux more closely than particle fluxes measured at high latitudes (Blake *et al.*, 1968; Bostrom, 1970).

An alternate route of particle entry into the region of latitudes around 65° to 70° appears to exist. By reference to Figures 10 and 12 it can be seen that the low latitude region ($< 30°$) in the 7.5 to $9R_e$ interval in the nightside of the earth can be directly populated from interplanetary space by longitudinal drift. Pitch angle scattering of protons at all local times near this altitude interval (Paulikas and Blake, 1970) seems efficient enough to drive particles into the atmospheric loss cone as well as into temporarily trapped (because of further scattering) trajectories. With such a mechanism (quasi-trapped orbits plus scattering) particle access to the low latitude portion of the polar caps should be rapid (< 10 min for ~ 5 MeV protons) without requiring field line connection near the earth. The high latitude region flux is presumed still to be dominated by particle propagation from the magnetotail.

We conclude that some scattering of low energy solar protons, but not necessarily of electrons, is required in order to account for the experimental observations of solar particles over the polar caps. Direct entry of particles through the flanks of the magnetosphere appears to play a role in supplying particles to the polar caps.

Acknowledgements

The measurements reported in this paper were conducted aboard various USAF-Office of Aerospace Research spacecrafts. We are grateful to the many officers and contractors of this organization for their efforts. This research was performed under U.S. Air Force Space and Missile System Organization Contract (F04701-69-C-0066).

References

Anderson, K. A. and Lin, R. P.: 1969, *J. Geophys. Res.* **74**, 3953.
Blake, J. B., Paulikas, G. A., and Freden, S. C.: 1968, *J. Geophys. Res.* **73**, 4927.
Bostrom, C. O.: 1970, this volume, p. 229.
Gall, R., Jimenez, J., and Camacho, L.: 1968, *J. Geophys. Res.* **73**, 1593.
Hudson, P. D. and Anderson, H. R.: 1969, *J. Geophys. Res.* **74**, 2881.
Kane, S. R., Winckler, J. R., and Hofmann, D. J.: 1968, *Planetary Space Sci.* **16**, 1381.
Leinbach, H.: 1967, *J. Geophys. Res.* **72**, 5473.
Montgomery, M. D. and Singer, S.: 1969, *J. Geophys. Res.* **74**, 2869.
Paulikas, G. A. and Blake, J. B.: 1970, *J. Geophys. Res.* (to be published).
Reid, G. C. and Sauer, H. H.: 1967, *J. Geophys. Res.* **72**, 4383.
Roederer, J.: 1967, *J. Geophys. Res.* **72**, 981.
Roederer, J.: 1969, *Rev. Geophys.* **7**, 77.
Taylor, H. E.: 1967, *J. Geophys. Res.* **72**, 4467.
Vampola, A. L.: 1969, *J. Geophys. Res.* **74**, 1254.
Williams, D. J. and Bostrom, C. O.: 1969, *J. Geophys. Res.* **74**, 3019.

ACCESS OF SOLAR PARTICLES TO
SYNCHRONOUS ALTITUDE

L. J. LANZEROTTI

Bell Telephone Laboratories, Murray Hill, N.J., U.S.A.

Abstract. When a source of solar-originated particles is present in interplanetary space, comparable fluxes of these particles with similar spectral characteristics are usually observed inside the magnetosphere at synchronous altitudes. The temporal and spectral changes in the access of these solar protons and α particles to synchronous altitude are discussed and reviewed. Possible consequences of the presence of substantial fluxes of low energy solar protons deep inside the magnetosphere are also discussed.

1. Introduction

A multitude of ground and satellite observations of solar proton fluxes at high latitudes have provided much convincing evidence that these protons have access to inner regions of the magnetosphere that would normally be prohibited by the Störmer theory (1955). These measurements and their relationship to theoretical particle trajectory calculations in model magnetospheres have been reviewed recently by Bostrom (1970), Paulikas (1970), and Hofmann and Sauer (1968).

Only recently have solar particle measurements been reported in the equatorial regions of the magnetosphere. Filius (1968) reported the access of solar protons (40–250 MeV) to 4 R_E at the equator and showed that the Störmer theory could be modified to agree with his observations. Lanzerotti (1968b) reported measurements of solar proton and alpha particle fluxes (1–20 MeV/nucleon) at synchronous altitude (6.6 R_E) and compared their temporal and spectral characteristics to simultaneous measurements of the interplanetary fluxes. The significance of these measurements was that substantial cutoff lowering was observed for ~1 MeV protons with near 90° pitch angles at the geomagnetic equator. Lanzerotti concluded that there must exist an effective diffusion process operative in the outer regions of the magnetosphere and/or the near regions of the geomagnetic tail that often preserves the spectral character and magnitude of the interplanetary fluxes after their penetration to the synchronous altitude. Paulikas and Blake (1969) have also reported the omni-directional intensity of solar protons at the synchronous altitude during a solar event in January, 1967. In addition to concluding that protons with energies greater than 21 MeV have essentially free access to synchronous altitude, they observed a diurnal effect in the fluxes of 5–21 MeV solar protons, with more protons observed near local midnight than near local noon. Lanzerotti (1968b) also observed such a diurnal effect in the 2.4 MeV proton fluxes.

This paper presents extensive measurements of the proton environment at synchronous altitudes. Most of the data are obtained from measurements made by the Bell Laboratories experiment on the ATS-1 satellite during the last half of 1967.

These measurements are compared to similar measurements obtained simultaneously by an instrument on the Explorer 34 (IMP F) satellite in interplanetary space. Since the ATS-1 satellite traverses all local times (solar-ecliptic longitudes) in a 24-hour period, and since the number of $\gtrsim 1$ MeV protons at synchronous altitude when interplanetary particles are not present is essentially zero, comparisons of the solar fluxes at this equatorial location with the interplanetary fluxes allows systematic investigations of the results of the particle access with other magnetosphere observations.

Due to the distortion of the magnetosphere by its interaction with the solar wind, the ATS-1 satellite does not remain on a line of constant magnetic field intensity B during its daily traversal of all local times. Rather the satellite is located on lines of lower B intensity at local midnight than at local noon. This fact, together with the observation that the satellite is always beyond the maximum of the outer electron zone gives rise to the familiar diurnal effect observed in the relativistic electron intensities (Lanzerotti et al., 1967; Paulikas et al., 1968). Also attributed to the distortion of the magnetosphere was the apparent diurnal effect (with the same sense as the electron fluxes) reportedly observed in the 0.3 to 10 MeV outer zone trapped proton fluxes by Armstrong and Krimigis (1968) using data from the elliptical orbit satellite, Explorer 33.

2. Experiments

The ATS-1 satellite (launched December 6, 1966) is a geostationary, spin-stabilized satellite (with the spin axis parallel to the local magnetic field) stationed at 158°W, 0°N and at a geocentric distance of 6.6 R_E. The Bell Telephone Laboratories (BTL) experiment flown on ATS-1 consists of a 6-element solid-state detector telescope oriented perpendicular to the spin axis of the satellite. By use of the particle dE/dx characteristics and appropriate logic circuitry, the experiment is capable of distinguishing between protons, α's, and electrons. The half-angle of the detector telescope collimator is 20°; the flux measured by the BTL experiment is the spin-averaged flux of those particles with pitch angles close to 90° (Lanzerotti, 1968a). ATS-1 local time is obtained by subtracting 10 hours from the universal time.

The IMP F satellite, launched May 24, 1967, is a spin-stabilized (with the spin axis perpendicular to the ecliptic plane) polar-orbit satellite with an apogee of approximately 34 R_E. The BTL experiment consists of a 4-element solid-state telescope oriented perpendicular to the spin axis. The half-angle of the detector telescope collimator is also 20°; the flux measured by the experiment is the spin-averaged flux of particles in the ecliptic plane. Protons and α's up to an energy of approximately 4 MeV/nucleon are distinguished by the energy deposited in the first two detectors of the telescope and subsequently measured in a 5-channel analyzer. Particle species above this energy are distinguished by the use of an on-board pulse multiplier. Electrons $\gtrsim 300$ keV are detected by their dE/dx characteristics and the appropriate logic circuitry (Lanzerotti et al., 1969a).

3. Observations

The first figure contains the 1 hour average temporal history of 1.9 and 2.4 MeV proton fluxes measured in interplanetary space and at synchronous altitudes, respectively, from day 318 to day 360, 1967. Also plotted at the bottom of the figure are the hourly average AE and Dst indices and the 3-hour average K_p index. The dark bars above the IMP data indicate periods when the Explorer 34 satellite was inside the magnetopause (D. H. Fairfield, private communication).

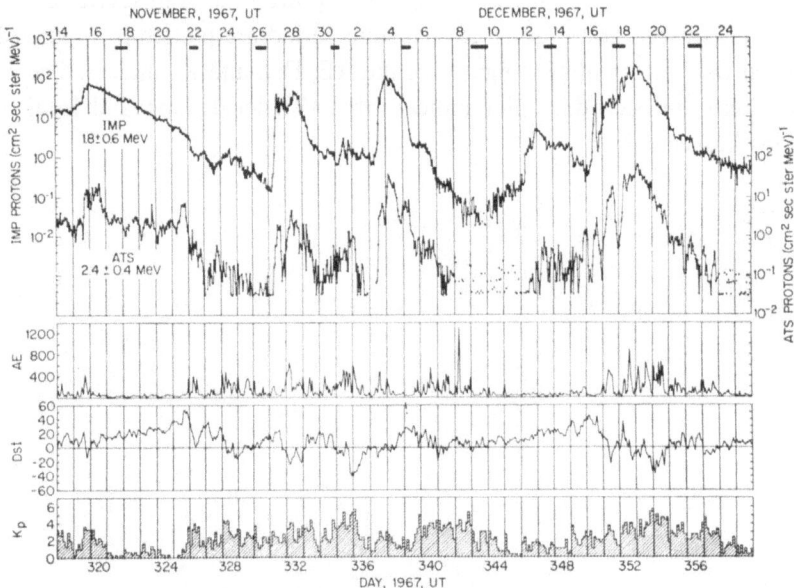

Fig. 1. 1-hour average Explorer 34 (IMP-F, interplanetary) and ATS-1 (synchronous altitude magnetosphere) proton fluxes from day 318 through day 359, 1967. The bars above the interplanetary data indicate periods when Explorer 34 was inside the magnetosphere.

This entire time period had appreciable fluxes of solar-origin interplanetary particles which are of interest in themselves. (In fact, the long period of interplanetary enhancements began on day 300, but continuous ATS coverage was not available during the entire period of day 300 to day 318.) This period was also moderately active geomagnetically as evidenced by the AE and K_p indices. However, although there were two sudden commencements reported, there were no large geomagnetic storms, the largest Dst negative excursion being less than 50 γ.

The overall, quite striking, impression obtained from this figure is that whenever there are interplanetary fluxes present, there are also proton enhancements inside the magnetosphere at synchronous altitudes. The enhancements have approximately the same temporal appearances and magnitudes as the interplanetary fluxes and do not persist after the interplanetary fluxes are absent.

There are also gross differences between the interplanetary fluxes and the synchronous altitude fluxes which will be discussed in detail later. The most striking of these are the decreases in the ATS fluxes that are often observed near local noon (2200 UT). The decreases are especially noticeable on days 331, 337, 338, 350, 351, 355, and 356, for example. Often the temporal histories of the flux decreases resemble PCA mid-day absorption recoveries (Leinbach, 1967).

A noticeable feature in the interplanetary data is the general absence of a modulation of the measured solar proton fluxes as the IMP satellite crosses the magnetopause. This feature of proton access to the magnetosphere is also discussed briefly later.

Expanded views of the 1-hour average ATS proton fluxes in several energy channels for a disturbed and a quiet geomagnetic period are plotted in Figures 2 and 3. Plotted at the bottom of each figure are the AE, Dst, and K_p indices. The open and closed circles on each data plot indicate the ATS local noon and local midnight locations, respectively.

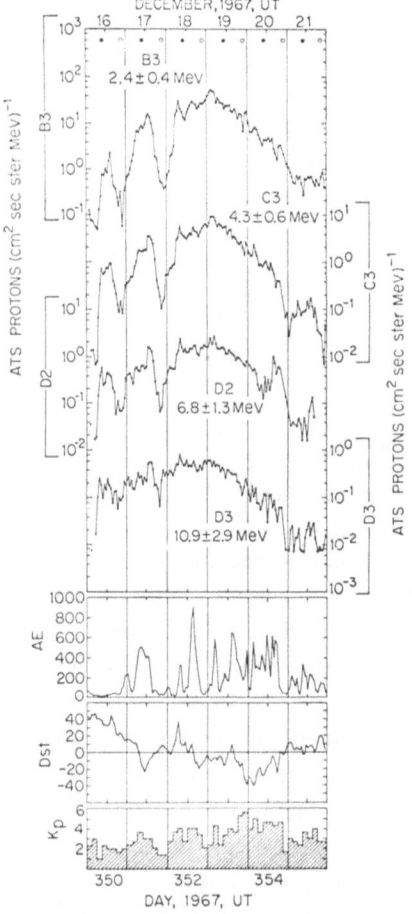

Fig. 2.   1-hour average ATS-1 proton fluxes from day 350 through day 355, 1967. This was a geomagnetically disturbed period.

During the geomagnetically disturbed period, days 350–355 (Figure 2), the diurnal variations in the proton fluxes were most pronounced at the lower energies. The diurnal minimum on day 351 was quite symmetrical about local noon in the 2.4 MeV proton channel, whereas both the symmetry and magnitude of the variations decrease at the higher energies.

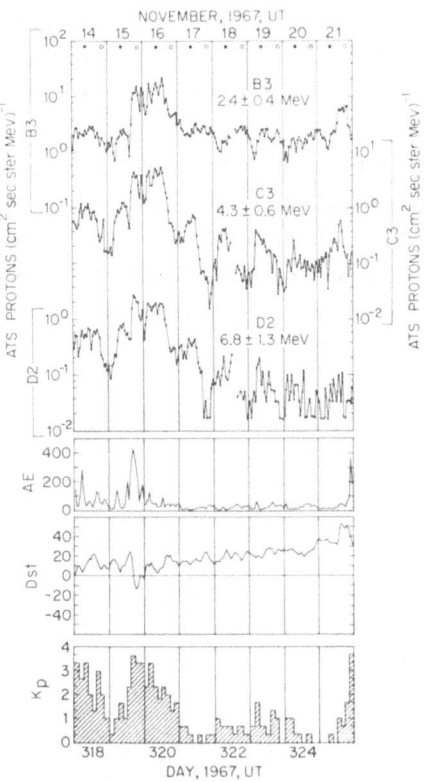

Fig. 3. 1-hour average ATS-1 proton fluxes from day 318 through day 325, 1967. The period including days 321–324 was a geomagnetically quiet period.

The temporal variations of the ATS proton fluxes during a quiet geomagnetic period, days 321–325 (Figure 3), were strikingly different than those observed in the data of Figure 2. The 2.4 MeV data show no decrease in the proton fluxes near local noon but rather small decreases near local evening on days 322–325. In contrast, the fluxes from the two higher energy proton channels plotted in the figure show significant decreases near local noon.

The hourly average electron fluxes for the disturbed and quiet periods are plotted in Figures 4 and 5. These data are shown in order to contrast the simultaneously measured electron diurnal variations (larger fluxes near local noon) with the proton diurnal variations seen in Figures 2 and 3 (larger proton fluxes near local midnight in general). It should be noted that even though no diurnal variations were observed in the proton fluxes on days 352–354, nevertheless, the electron flux intensities continued to reflect the distortion of the magnetosphere as the ATS satellite traversed all local times.

4. Results

A. DAILY VARIATIONS – DISTURBED CONDITIONS

Of great interest is the accessibility of the solar protons to synchronous altitude and the changes in the amount of accessibility with time, geomagnetic conditions, and energy of the protons. That indeed there are temporal changes in the accessibility was pointed out by Lanzerotti (1968b) and Paulikas and Blake (1969) and is seen clearly in Figures 1–3.

Although both the ATS-1 and IMP experiments measure proton fluxes in approximately the same energy ranges, the proton energy channels of the two experiments do not have the same energy widths nor do they have the same central energy values. In order to study the temporal variations of the ratio R of the synchronous altitude fluxes to the interplanetary fluxes for a common set of energies, the following proce-

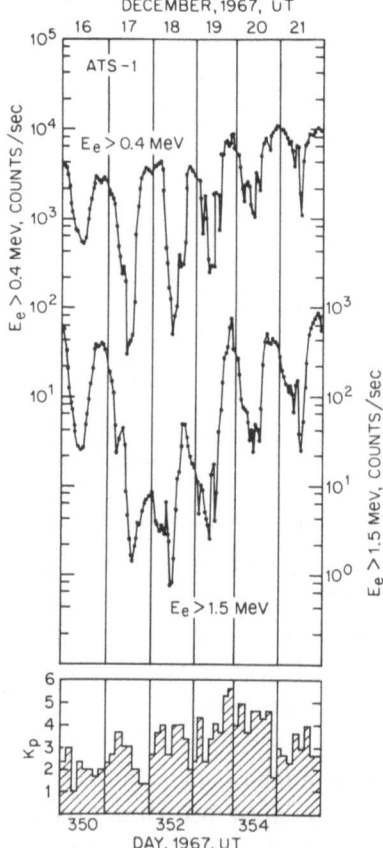

Fig. 4.   1-hour average data from two of the ATS-1 electron channels during the geomagnetically disturbed period, days 350–355, 1967.

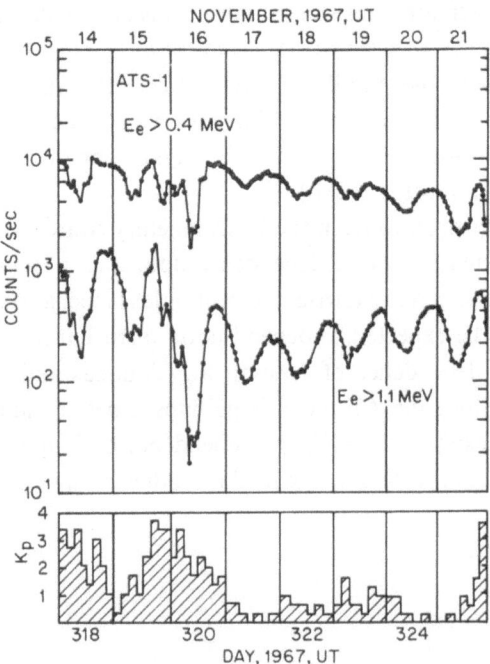

Fig. 5.   1-hour average data from two of the ATS-1 electron channels including the
geomagnetically quiet period, days 321–324, 1967.

dure was adopted. The differential-in-energy proton flux values at given energies for
each experiment were determined by linear interpolation between adjacent log (energy)
values of the hourly-averaged log (flux) values. These flux values were then used in
forming the ATS/IMP proton ratios.

It should be stressed that the ratio R does not necessarily measure the transmission
or accessibility of solar protons from the interplanetary source to the synchronous
altitude. There is a continued loss of the synchronous altitude protons as well during
the time the interplanetary source is present. Hence, R measures the result of the sum
of the transmission and loss processes present at synchronous altitude at any time.

Figure 6 contains five sets of equal energy ATS/IMP proton flux ratios for the
geomagnetically disturbed period, days 350–355. As stressed in the introduction, for
a symmetrically distorted dipole field, the ATS-1 satellite is effectively measuring
particles further out in the magnetosphere at local midnight than it is at local noon.
The effect of this distortion is seen on days 350–352 in the diurnal variation of R in
Figure 6. There are fewer protons at the ATS orbit at local noon than at local midnight. That is, there are fewer protons effectively further inside the magnetosphere.

The diurnal variation evident in Figure 6 is larger for the lowest energy protons.
The ratio R on day 351 is plotted in Figure 7 as a function of proton energy for ATS
local times of noon, morning, and midnight. The fluxes near local midnight are
essentially identical to the interplanetary fluxes at all energies. By local morning, the
ATS-measured fluxes at all energies have decreased. It is clear that there is an energy

dependence to the decrease in R at local noon; this energy dependence does not have a sharp cutoff.

After approximately the middle of day 352 UT, the value of R remains nearly constant at ~ 0.5 for all energies in Figure 6. This indicates very clearly that the proton spectra both inside and outside the magnetosphere are essentially the same at all local times (Lanzerotti, 1968b).

The access of solar protons from the interplanetary source region to the ATS orbit could well be expected to have a dependence upon the plasma and field conditions in the outer magnetosphere. Likewise, the proton loss mechanisms at the synchronous altitude would also be expected to depend upon disturbances in the outer magnetosphere. Hence, the dependence of R upon K_p (Paulikas and Blake, 1969) and Dst were investigated. Since the AE index is perhaps a better indicator of disturbances in the outer magnetosphere than K_p, the dependence of R upon AE was also studied.

It is quite clear from the data of Figure 2 that, due in particular to the large diurnal

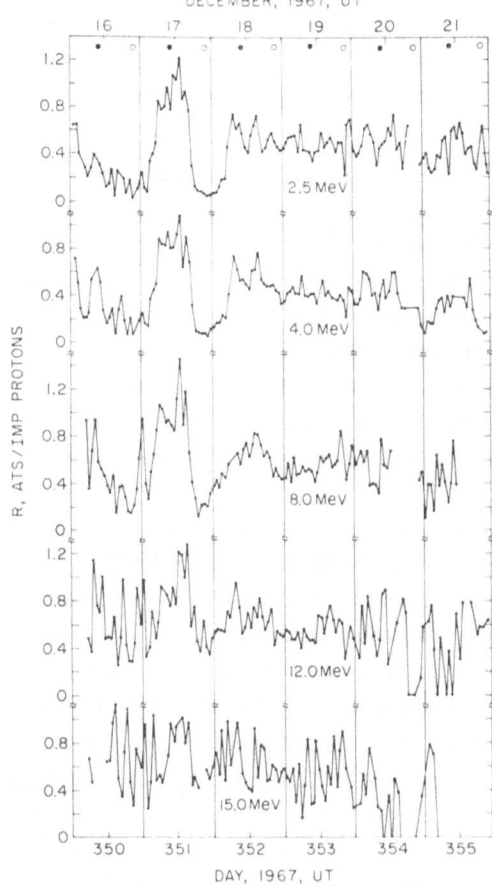

Fig. 6.   Temporal plot of the ratio R of the ATS-1 to Explorer 34 proton fluxes during the geomagnetically disturbed period, days 350–355, 1967.

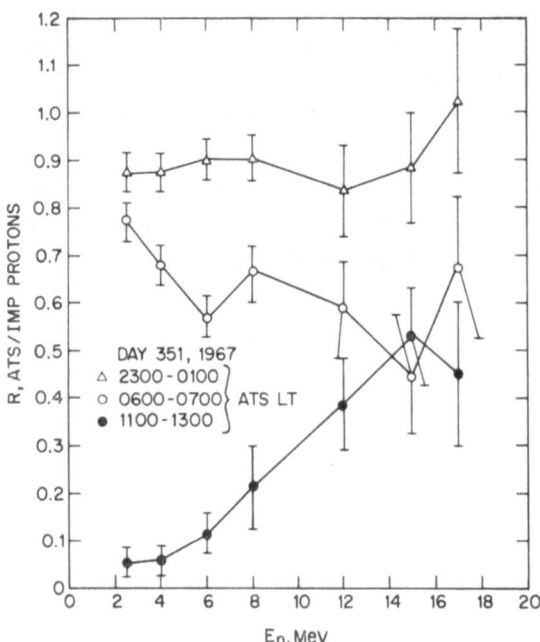

Fig. 7.   Ratio R as a function of energy and local time for day 351, 1967, a day when large diurnal variations in R were measured.

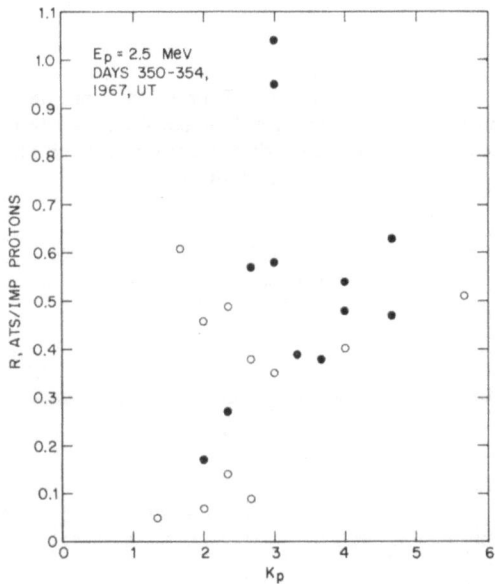

Fig. 8. R as a function of the 3-hour average K_p for 2.5 MeV protons – days 350–354, 1967. The open circles correspond to 3-hour average R values around ATS local noon (2100–2400 UT and 0000–0300 UT). The solid circles correspond to 3 hour average R values around ATS local midnight (0900–1200 UT and 1200–1500 UT).

variations, R does not have a simple correlation with the indices measuring the amount of geomagnetic disturbance. In Figures 8, 9, and 10a are the 2.5 MeV proton ratios plotted as a function of K_p, AE, and Dst for days 350–354, 1969. The ratios for several hours spanning local noon and local midnight on each day are denoted by open and closed circles, respectively. For large values of either K_p or AE, R is generally large, as could well be expected. Most of the R-values $\gtrsim 0.4$ are observed to occur for negative Dst values (Figure 10a). Other than these observations, R has little dependence upon the magnitude of the indices.

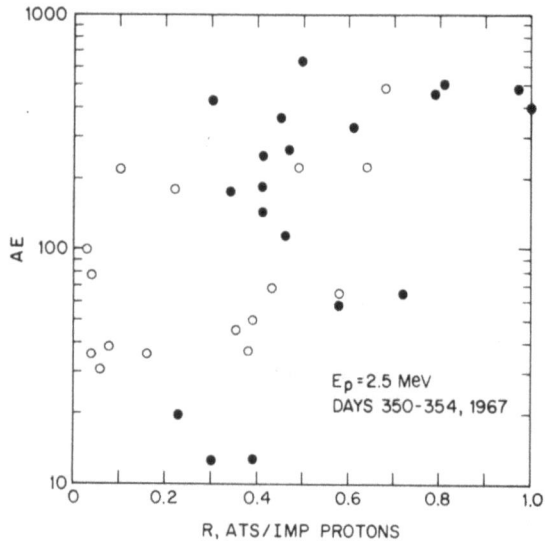

Fig. 9. R as a function of the hourly average AE for 2.5 MeV protons – days 350–354, 1967. The open circles correspond to the hourly average R values for 3 hourly periods around ATS local noon. The solid circles correspond to the hourly average R values for 3 hourly periods around ATS local midnight.

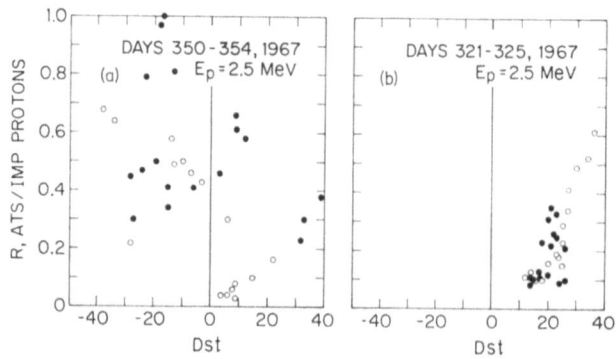

Fig. 10. R as a function of the hourly average Dst for 2.5 MeV protons. – (a) days 350–354, 1967. – (b) days 321–324, 1967. The open circles correspond to the hourly average R values for 3 hourly periods around ATS local noon. The solid circles correspond to the hourly average R values for 3 hourly periods around ATS local midnight.

B. DAILY VARIATIONS – QUIET CONDITIONS

The ratios, R, for 6 different proton energies for days 318–325, 1967, are plotted in Figure 11. The 'quiet-time' (days 321–324) ratios show the opposite temporal changes from the disturbed period ratios (Figure 6) in that a larger diurnal variation is generally observed at the higher energies.

   The distortion of the magnetosphere during these quiet-days can be estimated from the ATS magnetic field data, using the calculational results of Roederer (1969) for particle drift motion in a Williams-Mead model geomagnetic field. Since the magnetic field data were not available, estimates of the boundary stand-off distance and tail field strength were made by comparing the smoothly varying electron flux variations on day 321 (Figure 5) with similar electron daily variations in January, 1967, when the field data were available. Although this provides only a rough estimate, it was

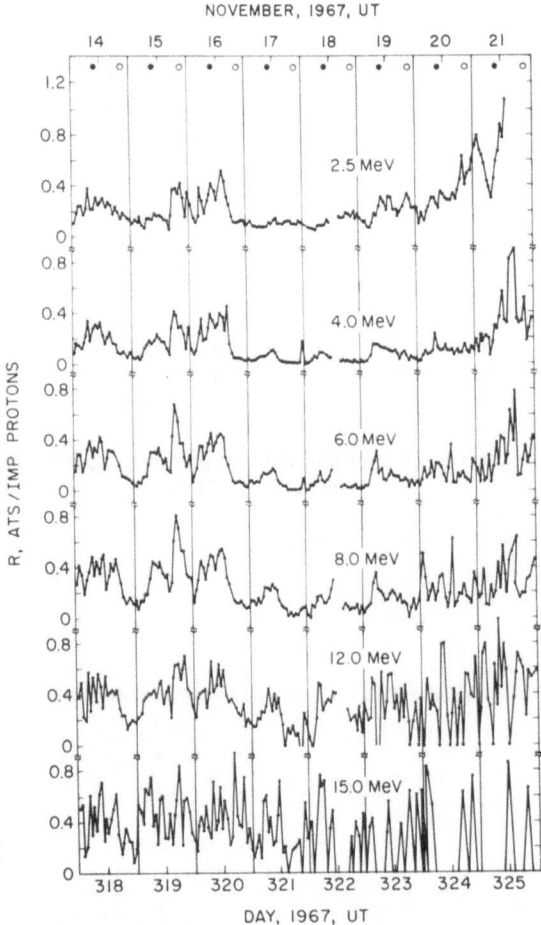

Fig. 11.   Temporal plot of the ratio R of the ATS-1 to Explorer 34 proton fluxes for a time period including the geomagnetically quiet period, days 321–324, 1967.

determined in this way that ATS was near $L \sim 6$ at local noon and near $L \sim 6.6$ at local midnight.

Using these results, a plot of R as a function of L is given in Figure 12 for four different proton energies. The radial gradient of the 2.5 MeV protons is constant over this L range (as was evident from the data of Figures 3 and 11) while the fluxes of higher energy protons all increase at higher L-values. The largest radial gradients are observed in the highest energy fluxes. A similar result of a radial gradient increasing with L was noted by Lanzerotti (1968b) for 2.4 MeV protons.

Plots of the ratio R as a function of K_p for 2.5 and 8.0 MeV protons during this period when $K_p \leqslant 1$ show no significant dependence of R upon the value of K_p (Figure 13). However, a striking temporal resemblance between the 2.5 MeV proton flux ratios and Dst during days 324–325 is observed in Figure 3. A plot of the quiet-time 2.5 MeV proton ratios and Dst in Figure 10b indicates an approximate linear

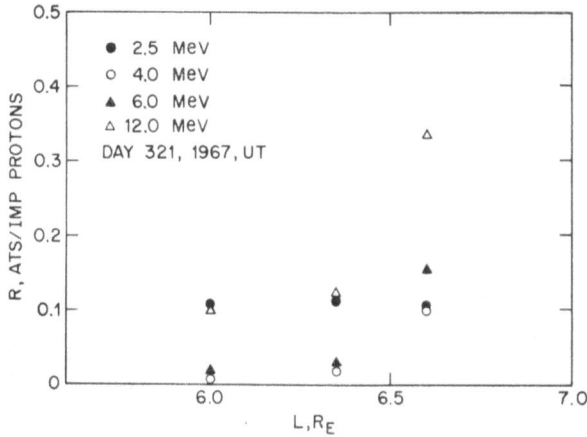

Fig. 12. Ratio R as a function of L for four different proton energies on the geomagnetically quiet day, day 321, 1967.

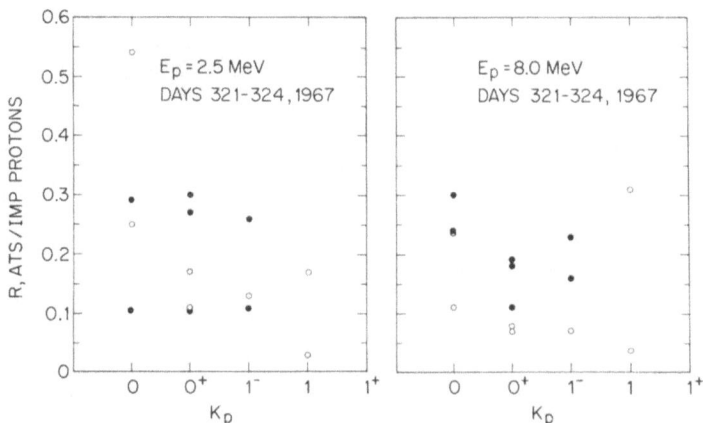

Fig. 13. R as a function of the 3-hour average K_p for 2.5 and 8.0 MeV protons – days 321–324, 1967. See caption of Figure 8 for time periods represented by open and closed circles.

relationship between R and the magnitude of the Dst positive excursion, particularly for the hours around local noon (open circles).

C. SOLAR PARTICLE ONSETS

It was quite evident from the data of Figure 1 that very frequently there is little or no delay between the observation of ~ 2 MeV solar protons in interplanetary space and their detection in the magnetosphere at synchronous altitude. Paulikas and Blake (1969) have also demonstrated this result for higher energy protons. Figure 14, taken from their paper, compares 21–70 MeV protons inside and outside the magnetosphere during the January, 1967, event.

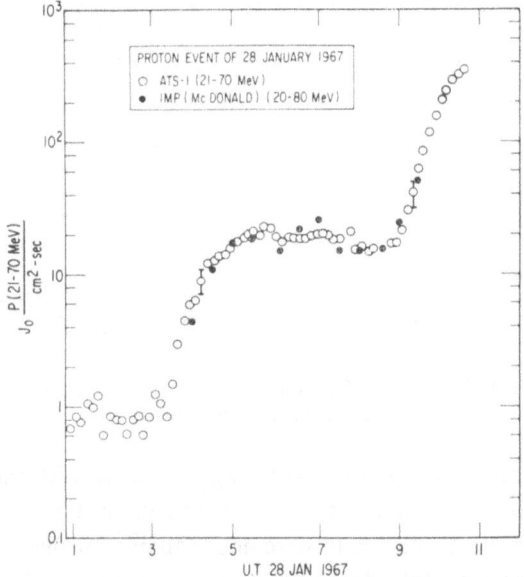

Fig. 14. 21–70 MeV protons observed at synchronous altitude and in interplanetary space on day 28, 1967. The K_D indices for the half-day were 2, 4⁻, 4⁻, and 3⁺. (From Paulikas and Blake, 1969.)

However, Lanzerotti (1968b) and Lanzerotti et al. (1970), also observed events when there were substantial delays before the low energy interplanetary protons were initially seen inside the magnetosphere. Figure 15, reproduced from Lanzerotti (1968b) shows such a delay very clearly. Once the solar particles are finally observed at synchronous altitudes, however, their intensities increase very rapidly to values comparable to those of the interplanetary fluxes.

Figure 16 plots the onset times of the interplanetary and magnetosphere particles as a function of proton energy for the event shown in Figure 15. The onset times were determined using 10-minute average data and determining the first significant flux increase above the background statistical fluctuations. There is approximately a 7-hour time delay between the onset of the 800 keV protons inside and outside the magnetosphere. There also appears to be a small decrease of the access time delay

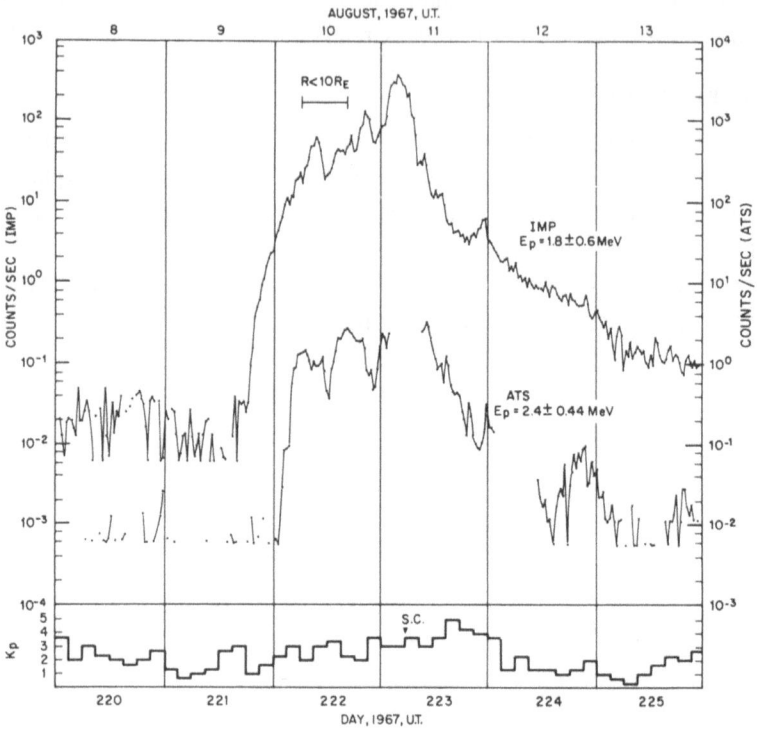

Fig. 15. Explorer 34 and ATS-1 proton fluxes during a solar event in August, 1967.
(From Lanzerotti, 1968b.)

for the higher energy particles inside the magnetosphere with the 2.4 MeV protons
being observed approximately $1\frac{1}{2}$ hours before the 800 keV protons.

Although ATS magnetic field data were not available for this period, the measured
ATS electron fluxes smoothly decreased in intensity during the first few hours of
day 222, indicating a decrease in the local magnetic field intensity. Just after the first
protons were observed, the electron fluxes decreased sharply, increased again, and
were disturbed for the next hour to hour and one-half. The nature of these fluctuations in the electron flux intensities indicate fluctuations in the local magnetic field
intensity (see Lanzerotti *et al.*, 1968, for a discussion of simultaneous field and particle changes). These disturbances of the local electron measurements indicate that
the proton onsets were probably due to the onset of magnetosphere-wide disturbances.
It is interesting to note, however, that the index AE had values of 350–400 during
most of the first 4 hours of day 222 (when no solar particles were seen at ATS) and
decreased to ~ 25 at ~ 0600 UT (when approximately full intensity solar fluxes were
observed).

An investigation of all ATS particle data for days 27 and 28 and days 221 and 222,
1967, showed that the electron count fluctuations on days 27 and 28 were much
more disturbed and had a much noisier character than did the electron fluxes on day
221 and the first few hours of day 222. This difference in the 'noisiness' of the electron

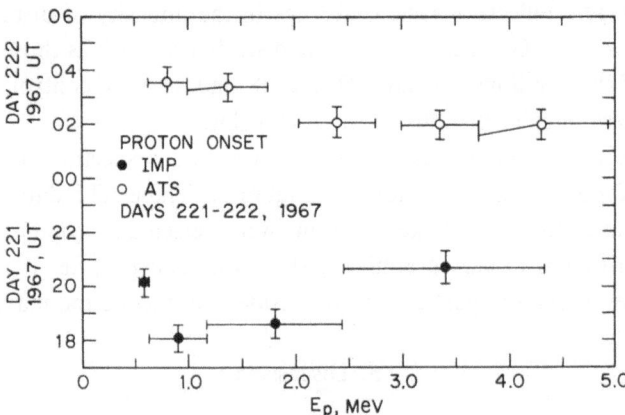

Fig. 16. Onset time as a function of proton energy for the interplanetary and synchronous altitude proton fluxes observed on days 221–222, 1967.

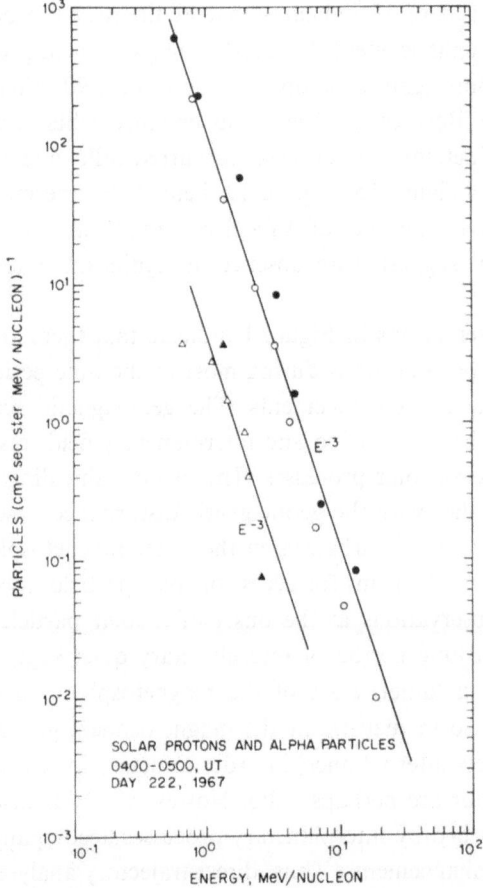

Fig. 17. Interplanetary and synchronous altitude proton and α spectra during hour 04, day 222, 1967. The circles denote protons; the triangles denote α-particles. Solid points are interplanetary fluxes; open points are magnetosphere fluxes.

fluxes undoubtedly reflects a basic difference in the intensity of disturbances in the outer magnetosphere. Outer magnetosphere disturbances such as those that produced the electron flux variations on days 27 and 28 undoubtedly control the initial and continued access of solar protons to the ATS orbit.

The interplanetary and magnetosphere proton and α spectra measured between 0400–0500 UT on day 222 are shown in Figure 17. These ATS data were measured only an hour after the first 800 keV protons were detected at synchronous altitudes. The spectra in Figure 17 very graphically show that shortly after the first magnetosphere particle onsets, the particle spectra inside and outside the magnetosphere are very similar.

5. Discussion

A. MAGNETOSPHERE SHIELDING

The foregoing observations and results, comparing near 90° pitch angle proton fluxes at 6.6 R_E geocentric radius to simultaneous measurements of interplanetary proton fluxes, indicate that solar protons generally have ready access to synchronous altitude. However, Figure 15 indicates that at times at the beginning of an interplanetary solar flux enhancement, the earth's magnetic field provides a very effective shield for several hours against the observation of near 90° pitch angle solar particles at the ATS altitude. Both of the 1967 solar enhancements in which significant synchronous altitude shielding was observed occurred following rather long periods of interplanetary quiet (Figure 15; Figure 19, below). Furthermore, as pointed out in the previous section, the absence of ATS electron flux variations prior to the day 221 proton enhancement suggested an absence of significant outer magnetosphere disturbances.

The extensive observations in Figure 1 indicate that there are at least moderately disturbed geomagnetic conditions during most of the time period accompanying the low energy solar particle enhancements. The geomagnetic activity is undoubtedly caused by the disturbed solar wind and interplanetary fields resulting from the energetic particle-producing solar processes. The proton 'shielding' observations in Figures 15 and 19, together with the geomagnetic disturbance observations in Figure 1, strongly suggest that the disturbances in the outer magnetosphere and/or the magnetotail provide the mechanisms for access of solar particles to synchronous altitude.

The 'shielding' observations at the onset of a solar particle enhancement which has occurred after a long period of interplanetary quiet suggest an important conjecture: there exists a 'quiet' state of the magnetosphere such that the trajectory calculations performed in realistic model magnetosphere geometries (using external fields as well as merely internal ones) in order to determine cutoffs for the low energy protons at the equator are perhaps valid. However, as was discussed, the magnetosphere is often disturbed by interplanetary processes accompanying or following the onset of solar flux enhancements. Thus, direct trajectory analyses that do not include outer zone noise and/or disturbances are probably unable to treat properly the problem of low energy solar particle access.

It is indeed interesting that during the 'quiet' geomagnetic period of days 321–324, although the fluxes observed inside the magnetosphere were low, they were non-zero. In fact, at the lowest energy, 2.5 MeV, the data suggest that the protons were stably trapped. The higher energy protons were also probably stably trapped (see next section). It is possible that the diffusion processes that move the interplanetary particles into the magnetosphere are still operative during a quiet period once protons have achieved access to the outer magnetosphere. Additionally, the proton loss processes may tend to zero intensity during a quiet period. It should be noted, however, that the radial gradient distributions of the higher energy protons were still opposite to those of the electrons, perhaps suggesting continued diffusion.

As has been pointed out by Lanzerotti (1968b), if solar protons can generally be found at synchronous altitudes during solar events, then they might be expected to be found everywhere in the outer magnetosphere, including the polar cap regions. Such a comparison has not been made. An important test of this hypothesis would be to compare polar cap and synchronous altitude data during a period when the magnetosphere is acting as a shield against the enhanced solar fluxes – such as day 222 (Figure 15) or day 261 (Figure 19, below).

B. GEOMETRY OF ACCESS

Low energy protons mirroring near the equator at the ATS altitude are normally stably trapped in a standard magnetosphere (stand-off boundary distance of 10 R_E; tail field of 20 γ). This can be seen from the results in Figure 18 (computer program courtesy of J. G. Roederer). Plotted here are the tail field and boundary position limits so that 70° and 90° pitch angle protons detected at 6.6 R_E altitude at local

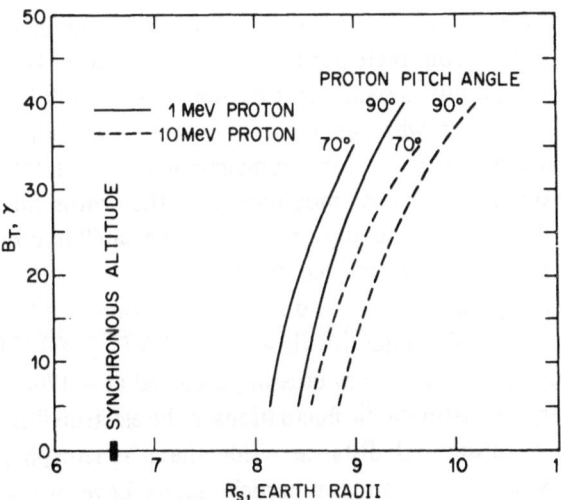

Fig. 18. Tail field (B_T) and stand-off boundary positions (R_S) as a function of proton energy and equatorial pitch angle such that a proton observed at local midnight would find the distant edge of its gyro-radius in the pseudo-trapping region. Magnetosphere parameters falling to the right and below a curve describe a magnetosphere configuration such that a proton with the given energy and pitch angle will complete a longitudinal drift period.

midnight would find the distant edge of their cyclotron orbits (~ 0.5 R_E for a 1 MeV proton in a 50 γ field) entering the pseudo-trapping region. Magnetosphere parameters falling to the right and below a curve in Figure 18 indicate that a proton with the given energy and pitch angle will complete a longitudinal drift period. If the magnetosphere configuration has parameters falling above and to the left of a curve, the proton will be lost out the dusk side of the magnetosphere.

The boundary and tail field limits on this plot seem merely to confirm the trajectory calculation findings that protons do not normally have access to the equatorial region of the magnetosphere at synchronous altitude. Some process must move their guiding centers by one or several gyro-radii in order that they be observed at the equator.

Theoretically, when the ATS satellite is at noon on the dayside, the cyclotron orbits of the near 90° pitch angle protons will not intersect a pseudo-trapping region during their longitudinal drifts. They will only become untrapped when the front boundary moves as close as one proton gyro radius to the satellite.

Smart *et al.* (1969), have recently shown, via trajectory calculations, that the vertically incident cut-off rigidities rapidly tend to zero for nightside latitudes in the region of the ATS field line. However, to get particles to mirror near the equator in order to be detected by the BTL ATS-1 experiment, scattering (or diffusion) of the pitch angles of the protons must take place. The means for accomplishing this is certainly not clear.

The question as to whether these solar particles are diffusing in everywhere across the magnetopause and/or through the magnetotail has not been satisfactorily answered yet. A recent paper by Williams and Bostrom (1969) suggested that the diffusion to synchronous altitude was predominantly via the neutral sheet and cusp region. This may indeed be a source for these particles. However, a recent data comparison by Lanzerotti *et al.* (1970), using ATS, IMP, and Vela magnetotail data has been made. Figure 19, taken from their paper, shows little relation between the Vela and ATS fluxes. However, a flux enhancement beginning at ~ 1900 UT at ATS, approximately 2 hours following what appears to be the corresponding interplanetary enhancement, is scarcely seen at all in the magnetotail (at ~ 1800 UT).

It was pointed out earlier that no modulations of the proton fluxes detected by the experiment on IMP F were generally observed as the satellite crossed the magnetopause. Although this might be expected because of the relatively large proton gyroradius, it is interesting that in the boundary crossing example shown in Figure 20 no diminution of the proton flux intensities is seen at ~ 3.5 R_E (2000 UT) inside the boundary at mid-latitudes. (The boundary crossing occurred at ~ 1700–1710 UT.) In contrast, however, large, quasi-periodic fluctuations in the electron fluxes ($E_e > 0.44$ MeV) are observed for a substantial distance inside the magnetosphere boundary. The implications of these observations for particle access have not yet been fully explored. It should be pointed out, however, that the data of Lanzerotti *et al.* (1970), Figure 19, show appreciable fluxes of solar protons in the outer regions of the magnetosphere long before there are any fluxes of near 90° pitch angle protons at synchronous altitude.

Data presented by Armstrong *et al.* (1969), was purported to show the exclusion of 0.82–1.9 MeV solar protons from the "ordered magnetic field of the near ($\lesssim 8\ R_E$) magnetosphere" near the equator on July 13, 1966. Krimigis (1969) expanded on this observation and stated that this apparent exclusion of 0.82 MeV protons from the geomagnetic field close to the equator showed that the radiation belts could not be directly populated via the infusion of particles of comparable energy from solar particle events. Although it is difficult, from the data presented by Armstrong *et al.*

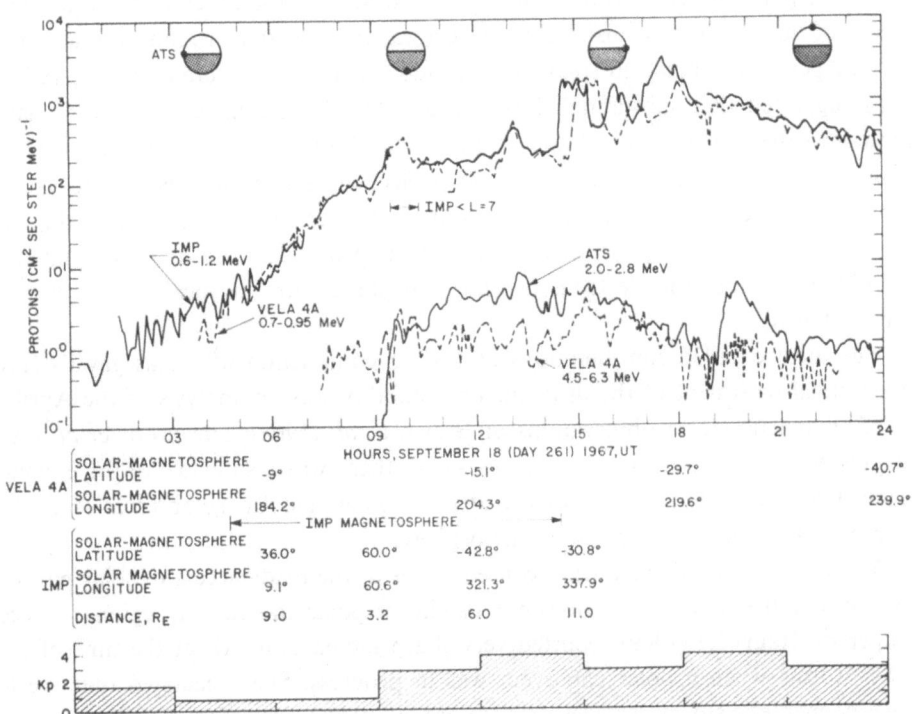

Fig. 19. Explorer 34, ATS-1 and Vela 4A magnetotail data during day 261, 1967. (From Lanzerotti *et al.*, 1970.)

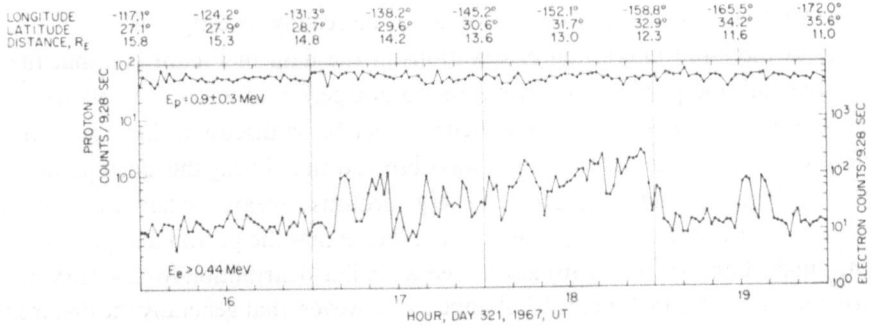

Fig. 20. Proton and electron data observed during the Explorer 34 magnetopause crossing (1700–1710 UT) on day 321, 1967. Geographic coordinates are shown above the data.

to tell when the magnetosphere boundary and trapping regions were entered, the data discussed here and presented by Lanzerotti (1968b) and Paulikas and Blake (1969) show conclusively that solar particles do frequently penetrate as deep as 6.6 R_E into the magnetosphere. In fact, as seen from the data of Figure 1, the exceptions are quite infrequent. The question of populating the radiation belts by these protons is discussed briefly below.

C. CONSEQUENCES OF SOLAR PARTICLES IN THE MAGNETOSPHERE

A recent paper by Lanzerotti *et al.* (1969b), described the observation of an occurrence of the drift mirror instability (Hasegawa, 1969) at synchronous altitude during a time period of enhanced solar fluxes inside the magnetosphere (day 177, 1967). It was suggested that substantial fluxes of 20–100 keV interplanetary protons having access to the synchronous altitude contributed to the occurrence of the instability. Figure 21, taken from their paper, shows electron and proton oscillations at synchronous altitude and electron heating during the period of occurrence of the instability. Their detailed analyses indicated that the protons and electrons oscillated out of phase and that the electrons oscillated in phase with the local ATS-1 magnetic field intensity.

The most detailed considerations of data in conjunction with Hasegawa's (1969) theoretical treatment of the drift mirror instability was an analysis of the April 18, 1965, magnetic storm observations of Brown *et al.* (1968). Lanzerotti *et al.* (1969) expressed the opinion that during this storm there was also evidence for enhanced interplanetary fluxes and that some evidence even existed for the observation of these protons at low latitudes in the magnetosphere.

There may be other significant consequences of the ready access of solar particles, not only to the equatorial regions but also in the polar and auroral regions. A recent paper by Haurwitz (1969) reported very sharp increases in AE at the time of large SSC storms when a polar cap event was in progress. She speculated that the low energy energetic storm particles that are present at the time of the SSC's are directly injected into the auroral zones via the magnetospheric tail and cause the bay activity. There were no large energetic storm particle enhancements at the times of the two sudden commencements of Figure 1. Furthermore, no large storm followed either SC, so that Haurwitz's conjectures could not be readily investigated.

One of the more interesting observations in the data of Figure 1 is that the synchronous altitude particle enhancements do not persist after the interplanetary particle source is gone. Hence, it is difficult to define or discuss a 'life-time' for these protons and α's in the outer zone. It was observed that during the 'quiet period', days 321–324 (Figures 3 and 11), the lowest energy protons appeared to have a diurnal distribution similar to the electrons. This could indicate that the proton loss processes were quite small. Lanzerotti (1968b) also noted a similar distribution for 2.4 MeV protons during a quiet time in June, 1967. It appears, however, that generally the disturbances in the outer magnetosphere are such as to cause continued losses of these protons so that no appreciable fluxes are trapped for any substantial length of time (~ 1 day).

The processes producing proton loss in the outer regions of the magnetosphere are unknown. The absence of a build-up of flux levels greater than the interplanetary fluxes at the ATS orbit suggests that there is a continual loss of protons back out of the magnetosphere during the period of an interplanetary enhancement. There is undoubtedly also a pitch angle scattering loss of these protons down the field lines. The lack of persistence of the synchronous altitude fluxes when the interplanetary source is gone suggests that 'back-diffusion' of the protons out of the magnetosphere is the predominant loss mechanism.

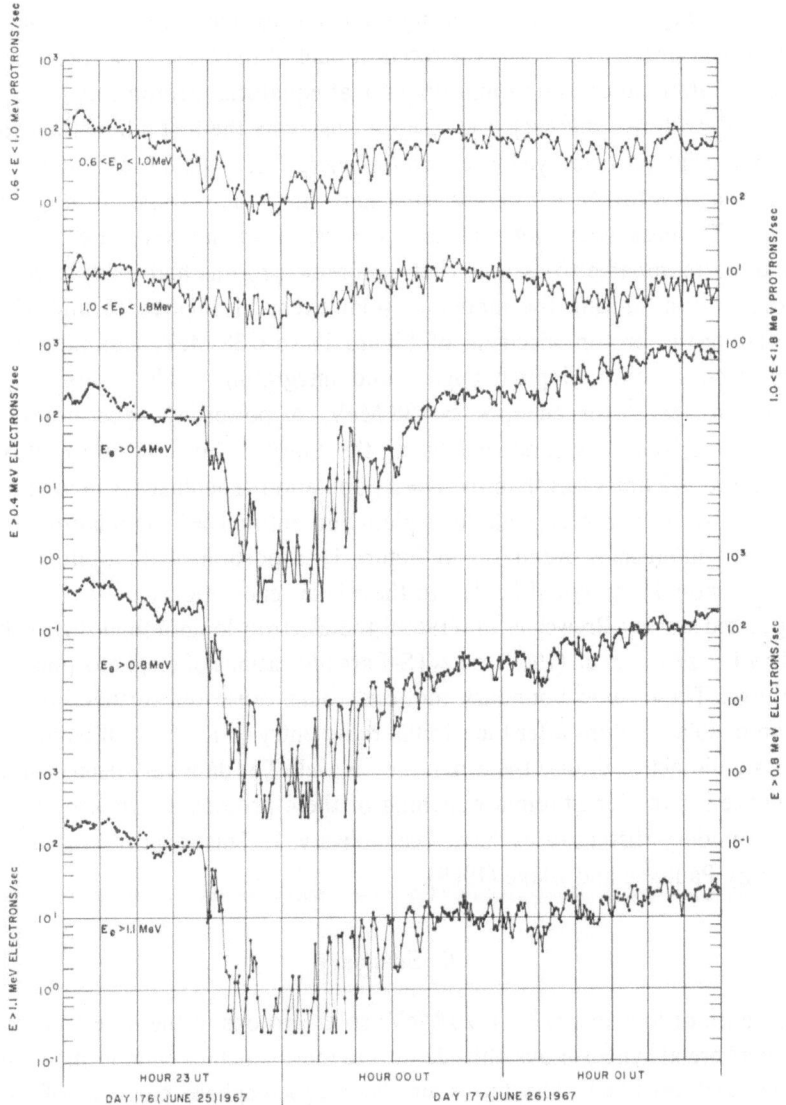

Fig. 21.   ATS-1 proton and electron data showing anti-correlation of the proton and electron oscillations and simultaneous electron heating during the substorm on day 177, 1967.
(From Lanzerotti *et al.*, 1969b.)

An interesting question that has not been answered yet concerns the possibility of a further diffusion of a small percentage of these solar protons deeper into the magnetosphere by the classic third-invariant violating diffusion mechanism first discussed by Kellog (1959). The population of the proton radiation belt by L-diffusion of protons from the magnetosphere boundary ($L = 10\ R_E$) has been extensively discussed, for example, by Nakada *et al.* (1965), and Nakada and Mead (1965), and reviewed by Hess (1968). For equatorial particles, the calculational results of Nakada and Mead (1965) appear to be convincing that cross-L diffusion of solar wind protons control the outer regions of the proton belt.

The possibility of partially populating the magnetosphere with these solar protons is nevertheless interesting. The most recent model of the 'inner belt' protons (Lavine and Vette, 1969) indicates an omni-directional equatorial proton flux ($E > 4.0$ MeV) of $\sim 4 \times 10^4$ protons·(cm^2 sec)$^{-1}$ at $L = 3$. Conserving the first adiabatic invariant of these equatorial particles yields a proton energy of ~ 0.38 MeV at $L = 6.6$ (and ~ 100 keV at $L = 10$). ($L = 3$ was selected here since the extrapolation of the fluxes from the ATS-measured energy values to 0.38 MeV was not very large.)

It is quite reasonable to assume that 0.38 MeV protons had ready access to synchronous altitude during the moderate solar event of Figures 15 and 17. Hence, extrapolating the proton spectrum of Figure 17 to 0.38 MeV, assuming a uniform proton spectrum over all pitch angles, and integrating yields a flux of $\sim 6 \times 10^3$ protons·(cm^2 sec)$^{-1}$ for energies > 0.38 MeV. A power-law spectrum remains a power-law with the same exponent during the L-drift. Thus, the proton fluxes from this event are ~ 7 times less than the $L = 3$ population. The fluxes from one moderate event are obviously not enough to explain the entire $L = 3$ population. However, with a large number of events (as in Figure 1), a portion of the quite stable inner proton belt population could be due to these solar energetic particles.

As pointed out by Brewer *et al.* (1969), the electron longitudinal dirft echoes observed by Lanzerotti *et al.* (1967), on ATS-1 are indications of magnetosphere boundary motions. These same boundary motions could cause the further inward radial diffusion of solar protons after their initial rapid entry to $L \sim 6$-7. Of course the same process should act to diffuse the α-particles as well. The data have been searched, but no evidence for the longitudinal bunching of these protons at synchronous altitude during boundary disturbances have been observed. This negative result was also reported by Paulikas and Blake (1969).

6. Summary

The existence of low energy (~ 1-20 MeV) solar protons in the outer regions of the magnetosphere at synchronous altitude during interplanetary enhancements has been discussed and reviewed. The fluxes of the magnetosphere particles often exhibit diurnal variations, with fewer particles seen at local noon than at local midnight. The access of these particles is only weakly correlated with such indices of geomagnetic activity as AE, K_p, and Dst. Infrequently, large (several hour) time delays are measured between the onset of the interplanetary and synchronous altitude enhancements. However, once access to 6.6 R_E has been achieved, the magnetosphere fluxes increase rapidly to the interplanetary flux values. The access of the interplanetary particles to the magnetosphere is probably due to a rapid diffusion (scattering) of the protons across the magnetic field lines. The protons are probably incident both across the magnetopause and through the geomagnetic tail. Two observations of a plasma instability in the magnetosphere have been identified during periods of solar proton enhancements. It was speculated that a fraction of the interplanetary particles would be further diffused deeper into the magnetosphere to contribute to the trapped radiation belt population.

Acknowledgements

I would like to thank Miss C. G. Maclennan for computational assistance. I would also like to thank Prof. P. J. Coleman and Mr. J. Barfield for use of their magnetic field data and Dr. W. L. Brown and Dr. G. A. Paulikas for numerous profitable discussions.

References

Armstrong, T. P. and Krimigis, S. M.: 1968, *J. Geophys. Res.* **73**, 143.
Armstrong, T. P., Krimigis, S. M., and Van Allen, J. A.: 1969, *Annals IQSY* **3**, 313.
Bostrom, C. O.: 1970, this volume, p. 229.
Brewer, H. R., Schulz, M., and Eviatar, A.: 1969, *J. Geophys. Res.* **74**, 159.
Filius, R. W.: 1968, *Annals de Geophys.* **24**, 821.
Hasegawa, A.: 1969, *Phys. Fluids* **12**, 2642.
Haurwitz, M. W.: 1969, *J. Geophys. Res.* **74**, 2348.
Hess, W. N.: 1968, *The Radiation Belt and Magnetosphere*, Blaisdell Pub. Co., 213–241.
Hofmann, D. J. and Sauer, H. H.: 1968, *Space Sci. Rev.* **8**, 850.
Kellog, P. J.: 1959, *Nature* **183**, 1295.
Krimigis, S. M.: 1969, *Annals IQSY* **3**, 457.
Lanzerotti, L. J., Roberts, C. S., and Brown, W. L.: 1967, *J. Geophys. Res.* **72**, 5893.
Lanzerotti, L. J.: 1968a, *Nucl. Instr. and Methods* **61**, 99.
Lanzerotti, L. J.: 1968b, *Phys. Rev. Letters* **21**, 929.
Lanzerotti, L. J., Brown, W. L., and Roberts, C. S.: 1968, *J. Geophys. Res.* **73**, 5751.
Lanzerotti, L. J., Lie, H. P., and Miller, G. L.: 1969a, *IEEE Trans. Nucl. Sci.* **NS-16** (1), 343.
Lanzerotti, L. J., Hasegawa, A., and Maclennan, C. G.: 1969b, *J. Geophys. Res.* **74**, 5565.
Lanzerotti, L. J., Montgomery, M. D., and Singer, S.: 1970, *J. Geophys. Res.* **75**.
Lavine, J. P., and Vette, J. I.: 1969, *Models of the Trapped Radiation Environment*, Vol. 5: *Inner Belt Protons*, 15.
Leinbach, H.: 1967, *J. Geophys. Res.* **72**, 5473.
Nakada, M. P., Dungey, J. W., and Hess, W. N.: 1965, *J. Geophys. Res.* **70**, 3529.
Nakada, M. P., and Mead, G. D.: 1965, *J. Geophys. Res.* **70**, 4777.
Paulikas, G. A., Blake, J. B., Freden, S. C., and Imamoto, S. S.: 1968, *J. Geophys. Res.* **73**, 4915.
Paulikas, G. A., and Blake, J. B.: 1969, *J. Geophys. Res.* **74**, 2161.
Paulikas, G. A.: 1970, this volume, p. 193.
Roederer, J. G.: 1969, *Rev. Geophys.* **7**, 77.
Smart, D. F., Shea, M. A., and Gall, R.: 1969, *J. Geophys. Res.*, **74**.
Störmer, C.: 1955, *Polar Aurora*, Oxford University Press.
Williams, D. J., and Bostrom, C. O.: 1969, *J. Geophys. Res.* **74**, 3019.

Discussion

Roederer: Two of the cases which you presented show delays in the penetration of protons to the synchronous altitude, when ATS was in the morning side, while no delays are observed in the other two cases, when ATS was in the evening side. Is this a regular feature?

Lanzerotti: Two examples are not statistically significant. This might be an important indication. However, I tend to think not, because the particles drift in longitude.

Singer: The Vela 4 satellite sees a good deal of examples of delays in the penetration into the tail of the magnetosphere. In no instances does Vela see an energy dependence in those delays. That might suggest that there is a difference between the mechanism by which the particles enter the tail and by which they penetrate to the synchronous altitude.

ENTRY OF LOW ENERGY SOLAR PROTONS
INTO THE MAGNETOSPHERE

CARL O. BOSTROM

The Johns Hopkins University, Applied Physics Laboratory, Silver Spring, Md., U.S.A.

Abstract. Some aspects of the temporal and spatial variations of low energy solar protons are examined by comparing data from the polar orbiting satellite 1963 38C with measurements in interplanetary space by Explorers 33 and 34. Selected periods during the events of September 1966 and May 1967 are used to show a variety of features in the polar region latitude profiles. The proton intensity in the low latitude 'rim' of the polar region appears to track the interplanetary flux of solar protons more readily than does the polar intensity. The rim straddles the energetic electron trapping boundary, and the rim intensity is frequently greater than the polar plateau flux. The intersatellite comparisons and latitude profiles are discussed in terms of model requirements.

1. Introduction

During the solar proton events of July 1961, Pieper *et al.* (1962) made the first satellite measurements of low energy protons in the polar regions and found that 1 to 15 MeV protons could apparently penetrate deep into the magnetosphere, particularly during magnetic storms. Measurements by low altitude satellites and riometers have demonstrated the complexity of the polar regions and have added considerably to our empirical knowledge of solar particle behavior (Stone, 1964; Blake *et al.*, 1968; Bostrom *et al.*, 1967; Zmuda *et al.*, 1963, 1967; McDiarmid and Burrows, 1969; Paulikas *et al.*, 1968; Evans and Stone, 1969; Reid and Sauer, 1967). Equatorial measurements have confirmed the low altitude observations of penetration by protons to very low *L*-values (Lanzerotti, 1968; Paulikas and Blake, 1969; Fillius, 1968). Simultaneous observations of particles inside and outside the magnetosphere have been made using numerous combinations of spacecraft locations (Krimigis *et al.*, 1967; Williams and Bostrom, 1967, 1969; Kane *et al.*, 1968; Montgomery and Singer, 1969), and these measurements have often been interpreted in terms of the magnetospheric models put forth by Michel and Dessler (1965) and by Dungey (1961). Considerable controversy has arisen because solar protons have not provided definitive support for one or the other of these models and one has the choice of emphasizing either the similarities or the differences between interplanetary and magnetospheric measurements. In this paper data from the polar-orbiting satellite 1963 38C are compared with Explorer 33 in September 1966 and with Explorer 34 in May 1967.

2. The Observations

The satellites and instruments have been described in detail previously (Bostrom *et al.*, 1967; Williams and Bostrom, 1969; Armstrong and Krimigis, 1968). Satellite 1963 38C is in a circular polar orbit at 1100 km altitude; Explorers 33 and 34 are in elliptical high altitude orbits so that these satellites spend a large fraction of their

V. Manno and D. E. Page (eds.), Intercorrelated Satellite Observations Related to Solar Events, 229–238. All Rights Reserved
Copyright © 1970 by D. Reidel Publishing Company, Dordrecht-Holland

orbital periods outside the magnetosphere. The detectors of interest in the comparisons are: (1) the 1963 38C proton spectrometer channels P_1 ($1.2 \leqslant E_p \leqslant 2.2$ MeV) and P_2 ($2.2 \leqslant E_p \leqslant 8.2$ MeV), (2) the Explorer 34 Solar Proton Monitoring Experiment (SPME) channel 4 ($1 \leqslant E_p \leqslant 10$ MeV), and (3) the Explorer 33 University of Iowa Experiment channel PN3 ($0.82 \leqslant E_p \leqslant 1.9$ MeV).

During the solar proton events of August 28 and September 2, 1966 about 100 high latitude passes were recorded from satellite 1963 38C at local times of 0100–0140 and 1300–1340. An interesting sequence of latitude profiles was recorded on September 3rd and is shown in Figure 1. The four passes are northern hemisphere nightside traversals of the invariant latitude range from 50° to >80°. The 2.2 to 8.2 MeV protons often show enhanced intensities near the low latitude edge of the polar cap. For the 0920 UT pass the enhanced region was between 60.5° and 72°; at 1110 UT the enhancement occurred between 58° and 76°; at 1258 UT the profile appears nearly flat from 60° to 82°. Qualitatively, this sequence of passes exhibits the behavior predicted by the long-tail model of the magnetosphere (Michel and Dessler, 1965). At 1446 UT the profile again shows an enhanced region but is quite different in that the intensity changes more slowly with latitude. The approximate location of the 280 keV electron trapping boundary is indicated on each curve. During this period the main phase of the magnetic storm was beginning (Dst ranged from -25γ to -50γ, and K_p was about 6) accounting for the depression of the boundary (Williams, 1967). If the trapping boundary on the night side is independent of energy (Williams and Mead, 1965), then the region of enhancement appears to straddle the boundary, favoring the high latitude side. This is not in agreement with the quiet time obser-

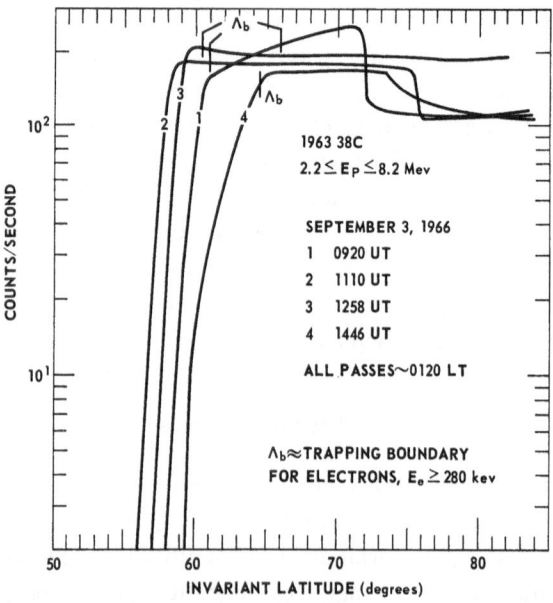

Fig. 1.   A sequence of nightside latitude profiles of solar protons ($2.2 \leqslant E_p \leqslant 8.2$ MeV) at 1100 km altitude on September 3, 1966.

vations of McDiarmid and Burrows (1969), from which they conclude that the enhancement occurs entirely within the trapping region. Although part of the disagreement may lie in the different definitions of boundaries, the enhanced region observed by Evans and Stone (1969) extends to $\sim 80°$ at times, well above the trapping region on both the day and night side of the earth.

In Figure 2 we show the temporal behavior on September 3 as observed by Explorer 33 and compared with the polar cap observations. The Explorer 33 data are $\frac{1}{2}$ hour sums (courtesy of Dr. S. M. Krimigis) and during the period shown the satel-

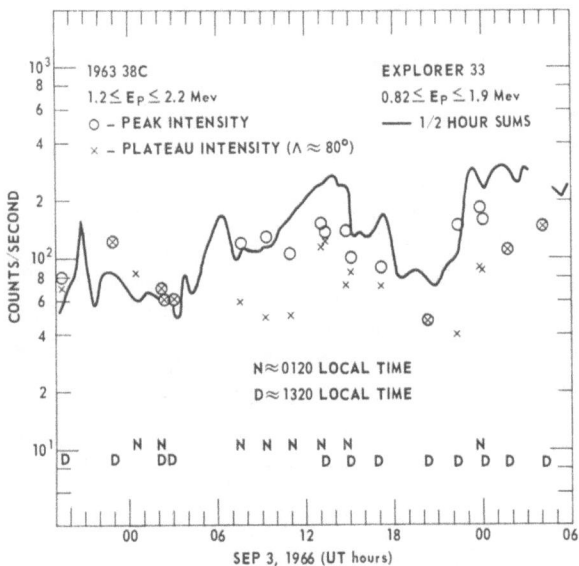

Fig. 2.   Low energy proton measurements by Explorer 33 (courtesy of S. M. Krimigis) and 1963 38C. Both peak and plateau intensities are given for the polar data.

lite was on the dusk side of the magnetosphere at distances ranging from ~ 12 to 37 R_E (Behannon *et al.*, 1968). Both peak intensities and plateau ($\Lambda \approx 80°$) values are shown for the low altitude polar data. The local time of each pass is indicated. Of the 19 passes shown having both values, seven display peak/plateau ratios of $\gtrsim 2$, seven have ratios of 1, and the rest are in between. The peak/plateau ratio is generally larger on the night side than on the day side for the cases where we have continuous data over the pole, although the largest ratio in this period was seen at 2210 UT on the day side. The ratio seems to increase following an increase in the interplanetary flux and then recover toward unity, i.e., the distribution becomes flat between the 'knee' and the pole. Another way to describe the behavior is to say that the interplanetary flux is tracked better by the peak intensities than by the plateau intensities. The implication, therefore, is that the protons have more ready access to the lower latitude peak region than to the polar region.

In Figure 3 we have compared the interplanetary fluxes observed by Explorer 34

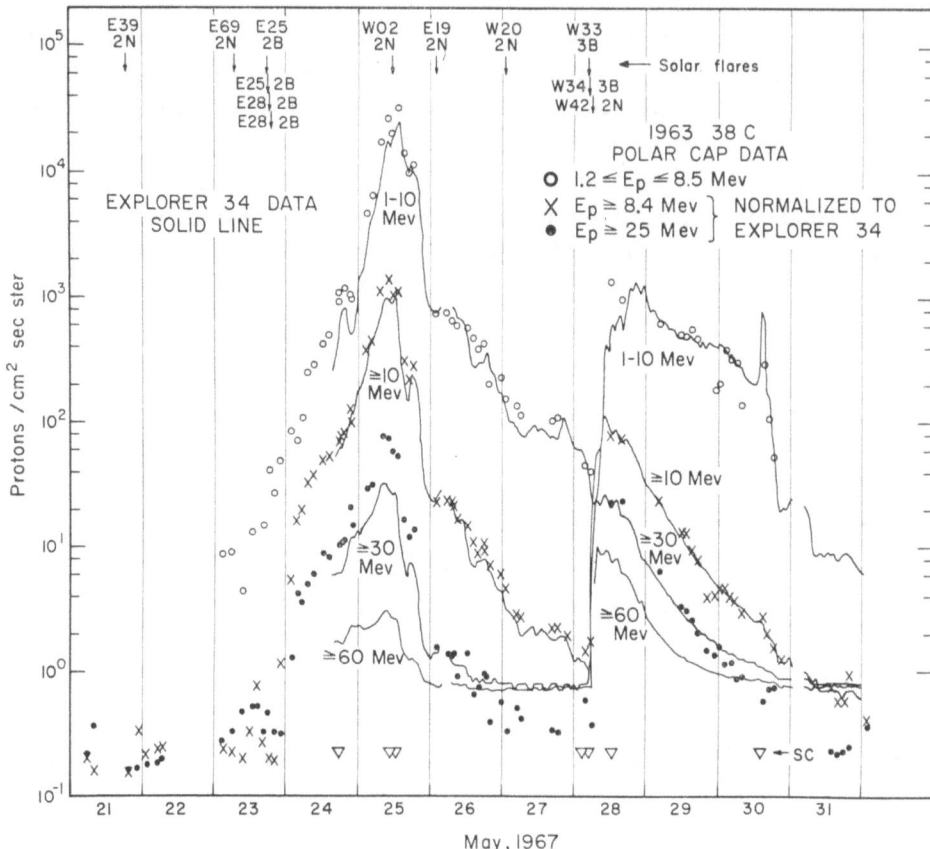

Fig. 3. The solar particle events of May 1967 as observed by the Explorer 34 Solar Proton
Monitoring Experiment and by satellite 1963 38C (polar cap averages).

with 1963 38C polar cap averages. The agreement can be considered to be quite
good, but there are several noticeable discrepancies. We have expanded the scale on
May 26 and show a more detailed comparison of the low energy channels in Figure 4.
While the time histories are similar it is worth noting that (a) the ratio of polar cap
to interplanetary fluxes varies by a factor of ∼3 during this period, and (b) the
significant decrease in the interplanetary fluxes occurring at ∼1120 UT is not observable over the polar cap except as a slower steady decrease. These data were
examined pass by pass by Williams and Bostrom (1969), and it was shown that the
structural features in the latitude intensity profiles are spatial rather than temporal.
It is clear, however, that these spatial features do change with time and very likely
are dependent on the variations in the magnetospheric configuration.

We show examples of an entirely different polar cap structure in Figures 5 and 6.
Figure 5 is a latitude profile recorded at 0232 UT, February 6, 1965. The detector
is sensitive to electrons ($E_e \geqslant 280$ keV) and protons ($E_p \geqslant 2$ MeV) and therefore the
proton cutoff is masked by the outer zone electrons. However, above 72° the response

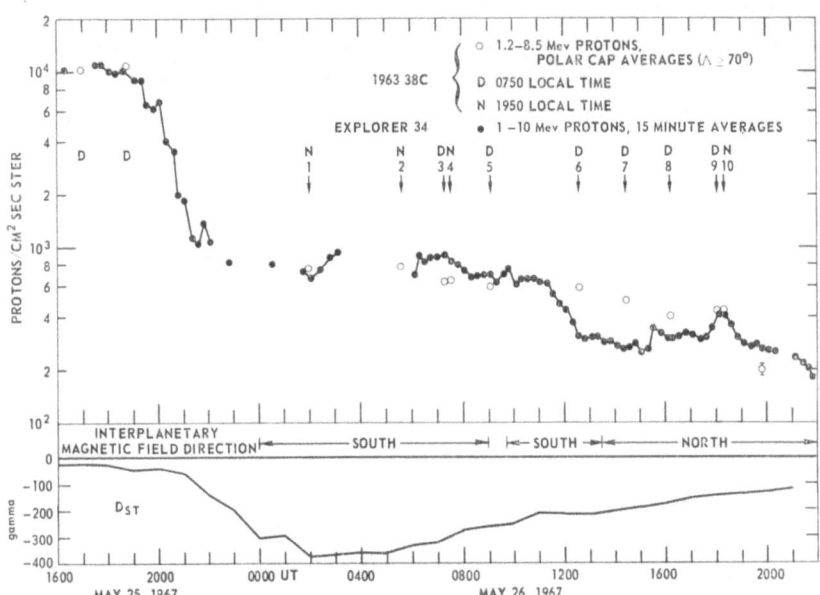

Fig. 4.   Time history of 1–10 MeV protons observed at high altitudes ($\geqslant 28.2\ R_3$) and 1.2–8.5 MeV protons observed at 1100 km over the northern polar cap on May 26, 1967. The orientation of the interplanetary magnetic field with respect to the ecliptic plane is also indicated (from Williams and Bostrom, 1969).

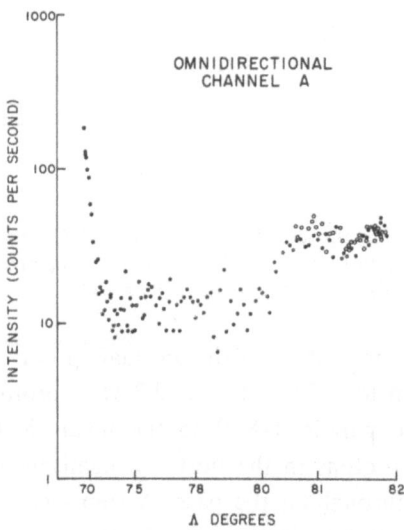

Fig. 5.   Counting rate of omnidirectional channel A ($E_e \geqslant 280$ keV, $E_p \geqslant 2.0$ MeV) vs. invariant latitude for the 1963 38C pass of 0232 UT February 6, 1965 (local time – 1536). The outer zone electron profile terminates at $\Lambda = 72°$. A high latitude proton enhancement begins at $\sim 80°$. The open circles show the descending portion of the pass (Williams and Bostrom, 1967).

is entirely due to protons and the observation of interest is a step increase by a factor of ~ 3 above $\Lambda \approx 80°$. On May 30, 1967 a similar high latitude increase was seen (Williams *et al.*, 1968; Williams, 1969). Figure 6 shows the counting rates of four detectors on 1963 38C and the Explorer 34 1–10 MeV data from 1501 to 1511 UT.

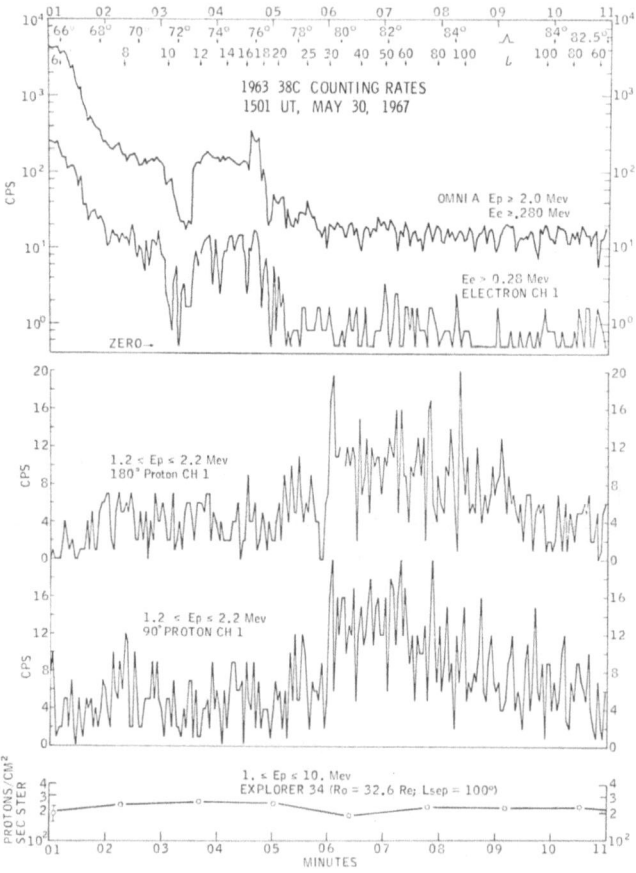

Fig. 6.   Counting rates vs. time for four detectors on 1963 38C during the 1501 UT pass on May 30, 1967 (local time – 0720). Both L and Λ are shown across the top of the figure and the simultaneous Explorer 34 measurements are also given.

This pass is also interesting because of the unusual appearance of energetic electrons at invariant latitudes up to $\sim 77°$. The 1.2–2.2 MeV proton flux is measured both perpendicular (90°) and parallel (180°) to the magnetic field. The 180° detector measures precipitating protons in the northern hemisphere, showing that the flux is essentially isotropic throughout the pass. A step increase by a factor of 3 to 4 occurs at about 79° while the simultaneous Explorer 34 data show no significant fluctuations occurring at this time outside the magnetosphere. Referring back to Figure 3, this period is seen to follow a sharp peak in the low energy proton intensity in interplanetary space. The time period around the peak is shown in Figure 7. The

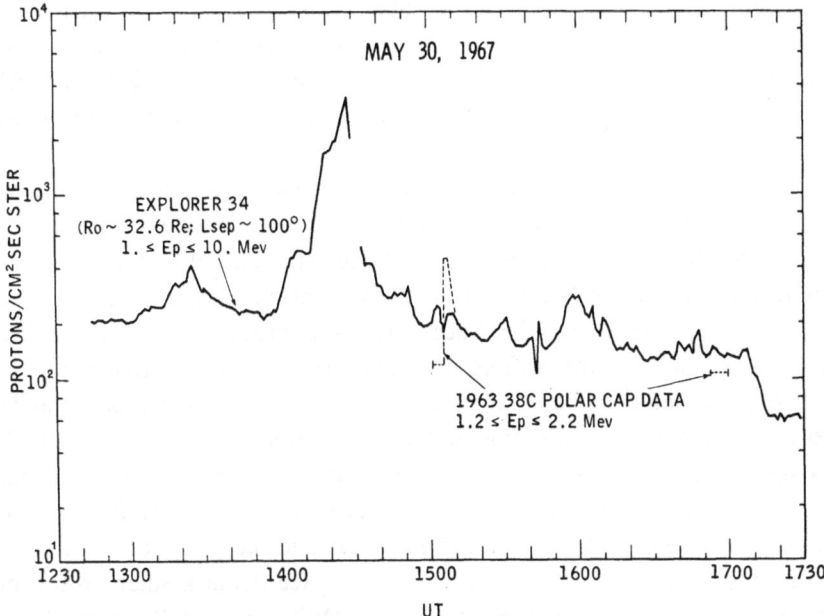

Fig. 7. Detailed time history of the sharp intensity increase on May 30, 1967. The 1963 38C passes at 1501 UT and 1650 UT are also plotted for latitudes $\geq 70°$.

1650 UT pass shows none of the structure evident in the earlier pass at 1501 UT. The absolute fluxes measured by the two satellites are in good agreement.

Both of the passes (Figures 5 and 6) exhibiting the polar enhancement were on the day side. The 1965 pass occurred at 1536 local time and the 1967 pass at 0720 local time. The location of the increases near $\Lambda = 80°$ leads one to consider that the increases may be associated with the neutral point or line common to all magnetospheric models. Indeed this may be the case, but Figure 7 could be interpreted as a further example of the low latitude edge of the polar region tracking the interplanetary flux more easily than the polar flux. That is, the polar flux may have increased (slowly?) in response to the interplanetary increase which began at ~ 1400 UT, but could not respond rapidly to the precipitous decrease at ~ 1430 UT. By 1650 UT the interplanetary flux had been relatively steady for about two hours and the latitude profile is essentially flat.

3. Discussion

From the data presented here and in the references, it can be stated that, at least for a large part of the time, the magnetospheric configuration does not permit direct and uniform access to solar protons over the polar caps. More than 50% of the low-altitude latitude profiles show an enhanced proton intensity at the low latitude edge of the polar region both during quiet (McDiarmid and Burrows, 1969) and disturbed times. It is clear that the region of enhancement can extend well inside the trapping region, but it appears to straddle the trapping boundary, i.e., the enhanced flux appears both on magnetic field lines which are closed and on those which extend into the magnetotail. However, these two kinds of field lines have the common topological property of being close to either the magnetopause or the neutral sheet. McDiarmid and Burrows (1969) propose that the polar intensity is representative of the interplanetary flux and the enhanced region occurs on closed field lines as a result of a scattering mechanism which favors scattering in from the tail over scattering out.

Our view is almost exactly opposite, in that we would consider the enhanced region fluxes to be more nearly representative of the interplanetary environment and the polar flux to be the result of a diffusion mechanism. Further, the isotropic fluxes found at all latitudes above the 'knee' and the apparently short proton (trapping) lifetimes argue against the drifting of temporarily trapped protons from the night to the day side and implies that protons penetrate the magnetopause at all local times and latitudes. If a scattering mechanism allowed rapid access to a depth inside the magnetopause of 2 to 3 R_E as suggested by Lanzerotti (1968), then Fairfield's model (Fairfield, 1968) would predict the dayside latitude cutoffs that have been observed. If, in addition, the tail current sheet and 'cusp' region on the nightside can be considered the same as interplanetary space (from a solar proton point of view), the nightside observations also follow (Williams and Bostrom, 1969). This approach does not require a scattering mechanism which favors inward scattering. The polar magnetic field lines are the least accessible of those which make up the magnetotail and temporal features observed outside the magnetosphere may be delayed or dispersed at the pole. Of course, this picture shares a common shortcoming with others in that it is an over-simplification. Reid and Sauer (1967) and Evans and Stone (1969) report significant and persistent asymmetry between the northern and southern polar caps, indicating that the northern and southern tail regions can be independent and may have very different properties. It is interesting, however, that the low latitude rim or enhanced region had very nearly the same intensity in both hemispheres throughout the period reported by Evans and Stone (1969).

Most observers have listed the restrictions and requirements placed on magnetospheric models by their observations. Much of the data cannot be interpreted in terms of charged particle motion in a static magnetic field, no matter how the field is distorted. For example, the storm-time presence of low energy solar protons at L-values as low as 3.0–3.5 R_E and the fact that the 'cutoffs' are essentially independent of energy (at least between 1.2 and 25 MeV) (Zmuda et al., 1967) are not reasonably explained in terms of any magnetic field configuration. What is apparently required is a mechanism for transporting these low energy protons through the magnetopause and into the outer magnetosphere with relative ease. The mechanism must have a weak energy dependence and be effective over a region whose inner boundary is around 6.5 R_E during quiet times and contracts in response to magnetic activity.

Although the data presented here have been selected to emphasize the features which cannot be explained by the direct interconnection of the interplanetary and the geomagnetic field, it should be recognized that we still hold the view expressed previously (Williams and Bostrom, 1967, 1969), namely that the details of proton entry into the magnetosphere depend on the magnetospheric configuration, which in turn depends on the boundary conditions imposed by the interplanetary medium.

Acknowledgements

I wish to thank Dr. S. M. Krimigis for the use of Explorer 33 data, and Drs. Krimigis and D. J. Williams for many discussions on this subject. This work has been supported in part by the National Aeronautics and Space Administration under DPR No. R21-009-004, and in part by the Naval Ordnance Systems Command, Department of the Navy, under contract NOw 62-0604-c.

References

Armstrong, T. P. and Krimigis, S. M.: 1968, 'Observations of protons in the magnetotail with Explorer 33', *J. Geophys. Res.* **73**, 143.

Behannon, K. W., Fairfield, D. H., and Ness, N. F.: 1968, 'Trajectories of Explorers 33, 34, and 35, July 1966 – July 1968', Goddard Space Flight Center Report X-616-68-372.

Blake, J. B., Paulikas, G. A., and Freden, S. C.: 1968, 'Latitude intensity structure and pitch angle distributions of low energy solar cosmic rays at low altitude', *J. Geophys. Res.* **73**, 4927.

Bostrom, C. O., Kohl, J. W., and Williams, D. J.: 1967, 'The February 5, 1965, solar proton event, 1. Time history and spectrums observed at 1100 kilometers', *J. Geophys. Res.* **72**, 4487.

Bostrom, C. O., Kohl, J. W., and Zmuda, A. J.: 1967, 'The solar proton events of August 29 and September 2, 1966', *Trans. Am. Geophys. Union* **48**, 178.

Dungey, J. W.: 1961, 'Interplanetary magnetic field and the auroral zones', *Phys. Rev. Letters* **6**, 47.

Evans, L. C. and Stone, E. C.: 1969, 'Access of solar protons into the polar cap: a persistent north-south asymmetry', *J. Geophys. Res.* **74**, 5127.

Fairfield, D. H.: 1968, 'Average magnetic field configuration of the outer magnetosphere', *J. Geophys. Res.* **73**, 7329.

Fillius, R. W.: 1968, 'Penetration of solar protons to four earth radii in the equatorial plane', *Ann. Geophys.* **24**, 821.

Kane, S. R., Winckler, J. R., and Hoffman, D. J.: 1968, 'Observations of screening of solar cosmic rays by the outer magnetosphere', *Planetary Space Sci.* **16**, 1381.

Krimigis, S. M., Van Allen, J. A., and Armstrong, T. P.: 1967, 'Simultaneous observations of solar protons inside and outside the magnetosphere', *Phys. Rev. Letters* **18**, 1204.

Lanzerotti, L. J.: 1968, 'Penetration of solar protons and alphas to the geomagnetic equator', *Phys. Rev. Letters* **21**, 929.

McDiarmid, I. B. and Burrows, J. R.: 1969, 'Relation of solar proton latitude profiles to outer radiation zone electron measurements', *J. Geophys. Res.* **74**, 6239.

Michel, F. C. and Dessler, A. J.: 1965, 'Physical significance of inhomogeneities in polar cap absorption events', *J. Geophys. Res.* **70**, 4305.

Montgomery, M. D. and Singer, S.: 1969, 'Penetration of solar energetic protons into the magnetotail', *J. Geophys. Res.* **74**, 2869.

Paulikas, G. A., Blake, J. B., and Freden, S. C.: 1968, 'Low-energy cosmic-ray cutoffs: diurnal variations and pitch-angle distributions', *J. Geophys. Res.* **73**, 87.

Paulikas, G. A. and Blake, J. B.: 1969, 'Penetration of solar protons to synchronous altitude', *J. Geophys. Res.* **74**, 2161.

Pieper, G. F., Zmuda, A. J., Bostrom, C. O., and O'Brien, B. J.: 1962, 'Solar protons and magnetic storms in July 1961', *J. Geophys. Res.* **67**, 4959.

Reid, G. C. and Sauer, H. H.: 1967, 'Evidence for non-uniformity of solar-proton precipitation over the polar caps', *J. Geophys. Res.* **72**, 4383.

Stone, E. C.: 1964, 'Local time dependence of non-Stormer cutoff for 1.2 MeV protons in quiet geomagnetic field', *J. Geophys. Res.* **69**, 3577.

Williams, D. J.: 1967, 'On the low-altitude trapped electron boundary collapse during magnetic storms', *J. Geophys. Res.* **72**, 1644.

Williams, D. J.: 1969, 'Solar proton observations, 1–10 MeV', NASA-GSFC Report X612-69-258, June 1969.

Williams, D. J. and Bostrom, C. O.: 1967, 'The February 5, 1965, solar proton event, 2. Low energy proton observations and their relation to the magnetosphere', *J. Geophys. Res.* **72**, 4497.

Williams, D. J. and Bostrom, C. O.: 1969, 'Proton entry into the magnetosphere on May 26, 1967', *J. Geophys. Res.* **74**, 3019.

Williams, D. J. and Mead, G. D.: 1965, 'Nightside magnetospheric configuration as obtained from trapped electrons at 1100 km', *J. Geophys. Res.* **70**, 3019.

Williams, D. J., Arens, J. F., Bostrom, C. O., and Kohl, J. W.: 1968, 'Explorer 34 and 1963 38C observations of low energy protons during the May 28, 1967 event', (abstract) Midwest Cosmic Ray Conference, University of Iowa, Iowa City, March 1–2, 1968.

Zmuda, A. J., Pieper, G. F., and Bostrom, C. O.: 1963, 'Solar protons and magnetic storms in February 1962', *J. Geophys. Res.* **68**, 1160.

Zmuda, A. J., Kohl, J. W., and Bostrom, C. O.: 1967, 'The solar proton event of September 2, 1966: day and night latitude profiles', *Trans. Am. Geophys. Union* **48**, 178.

Discussion

Axford: There are many boundaries you can fit to compare these things. You can fit the inner boundary to the trapping boundary or maybe one Larmor radius inside. You can fit the outer boundary to the various other boundaries that you can define from trapped to quasi trapped particles.

Bostrom: The passes shown are all on the night side. Data I have seen in the past seem to indicate that the night side electron trapping boundary is reasonably independent of energy, from 40 keV to relativistic energies. On the day side there is a 10° shift for the 40 keV and about 2° shift in the boundary for the relativistic electrons.

Reid: You suggest an access to the magnetosphere at all local times by some kind of scattering mechanism. This is awfully hard to reconcile with an asymmetry between the two polar caps.

Bostrom: The asymmetry between the polar caps occurs on the polar field line. The asymmetry does not occur in the rim section at all. The fluxes are identical north and south in the rim apparently but they are not identical in the lines near the pole, above 80°.

Hones: How rapidly does the rim track the interplanetary intensities?

Bostrom: The results presented here are not corrected for spectral effects. Generally I can say that the rim tends to track better than the polar region, within some tens of minutes.

Singer: The Vela results show no difference in the delay times as a function of spacecraft position inside the magnetosphere. Most observations have been in the so-called high latitude magnetotail. But in one or two cases one spacecraft was in the plasmasheet and the other in the interplanetary medium. There were detectable delays i.e. there was no immediate access to the plasmasheet.

Bostrom: This may have something to do with closing field lines through the plasmasheet also.

TRAPPED PARTICLE POPULATION CHANGES
ASSOCIATED WITH SOLAR EVENTS

JUAN G. ROEDERER

Dept. of Physics, University of Denver, Denver, Colo., U.S.A.

1. Physical Processes

Trapped particle population changes associated with solar events are caused by the effects of solar wind discontinuities on the magnetosphere. The most drastic sequence of such time variations occurs during a magnetospheric storm.

Some characteristic features of a typical storm are sketched in Figure 1. It shows the time dependence of the horizontal component H of the magnetic field as registered at an equatorial station, and the AE index (an indirect measure of the polar electrojet). In general, the arrival of a flare-initiated interplanetary shock is accompanied by a sudden compression of the magnetosphere (sudden commencement). After a few hours, a large decrease of the H-component is seen (main phase), signalling the buildup of a ring current. Prior to this decrease, a repetitive occurrence of magnetospheric substorms starts, with an associated display of repetitive auroral, ionospheric and polar magnetic disturbances (coincident with the peaks in the AE index). During its early stage, the ring current is highly assymmetric, mainly flowing in the midnight to evening sector of the magnetosphere. It is highly probable that this ring current, especially in its asymmetric stage, is caused by, or at least linked to, the repetitive action of magnetospheric substorms. Whether these substorms are triggered by the arrival of the driver gas behind the interplanetary shock wave (most distinctly marked by an increase of the He^{++}/H^+ ratio in the solar wind, Hundhausen, 1970),

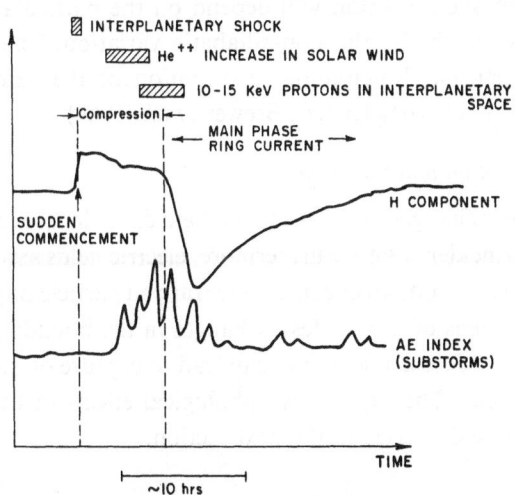

Fig. 1. Sketch of a magnetic storm.

V. Manno and D. E. Page (eds.), Intercorrelated Satellite Observations Related to Solar Events, 239–250. All Rights Reserved
Copyright © 1970 by D. Reidel Publishing Company, Dordrecht-Holland

and/or whether or not the formation of the ring current is associated with the recently detected enhancement of low-energy protons in interplanetary space in coincidence with the start of the main phase of a storm (Frank, 1969), are important questions to be settled. Typically, about six hours after the onset of the main phase, the substorm activity begins to decay, while the ring current develops for another few hours, becoming more symmetric. Then the field begins to recover; this is a slow process that may take days or even weeks.

During this sequence of events, important changes occur in the energetic particle population of the magnetosphere. We may classify them into the following categories:

A. ADIABATIC CHANGES

Magnetic field configuration changes that are slow as compared to a trapped particle drift period cause particle flux variations that are reversible – and well understood. This happens, for instance, during the recovery phase of a storm. McIlwain (1966) and Söraas and Davis (1968) have studied the adiabatic effect of the ring current decay on trapped particles. Also, slow compressions or expansions of the magnetosphere cause adiabatic changes in the particle population (Roederer, 1969). It should be pointed out here that a given magnetic field variation may be adiabatic for high energy particles (short drift periods), yet entirely irreversible (non-adiabatic) for lower energy particles (long drift periods).

B. NON-ADIABATIC EFFECTS

(a) *Drift bunching*

These are caused by impulsive, short term variations of the magnetic field configuration. Particles are driven by the electric field induced by the varying magnetic field, and their final drift shell position will depend on the particular longitude at which they were initially caught by the non-adiabatic variation. The final state will also depend on the particular longitudinal distribution of the field perturbation. This causes drift-bunching of particles (cf., Brewer *et al.*, 1969).

(b) *Particle acceleration and injection*

Severe non-adiabatic changes of the magnetic field cause local acceleration of particles (cf. Lezniak and Winckler, 1969). Furthermore, electric fields associated to tail plasma convection may cause profound effects on the trapped particle population, particularly to protons of a few tens of keV or less. Changes in the boundary of stable trapping and in the boundary of closed field lines can lead to capture of quasi-trapped particles into closed drift shells. The general morphological effects of these processes on the radiation belts will be discussed in the next section.

(c) *Diffusion*

Pitch angle diffusion and radial diffusion are considerably enhanced during a magnetospheric storm. Pitch angle diffusion occurs on a time scale that is comparable to early-phase storm variations and thus plays an important role in the development of the trapped particle perturbation. Radial diffusion occurs on a longer time scale (hours or days) and is important for trapped particle flux rearrangements at later times in the storm and throughout the post-storm period (cf. Söraas, 1969).

The physics of a magnetospheric storm and its effects on trapped particles is not yet fully understood. An important complication arises from the interplay between the trapping magnetic field, the trapped particle population (in particular, protons in the tens of keV range), the tail plasma (convective electric field), and electromagnetic and hydromagnetic waves. Figure 2 schematically illustrates this interaction.

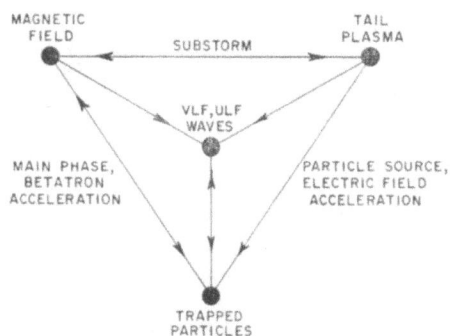

Fig. 2.   Diagram of the interactions between fields and particles during substorms.

2. Morphology of Trapped Particle Variations During Storms

Figure 3 reproduces Injun 3 data on the time-behavior of outer zone electrons of energy $\geqslant 1.6$ MeV (Craven, 1966), for different L-values. Data from different B-value intervals (representing different equatorial pitch angle intervals) are indicated. A qualitative examination of this figure leads to the following observations:

(i) Electron fluxes in this energy and L-value range are highly sensitive to geomagnetic activity, with large intensity increases in coincidence with geomagnetic storms.

(ii) The onsets of the increases occur over the whole range of L-values, with different responses at different L-values.

(iii) Flux increases never exceed a given value, indicating the existence of an asymptotic saturation level at each L-value.

(iv) Particles at different B-values for a given L behave similarly, indicating that at all times there is efficient pitch angle mixing.

(v) The decay curves are dissociated from each original storm, indicating that they are mainly controlled by diffusion (radial and pitch angle).

Figure 4 shows $\geqslant 280$ MeV/electron data obtained by Williams $et\ al.$ (1968) with satellites 1963 38C (low altitude) and Explorer 26 (nearly equatorial). These data reveal the following:

(i) At $L=3$, only large, well-defined perturbations are seen, associated to the largest storms (see Dst plot at bottom).

(ii) With increasing L, the variety of fluctuations increases. At $L = 5.5$, the flux is so agitated that the correlation with Dst gets lost.

(iii) The behavior of low altitude (small equatorial pitch angle) electrons is very similar to the behavior of near-equatorial ($\sim 90°$ pitch angle) particles. The only difference turns out to be due to the fact that in a distorted magnetosphere, two points with identical McIlwain L-values do not really represent identical particle shells (see Figure 6).

(iv) $\gtrsim 1$ MeV electrons (not shown in Figure 4) behave similarly, but low altitude

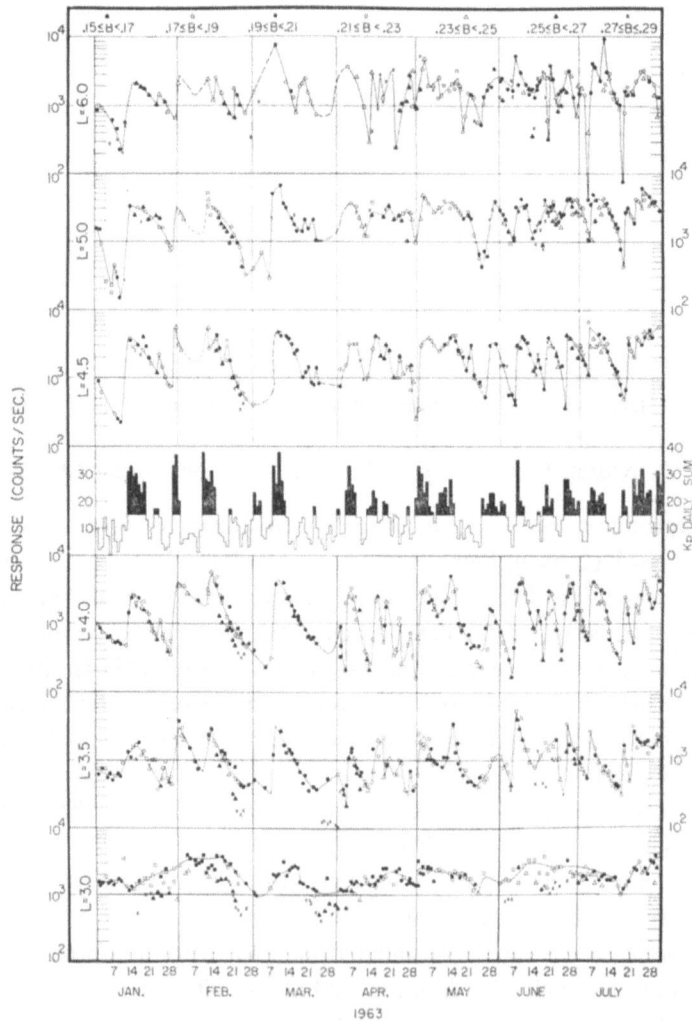

Fig. 3.   The responses of the Injun 3 213A GM tube due to the intensities of electrons ($E > 40$ keV) trapped in the outer radiation zone and mirroring at low altitudes, selected at $L = 3.0, 3.5, 4.0, 4.5,$ 5.0, and 6.0 ± 0.1 to display the temporal variations during the period from January 1 to July 31, 1963. The values of B are defined therein. Each datum represents an eight-sec average of the GM tube counting rate (Craven, 1966).

and equatorial particles do not achieve an equilibrium ratio so quickly. This suggests that pitch angle mixing is less effective at higher energies.

Williams *et al.* have analyzed the 'arrival half-time' (time interval until half of the peak value is reached) and the corresponding amplitude of the increase, as a function of L, for different storms. Figure 5 is an example. Other events look alike: particles appear *first*, and with *maximum intensity*, at a given L-value, and then diffuse radially away (inwards and outwards) from the initial acceleration region. The stronger the storm (the larger the Dst), the deeper in the magnetosphere is the acceleration region located. This is shown in Figure 6.

During the immediate post-storm period, the lifetime of the freshly accelerated $\gtrsim 300$ keV electrons (as graphically determined by fitting an exponential to the decay curve) is considerably shorter than during the recovery period (Figure 7). This is evidence of a considerable enhancement of the loss and particle transport mechanisms during the early phase of a storm. Notice in Figure 7 that high-altitude and low-altitude particles behave similarly during decay.

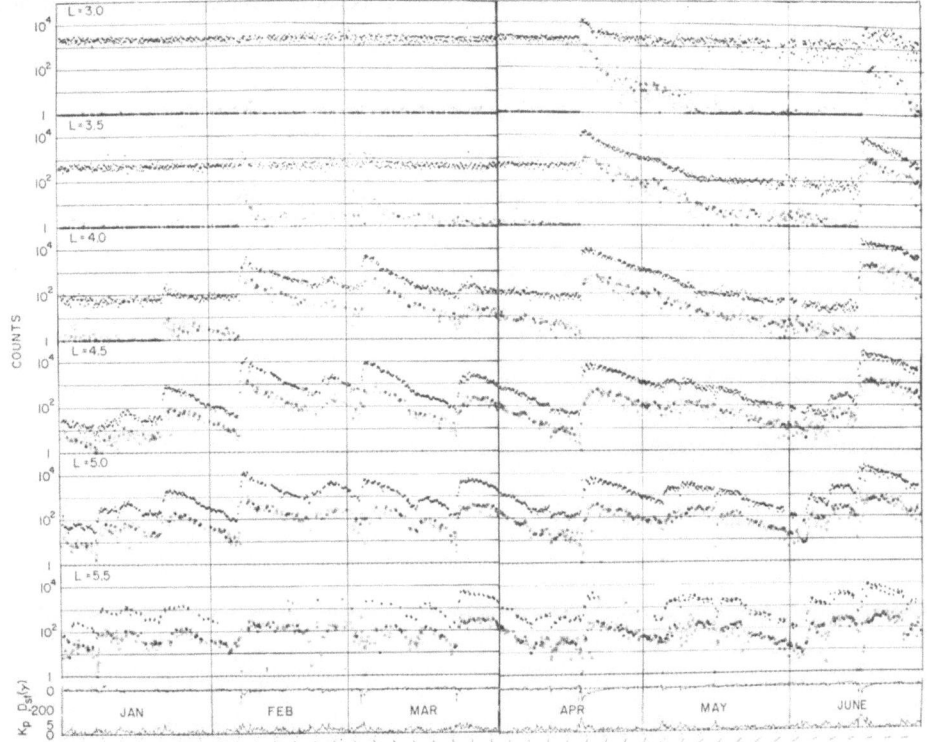

Fig. 4. Simultaneous electron data from the polar-orbiting satellite 1963 38C ($E > 280$ keV), denoted by □, and the near-equatorial satellite Explorer 26 ($E > 300$ keV), denoted by ×, for the period January 1, 1965, through June 29, 1965. The electron data at $L = 3.0, 3.5, 4.0, 4.5, 5.0,$ and 5.5 are shown in terms of counts for each of the experiments counting intervals. Below the electron data are plotted the hourly average Dst and the 3-hour average K_p index for the 6-month period (Williams *et al.*, 1968).

The behavior at lower L-values of trapped particles during geomagnetic perturbations has been discussed in a recent paper by Bostrom *et al.* (1969). Figure 8 shows data of high energy electrons ($\geqslant 1.2$ MeV) at low L-values during 5 years. These are mainly fission-debris electrons from the Starfish high altitude test, gradually decaying away. There is no visible response to magnetic storms. It is clear from Figure 8 that non-adiabatic effects are negligible in this range of L-values. Figure 9 depicts variations of $\geqslant 280$ keV electrons during the same period of time. Notice the response to geomagnetic storms, increasing with L. Figure 10 shows detailed behavior of these electrons during the May 1967 storm. Two distinct regions of particle behavior can be identified: (i) between $L=1.20$ and 1.40 the increase is immediate and rapid; (ii) beyond $L=1.80$ the increase is gradual, pointing to radial diffusion from a source located further out. Very likely, this is identical to the inward diffusion from an acceleration region as seen by Williams *et al.* (Figure 5). The two regions are separated by a transition region whose width decreases with increasing storm intensity. The acceleration mechanism acting in region (i) at very low L-values must be of a quite different nature than the one effective further out. It may be caused by recurrent field variations produced by ionospheric currents (Cladis, 1966), or it may be a non-adiabatic effect caused by the field decrease during the start of the main phase.

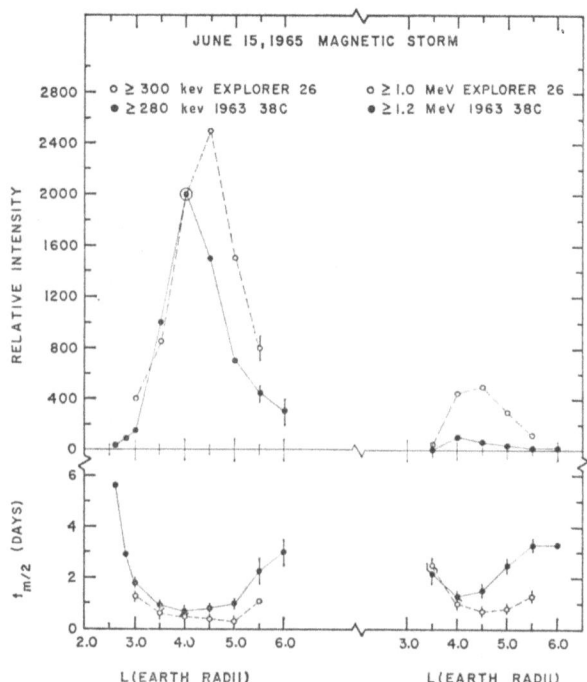

Fig. 5.   June 15, 1965, geomagnetic storm. The maximum storm-electron intensities and the time required for the intensities to reach half their peak values ($t_{m/2}$) are plotted vs. L for the high- and low-energy electrons observed on 1963 38C and Explorer 26. Explorer 26 $\geqslant 300$ keV and $\geqslant 1.0$ MeV data have been multiplied by 0.1. A very broad region ($L = 3.5$–5.0) of initial electron appearance was observed (Williams *et al.*, 1968).

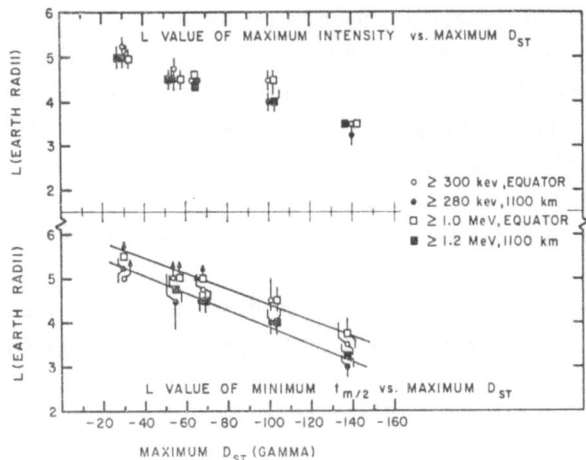

Fig. 6. (a) The L-value of maximum electron intensity plotted vs. the peak Dst value for five geomagnetic storms. Both energies at both low and high altitudes are included. (b) The L-value of the earliest $t_{m/2}$ plotted vs. the peak Dst value for the same five geomagnetic storms. A separation, attributable to a measure of the magnetospheric expansion, is observed between the high- and low-altitude electron data (Williams *et al.*, 1968).

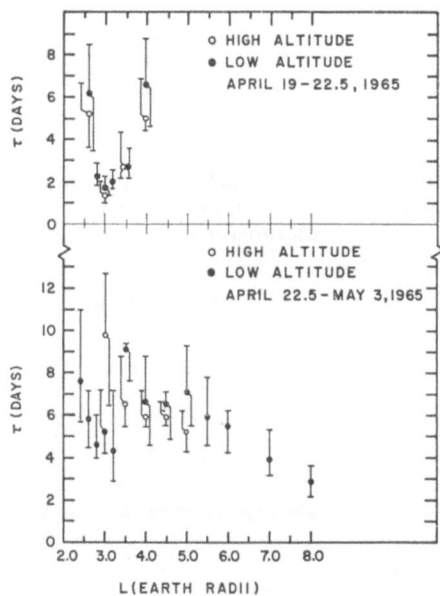

Fig. 7. Lifetimes (time to e^{-1} of initial value) for energetic electrons plotted vs. L for two periods after the April 17, 1965, geomagnetic storm. (a) Lifetimes for the immediate post-storm period, April 19 – April 22.5, 1965. Data beyond $L = 4.0$ were not available because outward electron diffusion was still operative during portions of this time period at the higher L-values, and no decay was observed. (b) Lifetimes for the post-storm period April 22.5 – May 3, 1965. The decay rates at the low L-values have increased by a factor of about 2 over the rates in (a) at $L = 2.6$–3.5. At all L-values, for both periods of time, the decay rates observed for these energetic electrons at high and low altitude are nearly identical (Williams *et al.*, 1968).

Protons behave in a more complicated way during storms. Whereas electrons *always* show intensity increases, proton fluxes may increase *or decrease* during the early phase of a storm. This is shown in Figure 11 for protons in the range 2.2–8.2 MeV (Bostrom *et al.*). Higher energy protons are considerably more stable and only show decreases at higher L-values during severe storms (McIlwain, 1965). Protons of energies below 2.2 MeV show a much more pronounced response than in Figure 11. In general, proton flux changes turn out smaller than those of the electrons, for any given event.

Figure 12 shows the L-profile of proton fluxes in different energy intervals, prior to and after the May 1967 storm. The general picture reveals an increase in a restricted

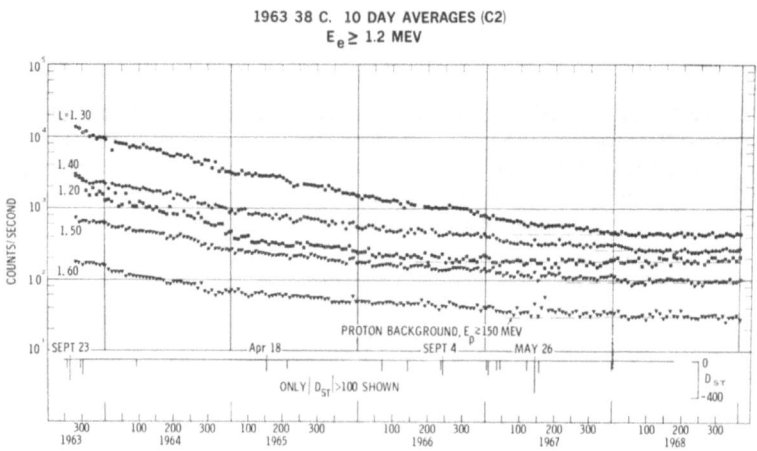

Fig. 8.   Ten-day average counting rates for ≥ 1.2 MeV electrons from October 1963 through December 1968 at $L = 1.2$, 1.3, 1.4, 1.5 and 1.6 R_E. The Dst scale indicates times of major magnetic storms (Bostrom *et al.*, 1969).

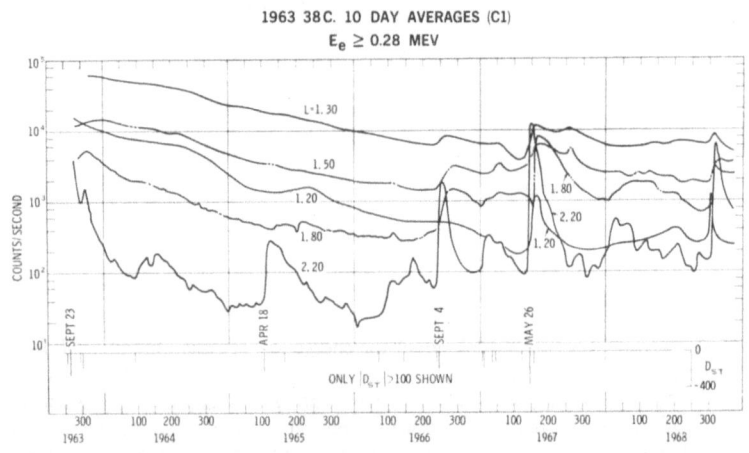

Fig. 9.   Curves derived from 10-day average counting rates for ≥ 280 MeV electrons from October 1963 through December 1968 at $L = 1.2$, 1.3, 1.5, 1.8 and 2.2 R_E. The Dst scale indicates times of major magnetic storms (Bostrom *et al.*, 1969).

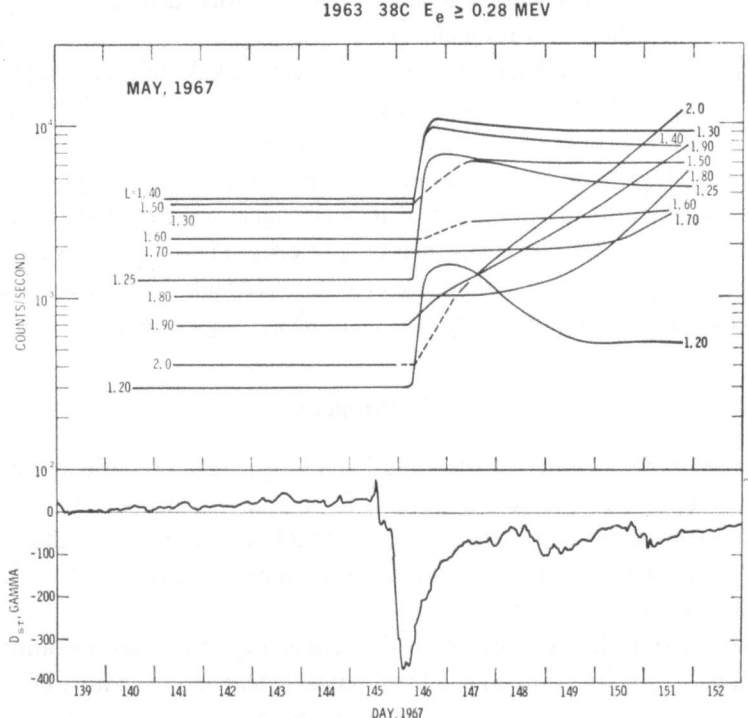

Fig. 10.   Counting rate of ≥ 280 keV electrons vs. time for $1.20 \leqslant L \leqslant 2.0\ R_E$ in May 1967. Dst values are plotted below the data (Bostrom *et al.*, 1969).

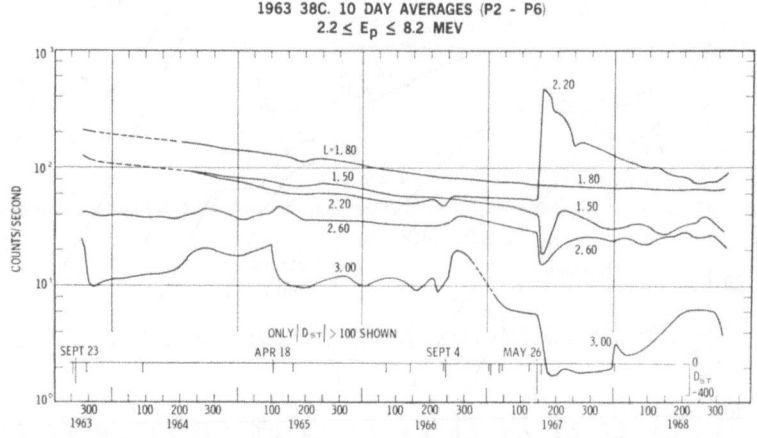

Fig. 11.   Curves derived from 10-day average counting rates for protons ($2.2 \leqslant E_p \leqslant 8.2$ MeV) from October 1963 through December 1968 at $L = 1.5$, 1.8, 2.2, 2.6 and 3.0 R_E. The Dst scale indicates times of major magnetic storms (Bostrom *et al.*, 1969).

L-interval (~ 2.2) and energy interval (<8 MeV), with depletion elsewhere, thus pointing to the action of a resonant mechanism.

Protons in the very low energy range of a few tens of keV play a key role in trapped particle dynamics. They are responsible for the ring current, they represent the upper end of the auroral proton spectrum, and, more generally, they represent particles in an intermediate stage between the plasma and the energetic, trapped population. They are most sensitive to the electric field in the magnetosphere which competes with the magnetic field in the control of their drift velocity. Figure 13 shows their behavior before, during and after the main phase of a storm (Frank, 1967), clearly identifying them as the particles responsible for the main-phase ring current.

3. Prospects

The detailed experimental and theoretical study of magnetospheric storm effects on trapped radiation is extremely important for radiation belt dynamics. Indeed, one may state that without storm effects, there probably would be no radiation belt at all (or just a tenuous 'halo' made up of decay protons and electrons from cosmic-ray albedo neutrons).

It is necessary to know more about the acceleration processes operating during a storm, in particular, their radial and longitudinal extent and the nature of the physical mechanisms involved. The key point for such a study would be *simultaneous measurements at different positions in the magnetosphere*, of the *pitch angle distribution* and the *energy spectrum* of the particles involved, as a function of time. In addition, since

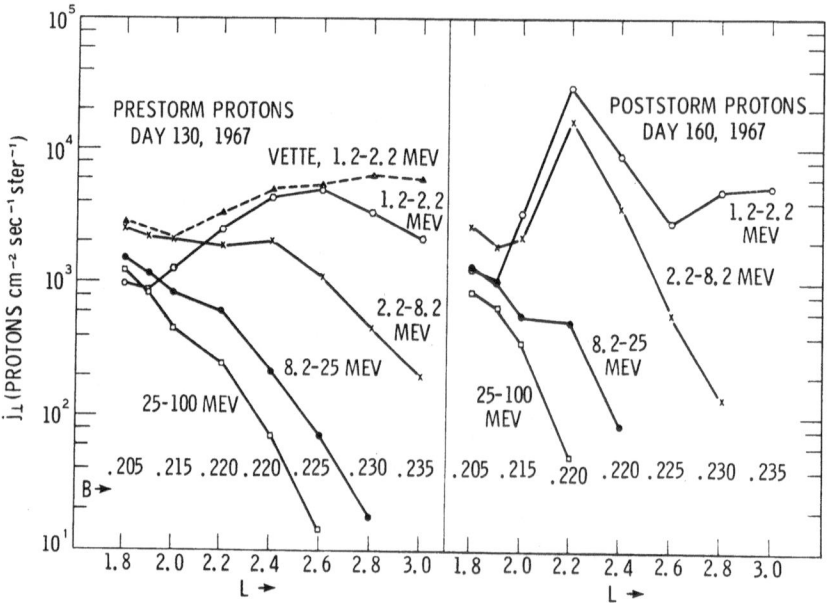

Fig. 12.   Proton flux vs. L for several energies showing the effect of the May 1967 storm on the proton distribution (Bostrom *et al.*, 1969).

several different phenomena are simultaneously involved in the control of the trapped particle population during storms (Figure 2), a *global study of field, plasma and waves is required, complemented with ground observations, especially at high latitudes.*

Finally, since the magnetospheric substorm seems to be a very fundamental process in the whole storm sequence, probably representing the action of a 'universal' instability of the system, it seems desirable to first *try to understand the mechanisms operating during an individual substorm.* Individual substorms occur in recurrent form at intervals of 3–4 hours even during magnetically quiet times, and are thus experimentally accessible in their 'purest' form at times other than during geomagnetic storms.

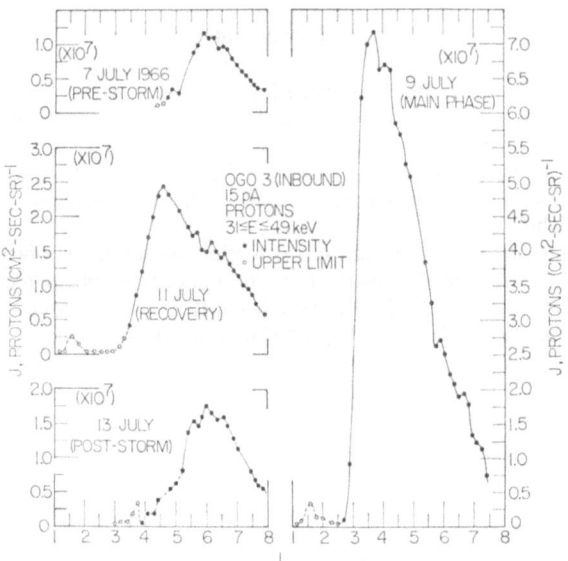

Fig. 13.   Directional intensities of protons ($31 \leqslant E \leqslant 49$ keV) as functions of L at low magnetic latitudes during the pre-storm, main phase, recovery phase and post-storm periods of the early July 1966 geomagnetic storm (Frank, 1967).

References

Bostrom, C. O., Beall, D. S., and Armstrong, J. C.: 1969, 'Time History of the Inner Radiation Zone October 1963 to December 1968', preprint of Applied Physics Laboratory, The Johns Hopkins University.

Brewer, H. R., Schulz, M., and Eviatar, A.: 1969, 'Origin of Drift-Periodic Echoes in Outer-Zone Electron Flux', *J. Geophys. Res.* **74**, 159–167.

Cladis, J. B.: 1966, 'Acceleration of Geomagnetically Trapped Electrons by Variations of Ionospheric Currents', *J. Geophys. Res.* **71**, 5019–5025.

Craven, J. D.: 1966, 'Temporal Variations of Electron Intensities at Low Altitudes in the Outer Radiation Zone as Observed with Satellite Injun 3', *J. Geophys. Res.* **71**, 5643–5663.

Frank, L. A.: 1967, 'On the Extraterrestrial Ring Current During Geomagnetic Storms', *J. Geophys. Res.* **72**, 3753–3767.

Frank, L. A.: 1969, 'On the Presence of Low Energy Protons, 5–50 keV, in the Interplanetary Medium', to be published in *J. Geophys. Res.*

Hundhausen, A. J.: 1970, this volume, p. 155.

Lezniak, T. W. and Winckler, J. R.: 1969, 'Magnetospheric Substorm Effects on Energetic Electrons in the Outer Van Allen belt', University of Minnesota, Cosmic Ray Group Tech. Report CR-137.

McIlwain, C. E.: 1965, 'Redistribution of Trapped Protons During a Magnetic Storm', *Space Res.* **5**, North-Holland Publ. Co., Amsterdam.

McIlwain, C. E.: 1966, 'Ring Current Effects on Trapped Particles', *J. Geophys. Res.* **71**, 3623–3628.

Roederer, J. G.: 1969, 'Quantitative Models of the Magnetosphere', *Rev. Geophys.* **7**, 77–96.

Söraas, F.: 1969, 'Comparison of Post-Storm Non-Adiabatic Recovery of Trapped Protons with Radial Diffusion', NASA-Goddard Space Flight Center Preprint X-612-69-241.

Söraas, F. and Davis, L. R.: 1968, 'Temporal Variations of the 100 keV to 1700 keV Trapped Protons Observed on Satellite Explorer 26 During First Half of 1965', NASA-Goddard Space Flight Center Preprint X-612-68-328.

Williams, D. J., Arens, J. F., and Lanzerotti, L. J.: 1968, 'Observations of Trapped Electrons at Low and High Altitudes', *J. Geophys. Res.* **73**, 5673–5696.

Discussion

Hones: Your results show that the flux of particles stays constant for a long time, and that the asymptotic value of the flux after the occurrence of a storm is lower than the precedent.

Roederer: This is certainly an important point, but it has not been looked at systematically.

Zhulin: Are the auroral particles precipitated on the auroral oval from the trapped region after injection from the magnetotail and subsequent longitudinal drift, or are they precipitated simultaneously along the whole oval?

Roederer: The key factor is the lifetime of the precipitated electrons, which seems to be shorter than the longitudinal drift time. Therefore, if any auroral phenomena are seen on the dayside, the electrons which produce them, should not survive that drift. This is an argument in favor of the ones who believe in acceleration and immediate precipitation at all local times.

GEOMAGNETIC STORMS AT ATS 1

PAUL J. COLEMAN, JR.

Institute of Geophysics and Planetary Physics*,
University of California, Los Angeles, Calif., U.S.A.

1. Introduction

A geomagnetic storm is accompanied by effects in the magnetic field at ATS 1 that are as complex as the storm-time effects in the surface field. Nevertheless, the comparison of field measurements taken at a geostationary satellite with those recorded simultaneously at the surface has yielded considerable information on the properties of the storm-time disturbance field. In addition, several types of oscillatory phenomena have been recorded at ATS 1 during storms.

This paper is intended as something of a progress report on our study of geomagnetic storms at ATS 1. Accordingly, we will first briefly review our earlier results and then summarize some more recent work.

To date, we have confined our attention to storms in the first six months of 1967. There were 18 storms during this interval. The onset times are listed in Table I. Data taken at ATS 1 during nine of these storms, as indicated in the table, will be discussed here.

TABLE I

Geomagnetic storm onsets for January through June, 1967 (reported by more than two stations)

Date	Time	Type	No. of stations reporting	ATS
0107	0758–0801	SC	17	*X*
0113	1202–1204	SC	17	*X*
0207	1636–1637	SC	11	*X*
0215	2348–2349	SC	17	*X*
0318	0300–0400	GC	3	*X*
0401	0806–0808	SC	8	
0404	0303–0305	SC	9	
0423	1400–1500	GS	7	
0501	1906–1908	SC	19	*X* (partial)
0507	0104–0106	SC	5	
0524	1724–1727	SC	6	
0525	1234–1237	SC	13	*X* (partial)
0527	19–	GC	3	
0528	0232	GC	3	
0530	1425–1427	SC	13	
0605	1914–1916	SC	16	
0625	0221–0224	SC	17	*X*
0626	1458–1459	SC	6	*X*

From: Principal Magnetic Storms *Solar-Geophysical Data.*

* Publication No. 774.

V. Manno and D. E. Page (eds.), Intercorrelated Satellite Observations Related to Solar Events, 251–279. All Rights Reserved
Copyright © 1970 by D. Reidel Publishing Company, Dordrecht-Holland

2. Trajectory and Coordinate Systems

ATS 1 was launched into the synchronous, equatorial orbit on December 6, 1966, from Cape Kennedy, Florida. The spacecraft is spin-stabilized at a nominal rotation frequency of $1.67\ sec^{-1}$. The spin axis is parallel to that of the earth.

The geocentric range of the synchronous orbit is a constant 6.6. R_E (1 R_E=radius of earth). ATS 1 is stationed in the equatorial plane at longitude 150.3°W. Thus, its coordinates are held fixed at geocentric range 6.6 R_E, latitude 0°, longitude 150.3°W.

In the discussions to follow, the magnetic field will usually be described in terms of its vector components in one of two coordinate systems denoted by X, Y, Z, and D, H, V. The XYZ system has its origin at the spacecraft, the Z axis is parallel to the earth's axis and positive northward, the X axis is in the direction of the projection of the spacecraft-to-sun line on the plane perpendicular to the spacecraft spin axis, which is also the equatorial plane, and the Y axis completes the right-handed system. The DHV system is also earth-centered and it rotates with the earth. Thus, H is in the direction of the earth's spin vector, V is positive upward along the local vertical or radially outward from the earth, and D, which is positive eastward, completes the right-handed system. The ATS 1 orbit and the two coordinate systems are shown in Figure 1.

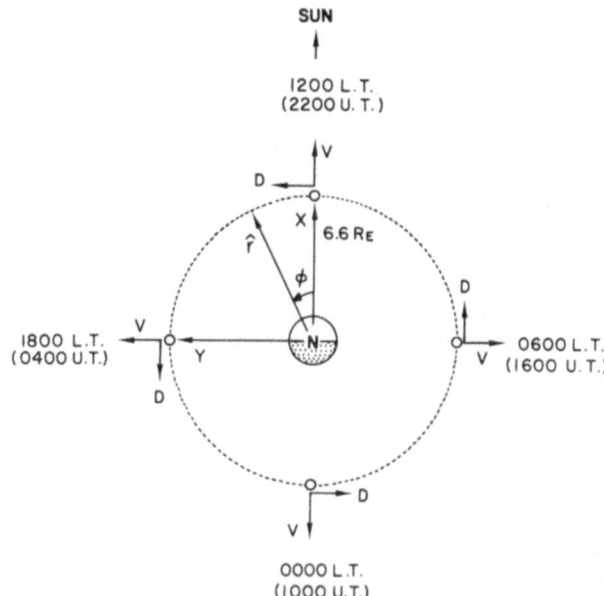

Fig. 1.   View from the north of the ATS 1 orbit in the equatorial plane showing the X and Y directions of the non-rotating XYZ system and the V and D directions of the rotating DHV system. The H and Z directions are normal to the equatorial plane and positive northward.

3. Instrumentation

The UCLA ATS 1 magnetometer is a biaxial, closed-loop fluxgate magnetometer which has been described in some detail by Barry and Snare (1966). The instrument measures the component of the magnetic field parallel to the spin axis of the space-craft and a single component transverse to the spin axis. The direction of this component is fixed relative to the spacecraft and therefore rotating in inertial space. The parallel component is sampled at a rate of 3.12 sec^{-1} and the transverse component is sampled at a rate of 6.24 sec^{-1}. The phase of the transverse component is measured relative to the pulse from a sun sensor with a field of view that sweeps across the sun during each revolution of the vehicle.

4. The Quiet-Day Field

A description of the quiet-day field at the synchronous, equatorial orbit will serve as a background for our discussion of the magnetic activity in the region. The ATS 1 data have been used to establish the properties of the quiet-day field during the first six months of 1967 (Cummings *et al.*, 1969b). The daily sum of the geomagnetic activity index, K_p, was used to determine the five quietest days in a given month. The days used in the quiet-day study, together with their corresponding K_p values, are listed in Table II.

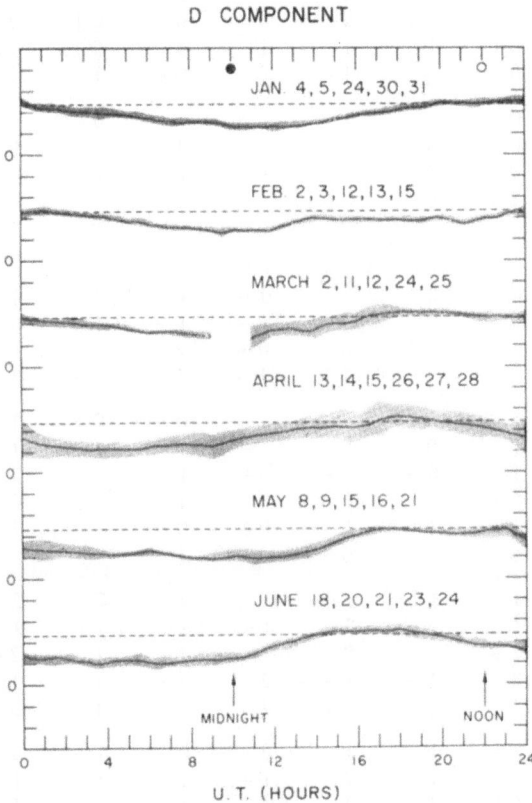

Fig. 2a. Mean values of D, H, and V for the quiet days of the indicated months. The shading indicates the envelope of one standard deviation. These results are derived from 6 min averages.

The ATS 1 record includes the five quietest days of every month except April. There are three gaps in Table II. A sudden commencement magnetic storm began on 2348 UT of February 15. Thus, the last 3-hour value of K_p for February 15 was $5-$, and the ATS 1 data for this interval have been omitted. The blank spaces for April 13 and 28 indicate intervals during which the ATS 1 magnetometer was turned off.

The data in Table II show that the quiet days of some months were much more magnetically active than the quiet days of others. For example, the values of $\sum K_p$ for May were more than twice the values for January.

TABLE II

Values of K_p and ΣK_p for the geomagnetically quiet days of the first six months of 1967

Month	Day	K_p Three-hour range indices								Sum
		1	2	3	4	5	6	7	8	
Jan.	4	0	0+	0+	0	1−	0+	1	1	4−
Jan.	5	1+	0+	0+	1−	1−	1−	0+	1−	5
Jan.	24	1	0+	0	0	0+	0+	0+	1+	4−
Jan.	30	0+	1−	1	1−	1−	0+	1−	1−	5
Jan.	31	0	0+	1−	0+	0+	1	1−	1−	4
Feb.	2	0	0	1−	1−	0+	0+	1+	1−	4
Feb.	3	0	0	0	2−	0+	0	1	1	4
Feb.	12	0+	0	0+	1−	1−	1−	0+	1+	4+
Feb.	13	1−	0	0+	1−	0+	2	2−	2−	7+
Feb.	15	1+	1	0+	0	0	0	1−		
March	2	0	1	1−	1+	2−	1+	1−	2−	8+
March	11	0	0	1	0+	1−	0+	0	0+	3−
March	12	0+	0	1−	1−	1−	1−	0+	0+	4−
March	24	1−	0+	0	0+	2+	1+	0+	0+	6−
March	25	0+	1	1+	1−	1+	1−	1	1−	7
April	13					0+	0+	1−	1+	
April	14	1−	0+	1+	1−	1−	1	1−	1−	6
April	15	1	0+	1−	0+	1	1	2−	1+	7+
April	26	0+	1	1−	1	1	0+	0+	0+	5
April	27	1	1−	1−	0+	0+	1−	1−	1	5+
April	28	0+	1	0+	0+	1−				
May	8	1	1	1	1	2	2	2	1+	11+
May	9	1	1+	2−	1+	1+	1+	1−	1	10−
May	15	3+	1+	0	1	1+	1	2+	2	12+
May	16	2	1+	1+	2−	1+	1	1+	2+	12+
May	21	1	1	0+	1	1−	1	3+	2+	11−
June	18	3−	1	1	0+	0+	1−	0+	1−	7
June	20	1−	1+	2	1−	1	1−	1−	1−	8−
June	21	0+	1	2	1	1	1	1	1+	9−
June	23	1−	1+	1	0+	0+	1	1	2	8−
June	24	1−	1−	1+	1−	1−	0+	1	2−	7

In this quiet-day study, 6 min averages of D, H, and V were used as the 'raw' data. For each month the magnetic fields for the five quiet days listed in Table II were averaged to yield an average quiet day for the month. The D components of these average quiet days are shown in Figure 2a as solid curves. The shaded regions in which the solid curves are centered represent the one standard deviation envelope, i.e., the width of the shaded region at any time point is equal to two standard deviations, $2\sigma_D$, where

$$\sigma_D = \sqrt{\sum_{i=1}^{n} \frac{(D_i - \langle D \rangle)^2}{n-1}}$$

and

$$\langle D \rangle = \frac{1}{n} \sum_{i=1}^{n} D_i.$$

Usually, $n=5$, but since some short gaps occur in the data, n is occasionally less than 5. Similar curves and envelopes are shown for H and V in Figures 2b and 2c, respectively.

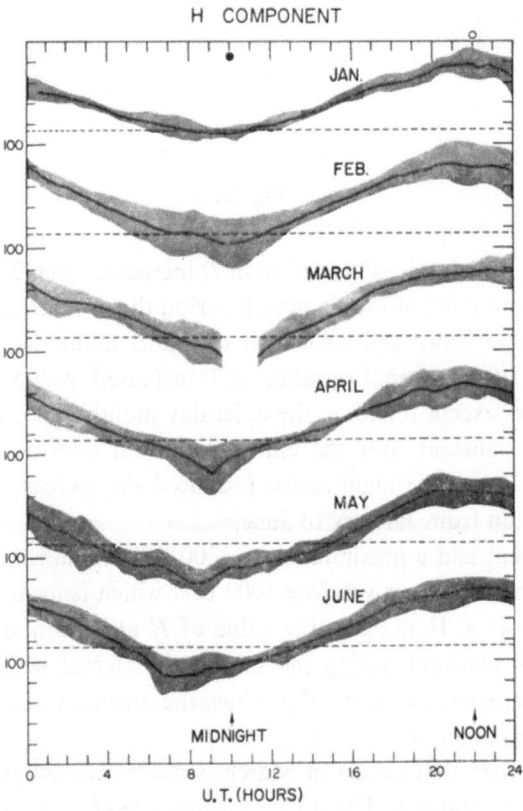

Fig. 2b.

V COMPONENT

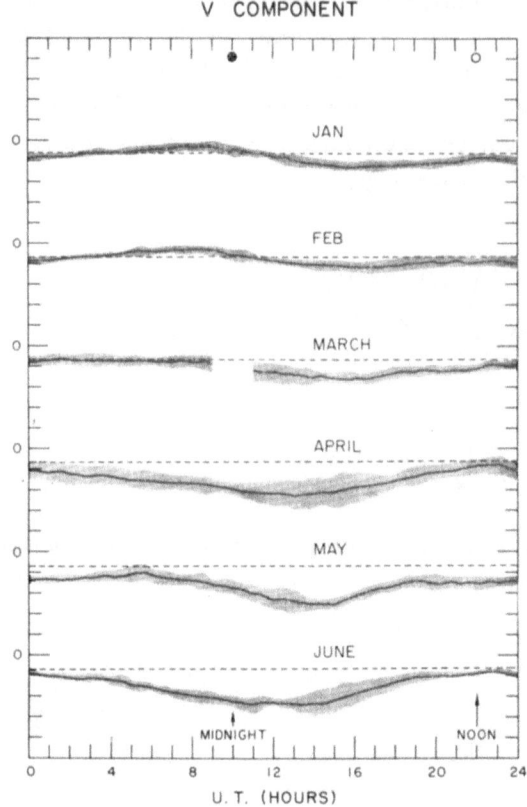

Fig. 2c.

The magnitude of the diurnal variation in D increased from 7 γ in January to 23 γ in June. Throughout most of the six month period the diurnal variation in D showed a peak shortly before dawn and a somewhat broader minimum shortly before dusk.

The magnitude of the diurnal variation in V increased from 5 γ in January to 20 γ in June. Note that, except for May, the quiet-day monthly mean value of V at local noon was nearly constant over the entire six-month interval. Thus, the seasonal change in the field over the night sector produced the increase in the amplitude of the diurnal variation from January to June.

The H component had a maximum near 2200 UT, which is local noon at ATS 1, and a minimum near or shortly before 1000 UT, which is local midnight. The total excursion was 35–45 γ. Here again the value of H at local noon on quiet days remained essentially constant during the six-month interval while the magnitude at local midnight decreased by about 10 γ. Thus, the diurnal variation in the quiet-day H increased by this amount.

The gap in the data that occurs in March is due to the eclipse of the sun by the earth near the spring equinox. The ATS 1 magnetometer data have been omitted for those approximately 1-hour eclipse intervals.

The earth's main field, as extrapolated from surface measurements (cf. Kahle, 1967), would appear as horizontal lines in Figures 2a, b, and c. This constant field has components $D_E = 23.5\,\gamma$, $V_E = -6.8\,\gamma$, and $H_E = 107.0\,\gamma$.

The cause of the January-to-June changes in the diurnal variation has not been definitely established. Olson and Cummings (1968) have found that a solar-wind flow inclined southward by 8° and constant over the year could produce such an effect. They assumed that the solar wind was specularly reflected at the magnetopause. Alternatively, since there was an increase in quiet-day activity of the interval studied, the changes may be associated with this increase.

5. The Storm-Time Disturbance Field

A. SUDDEN COMMENCEMENTS

Plots of ATS 1 magnetometer data taken during 3-day intervals covering nine geomagnetic storms of January–June 1967 are shown in Figures 3–10 (Coleman and Cummings, 1969). The ATS 1 data are 6 min averages. Also shown in each figure is a plot of 7.5 min samples of H at Honolulu (latitude 21°19′ N, longitude 158°00′ W). In some figures a similar plot of H at San Juan (latitude 18°07′ N, 66°09′ W) is also shown. The solid curve for each quantity is an average of the five QQ days of the appropriate month. No such curve is given for the H component at San Juan.

Eight of these storms exhibited sudden commencements. Field measurements from ATS 1 cover six of these commencements.

A comparison of the ATS 1 and Honolulu magnetograms during the sudden commencement of February 15, in Figure 6, indicates that the effects of a sudden commencement upon the distant geomagnetic cavity near local noon are just those expected for a simple compression. Specifically, the sudden commencement produces an increase in the field strength at ATS 1 and very little change in direction. Thus, there is an increase in H, since H is nearly parallel to the quiet-day field; a smaller increase in D since the quiet-day field has a small component in the positive D direction; and almost no change in V.

The sudden commencement of February 15 occurred when ATS 1 was only 1 hr 48 min past local noon. Similar behavior was exhibited by the field at ATS 1 during the sudden commencements of February 7 and June 25, Figures 5 and 10, even though the local times at ATS 1 were 0636 and 1621, respectively.

The other three sudden commencements covered by the ATS 1 data occurred when ATS 1 was in the night sector of its orbit. The data taken during these events are plotted in Figures 3, 4, and 10. The dates were January 7 and 13, and June 26.

In each event $|\delta V|/|V|$, the *relative* increase in the absolute value of V, was considerably greater than $|\delta H|/|H|$. The change in V was positive for the two SC's in January and negative for the one on June 26. This sign change is probably just the seasonal effect that one would expect under the assumption that ATS 1 was southward of the 'effective equatorial surface' of the cavity in the night sector during the (northern) winter, and northward of this surface during summer. The situation during the winter is sketched in Figure 11. Thus, one effect of these sudden commencements was to distort the lines of magnetic force into a configuration less dipolar and more like that of the geomagnetic tail.

From the behavior of the field at ATS 1 during the sudden commencement of January 7 we would argue that the tail current, as well as the surface current at the magnetopause, increased at the onset of the storm. If only the magnetopause current increased, one would expect an increase in the magnitude of the disturbance field from these currents. The *H* component of this disturbance field is positive and, during northern winter, the *V* component at ATS 1 should be positive. Thus, from the magnetopause current, one would expect a simultaneous compressional increase in *H* and *V* at ATS 1 during any sudden commencement that might occur when ATS 1 was near local midnight.

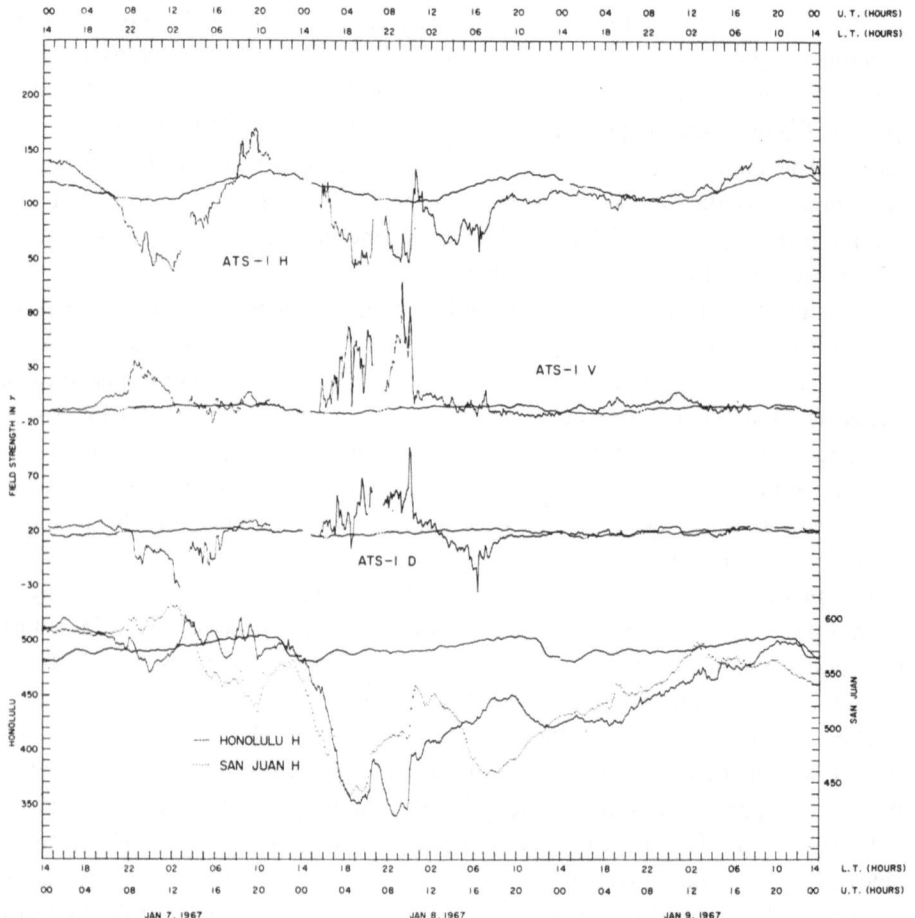

Fig. 3.   Magnetograms for the storm of January 7, 1967. The ATS 1 data are 6 min averages. Also shown are 7.5 min samples of *H* at Honolulu and San Juan. The solid curves are averages for the five QQ days of the month for *D*, *H*, and *V* at ATS 1 and *H* at Honolulu.

However, at the sudden commencement on January 7, there was no increase in H, although the record shows a decrease in the magnitude of the time derivative. Thus, the tail current apparently increased simultaneously and produced a decrease in H that compensated for the compressional increase.

The sudden commencement on January 13 was a more complicated event. A substorm expansion occurred simultaneously with the SC. Thus, it is likely that the increase in H at ATS 1 was produced in part by the SC compression and in part by the increase in H that accompanies most substorm expansions (cf. Cummings and Coleman, 1968a; Cummings et al., 1968).

The sudden commencement of June 26 produced increases in D and H and a decrease in V. The magnitude of the change in D was relatively rather small, however. Since ATS 1 was only 1 hour from the dawn meridian, one would expect the com-

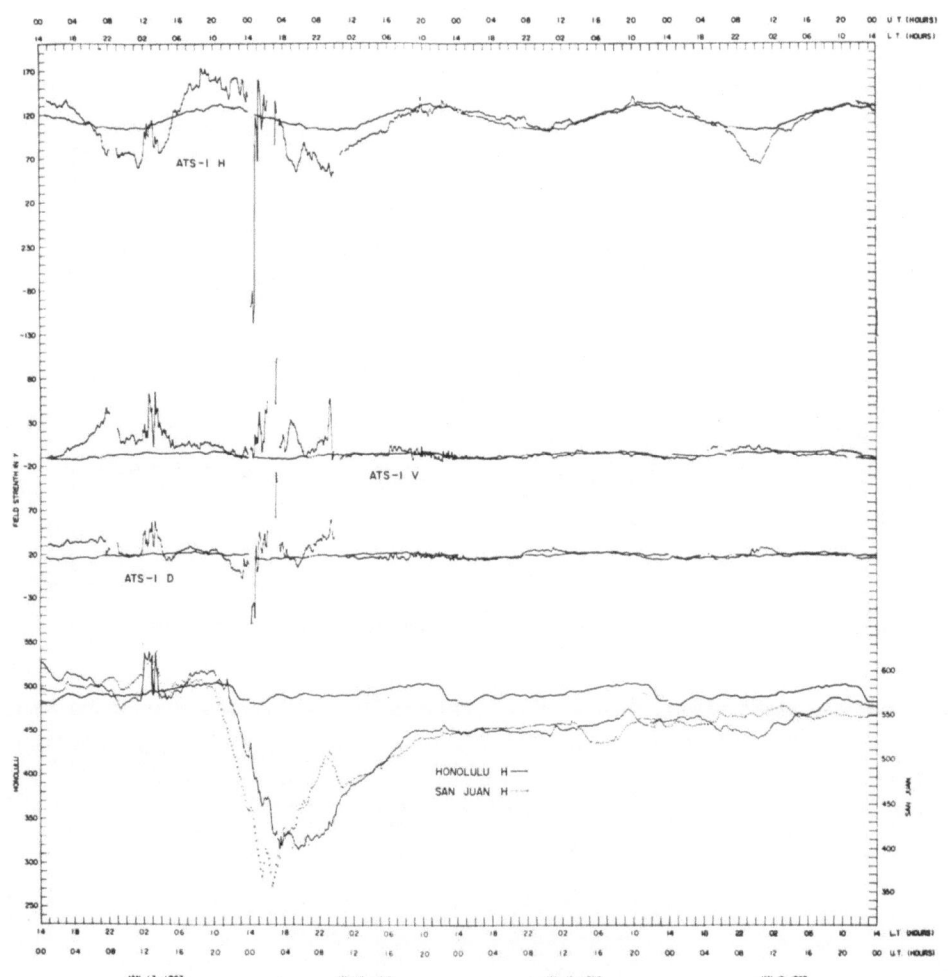

Fig. 4.   Magnetograms for the storm of January 14, 1967.

pressional effects of the SC to increase H and to increase D to a lesser extent, so long as ATS 1 was above the effective equatorial surface. This location is likely in the night sector during the summer.

However, the behavior of V cannot be explained in terms of compressional effects alone. The change in V could have been produced by an increase in a westward current in the dawn sector if the plane of symmetry of the current system was southward of ATS 1. In this case, the V component of the disturbance field at ATS 1 would be in the negative V direction. Since H increased at the SC, this interpretation requires either a compressional increase in H greater than any simultaneous decrease produced by the azimuthal current, or an azimuthal current far enough inside 6.6 R_E to have had a negligible or positive effect on H. The behavior could have been produced also by the neutral sheet current if the inner edge of the neutral sheet were 'U'-shaped and extended to the dawn meridian. Here again, the behavior of H would require a compressional increase greater than the decrease produced by the sheet

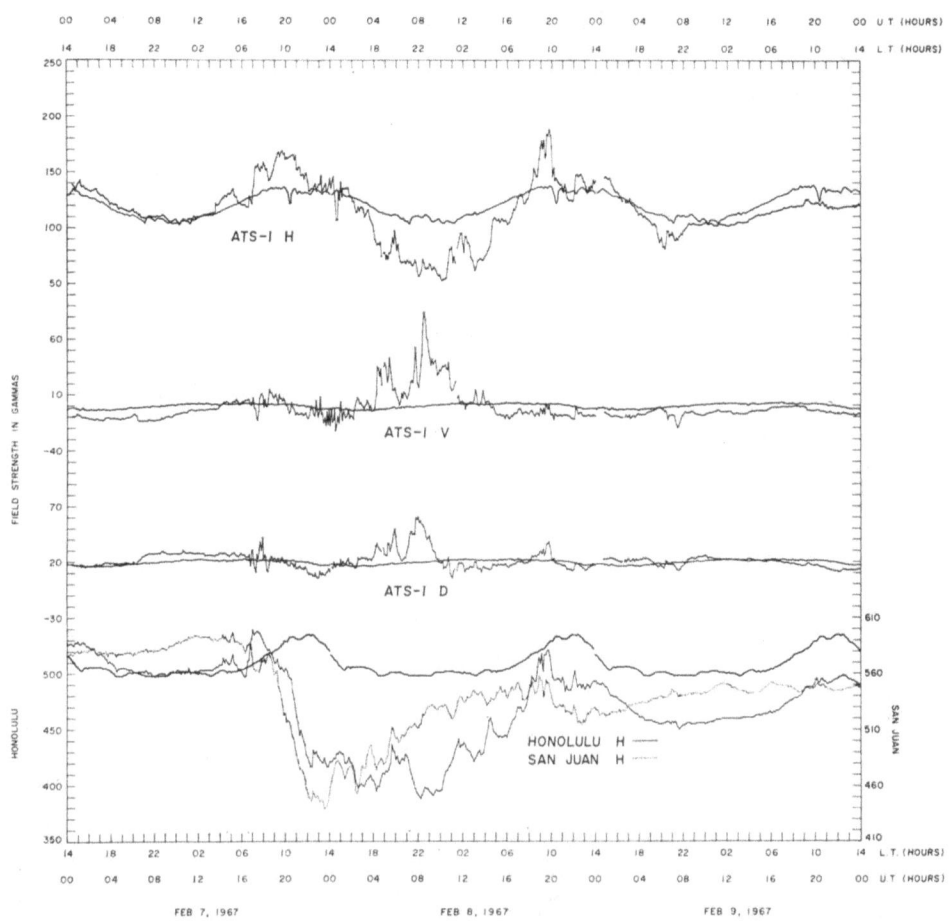

Fig. 5.   Magnetograms for the storm of February 7, 1967.

current. However, the effects of this SC may have been anomalous in a number of ways since it occurred during the recovery phase of the storm that had begun only about 36 hours before. The Honolulu magnetogram in Figure 10 indicates that there was an appreciable recovery-phase ring current flowing at the time of this SC. Thus, the effect of the SC on the V component of the disturbance field may have resulted from the effects of the SC on this ring current.

B. THE INITIAL PHASE

During the initial phases of these storms, the H component of the field at ATS 1 was usually greater on the day side and smaller on the night side than it was during quiet times. This behavior might have been anticipated in view of the changes that occurred at the sudden commencements.

Day sector increases are apparent in Figures 3, 4, 5, 6, 9, and 10. Night sector de-

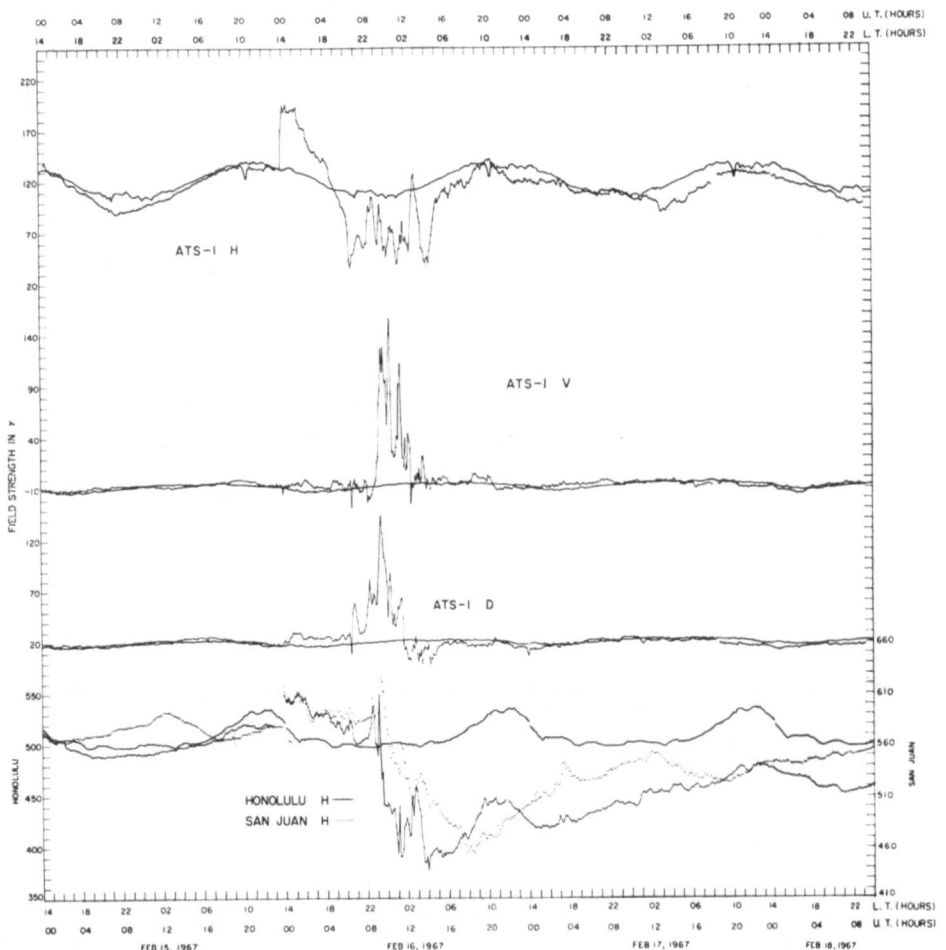

Fig. 6. Magnetograms for the storm of February 15, 1967.

creases are shown in Figures 3 and 4 near the beginning of the initial phases; in Figure 6 before the main phase decrease; in Figure 8 near 1000 UT (2400 LT) on May 2; and in Figure 10 during the entire night sector traversal following the SC of June 25. These results indicate that the geomagnetic cavity remains compressed and that the tail current remains enhanced during the initial phase of a storm.

However, the night sector decreases in H during the initial phases were often no greater than those that precede major substorm activity. Thus, our conclusion that the greater tail current indicated by the disturbance field during the initial phase is a storm-associated phenomenon should be considered somewhat tentative.

C. MAIN PHASE DECREASE

During several main phase decreases, the cavity remained compressed near local noon and distorted in a tail-like geometry near local midnight. The compression near local noon is apparent in Figures 4, 8, and 9. During the main phase decreases shown in Figures 4 and 9 the compression became so great at times that the mag-

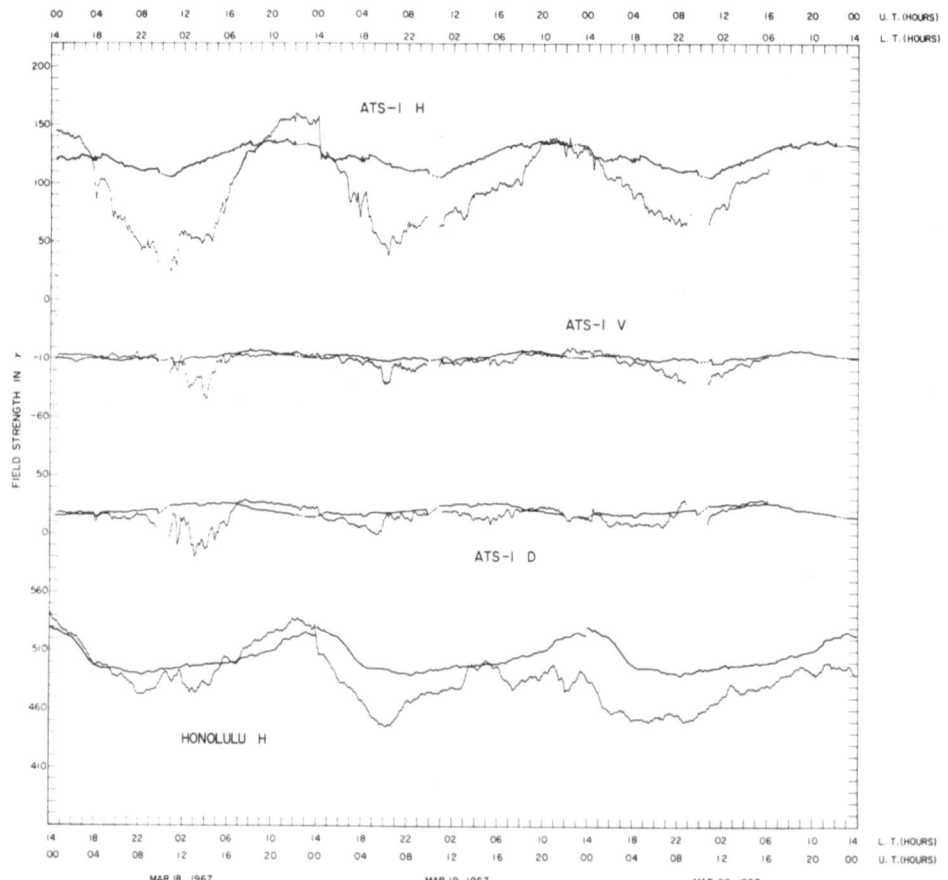

Fig. 7.   Magnetograms for the storm of March 18, 1967.

netopause near the subsolar point was inside the orbit of ATS 1. The boundary crossing during the first storm has been described in detail by Cummings and Coleman (1968b).

On the other hand, the main phase decrease in the storm that began on June 25, shown in Figure 10, was clearly accompanied by a decrease in H at local noon at ATS 1. Simultaneous decreases are also suggested in the data from the second storm in Figure 10 and from the storm of February 7, shown in Figure 5. It would appear therefore that the behavior near local noon is so dependent on the locations of the developing ring current and the magnetopause that the field near local noon at ATS 1 during main phase decrease can be either increased or decreased.

The evidence for tail-like distortion of the field at local midnight during a main phase decrease was obtained during the storm of February 15 shown in Figure 6. The main phase decrease occurred rather abruptly between 0900 and 1100 UT on February 16. An expanded section of the ATS 1 record for the interval is shown in Figure 12 along with data from the University of Minnesota electron spectrometer (courtesy of J. R. Winckler, T. W. Lezniak, and G. K. Parks) also on board ATS 1. It is evident from the behavior of the V component that the field was often in a

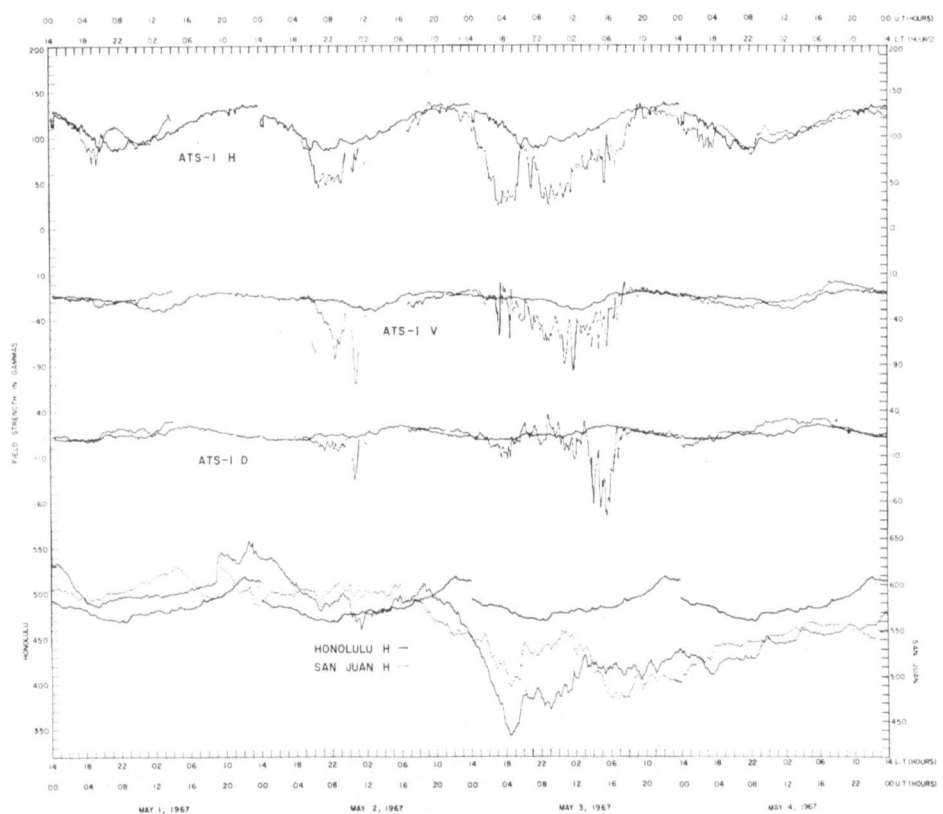

Fig. 8.   Magnetograms for the storm of May 1, 1967.

tail-like topology, i.e., there was a strong radial component during this interval. Decreases in the electron count rates accompanied these tail-like fields.

We would comment here that the abrupt changes of the field in the night sector between the quiet-day orientation and a more tail-like orientation seem to occur throughout the main phase of a storm. Since changes of this type are usually accompanied by substorms, at least when there is no storm in progress, this behavior in the field at ATS 1 is consistent with the statistical relationship between polar sub-

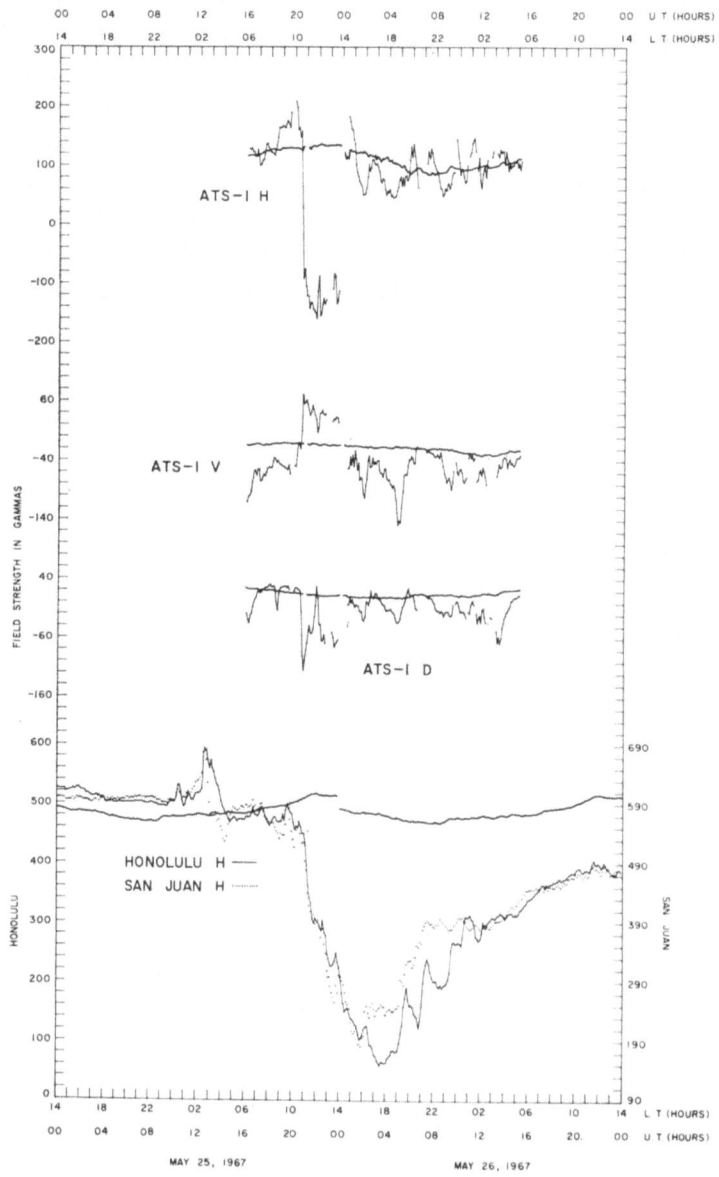

Fig. 9.   Magnetograms for the storm of May 25, 1967.

storm activity and ring current growth discussed by Davis and Parthasarathy (1967).

D. THE MAIN PHASE MINIMUM

During the storms that were of sufficient magnitude to enable one to identify main phase minima with some confidence, the minima occurred when ATS 1 was at least 5 hours from local noon. Thus, we are unable to determine the state of the field near local noon at the time of a strong minimum. The situation on the night side is apparent, however, from nearly all the storms, i.e., the field was usually in a more tail-like topology. However, as just mentioned, the field switched back and forth rather often during this phase of the storms.

E. THE RECOVERY PHASE

Figures 3, 4, and 6 show that the *H* component of the field near local noon at ATS 1 had decreased to its quiet day value or to smaller magnitudes by the time the *H* component at the surface had recovered about half the main phase decrease. Similarly, the tail current had returned to a quiet day state.

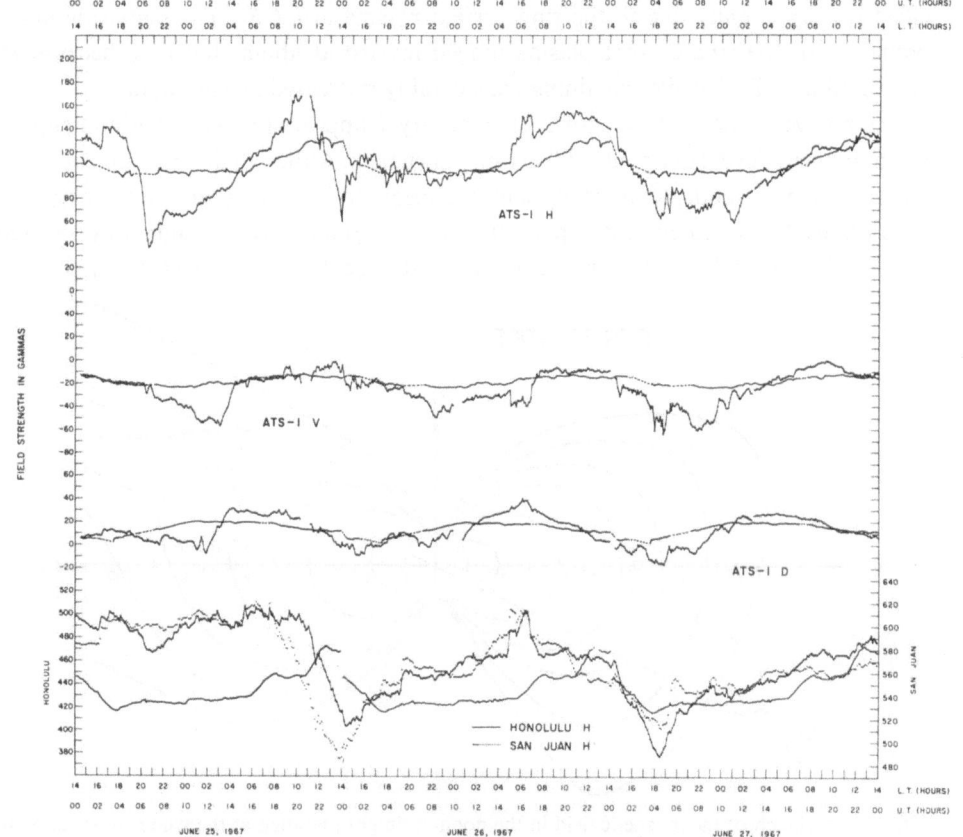

Fig. 10.   Magnetograms for the storm of June 25–26, 1967.

In Figure 3, the H component at ATS 1 was still decreased at 2200 UT (1200 LT) on January 8. This decreased H was accompanied by a slightly decreased V, and this relationship between the H and V components is just that expected for a ring current when the effective equatorial surface in which the current flows is southward of ATS 1. That there was still a strong azimuthal current at this point in the recovery phase is evident from the Honolulu and San Juan records.

In Figure 4 there was a recovery in H over the day sector at ATS 1 during January 15, but there was very little progress in the recovery at the surface. We interpret this behavior as evidence that the ring current plasma moved inward during this recovery.

Consider now the timing of the return of the tail current to a quiet-day state. In Figure 3, the behavior of the H and D components indicates that there was still an appreciable tail current at 1800 UT (0800 LT) on January 8. On January 9, the tail current had returned to its quiet-day state. However, the H components at Honolulu and San Juan were still considerably decreased.

In Figure 4, the behavior of the D component indicates that the tail current returned to normal at about 1100 UT (0100 LT) on January 14. At this time, H at Honolulu had recovered about half of the main phase decrease. However, the depression in H persisted at ATS 1 until about 0030 UT (1430 LT) on January 15. This effect was apparently due to the ring current plasma at $6.6 R_E$. Evidently, the inner boundary of the ring current plasma moved inward at about that time because H recovered at ATS 1 while remaining considerably decreased at Honolulu.

The recovery phase of the storm of February 7 appears to have been interrupted by an impulsive event at about 1900 UT on February 8. Although this phenomenon was not designated a sudden commencement, it appears to have been quite similar to one.

In Figure 6, the tail current apparently returned to its quiet-day state around 1600 UT (0600 LT) on February 16. Again, H continued to be depressed from time to

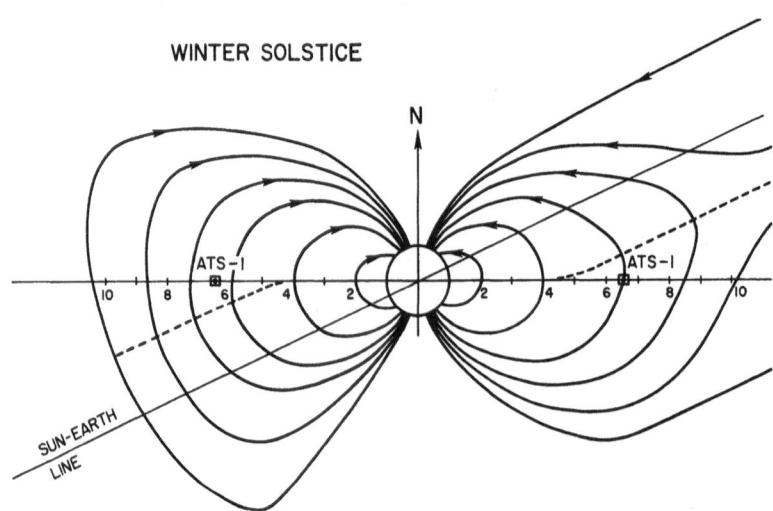

Fig. 11.   A sketch of the magnetic field in the noon-midnight meridian at the winter solstice. Note that on the night side ATS 1 is below the ecliptic plane and thus it is probably below the plane of the surface of symmetry for any ring current or tail sheet current.

time, indicating the effects of ring current plasma. In Figure 8 the behavior of the field during the prolonged recovery indicates that the tail current was back to normal by 1800 UT (0800 LT) on May 3, by which time H had recovered about half of its main phase decrease.

These observations suggest that the tail-current system returns to a quiet-day state at about the time of the beginning of the recovery phase. We have not yet been able to determine whether the external pressure on the cavity decreases at this same time. However, during several of the storms the cavity was still compressed near the end of the main phase decrease so that one might speculate that the recovery phase begins when the pressure decreases.

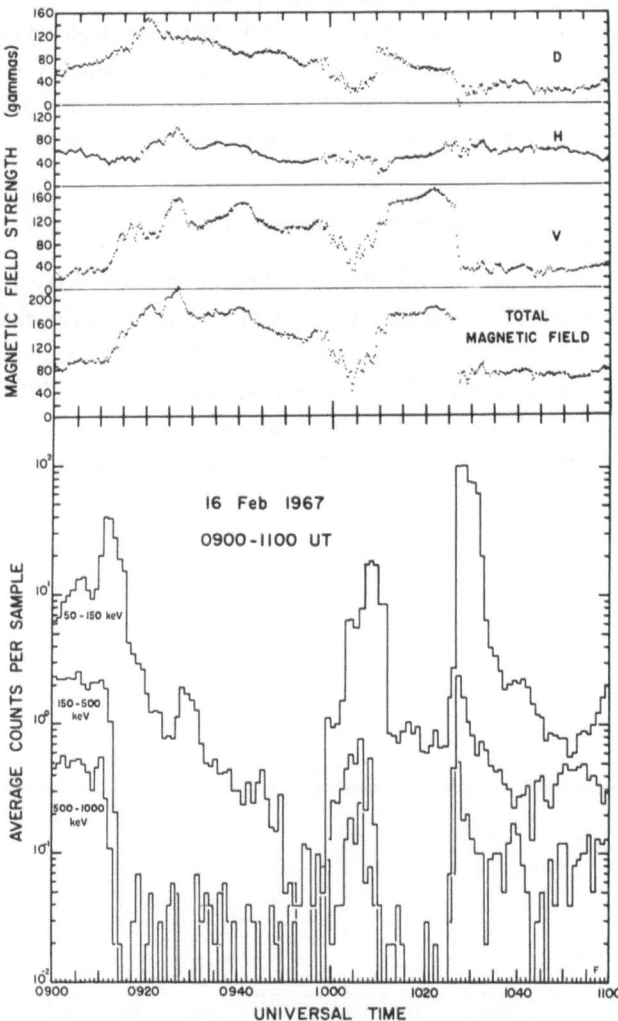

Fig. 12.   The field at ATS 1 during the main phase decrease of the storm of February 15, 1967. Also shown are the count rates from the University of Minnesota electron spectrometer on board ATS 1.

F. THE D COMPONENT

The observed storm-time behavior of the D component at ATS 1 merits some additional discussion. An appreciable D component was often present near midnight when there was an enhanced V component or decreased H component. The systematic appearance of this D perturbation was mentioned in the discussion of substorm related effects at ATS 1 (Cummings *et al.*, 1968). This component of the disturbance field was positive in the winter (Figures 3, 4, 6) and negative in the summer (Figures 8 and 10). We have argued that the H and V components of the disturbance field are produced at such times by the tail current. A severely skewed current system, at least near midnight at 6.6 R_e, would be required to account for the D component. Skewing of this type might be an effect of the earth's rotation.

A field-aligned current is another possibility. We have neglected such currents in our interpretation throughout. However, the foregoing difficulty with the interpretation of the behavior of the D component of the disturbance field suggests that field-aligned currents may be an important source of the disturbance field at ATS 1.

6. Storm-Associated Pulsations

Quasi-sinusoidal fluctuations of relatively large amplitudes have been recorded in the magnetic field at ATS 1 during several geomagnetic storms (Barfield and Coleman, 1969). Field fluctuations of a somewhat similar nature were observed during the geomagnetic storm of April 17, 1965, as reported by Cahill (1966), Brown *et al.* (1968), Davis and Williamson (1966), and others.

Figure 13 is a section of the ATS 1 magnetogram for a 1 hr interval during which

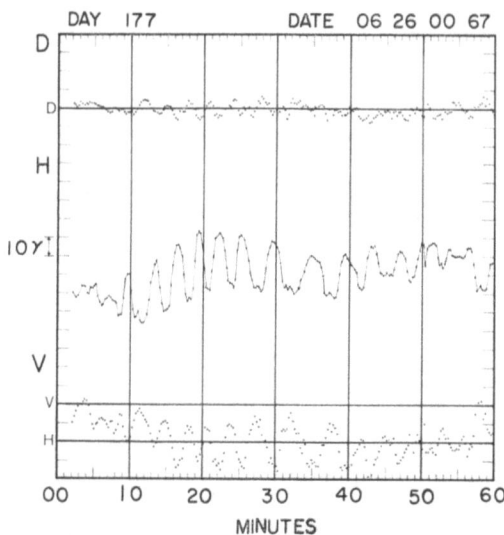

Fig. 13. ATS 1 15-sec average magnetogram for 0000–0001 UT, June 26, 1967, during the main phase of the first storm shown in Figure 10.

quasi-sinusoidal fluctuations were detected. The data cover the first hour (UT) of June 26, 1967, during the main phase of the geomagnetic storm that began at 0222 UT on June 25, 1967. Magnetograms from ATS 1, Honolulu, and San Juan for June 25–27 are shown in Figure 10. The field oscillations had a maximum amplitude of about 20 γ and an average period of about 4 min. For this particular event the duration of the oscillations was about four hours. The oscillations were principally in the H and V directions, i.e., in the meridional plane.

We have examined the ATS 1 records for six sudden commencement storms that occurred during the first six months of 1967. Obviously similar fluctuations were recorded during four of these storms. In each instance the oscillations were detected during or after the main phase minimum. The oscillation period varied from 2 to 15 min with 4 to 5 min most common. Thus, the periods were fairly well confined to the Pc 5 band. The maximum fluctuation amplitude was 20 γ, while the average was closer to 8 γ. The polarization of the fluctuations was nearly linear during each event. The direction of the polarization varied somewhat, but the fluctuations in H and V were 180° out of phase, while those in D and V were in phase. The fluctuation amplitude parallel to the average field was usually comparable to that in the transverse direction. The mean value of V was less than zero during each event. Thus, one might expect the H and V fluctuations to be in phase when $V > 0$.

An expanded subsection of the interval covered in Figure 13 is shown in Figure 14. The expanded record shows that the 3 to 4 min oscillations were accompanied by fluctuations with much higher frequencies.

These higher frequency oscillations had periods of about 5 sec and amplitudes of 1–2 γ. They were left-hand elliptically polarized, but the direction of the major axis of the polarization ellipse varied during the event. Further, they were amplitude-modulated with amplitude maxima at the minima and maxima of 3–4 min oscillations.

Also, during this event the intensities of energetic electrons and protons fluctuated coherently with the 3 to 4 min field oscillations. The records from the particle detectors on board ATS 1 showed that the fluctuations with energies greater than 50 keV were in phase with the 3 to 4 min oscillations in the Z, or H, component of the field (Parks et al., 1969). The fluctuations in the intensity of protons with energies between 0.6 and 1.0 MeV were out of phase with these H oscillations by 180° (Lanzerotti et al., 1969). Also, the electron spectrum was observed to harden as the field oscillations progressed (Parks et al., 1969).

The record from the electron detector has been examined for intensity fluctuations with periods near 5 sec, the period of the higher frequency field fluctuations. No such fluctuations have been found (Parks et al., 1969).

The average flux of 0.6–1.0 MeV protons during this event was considerably greater than that for comparatively quiet times. The enhanced intensity was apparently due to solar flare protons that had begun to arrive in the vicinity of the earth on June 25 and had almost direct access to the region of the geomagnetic cavity at 6.6 R_E (Lanzerotti, 1968).

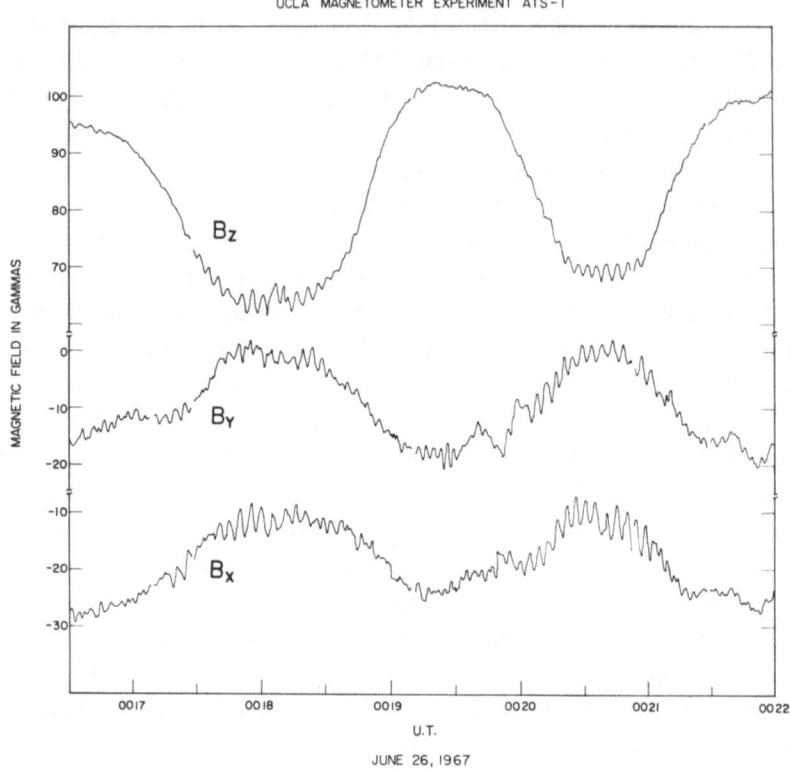

Fig. 14.   ATS 1 1-sec average magnetogram for the part of the event shown in Figure 13.

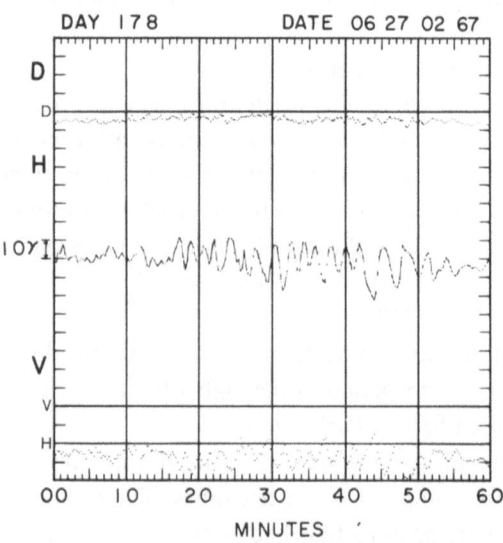

Fig. 15.   ATS 1 15-sec average magnetogram for 0200–0300 UT, June 27, 1967, during the main phase decrease of the second storm shown in Figure 10.

Fluctuations recorded during the main phase of the second storm in Figure 10 are shown in Figure 15. In this case, the periods are about 2 min. The pulsations were recorded from 0200–0330 UT on June 27. Otherwise, the behavior of the field is quite similar to that recorded during the first storm. The 5 sec field fluctuations again accompanied these longer period oscillations and the electron and proton intensities varied coherently with them in the manner just described. Further, the solar flare protons were still in evidence.

During these two events, ATS 1 was in the afternoon sector. Further, the H component of the average field was severely depressed below its quiet day value. Thus, the particle pressure was probably unusually high throughout both events.

In order to determine whether or not the occurrence of the shorter period oscillations depends upon the presence of the longer period fluctuations, we studied a period when the intensity of solar flare protons at ATS 1 was high, and the longer period oscillations were not present. During this period, the shorter period oscillations were observed. Thus, it appears that although both types of oscillations were observed during periods of intense solar flare protons, it is not necessary that they occur simultaneously.

Our preliminary studies of these storm-associated oscillations may be summarized as follows:

(1) During four of the six geomagnetic storms studied to date, the relatively large amplitude, quasi-sinusoidal oscillations have been detected following the main phase minimum.

(2) During these two events in which the H component of the field was unusually weak and the intensity of solar flare protons was unusually strong at ATS 1, the higher frequency fluctuations occurred simultaneously.

At this point, we can only speculate on the sources of these two types of oscillations. Lanzerotti *et al.* (1969) have shown that the phase characteristics of the proton flux oscillations imply that the drift mirror instability of Hasegawa (1969) existed during the storm event of June 25–27, 1967. The phase relationship between the electrons and the magnetic field, as reported by Parks *et al.* (1969), the progressive hardening of the electron spectrum during the magnetic field oscillations, and the phase characteristics of the longer period magnetic field oscillations support this indication.

The shorter period magnetic field oscillations fall within the frequency range of the ion cyclotron resonance. However, the signal is rather complicated, and further analysis is required before we can make a definitive classification of this oscillation.

Pulsations of a much more irregular character have also been recorded during the main phase minimum of a storm. Figures 16 and 17 are 15 sec average ATS 1 magnetograms for two separate 1 hour intervals during the minimum in the main phase of the geomagnetic storm that began at 1636 UT on February 7, 1967 (Figure 5). Irregular field fluctuations were recorded at ATS 1 almost continuously from 2200 UT on February 7 to 0300 UT on February 8.

The spacecraft covered the local-time sector from 1200 to 1600 during this interval and the components of the mean field were $\langle D \rangle \simeq 15\,\gamma$, $\langle H \rangle \simeq 125\,\gamma$, and $\langle V \rangle \simeq -5\,\gamma$.

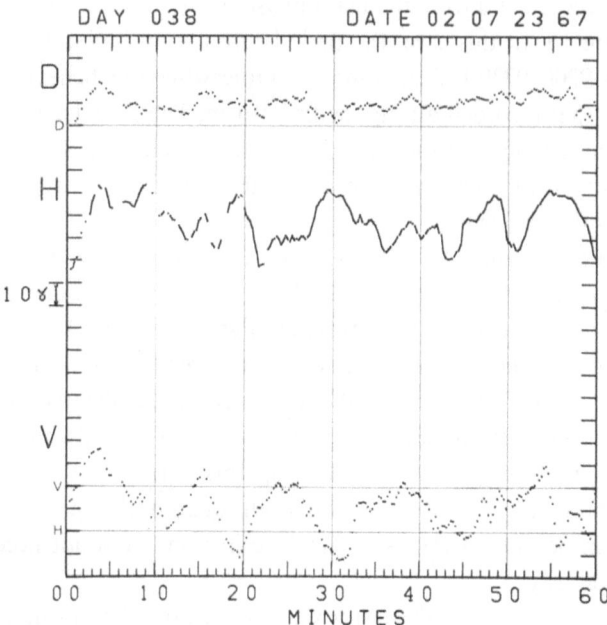

Fig. 16.   Field pulsations recorded at ATS 1 during the main phase of the storm shown in Figure 5.

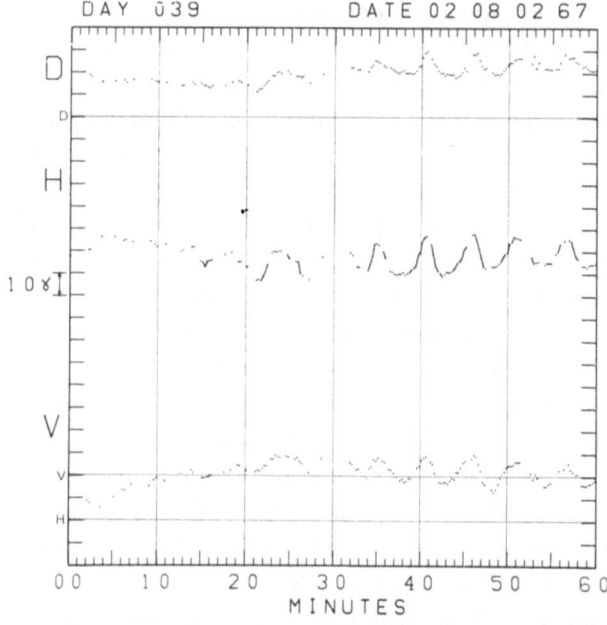

Fig. 17.   A later section of the record of the pulsation event of Figure 16.

The first of the two figures, Figure 16, shows the rather complicated fluctuations recorded early in the interval, 2300–2400 UT. We have not yet analyzed these variations in any detail. However, it is apparent that the dominant period was 10–15 min. The fluctuation amplitudes in H and V were about 15 γ while that in D was more like 5 γ. The phase relationships between the vector components of the fluctuating field are also somewhat complicated. However, it is clear that during some intervals the fluctuation in H may be described as a rectified version of a somewhat implied fluctuation in V. This relationship is particularly apparent between 2310 and 2330. The H and V fluctuations were in phase when V is greater than about $-10\,\gamma$ and they were 180° out of phase when V is less than this value.

The magnetogram for an hour near the end of the event is shown in Figure 17. This shows a more regular oscillation that appeared from time to time during the event. For this phenomenon the fluctuations in D, H, and V were all in phase. The maximum amplitudes were 6, 8, and 7 γ, respectively. The period was 5–6 min.

These pulsations may well be of the same type as those shown in Figures 13 and 14. The essential difference is the polarization. During the previously described events, the H and V fluctuations were 180° out of phase. In this case they were in phase, but this behavior may have resulted from the fact that $\langle V \rangle > 0$. Note that the oscillations in H are obviously distorted in a similar fashion in Figures 13 and 17. In both cases the maxima were sharper than the minima. This behavior may indicate a tendency toward rectification in the H fluctuation. In any case, our present opinion is that the fluctuations shown in Figure 17 are simply the smaller amplitude, shorter period, more regular versions of those shown in Figure 16. The fluctuations plotted in Figures 13 and 15 may have been of this type also.

Pulsation events have been found in the records obtained with the ATS 1 magnetometer during the recovery phases of two geomagnetic storms. Here again we have not as yet performed a systematic search of the records for similar events.

Field fluctuations recorded at ATS 1 during the recovery phase of the sudden commencement storm that began at 1203 UT on January 13, 1967 (Figure 4), are shown in Figures 18 and 19. These pulsations were recorded almost continuously from 1430 UT on January 14 until the first hour of January 15. In this interval the spacecraft moved through the local-time sector from 0430 to 1400.

The H component at Honolulu had recovered about half of the main phase decrease by the time these pulsations began. At ATS 1 the mean field changed gradually from $\langle D \rangle = 17\,\gamma$, $\langle H \rangle = 85\,\gamma$, $\langle V \rangle = -7\,\gamma$ to $\langle D \rangle = 15\,\gamma$, $\langle H \rangle = 120\,\gamma$, $\langle V \rangle = -10\,\gamma$ while the pulsations were being recorded. Over this same interval the H component at Honolulu recovered to about one quarter of the maximum decrease, where it remained for the next 24 hours.

There were two obvious components in the recorded fluctuations. The low frequency component varied in period from 5 to 40 min. The maximum amplitude in the V direction was about 8 γ. The fluctuations in the DV plane were right-hand elliptically polarized.

Perhaps the most striking property of the low frequency fluctuations is the fact

PAUL J. COLEMAN, JR.

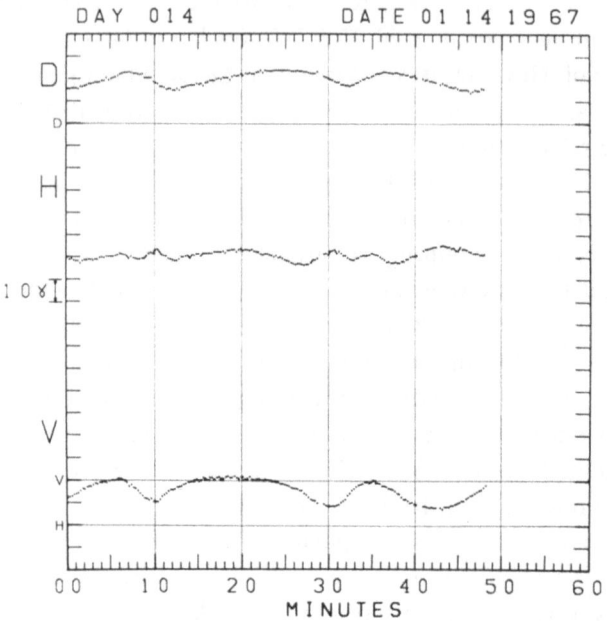

Fig. 18.   Field pulsations recorded at ATS 1 during the recovery of the storm shown in Figure 4.

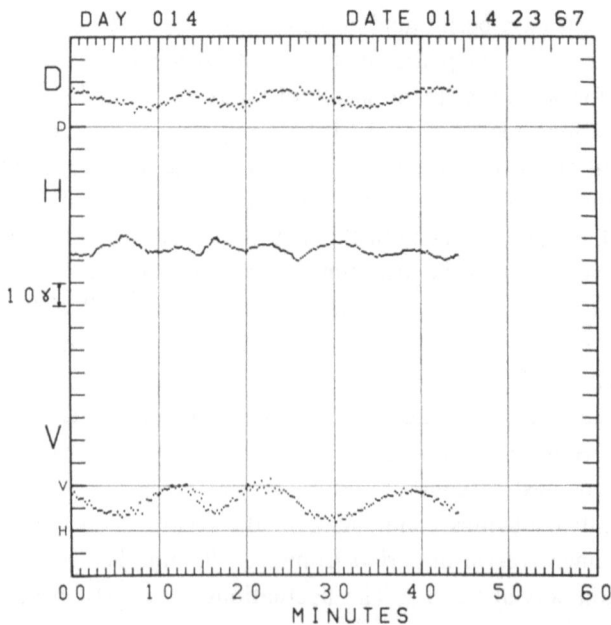

Fig. 19.   A later section of the record of the pulsation event of Figure 18.

that the fluctuation in H was simply a rectified version of the fluctuation in V. In Figure 18 the H and V fluctuations were in phase when V was greater than $-5\,\gamma$ and 180° out of phase when V was less than $-5\,\gamma$. In Figure 19, which covers the interval 2300–2400 UT or 4 hours later than the interval 1900–2000 UT of Figure 18, the phase of the V and H fluctuations reverses at $V \simeq -10\,\gamma$.

Superimposed on the long period pulsations was a short period fluctuation. The period was roughly constant at about 50 sec. These oscillations were primarily transverse to the mean field and they reached a maximum amplitude of $2\,\gamma$ in V. They were linearly polarized in the DV plane and the D and V fluctuations were in phase.

A more complicated event occurred during the recovery of the sudden commencement storm that began at 0800 UT on January 7, 1967 (Figure 3). Pulsations were recorded almost continuously from 1830 on January 8 to 0400 UT on January 9. During this interval the spacecraft covered the local-time sector from 0830 to 1800.

As in the later storm just discussed, the pulsations began at about the half-way points in the recovery. At ATS 1 the components of the mean field varied slowly over the following ranges: $\langle D \rangle$, 10 to 20 γ; $\langle H \rangle$, 100 to 115 γ; and $\langle V \rangle$, -10 to $-15\,\gamma$.

The ATS 1 15-sec average magnetogram in Figure 20 covers a 1-hour interval early in the event. The pulsations recorded during this interval are similar in many respects to the low frequency pulsations recorded during the storm of January 13. Thus, the fluctuations in the DV plane were right-hand elliptically polarized. The

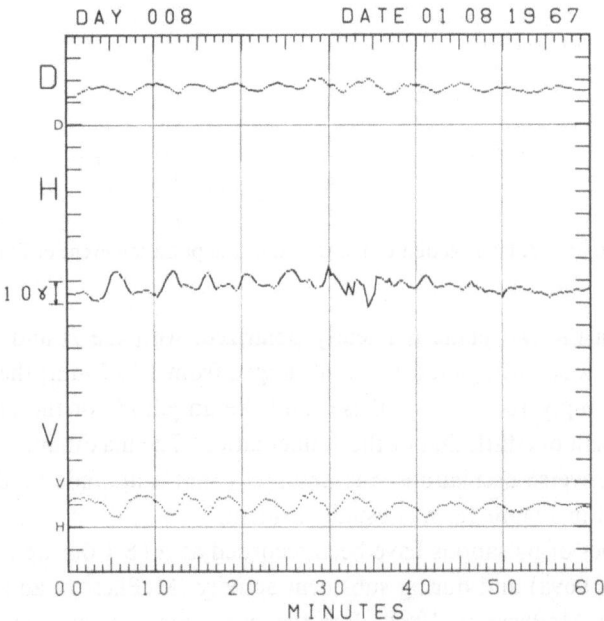

Fig. 20.   Field pulsations recorded at ATS 1 during the recovery of the storm shown in Figure 3.

maximum amplitude of the fluctuations in V was about 6 γ and the fluctuation in H was a rectified version of that in V. In this instance the phase of H reversed at $V \simeq -8\,\gamma$. As before, the H and V fluctuations were in phase when V was greater than the value at the phase reversals. These fluctuations exhibited periods of 4–7 min, which is the short period end of the spectrum of those recorded during the recovery of the January 13 storm.

Figure 21 is a similar magnetogram for an hour about half way in the interval under discussion. This shows a shorter period fluctuation of a type that appeared at various times throughout the interval. We have tentatively concluded that these oscillations are basically different from the longer period pulsations that were dominant earlier in the interval. This conclusion is based on the following points: The

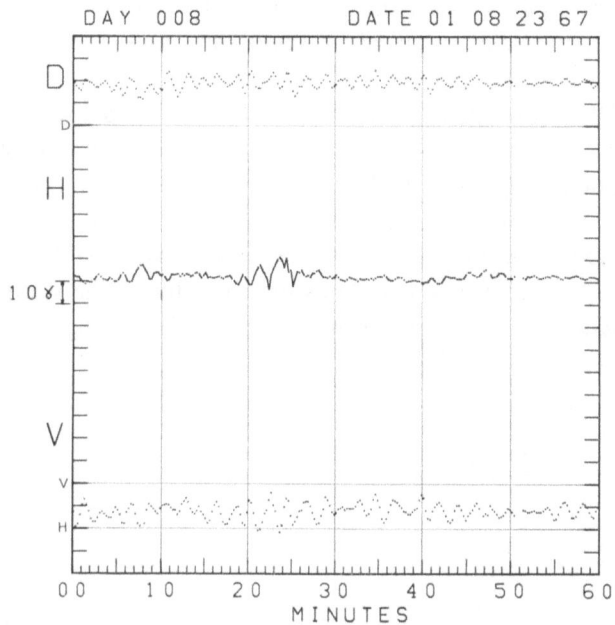

Fig. 21.   A later section of the record of the pulsation event of Figure 20.

component in the DV plane is linearly polarized, with the D and V oscillations in phase; the periods throughout the event ranged from 1 to 2 min; the H and V fluctuations were simply 180° out of phase; and the amplitude of the H fluctuation was usually less than one-fifth that of the V fluctuation. The maximum amplitude of these primarily transverse oscillations was about 6 γ, but amplitude modulation, such as that exhibited in Figure 18, was usually evident.

Several types of pulsations have been recorded at ATS 1 during quiet times (Cummings et al., 1969a) and during substorm activity (McPherron and Coleman, 1969; Coleman and McPherron, 1969). For the most part, these phenomena differ significantly from those recorded during storms.

7. Summary

From our studies of magnetic storms at ATS 1, we have concluded that the effects of a sudden commencement usually include an increase in the tail current along with the increase in the compression of the geomagnetic cavity, i.e., the increase in the surface currents at the magnetopause. The tail current increase may be delayed somewhat relative to the SC. The increased compression persists and the tail current remains enhanced during the initial phase. During the main phase decrease the increased compression is still present at least part of the time, although we have not yet determined whether the increased compression becomes intermittent during this phase of a storm. The tail current intermittently recovers (toward its quiet-day state) and increases as the main phase decrease proceeds. These recoveries, as indicated by the behavior of the disturbance field in the night sector of the ATS 1 orbit, are similar in most respects to those associated with substorms. At the same time the disturbance field at ATS 1 exhibits changes consistent with the growth of a ring current. During the storms that occurred near the solstices, the effects of the ring current were apparent in both H and V. At the minimum of the main phase the tail current is still intermittently enhanced. Thus, throughout the main phase, the tail current is evidently quite variable. However, the separation of the ring current and tail current effects at ATS 1 is an uncertain process. This difficulty leads us to suggest that the neutral sheet may move inward and partially or completely encircle the earth during the early part of the main phase and that the ring current and neutral sheet plasmas are contiguous, at least intermittently, during the main phase.

By the time the recovery starts, or shortly thereafter, the compression and the tail currents have returned to their respective quiet-day states. As the recovery proceeds, the ring current continues to flow, according to the field measurements at the surface of the earth, and, if our assumption of a connection between ring current and neutral sheet plasmas is correct, the connection must be broken early in the recovery. Further, the recovery of the field at ATS 1 from the ring current effects proceeds much faster than the recovery of the field at the surface. Thus, the ring current probably moves inward in the course of the recovery.

We have established the existence of several different storm-time pulsation phenomena at the synchronous, equatorial orbit. In the Pc 1–2 band are the 5 sec oscillations that were recorded during the main phases of several storms. In the Pc 4 band are the transverse oscillations, with periods from 50 to 120 sec, which were recorded during the recovery phases of two storms.

In the Pc 5 band are the fluctuations recorded during the main phase decrease of one storm and the main phase minimum of another and accompanied on both occasions by oscillations in the Pc 2 band. Also, in the Pc 5 band and beyond, toward longer periods, are the storm-associated pulsations that are right-hand elliptically polarized in the DV plane and exhibit a fluctuation in H that appears to be a rectified version of the fluctuation in V. These pulsations have been detected during the recovery phases of two storms. (In each case they were accompanied by or alternated with the previously mentioned transverse oscillations in the Pc 4 band.) It is also possible that fluctuations of this type were recorded during the main phase minimum of another storm. The periods ranged from 4–40 min. This last phenomenon may be related to the irregular pulsations in the magnitude of the total field at the surface that have been detected in the band from 10–110 min (Herron, 1967).

Acknowledgements

This work was supported in part by the National Aeronautics and Space Administration under Research Grant NGR 05-007-004.

References

Barfield, J. N. and Coleman, Jr., P. J.: 1969, 'Storm-related wave phenomena observed at the synchronous, equatorial orbit', submitted to *J. Geophys. Res.*

Barry, J. D. and Snare, R. C.: 1966, 'A fluxgate magnetometer for the Applications Technology Satellite', *IEEE Trans. Nuc. Sci.* **NS-13**, 326.

Brown, W. L., Cahill, L. J., Davis, L. R., McIlwain, C. E., and Roberts, C. S.: 1968, 'Acceleration of trapped particles during a magnetic storm on April 18, 1965', *J. Geophys. Res.* **73**, 153.

Cahill, L. J., Jr.: 1966, 'Inflation of the inner magnetosphere during a magnetic storm', *J. Geophys. Res.* **71**, 4505.

Coleman, P. J., Jr. and Cummings, W. D.: 1969, 'The storm-time disturbance field at ATS 1', Publication No. 771, Institute of Geophysics and Planetary Physics, University of California, Los Angeles.

Coleman, P. J., Jr. and McPherron, R. L.: 1969, 'Fluctuations in the distant geomagnetic field during substorms: ATS 1', Publication No. 768, Institute of Geophysics and Planetary Physics, University of California, Los Angeles.

Cummings, W. D. and Coleman, Jr., P. J.: 1968a, 'Simultaneous magnetic field variations at the earth's surface and at synchronous, equatorial distance, 1, Bay-associated events', *Radio Science* **3**, 758.

Cummings, W. D. and Coleman, Jr., P. J.: 1968b, 'Magnetic fields in the magnetopause and vicinity at synchronous altitude', *J. Geophys. Res.* **73**, 5699.

Cummings, W. D., Barfield, J. N., and Coleman, Jr., P. J.: 1968, 'Magnetospheric substorms observed at the synchronous orbit', *J. Geophys. Res.* **73**, 6687.

Cummings, W. D., O'Sullivan, R. J., and Coleman, Jr., P. J.: 1969a, 'Standing Alfvén waves in the magnetosphere', *J. Geophys. Res.* **74**, 778.

Cummings, W. D., Coleman, Jr., P. J., and Siscoe, G. L.: 1969b, 'The quiet-day magnetic field at ATS 1', Preprint, Publication No. 752, Institute of Geophysics and Planetary Physics, University of California, Los Angeles.

Davis, T. N. and Parthasarathy, R.: 1967, 'The relationship between polar magnetic DP and the growth of the geomagnetic ring current', *J. Geophys. Res.* **72**, 5825.

Davis, L. R. and Williamson, J. M.: 1966, 'Outer zone protons', in *Radiation Trapped in the Earth's Magnetic Field* (ed. by B. M. McCormac), D. Reidel Publ. Co., Dordrecht, Holland.

Hasegawa, A.: 1969, 'Drift mirror instability in the magnetosphere', Bell Telephone Laboratories, Murray Hill, N.J.

Herron, T. J.: 1967, 'An average geomagnetic power spectrum for the period range 4.5 to 12900 sec', *J. Geophys. Res.* **72**, 759.

Kahle, A. B.: 1967, 'Harmonic analysis for a spheroidal earth', Rep. No. P-3684, Rand Corp., Santa Monica, Calif.

Lanzerotti, L. J.: 1968, 'Penetration of solar protons and alphas to the geomagnetic equator', *Phys. Rev. Letters* **21**, 929.

Lanzerotti, L. J., Hasegawa, A., and MacLenna, C. G.: 1969, 'Drift mirror instability in the magnetosphere: Particle and field oscillations and electron heating', *J. Geophys. Res.* in press.

McPherron, R. L. and Coleman, Jr., P. J.: 1969, 'Magnetic fluctuations during magnetospheric substorms, 1, Expansion phase', Preprint, Publication No. 772, Institute of Geophysics and Planetary
  Physics, University of California, Los Angeles.
Olson, W. P. and Cummings, W. D.: 1968, 'A hinged model of the current system in the magnetospheric tail', Abstract, *Trans. Am. Geophys. U.* **49**, 742.
Parks, G. K., Winckler, J. R., Cummings, W. D., Barfield, J. N., and Lanzerotti, L. J.: 1969, 'Periodic modulations observed in energetic electrons and the magnetic field at the ATS 1 during a
  magnetospheric substorm', Abstract, *Trans. Am. Geophys. U.* **50**, 284.

Discussion

Parks: I would like to make a brief comment for the ones who are interested in correlations between particle fluxes and magnetic field variations.

We have looked into a particular event and find that the 4 to 5 min variations in the magnetic field tie up with variations in particle flux with the electrons in phase and the protons out of phase. However, we do not find any correlation of 50-KeV electrons with fast field oscillations.

MAGNETOSPHERIC ELECTRIC FIELDS

H. VÖLK and G. HAERENDEL

Max-Planck-Institut für Physik und Astrophysik, Institut für extraterrestrische Physik,
Garching bei München, Germany

1. Introduction

High-latitude, large-scale electric fields, their implication for magnetospheric convection, their relation to ionospheric currents and their ultimate connection with
magnetic perturbations observed at ground level have been the subject of extensive
discussion during the last decade.

It is difficult to measure these electric fields, because they are quasi-DC fields of the
order of a few millivolts/m to generally some tens of mV/m. For this reason indirect
evidence is used extensively. Indeed, the first indications of the magnetospheric
convection pattern and the associated electric fields were derived by Axford and Hines
(1961) through relating the magnetic perturbations on the ground to overhead Hall
currents.

In the last few years, however, a number of more or less direct field measurements
have been done, with various techniques and in various regions of the ionosphere and
the magnetosphere. The aim of this article is a discussion of recent observational
results in their connection with the magnetosphere. Thus we concentrate on experiments done in high latitudes and we shall omit altogether the very important observations connected with ionospheric dynamo fields, in particular in the equatorial
region.

In Section 2 we shall briefly discuss the problem of mapping ionospheric fields into
the magnetosphere. The following Section 3 first of all refers to some recent electrostatic probe measurements with sounding rockets. Here, also interesting field phenomena connected with auroral displays have been observed. Some such probes have
been flown on satellites and show the onset of an intense irregularity structure poleward from the auroral zones. Also the plasmapause seems to have been detected.
With respect to the latter extensive ground-based Whistler studies have been reported
in particular changes in the plasmapause position presumably due to the convective
action of electric fields.

Coming back to satellite observations, low energy ion flux anisotropies have been
detected using retarding potential analyzers.

The section concludes with a review of a balloon experiment with electrostatic
probes in the auroral zone. Such an experiment is capable at least potentially of
giving a synoptical view of the local electric field distribution for a whole day.

Measurements with ion clouds are discussed in Section 4 where a synoptical view,
now from the results of 17 barium experiments in the auroral ionosphere, will be
presented.

Recently also a first ion cloud experiment in the distant magnetosphere has been
performed from the ESRO-satellite HEOS I. It will be described in the last section.

V. Manno and D. E. Page (eds.), Intercorrelated Satellite Observations Related to Solar Events, 280–296. All Rights Reserved
Copyright © 1970 by D. Reidel Publishing Company, Dordrecht-Holland

2. Mapping of Ionospheric Electric Fields into the Magnetosphere

Many of the measurements have been performed in the ionosphere and thus the problem of mapping the fields into the magnetosphere arises. Usually this problem is solved by considering the magnetic field lines as equipotential lines of the electrostatic field and by assuming some static magnetic field model. These assumptions imply an infinite conductivity (frozen in magnetic field) and in addition a negligible distortion of the magnetic field due to the convective motion of the plasma or – even more stringent – a steady state convection. Under these conditions the plasma motions are transverse to **B** along electric equipotential surfaces; the perpendicular drift velocity is then given by:

$$\mathbf{v}_\perp = c/B^2 [\mathbf{E}, \mathbf{B}]$$

which relates $\mathbf{v}_\perp$ to the transverse component $\mathbf{E}_\perp$ of the electric field.

This relation is assumed to hold everywhere above the ionospheric F-layer. It also forms the basis of electric field measurements through the observation of ionospheric drifts for instance by artificially injected ion clouds.

In order to insure the condition of infinite conductivity the spatial scales of the electric field in the ionosphere have to exceed several kilometers, because the finite Pederson conductivity in the lower ionosphere tends to short out small scale electric fields. Another problem is the possible occurrence of turbulent resistivity in the presence of strong field-aligned currents as for instance predicted by Swift (1965). Temporal changes of the convective motions are considered as being quasistatic in the above scheme. Thus the timescales involved must be long compared to the time of propagation of an Alfvén wave through the magnetosphere between conjugate points, i.e. several minutes.

Finally, the use of a particular static **B**-field model involves in general a difficult decision. This is true in particular, if we try to fit observations performed under conditions of different magnetic agitation and season into a single synoptical picture. We will come back to this point in Section 4.

In the following sections we shall deal with the experimental results almost exclusively. The theoretical implications and their relation to the various convection models (Axford and Hines, 1961; Taylor and Hones, 1965; Nishida, 1966; and Brice, 1967) are discussed extensively in the excellent review article by Axford (1969).

3. Recent Field Measurements Other than With Ion Clouds

3.1 SOUNDING ROCKET EXPERIMENTS IN THE IONOSPHERE WITH ELECTROSTATIC PROBES

The measurement of potential differences with electrostatic probes employing cylindrically or spherically ending booms is certainly the most direct approach to the electric field measurement. One can adapt it to rockets, satellites and balloons at all local times. There are difficulties involved, however, because the potential difference between the probe and the plasma is much greater than the voltage to be measured.

Thus the booms have to be as long as possible which is hard from an engineering point of view and the payload must be very symmetric. Also the [**v**, **B**]-electric field seen by the satellite or rocket is generally much larger than the field to be measured and must be subtracted. Differences in work function of the two spheres are a major problem, although it can be partly eliminated by coating the two spheres with carbon thus producing surfaces with work function variations less than a few millivolts (Fahleson *et al.*, 1968).

Several rocket flights have been performed in the auroral ionosphere with double probes. Transverse electric fields as large as 65 mV/m were reported with directions generally in agreement with the motion of auroral forms (Mozer and Bruston, 1967; Mozer and Fahleson, 1968). These authors also reported a surprisingly large field

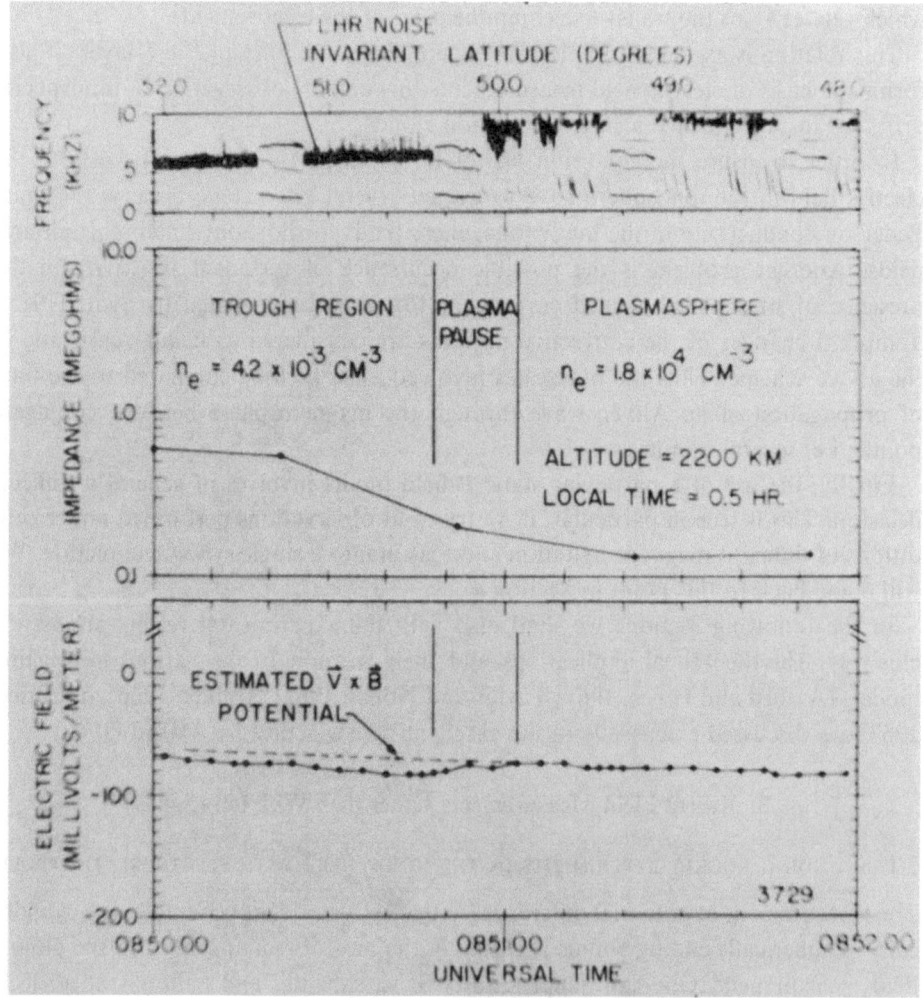

Fig. 1.   (After Gurnett and Cauffman, 1969.)

component parallel to **B** up to 20 mV/m and saw no strong variation of the mean transverse fields in crossing an auroral arc. This latter result is in contrast to the GSFC rocket experiments, summarized by Aggson (1969), which show a strong anti-correlation of the measured transverse electric field with auroral intensities. The field inside an arc decreases by a factor of 2 to 4 below its value outside. It is, however, not yet clear, whether this interesting ionospheric field structure on a small spatial scale has an equivalent effect upon the convection in the magnetosphere (see Section 2). Other probe experiments on rockets have been flown by Bernstein *et al.* (1969).

3.2 SATELLITE OBSERVATIONS AND WHISTLER MEASUREMENTS

On satellites, electrostatic probe measurements have been performed too. The polar orbiting satellites OVl-10 (Maynard and Heppner, 1969; Heppner, 1969) and Injun V (Gurnett and Cauffman, 1969) found an electric irregularity boundary distinct in latitude and time on every polar crossing. The low frequency (e.g. 3–30 Hz) irregu-larities are interpreted (Heppner, 1969) as a consequence of the satellite moving at a

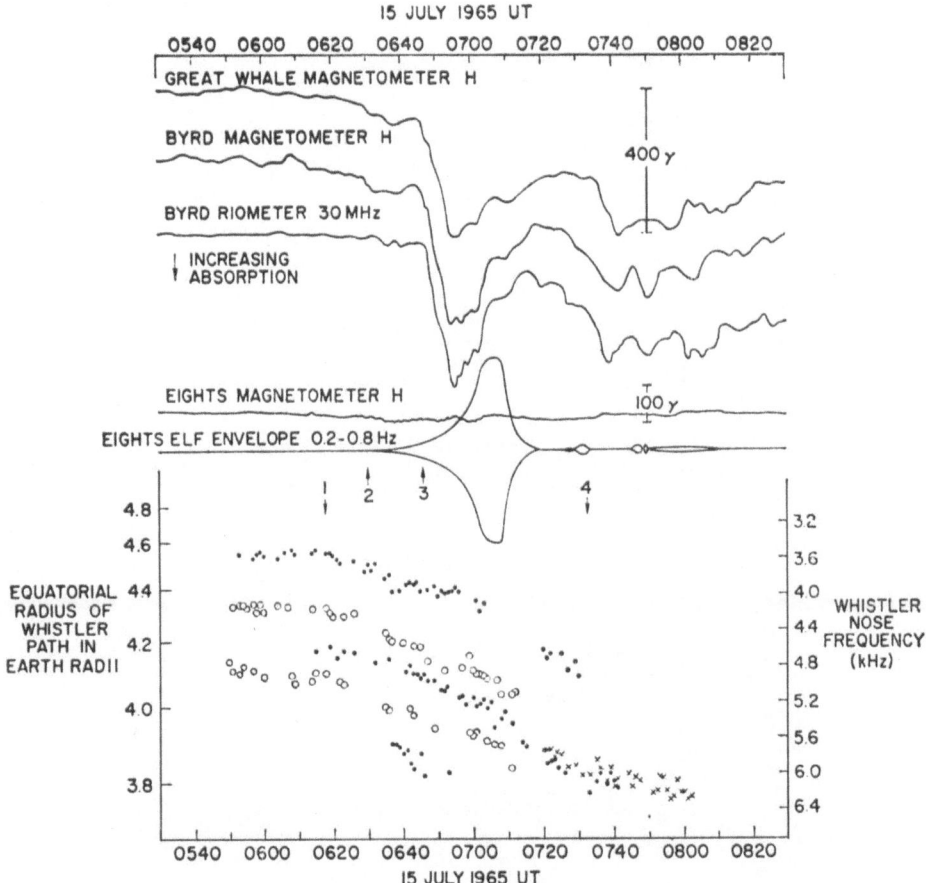

Fig. 2.   (After Carpenter and Stone, 1967.)

high speed across the spatial irregularity structure. The southern boundary of this structure resembles the form of the auroral belt at slightly lower latitudes and is interpreted as the boundary between weak and strong DC electric fields. Recent measurements on the OGO 6 satellite seem to confirm this picture (Maynard and Heppner, 1969). It is also in accord with the results from Injun V, which yield irregularity peak amplitudes typically of order 100 mV/m and large-scale disturbances near the auroral zone, of amplitudes around 200 mV/m.

On the latter satellite (Gurnett and Cauffman, 1969) also the behaviour of the electric field at the plasmapause transition was examined (Figure 1). The boundary crossing as indicated by the 'lower hybrid resonance (LHR) breakup' effect (Carpenter et al., 1968), the electron density change measured on board and the characteristic sheath resistance increase, was accompanied by a smooth transition from a corotation field inside (only [V, B]-field, V measured relative to the rotating earth) to a roughly 15 mV/m convection field outside. The convection velocity is then of the order of 0.4 km/sec, a result typical also for other investigated crossings.

With respect to the plasmapause (Carpenter, 1962, 1963; Gringauz, 1963) impor-

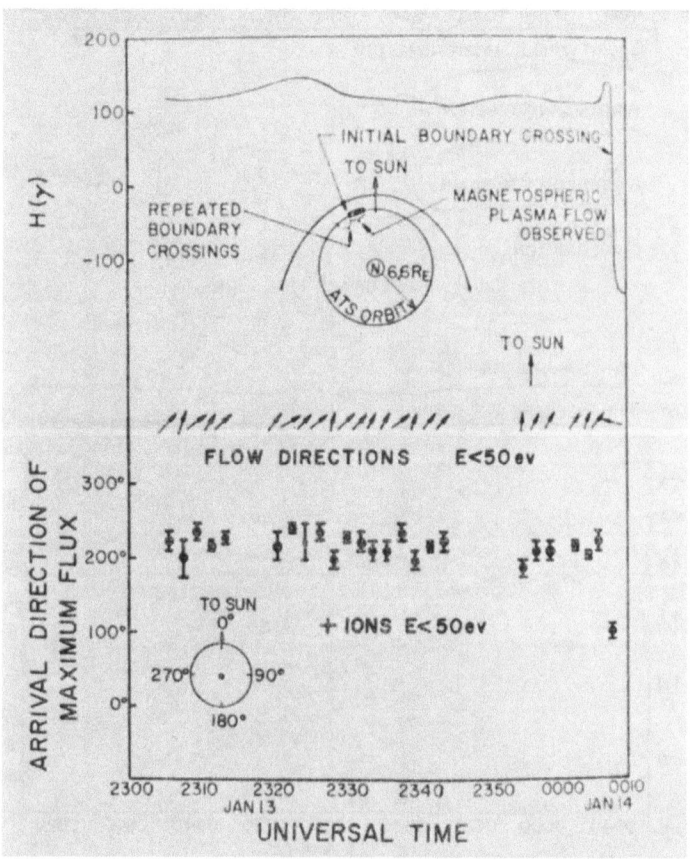

Fig. 3.   (After Freeman et al., 1968.)

tant results about the radial motions of this boundary and the associated electric fields have been obtained by tracing the radial motions of whistler ducts inside the plasmasphere using ground-based receivers (Carpenter and Stone, 1967, 1968; Carpenter, 1968). Figure 2 from Carpenter and Stone (1967) shows such an event during a polar substorm. The plasmapause starts to contract about half an hour before the onset of the magnetic perturbations (H-component) at the $L = 7$ stations Great Whale and Byrd. The inward speed of the whistler ducts is about 700 m/sec or 0.4 R_E/hour corresponding to a westward electric field $E \sim 0.3$ mV/m or 2 kV/R_E.

Returning to satellite observations, anisotropies of the low energy ion flux have been measured at synchronous altitude on ATS 1 using retarding potential analyzers (Freeman, 1968, 1969; Freeman et al., 1968). During several magnetic storms strong directional ion fluxes were observed in the 0–50 eV energy range.

In one particular event, on Jan. 13–14, 1967, the magnetopause became pushed inwards beyond 6.6 R_E and the satellite was repeatedly in the magnetosheath in the noon to dusk quadrant. As long as the satellite was within the magnetosphere a roughly sunward flow was observed (Figure 3) with a velocity of about 27 km/sec, corresponding to an electric field of about 5 mV/m directed roughly from dawn to dusk. The number of α-particles assumed to be observed is rather high and their energies too low (~ 10 eV) for being solar wind particles. Thus this event may represent rather extreme conditions (Axford, 1969).

Quite near the magnetopause, but still inside, there seemed to exist a return flow (Figure 3), represented diagrammatically in Figure 4, similar to that postulated originally by Axford and Hines (1961), on the basis of viscous interaction between the solar wind and the magnetospheric plasma. This last result would indicate that the magnetopause is an equipotential surface, at least under such severe conditions.

Other events and also statistical results from magnetically quiet days tend to support the pattern of roughly sunward flow in the noon to dusk quadrant (Freeman, 1969). No anisotropy observations were reported at other local times.

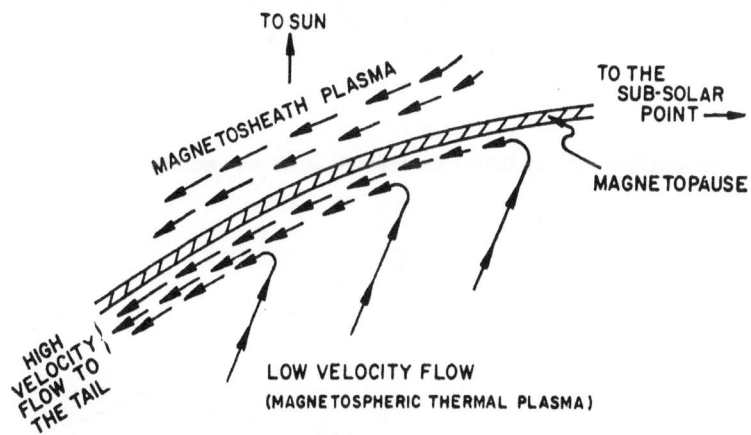

Fig. 4.   (After Freeman et al., 1968.)

3.3 BALLOON OBSERVATIONS

We shall conclude this section with a note on a recent balloon experiment with electrostatic probes (Mozer and Serlin, 1969). Originally the method was suggested by Kellog and Weed (1969). In such an experiment the [V, B] electric field is very small. Among non-satellite experiments it has in addition the unique feature of being able to cover a 24 hour period in one single flight. Therefore, a synoptical view of the electric field pattern can potentially be obtained. At balloon heights, there is, however, the problem of disentangling weather produced electric fields from ionospheric fields, which leads to some uncertainty in the interpretation of the results.

Two balloons were launched at Ft. Churchill, Canada, on August 5 and 6, 1968. Figure 5 contains a plot of half hour averages of the electric field vectors, projected into the equatorial plane using the magnetic field model of Fairfield, 1968. The field when there are bays somewhere in the world are labelled B. Distances are given in units of earth radii.

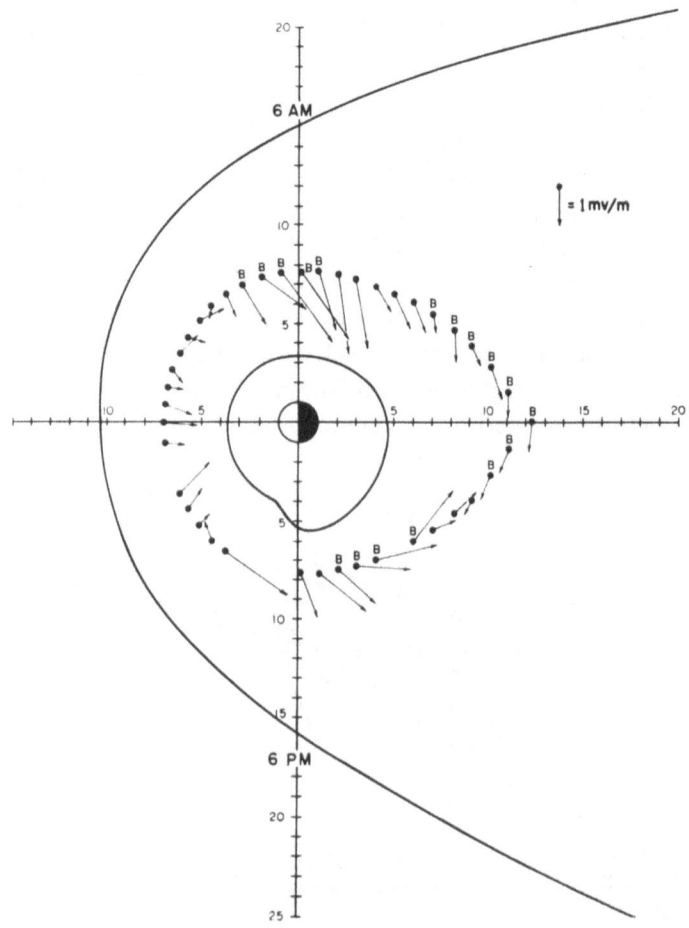

Fig. 5.   (After Mozer and Serlin, 1969.)

The general conclusions drawn by Mozer and Serlin (1969) are: the convection velocity is radially inward near local midnight, has about equal radial and azimuthal components at 6.00, is azimuthal with about the corotation velocity at noon and is directed outwards in the afternoon and evening hours.

There was a negative bay with $B_x^{max} \sim -200\,\gamma$ at Ft. Churchill, starting at about 2300 hours local time. The electric fields in the equatorial plane are between 0.6 and 1 mV/m. The results are grossly consistent with the ATS 1 and barium cloud data.

4. A Synopsis from Ion Cloud Experiments in the Auroral Ionosphere

The ion cloud technique provides artificial irregularities, whose motion is used to measure the electric field (Föppl et al., 1968; Haerendel and Lüst, 1968; Haerendel et al., 1969; Wescott et al., 1969a, b; Haerendel and Lüst, 1970). It has been shown (Haerendel et al., 1967) that this drift is practically the [**E, B**]-drift of a single ion provided that the perturbation of the integrated Pedersen conductivity is small. Thus one has to use small amounts of barium at altitudes $\geqslant 200$ km. The technique is applicable at all latitudes but limited by the conditions of fair weather, twilight time and short observation times. Still the set of observations is the largest yet available.

Here we will give a brief synopsis from 17 ion cloud experiments in the auroral zone from the point of view of magnetospheric convection and electric fields. Looking at these data, however, one should always keep in mind the problems and extrapolations inherent in the field mapping procedure mentioned in Section 2 above.

Table I gives a list of the 17 experiments, performed by the Garching group and

TABLE I

List of barium ion cloud experiments

No.	Date	Launching range	Time of cloud appearance (UT)	Altitude (km)	K_p
22.1	Aug. 17, 66	Ft. Churchill	3:44:42	258 ⎱	1
22.2	Aug. 17, 66	Ft. Churchill	3:46:26	376–230 ⎰	
26	Apr. 7, 67	Kiruna	20:01	231	2+
27	Apr. 8, 67	Kiruna	20:07	237	1+
28	Apr. 9, 67	Kiruna	01:04	243	3
29	Apr. 10, 67	Kiruna	20:17:59	230	2+
30	Apr. 11, 67	Kiruna	20:25	237	1+
W1.1	Aug. 31, 67	Andenes	22:08	200–300	2+
W2.1	Sept. 2, 67	Andenes	20:33	200–300	2−
W3.1	Sept. 12, 67	Andenes	20:59	200–300 ⎱	2−
W3.3	Sept. 12, 67	Andenes	21:02	200–300 ⎰	
33	Oct. 23, 67	Kiruna	16:08:52	217	1−
34	March 20, 68	Kiruna	18:26:52	215	2
35	March 23, 68	Kiruna	18:38:53	200–210	3
42	March 15, 69	Kiruna	17:58:48	211	3
43	March 17, 69	Kiruna	18:13:51	255	4+
45	April 17, 69	Andenes	21:55:29	250	3+

W = Exp. of Wescott et al. (1969a).

GSFC. Experiments with a W in front are from Wescott *et al.* (1969a). Obviously a wide variety of geomagnetic conditions was met, K_p ranging from 1_- to 4_+. The clouds happened to be inside as well as outside auroral activity. Experiment No. 30 was inside the plasmapause.

Figure 6 shows the X-component of the magnetic field H at Kiruna and Andenes for the Garching experiments. The corresponding magnetograms of the GSFC experiments are given in the article by Wescott *et al.* (1969a). During the Churchill experiments (No. 22) the field was very quiet ($H \sim 20$–$30\ \gamma$).

One of the prominent features is the close correlation of positive perturbations of H with westward drifts in a frame rotating with the earth and negative perturbations with eastward drifts (Table II). This correlation is also known for auroral forms (Harang and Tröim, 1961) and is consistent with magnetic perturbations from overhead Hall currents.

X-component in Kiruna

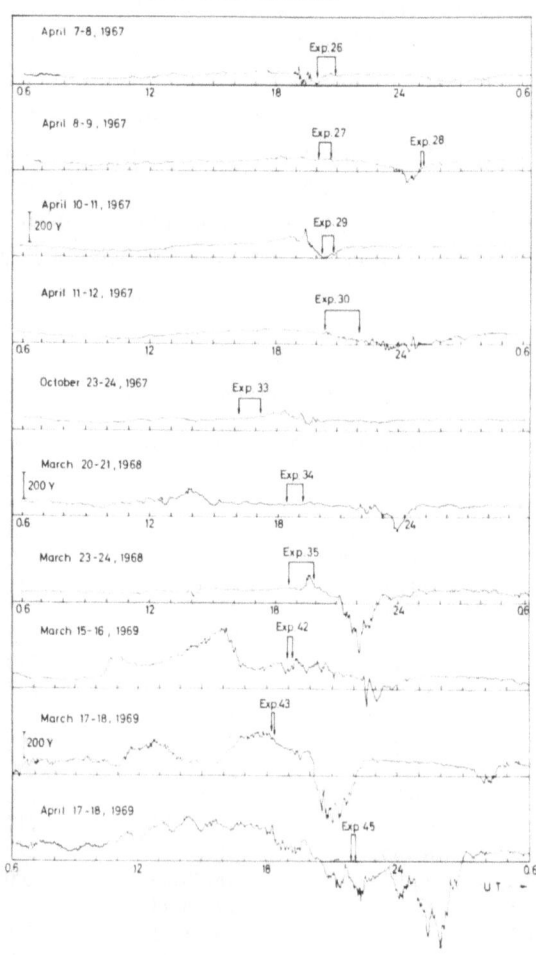

Fig. 6.

TABLE II

Magnetic perturbation	Drift in rotating frame towards		Drift in non-rotating frame towards	
H	West	East	Dusk	Dawn
positive	10	2	8	5
negative	1	6	0	7

In one of the exceptional cases (W3) the electric field was very irregular as judged from the observation of three simultaneous clouds. In the other case (No. 34) the drifts were slow and ΔX small.

In the one case (No. 30) where a negative bay was formed, although the drift was directed westward, the cloud was just in the plasmasphere according to OGO 2 measurements (Carpenter, private communication). The motion of the cloud was almost corotational. The magnetic perturbations at Kiruna ($H = -33$, $Z = -28$), Abisko ($H = -47$, $Z = -25$) and Tromsö ($H = -101$, $Z = +7$) being located each in turn further northward from the cloud's path show that slightly north of the cloud a stronger eastward drift occurred which caused the negative magnetic perturbation. Thus an electric field discontinuity seems to exist at the plasmapause in consistence with the Injun V measurements, referred to in the last section.

By subtracting the corotation, i.e. transforming to a stationary observer, some of the clouds reverse their direction (Table II). Presenting the data in this way a very pronounced correlation appears between velocity components directed towards dawn (eastward) with southward components and between those directed towards dusk (westward) with northward components (Table III, Figure 7). In Experiment 33 no

TABLE III

Drift in non-rotating frame towards	Meridional component of drift		Local magnetic time before	after
	N	S	2300 hours	
dusk	7	0	5	3
dawn	0	11	2	9

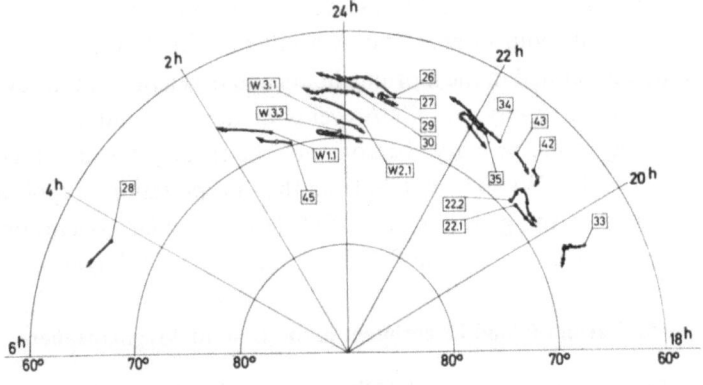

Fig. 7.

clear eastward component, in Experiment 22.1 no clear north-south component could be assigned. Figure 7 demonstrates furthermore that motions toward the east occur mostly after 2300 MLT and westward motions predominantly before 2300 MLT. A similar result was derived by Heppner (1969) from magnetic perturbations.

If we convert to corresponding L-values we get an impression of how the convection might look in the magnetosphere (Figure 8). For this projection the simple dipole model was used for the geomagnetic field. This, of course becomes questionable beyond $L=7$, especially in the tail. However, Figure 8 comprises experiments from rather different magnetic conditions, where the static **B**-field can be significantly distorted. Thus we have refrained from using a better field approximation.

Although everything happened in entirely different events it is tempting to understand the main characteristics of the motions of the clouds, given above, as applicable to an average substorm: Positive and negative phases seem to be clearly distinguished by nearly opposite drift directions; the plasmasphere seems to be a sharp boundary in the convection pattern.

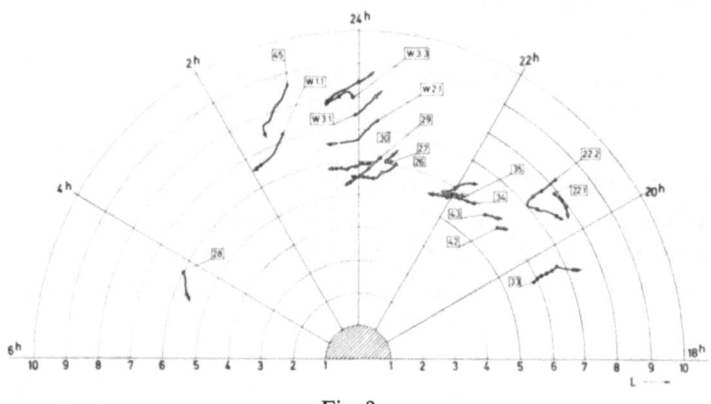

Fig. 8.

These results roughly agree with the corresponding ones described in Section 3, as far as a comparison is possibile. Still, however, the data are certainly not yet sufficient to derive definite conclusions about the overall topology of auroral fields and currents.

The observed antisymmetry in the east-west components of the electric field might be related to the behaviour of energetic electrons of a few keV to a few tens of keV. They would be accelerated in the morning hours and decelerated in the evening hours. The predominance of X-ray and CNA-events in the post-midnight hours and the general hardening of the primary electrons may be regarded as a consequence. It is interesting to note that Winckler (1970) from the interpretation of typical changes in the electron fluxes at the geostationary orbit derives similar conclusions about the east-west component of the electric field before and after midnight.

5. Barium-Cloud Experiment in the Distant Magnetosphere

Releasing a barium cloud in the magnetosphere leads to an experiment which is in some respect quite different from the well known ionospheric ion cloud experiments. In the ionosphere the total momentum of the cloud is lost through collisions with the neutral atmosphere in a matter of seconds. From then on diffusion and $[\mathbf{E}, \mathbf{B}]$-drift determine the cloud's shape and bulk motion.

In the magnetosphere, an ion cloud may also be used as an electric field tracer. Due to the collisionless nature of the environment, however, the high particle density in the cloud – at least in the beginning – and the weak magnetic fields, the time-scale of adjustment of the cloud to the speed of the ambient plasma is very much prolonged. Indeed, the study of this transient process provides a novel and most interesting aspect of this experiment. In addition, since the energy density in the cloud is initially much higher than in the ambient plasma and magnetic field, there will be an initial expansion leading to a diamagnetic cavity into which only later the B-field penetrates again due to collisions or turbulence. This will affect the initial shape and distort the ambient $\mathbf{B}$-field.

In this section we shall discuss the preliminary results from the first magnetospheric ion-cloud experiment.* The ion cloud was released on March 18, 1969, at 0720.52 UT from a capsule ejected about 10^4 sec prior by ground command from the ESRO-satellite HEOS I. The release distance from the earth was 12.5 R_E, geographic latitude = 31.5°, geographic longitude = 319°; magnetic longitude = 42°, MLT = 0430. The cloud was observed for about 25 min from stations in Arizona and Chile and several other sites. It contained about 4.4×10^{23} Barium ions. At the end of the observation time it had reached a length of about 5000 km and a width of about 100 km. Figure 9 shows the cloud as photographed through the 48 inch Schmidt telescope at Mt. Palomar by G. Munch. The three exposures, taken at 0724.00, 0725.30 and 0727.00 UT with respective exposure times of 20, 40 and 120 sec (from bottom to top of Figure 9) are superimposed on the same plate. They clearly show some magnetic field aligned striations. The striation farthest to the left is imbedded in a short diffuse tail, which separates slowly from the sharp-edged main structure indicating the interaction with the ambient plasma flow. The direction of elongation of the cloud is consistent with the magnetometer data from HEOS I (Hedgecock and Sear, 1969). The field has a magnitude of 49 γ. Its direction is roughly antisolar, consistent with Fairfield's (1968) model from which Figure 10 is constructed. The point with an arrow in this meridional plot shows that the cloud was most probably above the plasma sheath at the beginning of the geomagnetic tail. Estimates of the ambient plasma density, n_a, are rather uncertain. Assuming $n_a = 0.01 \, \text{cm}^{-3}$ or $n_a = 1 \, \text{cm}^{-3}$ one obtains Alfvén velocities $v_A = 1.1 \times 10^9$ cm/sec and $v_A = 1.1 \times 10^8$ cm/sec, respectively. Ambient temperatures might have been of the order of a few kV.

The neutral Ba cloud expands with a mean velocity of 1.2 km/sec due to the

* These results (as well as some of the material in Section 4) have been reported for the first time about 2 months prior to the writing of this résumé by Haerendel and Lüst (1970) at the Summer Advanced Study Institute 'Earth's Particles and Fields' at U.C. Santa Barbara, Aug. 1969. In order to avoid unnecessary duplication of this material we shall omit some details here and concentrate on the topics primarily relevant to the general line of the present article.

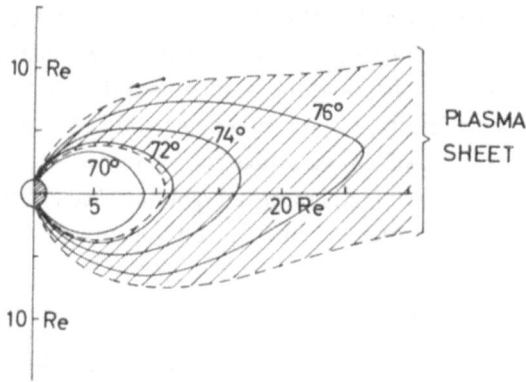

Fig. 9.

Fig. 10.   (After Fairfield, 1968.)

chemically produced kinetic energy of about 1 eV per atom. During the photo-ionization the electrons gain an average thermal energy of the order of 0.5 eV. Initially the energy density in the plasma cloud is much higher than in the ambient magnetic field and thus the plasma cloud will expand against the B-field (Pilipp, 1969). In fact, the plasma cloud expands more rapidly than the neutral cloud because of the energy gain of the electrons and will thus form a diamagnetic cavity. The motion transverse to B will be stopped eventually when the plasma pressure is balanced by the field pressure and stresses. This stage is reached after about 17 sec at a cavity radius of about 20 km. Since the time-scale of photoionization is 20 sec, there will be still neutral atoms left which now can overshoot the diamagnetic core and become ionized in a stronger B-field. Thus the central part of the cloud will be surrounded by a region of rapidly decreasing plasma density and increasing magnetic field strength. The residual diamagnetic effect of atoms ionized at the satellite's position has been observed by the on-board magnetometer to be about -2γ (Hedgecock and Sear, 1969).

The striations are most probably formed due to a Rayleigh-Taylor instability during the deceleration phase of the expansion of the central plasma cloud (Pilipp, 1969; Haerendel and Lüst, 1970).

The density and magnetic field profile discussed above has important consequences for the momentum exchange with the ambient plasma (Scholer, 1969). The dominant effect is the propagation of the electric field, seen by the injected plasma, with Alfvén velocity along the unperturbed magnetic field. Initially the barium plasma will have a bulk velocity equal to that of the satellite which at the release instant amounted to 2.3 km/sec with a strong radial component. If this velocity is different from that of the ambient plasma, the cloud will deform the B-field and force the background plasma in the same flux tube to participate in its motion. When the front of this distortion has passed over a mass comparable to that of the injected plasma, i.e. after a time $\tau = M/F \cdot 2v_A \cdot \varrho_a$, the momentum exchange is essentially completed (M = mass of injected plasma, F = cross-section of magnetic flux tube occupied, ϱ_a = ambient mass density). Taking the total mass of barium atoms released and dividing by the cloud's cross-section we get $\tau \approx 10$ hours for $n_a = 10^{-2}$ cm^{-3} and $\tau \approx 1$ hour for $n_a = 1$ cm^{-3}.

The above deceleration time neglects any viscous kind of interaction of the cloud's flux tubes with the ambient plasma. The effects of the ionosphere and of the inhomogeneity of the magnetospheric plasma, on the other hand, can readily be taken into account (Scholer, 1969). Thus the time-scale τ will in fact be shorter and the formula above gives only an upper limit.

Applying the formula for τ locally it is clear that the dilute parts of the cloud will adjust more quickly to the ambient flow velocity than the denser and more diamagnetic central parts. Thus a tail will form much like a comet tail. The tail will, however, be so dilute at larger distances from the core that only its nearby portions are visible. They will not yet have reached the ambient flow speed. Still, aside from aberration, the tail will point in the direction of the ambient flow, thereby indicating the direction of the ambient electric field. Preliminary results show that the tail pointed roughly azimuthally towards midnight. On the other hand, the main part of the cloud practi-

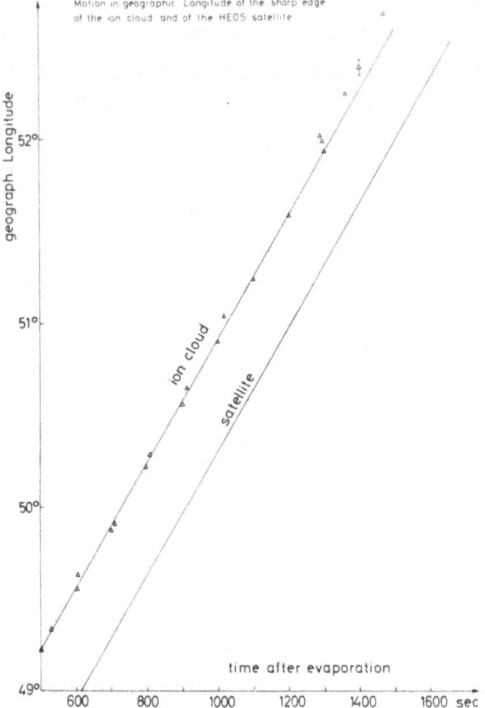

Fig. 11.

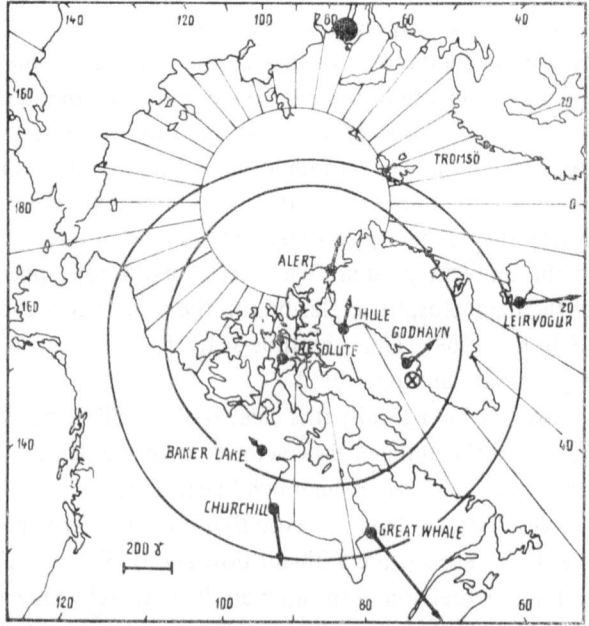

Fig. 12.

cally follows the motion of the satellite (Figure 11). Only towards the end of the observation time a small deviation seems to show up.

One very interesting observation can be made by comparing the inferred drift direction with magnetic perturbation vectors on the ground. The experiment took place in the middle of a well-developed negative bay. The field line through the cloud meets the ionosphere near Godhavn (Greenland) (Figure 12). From the tail direction we expect an ionospheric drift towards magnetic west near Godhavn with a northward component if aberration is not negligible.

The magnetic perturbation vectors, indicated in Figure 12, however, show that this drift direction is inconsistent with the assumption of either overhead Hall or Pedersen currents causing the perturbations. Similar discrepancies have been reported by Wescott *et al.* (1969b) from barium releases north of the auroral oval. Thus it seems that in the polar cap the magnetic perturbations are not caused by horizontal ionospheric currents as in the auroral zone (see Section 4). In fact Akasofu's (1970) model assuming strong field aligned currents in the polar cap region seems to represent the situation rather well, although more experiments are needed to prove this point.

6. Final Remarks

The measurements discussed here do not comprise all the methods applied up to now to measure electric fields. For instance in the distant magnetotail an upper limit to the electric field of 0.5 mV/m was derived by Van Allen and Ness (1969) from an interpretation of the shadow effect of the moon on solar electrons (see also Van Allen, 1969). By a similar method, Anderson (1969) deduced 0.3 mV/m as an upper limit. There also exists a preliminary experiment for a potentially very useful modification of the field mill-technique (Knott, 1969). And the incoherent scatter technique (Woodman and Balsley, 1969) will hopefully be applied some time in the auroral zone.

Altogether the experimental evidence is not yet conclusive enough to provide a detailed picture of the electric field distributions during magnetospheric substorms or events induced by solar disturbances. Still, however, the general magnitude of the fields and some special features of the magnetospheric convection pattern can well be deduced. In particular the limitations of a supposed quantitative and – in the polar cap – even only qualitative correspondence between magnetic perturbations on the ground and the magnetospheric electric fields are quite apparent.

References

Aggson, T. S.: 1969, in *Atmospheric Emissions* (ed. by B. M. McCormac and A. Omholt), Reinhold Book Co., New York, p. 305.
Akasofu, S. I.: 1970, in *Particles and Fields in the Magnetosphere* (ed. by B. M. McCormac), D. Reidel Publ. Co., Dordrecht-Holland, p. 34.
Anderson, K. A.: 1969, Abstract IAGA General Sci. Assemblies, Madrid.
Axford, W. I. and Hines, C. O.: 1961, *Can. J. Phys.* **39**, 1433.
Axford, W. I.: 1969, *Rev. Geophys.* **7**, 421.

Bernstein, W., Innouye, G. T., Sanders, N. L., and Wax, R. L.: 1969, Abstract, Conf. on Electric Fields in the Magnetosphere, Rice University.

Brice, N. M.: 1967, *J. Geophys. Res.* **72**, 5193.

Carpenter, D. L.: 1962, Stanford Univ. Radio Sci. Lab. Rept. SEL-62-059.

Carpenter, D. L.: 1963, *J. Geophys. Res.* **68**, 1675.

Carpenter, D. L.: 1968, *Radio Sci.* **3**, 719.

Carpenter, D. L. and Stone, K.: 1967, *Planetary Space Sci.* **15**, 395.

Carpenter, D. L. and Stone, K.: 1968, Abstract 7-12, Int. Symp. on the Physics of the Magnetosphere, Washington D.C.

Carpenter, D. L., Walter, F., Barrington, R. E., and McEwen, D. J.: 1968, *J. Geophys. Res.* **73**, 2929.

Fahleson, U. V., Kelley, M. C., and Mozer, F. S.: 1968, U.C. Berkeley, preprint.

Fairfield, D. H.: 1968, *J. Geophys. Res.* **73**, 7329.

Föppl, H., Haerendel, G., Haser, L., Lüst, R., Melzner, F., Meyer, B., Neuss, H., Rabben, H., Rieger, E., Stöcker, J., and Stoffregen, W.: 1968, *J. Geophys. Res.* **73**, 21.

Freeman, Jr., J. W.: 1968, *J. Geophys. Res.* **73**, 4151.

Freeman, Jr., J. W.: 1969, *Science* **163**, 1061.

Freeman, Jr., J. W., Warren, C. S., and Maguire, J. J.: 1968, *J. Geophys. Res.* **73**, 5719.

Gringauz, K. I.: 1963, *Planetary Space Sci.* **11**, 281.

Gurnett, D. A. and Cauffman, D. P.: 1969, University of Iowa Report 69-45.

Haerendel, G. and Lüst, R.: 1968, in *Earth's Particles and Fields* (ed. by B. M. McCormac), Reinhold Publ. Comp. New York, p. 271.

Haerendel, G. and Lüst, R.: 1970, in *Particles and Fields in the Magnetosphere* (ed. by B. M. McCormac), D. Reidel Publ. Co., Dordrecht-Holland, p. 213.

Haerendel, G., Lüst, R., and Rieger, E.: 1967, *Planetary Space Sci.* **15**, 1.

Haerendel, G., Lüst, R., Rieger, E., and Völk, H.: 1969, in *Atmospheric Emissions* (ed. by B. M. McCormac and A. Omholt), van Nostrand Reinhold Comp. New York, p. 293.

Harang, L. and Tröim, J.: 1961, *Planetary Space Sci.* **5**, 33.

Hedgecock, P. C. and Sear, J.: 1969, preprint to be published.

Heppner, J. P.: 1969, in *Atmospheric Emissions* (ed. by B. M. McCormac and A. Omholt), Reinhold Book Co., New York, p. 251.

Kellog, P. J., and Weed, M.: 1969, in Planetary Electrodynamics (ed. by S. C. Coroniti and J. Hughes), Gordon and Breach, N.Y., Vol. 2, p. 431.

Knott, K.: 1969, Preprint University of Heidelberg (to be published in *Space Research*).

Maynard, N. C. and Heppner, J. P.: 1969, GSFC Report X-612-69-374.

Mozer, F. S. and Bruston, P.: 1967, *J. Geophys. Res.* **72**, 1109.

Mozer, F. S. and Fahleson, U. V.: 1968, U.C. Berkeley, preprint.

Mozer, F. S. and Serlin, R.: 1969, *J. Geophys. Res.* **74**, 4739.

Nishida, A.: 1966, *J. Geophys. Res.* **71**, 5669.

Pilipp, W.: 1969, Ph.D. Thesis, University of Munich (to be published).

Scholer, M.: 1969, Ph.D. Thesis, University of Munich (to be published).

Swift, D.: 1965, *J. Geophys. Res.* **70**, 3061.

Taylor, H. E. and Hones, E. W.: 1965, *J. Geophys. Res.* **70**, 3605.

Van Allen, J. A.: 1969, University of Iowa report 69-29.

Van Allen, J. A. and Ness, N. F.: 1969, *J. Geophys. Res.* **74**, 71.

Wescott, E. M., Stolarik, J. D., and Heppner, J. P.: 1969a, *J. Geophys. Res.* **74**, 3469.

Wescott, E. M., Stolarik, J. D., and Heppner, J. P.: 1969b, GSFC Report X-612-69-411.

Winckler, J. R.: 1970, in *Particles and Fields in the Magnetosphere* (ed. by B. M. McCormac), D. Reidel Publ. Co., Dordrecht-Holland, p. 332.

Woodman, R. F. and Balsley, B. B.: 1969, *J. Atmos. Terr. Phys.* **31**, 865.

5A. THE MAGNETOTAIL

EXPERIMENTAL OBSERVATIONS IN THE MAGNETOTAIL
DURING AN INTERPLANETARY DISTURBANCE

EDWARD W. HONES, JR.

University of California, Los Alamos Scientific Laboratory, Los Alamos, N.M., U.S.A.

1. Introduction

In the few years since it first became possible to observe the solar wind experimentally many studies have been made with the purpose of determining the degree to which each of the observable properties is responsible for disturbances within the earth's magnetosphere. In many of these studies the parameter characterizing the degree of disturbance of the earth's magnetosphere has been some index of disturbance in the surface magnetic field such as the K_p index. Snyder *et al.* (1963) showed that K_p was correlated with the velocity of the solar wind. Wilcox *et al.* (1967) and Schatten and Wilcox (1967) found that the K_p index correlated with the magnitude of the interplanetary magnetic field, with the direction of its north-south component, B_z, and with the Alfvén Mach number and energy flux of the solar wind particles. Fairfield and Cahill (1966) and Fairfield (1968) observed that a southward pointing magnetic field in the magnetosheath was associated with high-latitude geomagnetic disturbances. Rostoker and Falthammer (1967) have supported these results. Nishida (1968) has shown that the *DP*2 geomagnetic variations, apparent at polar and equatorial stations, are coherent with variation in the north-south component of the interplanetary magnetic field. In a very recent work Hirshberg and Colburn (1969) have confirmed the correlation of K_p with a southward pointing B_z, with the intensity of B and with the variance of B. They found further, that the long initial phase of the Feb. 15, 1967 storm was terminated and the main phase began within an hour after the interplanetary magnetic field turned from northward to southward pointing. Hirshberg and Colburn (1969) conclude that the main phase of a magnetic storm may be initiated by arrival at the earth of the 'driver gas' whose shock wave, arriving perhaps many hours earlier, caused the storm sudden commencement. This view is supported by the observation (Hirshberg *et al.*, 1969) that the solar wind suddenly became helium-enriched at the start of the main phase of the Feb. 15, 1967 storm. The appearance of helium coincided, too, with the onset of the main phase of the April 17, 1965 storm (Gosling *et al.*, 1967). It has also been noted that helium appeared during the main phases of two other storms (Ogilvie *et al.*, 1968; Bame *et al.*, 1967b). It is not clear what properties of the piston of driver gas (e.g., increased dynamic pressure, increased velocity, change of composition, change of magnetic field) might be most influential in starting the storm main phase, though the studies of Hirshberg and Colburn (1969) and others referred to above suggest the direction of B_z may be of particular importance in this respect.

V. Manno and D. E. Page (eds.), Intercorrelated Satellite Observations Related to Solar Events, 299–308. All Rights Reserved
Copyright © 1970 by D. Reidel Publishing Company, Dordrecht-Holland

This paper presents observations made concurrently by two Vela satellites – one in the solar wind and one in the plasma sheet of the magnetotail – before, and during the first seven hours of, a weak s.c. storm that started May 1, 1967. In this event a sudden compression of the plasma in the plasma sheet was observed a few minutes after the s.c. at the earth. A sudden moderate helium enrichment of the solar wind, occurring $3\frac{1}{2}$ hours after the s.c., was followed within minutes by the sharp onset of negative bays in the midnight sector of the auroral zone and by sudden diminution of plasma intensity in the plasma sheet.

Before describing the event in detail I shall discuss some Vela satellite observations, not specifically related to interplanetary disturbances, which are, nevertheless, relevant to the observations during this event.

2. Observations

The Vela satellites are in nearly circular orbits of radius $\sim 18 \, R_E$. The orbits of all of the satellites are nearly co-planar and inclined about 60° to the ecliptic plane. Because of the earth's motion around the sun the inclined circular orbits, fixed in initial space, trace out a truncated spherical surface around the earth each year as illustrated in Figure 1. The satellites carry electrostatic analysers which measure energy spectra of protons and electrons of energies between ~ 100 eV and 20 keV. (See Hundhausen *et al.*, 1967, for more detailed description of the electrostatic analysers.) They also carry Geiger counters, and solid-state detectors which measure the fluxes of more energetic electrons and protons above several energy thresholds.

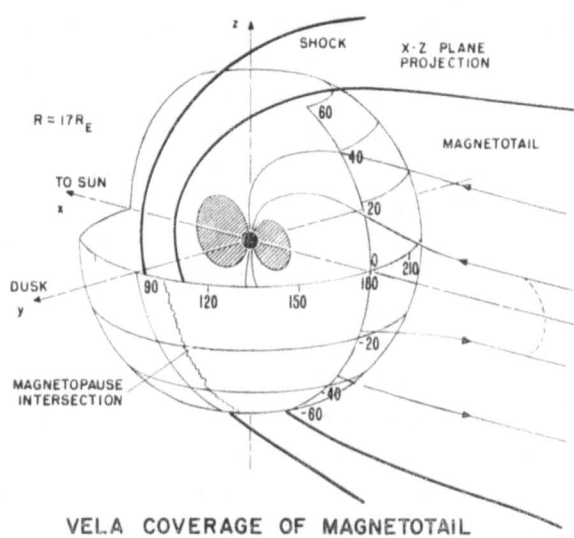

VELA COVERAGE OF MAGNETOTAIL

Fig. 1.   Representation of the approximately spherical surface traced out by the 60° inclined circular orbits of the Vela satellites each year. Indicated schematically are the magnetopause, the bow shock and the Van Allen radiation belts. The gray portion of the sphere indicates where it cuts the plasma sheet in the magnetotail (Bame *et al.*, 1967a).

A. PLASMA PRESSURE NEAR THE NEUTRAL SHEET

During the period July 1965 to June 1967 Vela 3A and 3B were about 180° apart in their orbits and thus provided a number of opportunities to make concurrent measurements of the solar wind and of the plasma in the magnetotail. About 10 000 of the solar wind spectra taken by both 3A and 3B during the two year interval were carefully analyzed in the manner described by Hundhausen *et al.* (1967) to give *proton* densities and temperatures as well as the bulk velocity. This set of measurements was searched to find times when that one of the two satellites which was not in the solar wind was measuring the plasma properties within $\pm 1R_E$ of the neutral sheet in the magnetotail. The formula of Russell and Brody (1967) was used to estimate the position of the neutral sheet. Time delays up to 3000 sec from the time of a given solar wind measurement to the time(s) of the correlated tail plasma measurement(s) were allowed in this comparison. For each set of corresponding solar wind and tail plasma measurements the flow kinetic energy density of the solar wind *protons*, $u_w = \frac{1}{2}\rho_w v_w^2$, and the kinetic energy density of the plasma sheet electrons or protons were compared. Figure 2 shows the energy density of the *protons* in the tail plasma, $u_{pt} = \frac{1}{2}\rho_{pt} v_{pt}^2$

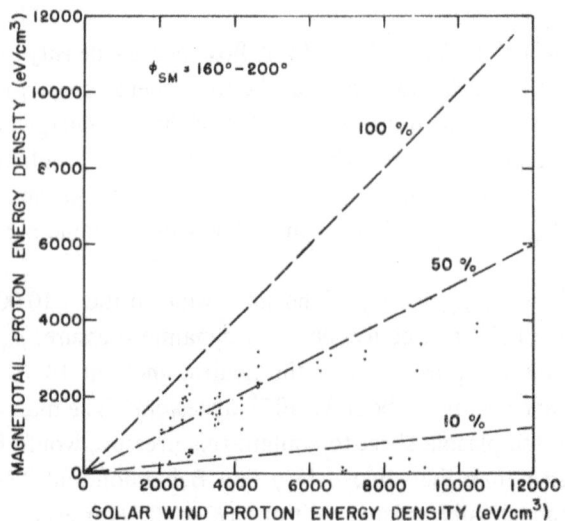

Fig. 2. Proton energy density in plasma sheet vs. solar wind proton flow energy density. All data points were obtained when the Vela satellite in the tail was within ± 1 earth radius of the (estimated) neutral sheet position and in the solar magnetospheric longitude range 160° to 200°.

plotted against u_w. u_{pt} ranges up to $\sim 100\%$ of u_w and averages $\sim 50\%$ of it. Figure 3 shows the same sort of comparison between the energy density of the *electrons* in the tail plasma, u_{et}, and u_w. u_{et} ranges up to $\sim 20\%$ of u_w and averages $\sim 10\%$ of it. Thus, the ordinates of the points in Figure 2 represent only $\sim 80\%$ to $\sim 90\%$ of the total energy density of the plasma sheet plasma. And, since helium typically constitutes $\sim 4\%$ of the particle number density of the solar wind, the abscissae of the points

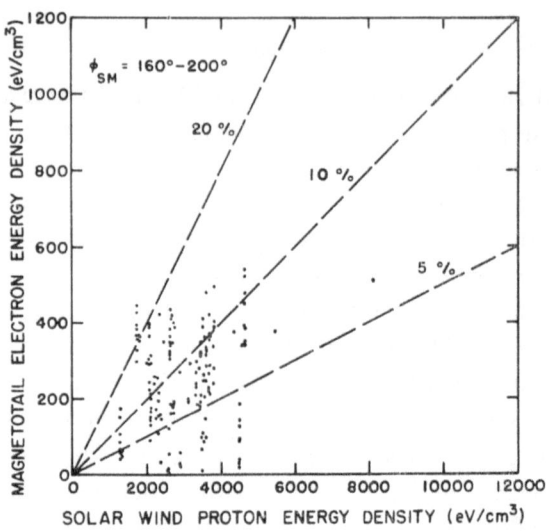

Fig. 3. Electron energy density in plasma sheet vs. solar wind proton flow energy density. All data points were obtained when the Vela satellite in the tail was within ± 1 earth radius of the (estimated) neutral sheet position and in the solar magnetospheric longitude range 160° to 200°.

represent only $\sim 80\%$ to 90% of the total flow energy density of the solar wind. Essentially, then, Figures 2 and 3 say that the total energy density of the plasma near the neutral sheet at $\sim 18\ R_E$ in the tail is $\sim 50\%$ of the *flow* energy density of the solar wind. In terms of pressure, since the solar wind flow is unidirectional while the particle flux in the plasma sheet is generally isotropic, the plasma pressure near the neutral sheet at $18\ R_E$ is about 16% of the solar wind dynamic pressure (momentum flux).

The average flow energy density of the solar wind in the $\sim 10\,000$ spectra used in this study was ~ 6 keV/cm^3, equivalent to a dynamic pressure, $\rho_w v_w^2$, of $\sim 2 \times 10^{-8}$ dynes/cm^2. The plasma pressure near the neutral sheet at $18\ R_E$, being $\sim 16\%$ of this, must then have averaged about 3×10^{-9} dynes/cm^2. The magnetic field required above and below the plasma sheet to contain this pressure would be $\sim 27\ \gamma$. This is considerably higher than the value $\sim 18\ \gamma$ that Behannon and Ness (1968) show at $18\ R_E$ for geomagnetically quiet conditions ($K_p \leqslant 2$). However, our data included many magnetically active periods and the data of Behannon and Ness (1968) indicate that the tail field within $31\ R_E$ exceeded $27\ \gamma\ \sim 25\%$ of the time for $K_p \geqslant 2$.

The tail is confined by the normal component of the pressure of the magnetized plasma flowing in the magnetosheath and this has not been evaluated in any systematic fashion experimentally. Detailed gas-dynamic calculations by Spreiter *et al.* (1968), which do not, however, include the effects of the magnetic field on the magneto-sheath, suggest that the pressure normal to the magnetopause 10 to 12 R_E down-stream from the earth is $\sim 5\%$ of the solar-wind-dynamic pressure. The simple Newtonian pressure formula, $p_\perp = p_w \cos^2 \psi$ where ψ is the angle between the solar wind flow direction and the surface normal, gave $p_\perp \approx 0.03 p_w$ for the same model of the magnetosphere. For the same model, $p_\perp$ at 18 R_E downstream from the earth, the distance of the Vela satellites, would be less than these values due to the increased value of ψ. Values given by Spreiter et al. (1968, Figure 24) for the magnetic field in the magnetosheath relative to that in the undisturbed solar wind, together with the fact that the pressure of the interplanetary magnetic field (~ 5 γ) is typically only $\sim 1\%$ of the solar-wind-dynamic pressure suggest that the magnetic field in the magnetosheath might double $p_\perp$, making it, perhaps $\sim 8\%$ of p_w. Also, $p_\perp$ is very sensitive to the angle, ψ, varying about as $(\pi/2 - \psi)^2$, and thus to the detailed shape of the magnetosphere. To conclude this discussion of the measured pressure of the plasma near the neutral sheet and its relation to the tail magnetic field strength and, ultimately, to the solar wind dynamic pressure, we wish to emphasize (a) that the ratio u_p/u_w is found to average $\sim \frac{1}{2}$; (b) that the average plasma pressure near the neutral sheet is not in serious disagreement with tail magnetic field strengths measured by Behannon and Ness (1968) and (c) that the ratio is considerably higher than one would expect on the basis of Spreiter's et al. (1968) calculations and suggests that the magnetic pressure in the magnetosheath and/or the flaring angle of the magnetopause ~ 18 R_E down the magnetotail may be, on the average, larger than those found in Spreiter's work.

B. VARIATIONS OF PLASMA SHEET RELATED TO MAGNETOSPHERIC SUBSTORMS

At 18 R_E in the magnetotail the plasma sheet (i.e., that region encompassing the neutral sheet, where in the plasma energy density equals or exceeds the magnetic field energy density) becomes thinner as a negative bay develops in the night-side auroral zone. It is re-inflated with plasma hotter than that previously present, usually during the recovery phase of the bay (Hones et al., 1967; Hones, 1968; Hones et al., 1968a, b). This phenomenon is illustrated in Figure 4, where data taken by Vela 2A during a pass through the magnetotail on October 12, 1964, are presented. The Ae index (negative and positive envelopes plotted separately) show two large substorms whose negative peaks occur ~ 0700 UT and ~ 1200 UT. The average energy and energy density of the plasma electrons (~ 150 eV $< E_e < 20$ keV) and the counting rate of the Geiger counter ($E_e > 40$ keV) all decrease during the development of the two substorms. The average energy of the plasma electrons and, particularly, the counting rate of the Geiger counter then jump to values higher than the pre-substorm values during the later phases of both substorms. The degree of *thinning* of the plasma sheet is well illustrated in the first substorm when the plasma electron energy density drops from normal values to $\sim 1\%$ of these values though the satellite is only a few tenths of an earth radius from the estimated position of the neutral sheet. The degree of *thickening* of the plasma sheet is best illustrated by the second substorm; at about the peak of the substorm the plasma and energetic electrons return to the satellite which is now ~ 4.5 R_E from the neutral sheet. We have shown, more conclusively, that this phenomenon is indeed, a *thinning* and *thickening* of the plasma sheet and *not* a flapping motion, by observing that two Vela satellites, obviously on opposite sides of the neutral sheet, sense the deflation and re-inflation of the plasma sheet more or less together (Hones, 1969).

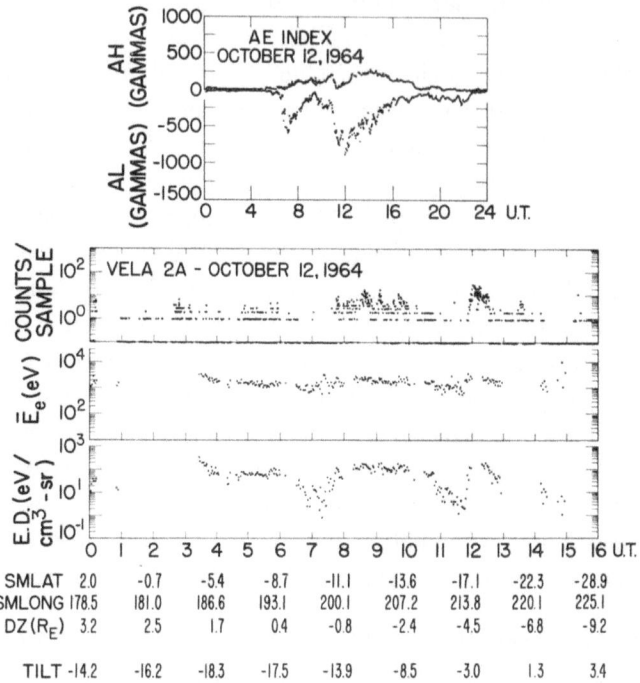

Fig. 4. *Ae* index and data from Vela 2A in the magnetotail, October 12, 1964. Top curve of Vela data is counting rate of Geiger counter sensitive to electrons ($E_e > 40$ keV). Bottom two curves are average energy, $\bar{E}_e$ and energy density (E.D.), of electrons (~ 150 eV $< E_e < 20$ keV) measured with electrostatic analyzer. At bottom are solar magnetospheric latitude and longitude of the satellite, its perpendicular distance (in R_E) from the neutral sheet, and the tilt of the earth's dipole axis. (Positive values mean north pole is tilted toward the sun.)

C. VELA SATELLITE OBSERVATIONS DURING AN INTERPLANETARY

DISTURBANCE – MAY 1–2, 1967

A sudden impulse was sensed by magnetometers at the earth at 1906 UT May 1, 1967 (Figure 5). Within $\sim \frac{1}{2}$ hour a weak negative bay began at Kiruna, recovering ~ 2100 UT. At 2240 large negative bays started suddenly at Kiruna and Leirvogur (magnetic local times (MLT)~ 0140 and ~ 2240, respectively). Throughout this series of events Vela 3A was in the solar wind near the noon meridian and Vela 3B was in the magnetotail near the neutral sheet and near local midnight (Figure 6). The proton spectra from Vela 3A show that the solar wind velocity jumped from 355 km/sec to 460 km/sec some time between 1900 and 1907 UT, the result of an interplanetary shock wave passing the satellite. Measurements in real-time mode about 1 hour before the s.i. and 1 hour after it showed proton densities of 5/cm^3 and 15/cm^3 respectively. The proton flow energy density before and after the shock was ~ 3700 eV/cm^3 and ~ 19000 eV/cm^3 respectively. At Vela 3B, ~ 0.3 R_E from the neutral sheet and ~ 1 hour before local midnight, the electron energy density jumped from 300 eV/cm^3 to 1200 eV/cm^3 some time between 1914 and 1921 UT. From proton measurements taken in real-time mode at ~ 1400 UT and ~ 2130 UT it is found that the proton energy density in the plasma sheet was ∼8 times that of the electrons. Thus, we estimate that the total particle kinetic energy density near the neutral sheet jumped from ∼2700 eV/cm³ to ∼11 000 eV/cm³, up a factor of 4, the change occurring between 8 and 15 min after the s.i. at the earth. The time delay (8 min $< \Delta t < 15$ min) is much longer than the 2 to

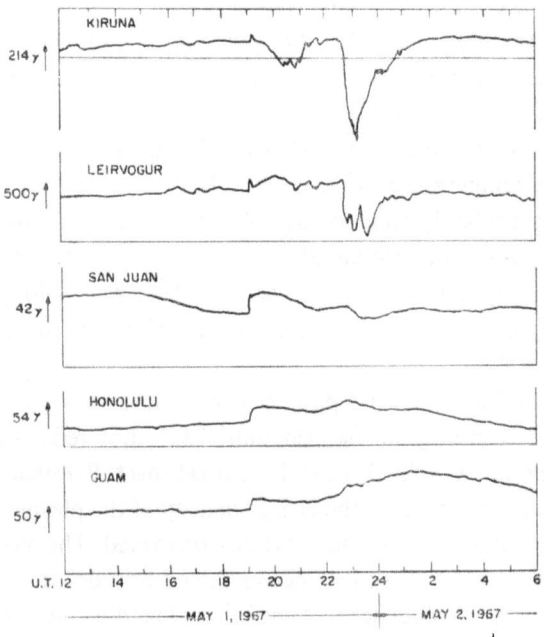

Fig. 5.   Magnetic H- or X-components from several stations on May 1–2, 1967.

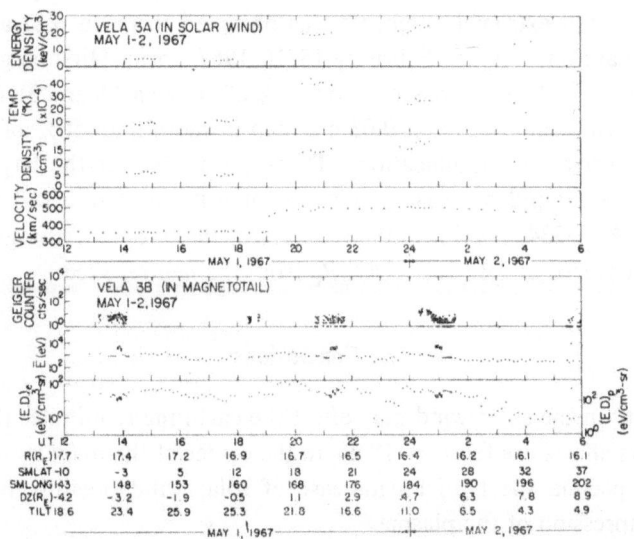

Fig. 6.   Data from Vela 3A and 3B, May 1–2, 1967. The X's in the 3B data are proton values and the dots are electron values. The scale for proton energy density is given at the right side of the bottom box·

3 min that one would expect for transmission of a signal from the front of the magneto-sphere at the Alfvén speed. The pre-shock and post-shock solar wind proton density measurements yield an estimated shock speed of 500 km/sec (4.7 R_E min). The shock in the solar wind should thus have reached a point 16.8 R_E down the tail from the earth (the distance to Vela 3B) no later than 1910 UT. The pressure increase at the center of the tail was felt 4 to 11 min after this time.

Figure 6 shows that the energy density change of the tail plasma was due largely to a change of particle number density. The electron average energy increased only a factor ~ 1.5 while the number density increased a factor ~ 2.7. In an adiabatic compression of a monatomic gas ($\gamma = \frac{5}{3}$) a particle density increase by 2.7 would be accompanied by a temperature increase of $(2.7)^{2/3} = 1.9$. Within the accuracy of the measurement it is probably safe to say that the initial pressure increase near the neutral sheet was consistent with an adiabatic compression of the plasma. Before the s.i. the particle energy density near the neutral sheet was $\sim 70\%$ of the solar wind energy density; after the s.i. and after the increase of tail plasma pressure had taken place the ratio was 60%.

The properties of the solar wind plasma as measured by Vela 3A, particularly the velocity, remained relatively unchanged until 2231, not reflecting in any way the significant change seen by Vela 3B near the neutral sheet. Parallelling the weak bay at Kiruna (but lagging by ~ 15 min) the energy density of the plasma 1 to 3 R_E from the neutral sheet fell to pre-s.i. levels then partially recovered. The Vela data taken alone lead one to believe that this weak bay, accompanied by a deflation and re-inflation of the plasma sheet was a consequence of some internal readjustment within the magneto-sphere.

Some time between 2231 and 2239, however, the α-to-proton ratio, a/p at Vela 3A in the solar wind rose from 0.03 to 0.07 and remained so until well past 2400. An a/p value of 0.07 is not extremely high. (It was observed to reach 0.22 about the time of the main phase onset of the February 15–16, 1967 storm (Hirshberg et al., 1969).) However, a/p ratios of 0.07 were seen in only 4% of the Vela 3A and 3B data from July 1965 to June 1967 while ratios of 0.03 are seen in more than 50% of these spectra (Hundhausen, private communication). The sudden onset of the bays at 2240 was accompanied by a rapid decrease of plasma intensity at Vela 3B, 3.8 R_E from the neutral sheet. At ~ 2340, well into the recovery of the bays at Leirvogur and Kiruna the plasma recovered at 3B (now ~ 4.5 R_E from the neutral sheet).

3. Conclusions

A discontinuity in the solar wind may affect the earth many minutes (8 to 15 in this case) before its effects are felt at ~ 18 R_E in the center of the magnetotail. The initial effect, in the plasma sheet, of an increase of solar wind pressure is evidently an adiabatic compression of the plasma.

The large increase in dynamic pressure of the solar wind (by a factor of 5) was followed within a half hour only by *weak* negative bay activity and the typically associated deflation of the plasma sheet. However, a relatively minor increase in the solar wind a/p ratio, occurring $3\frac{1}{2}$ hours after the s.i. was followed within minutes by the sudden onset of large magnetic bays. It is doubtful that the dynamic pressure increase ($\sim 16\%$) associated with the added helium would have produced such a striking effect. If, indeed, the sudden bay onset *was* related to the change in solar wind composition, then it may have resulted from the composition change itself or from some feature (e.g., a magnetic field variation) not detected by the Vela satellite.

The coincidence of the plasma sheet recovery and the recovery of the large bays at Leirvogur and Kiruna reinforces our view that re-inflation of the plasma sheet is causally related to the termination or recovery of auroral zone negative bays.

Acknowledgements

The Vela nuclear test detection satellites have been designed, developed and flown as part of a joint program of the Advanced Research Projects Agency of the U.S. Department of Defense and the U.S. Atomic Energy Commission. The program is managed by the U.S. Air Force.

References

Bame, S. J., Asbridge, J. R., Felthauser, H. E., Hones, E. W. and Strong, I. B.: 1967a, *J. Geophys. Res.* **72**, 113.

Bame, S. J., Asbridge, J. R., Hundhausen, A. J., and Strong, I. B.: 1967b, *J. Geophys. Res.* **73**, 5761.

Behannon, K. W. and Ness, N. F.: 1968, in *Physics of the Magnetosphere* (ed. by R. L. Carovillano, J. F. McClay, and H. R. Radoski), Reidel, Dordrecht, The Netherlands.

Fairfield, D. H.: 1968, *Space Research VIII*, North-Holland Publ. Co., Amsterdam.

Fairfield, D. H. and Cahill, L. J., Jr.: 1966, *J. Geophys. Res.* **71**, 155.

Gosling, J. T., Asbridge, J. R., Bame, S. J., Hundhausen, A. J., and Strong, I. B.: 1967, *J. Geophys. Res.* **72**, 1813.

Hirshberg, J. and Colburn, D. S.: 1969, *Planetary Space Sci.* **17**, 1183.

Hirshberg, J., Alksne, A., Colburn, D. S., Bame, S. J., and Hundhausen, A. J.: 1969, to be published.

Hones, E. W., Jr.: 1968, in *Physics of the Magnetosphere* (ed. by R. L. Carovillano, J. F. McClay, and H. R. Radoski), Reidel, Dordrecht, The Netherlands.

Hones, E. W., Jr.: 1969, Paper presented at the Summer Institute, Earth's Fields and Particles, University of California, Santa Barbara, Calif.

Hones, E. W., Jr., Asbridge, J. R., Bame, S. J., and Strong, I. B.: 1967, *J. Geophys. Res.* **72**, 5879.

Hones, E. W., Jr., Bame, S. J., Singer, S., and Brown, R. R.: 1968a, *J. Geophys. Res.* **73**, 6189.

Hones, E. W., Jr., Singer, S., and Rao, C. S. R.: 1968b, *J. Geophys. Res.* **73**, 7339.

Hundhausen, A. J., Asbridge, J. R., Bame, S. J., Gilbert, H. E., and Strong, I. B.: 1967, *J. Geophys. Res.* **72**, 87.

Nishida, A.: 1968, *J. Geophys. Res.* **73**, 5549.

Ogilvie, K. W., Burlaga, L. F., and Wilkerson, T. D.: 1968, *J. Geophys. Res.* **73**, 6809.

Rostoker, G. and Falthammar, C. G.: 1967, *J. Geophys. Res.* **72**, 5853.

Russell, C. T. and Brody, K. I.: 1967, *J. Geophys. Res.* **72**, 6104.

Schatten, K. H. and Wilcox, J. M.: 1967, *J. Geophys. Res.* **72**, 5185.

Snyder, C. W., Neugebauer, M., and Rao, U. R.: 1963, *J. Geophys. Res.* **68**, 6361.

Spreiter, J. R., Alksne, A. Y., and Summers, A. L.: 1968, *Physics of the Magnetosphere* (ed. by R. L. Carovillano, J. F. McClay, and H. R. Radoski), Reidel, Dordrecht, The Netherlands.

Wilcox, J. M., Schatten, K. H., and Ness, N. F.: 1967, *J. Geophys. Res.* **72**, 19.

Discussion

Parks: Can you say something on the origin of the electrons which you see in the tail during the recovery phase of the magnetic storm?

Hones: Our view is that what you see upon recovery of a bay is a relaxation of the contained plasma out into the tail, carrying with it some of the energetic electrons which are engendered during break up.

TEARING INSTABILITIES IN THE MAGNETOSPHERE

K. SCHINDLER

European Space Research Institute, (ESRIN) of the European Space Research Organisation (ESRO)
Frascati (Rome), Italy

1. Introduction

This is a brief description of studies concerning plasma instabilities which involve topological changes of the magnetic field configuration.

There is evidence that magnetic field loops are formed in the magnetospheric tail (Mihalov *et al.*, 1968) as predicted by Coppi *et al.* (1966). In this paper we shall emphasize that similar processes might also occur in other parts of the magnetosphere. Consequently, we shall treat the neutral sheet only as one region in a series of others, where 'tearing' instabilities might play a role.

We can detect those regions easily for a two-dimensional magnetosphere model, where the earth is represented by a long rod along which all quantities are constant. We then can use the two-dimensional tearing theory for which there is a variational principle derived from the Vlasov equations, which gives sufficient stability criteria, depending on a certain 'potential' V.

It will be interesting to note that the search for unstable equilibria will automatically lead to antiparallel field situations, pointing towards field line reconnection.

2. The Negative V Regions

Let the equilibrium current flow along the z-direction (along the rod). The equilibrium distribution functions for electrons and ions are chosen to be

$$f_0 = F_0(H, P_z) \tag{1}$$

$$H = \frac{m}{2}(v_x^2 + v_y^2) + \frac{1}{2m}(P_z - eA_0(x, y))^2 + e\phi_0(x, y) \tag{2}$$

$$P_z = mv_z + eA_0(x, y) \tag{3}$$

where A_0 is the z-component of the vector potential (the other components vanish) and ϕ_0 is the electric potential.

Clearly, energy H and generalized momentum component P_z are constants of motion and consequently f_0 is a solution of the stationary Vlasov equation. Note that f_0 is isotropic in the v_x-v_y plane, so that there is an isotropic pressure p_0. We mpose the condition

$$\partial F_0/\partial H < 0 \tag{4}$$

which for example holds in the case of a local Maxwellian, this excludes several kinds of microinstabilities, which can perhaps be treated separately.

V. Manno and D. E. Page (eds.), Intercorrelated Satellite Observations Related to Solar Events, 309–315. All Rights Reserved
Copyright © 1970 by D. Reidel Publishing Company, Dordrecht-Holland

The velocity moments depend on space only via the potentials A_0 and ϕ_0. It is then possible to derive a variational principle for example by the method which was used by Schindler and Soop (1968) for one-dimensional equilibria. Assuming quasi-neutrality

$$n_i(A_0, \phi_0) = n_e(A_0, \phi_0) \tag{5}$$

and confining the discussion to two-dimensional perturbations, being constant along the direction of equilibrium current ($\partial/\partial z = 0$) one finds a quadratic form W

$$W \equiv \int \left[(\nabla A_1)^2 + V(x, y) A_1^2 \right] dx\, dy \tag{6}$$

such that the equilibrium is stable (against the two-dimensional perturbations) if

$$W > 0 \quad \text{for all } A_1 \tag{7}$$

(sufficient stability criterion).

The 'potential' $V(x, y)$ is computed in the following way: the total pressure p_0 can be written as

$$p_0 = p_0(A_0)$$

after elimination of ϕ_0, using (5). Then

$$V = -\mu_0\, d^2 p_0 / dA_0^2. \tag{8}$$

The term $-\int V(x, y)\, A_1^2\, dx\, dy$ may be called the 'free energy' available for driving the instability. Clearly no instability is possible for $V > 0$.

The regions in which $V < 0$, are possible candidates for a tearing instability. These regions are however not necessarily unstable; a much more detailed study is necessary to prove instability (or stability) in any particular case. If the instability is present it is roughly localised to the regions of negative V. Regions with $V > 0$ are stable.

It is sometimes useful to note that (6) is formally coincident with the variational integral used in quantum mechanics, and that the minimising function A_1 obeys a Schrödinger equation $-\Delta A_1 + V A_1 = \lambda A_1$, where λ is the eigenvalue. The unstable mode is therefore localised in the same way as a quantum mechanical wave function of a bound state is localised in the region of negative potential.

In order to determine the regions of negative V we have to make a guess about the pressure distribution in our model magnetosphere. Figure 1 shows a possible profile in arbitrary units. We have introduced a plasma sphere, a plasma sheet and a neutral sheet. Radiation belts will be discussed separately. The field lines are labelled by their near earth latitude θ; one easily transforms to the vector potential A_0 as a variable (lower part). Regions of negative V follow to be connected with the plasmapause, with the edges of the plasma sheet and with the neutral sheet. Figure 2 is a sketch of these regions in the magnetosphere. For the further discussion we distinguish between the plane regions (neutral sheet and outer edge of plasma sheet in the far tail) and the regions with field line curvature.

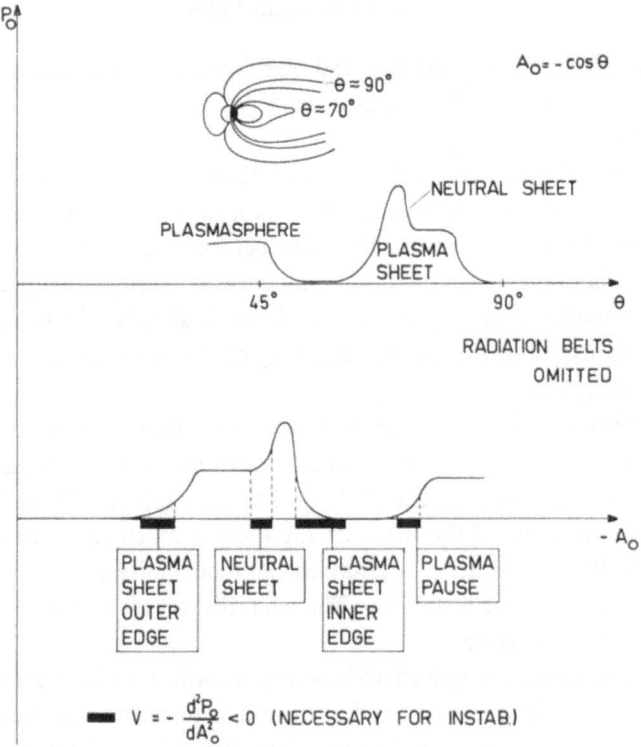

Fig. 1.   Pressure p_0 as a function of latitude θ and of the azimuthal component of the vector potential,
A_0, for a two-dimensional model magnetosphere.

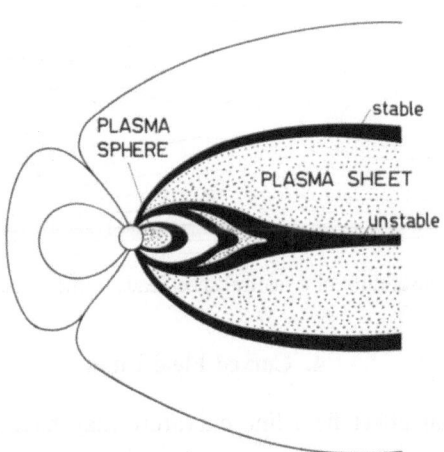

Fig. 2.   The regions of negative V (black) as obtained from Figure 1. In these regions the tearing
instability might occur ($V < 0$ is necessary but not sufficient for instability). As indicated, the far
outer edge of the plasma sheet is stable whereas the neutral sheet is unstable.

3. Straight Field Lines

The case of a unidirectional magnetic field, varying in one direction perpendicular to the field, has been treated for general distribution functions $F(H, P_z)$ by Schindler and Soop (1968).

The necessary and sufficient condition for stability against tearing is simply that the magnetic field B must not go through zero. That means that if B vanishes somewhere, the tearing instability must occur. Thus the neutral sheet is unstable whereas the (plane part of the) outer edge of the plasma sheet is tearing stable, as indicated in Figure 2. The geometry of the tearing instability is sketched in Figure 3. There is considerable discussion of the tearing instability of the neutral sheets in the literature (see e.g. Schindler, 1968).

It will therefore be sufficient to mention only a very recent result (to be published).

It seems that the tearing instability, even though active microscopically, does not lead to gross destruction of a neutral sheet. The reason is that there is strong non-linear stabilization, active if the width of the sheet is much larger than the electron Larmor radius. If one assumes a random tearing mode noise, an anisotropy develops quickly which stabilizes the instability. The neutral sheet turbulence is of interest for particle acceleration processes.

The absence of a macroscopic effect is consistent with the observation of a neutral sheet (in a macroscopic sense) centered in the magnetospheric tail (Ness, 1965).

Similarly, this may explain why the sector boundaries in the interplanetary medium do not undergo gross tearing. The inconsistency of the observations with unlimited growth of the loops was pointed out by Dessler (1967).

The absence of gross tearing effects is closely related to the fact that in a neutral sheet the equilibrium field lines are approximately straight.

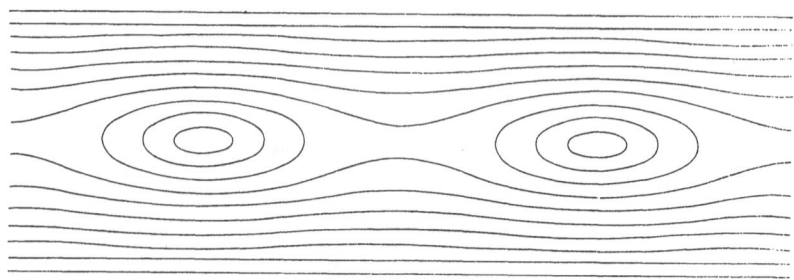

Fig. 3.   The tearing instability of a neutral sheet leads to the formation of field line loops.

4. Curved Field Lines

An indication of what effect field line curvature may have can be obtained in the following way.

Let's assume that B_0 does not vanish. Then one can substitute in (6),

$$\phi = A_1/B_0$$

obtaining

$$W = \int \left[(\nabla \phi)^2 - (\nabla \alpha)^2 \right] B_0^2 \, dx \, dy \tag{9}$$

where

$$\alpha = \arctan(B_{0y}/B_{0x})$$

is the angle between the field line and the x-direction.

For $\nabla \alpha = 0$ (straight lines) one recovers the stability part of the general criterion discussed in the preceding section. Clearly $\nabla \alpha$ has a destabilizing effect. The field line curvature must however be sufficiently pronounced in order to lead to an instability. For instance, it is easy to verify that vacuum fields are stable.

A quantitative (sufficient) stability criterion is obtained in the following way.

Suppose one can find Cartesian coordinates such that everywhere $B_{0x} \neq 0$. Then substituting into (6)

$$\Psi = A_1/B_{0x}$$

one obtains

$$W = \int B_{0x}^2 (\nabla \Psi)^2 \, dx \, dy > 0$$

i.e. the situation is stable.

Figure 4 shows a case where this criterion is satisfied (a) and where it is violated (b). The tendency to create a weak field region in case (b) is obvious.

Furthermore, the field loops (b) are developed to the extent that an antiparallel field situation is created. This points towards reconnection of field lines similar to tearing of a neutral sheet (Figure 3).

The discussed properties suggest that the tearing instability will be most pronounced in regions where V is negative over large areas in which the field lines are strongly curved.

Therefore, in our model magnetosphere the tearing will be most pronounced in the region where the neutral sheet connects onto the closed-field-line region. In fact, observations seem to suggest that there is an instability of macroscopic nature originating from that region (e.g. Roederer, this volume, p. 239). The present con-

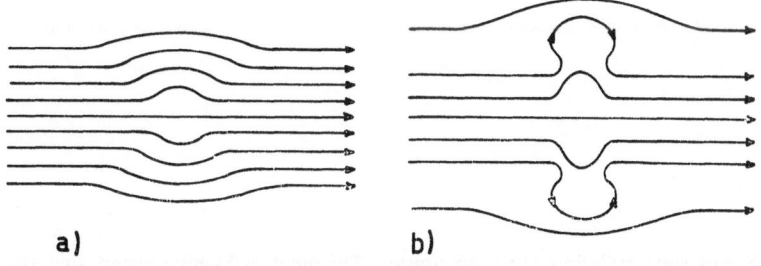

a)   b)

Fig. 4. – (a) There is a direction such that the corresponding component of the magnetic field is positive everywhere: configuration stable. – (b) There is no such direction: necessary for tearing instability. Note formation of antiparallel field situation.

siderations might also have some bearing on the mechanism of solar flares (Axford, this volume, Summary Lecture, p. 7).

Clearly, further study is necessary to find out what role tearing plays in these processes. So far we might say that we have a promising candidate.

5. Radiation Belts

It seems to be generally accepted that radiation belts undergo instabilities of microscopic nature, leading to particle diffusion (Kennel, 1969).

A macroscopic instability which might be of importance for radiation belts is the exchange mode (Swift, 1967). This mode would depend on r and ϕ (using cylindrical coordinates r, ϕ, z).

The tearing mode is an alternative possibility. Being constant along the direction of equilibrium current (ϕ-direction) it describes perturbations lying in the r-z-plane. It will be of particular importance for strongly inflated belts with $\beta \gtrsim 1$, where β is the maximum pressure to dipole magnetic energy density ratio. The first task is of course to find the corresponding equilibria by prescribing the distribution function

$$f_0 = F_0(H, P_\phi) \tag{10}$$

in analogy to (1).

This problem was solved by K. Lackner (unpublished).* Figure 5 shows the field lines of a case with $\beta = 27$. This is an extreme case. Field reversal occurs already at considerably lower values of β.

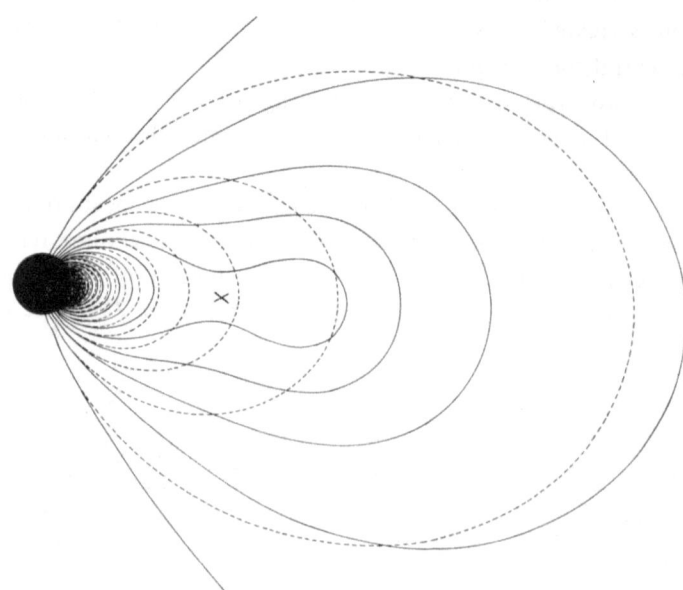

Fig 5.   Steady state radiation belt configuration. The dipole is highly inflated, such that a neutral line ($\times$) appears (computed by K. Lackner).

* Will be published in *J. Geophys. Res.*

In the present context we confine ourselves to giving a rough estimate which indicates under what conditions one might expect the tearing instability to occur. The sufficient stability criterion (7) holds analogously also for a rotationally symmetric case with $\partial/\partial\phi=0$. An equilibrium is stable if for all A_1

$$W = \int \left[(\nabla \times \mathbf{A}_1)^2 + V A_1^2 \right] r \, dr \, dz > 0$$

where

$$\mathbf{A}_1 = (0, A_1, 0) \quad \text{and} \quad V = -\mu_0 \frac{\partial j_0(A_0, r)}{\partial A_0}.$$

In the limiting case ($W_{\min}=0$) the minimizing function A_1 satisfies a Schrödinger-type equation, with vanishing eigenvalue

$$-\varDelta A_1 + \frac{A_1}{r^2} - \mu_0 \frac{\partial j_0}{\partial A_0} A_1 = 0.$$

A rough order of magnitude estimate gives

$$\mu_0 \frac{\partial j_0}{\partial A_0} \approx \mu_0 \frac{\delta j_0}{\delta A_0} \approx \frac{B_p/L}{BL} \approx \frac{B_p}{B} \cdot \frac{1}{L^2}$$

where B_p is the perturbed field, B is the total field, and L is the characteristic length of the perturbation. Since $\varDelta \approx 1/L^2$ we find that the field perturbation has to be of the same order as the total field. This is roughly the same as saying that

$$\beta \gtrsim 1.$$

Low β radiation belts are tearing stable. The physical reason of course is that there is not sufficient energy in the plasma to provide the magnetic energy necessary for field deformation.

A detailed tearing stability analysis seems desirable because observations show that β can be larger than one (Frank, 1967).

Acknowledgements

It is a pleasure for me to thank my colleagues at ESRIN for many stimulating discussions. In addition, I should like to thank Dr. K. Lackner for providing Figure 5.

References

Coppi, B., Laval, G., and Pellat, R.: 1966, *Phys. Rev. Letters* 16, 1207.
Dessler, A. J.: 1967, *Rev. Geophys.* 5, 1.
Frank, L. A.: 1967, *J. Geophys. Res.* 72, 3753.
Kennel, C. F.: 1969, *Rev. Geophys.* 7, 379.
Mihalov, J. D., Colburn, D. S., Currie, R. G., and Sonnett, C. P.: 1968, *J. Geophys. Res.* 73, 943.
Ness, N. F.: 1965, *J. Geophys. Res.* 70, 2989.
Schindler, K. (ed.): *The Stability of Plane Plasmas*, ESRO SP-36.
Schindler, K. and Soop, M.: 1968, *Phys. Fluids* 11, 1192.
Swift, D. W.: 1967, *Planetary Space Sci.* 15, 1225.
Wilcox, J. M. and Ness, N. F.: 1965, *J. Geophys. Res.* 70, 5793.

6. SOME IONOSPHERIC EFFECTS FOLLOWING SOLAR EVENTS

CURRENT PROBLEMS IN POLAR-CAP ABSORPTION

GEORGE C. REID

*Cooperative Institute for Research in the Environmental Sciences, University of Colorado, U.S.A.
and ESSA Research Laboratories, Boulder, Colo., U.S.A.*

1. Introduction

Polar-cap absorption was first recognized as a distinct ionospheric phenomenon after the great cosmic-ray flare of 23 February 1956 and the somewhat less dramatic proton flares that followed during the early part of the International Geophysical Year (Bailey, 1957; Bailey, 1959; Reid and Collins, 1959; Hultqvist and Ortner, 1959; Reid and Leinbach, 1959). During the intervening 12 years or so, a large amount has been learned about its characteristics, and this knowledge has contributed in a significant way to our understanding of solar-terrestrial physics. This paper will outline some of the more important features of PCA, describe a few areas in which our understanding is still far from adequate, and comment on ways in which coordinated satellite and ground-based observations can contribute.

PCA represents the ionospheric response to the large fluxes of solar protons and heavier nuclei that are generated by certain flares. These particles create abnormally large amounts of ionization in the lower part of the ionosphere, giving rise to radio-propagation effects that can be readily detected by a wide variety of techniques. The most noticeable effect is the occurrence of intense absorption of radio waves in the HF and lower VHF bands, creating severe communications problems at high latitudes, and also being responsible for the name 'polar-cap absorption'. Events that are too weak to be recorded through HF absorption can be detected at frequencies in the LF and VLF bands through changes in the phase and amplitude of signals propagating across the polar caps (Eriksen and Landmark, 1961; Egeland et al., 1961; Belrose and Ross, 1961; Bates, 1962; Westerlund et al., 1969). Undoubtedly these latter observations provide the most sensitive ground-based technique available for detecting solar-proton fluxes in the ionosphere, but unfortunately our understanding of the propagation mechanisms involved is not yet well enough developed to allow any meaningful quantitative conclusions to be drawn from them. In fact, the bulk of the quantitative data obtained on PCA has come from the riometer technique, which monitors the strength of cosmic-noise signals propagating roughly vertically through the ionosphere, and this paper will lean heavily on the results of riometer measurements, and their relation to satellite observations of solar-proton fluxes.

In particular, PCA shows marked variations with both local time and position on the earth which are unrelated to variations in the flux and spectrum of the incident solar protons. These temporal and spatial variations are perhaps the most interesting features of a PCA event, and have been shown to be a potential source of a great deal

V. Manno and D. E. Page (eds.), Intercorrelated Satellite Observations Related to Solar Events, 319–334. *All Rights Reserved*
Copyright © 1970 by D. Reidel Publishing Company, Dordrecht-Holland

GEORGE C. REID

of information about the structure of the magnetosphere and about the physical
properties of the lowest region of the ionosphere; and the remainder of the paper will
be devoted to a discussion of the present status of our knowledge of them, and the
direction in which further studies might be aimed.

2. Diurnal Variation of PCA

One of the most pronounced features of PCA as observed by HF-absorption or
VHF-forward-scatter observations is the existence of a strong diurnal variation,
which has excited a considerable amount of attention since the earliest observations.
The variation is not continuous throughout the day, like the diurnal variation in the
higher regions of the ionosphere, but consists of a relatively rapid transition during the
twilight period between a constant daytime value and a constant nighttime value. The
nighttime absorption measured by a riometer is typically less than the daytime absorption by a factor of from 3 to 6 (in decibels), and a plot of absorption as a function
of local time in the presence of a constant flux of protons thus has a trapezoidal shape.
At high geographic latitudes near the solstices, when there is either continuous day or
continuous night, the diurnal variation disappears entirely.

The magnitude of the absorption observed during a PCA event obviously depends
on the complicated chemical reactions that determine the steady-state electron density
maintained by the ionizing protons in the lower ionosphere, and measurements of
PCA, in particular its twilight variation, are a powerful source of information about
these reactions. In the work carried out to date, most effort has gone into attempting

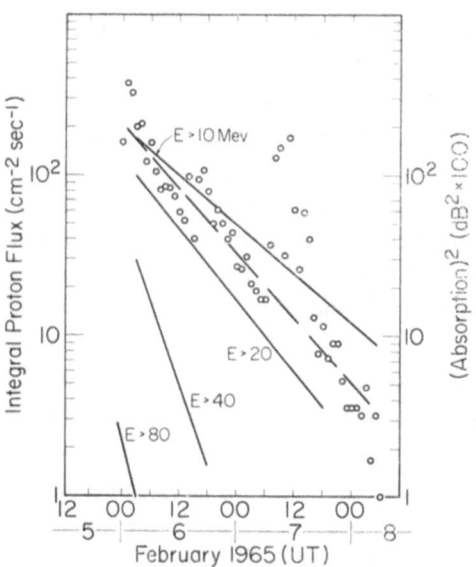

Fig. 1.   Decay of the flux of protons in four energy channels during the solar-proton event of February
1965, compared with the decay of the 30-MHz absorption recorded at the South Pole.
The proton flux data is taken from Paulikas *et al*. (1966).

to construct models that will quantitatively predict the magnitude of the daytime absorption, and much less attention has been devoted to the equally important problems of the steady nighttime absorption and the detailed twilight variation.

Observationally, a riometer operating well within the polar cap at a frequency of 30 MHz can be considered as responding to the integral flux of protons with energy above about 20 MeV during daytime conditions in the lower ionosphere. This is illustrated in Figure 1, which shows data obtained during the solar-minimum PCA of February 1965. The four straight lines show the approximately exponential decay with time of the fluxes of protons above 10, 20, 40, and 80 MeV as recorded on the satellite 1964-45A over the polar caps by Paulikas *et al.* (1966). The individual points show the square of the vertical absorption recorded by a 30-MHz riometer at the South Pole, which was in continuous sunlight; the broken line is an attempt to fit a straight line to the periods when the PCA was not obscured by superimposed auroral absorption. The square of the absorption was used since the ambient electron number density is expected to be proportional to the square-root of the rate of electron production, and hence the square-root of the proton flux. It can be seen that a relation of this kind is a fairly good approximation, and that the decay of the absorption during this event followed the decay of the proton flux above 20 MeV quite closely. The numerical relation derived from this data is

$$J(> 20 \text{ MeV}) = 61 \ A^2 \tag{1}$$

where J is the 2π-omnidirectional flux in protons cm^{-2} sec^{-1} and A is the vertical absorption in decibels observed at a frequency of 30 MHz in the sunlit polar cap.

In principle, the energy response of the riometer can be changed by changing the frequency, since lower frequencies respond to ionization at higher altitudes produced mainly by protons of lower energies. In practice, however, it is not possible to alter the response much in this way. For example, to lower the effective response from 20 MeV to 2 MeV would mean lowering the frequency from 30 MHz to about 0.5 MHz, which is well below the F-region penetration frequency; raising the response to 200 MeV would mean increasing the frequency to about 1800 MHz, where ionospheric absorption effects are completely unmeasurable by existing techniques. During an 'average' PCA event, then, the riometer technique is capable of monitoring from the ground the flux of protons above about 20 MeV under daytime conditions, and gives little information about the flux at energies much higher or lower than this.

During the night, however, the situation is apparently radically changed. Figure 2 shows the 30-MHz absorption recorded by several riometers under both daytime and nighttime conditions during the event of September 1966 (Reid, 1969). Note first that the nighttime values are considerably lower than the daytime values, reflecting the diurnal variation mentioned above. Of more importance, however, is the fact that the shape of the two sets of points is quite different, implying that the nighttime absorption is not responding to the same protons as the daytime absorption. This has been verified by correlating the absorption with the fluxes of protons in different energy ranges as recorded by the satellite 1963-38C (Bostrom, private communication) over the polar caps. Figure 3 shows the correlation coefficient as a function of proton energy for both daytime and nighttime conditions, and demonstrates that during the night the riometer is apparently more sensitive to protons with energies between about 3 and 10 MeV than to the fluxes above 20 MeV that cause the bulk of the daytime effect.

This result suggests that the bulk of the daytime absorption comes from a very low region in the ionosphere that is heavily ionized by protons with energy greater than 20 MeV (vertical penetration altitude 54 km), while most of the nighttime effect is due to a considerably higher region ionized chiefly by protons with energy less than

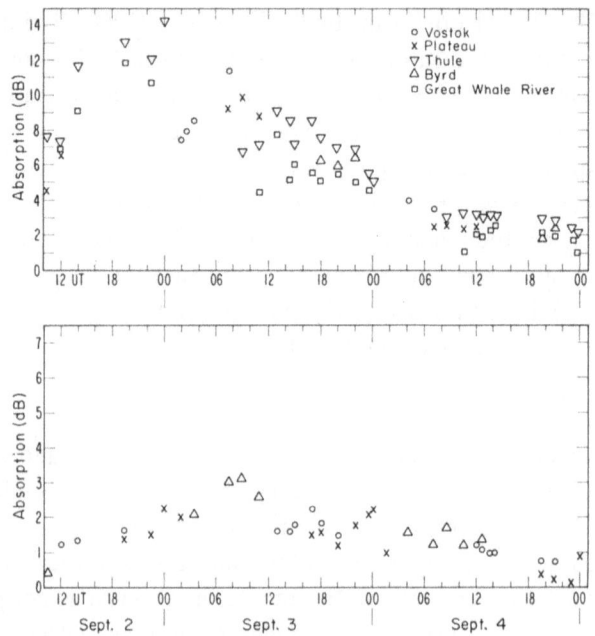

Fig. 2.   30-MHz absorption recorded by several high-latitude riometers during the PCA event of September 1966. The upper diagram shows daytime observations, and the lower nighttime observations (Reid, 1969).

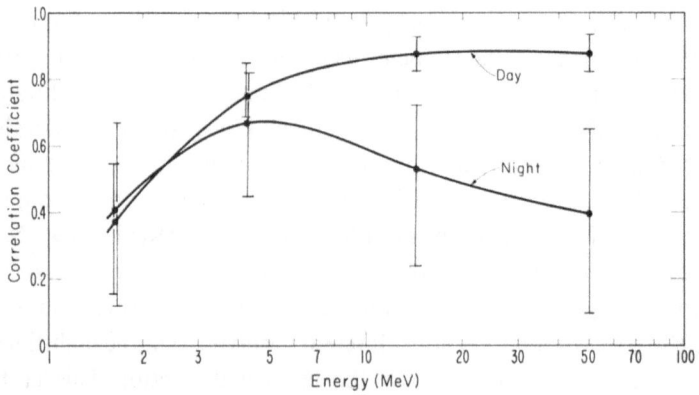

Fig. 3.   Linear correlation coefficient between 30-MHz absorption and the square root of the proton flux for daytime and nighttime conditions (Reid, 1969).

about 10 MeV, and possibly in the vicinity of 5 MeV (vertical penetration altitude 73 km). This result will be discussed later, after a brief look at the twilight variation.

The twilight variation shows us the detailed way in which the lower ionosphere makes the transition between the two steady-state conditions discussed above, and is well worth studying from this point of view. It was realized at an early stage that the key to the twilight variation probably lay in the formation of negative ions by electron attachment to neutral atoms and molecules, and that the chief difference between day and night conditions was the existence of photodetachment reactions in sunlight and their absence during the night (Bailey, 1957). An anomaly soon became evident in the observations, however, since the onset of the twilight variation at sunset was observed in several events to occur at a solar zenith angle of close to $90°$; this corresponds to ground sunset, when the lower ionosphere is still in essentially unattenuated visible sunlight. The negative-ion species that might have been expected to be the dominant one, O_2^-, however, is known to be photodetached efficiently by the visible spectrum (Smith et al., 1958), and twilight would not have been expected to begin until the visible-light shadow reached the base of the ionized region, i.e., until a solar zenith angle of about $93°$. Thus it became evident that, at least at the base of the ionized region, which is first affected at sunset, the dominant negative ion probably required ultraviolet light for efficient photodetachment, and the effective shadow was not that of the earth itself, but corresponded more nearly to that of the ozone layer with an effective top in the neighborhood of 35 km (Reid, 1961).

Since then, the effectiveness of ultraviolet light, and the relative inefficiency of visible light, have been shown quite clearly in the case of the quiet D region by rocket observations (Bowhill and Smith, 1966), and a great deal of attention has been devoted to a study of the possible negative-ion reactions that might occur in the mesosphere. One of the most important results to emerge has been the overwhelming importance of associative-detachment reactions such as

$$O_2^- + O \rightarrow O_3 + e$$

in which O_2^- is destroyed by atomic oxygen (Fehsenfeld et al., 1967), producing a free electron, and this result opened up the possibility that the diurnal variation in PCA might not be due to photodetachment at all, but might simply reflect the diurnal variation of atomic-oxygen concentration in the mesosphere. Since the two main sources of O in the altitude region of concern are photo-dissociation of O_3 and O_2 in the ultraviolet Hartley and Herzberg continua respectively, and since the time-constant for decay of O at sunset below about 75 km is very short, the observed ultraviolet-shadow effect could be explained without invoking any photodetachment. This mechanism could also explain qualitatively the altitude difference between the day-time and nighttime absorption effects mentioned above, since above 75 km the lifetime of O rapidly becomes long compared with the duration of the night (except possibly during the long polar nights), and there should be little diurnal variation. The relative importance of oxygen dissociation vs. ultraviolet photodetachment in the PCA region remains an unanswered question.

Information about the top of the negative-ion region ought to be available from observations of the onset of twilight at sunrise, since the upper regions are the first to be affected. The evidence on this is not so clear or consistent as in the case of the sunset onset, but in several cases a rather sharp sunrise onset has been observed at a solar zenith angle of about 99°. An example is shown in Figure 4, where the 30-MHz absorption observed at Dumont d'Urville, Antarctica, is shown for the period May 24–25, 1967 (Lavergnat, private communication), together with the absorption measured at the South Pole, which was in continuous darkness. Dumont d'Urville experienced a brief period of daytime absorption, which began and ended rather abruptly at a solar zenith angle of 99°. This is shown more clearly in Figure 5, in which the ratio of the absorption at Dumont d'Urville to that at South Pole is plotted as a function of solar zenith angle at Dumont d'Urville for sunrise and sunset. The average ratio is close to unity for zenith angles in excess of 99° both before sunrise and after sunset. The visible-light shadow for a zenith angle of 99° lies at about 80 km altitude; refraction tends to lower this height, but attenuation of the solar radiation in passing through the lower atmosphere tends to raise the effective shadow height again, so that the effective shadow for visible light probably lies in the vicinity of 80

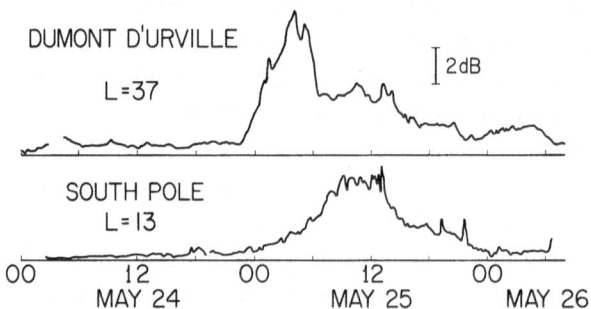

Fig. 4.   30-MHz absorption recorded at Dumont d'Urville (Lavergnat, private communication) and South Pole during the PCA event of May 1967.

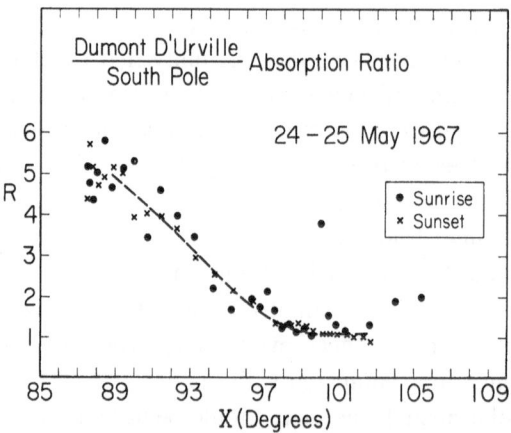

Fig. 5.   Ratio (R) of absorption recorded at Dumont d'Urville and South Pole as a function of solar zenith angle at Dumont d'Urville during 24–25 May 1967.

km. The ultraviolet shadow, however, lies above 110 km, so that it is unlikely to be having any appreciable influence on the negative-ion region at this time. The small amount of evidence that exists thus suggests that the top of the negative-ion region may be sharply bounded (as revealed by the suddenness of the onset and ending of nighttime conditions) and may contain ions that can be photodetached by visible sunlight; O_2^- is one obvious possibility.

Existing observational evidence thus suggests that the negative ions near the base of the PCA region of the ionosphere (about 30–40 km) are tightly bound, requiring ultraviolet light for photodetachment, while the negative ions near the top of the negative-ion layer (about 75–80 km) may be less tightly bound, being destroyed by visible light.

Detailed models of the ion chemistry of the mesosphere have been proposed (Adams and Megill, 1967; LeLevier and Branscomb, 1968), based on the reaction rates measured in the laboratory and estimates of the concentrations of various minor neutral species. While these models have shown a certain amount of agreement with observation, a great deal of work remains to be done. The ideal way to attack the problem would be to measure the electron-density profile and the ion mass spectrum, together with the proton flux and spectrum as a function of altitude, by rocket techniques. To obtain a satisfactory picture, however, such measurements would have to be carried out during a large number of PCA events, at a variety of locations, and dur ng several local times. In practice, rocket probing of PCA events has proven to be a frustrating task, largely due to the comparative rarity of the events, even near the peak of the solar cycle, and to the consequent logistic difficulties of waiting in readiness for long periods at remote rocket facilities within the polar cap. Less satisfactory, but perhaps more likely to yield early results, is the combination of ground-based observations of the ionosphere with satellite measurements of proton fluxes and spectra within the loss cone over the ground base. Riometer observations of integrated absorption have the advantage of yielding long-term synoptic data with relatively little difficulty, but the disadvantage of providing only a height-integrated measurement of the mesospheric ionization. An estimate of the electron-density profile can be made from multi-frequency riometer measurements (Parthasarathy et al., 1963), and the results of one such experiment have been combined with satellite observations to produce an effective electron-loss coefficient for the lower ionosphere (Adams and Masley, 1965). Recently, however, more direct measurements of electron-density profiles during PCA events have been carried out at Resolute Bay using the partial-reflection technique (Hewitt, 1969), and a combination of measurements of this kind with satellite measurements of proton flux appears to be a potentially valuable source of information on the complex physical and chemical mechanisms of the mesosphere.

3. Spatial Variation of PCA

The confinement of PCA to the polar-cap region of high magnetic latitudes is qualitatively the kind of effect that would be predicted by the classical Störmer theory of the motion of charged particles in a dipole magnetic field, and its later refinements to include higher-order terms in the spherical-harmonic expansion of the geomagnetic field (Quenby and Webber, 1959; Sauer, 1963). Störmer theory predicts that 100-MeV protons should have access to the earth only at geomagnetic latitudes higher than about 68°, and this is roughly in accord with the observation that PCA effects are normally seen only at latitudes higher than that of the auroral zones. More detailed examination of the shape of the edge of the PCA region, however, reveals marked departures from the predictions of Störmer theory. For example, a station such as Thule, with a geomagnetic latitude close to 90°, ought to have essentially zero cutoff energy for protons, while College, at a geomagnetic latitude of about 65°, should lie outside the PCA region almost entirely, with a cutoff energy in the vicinity of 120 MeV. Observationally, however, Leinbach et al. (1965) found that the ratio of PCA at College to that at Thule was about 0.8 under quiet magnetic conditions. Furthermore, observations at Farewell, only about 3° south of College, showed that the absorption there was only about one-sixth of that at College. The observations thus showed that the absorption was much more uniform over the polar cap, and showed a much more abrupt edge than Störmer theory would have predicted.

The solution to this problem soon became apparent after the discovery of the

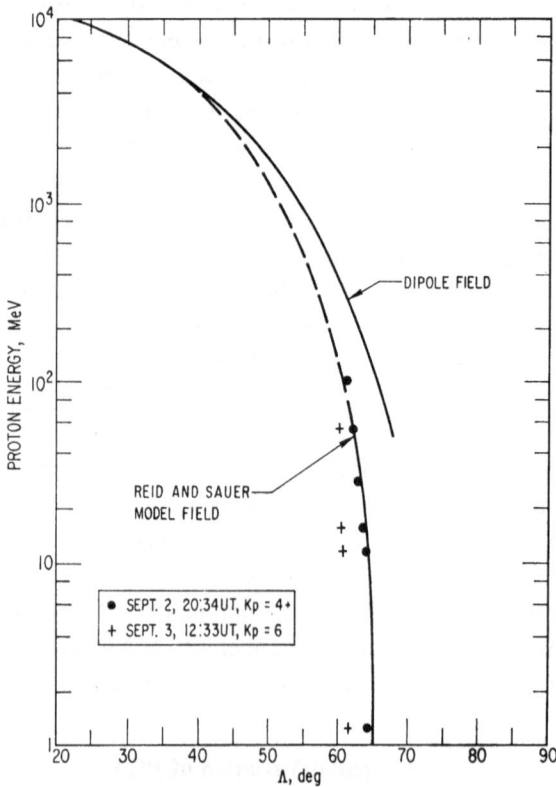

Fig. 6. Geomagnetic cutoffs near the midnight meridian predicted by Störmer theory for a dipole field and by a simplified theory including the effect of a geomagnetic tail. The individual points represent satellite measurements (from Paulikas et al., 1968).

geomagnetic tail (Ness, 1965). The geomagnetic field lines from the polar caps ultimately form the tail, which is apparently a sufficiently 'open' structure that it can be
permeated by nearly all the protons responsible for PCA, thus accounting for the flat
'plateau' appearance of PCA across the polar caps. Calculations based on a rather
simple model for the geomagnetic field immediately inside the tail region, and
assuming a completely accessible tail, yielded the proton-cutoff vs. latitude relation
shown in Figure 6 (Reid and Sauer, 1967a), in which the Störmer predictions are also
shown, together with some experimental measurements of Paulikas *et al.* (1968). The
effect of the tail is to produce a large reduction of cutoff down to auroral latitudes, and
an extremely rapid increase in cutoff with decreasing latitude through the subauroral
region, accounting for the sharp edge observed in the PCA region. The predicted
cutoffs for selected proton energies over North America are shown in Figure 7; as
mentioned above, the protons responsible for PCA mainly lie between energies of 1
and 100 MeV, and thus the edge of the PCA plateau is mapped out roughly by the
region lying between the 1 and 100-MeV contours in Figure 7, which corresponds to a
linear distance of some 500 km.

The calculation on which these predictions were made is valid only for the local-
time period close to magnetic midnight, when the station is facing in the direction of
the tail. Observationally, however, the edge of the PCA region does not vary much
with local time, except under special circumstances that will be mentioned below. The
magnetic field lines delineating this edge, corresponding to L-values between about
4 and 5.5, certainly lie well within the domain of closed field lines on the day-side of
the magnetosphere. Thus the protons responsible for daytime PCA probably reach

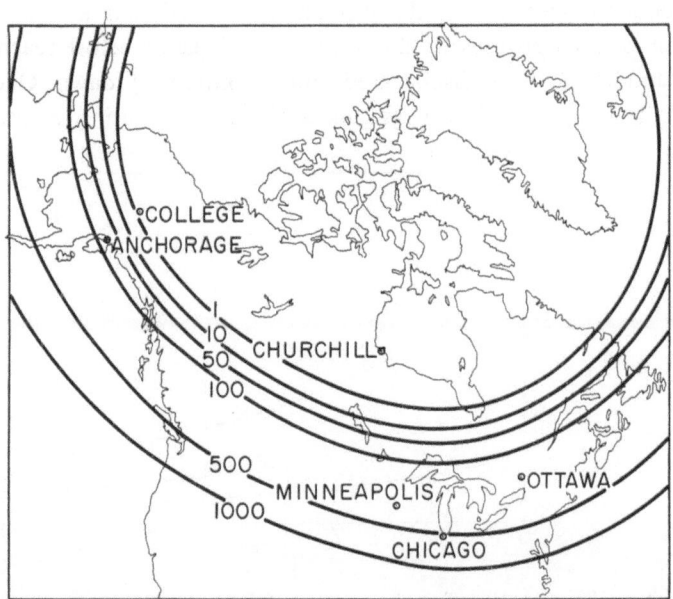

Fig. 7.   Predicted geomagnetic cutoffs near local midnight over North America for moderately
disturbed conditions. The labels on the contours are proton energies in MeV (Reid and Sauer, 1967).

these latitudes by performing rather complicated orbits that carry them around from the tail. Detailed orbit calculations in a model field have been carried out by Taylor (1967) and by Gall *et al.* (1968), and have shown that orbits of this kind must certainly exist. At the lower energies the protons on the dayside of the magnetosphere will be 'pseudo-trapped', in the sense described by Roederer (1967), i.e., the dayside segment of their orbit will be very similar to the orbit of a trapped proton bouncing back and forth between mirror points and drifting in longitude. Roederer (1967) has shown that the mirror-point domain for pseudo-trapped protons injected on the midnight meridian lies at rather high latitudes on the dayside, and thus contains protons with rather small equatorial pitch-angles. The protons recorded by PCA measurements, of course, lie within the atmospheric loss-cone, and are thus well within the pseudo-trapping domain.

This rather simple explanation of the appearance of PCA at L-values as low as 5.5 on the dayside of the magnetosphere does not seem to provide the whole answer, however. The pseudo-trapped protons bounce back and forth between mirror points as they drift around from the dusk to the dawn side of the magnetosphere; in the case of the protons that produce PCA the mirror points must lie in the atmosphere, and it is difficult to see how a proton can survive more than one or two bounces. Using a dipole-field geometry (an adequate approximation for the dayside of the magneto-sphere well inside the magnetopause), the number of bounces can be readily calculated from existing expressions for the bounce period and the drift period (Hamlin *et al.*, 1961). Figure 8 shows the number of bounces, N, made by a proton of energy 5, 20, and 80 MeV in drifting through $180°$ of longitude on L-shells varying from 5 to 10. At $L=6$, a 20-MeV proton (corresponding to the average response of a sunlit riometer) bounces about 10 times between the dusk and dawn meridians, and one would expect that the flux of such protons would be seriously depleted before reaching the noon meridian, and would have disappeared almost entirely by dawn. One would then expect PCA to have a very low magnitude at this L value during the pre-noon hours,

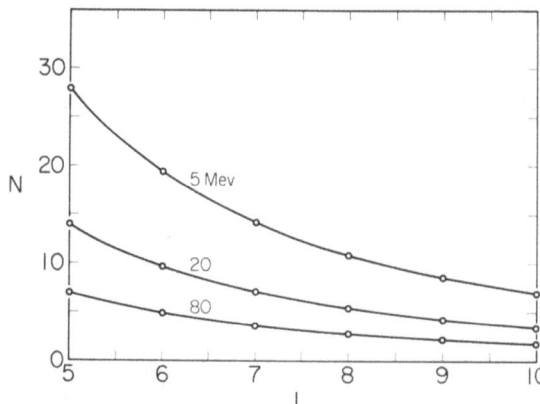

Fig. 8.   Number of bounces (N) suffered by a proton of the energy indicated in the course of drifting through $180°$ of longitude from the dusk to the dawn meridian in a dipole field. The abscissa is the L-value of the magnetic shell on which the proton drifts.

gradually increasing through the afternoon to a constant high nighttime value (forgetting for the moment the additional complication of the diurnal ionospheric response), and dropping rather abruptly to a very low value again near dawn. The situation is still further complicated by the 'shell-splitting' predicted by the calculations of Roederer (1967). Particles injected on the same magnetic shell near the midnight meridian with different equatorial pitch-angles will drift to somewhat different shells in the day-side of the magnetosphere, and their equatorial pitch-angles will also change, modifying the rate of precipitation into the atmosphere with both latitude and local time.

Observationally, PCA is often found to be quite uniform in local time at a given location on the sunward side of the magnetosphere, and this uniformity is difficult to explain simply on the basis of the drift paths of pseudo-trapped protons in a magnetosphere that is symmetrical about the sun-earth axis. On other occasions, however, a pronounced decrease in absorption is observed in the middle of the day at locations near the low-latitude boundary of the PCA region. This 'midday recovery' has been discussed in detail by Leinbach (1967), who suggests that it may be related to the development of an anisotropic pitch-angle distribution among the protons drifting around the sunward side of the magnetosphere, rather than to a simple increase in cutoff energy, as had been suggested earlier (Leinbach, 1962; Reid and Sauer, 1967a). It is possible that such an anisotropy might be associated with the shell-splitting of pseudo-trapped protons mentioned above, and the accompanying rise in mirror-point altitudes as the protons drift from the dusk meridian toward noon. Beyond noon, the mirror points should drop down again, and protons that were able to avoid precipitation in the evening sector by having their mirror points raised might be precipitated in the morning sector as their mirror points descend again. The details of this mechanism for explaining the midday recovery phenomenon are presently being investigated.

The midday recovery shows some additional features, however, that remain unexplained. In particular, only some 20% of PCA events have shown any midday recovery at all (Leinbach, 1967), whereas mechanisms of the type discussed above would predict its appearance in every event. Furthermore, an event that lasts for several days rarely displays a midday recovery on more than one day, and no other feature has yet been found that correlates with the appearance or otherwise of a midday recovery. This is an area in which satellite observations can play a key role, in particular in searching for anisotropies in the proton pitch-angle distribution near the edge of the polar-cap plateau on the sunward side of the magnetosphere. Since the riometer observations during day-time probably reflect the fluxes of protons above 20 MeV as indicated above, satellite measurements in this energy range should be particularly valuable.

4. Magnetic-Storm Effects

During magnetic storms, the PCA region is often observed to expand rather rapidly to lower latitudes, and then retreat more slowly back to its original location. This is the ionospheric counterpart of the magnetic-storm depression of cosmic-ray cutoffs that has been discussed by several authors. An example of this effect is shown in Figure 9, where riometer observations made at a chain of stations in Alaska are shown for a portion of the major PCA event of May 23–26, 1967. Also shown are the H-component magnetograms from Honolulu and College (the latter on a reduced vertical scale). Two storm sudden commencements occurred during the period shown, one at 1021 UT and the other at 1235 UT. In association with this magnetic disturbance, the edge of the PCA region obviously swept down to considerably lower L values than before. The recovery is not shown here, and is complicated by the fact that in this event the absorption did show a marked midday recovery at all the southerly locations, centered at about 2000 UT and both widening and deepening with decreasing latitude.

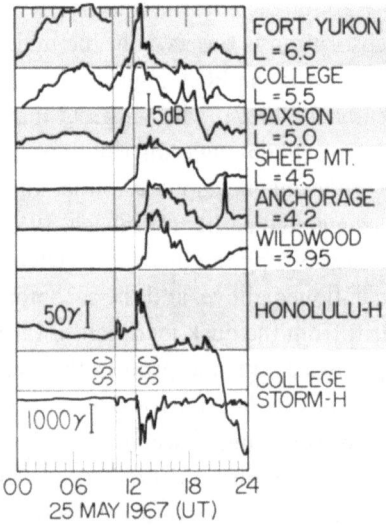

Fig. 9.   30-MHz absorption recorded at a chain of stations in Alaska during the PCA event of May
1967. The horizontal component of the geomagnetic field recorded at Honolulu
and College is also shown.

The onset of the storm-associated motion of the PCA edge is not clearly correlated with any unique feature of the magnetograms; there was no pronounced main phase (reflecting the build-up of a ring current) until well after the cutoff-decrease appeared, and a polar substorm at College had subsided long before the cutoff-decrease recovered.

This lack of correspondence with features on magnetograms may not be too surprising. As discussed above, it is now fairly well established that the permanent lowering of cutoffs below the Störmer values is due to the permanent existence of a geomagnetic tail that allows fairly easy access to solar protons. The simplest interpretation of the storm-reduced cutoffs is to assume that they reflect the opening up of more field lines, and the consequent intrusion of the influence of the tail to lower latitudes than usual. Since the existence of the tail is a consequence of the pressure of the solar wind on the magnetosphere, any increase in this pressure ought to result in the stretching of more field lines into the tail. This phenomenon cannot readily be detected by groundbased magnetometers, however, as has been shown by the rather poor correlation between such parameters as K_p and the solar-wind flux or energy density or the standoff distance of the magnetopause. It may be that the reduction in cutoff associated with magnetic storms should be correlated directly with the solar-wind parameters or with the standdoff distance, but no such correlation has yet been carried out. Here again there is an opportunity for satellite observations to play an important role in carrying out such a correlation. The various parameters of the solar wind can be monitored continuously by satellites located outside the magnetopause or bow shock, and the position of the geomagnetic cutoff for solar protons of different energy ranges can be observed by polar-orbiting satellites as they enter the polar caps; in the case of the more intense events, this information can be at least partly provided by a riometer chain. It would be of considerable interest to see whether such a correlation exists, and if so, which are the most important solar-wind parameters. It would also be of interest to observe the change in cutoff at different local times, with the hope of obtaining some information about the nature of the transient response of the magnetosphere to sudden changes in the solar wind.

A more uncommon type of magnetic-storm variation in PCA is shown in Figure 10, where riometer recordings from three stations are shown for the same event as that in Figure 9. Byrd and Great Whale River are magnetically conjugate locations; during the period shown, the lower ionosphere at Byrd was in continuous darkness, while Great Whale River was mainly in sunlight, but with a short period of darkness shown by the pronounced decrease in absorption between 0100 and 0800 UT on 25

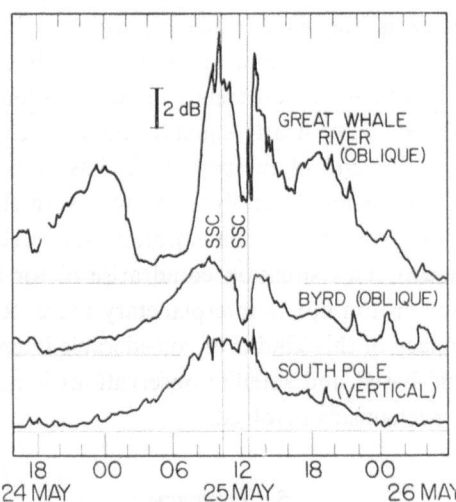

Fig. 10.   30-MHz absorption recorded at three stations during the PCA event of May 1967. The riometer antennas at Byrd and Great Whale River are directed obliquely at an angle of 45° to the vertical.

May. The record from South Pole, also in continuous darkness, is shown at the bottom.

Following the first of the two sudden commencements, at 1021 UT, there is a very marked decrease in absorption at both Great Whale River and Byrd, and the absorption is rapidly restored to near its original level following the second sudden commencement at 1235. That this decrease does not reflect a decrease in the interplanetary flux of protons is shown by its absence at South Pole; satellite measurements also give no evidence of any major decrease at this time, and we are forced to the conclusion that the decrease at the two conjugate stations represents some kind of increase in local cutoff, which must have been restricted in local time, since there is no trace of it in the Alaskan data shown in Figure 9. Protons with energies at least as high as 20 MeV were probably affected, since we have seen that this is the energy domain to which the sunlit riometer at Great Whale River was probably responding.

Similar effects have been reported previously by Ortner *et al.* (1961), and have been associated with the initial phase of magnetic storms. Qualitatively one might expect that the compression of the lines of force associated with the initial phase of a storm would increase field strengths on the sunward side of the magnetosphere and thus increase cutoffs, but this interpretation implicitly assumes that protons are entering this region of increased field directly from interplanetary space. If we accept the concept of the dayside protons executing pseudo-trapped orbits originating in the tail, the consequences of an increased field strength on the sunward side are not so clear. Furthermore, in the example shown here the magnitude of the initial phase following the first SC was much less than that of the initial phase accompanying the second SC, when the cutoffs were apparently restored to their pre-storm levels.

Figure 10 shows another anomaly that is not immediately explainable. During the period 0200 to 0700 UT on 25 May, the Great Whale River riometer is recording nighttime absorption, as is the Byrd riometer, which is in nighttime conditions throughout the event. These two stations are approximately conjugate magnetically, yet there is an obvious and major difference in their response. The absorption at Byrd climbs steadily from a value nearly equal to that at Great Whale River to more than twice that value at the end of the period. As has been mentioned above, the nighttime response of a riometer is mainly to protons with energy somewhat below 10 MeV, and the data thus suggest that these protons were precipitating preferentially in the southern polar cap. This situation could arise if, for instance, there were a strong anisotropy in the proton flux in interplanetary space (Reid and Sauer, 1967b). A breakdown in conjugacy of this kind is of considerable interest, and here again the coordination of ground-based and satellite observations is necessary to a complete understanding of the mechanisms involved.

5. Summary

Ground-based observations of polar-cap absorption form a potentially valuable source of information on the manner in which solar protons enter the magnetosphere, and on the processes responsible for the loss of electrons in the lower ionosphere. This potential can only be realized, however, through the existence of simultaneous observations of the solar protons themselves by satellite techniques. The concept involves the use of a network of high-latitude riometers as a synoptic monitor of proton precipitation, giving an indication of the occurrence of unusual conditions, such as sudden apparent reductions or increases in magnetic cutoff, midday recoveries, or abnormal twilight variations. The satellite data could then be used to provide the basis for a detailed study of these effects, hopefully providing the evidence necessary for verification or rejection of some of the current ideas about them.

To achieve this objective, the satellite instrumentation should include detectors of protons with energies near 20 MeV and near 5 MeV, the former to correspond to the response of a riometer in daytime, the latter for the nighttime situation. A directional capability should be included, since several of the important questions hinge on pitch-angle distributions at the top of the atmosphere, or anisotropies in interplanetary space. The possible role of α-particles, in addition to protons, as contributors to the ionospheric effects, also needs investigation.

What is needed, then, is a satellite experiment that consciously uses the excellent potential of the riometer as a synoptic monitoring tool to aid in the identification of periods of special interest. The riometer, of course, has severe limitations as a quantitative tool, and there is also a great need for the measurement of electron-density profiles in PCA conditions. Rocket experiments capable of measuring ion concentration, as well as electron density, would probably be the most fruitful in this respect, but as mentioned above there have proven to be great logistic difficulties. Alternative approaches, such as the use of partial-reflection observations to measure the electron-density profile, should be exploited as far as possible.

The general problem of the interaction of energetic solar particles with the earth's environment is one of considerable importance, and hopefully a combination of all the techniques mentioned will lead to a significant advance in the state of our knowledge.

References

Adams, G. W. and Masley, A. J.: 1965, 'Production rates and electron densities in the lower ionosphere due to solar cosmic rays', *J. Atmos. Terr. Phys.* **27**, 289.

Adams, G. W. and Megill, L. R.: 1967, 'A two-ion *D*-region model for polar cap absorption events', *Planetary Space Sci.* **15**, 1111.

Bailey, D. K.: 1957, 'Disturbances in the lower ionosphere observed at VHF following the solar flare of 23 February 1956 with particular reference to auroral-zone absorption', *J. Geophys. Res.* **62**, 431.

Bailey, D. K.: 1959, 'Abnormal ionization in the lower ionosphere associated with cosmic-ray flux enhancements', *Proc. IRE* **47**, 255.

Bates, H. F.: 1962, 'Very-low-frequency effects from the November 10, 1961, polar-cap absorption event', *J. Geophys. Res.* **67**, 2745.

Belrose, J. S. and Ross, D. B.: 1961, 'Observations of unusual low-frequency propagation made on 12 November, 1960', *Canad. J. Phys.* **39**, 609.

Bowhill, S. A. and Smith, L. G.: 1966, 'Rocket observations of the lowest ionosphere at sunrise and sunset', in *Space Research VI* (ed. by R. L. Smith-Rose), 511, Spartan Books, Washington.

Egeland, A., Hultqvist, B., and Ortner, J.: 1961, 'Influence of polar cap absorption events on VLF propagation', *Ark. Geofys.* **3**, 481.

Eriksen, K. W. and Landmark, B.: 1961, 'Some results concerning the behaviour of long distance VLF circuits during polar cap absorption events', *Ark. Geofys.* **3**, 489.

Fehsenfeld, F. C., Schmeltekopf, A. L., Schiff, H. I., and Ferguson, E. E.: 1967, 'Laboratory measurements of negative ion reactions of atmospheric interest', *Planetary Space Sci.* **15**, 373.

Gall, R., Jimenez, J., and Camacho, L.: 1968, 'Arrival of low-energy cosmic rays via the magnetospheric tail', *J. Geophys. Res.* **73**, 1593.

Hamlin, D. A., Karplus, R., Vik, R. C., and Watson, K. M.: 1961, 'Mirror and azimuthal drift frequencies for geomagnetically trapped particles', *J. Geophys. Res.* **66**, 1.

Hewitt, L. W.: 1969, 'Ionization increases associated with the small solar proton events of 5 February 1965 and 16 July 1966', *Canad. J. Phys.* **47**, 131.

Hultqvist, B. and Ortner, J.: 1959, 'Strongly absorbing layers below 50 km', *Planetary Space Sci.* **1**, 193.

Leinbach, H.: 1962, 'Interpretations of the time variations of polar cap absorption associated with solar cosmic ray bombardments', Sci. Rept. No. 3, NSF Grant No. G14133, Geophysical Institute, University of Alaska.

Leinbach, H.: 1967, 'Midday recoveries of polar cap absorption', *J. Geophys. Res.* **72**, 5473.

Leinbach, H., Venkatesan, D., and Parthasarathy, R.: 1965, 'The influence of geomagnetic activity on polar cap absorption', *Planetary Space Sci.* **13**, 1075.

LeLevier, R. E. and Branscomb, L. M.: 1968, 'Ion chemistry governing mesospheric electron concentrations', *J. Geophys. Res.* **73**, 27.

Ness, N. F.: 1965, 'The earth's magnetic tail', *J. Geophys. Res.* **70**, 2989.

Ortner, J., Leinbach, H., and Sugiura, M.: 1961, 'The geomagnetic storm effect on polar cap absorption', *Ark. Geofys.* **3**, 429.

Paulikas, G. A., Freden, S. C., and Blake, J. B.: 1966, 'Solar proton event of February 5, 1965', *J. Geophys. Res.* **71**, 1795.

Paulikas, G. A., Blake, J. B., and Freden, S. C.: 1968, 'Low-energy solar-cosmic-ray cutoffs: diurnal variations and pitch-angle distributions', *J. Geophys. Res.* **73**, 87.

Parthasarathy, R., Lerfald, G. M., and Little, C. G.: 1963, 'Derivation of electron-density profiles in the lower ionosphere using radio absorption measurements at multiple frequencies', *J. Geophys. Res.* **68**, 3581.

Quenby, J. J. and Webber, W. R.: 1959, 'Cosmic ray cutoff rigidities and the earth's magnetic field', *Phil. Mag.* **4**, 90.

Reid, G. C.: 1961, 'A study of the enhanced ionization produced by solar protons during a polar cap absorption event', *J. Geophys. Res.* **66**, 4071.

Reid, G. C.: 1969, 'Associative detachment in the mesosphere and the diurnal variation of polar-cap absorption', *Planetary Space Sci.* **17**, 731.

Reid, G. C. and Collins, C.: 1959, 'Observations of abnormal VHF radio wave absorption at medium and high latitudes', *J. Atmos. Terr. Phys.* **14**, 63.

Reid, G. C. and Leinbach, H.: 1959, 'Low-energy cosmic-ray events associated with solar flares', *J. Geophys. Res.* **64**, 1801.

Reid, G. C. and Sauer, H. H.: 1967a, 'The influence of the geomagnetic tail on low-energy cosmic-ray cutoffs', *J. Geophys. Res.* **72**, 197.

Reid, G. C. and Sauer, H. H.: 1967b, 'Evidence for nonuniformity of solar-proton precipitation over the polar caps', *J. Geophys. Res.* **72**, 4383.

Roederer, J. G.: 1967, 'On the adiabatic motion of energetic particles in a model magnetosphere', *J. Geophys. Res.* **72**, 981.

Sauer, H. H.: 1963, 'A new method of computing cosmic-ray cutoff rigidity for several geomagnetic field models', *J. Geophys. Res.* **68**, 957.

Smith, S. J., Burch, D. S., and Branscomb, L. M.: 1958, 'Experimental photodetachment cross section and the ionospheric detachment rate for O^-_2', *Ann. Geophys.* **14**, 225.

Taylor, H. E.: 1967, 'Latitude local-time dependence of low-energy cosmic-ray cutoffs in a realistic geomagnetic field', *J. Geophys. Res.* **72**, 4467.

Westerlund, S., Reder, F. H., and Abom, C.: 1969, 'Effects of polar cap absorption events on VLF transmissions', *Planetary Space Sci.* **17**, 1329.

INTERDISCIPLINARY INVESTIGATIONS OF THE
AURORAL PHENOMENA

I. A. ZHULIN

Institute of Terrestrial Magnetism, Ionosphere and Radiowave Propagation,
Academy of Sciences, Moscow, U.S.S.R.

Abstract. Problems and difficulties of the interdisciplinary analysis of auroral phenomena can be discussed on the example of investigations in magnetically conjugate points. One such attempt was undertaken within the frame of Soviet-French cooperation in the field of space research. In March-April, 1968, a series of balloon launchings was performed in the northern (Arkhangelsk region) and southern (Kerguelen Island) conjugate subauroral zones ($L \approx 3.85$). An interdisciplinary program was organized to observe the geophysical phenomena at the network of expeditionary stations in the environments of the conjugate points. During this experiment a better conjugacy was achieved than during all the other known flights of balloons (approximately 50 km).

On the basis of calculations made by Roederer, the influence upon conjugacy has been estimated of the daily migration of conjugate points due to the change of spatial position of the geomagnetic-field axis.

On the 1st of April, 1968, three events of auroral-electron precipitation were recorded. In the energetic spectrum of these electrons the presence of soft and hard components was revealed (characteristic energies 6 and 30 keV). In one of the events the presence of the maximum in the exponential spectrum near 90 keV (nearly monoenergetic flux of electrons) was revealed. Possibilities for the intercorrelated analysis of obtained data in comparison with satellite data (especially from ESRO-1) are discussed.

The investigations of space disturbances, which accompany active phenomena on the sun, is a unique possibility to study complicated dynamic processes in the interrelation of charged particles with electric and magnetic fields, causing acceleration of particles and their precipitation into the earth's atmosphere at high latitudes. Interdisciplinary analysis of a concrete disturbance development (magnetospheric storm) is the most promising approach to realize this possibility. Of course, concrete disturbances are very different, which does not allow generalizations based on one or several cases. Nevertheless, while making the interdisciplinary analysis of a concrete event, one may reveal its most important peculiarities, inherent in the 'ideal' disturbance.

The interdisciplinary analysis of the magnetospheric storm development is a rather complicated task, because it needs the simultaneous work of a large complex of both the ground and flight apparatus. The greatest difficulties may be met when organizing interdisciplinary experiments on the investigation of auroral phenomena in magnetically conjugate points of the earth's northern and southern hemispheres. At the same time, just such experiments give the most complete diagnostics of the processes in the earth's magnetosphere.

One such attempt to make interdisciplinary investigations was undertaken within the frame of Soviet-French cooperation in the field of space research. To investigate the auroral X-rays, as well as to obtain the information on the energetic spectrum of auroral electrons, in March-April 1968 a series of simultaneous launching of balloons was carried out in the northern (Arkhangelsk region) and in the southern (Kerguelen Island in the Indian Ocean) conjugate subauroral zones, for which the MacIlwain parameter L equals 3.85. The most interesting results were obtained in the energy range from 15 to 150 keV, with the aid of the auroral X-rays spectrometers, developed and manufactured by Centre d'étude Spatiale des Rayonnements (Astor, 1968).

Preliminary results of this experiment were given by Cambou et al. (1969). In February-March 1969 one more series of balloon launchings was also carried out in the Arkhangelsk region; the results of measurements are now being studied.

During both the above-mentioned periods in the Arkhangelsk region and on Kerguelen, an interdisciplinary program was organized to observe the geophysical phenomena at the network of expeditionary stations in the environments of the conjugate points (recording of magnetic variations and VLF-emissions, vertical

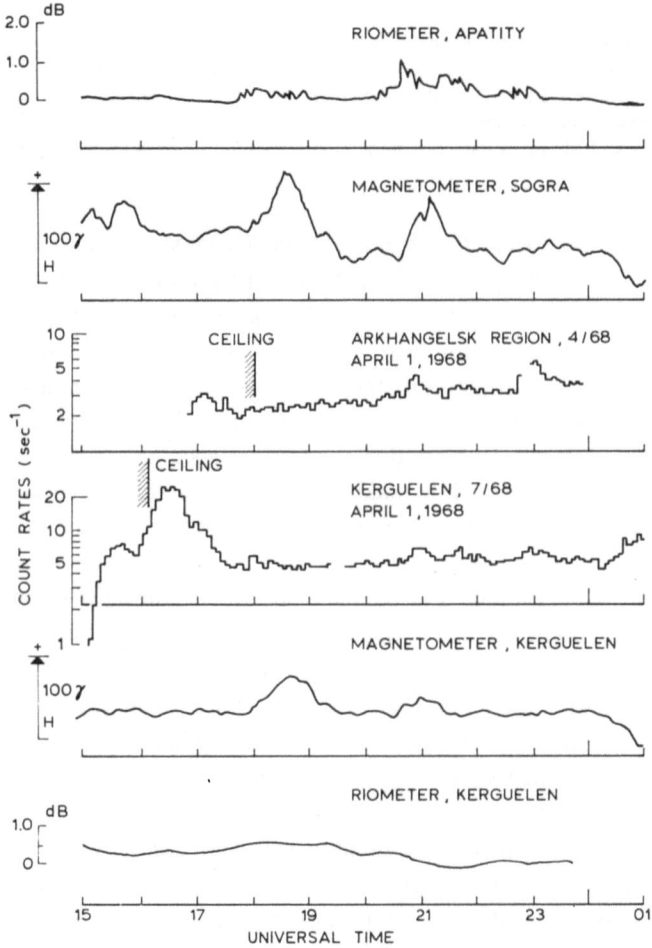

Fig. 1.   Riometer data, magnetograms and count rate of X-ray detectors vs. universal time, April 1, 1969.

ionospheric sounding, observations of aurorae with all-sky cameras, measurements of radiowave absorption, etc.)

To co-ordinate the launchings of balloons, based on the estimation of geophysical conditions in magnetically conjugate points, the direct intercommunication Ker-guelen-Paris-Moscow-Arkhangelsk region was in operation during all the time of the experiment.

Interesting results were obtained on the 1st of April, 1968, when the best conjugacy was achieved in the flights of balloons. Figure 1 gives the one-minute averaged values of count rate vs. universal time for launchings both from Kerguelen Island, and from Arkhangelsk region. Vertical lines with an inclined hatching correspond to the moments of balloons going up to the 'ceiling'. For approximate estimate of the geophysical condition some riometric and geomagnetic data for conjugate points are also given. Table I presents geomagnetic coordinates of base stations.

TABLE I

Name of station	Geomagnetic latitude	longitude
Kerguelen	57°. 3 S	128°. 4 E
Apatity	63° N	125° E
Sogra	56°. 3 N	129°. 8 E

For most detailed presentation, a part of the data for April 1, given in Figure 1, is repeated in Figure 2 at a somewhat changed scale. The lower part of Figure 2 presents

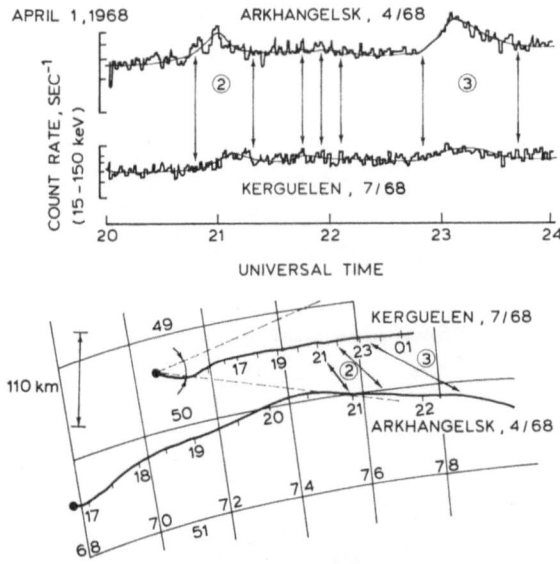

Fig. 2.   Upper part – count rate of X-ray detectors vs. universal time, April 1, 1969. Lower part – trajectory of balloon launched at Kerguelen and the line geomagnetically conjugated to the real trajectory of balloon launched at the Arkhangelsk region (both in geographic coordinates).

in geographical coordinates the flight trajectory of the balloon launched from Kerguelen. On the basis of the geomagnetic field model (Cain, 1968) the flight trajectory
of the balloon launched in the Arkhangelsk region, determined with high precision,
was plotted in geomagnetic coordinates, projected along the lines of force into the
southern hemisphere, and then re-plotted in geographical coordinates. Apparently,
for the precipitation observed at about 21 h UT, the conjugacy was about 50 km. It
should be noted, that during this experiment a better conjugacy was achieved, than
during all the other flights of balloons in the well-known experiments performed on
Kotzebue and Macquarie (Brown, 1968), when the longitudinal discrepancy was
about 7° (about 700 km).

During similar experiments both daily and seasonal migrations of conjugate points
are to be considered. Illustrative calculations of such migrations have been recently
given by Barish and Roederer (1969). In our case, it would be of interest to see how
the conclusions on conjugacy of X-ray bursts would change with the account of daily
variations of position of the points conjugate to the points along the trajectory of the
balloon launched from the Arkhangelsk region. Figure 3 reproduces the calculated
data by Barish and Roederer (1969) on daily variation of position of the point conjugate to Arkhangelsk for the case of moderately compressed magnetosphere (the
stand-off distance $Rs = 8$ earth's radii, the field intensity at the tail $B_T = 30$ γ's for
March 21, which approaches the case of April 1st under consideration). Along the

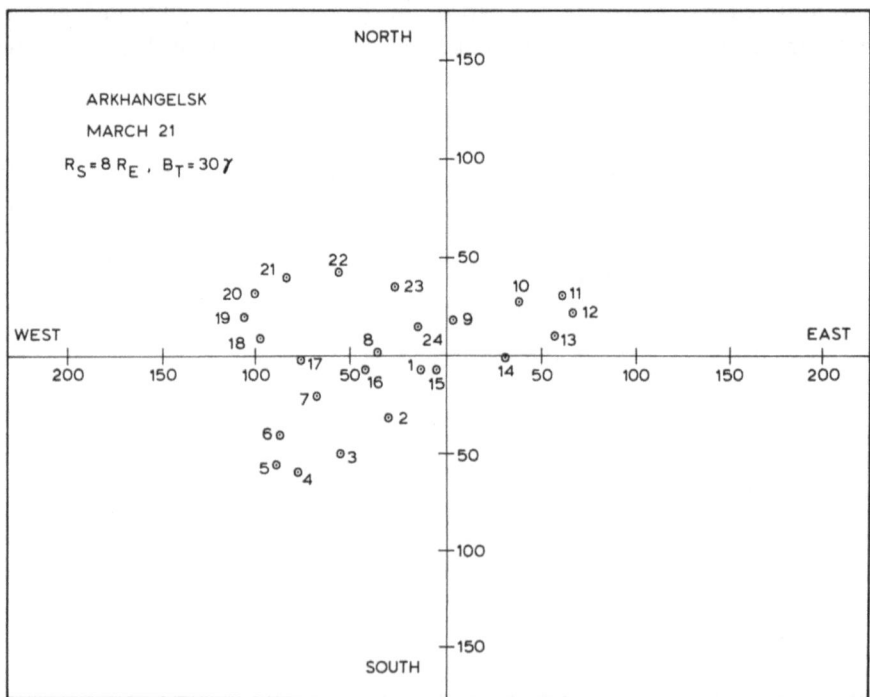

Fig. 3.   Diurnal variation of the point magnetically conjugated to Arkhangelsk (March 21).

axes deviations are plotted in the north–south and east–west directions from the point conjugate to Arkhangelsk according to the data on the main geomagnetic field computed by Cain (1968). It is obvious, that when introducing some corrections into Figure, 2 corresponding to the daily variation of position of the points conjugate to the points along the balloons' trajectory, the conjugacy of balloons' positions during X-ray bursts will appear to be still closer to an ideal one.

Energetic spectra of the auroral electrons, responsible for the recorded X-ray bursts, may be presented as follows:

$$I_e(T) = A \left[\exp\left(-T/T_0\right) + R \exp\left(-T/T_1\right) + S\delta\left(T - T_2\right) \right],$$

where $I_e(T)$ and A=intensities of electron flux, T=electron energy, T_0, T_1 and T_2=characteristic energies of spectral components, R and S=dimensionless values determining relative intensities of different components of the spectrum. $\delta(T - T_2)=$ δ-function, describing monoenergetic spectrum of the electrons with energies T_2. For the three bursts recorded on April 1st, in southern hemisphere, the values of parameters characterizing the energetic spectra, are the following ones:

Spectrum 1. $(16^h30^m - 17^h25^m \text{ UT})$
$\quad A = 1 \times 10^6 \text{ cm}^{-2} \text{ sec}^{-1} \text{ sterad}^{-1} \text{ keV}^{-1}$
$\quad R = 1.6 \times 10^{-2}, \ S = 0, \ T_0 = 6 \text{ keV}$
$\quad T_1 = 30 \text{ keV}$

Spectrum 2. $(20^h54^m - 20^h57^m \text{ and } 21^h05^m - 21^h11^m \text{ UT})$
$\quad A = 1.5 \times 10^5 \text{ cm}^{-2} \text{ sec}^{-1} \text{ sterad}^{-1} \text{ keV}^{-1}$
$\quad R = 2.2 \times 10^{-2} \quad S = 0$
$\quad T_0 = 6 \text{ keV} \quad T_1 = 30 \text{ keV}$

Spectrum 3. $(22^h55^m - 23^h13^m \text{ and } 23^h20^m - 23^h29^m \text{ UT})$
$\quad A = 6.7 \times 10^5 \text{ cm}^{-2} \text{ sec}^{-1} \text{ sterad}^{-1} \text{ keV}^{-1}$
$\quad R = 0 \quad S = 5.4 \times 10^{-3}$
$\quad T_0 = 5 \text{ keV} \quad T_2 = 90 \text{ keV}$

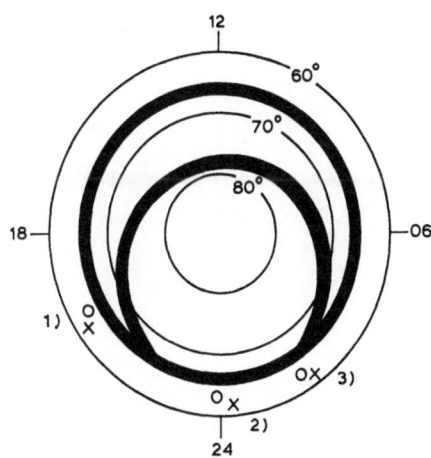

Fig. 4. Auroral oval and auroral zone (schematically) and approximate positions of Kerguelen balloon (cross) and Apatity riometer (circle) for three cases of precipitation shown on Figure 2.

A possible error in defining the electron flux is equal to $\approx 20\%$, in defining the characteristic energies – about 1 keV. An error in defining T_2 (Spectrum 3) equals ≈ 5 keV. With the account of the latter, and the computing method used (Maral, 1969), one cannot certainly speak of the revealing of strictly monoenergetic flux of precipitating electrons. Nevertheless, one may consider as proved the presence of the maximum in the exponential spectrum near 90 keV.

Approximate positions of balloons during the three precipitations of auroral electrons as to the auroral oval and zone of aurorae, are presented in Figure 4. The circles show the position of the riometer at Apatity.

It would be of considerable interest to present the comparison of the data obtained during the described interdisciplinary experiments, with the satellite data (especially, with the data of the ESRO-I-satellite).

References

Astor, J. L.: 1968, Thèse de Spécialité, Nr. 632, Université de Toulouse.

Barish, F. D. and Roederer, J. G.: 1969, Influence of Magnetosphere Cavity Model Field on Geomagnetic Conjugacy of High Latitude Stations, IAGA General Assembly, Madrid, Spain, September.

Brown, R. R.: 1968, *Radio Sci.* 3, 171.

Cain, J. G.: 1968, *Radio Sci.* 3, 149.

Cambou, F., Maral, G., Zhulin, I. A., and Kopylov, Yu. M.: 1969, *Kosm. Issled.* [Space Researches] 7,

Maral, G.: 1969, *J. Geophys. Res.* 74.

ASPECTS OF AURORAL MORPHOLOGY

T. NEIL DAVIS

Geophysical Institute, University of Alaska, College, Alaska, U.S.A.

Abstract. The temporal relationship of the auroral substorm to increases in the total particle kinetic energy within the trapping region and also the characteristics of auroras near the equatorward boundary of the typical display indicate a close physical connection between acceleration processes causing auroras and growth of the ring current. Conjugate studies of auroras during disturbed epochs indicate sizable distortions in the geomagnetic field at the higher latitudes; these distortions may be caused by meridianal asymmetry in the ring current as well as by magnetopause and neutral sheet currents.

Recent extensive reviews on auroral morphology are given by Akasofu *et al.* (1966), Hultqvist (1969) and Feldstein (1969). In part review, this paper attempts to cover topics of special interest to those attending the ESLAB/ESRIN Symposium. Also there is some discussion of work underway by the group at the University of Alaska.

Just as the *L*-parameter is a concept critical to the ordering of particle distributions in the geomagnetic field, the auroral substorm is a concept necessary to the understanding of the temporal behavior of the aurora. Lasting a few tens of minutes to perhaps over an hour, the auroral substorm apparently is coincident with the Dp (polar magnetic disturbance) substorm, the ionospheric substorm, etc., all of which are considered to be manifestations of the general magnetospheric substorm. The sequence of events within each type of substorm and the exact nature of the interrelations between substorm types are not yet completely known, but these substorms, all of which are the result of particle precipitation into the atmosphere, exhibit a temporal correlation with changes in the particle energy content of the magnetosphere and the distributions of particles contained there. Thus it seems essential that the investigator of magnetospheric particle populations and magnetosphere fields consider the substorm time parameter as well as spatial parameters. Study of the temporal variations observed within the magnetosphere (including the ionosphere) and time-varying parameters of the interplanetary medium is the chief means being used to answer the problem of how magnetospheric substorms originate. These may occur in direct response to solar wind variations or be the result of an internal mechanism involving energy storage and then spontaneous or solar-wind triggering of the burst-like substorm.

The constraints on auroral observation are so severe that the tendency has been to utilize magnetic variations for most studies and then to apply the results to auroral morphology by using knowledge of the association between auroras and magnetic variations. Interesting results in this respect have been obtained by Nishida (1968a, b). Following a classification suggested by Obayashi (1967), Nishida considered the polar magnetic disturbance Dp to consist of Dp 1 and Dp 2. Dp 1 is strongest at the auroral zones and is the component considered to undergo the magnetic substorm variation (coincident with the auroral substorm). Dp 2 has a quasi-periodic burst-like behavior

V. Manno and D. E. Page (eds.), Intercorrelated Satellite Observations Related to Solar Events, 341–350. *All Rights Reserved*
Copyright © 1970 by D. Reidel Publishing Company, Dordrecht-Holland

not unlike Dp 1 but is generally weaker and is observed both in equatorial and polar cap regions. Nishida finds the Dp 2 variations to be strongly correlated with the north–south component of the interplanetary magnetic field, irrespective of whether this component is directed north or south. Nishida notes that the seat of the Dp 1 and other related processes, probably lies in the nightside magnetosphere and suggests that the source of Dp 2 is a deeply penetrating interplanetary electric field entering the magnetosphere probably near the dawn and dusk meridians. Nishida's findings, together with a general failure to find more than a weak correlation between variation in various solar-wind parameters and substorm phenomena (time scale 10 min to an hour or two), does suggest that the seat of the processes giving rise to the substorm is indeed internal to the magnetosphere (see also Axford, 1969). Seemingly of significance also are the results of Davis and Parthasarathy (1967), Pudovkin *et al.* (1968) and Davis (1969) which indicate that the particle precipitation causing auroras, Dp 1 and similar substorm phenomena is essentially coincident with increases of the particle kinetic energy content of the trapping region of the magnetosphere (see Figure 1).

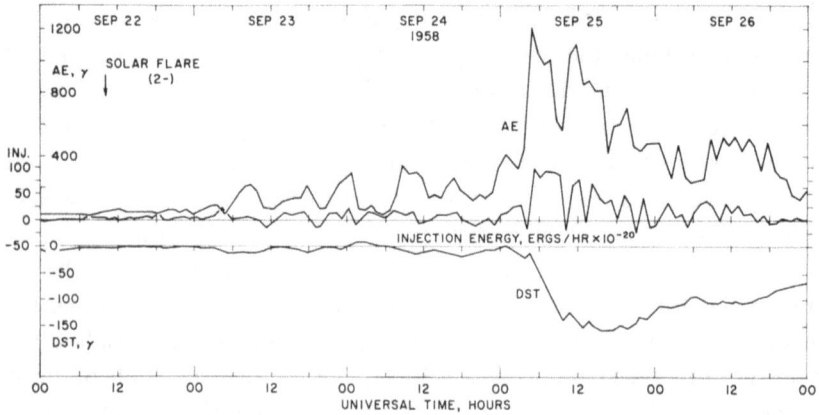

Fig. 1.   The top curve *AE* is a global measure of Dp 1 and illustrates the substorm behavior before and during a moderate magnetic storm. The middle curve represents injection of energy into the ring current and reflects the substorm behavior seen in Dp 1. *Dst*, the lower curve, is a measure of the total kinetic energy content in the trapping region. After Davis (1969).

Thus it seems likely that there is a single mechanism or group of closely interrelated mechanisms developing the kinetic energy responsible for auroras and the geomagnetic ring current. These processes are not steady state even at quiet times; their behavior is quasi-periodic. The magnetospheric substorm is always burst-like but the duration and structure appears to vary from one magnetic storm to another. This suggests that even though the mechanism may be internal to the magnetosphere, it is not indepen-dent of the solar wind parameters. Individual substorms have durations of a few tens of minutes to more than one hour. During magnetic storms the substorms become larger and more frequent, perhaps even overlapping one another. On such occasions, extensive, complex auroral displays appear near the auroral zone for many hours.

The concept of the instantaneous auroral oval (Feldstein, 1966; Akasofu, 1966) and of the auroral substorm (Akasofu, 1964) are major advances toward a full understanding of the detailed and global nature of phenomena related to the precipitation of auroral particles. Recently Montbriand (1969) has enlarged upon the auroral substorm descriptions given by Akasofu and coworkers by incorporating a description of the patterns of auroral hydrogen emission and auroral absorption. Part of Montbriand's revised substorm diagrams showing the boundaries of hydrogen aurora in relation to electron produced aurora are contained in Figure 2. At present the morphology of hydrogen emission is of particular interest because there is indication that the proton energy and spatial distributions radically change near the time of the auroral break-up, and there is the suspicion that these variations may play a key role in the break-up process. Also it is now generally recognized that the configuration and location of the auroral oval is determined by the outer boundary of trapping in the magnetosphere. However, the exact relationship between the trapping boundary and the regions of electron and proton precipitation is somewhat uncertain in that at least part of the precipitating particles may derive from the trapping region or perhaps they all come from outside its boundary.

Closely related to this question is that of the configuration of the Dp (Dp 1) in

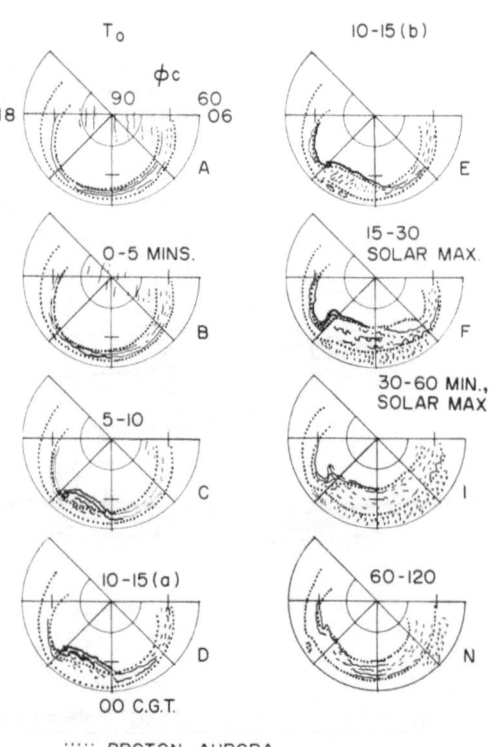

Fig. 2. A revision of Akasofu's auroral substorm diagrams to include regions of proton precipitation. Redrawn after Montbriand (1969).

Obayashi's classification currents. Three different schematic representations of two-dimensional current systems are given in Figure 3. Although there is not complete agreement on the matter it appears that configuration C is most nearly correct. It seems likely that the temporal variations of the eastward and westward electrojets are somewhat dissimilar, and therefore the two may have different sources. Yet, in at least a gross sense, the temporal variations are similar in the two electrojets. An argument for some physical connection between the two electrojets is seen in the auroral form and disturbance vector relationships shown in Figure 4 (Davis and Kimball, 1962). Recently, Meng and Akasofu (1969) have found semi-quantitative

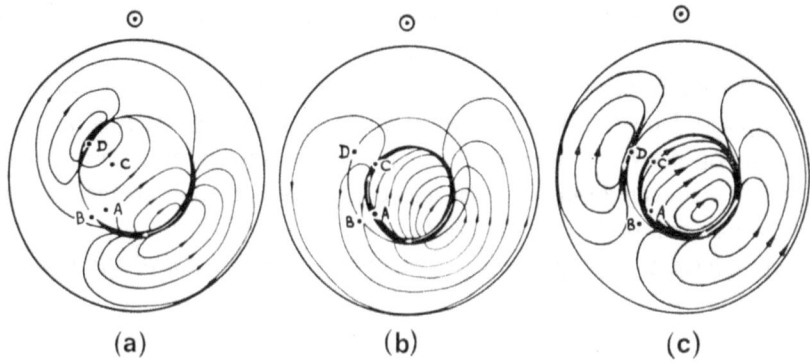

Fig. 3.   Schematic representation of the current system of Dp disturbance in the Northern Hemisphere: (a) classical system; (b) single-vortex system with westward electrojet along the auroral oval; (c) double-vortex system with westward electrojet along the oval. The solar direction is at top.
After Feldstein (1969).

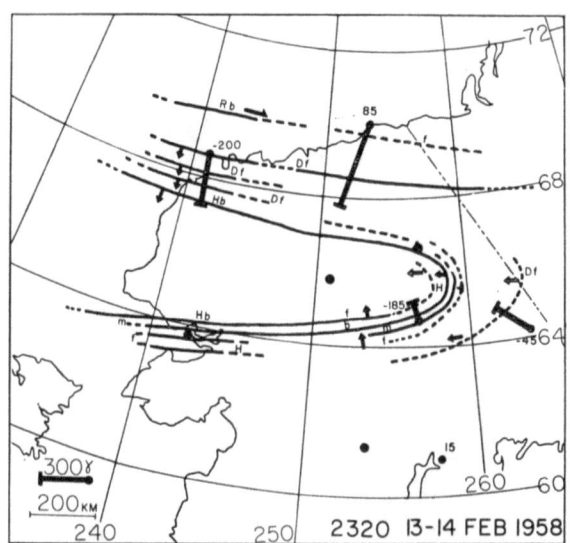

Fig. 4.   A system of stable west-opening loops observed for over an hour on the night of February 14, 1958. Heavy bars are horizontal disturbance vectors which were typically perpendicular to the auroral configurations, indicating current directed along the auroras. After Davis and Kimball (1962).

agreement between magnetic variation measurements and a three-dimensional current system model involving an asymmetric ring current in the day-evening sector and current along the magnetic field lines connecting to the westward electrojet in the post midnight sector.

This past decade has seen a remarkable improvement in the understanding of auroral morphology, the single most important development being the substorm concept. However, there still remains a number of problems of strictly a descriptive nature. The relationship of the approximately sun–earth aligned auroras in the polar cap to auroras in the oval is still uncertain; they probably represent a separate entity since these auroras evidently occur only during quiet times and then during major substorms are replaced by auroras more typical of those seen in the auroral zone (Stringer, 1966). Montbriand (1969) having just completed a major compilation of auroral data, notes that the large, stable westward-opening auroral loops reported earlier by Nichols (1959) and Davis (1962) (and as shown in Figure 4) do not seem to fit properly into the substorm descriptions now available. I speculate that a complete substorm description must allow for a development phase occurring prior to substorm time $T = 0$. This development phase, probably lasting near 30 min, is suggested by the observation that one often sees the slow increase in number and intensity of quiet arc-like forms to the stage shown for $T = 0$ in Figure 2. If such a development stage actually does exist, it quite likely can be detected also in magnetic field and particle population or kinetic energy distribution measurements in the night side magnetosphere.

There are several indications of possible longitudinal variations in auroras; such variations could arise from the daily variation in the approach direction of the solar wind relative to the geomagnetic field, or from gross asymmetry in the geomagnetic field which affects the mirror heights of trapped or quasi-trapped particles at auroral near $L = 6$–10 (see Davis, 1967 for a discussion of this point). The question is important because major longitudinal asymmetries might affect auroral morphology to the extent that observations at one longitude cannot be reliably extended to other parts of the globe. The matter has not yet been resolved because what might be interpreted as a longitudinal effect may in reality only be a seasonal variation.

Figure 5, compiled from ASCAPLOT data (AIGY, 1962), indicates a greater auroral occurrence near the Alaska sector of the northern auroral zone than in the Scandinavian sector. In view of the crudeness of these data and the possibility that the various all-sky cameras were not equally sensitive or were scaled differently, the plot in Figure 5 is suspect. However, the occurrence rate of aurorally related radio-wave absorption also appears to be greater in Alaska than in Canada or Norway; Waite (1965) suggested that the apparent longitudinal discrepancy may be the result of seasonal or solar cycle effects. Another possible indicator of longitudinal effects is the repetitive pattern of Dp magnetic variations at the same universal time on successive days reported by Heppner (1967). The greater average intensity of auroras over Alaska as compared to the conjugate region during a disturbed epoch of March 1968 (Belon et al., 1969) may be a seasonal effect or a longitudinal effect, the longitudinal effect presumably arising through differences in mirror altitudes.

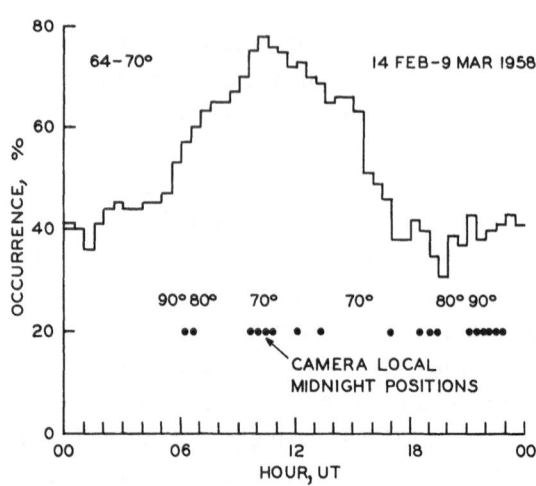

Fig. 5. Percentage of clear, dark hours when auroras occurred over all-sky cameras in the dipole latitude range 64–70°N at positions indicated. (Note: West longitude equals UT hour times 15.) The numbers 70°, 80° and 90° show the pitch angle at 200 km altitude of particles mirroring at altitude 200 km in the Southern Hemisphere. Data shown here are most reliable over the camera midnight positions.

Investigation of auroral conjugacy with camera-equipped aircraft flying along dipole longitude 256° during a magnetically quiet period in March 1967 showed a high degree of conjugacy of auroras observed between dipole latitudes 64° and 71° (Belon *et al.*, 1969). These results indicate symmetrical precipitation in the two hemispheres of both protons and electrons along stable geomagnetic field lines relatively undistorted by external magnetic fields. A second series of flights in March 1968 were undertaken during a disturbed period (College $K=5$ and 6) with quite different results. Major relative displacements of conjugate auroras were observed as were complete failures in conjugacy at the higher latitudes. Once a displacement in conjugacy occurs, the data become very difficult to analyze owing to the high aspect dependence in auroral observations. This problem is so severe that we have experienced difficulty in identifying a given auroral feature in image orthicon televison observations taken in one hemisphere at two sites only 50 km apart. The apparent intensities of auroras as well as their apparent geometrical shapes are critically dependent upon the view direction relative to the local magnetic field direction. Consequently, even if an aurora in one hemisphere has an exact copy in the other, any shift in relative conjugacy makes the two auroras appear differently to observers at the two ends of a particular magnetic field line. Because of this problem, analysis of the data is painstaking and is proceeding slowly. Near the equatorward boundaries of the conjugate displays the hydrogen arcs with the associated electron generated emissions within their poleward edge appear quite similar in intensity and shape although there may at times be a very slight meridianal displacement of one hemisphere's aurora relative to the other's. Proceeding poleward past the aurora within approximately 100 km of the equatorward boundary sizable displacements in conjugacy are evident. Examples analyzed so far show a relative equatorward and eastward displacement of the Southern Hemisphere aurora, one is shown in Figure 6. Relative displacements measured on March 29, 1967 between the times 0910 and 1043 UT, are plotted in Figure 7 together with a plot of daily conjugate variations calculated by Barish and Roederer (1969) assuming field distortion by magnetopause and neutral sheet currents. The calculated displacements are dependent upon the degree of magnetosphere compression and tail field intensity, so the fact that the actual points fall outside the locus of calculated points is of no consequence. For complete agreement, the data points taken during an hour need only lie in that hour's sector of the diagram. The observations indicate a greater meridianal displacement than allowed by this model.

Distortion in this direction could be caused by a ring current asymmetric to the geomagnetic equatorial plane. In view of the high mobility of the conjugate displacements poleward of the equatorward boundary of the aurora, the near lack of displacement at the position of the equatorward hydrogen arc is surprising. Even for quiet magnetosphere conditions, the model of Barish and Roederer predicts for the Alaska sector a meridianal displacement of 20–30 km at 1000 UT in March, yet the observed displacement appearing on several disturbed nights was not more than 10 km. There appears to be a rather definite boundary lying just poleward of the evening hydrogen arc, between the stable field lines and those easily distorted. Very likely this is the boundary of stable trapping.

The auroras in the region near the equatorward boundaries of moderate to major auroral displays differ from the higher-latitude auroras in that they exhibit more quasi-periodic temporal variation in the frequency range 0.01–10 Hz (Cresswell and Davis, 1966). The affinity of the pulsating auroras for the lower latitudes is suggestive that occurrence on closed, relatively undistorted field lines is a prerequisite for this type of quasi-periodic behavior. Observations during March 1968 of conjugate pulsatings with television systems having a time resolution of 0.03 sec showed several clear instances where conjugate pulsatings of period 1–3 sec occurred simultaneously in the two hemispheres, the simultaneity being in error by at most 0.1 sec (Davis, 1969). The simplest explanation for the simultaneity is a source mechanism in the equatorial plane. If separate sources in the two hemispheres are responsible, the sources must be connected through the geomagnetic field by an agent travelling near the speed of light or be initiated by a slower wave originating at the equatorial plane.

Another periodic variation quite unlike the normal pulsating aurora has been observed at various locations, including conjugate points, just at or just prior to the time of auroral break-up (Beach et al., 1968). Of interest because it may be a critical part of the break-up process, this variation has a fixed frequency at 10 ± 3 Hz.

In Alaska, use is being made of television techniques to examine the characteristics of auroral micro-morphology. One of these investigations (Hallinan, 1969) shows that most or all auroral rays result from small-scale folds and curls in auroral arcs. The curls exhibit a remarkable spiralling identical to those produced with a laboratory electron sheet beam (Webster, 1957) and produced by a computer model investigation of the long-wave diocotron instability in a low-density electron beam (Levy and

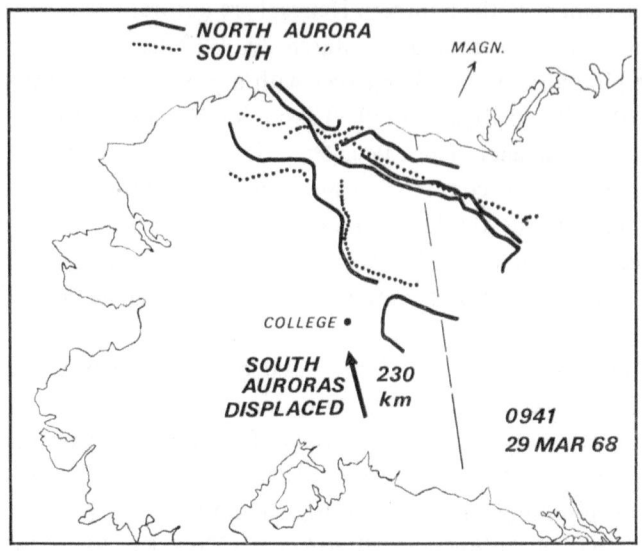

Fig. 6. Conjugately displaced auroras. The aurora in the Southern Hemisphere has been mapped through the internal geomagnetic field to the Northern Hemisphere and shifted 230 km, as indicated on the diagram, in order to match the northern auroras.

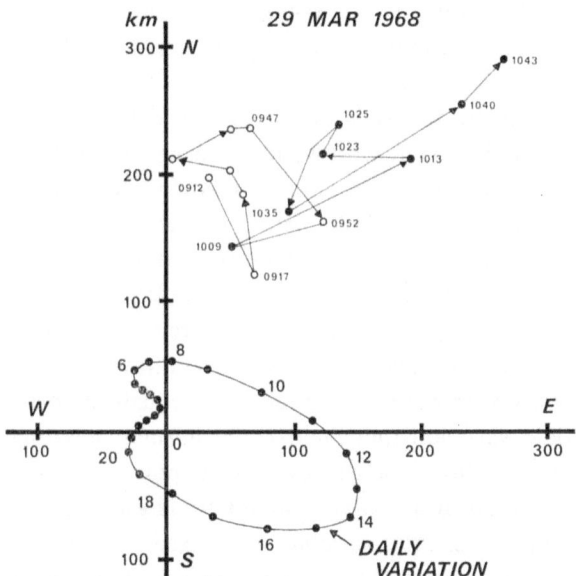

Fig. 7. Measured displacements in auroral conjugacy plotted on diagram of daily variation calculated by Barish and Roederer (1969) by taking into account magnetopause and tail fields during a quiet epoch in March. The daily variation is calculated for Kotzebue, just west of the region of measurement. Open-circle points are 5-min averages about the times indicated; closed circle points are from single measurements. The basic diagram shows the direction and magnitude in the Southern Hemisphere of displacement of a field line originating in the Northern Hemisphere.

Hockney, 1968). The scale size of the folding (2–5 km), the growth time (~ 0.3 sec) and the other observed characteristics such as the sense of rotation are compatible with the electron sheet beam instability. If the sheet beam instability model is correct, the observed growth characteristics indicate that the instability originates well above the ionosphere.

Hess (1969) reported the success of an experiment to produce artificial aurora with a rocket-borne accelerator beaming electrons downward into the atmosphere. Prior to the experiment, it was suggested that the electron beam might become unstable and then fail to produce observable auroras. Had this actually happened, it would be necessary to seek an auroral acceleration mechanism which can maintain low-density particle beams down to E-region height. The success of the artificial aurora experiment and the results of studies of auroral microstructures are compatible with the idea that the complex, small-scale auroral morphology is of magnetospheric rather than ionospheric origin.

Acknowledgements

Hans Nielsen has performed most of the analysis of conjugate data reported here. The preparation of this manuscript was supported by National Science Foundation, Atmospheric Sciences Section Grant No. GA-3947.

References

AIGY, I.G.Y. ASCAPLOTS: 1962, *Annals of the International Geophysical Year* **20**, 147 pp.

Akasofu, S.-I.: 1964, 'The Development of the Auroral Substorm', *Planetary Space Sci.* **12**, 273–282.

Akasofu, S.-I.: 1966, 'The Auroral Oval, the Auroral Substorm and their Relation with the Internal Structure of the Magnetosphere', *Planetary Space Sci.* **14**, 587–595.

Akasofu, S.-I., Chapman, S., and Meinel, A. B.: 1966, 'The Aurora', in *Handbuch der Physik*, vol. 49, pp. 1–158.

Axford, W. I.: 1969, 'Magnetospheric Connection', *Rev. Geophys.* **7**, 421–459.

Barish, F. D. and Roederer, J. G.: 1969, Influence of a Magnetospheric Cavity Model Field on Geomagnetic Conjugacy at High Latitude Stations. Presented at IAGA General Assembly, Madrid, Spain, September.

Beach, R. F., Cresswell, G. R., Davis, T. N., Hallinan, T. J., and Sweet, L. R.: 1968, 'Flickering, a 10-cps Fluctuation within Bright Auroras', *Planetary Space Sci.* **16**, No. 12, 1525–1529.

Belon, A. E., Maggs, J. E., Davis, T. N., Mather, K. B., Glass, N. W., and Hughes, G. F.: 1969, 'Conjugacy of Visual Auroras during Magnetically Quiet Periods', *J. Geophys. Res.* **74**, 1–28.

Cresswell, G. R. and Davis, T. N.: 1966, 'Observations on Pulsating Auroras', *J. Geophys. Res.* **71**, 3155–3163.

Davis, T. N.: 1962, 'The Morphology of the Auroral Displays of 1957–58, Part 2', *J. Geophys. Res.* **67**, 75–110.

Davis, T. N.: 1967, 'Worldwide Auroral Morphology', in *Aurora and Airglow* (ed. by B. M. McCormac), Reinhold Publ. Corp., New York, pp. 41–58.

Davis, T. N.: 1969, 'Television Observations of Auroras', in *Atmospheric Emissions* (ed. by B. M. McCormac and A. Omholt), Van Nostrand Reinhold Co., New York, pp. 27–35.

Davis, T. N.: 1969, 'Temporal Behavior of Energy Injection into the Geomagnetic Ring Current', *J. Geophys. Res.* (in press).

Davis, T. N. and Kimball, D. S.: 1962, The Auroral Display of February 13–14, 1958. Rept. UAG-128, Geophys. Inst., College, Alaska.

Davis, T. N. and Parthasarathy, R.: 1967, 'The Relationship between Polar Magnetic Activity Dp and Growth of the Geomagnetic Ring Current', *J. Geophys. Res.* **72**, 5825–5836.

Feldstein, Y. I.: 1966, 'Peculiarities in the Aurora Distribution and Magnetic Disturbance Distribution in High Latitudes Caused by the Asymmetrical Form of the Magnetosphere', *Planetary Space Sci.* **14**, 121–130.

Feldstein, Y. I.: 1969, 'Polar Auroras, Polar Substorms, and their Relationships with the Dynamics of the Magnetosphere', *Rev. Geophys.* **7**, 179–218.

Hallinan, T. J.: 1969, The Morphology of Small-Scale Folds and Curls in the Aurora. M.S. Thesis, Univ. of Alaska, College, Alaska, May.

Heppner, J. P.: 1967, 'High Latitude Magnetic Disturbances', in *Aurora and Airglow* (ed. by B. M. McCormac), Reinhold Publish. Corp., New York, pp. 75–92.

Hess, W. N.: 1969, 'Generation of an Artificial Aurora', *Science* **165**, 1512–1513.

Hultqvist, B.: 1969, 'Auroras and Polar Substorms: Observations and Theory', *Rev. Geophys.* **7**, 129–177.

Levy, R. H. and Hockney, R. W.: 1968, 'Computer Experiments on Low-Density Crossed-Field Electron Beams', *Phys. Fluids* **11**, 766–771.

Meng, C.-I. and Akasofu, S.-I.: 1969, 'A Study of Polar Magnetic Substorms. 2, Three-Dimensional Current System', *J. Geophys. Res.* **74**, 4035–4053.

Montbriand, L. E. J.: 1969, Morphology of Auroral Hydrogen Emissions during Auroral Substorms. Ph.D. thesis, Univ. of Saskatchewan, Saskatoon.

Nichols, B.: 1959, 'Auroral Ionization and Magnetic Disturbance', *Proc. Inst. Radio Engrs.* **47**, 245–254.

Nishida, A.: 1968a, 'Geomagnetic Dp 2 Fluctuations and Associated Magnetospheric Phenomena', *J. Geophys. Res.* **73**, 1795–1803.

Nishida, A.: 1968b, 'Coherence of Dp 2 Fluctuations with Interplanetary Magnetic Variations', *J. Geophys. Res.* **73**, 5549–5559.

Obayashi, T.: 1967, 'The Interaction of Solar Plasma with Geomagnetic Field, Disturbed Condition', in *Solar Terrestrial Physics* (ed. by J. W. King and W. S. Newman), Academic Press, New York.

Pudovkin, M. I., Shumilov, O. I., and Zaitzeva, S. A.: 1968, 'Polar Storms and Development of the DR-Currents', *Planetary Space Sci.* **16**, 891–898.

Stringer, W. J.: 1966, Aspects of Auroral Morphology during the International Quiet Sun Year. M.S. Thesis, Univ. of Alaska, College, Alaska, May.

Waite, C. W.: 1965, 'Auroral Phenomena in an Integral Invariant Coordinate System', *Can. J. Phys.* **43**, 2319–2330.

Webster, H. F.: 1957, 'Structure in Magnetically Confined Electron Beams', *J. Appl. Phys.* **28**, 1395–1397.

ACCELERATION AND PRECIPITATION OF VAN ALLEN ELECTRONS DURING MAGNETOSPHERIC SUBSTORMS*

GEORGE K. PARKS

*Université de Toulouse, Centre d'Etude Spatiale des Rayonnements
(Equipe de Recherche Associée au C.N.R.S.), Toulouse, France*

Abstract. This paper will present a representative portion of data from the more extensive file of the ATS-1 auroral particle correlation experiment. The main purpose is to show that magnetospheric and auroral processes are intimately related and that Van Allen electrons of $E > 50$ keV are accelerated during Magnetospheric Substorms. In particular, it is shown that (1) intense auroral precipitation does not occur randomly, but only during periods when the outer zone electron fluxes increase (2) electron pitch-angle distribution in the equatorial plane during substorms is peaked toward 90° (3) the pitch-angle distribution changes in time and depends on the flux and (4) life-times of freshly energized Van Allen electrons are considerably shorter than their drift periods. It will be shown that our observations can be explained consistently in terms of energy and pitch-angle diffusion processes accompanying whistler-cyclotron resonance interaction.

1. Introduction

Observations of the behavior of trapped magnetospheric energetic electrons by the University of Minnesota electron spectrometer experiment on-board the ATS-1 geostationary satellite together with simultaneous auroral measurements of precipitated energetic electrons near the magnetic conjugate of the ATS-1 have provided extensive information concerning the origin of auroral and Van Allen electrons. This article is intended to bring together the main results of our experiments and interpret them in a coherent way, employing the present knowledge of plasma theory. It is hoped that our study will contribute to further understanding of the relation of waves and particles in the magnetosphere responsible for energization and removal of the particles from the magnetic lines of force. Previous publication of the ATS-1 results and the results of the auroral equatorial plane correlation are contained in Lezniak *et al.* (1968), Parks *et al.* (1968), Parks and Winckler (1968, 1969), Arnoldy and Chan (1969), Pfitzer and Winckler (1969), Winckler (1969), and Parks (1969).

2. Instrumentation

The Minnesota experiment consisted of a narrow aperture magnetic electron spectrometer which measured electrons of energies from 50 keV to 1 MeV in three energy ranges: 50–150 keV, 150–500 keV, and 500–1000 keV. The factors converting the 'counts/sample' to directional flux in $(cm^2\text{-sec-ster-keV})^{-1}$ are 4.5×10^2, 7.79×10^1, and 4.03×10^1, respectively. Each energy channel and the background radiation were sampled every 40 millisec sequentially. The full view angle of the spectrometer

---

* This study was conducted while the author was in the School of Physics and Astronomy, University of Minnesota. This manuscript was written in Toulouse, France.

V. Manno and D. E. Page (eds.), Intercorrelated Satellite Observations Related to Solar Events, 351–363. All Rights Reserved
Copyright © 1970 by D. Reidel Publishing Company, Dordrecht-Holland

was ~11°. Nominally, particles with equatorial pitch-angles 60–90° were detected.

The auroral measurements were conducted from College, Alaska (64.5°N, 149.5°W), approximately 600 km west of the calculated magnetic conjugate region of the ATS-1 in the auroral zone. According to the geomagnetic field model of Cain *et al.* (1964), the region of magnetic conjugacy of the ATS-1 is centered around 62.8°N, 128°W, in the Yukon Territories, Canada. Auroral detectors were NaI (Tl) coupled to photomultipliers carried to 2–3 g/cm² atmospheric depths by high-altitude balloons, detecting the atmospheric bremsstrahlung X-rays from precipitated energetic electrons. The geometry factor of the X-ray detectors was ≃ 35 cm²-ster.

3. Behavior of Energetic Electrons at Synchronous Altitudes

The geostationary satellite (located at 6.6 R_e equator) is stationary with respect to a point on the earth's surface and, to a large extent, stationary in the magnetosphere. The direct measurement on-board the satellite of the local magnetic field gives information on the relative motion of L-shells due to the distortion of the magnetosphere. Consequently, the variations of electron fluxes on a fixed shell with time may now be separated in principle from changes caused by drift shells passing the satellite, and temporal behavior of energetic electrons in the equatorial plane of the magnetosphere can be studied.

Figure 1 shows the count-rate time profile (15 min averages) of electron fluxes for the three energy ranges as observed at synchronous altitudes in the course of a typical geomagnetic quiet day. In interpreting this figure, one recognizes that due to the

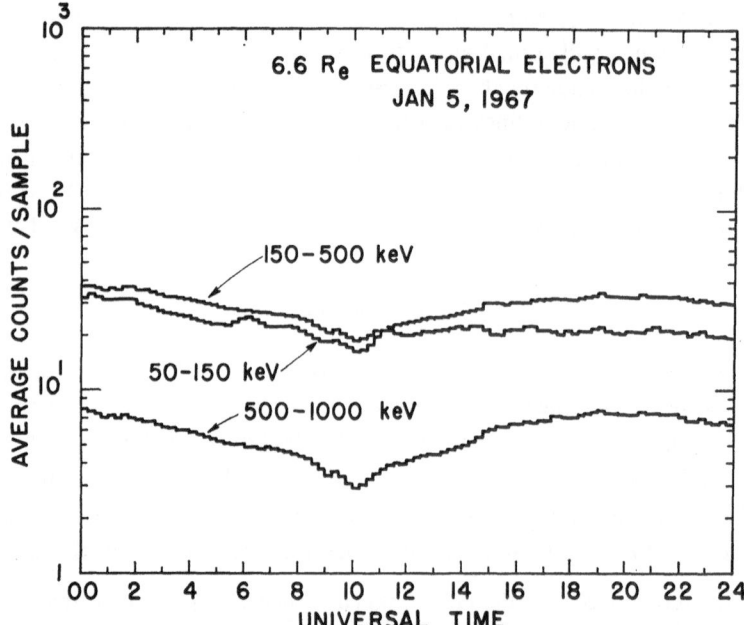

Fig. 1.   A quiet time radiation belt as observed in the geostationary reference frame in the course of one day. The data represent counts averaged over 15 min. Local Time = Universal Time − 10 hours.

solar wind distortion of the magnetosphere, the ATS-1 is not geomagnetically fixed. Consequently, the circular orbit of the satellite will sample particles from a range of *L*-shells in the course of a day. The diurnal variation shown in Figure 1 arises because there exists particle gradient in *L* and is a measure of the distortion of the magnetosphere in the geographic reference frame. This interpretation is substantiated by the existence of similar diurnal variation in the locally measured magnetic field at the ATS-1 orbit (Cummings *et al.*, 1968). Paulikas *et al.* (1968) and Lanzerotti *et al.* (1967) have shown that the particle variations during quiet periods are ordered fairly well by the local magnetic field and they have ascribed the diurnal variations to adiabatic causes. It has been shown by Pfitzer *et al.* (1969) that the quiet time radiation belt particles obey the kinematics as described according to Roederer (1968).

Accompanying moderate geomagnetic activity, electron flux-time profiles at synchronous altitudes have superposed on the regular diurnal variation large *non-adiabatic* increases of electron fluxes, typically of 30–60 min duration. Figure 2 shows an example of the count-rate time profile for the three energy ranges (15 min averages). Also included in the figure are locally measured magnetic field intensities and the

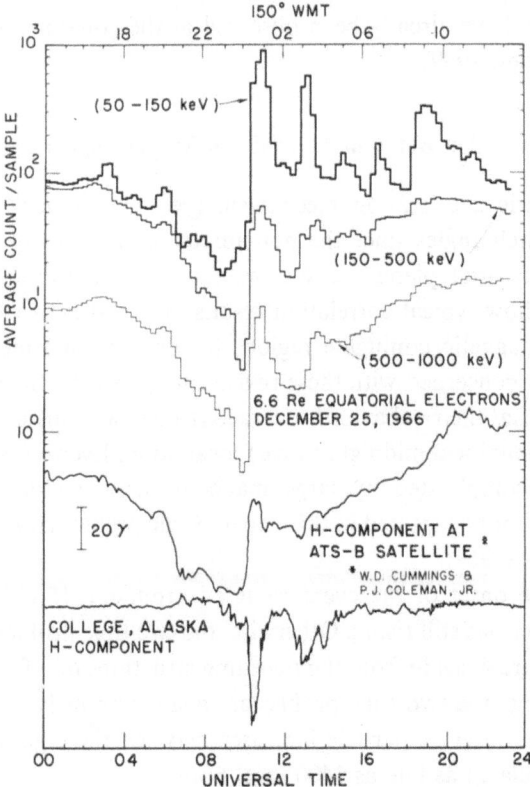

Fig. 2.   A typical time profile of electron fluxes observed during moderate geomagnetic activity The large increases of electron fluxes are nonadiabatic and believed freshly accelerated Van Allen electrons.

auroral zone magnetic record of College, Alaska. For particles, the most interesting feature is the order of magnitude increases of electron fluxes observable throughout the entire day. Note, however, that peak intensities are considerably larger between local midnight and local noon than between local noon and midnight, giving rise to the 'asymmetric' pattern. The onset of electron flux increases in the morning sector are correlated with the *expansion* phase of the auroral substorm (Akasofu, 1964). Here, we have direct evidence showing that electrons of energies > 50 keV are accelerated in the magnetosphere during auroral substorms. Parks and Winckler (1968, Figure 4) recently showed that the magnetospheric acceleration mechanism may be 'quasi-periodic' and that the substorm correlated increase of energetic electron fluxes are probably the source of energetic electrons for the outer regions of the Van Allen radiation belts.

The behavior of magnetic field fluctuations at synchronous altitude during substorms has been studied by Coleman and McPherron (1969) and the paper by these authors should be consulted for details. Briefly, the H-component in the equatorial plane *recovers* during the expansion phase of the auroral substorm. Often, Pi-1 type fluctuations are observed at the equator accompanying the recovery, similar to those observed on the ground in the auroral zone. Correlated fluctuations of ~ 5–40 sec in energetic electrons have already been observed at the equator and the auroral zone (Parks and Winckler, 1969).

4. Correlation Between Auroral and Magnetospheric Electrons

The magnetospheric acceleration mechanism (yet unidentified) evidently affects a broad range of pitch-angles since the non-adiabatic increases of electron fluxes observed in the equatorial plane are accompanied by intense auroral precipitation. Figures 3 and 4 show typical correlation results obtained between synchronous altitudes and their magnetic conjugate regions in the auroral zone. Although in this article we are only concerned with these two events (April 19 and August 11, 1967) it is worthy of note that their behavior is characteristic of all substorm correlated events occurring from about local midnight to past local noon. Events occurring in the afternoon sector are complicated by large magnetospheric distortions accompanying substorms and will not be treated in this paper. Some preliminary examples have been indicated by Parks (1969).

In Figure 3, the onset of the event centered around 1515 UT was not observed because the balloon was still rising. Otherwise, the two time profiles are extremely well correlated. In Figure 4, aside from the pulsating structures of a few minutes period in the equatorial plane, the two time profiles are again extremely well correlated. Here one notes that the onset of particle increases between the equator and the auroral zone are well correlated as late as 1400 local hours.

The peak electron fluxes shown in Figure 3 are ~ 6–7×10^6 $(\text{cm}^2\text{-sec-ster})^{-1}$ for energies between 50–150 keV. For this event > 95% of the electron fluxes were observed in this energy range, and < 5% for electron energies > 150 keV. Consequently, the electron energy spectrum for these increases was very soft. Assuming exponential type spectrum, the electron *e*-folding energy was ~20 keV. This *e*-folding energy is quite similar to the energies of electrons observed in the auroral zone. If we assume that the bremsstrahlung X-rays are results of primary electron energy spectrum of an exponential form, the thick-target bremsstrahlung calculation gives for the precipitated *primary* electron energy spectrum an *e*-folding energy $\simeq 15$–20 keV. The simi-

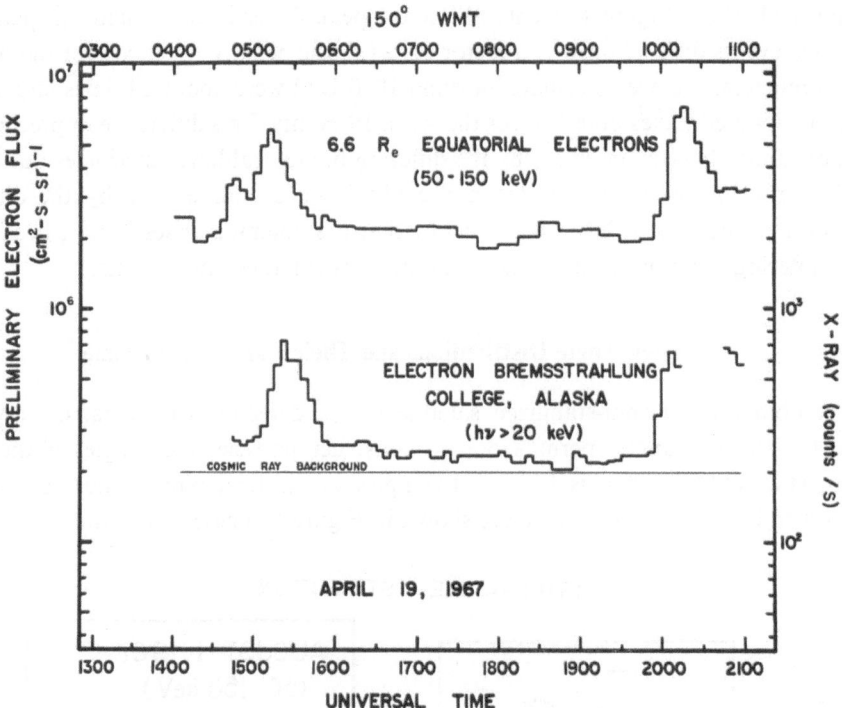

Fig. 3. Correlated precipitated and increases of trapped electron fluxes observed between synchronous altitude and the auroral zone. Data represent 5-min averages.

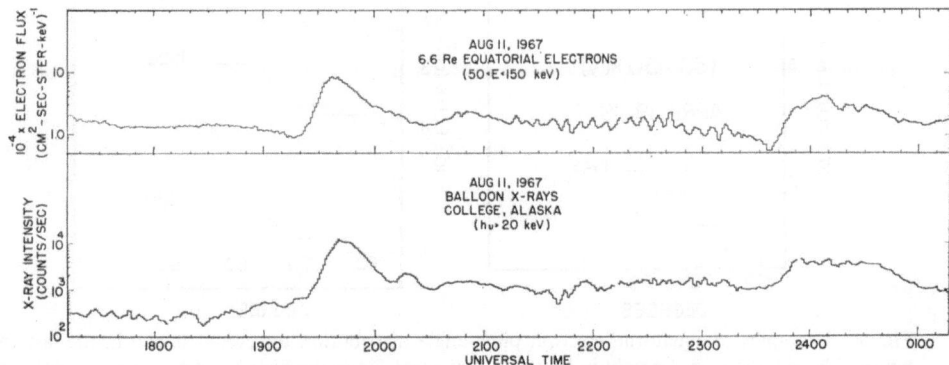

Fig. 4. Same as Figure 3. Data represent 1-min averages.

larity of precipitated and trapped electron energy spectra has already been observed and reported by McDiarmid and Budzinski (1964). The soft energy spectral characteristic for substorm-correlated electrons has also been observed on OGO-5 by Farley (private communication).

The energy spectrum characteristics for the events shown in Figure 4 were essentially similar to those of Figure 3. The derived spectra were again relatively steep in both auroral and equatorial regions, with e-folding energies of $\simeq 20$ keV.

There is one significant difference between the April 19, 1967 (Figure 3) and the August 11, 1967 (Figure 4) events. While the peak fluxes in the equatorial plane for the two events differed only by a factor of $\simeq 1.5$, the peak X-ray fluxes for the August 11 events (e.g. the events centered around 1940 UT) were about 20 times larger than the precipitated fluxes observed for the April 19 events. This difference in precipitated fluxes cannot be accounted for by the differing bremsstrahlung efficiencies due to the differing primary energy spectra nor could it be accounted for by the differing balloon altitudes. It will be shown below that the dilemma arises in part because of the differing equatorial pitch-angle distributions for these two events.

5. Pitch-Angle Distributions and Their Variations in Time

A combination of spin-stabilized satellite and a detector with a narrow aperture ($\simeq 11°$ full view angle) permits one to construct particle pitch-angles if the local magnetic field direction is known. Examples of electron pitch-angle distributions observed in the *equatorial plane* are shown in Figure 5. These distributions were con-

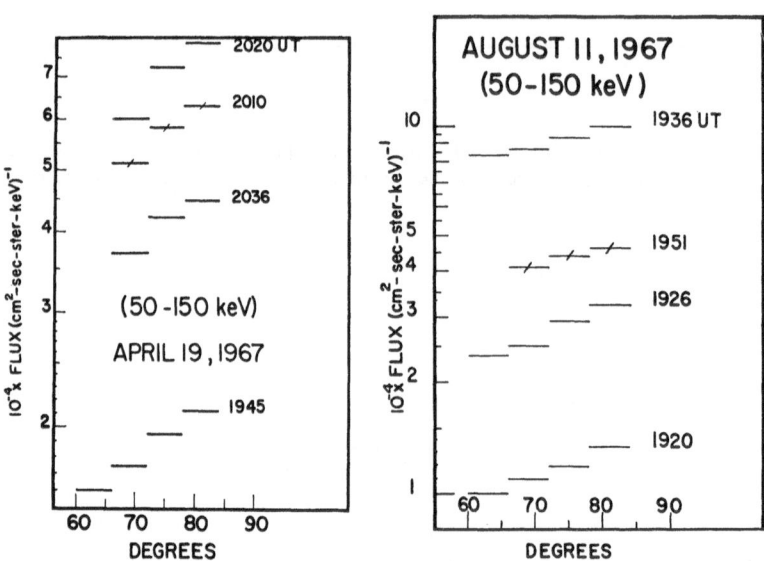

Fig. 5. Examples of equatorial electron pitch-angle distribution for events shown in the last two figures. The numbers on the right indicate the Universal Times the distributions represent. Note that the distributions are peaked toward 90° pitches.

structed using the satellite measured magnetic field direction (courtesy of Coleman and Cummings, UCLA). The pitch-angles were averaged over 6° intervals and electron fluxes were averaged over 72 sec. A number of distributions is shown in each case to illustrate their dependence on time as well as on the electron flux. An important feature indicated by these curves is that the Van Allen electrons are accelerated at all pitch-angles but more preferentially at 90°. *The pitch-angle distribution is anisotropic and peaked toward 90° at all phases of the acceleration during substorms.*

Since a broad range of pitch-angles is sampled, it is possible to study the behavior of pitch-angle anisotropy with time. Our studies indicate that the pitch-angle anisotropy either increases with increasing electron fluxes or it decreases with increasing electron fluxes. Figure 6 shows examples of these two cases, where we have also in-

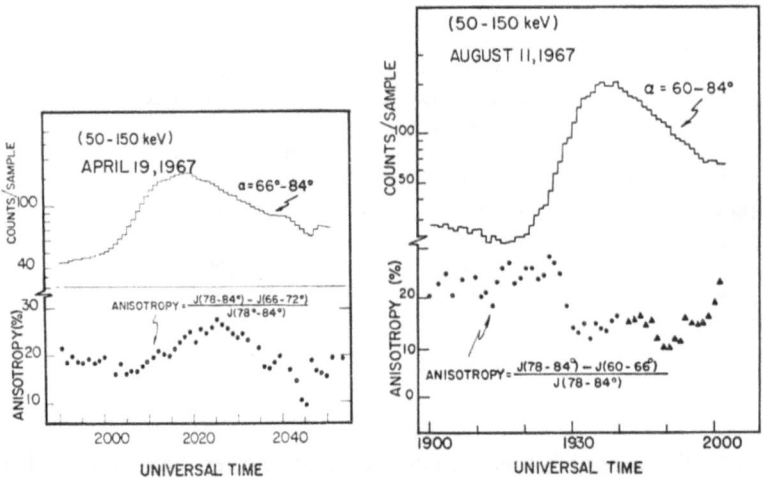

Fig. 6. Behavior of anisotropy with time for the two events discussed in the last figure. See text for details. Note that the anisotropy is measured in percent.

dicated the definition of the pitch-angle anisotropy. Note that the definition may differ from one event to the next because the range of pitch-angles detected depends on the local magnetic field direction. For the April 19, 1967 acceleration event (left curve) one notes that the anisotropy-time curve generally increased and decreased with the flux-time profile. The time-lag between these two curves is a real effect but the significance is not understood at present. In contrast to the April 19, 1967 event, the August 11, 1967 event shows that the pitch-angle anisotropy decreased with increasing electron fluxes (right curve). After about 1940 UT, the range of pitch-angles sampled changed and consequently the anisotropy factor was redefined as $1 - J(66-72°)/J(78-84°)$, and these are shown as triangles. Note that while the pitch-angle anisotropy decreased, the pitch-angle distribution was still peaked toward 90°. The unusually intense precipitated fluxes observed on August 11 can now be explained in part by the behavior of the pitch-angle distribution of this event (Figure 4).

6. Life Times of Freshly Accelerated Van Allen Electrons

The thick-target bremsstrahlung theory provides a means for obtaining the primary precipitation electron energy spectrum and fluxes from observed X-rays. Since *simultaneous* observations of both trapped and auroral precipitation fluxes are available, life-times of these freshly accelerated Van Allen electrons can be calculated. Table I summarizes the results of the life-time calculations for 7 separate events. These calculations are based on formulas given in Hess (1968) and O'Brien (1962). In deriving the average electron fluxes, we assumed the electron energy spectrum to be exponential in form. The life-times of the freshly energized electrons varied from about 200–2000 sec. Electrons in a typical event have life-times of about 1000 sec.

TABLE I

Life-time of substorm electrons

Date	Average equatorial electron flux $(cm^2\text{-sec-ster})^{-1}$	Average auroral precipitated electron flux $(cm^2\text{-sec-ster})^{-1}$	Life-time (sec)
April 2, 1967	8×10^6	10^6	1930
April 19, 1967	4×10^6	10^6	970
	5×10^6	10^6	1200
April 21, 1967	8×10^6	1.6×10^6	1300
August 11, 1967	6×10^6	7×10^6	200
	3×10^6	3.5×10^6	200
August 17, 1967	6×10^6	10^6	1400

One notes that these life-times are considerably shorter than electron drift periods in a geomagnetic field. For electrons of energy about 70 keV and pitch-angles $\simeq 90°$, the drift period at $L \simeq 6.5$ is about 6000 sec assuming a dipole field and absence of electric fields. The short life-time is direct evidence that the satellite was detecting *locally accelerated* electrons created in the near vicinity of the satellite. In fact, the short life-times argue against the magnetotail region as the primary acceleration region. Instead, the short life-times together with observations of acceleration and correlated precipitation in a broad local time region suggest that the magnetospheric acceleration is *a macroscopic and a large-scale phenomenon*. Similar conclusions have been reached by McDiarmid *et al.* (1969).

7. Summary and Discussion

There is a large body of information available concerning magnetospheric and auroral particle fluxes, spatial distribution, and energy spectra (e.g. Frank, 1965; O'Brien, 1964). Many of the results presented in this article are consistent with conclusions and speculations derived from these earlier studies. For example, the original 'splash catcher' model of O'Brien is qualitatively correct even in the equatorial plane. The rapid flux variations observed by Frank (1965) which were thought to be associated with some acceleration process are consistent with the ATS-1 observations. Although Frank could not distinguish whether his observations were due to spatial or temporal effects, is it now clear that the magnetospheric electron fluxes undergo large temporal variations (see Figure 2).

The observations presented in this article show that precipitation and acceleration are strongly coupled since the time profiles and energy spectra in the equator and the auroral zone are correlated. Moreover, the mechanism responsible for these observations produces particles at all pitch-angles, but the pitch-angle distribution is anisotropic and peaked toward $90°$. If we assume the distribution to have a $\sin^n \alpha$ form, n varies between 2 and 3. The anisotropy in pitch-angle distribution is not fixed but changes with time. As was shown, the anisotropy factor can increase or decrease with increasing electron flux, but one notes that the peak in distribution was at all times in the $90°$ direction. Finally, estimates of life-times show that the freshly accelerated energetic electrons are short-lived. Most of these electrons are lost before they can traverse the magnetosphere by the drift motion.

The excellent correlation shown above between acceleration and precipitation raises an interesting question concerning the origin of these particles. Recently, Mozer and Bruston (1966) and Evans (1967) suggested that the earth's ionosphere might be a source region for the $\geqslant 40$ keV Van Allen particles and there is theoretical work of Perkins (1968) showing that 'loss cone' instabilities in the ionosphere under suitable conditions could lead to production of very high energy electrons. But then, *why* and *how* do these particles appear as predominantly $90°$ pitch-angle particles in the equatorial plane? Why is there a general agreement between the energy spectra of precipitated and trapped electrons (McDiarmid and Budzinski, 1964)?

On the other hand, if we assume that the equatorial plane is where complicated wave-particle interactions occur to accelerate and precipitate particles (since it appears that most of the particles in a flux tube are affected by acceleration and precipitation), then one must ask whether the precipitated particles represent freshly accelerated particles in the loss cone or whether they are particles accelerated initially with large pitch-angles which are subsequently scattered into the loss cone and precipitated. We are certain of one feature and that is that the acceleration and precipitation processes must be strongly coupled.

A complete theory for particle acceleration and precipitation is not available at present. However, the literature provides us with numerous models (Axford and Hines, 1961; Swift, 1965; Fälthammer, 1965; Hasegawa, 1969; Kennel, 1969). These models are all capable of producing pitch-angle distributions with peak toward $90°$, as our observation requires. However, from the energies of particles observed and the time profiles, it does not appear that the convective flow (Axford and Hines, 1961) or the radial inward diffusion (Swift, 1965; Fälthammer, 1965) are capable of explaining the present results: the maximum energy gained in a convective flow model is $\lesssim 50$ keV, corresponding to the potential across the tail; radial diffusion times are not only longer than substorm durations, but radial diffusion cannot account for the nearly simultaneous *onset* of precipitation observed across many *L*-shells (Winckler *et al.*, 1962).

The electron acceleration models proposed by Hasegawa (1969) and Kennel (1969) both depend on wave absorption, the coupling occurring at hybrid and cyclotron frequencies, respectively. Instabilities at the hybrid resonance frequency as Hasegawa has described is efficient for accelerating protons and low energy electrons (Hasegawa, 1969). For electrons of energies considered here, cyclotron resonance is probably more efficient. Below, we examine some consequences of cyclotron resonance interactions.

It has been shown by Kennel (1969) that particles can undergo both energy and pitch-angle diffusion in the whistler-mode interaction. Assuming that large-amplitude whistler spectrum is present in the magnetosphere (produced for example by 1–10 keV magnetospheric electrons), particles interacting with the whistler waves can undergo both energy and pitch-angle diffusion. Evidence that the whistler interaction occurs during precipitation is provided by the correlation between enhancement of VLF chorus emission and precipitation (Oliven and Gurnett, 1968). Recent observations of whistler noise spectrum during substorms also indicate that large-amplitude whistlers are present in the magnetospheric equator (Brody *et al.*, 1969). Consequently, we assume that the whistler interaction is probably responsible for precipitation and acceleration. Although time-dependence and coupling of pitch-angle and energy diffusion are not completely understood, it will be shown that the whistler theory can explain numerous features of our observation.

The ambient electron flux at the ATS-1 orbit for $\simeq 50$ keV electrons is $\simeq 3 \times 10^7$ (cm^2-sec)$^{-1}$, corresponding to the stably trapped limit value predicted by Kennel and Petschek (1966). When the acceleration creates new particles, the magnetosphere flux will exceed the stable limit and therefore plasma waves will be generated. Since particles are produced anisotropically ($J_\perp > J_\parallel$), particles must be eventually scattered into the loss cone and precipitated. In the whistler theory, the precipitation rate is proportional to the diffusion rate and the diffusion coefficient proportional to the whistler spectrum. If the acceleration preferentially produces a pancake distribution, the whistler growth-rate is increased and leads to enhancement in precipitation rate. According to our life-time calculations, the ATS-1 electrons are in the region between weak and strong diffusion, with the exception of the August 11 event when the electrons appear to be in strong diffusion. Note that the 200-sec life-time is in good agreement with the theoretical prediction for electrons in strong diffusion (Kennel, 1969). Accordingly, one can also understand the behavior of pitch-angle anisotropy for the August 11, 1969 event. In the limit of strong diffusion, the pitch-angle anisotropy tends to approach isotropy, as was noted for the August 11 event.

To understand the anisotropy behavior for the April 19, 1969 event, let us consider here the time-dependent situation and, for simplicity, that electrons are near strong diffusion. We also define an anisotropy factor as $A = 1 - J(\alpha < 90°)/J(\alpha = 90°)$; where $J(\alpha)$ is the electron flux at pitch-angle α, and in our case, α is usually $< 90°$. It should be evident that the $\alpha < 90°$ particles diffuse more rapidly toward the loss cone than the

$\alpha = 90°$ particles, since the life-times of $\alpha < 90°$ particles are shorter than the $\alpha = 90°$ particles. The reason for this is that the resonance condition in whistlers $\omega - k_{\parallel} v_{\parallel} - |\Omega_-| = 0$ implies that $\omega \simeq |\Omega_-|$ for $v_{\parallel} = 0$ particles. These waves will be strongly damped by the thermal electrons, and since the diffusion coefficient is proportional to the spectral wave energy, $D(\alpha) \simeq |\Omega_-|(B_K/B)^2$, the diffusion coefficient, $D(\alpha) \simeq 0$ when $\omega \simeq |\Omega_-|$. Consequently, particles with near $90°$ pitches only weakly interact with whistlers and therefore their life-times are long. Their life-times are not infinite since $90°$ pitch particles can be scattered in pitch-angles by the bounce resonance mechanism provided a suitable wave spectrum exists (Roberts and Schulz, 1968). The important conclusion is that $90°$ pitch particles have longer life-times than particles with pitches $< 90°$. Since $\alpha < 90°$ particles are rapidly diffusing toward the loss cone and gaining energy (recall that we are near strong diffusion), the quantity $J(\alpha < 90°)/J(\alpha = 90°)$ should initially decrease with time. Therefore, the anisotropy factor, A, increases. The anisotropy A should finally return to the equilibrium value after a time of the order of the life-times of $90°$ pitch particles. These qualitative arguments presented above can account for the observed similarities in the precipitation and trapped flux-time profiles as well as the pitch-angle anisotropy behavior. We note that in strong pitch-angle diffusion, the acceleration and precipitation are strongly coupled. The acceleration and pitch-angle diffusion times are related through $T_E \simeq T_\alpha (E/E_c)^2$ where T_α and T_E are pitch-angle diffusion and acceleration times, E is the energy, and E_c the critical energy, $B^2/8\pi N$. In the outer radiation zones, whistler waves amplitudes in the equatorial plane are often as large as 70 milligammas during substorms (Brody et al., 1969; Russel et al., 1969). If 50 milligamma whistler field is used and assuming that $E/E_c \simeq 2$ ($E_c \simeq 20$–30 keV in the outer zone) the corresponding acceleration time is about 10 min. If $E/E_c \simeq 1$, $T_E \simeq 200$ sec. This means that electrons can gain 20–30 keV energy in a few minutes of time during strong diffusion.

Since the diffusion coefficient and the whistler wave spectrum are related by $D^* \simeq k^* \Omega_- (B_k/B_0)^2$, we can estimate from our observations the whistler wave amplitude near resonance. If the observed diffusion coefficient is used where $D^* \simeq 1/T_L = 10^{-3}$ (sec)$^{-1}$, and Ω_- is about 2×10^4 rad/sec, we get a wide-band wave intensity $B' \simeq 10^{-2}$ gammas if 100 gamma ambient flux value is used. This value agrees with the recent OGO-3 and OGO-5 observations (Russel et al., 1969; Brody et al., 1969) of whistler noise and chorus in the equatorial plane which showed RMS values of about 10^{-2} gammas during substorms. Finally, the wave growth rate is roughly $\simeq \Omega_- \eta A$. The observed anisotropy (A) is about $\frac{1}{3}$, Ω_- is about 2×10^4 rad/sec, and the number of observed resonant particles (η) is $\simeq 3 \times 10^{-3}$ (cm^3)$^{-1}$, giving a growth rate of about 20 rad/sec. This growth rate is roughly 10 times the growth factor for electrons in weak diffusion (Kennel and Petschek, 1966).

We have shown that it is possible to understand our data in the frame work of energy and pitch-angle diffusion accompanying whistler cyclotron resonance. But this does not deny that in the magnetosphere, numerous different processes could act collectively to accelerate particles. For example, the low energy electrons needed to produce the large-amplitude whistlers could have been accelerated initially by the convective flow (Axford and Hines, 1961) in the magnetotail region by energizing the solar wind type plasmas in transit from the neutral sheet to the auroral zone, or by energizing the magnetospheric plasmas in the manner described by Hasegawa (1969). Finally one notes that if cross-L diffusion process is operating in the magnetosphere (Swift, 1965; Fälthammer, 1965), pitch-angle and energy diffusion will be enhanced since cross-L diffusion energizes predominantly the perpendicular energy leading to larger anisotropies in the pitch-angle distribution.

The foregoing has been a study of the magnetospheric dynamics based on information obtained mainly from electrons of energies > 50 keV. It remains to be seen what important roles the electrons of energies $\simeq 10$ keV (the predominant component of visual auroras), the keV type protons (ring current protons), and electromagnetic waves over broad frequencies play in the triggering of the catastrophic magnetospheric substorms to produce the extensive correlated auroral and magnetospheric effects.

Acknowledgements

This text was written at the Faculté des Sciences, Université de Toulouse, France.

The work in Minnesota was in part supported by the National Aeronautics and Space Administration contract NAS 5-9542 and the National Science Foundation Grant NSF GA-487.

References

Akasofu, S.-I.: 1964, *Planetary Space Sci.* **12**, 273.
Arnoldy, R. L. and Chan, K. W.: 1969, Univ. of New Hampshire report; also, 1969, *Trans. Am. Geophys. Union* **50** (abstract).
Axford, W. I. and Hines, C. O.: 1961, *Can. J. Phys.* **39**, 1433.
Brody, K. I., Russel, C. T., Holzer, R. E., Kennel, C. F., Fredericks, R. W., and Scarf, F. L.: 1969, *Trans. Am. Geophys. Union* **50**.
Coleman, P. J. and McPherron, R. C.: 1969, University of California, Los Angeles, Planetary and Space Science Dept., preprint No. 768.
Cummings, W. D., Barfield, J. N. and Coleman, P. J.: 1968, *J. Geophys. Res.* **73**, 6687.
Evans, D. S.: 1967, *J. Geophys. Res.* **72**, 4281.
Fälthammer, C.-G.: 1965, *J. Geophys. Res.* **70**, 2503.
Frank, L. A.: 1965, *J. Geophys. Res.* **73**, 6189.
Hess, W. N.: 1968, *The Radiation Belt and Magnetosphere*, Blaisdell, Waltham, Mass.
Kennel, C. F.: 1969, *Rev. Geophys.* **7**, 379.
Kennel, C. F. and Petschek, H. E.: 1966, *J. Geophys. Res.* **71**, 1.
Lanzerotti, L. J., Roberts, C. S., and Brown, W. L.: 1967, *J. Geophys. Res.* **72**, 5893.
Lezniak, T. W. and Winckler, J. R.: 1969, Univ. of Minnesota Cosmic Ray Technical Report CR-137, to be published.
Lezniak, T. W., Arnoldy, R. L., Parks, G. K., and Winckler, J. R.: 1968, *Radio Science* **3**, 710.
McDiarmid, I. B. and Budzinski, E. E.: 1964, *Can. J. Phys.* **42**, 2048.
McDiarmid, T. B., Burrows, J. R., and Wilson, M. D.: 1969, *J. Geophys. Res.* **74**, 1749.
Mozer, F. S. and Bruston, P.: 1966, *J. Geophys. Res.* **71**, 4461.
O'Brien, B. J.: 1962, *J. Geophys. Res.* **67**, 3687.
O'Brien, B. J.: 1964, *J. Geophys. Res.* **69**, 13.
Oliven, M. N. and Gurnett, D. A.: 1968, *J. Geophys. Res.* **73**, 2355.
Parks, G. K.: 1969, Univ. of Minnesota Technical Report No. CR-138.
Parks, G. K. and Winckler, J. R.: 1968, *J. Geophys. Res.* **73**, 5786.
Parks, G. K. and Winckler, J. R.: 1969, *J. Geophys. Res.* **74**, 4003.

Parks, G. K., Arnoldy, R. L., Lezniak, T. W., and Winckler, J. R.: 1968, *Radio Science* 3, 715.
Paulikas, G. A., Blake, J. B., Freden, S. C., and Imamoto, S. S.: 1968, *J. Geophys. Res.* 73, 4915.
Perkins, F. W.: 1968, *J. Geophys. Res.* 73, 6631.
Pfitzer, K. A. and Winckler, J. R.: 1969, submitted to *J. Geophys. Res.* publication, 1969.
Pfitzer, K. A., Lezniak, T. W., and Winckler, J. R.: 1969, preprint Univ. of Minnesota Cosmic Ray Report.
Roberts, C. W. and Schulz, M.: 1968, *J. Geophys. Res.* 73, 7361.
Roederer, J.: 1968, *Earth's Particles and Fields*, Reinhold Book Co., New York, p. 193.
Russel, C. T., Holzer, R. E., and Smith, E. J.: 1969, *Trans. Am. Geophys. Union* 50.
Swift, D. W.: 1965, *J. Geophys. Res.* 70, 2529.
Winckler, J. R.: 1969, paper presented at Leningrad, June 3–7, 1969.
Winckler, J. R., Bhavsar, P. D., and Anderson, K. A.: 1962, *J. Geophys. Res.* 67, 397.

Discussion

Lanzerotti: Some of the life-times you calculate seem to be irreconcilable with the conclusion of local acceleration as late as 02.00 in the afternoon.

The graph of the April 21st event shows an enhancement of 150 keV electrons, occurring well before that of the low energy electrons. This could be interpreted as first a drift and then a continuous precipitation after the initial electrons arrived.

Parks: Detailed study shows the onset times to be really the same. The peak in the high energy channel is observed considerably earlier but that can be explained in terms of whistler diffusion.

Lanzerotti: I think it is necessary to be very careful in interpreting what is the onset time.

Parks: Okay, but the point is: where are the particles observed on the field line at 2 o'clock in the afternoon produced?

Lanzerotti: It would seem that the onset of the precipitation on the dayside is consistent with an initial gradient drift on the nightside at any local time and then perhaps some other process is taking place.

I don't think that acceleration and precipitation on the dayside coincides with the onset of a substorm on the nightside. How the life-times and length of time of precipitation fit into the picture is very complex. Probably there is a combination of processes.

Parks: There could be time delays but these can also be accounted for by a large acceleration region which has a time dependence in the intensity.

7. SATELLITE AND ROCKET RESULTS FOR THE FEBRUARY 25, 1969 EVENT

INTRODUCTION TO THE ESRO SPACECRAFT
ESRO-II/IRIS, ESRO-I/AURORAE AND HEOS-I

D. E. PAGE

Space Science Department (ESLAB), Noordwijk, The Netherlands

ESRO-II was launched in May 1968, ESRO-I in early October 1968, and HEOS-1 in early December 1968. The solar event of February 25, 1969 was the first to be seen by all three satellites. For that reason, and not because of any particularly exciting solar features, was the event chosen for discussion. Choice of a later event would have given even less time for data analysis before the Symposium and there was no guarantee that at that later time all three satellites would still be operational. Indeed, there was no guarantee that another convenient solar event would arrive in time. The February 25, 1969 event was made more attractive by the fact that several ESRO rockets were launched from the Northern auroral zone at that time as part of a PCA campaign.

The main features of the three satellites and the experiments are described below.

1. ESRO-II/IRIS Satellite

ESRO-II/IRIS, the first ESRO-satellite, was successfully launched into orbit by NASA on May 17, 1968 and since that date has operated very satisfactorily.

A. SCIENTIFIC MISSION

The seven experiments study:

(a) X-radiation from the sun in order to reach a better understanding of processes on the sun itself and to correlate solar X-ray changes with heating and ionization in the earth's atmosphere.

(b) Corpuscular radiation from the sun and the galaxy, including that trapped in the Van Allen belts.

(c) Electron component of the primary cosmic radiation.

Orbit: To fulfil the requirements of the scientific mission the orbit was chosen to be near-polar, 98° retrograde (always sunlit) with apogee about 1100 km and perigee about 350 km.

Attitude and Stabilization: The spacecraft, for stability, spins at about 40 revolutions per minute. The spin axis is held in a plane perpendicular to the solar direction. (Spin-up is achieved using gas jets, while the spin axis is realigned by means of a magnetotorquer.) Thus experiments in the satellite equator point at the sun (and away from the sun) 40 times each minute.

B. SPACECRAFT DETAILS

The 'exploded' diagram (Figure 1) shows the layout of the satellite, and Table I gives brief details of the experiments.

V. Manno and D. E. Page (eds.), Intercorrelated Satellite Observations Related to Solar Events, 367–381. All Rights Reserved
Copyright © 1970 by D. Reidel Publishing Company, Dordrecht-Holland

Structure: 12-sided cylindrical, 86 cm long, 78 cm diameter.
Power Source: 3456 n-on-p solar cells, each 20.5 mm^2.
Battery: Nickel-cadmium 16-cell, 3 amp-hour.
Weight: With Scout E-section, 84.6 kg (including 20.4 kg experiments).
Telemetry:

Real-Time Continuous	Carrier	136.89 Mc/s
	Modulation	PCM/FM/PM
	Rate	128 bits/sec

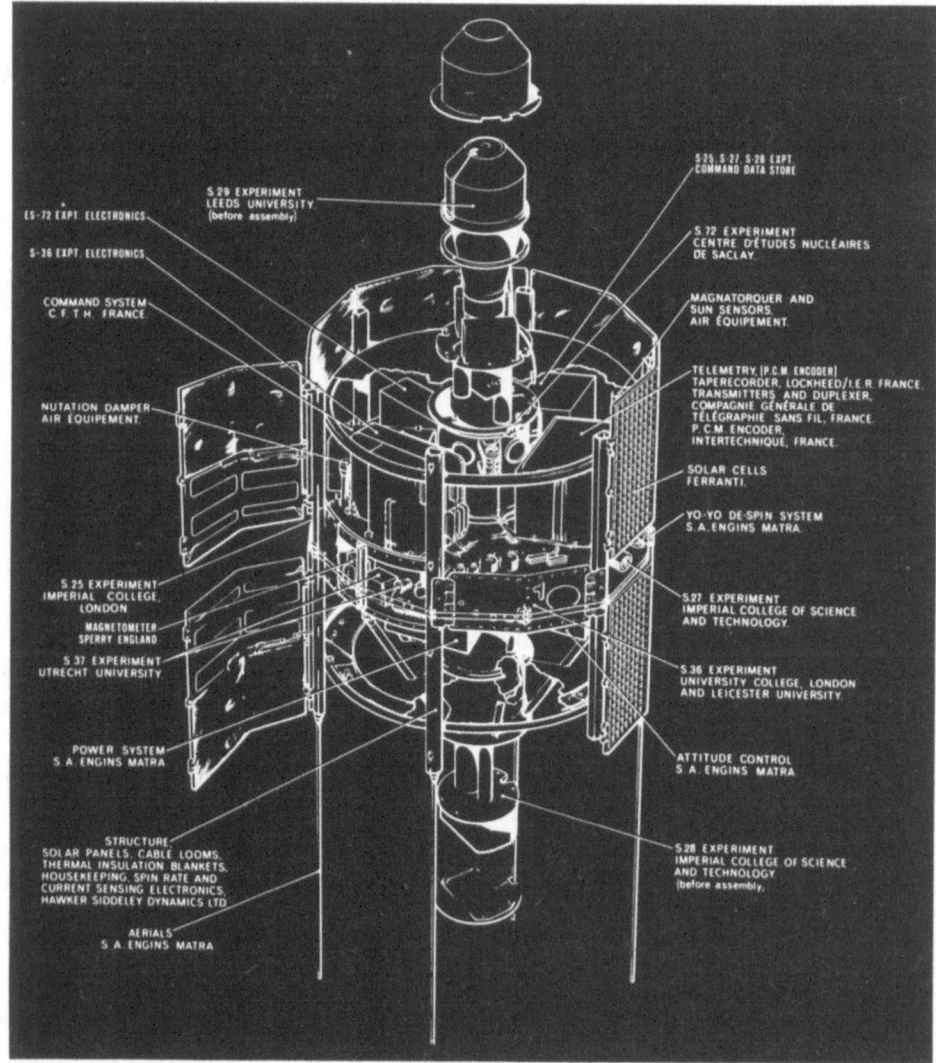

Fig. 1.   Interior configuration of ESRO-II/IRIS.

Playback	Carrier	136.05 Mc/s
	Modulation	PCM/PM
	Rate	4096 bits/sec
Tape Recorder	Records	128 bits/sec for 110 min
	Endless tape	
Command	Carrier	148.25 Mc/s
	Standard	tone-digital
	GSFC	

TABLE I

Experiment	Title	Group	Group leader
S-25	Monitor of energetic particle flux	Imperial College, London	Prof. H. Elliot
S-27	Solar and Van Allen belt protons	Imperial College, London	Prof. H. Elliot
S-28	Solar and galactic α-particles and protons	Imperial College, London	Prof. H. Elliot
S-29	Primary cosmic ray electrons	University of Leeds	Dr. P. Marsden
S-36	Hard solar X-rays	University College, London	Prof. R. L. F. Boyd
S-37	Soft solar X-rays	Laboratorium voor Ruimte-Onderzoek, Utrecht	Prof. C. de Jager
S-72	Flux and energy spectrum of solar and galactic particles	C.E.N.S., France	Prof. J. Labeyrie

C. EXPERIMENT DESCRIPTION

All seven experiments have performed satisfactorily for most of the time and the mission can certainly be regarded as having been successful. One experiment displayed what appears to have been a corona break-down after six weeks in orbit but it cured itself again after a further two months. Another for some time suffered a high-voltage oscillation in phase with the spin of the spacecraft.

S-25: Monitor of Energetic Particle Flux (Imperial College, London)

The experiment measures the particle populations in the Van Allen radiation belts and how these change with time. The Geiger-Müller detectors used are very stable over long periods and enable comparisons to be made with measurements obtained by earlier satellites, particularly ARIEL 1. They also provide a valuable check on the stability of other experiments in IRIS.

The two Geiger-Müller counters are:

	Electron threshold	Proton threshold	Max. count rate
Anton 302	3 MeV	20 MeV	10^4 sec^{-1}
Anton 112	1 MeV	15 MeV	2×10^3 sec^{-1}

S-27: Solar and Van Allen Belt Protons (Imperial College, London)

The experiment aims to monitor solar and Van Allen belt protons between 1 and 100 MeV and α-particles between 5 and 70 MeV. Spectral and intensity variations are being studied in association with solar and geomagnetic activity.

The detector is a telescope of four solid-state detectors separated by small amounts of absorber and shielded at one end. By choosing certain combinations of detectors and suitably setting electronic discriminator levels particles are sorted into different energy ranges.

S-28: Solar and Galactic α-Particles and Protons (Imperial College, London)

The experiment measures the ratio of solar α-particles to solar protons in the magnetic rigidity range 0.4–0.8 GV. The time variations of this ratio help to explain the cosmic ray modulation mechanisms in interplanetary space. In addition, one channel counts fast heavy particles with atomic number greater than 2.

The sensor is a telescope consisting of two scintillators, two proportional counters and a perspex Čerenkov detector. Suitable electronic coincidence arrangements select protons or α's.

S-29: Primary Cosmic Ray Electrons (University of Leeds)

The experiment measures the flux and energy distribution of cosmic ray electrons in the energy range between about 1 and 5 GeV. Such data are relevant to the theories of origin and acceleration of cosmic ray particles and, when combined with radio noise observations, provide a means of estimating the strength of the galactic magnetic field.

The detector is a gas Čerenkov counter, with a threshold (20 GeV) for protons high enough to reduce the contaminating proton background by a factor of 20, followed by lead/scintillator sandwiches where electrons are identified by the particle cascades produced. By suitable electronic selection the integral flux of protons of about 250 MeV is also obtained. The three photomultipliers in the detector are calibrated frequently by solid-state light flashers. Assuming that the flashers remain constant it is thus possible to continuously monitor the gains of each of the scintillator/photomultiplier combinations.

S-36: Hard Solar X-Rays (University College London and University of Leicester)

The experiment measures the flux and energy spectrum of solar X-rays in the wavelength range 1–20 Å. Study of the more energetic part of this radiation, which is closely related to solar flares and exhibits great variation with time, contributes to the knowledge of solar flare phenomena and their effects on the ionosphere.

The data obtained from experiment S-36 are being used to continue the investigation in which the two scientific groups concerned with this experiment have been involved since the launch of a similar X-ray experiment on the OSO-IV spacecraft in October 1967. This earlier experiment functioned continuously until just before the ESRO-II launch.

Five gas-filled proportional counters with different window sizes and wavelength responses look out of the satellite waistband so that each sees the sun once per satellite spin. The counters and counting electronics are grouped such that automatic switching can accommodate X-ray fluxes varying by six orders of magnitude. In-flight calibration is achieved using Fe^{55} sources of 2.1 Å X-radiation. At any time eight-channel pulse-height analysis on each of three counters being used (high sensitivity or low sensitivity) yields a 24-point spectrum.

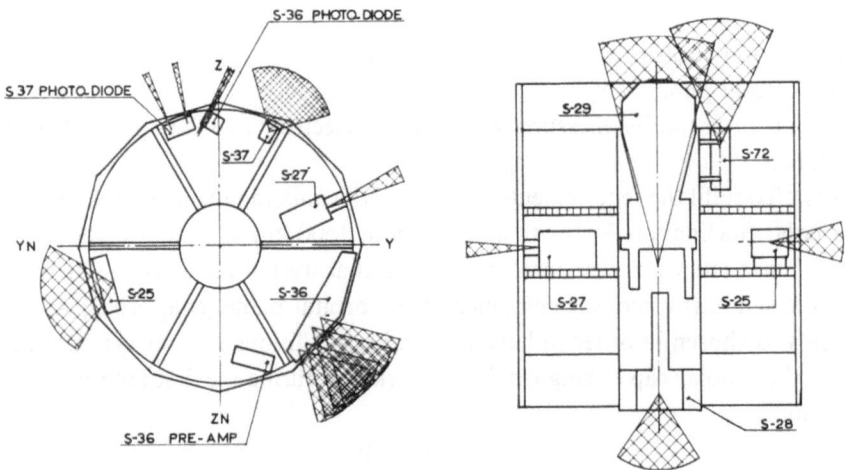

Fig. 2. Viewing angles and installation of experiments, ESRO-II.

S-37: Soft Solar X-Rays (Laboratorium voor Ruimte-Onderzoek, Utrecht)

The experiment measures the flux of solar X-rays in the wavelength band 44–60 Å. As in the case of S-36, the aim is to study the X-radiation during solar outbursts and associated ionospheric phenomena. Two mylar-window proportional counters look out of the satellite equator at opposite ends of a satellite diameter. Thus, while one detector scans the sun the other measures the flux in a direction away from the sun.

S-72: Flux and Energy Spectrum of Solar and Galactic Cosmic Ray Particles (Centre d'Études Nucléaires de Saclay)

Solar and galactic protons between 35 MeV and 1 GeV and α-particles between 140 and 1200 MeV are being measured to help explain solar flares, acceleration processes on the sun and the modulation of the galactic cosmic radiation in interplanetary space. At low geomagnetic latitudes, where only minimum ionizing particles arrive, it should be possible to differentiate between nuclei with atomic numbers between 1 and 6.

2. ESRO-I/AURORAE Satellite

ESRO-I/AURORAE, the second ESRO satellite, was successfully launched into orbit on October 3, 1968 by a NASA Scout rocket from Western Test Range, Calif. Since that date all satellite subsystems and experiments have operated very satisfactorily.

A. SCIENTIFIC MISSION

The aim of the ESRO-I programme is to study the northern polar ionosphere and aurorae. The main emphasis of the eight experiments (see Table II) aboard the satellite is on the investigation of the fine structure of the aurora borealis, and on correlation studies between auroral particles, auroral luminosity and ionospheric composition and heating effects. The experiments can conveniently be considered in three basic groups:

(a) Auroral electron and proton measurements;

(b) Auroral photometry;

(c) Langmuir probe measurements on the electron and ion population of the ionosphere.

Orbit: To fulfil the requirements of the scientific mission the orbit was chosen to be near-polar (inclination 94°) with apogee about 1500 km and perigee about 275 km. ESRO-I was launched near local noon, such that during the early life of the spacecraft the sun–earth line was contained in the orbital plane (cold orbit). An autumn launch was chosen in order to have favourable conditions for viewing aurorae, with the northern polar cap having maximum darkness during the first months of satellite operation.

TABLE II

Experiment	Title	Group	Group leader
S-32	Absolute auroral luminosity	Norwegian Inst. of Cosmic Physics, Oslo	Prof. A. Omholt
S-44	Electron temperature and density	University College, London	Prof. R. L. F. Boyd
S-45	Positive-ion composition and density	University College, London	Prof. R. L. F. Boyd
S-71A	Electron flux and energy spectra in the range 50–450 keV	Radio & Space Research Station, Slough	Mr. R. Dalziel
S-71B	Low-energy electrons and protons in the 1–13 keV range	Kiruna Geophysical Observatory	Prof. B. Hultqvist
S-71C	Auroral proton energy spectra in the range 0.1–6 MeV	University of Bergen and Danish Space Research Institute, Lyngby	Prof. B. Trumpy Prof. B. Peters
S-71D	Monitor of energetic particle flux	Norwegian Defence Res. Establ., Kjeller and Danish Space Research Institute, Lyngby	Dr. B. Landmark Prof. B. Peters
S-71E	Proton energy spectra between 5 and 30 MeV	Radio & Space Research Station, Slough	Mr. R. Dalziel

Attitude and Stabilization: The ESRO-I satellite is passively stabilized along the geomagnetic field lines. The magnetic stabilization enables the particle experiments to measure particles impinging at various angles with the local geomagnetic field vector and thus to discern between particles precipitated along the magnetic field lines and trapped particles moving predominantly perpendicular to the magnetic field. As the satellite passes over the northern polar region two photometers look down to earth and measure the absolute auroral luminosity. Figure 3 shows the orientation of the ESRO-I satellite and of its experiment sensors above the northern polar region.

To achieve magnetic stabilization the satellite had to be despun. The residual spin is about 2 revolutions per orbit.

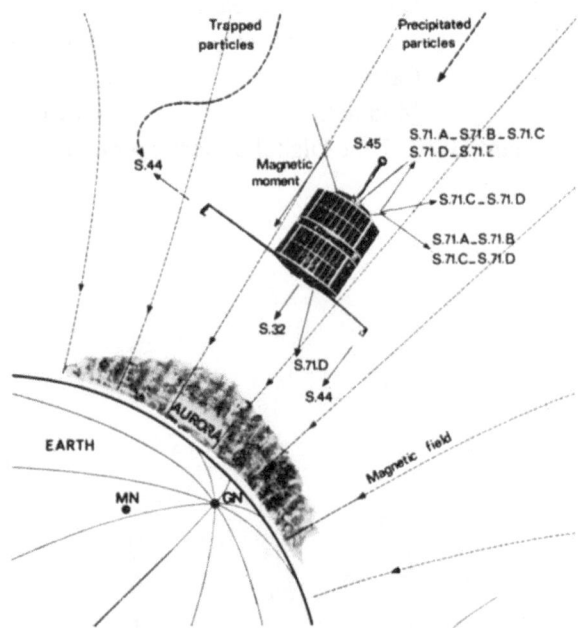

Fig. 3.   Orientation of ESRO-I and its experiment sensors above the northern polar region.

B. SPACECRAFT DETAILS

Figure 4 shows a typical section of the spacecraft.

Power Supply: 6990 n-on-p solar cells, each 2 cm². 16-cell, nickel-cadmium battery, 3 Ah.

Weight: 86 kg, including Scout E section and 22 kg of experiments.

Telemetry: Two separate telemetry systems:

Low-speed data:	320 bits/sec, continuously in real time;
	stored by magnetic tape recorder;
	play-back ratio 32:1 (10240 bits/sec).
	Low-speed format 12.8 sec.
High-speed data:	5120 bits/sec, real time, on command.
	High-speed format 800 msec.

Transmission:

Low-power transmitter:	Operation	continuously
	Effective radiated power	0.2 W
	Data rate	320 bits/sec
	Carrier	136.17 MHz
	Modulation	PCM/PSK/PM
High-power transmitter:	Operation	on command
	Effective radiated power	1.2 W
	Data rate: HS real time	5120 bits/sec
	LS play-back	10240 bits/sec
	Carrier	136.95 MHz
	Modulation	PCM/PM

Telecommand: Standard GSFC Tone Digital System operating at 148.25 MHz with 36 commands.

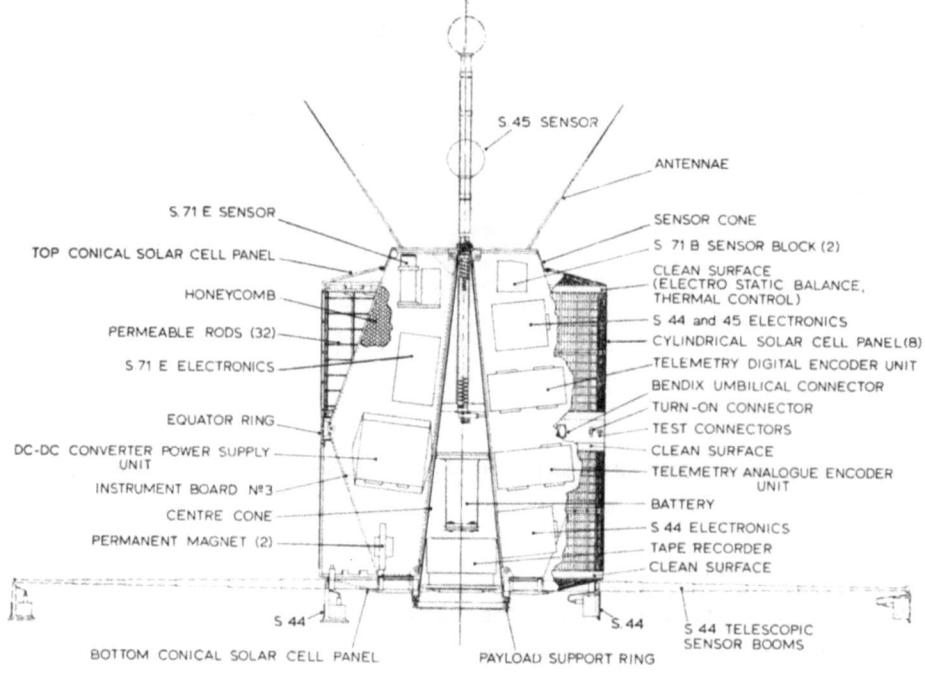

Fig. 4.   ESRO-I satellite – typical section.

C. EXPERIMENT DESCRIPTION

Introduction. All eight experiments were switched on for the first time within 2–10 days after launch. No problems were encountered and the experiments have performed very satisfactorily since then. Six experiments are operating continuously, while the remaining two were designed for highspeed operation only and are switched on by command during high-speed real-time transmissions, mainly above the telemetry station at Tromsö, Norway.

Correlation of the results from all experiments aboard the ESRO-I/AURORAE satellite is of prime importance for the achievement of its aim, the investigation of aurorae and related phenomena. The ultimate success of the scientific mission will depend largely on the close cooperation of all participating scientific groups during the time of data analysis. Furthermore, correlation of the satellite data with magnetic, ionospheric and auroral ground observations is of great importance. Measurements by ESRO-I are concentrated mainly over Northern Europe and close cooperation with the Geophysical Observatory at Kiruna, Sweden, and the Auroral Observatory at Tromsö, Norway, has been assured. Other European institutes as well as stations in Greenland, Iceland and Alaska, have expressed their interest in cooperation and exchange of data with ESRO-I experimenters and special observation periods have been defined when more frequent and intensified ground measurements are taken during ESRO-I overhead passes.

S-32: Auroral Photometry (Norwegian Institute of Cosmic Physics, Oslo)

Two photometers looking downwards to earth when the satellite passes over the northern polar cap measure the absolute auroral luminosity. The experiment aims to map the occurrence of aurorae and in particular to study the relation between proton-excited and electron-excited aurorae.

One photometer measuring the 4278-Å band emitted from ionized nitrogen (N_2^+) gives a measure of the total energy input by fast particles into the ionosphere and of the total rate of ionization. Apart from its specific optical investigations this photometer can therefore serve as a particle monitor and allows a cross-check with data from the particle detectors.

The other photometer measures the 4861-Å ($H\beta$) line of neutral hydrogen. The excited emitting hydrogen atoms are produced from fast protons impinging upon the atmospheric atoms and molecules. Although the relation between this emission and the flux of protons is rather complicated, the photometer in fact maps the regions where protons are a significant constituent of the beam of primary particles producing the aurora.

S-44: Electron Temperature and Density (University College, London)

The experiment consists of two Langmuir plasma probes to measure the electron density and temperature. The aim is to obtain global coverage of these parameters by storing data on the on-board tape recorder, so as to observe variations with latitude, local solar time and altitude.

Important to the ESRO-I/AURORAE mission will be the correlation of the direct ionospheric measurements with data from the auroral photometers and the particle experiments. More incidental to the overall project are probe studies for their own sake, for example a study of variations in spacecraft potential with respect to the plasma, and observations on the satellite wake.

The two electron probes are plane, circular discs, mounted on two 80-cm booms extending from the bottom of the spacecraft perpendicular to its spin axis. One sensor is oriented perpendicular to the magnetic field while the other is pointed anti-parallel to the magnetic field vector. An attempt will be made to measure magnetic field aligned effects on ionospheric electrons.

S-45: Ion Composition and Temperature (University College, London)

The sensor of the ion probe consists of a rhodium-plated sphere, 9 cm in diameter, surrounded by a concentric spherical grid 10 cm in diameter, with a transparency of 30%. The probe is mounted on a boom 0.5 m long, extending from the top of the satellite along its spin axis.

The aim of this experiment and the measuring technique used are very similar to those of the electron-probe experiment, which is supplied by the same scientific group. The ion probe is also referred to as an ion mass spectrometer because of its ability to resolve different ion species. This feature depends on the fact that the spacecraft velocity largely exceeds the mean thermal velocity of the ions, which is not the case for the electrons. The ion energy is provided mainly by the ion drift velocity due to the spacecraft motion. By increasing the positive potential applied to the probe, ions of increasing mass are retarded and may be identified as peaks in the second derivative of the probe current/voltage characteristic. A signal proportional to the logarithm of the second derivative is transmitted in low speed and also stored on the on-board tape recorder.

S-71A: Electron Flux and Energy Spectra (Radio & Space Research Station, Slough)

The experiment uses plastic scintillators on the face of photomultipliers, with subsequent pulse-height analysis of the output to investigate differences in the flux and energy spectrum of precipitated and trapped electrons in the energy range 50–450 keV. Precipitated electrons are observed by one detector pointing approximately in the direction of the earth's magnetic field, and trapped electrons are observed by a second detector perpendicular to the magnetic field. Background radiation is measured by two other detectors of the same construction but which contain magnets in their collimators to reject electrons below 450 keV.

The experiment operates in a low-speed and a high-speed mode. In low speed the energy spectrum is obtained once every 12.8 sec and integral intensities may be measured between 2.10^4 and 2.10^8 electrons/cm^2 sec ster. In the high-speed mode the integral intensities of the electrons are measured every 100 msec.

S-71B: Low-Energy Electrons and Protons (Kiruna Geophysical Observatory)

The purpose of the experiment is to measure electron and proton spectra in the 1–13 keV range in two directions (10° and 80°) relative to the earth's magnetic field lines. From recent measurements it may be expected that these particles are the main contributors to the excitation of aurorae.

The energies of the incoming particles are defined by means of cylindrical, electrostatic analysers. Particles which have passed the analyser are detected and multiplied by windowless continuous-dynode electron multipliers ('channeltrons'). Twelve sensors are used in two blocks of six. Each block contains one reference sensor counting radiation from a built-in Ni-63 radioactive source to monitor the long-term behaviour of the channel multipliers in orbit. It is known that the multipliers degrade with time and accumulated counts, and corrections have to be made for this effect.

The experiment operates in the high-speed mode only and is switched on during real-time passes above the three telemetry stations in the northern auroral zone (mainly Tromsö, but also Fairbanks and Spitzbergen).

S-71C: Medium-Energy Proton Experiment (University of Bergen & Danish Space Research Institute, Lyngby)

Solid-state detectors are used to measure the energy spectra of auroral protons in the energy range 100 keV to 6 MeV at two different orientations (0° and 90°) with respect to the satellite reference axis. A third detector at 45° is shielded by 0.3 mm of aluminium and measures the high-energy particle background. All sensors are magnetically shielded in order to deflect electrons with energies less than 500 keV. The response of the two main detectors is linear for protons with energies less than 5.6 MeV. The pulses from the detectors are fed through pre-amplifiers and amplifiers to pulse-height analysers with 6 channels.

The instrument can operate in either a low-speed or a high-speed mode. In the low-speed mode it takes 3.6 sec to cycle through all three detectors, six channels on each. In the high-speed mode the corresponding cycle time is reduced to 0.225 sec.

S-71D: Integral Particle Flux at Different Orientations (Norwegian Defence Research Establishment, Kjeller and Danish Space Research Institute, Lyngby)

Four Geiger-Müller tubes are used to obtain information on the angular distribution of energetic particles impinging on the upper atmosphere. Three detectors are oriented in the usual directions with respect to the geomagnetic field, parallel to measure precipitated particles, at 45°, and perpendicular to detect trapped particles. The fourth detector is oriented at 160° and measures back-scattered particles moving up along the magnetic field lines when the satellite is above the northern polar region.

The Geiger-Müller counters measure the integral particle flux of electrons with energies > 40 keV and of protons > 0.5 MeV. By correlation with the proton measurements of experiment S-71C, the electron flux alone may be obtained.

An outstanding feature of the experiment is its high time-resolution. The experiment was originally designed for high-speed operation only and the basic sampling period during real-time transmission is 17 millisec. Following a recommendation by NASA that the integral particle flux should be monitored throughout the satellite orbit rather than only in the auroral zone, it was decided to operate the experiment continuously and to connect a ratemeter to the Geiger-Müller tube measuring at right angles to the magnetic field. This ratemeter was designed and built by ESLAB.

S-71E: Proton Energy Spectra (Radio & Space Research Station, Slough)

This experiment measures the energy spectra of protons over the range 5–30 MeV, particularly during polar cap absorption (PCA) events. The spectra are obtained by pulse-height analysis of the outputs from a telescope consisting of a semi-conductor detector and a caesium-iodide crystal scintillator which is viewed by a photomultiplier. The telescope accepts protons arriving within a cone of 30° half-angle about its axis, which is aligned with the Z-axis of the satellite.

The experiment operates in low-speed only. A complete spectrum is obtained every 12.8 sec. The dynamic range of the experiment is from 5 to 4.10^4 protons/cm² sec ster.

3. HEOS-I Satellite

A. SCIENTIFIC MISSION

The HEOS-1 spacecraft (Highly Eccentric Orbit Satellite) carries experiments to investigate the environment of interplanetary space and the distant magnetosphere at the time of the present maximum of the solar cycle. It was launched from Cape Kennedy on December 5, 1968, by a thrust-augmented Improved Delta into an orbit with apogee 223 000 km and inclination 28.2° (Figure 5). The launch time was such that the apogee remained in interplanetary space, outside the earth's bow shock, for the first 150 days of life before traversing the tail region of the magnetosphere. The apogee re-emerged into interplanetary space after 240 days. For the initial orbits 85% of the time was spent outside the bow shock.

Besides the improvement in our understanding of interplanetary physics through the HEOS-1 mission, the possibility exists to make correlations with the results from ESRO's first two spacecraft IRIS and AURORAE in order to study the wider field of solar-terrestrial relationships.

The HEOS-A2 mission is a follow-up to HEOS-1 and for the A2 spacecraft, the maximum use of the existing design will be made.

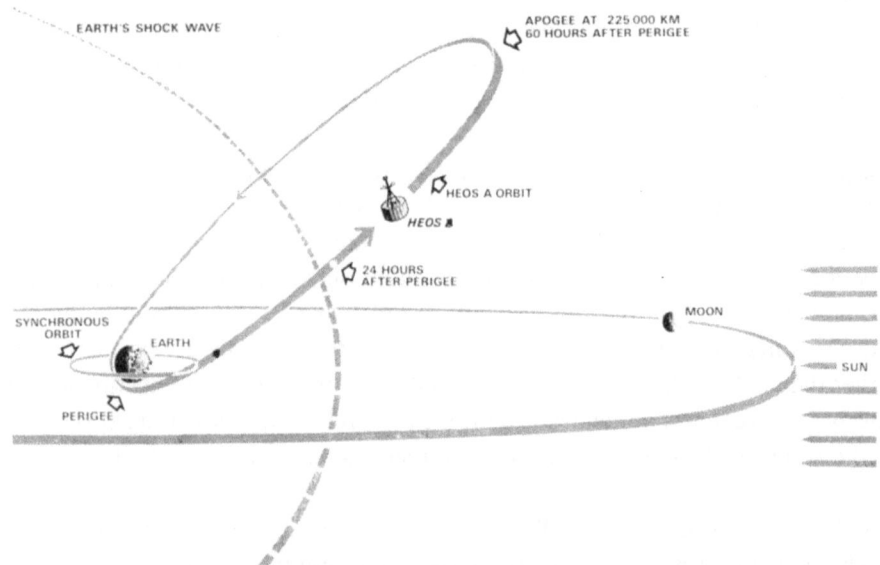

Fig. 5.   Orbit of HEOS-I.

B. SPACECRAFT

Spin stabilization (at 10 r.p.m. nominal) was chosen for HEOS for technical simplicity. A cold-gas reorientation system is included, so that the spin axis can be oriented in any desired direction. These features allow variation in the scanning plane of the experiment sensors S-24 A/B/C and S-58/73, so that intercalibration of sensors can be performed and measurements made, for example, in the ecliptic plane and perpendicular to this plane.

Since launch, four reorientation manoeuvres have been carried out such that approximately equal times have spent scanning in and perpendicular to the ecliptic plane.

The spacecraft attitude can be determined to within $\pm 2°$ through the use of a solar aspect sensor and an albedo sensor. Nine sun sensors deliver pulses to the experiments as the sun passes through nine different meridian planes of the spacecraft, enabling angular measurements to be made.

A cut-away view of the spacecraft (Figure 6) shows the main structural details, i.e. the adaptor which was mated to the launch vehicle, the octagonal-tube main structure, which supported the majority of boxes, and the outriggers supporting the 16-sided solar array.

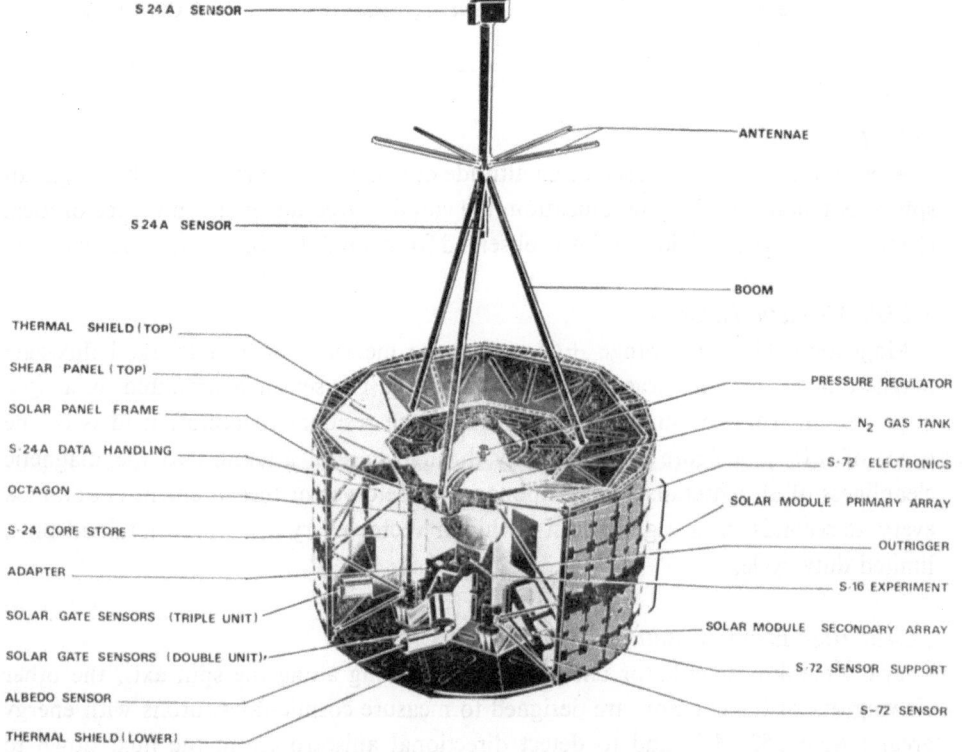

Fig. 6. Perspective view of H EOS-I.

The data transmission rate of 12 bits per second was determined by apogee height and limited transmitted power. No tape recorder is carried and so all data are transmitted in real time. However, typically 98% of every orbit can be covered by the ground stations.

C. EXPERIMENT DESCRIPTION

Table III gives a list of the experiments and the scientific groups responsible.

TABLE III

Experiment	Title	Group	Group leader
S-16	Ion cloud	Max-Planck-Institute for Extraterrestrial Physics, Garching	Prof. Lüst
S-24A	Magnetic fields	Imperial College, London	Prof. Elliot
S-24B	High-energy cosmic ray anisotropy	Imperial College, London	Prof. Elliot
S-24C	Low-energy solar protons	Imperial College, London	Prof. Elliot
S-58/73	Solar wind	Universities of Florence, Rome, Brussels	Professors Bonetti, Pizzella, Coutrez
S-72	Flux and spectrum of cosmic rays	C.E.N. Saclay	Prof. Labeyrie
S-79	Primary cosmic ray electrons	C.E.N. Saclay, University of Milan	Professors Labeyrie & Occhialini-Dilworth

S-16: Ion Cloud

A barium cloud was released at an altitude of some 75 000 km, inside the magnetosphere boundary. Following ionization, the cloud moved under the influence of local electric and magnetic fields, and was observed from a number of ground stations.

S-24A: Magnetic Fields

Magnetic fields in the range ± 64 γ's can be measured with a tri-axial flux-gate magnetometer to an accuracy of better than 0.5 γ. The sensor is mounted on a rigid boom 1.7 m from the main structure, where the residual spacecraft field is of the order of 0.15 γ, a figure achieved only through careful attention to the magnetic cleanliness of all materials and components. A field vector measurement is obtained every 48 seconds on a regular basis, while a vector every 1.5 sec is obtained on a limited duty cycle.

S-24B: High-Energy Cosmic Ray Anisotropy

Two Čerenkov/scintillator telescopes, one viewing along the spin axis, the other viewing out of the equator, are designed to measure cosmic-ray protons with energy greater than 350 MeV and to detect directional anisotropies in the flux, down to 0.25%. The geometric factor of the telescopes is of the order of 5 cm^2 ster.

S-24C: Low-Energy Solar Protons

Two range/energy telescopes, each employing four surface barrier solid-state detectors (located on the spacecraft similarly to S-24B), will measure the flux, energy spectra and arrival direction of protons from 1 to 20 MeV. The energy range is split up into four windows by coincidence/anticoincidence pulse-height conditions on the detectors. The geometric factor of the telescopes is of the order of 10^{-2} cm^2 ster. Correlations of magnetic-field data and particle-arrival-direction data should yield fundamental information on the propagation mechanisms of solar protons.

S-58/73: Solar Wind

S-73 can measure the energy spectrum and S-58 the angular distribution of the positive component of the solar wind in the energy range 0.15–15 keV, using an electrostatic analyser and Faraday cup. The experiment scans in the equatorial plane of the satellite. A measurement of the 28-step energy spectrum is obtained every 6.4 min. The Faraday cup is capable of measuring particle fluxes down to 10^6–10^7 particles/cm^2/sec. The field of view defined by the electrostatic analyser is 12° in azimuth and 100° in elevation.

S-72: Flux and Spectrum of Cosmic Rays

A four-element solid-state detector telescope is used to measure the flux and energy spectra of protons from 3.7 MeV to relativistic energies, relativistic electrons and α-particles in the range 40–440 MeV/nucleon. Particle energy and selection is determined by coincidence/anticoincidence pulse-height conditions on the detectors. The experiment scans through the equatorial plane but no directional measurements are performed.

S-79: Primary Cosmic Ray Electrons

The flux and energy spectrum of primary electrons in the range 10–600 MeV is measured by a four-element telescope including a gas Čerenkov detector to eliminate protons below 15 GeV, and a lead-glass Čerenkov shower detector for energy determination. A solid-state detector is incorporated to define more closely the acceptance cone and a geometric factor of 0.4 cm^2 ster is achieved. Particles which penetrate the lead glass detector are observed by a CaF scintillator.

Discussion

Pieper: What were the launch dates of the three satellites?
Page: ESRO II: May 17, 1968; ESRO I: Oct. 2, 1968; HEOS I: Dec. 5, 1968.

GROUND OBSERVATIONS OF THE SOLAR EVENT
OF FEBRUARY 25, 1969

K. G. LENHART
ESOC, Darmstadt, Germany

Abstract. In order to give a broad basis for correlation of satellite data with ground-based measurements for the solar event of February 25, 1969, a collection of ground-based observations centred on this period has been made. The data presented include solar measurements (magnetic fields, radio emission, sunspots) and effects caused on the earth (ionospheric effects, magnetic measurements and neutron monitor data).

1. Introduction

The purpose of this paper is to present data of ground-based measurements in relation to the solar event of February 25, 1969. The intention is to give information on many different measurements in order to have a broad basis for correlation with the satellite data. Because of the limited space available only some typical data can be shown, but a special paper has been published which contains most of the collected data.

2. Summary of Solar Activity at End of February, 1969

The complex region (McMath plage region 9946), which was responsible for the high solar activity during end of February, 1969, started a period of enhanced activity (see Table I) with an importance 1B class M flare on February 23, 1969, at 04 41 UT (location N12 W09), followed by a large class M event of importance 2B on February 24, 1969, at 23 06 UT (location N12 W32). Ten hours later on February 25,

TABLE I

Time (UT)		Importance	Class	Location
1969, Feb. 23	04.41	1B	M	N12 W09
24	23.06	2B	M	N13 W32
25	09.03	2B	X	N13 W37
25	19.37	1B	M	N13 W43
26	04.19	2B	X	N13 W46
26	05.50	1B	M	N13 W46
27	13.52	2B	X	N13 W65

Definitions:
   Class M – Solar events which are accompanied by significant X-ray production, greater than 10^{-2} ergs cm^{-2} sec^{-1} in 0–8 Å band, or 10^{-3} ergs cm^{-2} sec^{-1} in 0.5–5 Å band, comparable SID (SWF or SPA), or by a 10 cm radio-noise outburst of at least 75 flux units over background and of duration >1 min.
   Class X – Solar events which are accompanied by great X-ray production, greater than 10^{-1} ergs cm^{-2} sec^{-1} in 0–8 Å band, or 10^{-2} ergs cm^{-2} sec^{-1} in 0.5–5 Å band, comparably great SID, or by a 10 cm radio-noise outburst of more than 1000 flux units over background and duration greater than 10 min.

1969, at 09 03 (location N13 W37) another importance 2B flare was observed and classified as class X. It produced an intense complex radio burst, severe ionospheric disturbances and ground level increase on neutron monitors. A considerable X-ray flux (peak ratio to quiet sun of 90) and high energy solar protons were observed by satellites. The relatively short-lived event reached maximum for the X-ray flux at about 09 18 UT and for the protons between 11 00 and 13 00 UT.

The same region produced later several other major flares:

Feb. 25, 19 37 UT an importance 1B class M flare (location N13 W43)
Feb. 26, 04 19 UT an importance 2B class X flare (location N13 W46)
accompanied by a complex 10 cm radio burst (620% increase)
Feb. 26, 05 50 UT an importance 1B class M flare (location N13 W46)
Feb. 27, 13 52 UT an importance 2B class X flare (location N13 W65)

with only minor type IV radio emission (290% increase at 10 cm). (For summary see Table I.)

The flares of February 23 and 24, 1969 produced, according to Kiruna Geophysical Observatory, two weak PCA (Polar Cap Absorption) events and the flare of February 25, 1969 was reported to have been followed by a medium PCA. (The main source of information was phase and amplitude records of VLF radio signals.) For the flare of February 27, 1969 there were conflicting reports of the ionospheric detection of a PCA.

This paper reports mainly on the effects caused by the flare of February 25, 1969.

3. General Solar Terrestrial Conditions

An inspection of the general solar activity during the first quarter of 1969 is useful for establishing the background. In Figure 1 four histograms of frequently used daily measurements are drawn (the horizontal coordinate is the time in days).

(a) the relative sunspot number Rz from Zürich (provisional values). A very pronounced maximum (reaching more than 200) appears at the time of interest (end of February);

(b) still better to see and centred round February 25 is the maximum of the daily solar flux as recorded at 2800 MHz in Ottawa. (The adjusted flux to 1 AU is given.) The units (scale to the right) are $10^{-22} \mathrm{Wm}^{-2} \mathrm{Hz}^{-1}$;

(c) the next histogram shows the daily averaged neutron monitor counts from Deep River (also referred to as cosmic ray indices). The remarkable maximum of the daily values on February 25, 1969 indicates the arrival of high energy solar cosmic rays following the solar event. A well-developed Forbush decrease occurring in several steps (see also Figure 15) coincides with geomagnetic disturbances of relatively small size;

(d) these small disturbances can be seen in the increase of the daily average index A_p (derived from the K_p indices) on February 26 and 27, 1969 from a quiet level during the week before this event.

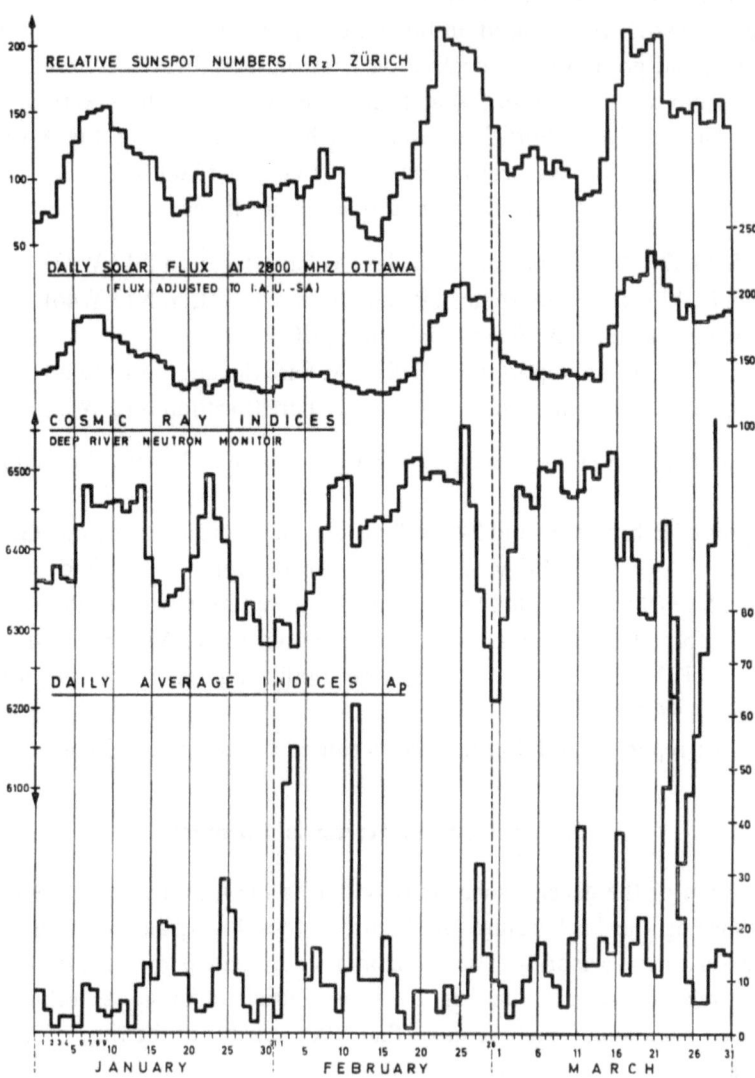

Fig. 1.   Solar activity during the first quarter of 1969.

From this general impression one could already define the event as accompanied by strong radio emission, producing sufficient high-energetic protons to create a pronounced ground level effect (GLE) but only small numbers of geomagnetic storm particles that hit the earth.

We have two different types of measurements, if we wish to distinguish between the origin:

(a) measurements directed towards the sun (see Section 4);

(b) measurements of effects near or on the earth caused by the sun's activity (see Section 5).

4. Observations of the Sun

Figures 2 show the development of the flare of February 25, 1969, at various universal times.

Coincident with the time of Figure 2b, when a split of the flare is recognisable, were the highest emissions of microwave and radio fluxes, indicating that some process of acceleration was taking place.

(The photographs have been provided by courtesy of Dr. P. Simon of the Observatory of Meudon.)

4.1. SOLAR MAGNETIC FIELDS

The measurement of the solar magnetic fields associated with sunspots is particularly important for the understanding of the sources of energy, which is released during flares. Figures 3–5 showing these data demonstrate the complex structure of the magnetic fields. In Figure 3 the measurements from Crimea are drawn, containing an overall picture of the sun (at the left side) for February 25, 1969, and details of various regions which are marked by Roman numerals. Figure 4 gives more details of the same measurements for the interesting region from which the flare originated. N means north seeking polarity and S south seeking polarity. The numbers following the letters are the field intensities in units of 100 G. The spots can be divided into

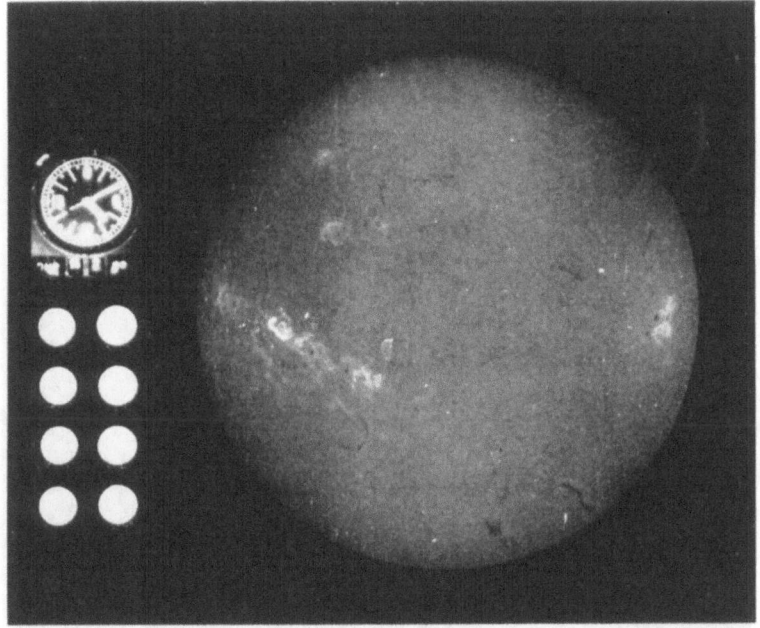

Fig. 2a.   UT 0910

The development of the solar flare of February 25, 1969. The flare was photographed in Hα light The coordinates of the flare are N 13 and W 37.

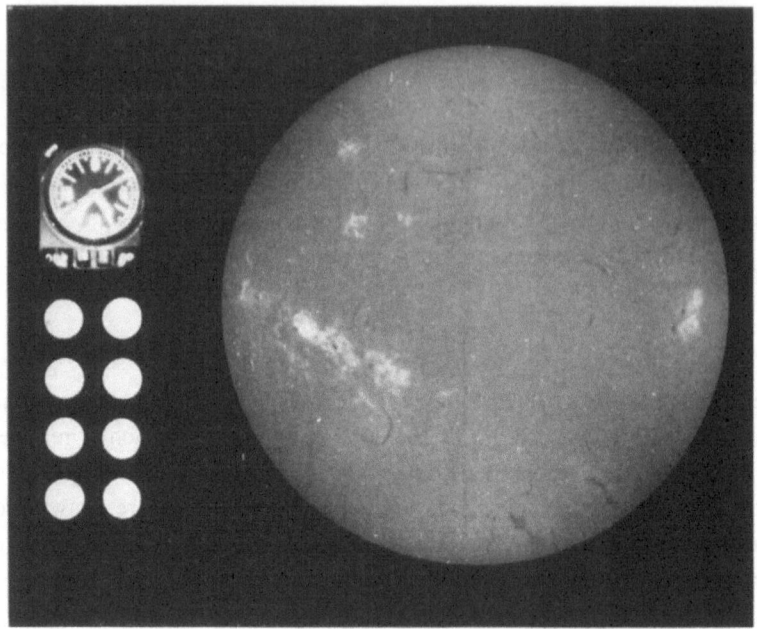

Fig. 2b. UT 0919

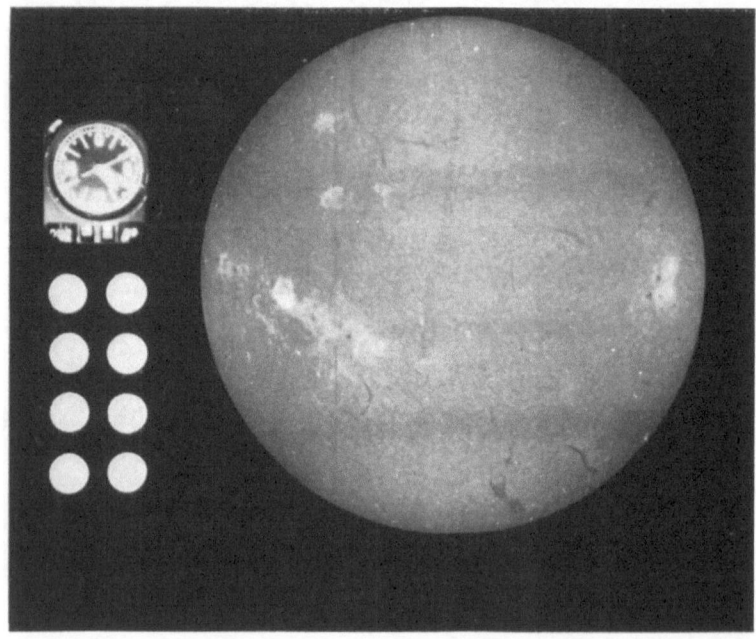

Fig. 2c. UT 0937

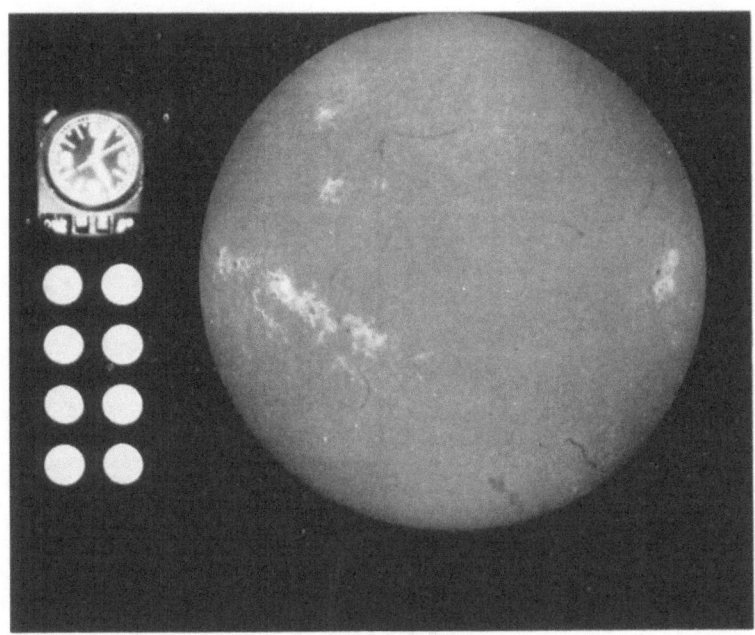

Fig. 2d.   UT 1004

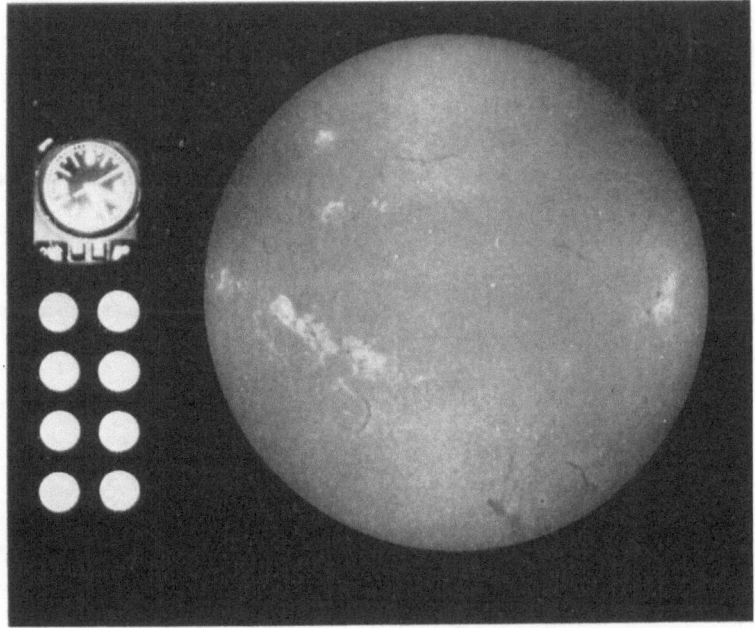

Fig. 2e.   UT 101 9

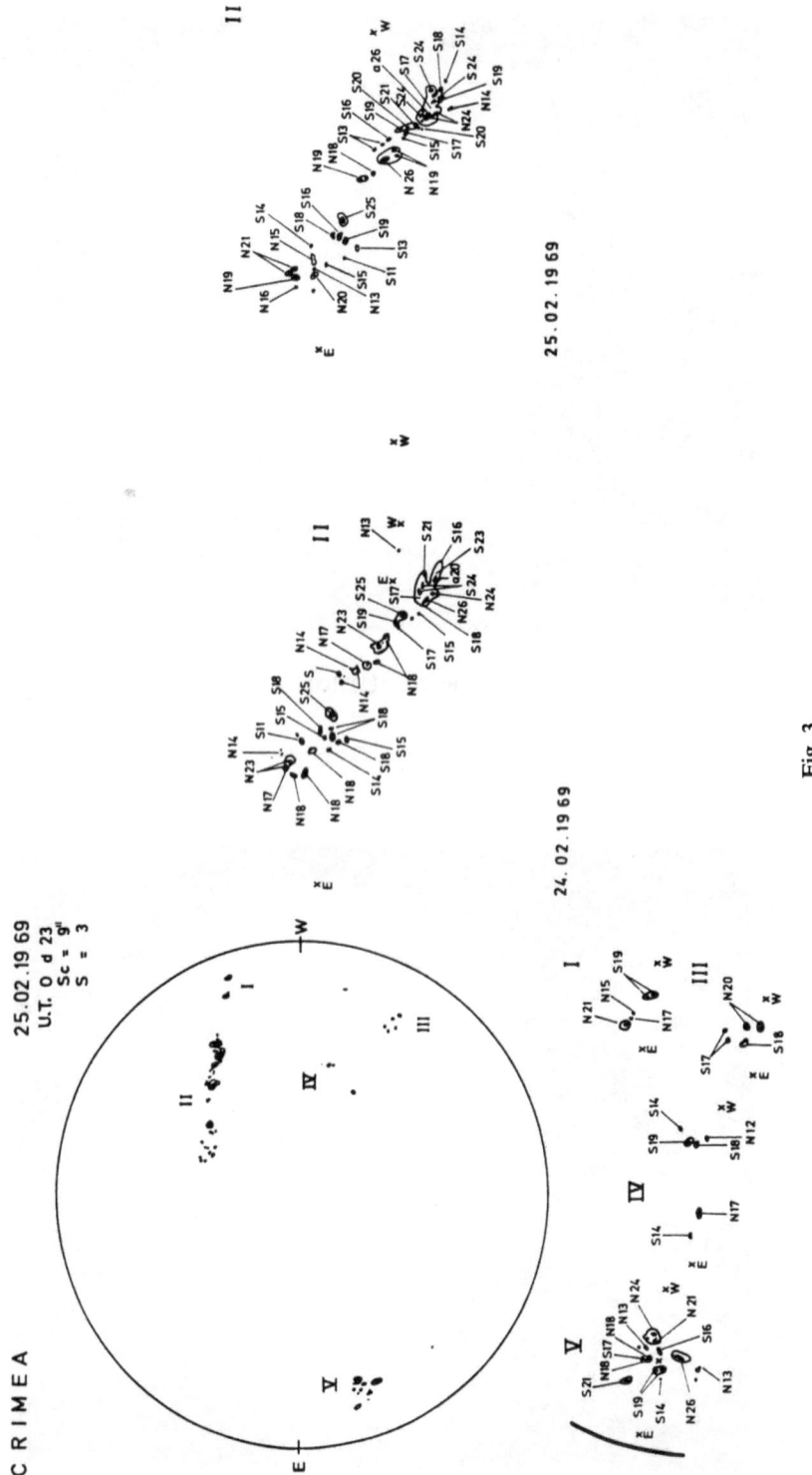

Fig. 3.

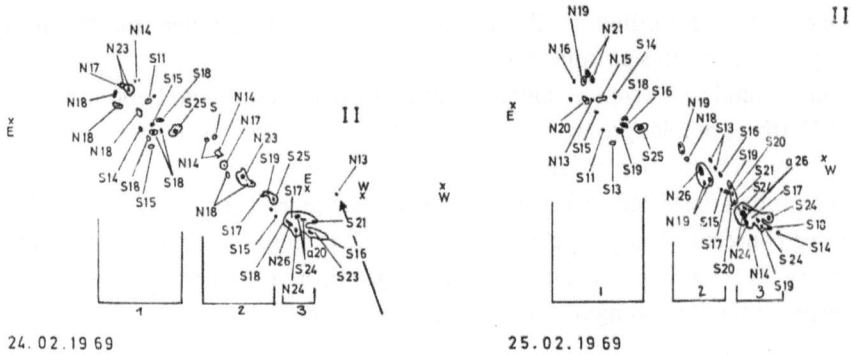

24.02.1969   25.02.1969

Fig. 4.   Bipolar spot groupings, Crimea.

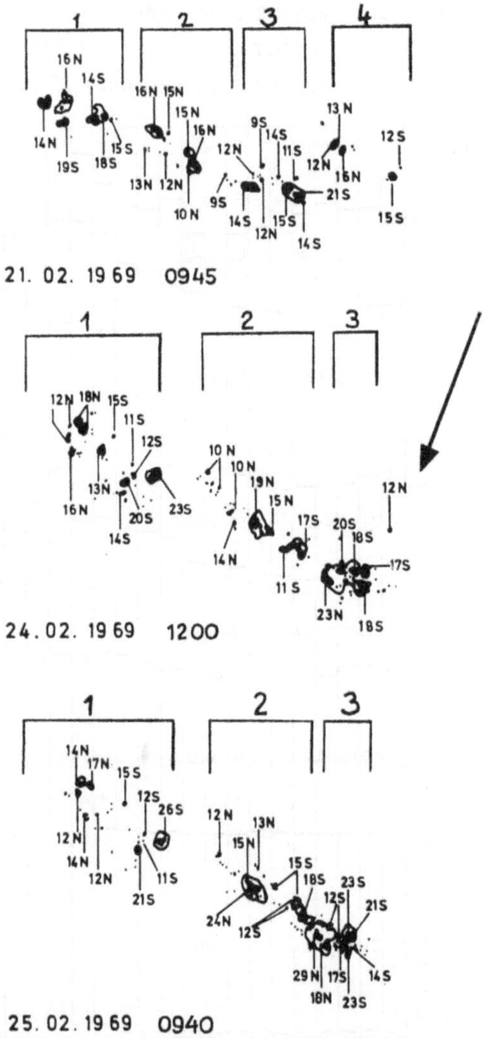

21. 02. 1969   0945

24.02.1969   1200

25.02.1969   0940

Fig. 5.   Bipolar spot groupings, Rome.

bipolar groups according to the normal arrangement expected for the northern hemisphere in the present solar cycle.

A clear tendency from a more complex field structure on February 24, 1969 (0 h UT) (three bipolar groups and one unipolar spot) to a simpler one on February 25, 1969 (only three bipolar groups) (0 h UT) can be seen. This configuration change is still better demonstrated in Figure 5, solar magnetic field recordings made in Rome on three consecutive days (February 23–25, 1969). Here the configuration starts on February 23 with four bipolar spot groupings, February 24 and 25 show then the increasingly simpler configuration similar to Figure 4.

4.2. SOLAR RADIO EMISSION

Figure 6 shows the solar radio fluxes (daily averages in units of 10^{-22} Wm^{-2} Hz^{-1}) for seven different frequencies ranging from 100 MHz to 9100 MHz, during the period before the event and shortly after it (February 10–28).

The buildup of the flux level starting with February 15 is particularly impressive

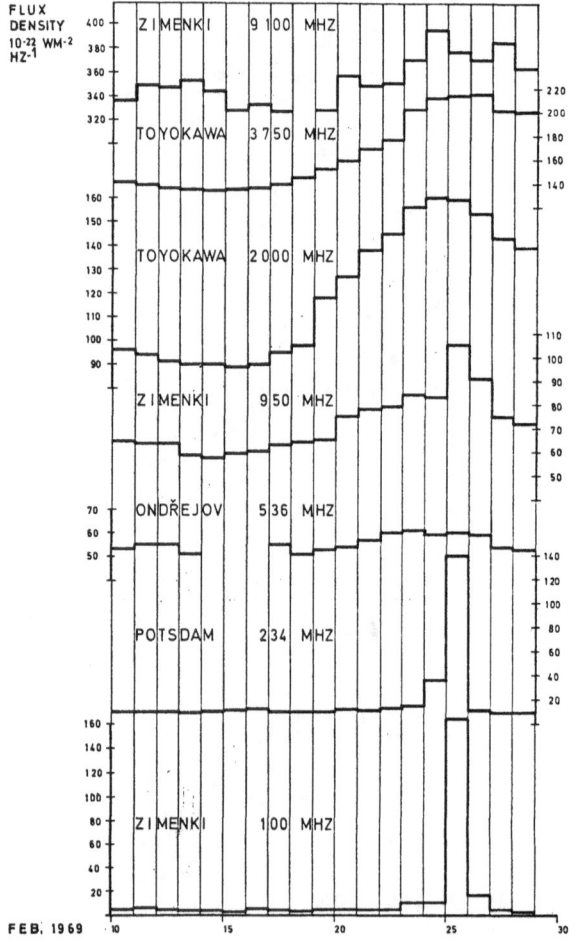

Fig. 6.   Solar radio emission. Daily data of single frequency observations.

for the higher frequencies (950–9100 MHz), whereas the lower frequencies (234 and 100 MHz) have a very sharp peak on February 25 itself.

An important type of information for a solar event consists in the radio spectrum emitted. In order to give an impression of the complexity, 9 representative frequencies varying from 100–71000 MHz have been plotted against time (Figure 7). The mean and peak flux density values are given (data are taken from *Solar Geophysical Data*). The crosses mark the time when the maximum occurred. The emission times were mainly of the order of 1 h, in some lower frequencies lasting up to 215 min.

Most of the typical criteria for a proton flare were met for this event, e.g. complex structure and high intensity and strong gradients of the magnetic fields, maturity (it was the second rotation of the region), location and solar radio emission.

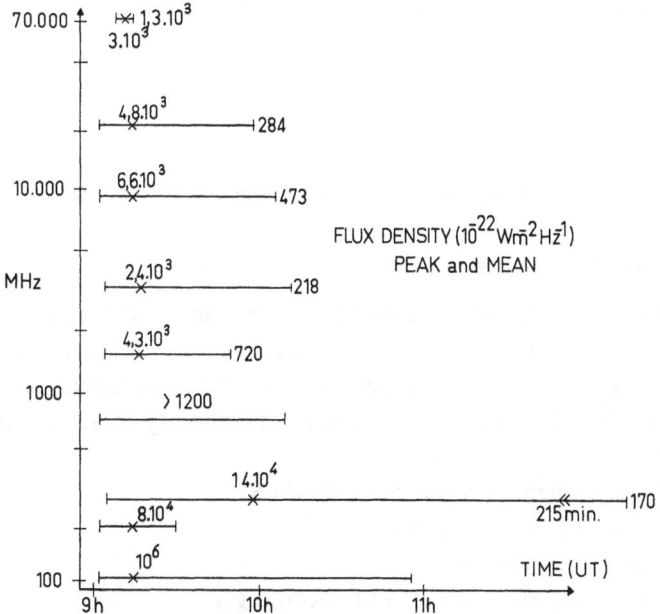

Fig. 7.   Solar radio emission. Outstanding occurrences, February 25, 1969.

5. Ground Effects

Figure 8 summarises the ground effects caused by the solar flare on February 25. The characteristic measurements, examples of which are given later, are drawn schematically in the time sequence as they were actually recorded. A distribution was made into three groups: simultaneous effects caused by the electromagnetic radiation (see Figures 9, 10, 11, 12 and 20), the effects caused by the high energetic particles i.e. neutron monitor count increase (see Figures 13, 14, 15) and ionospheric absorption near to the poles (polar cap absorption: PCA) (see Figures 16, 17, 18) and the effects caused by the arrival of the ejected magnetic storm particles i.e. storm sudden commencements (SSC) (see Figures 10, 19, 20) and Forbush decrease (Figure 15).

The following part of the paper gives typical recordings of these ground effects.

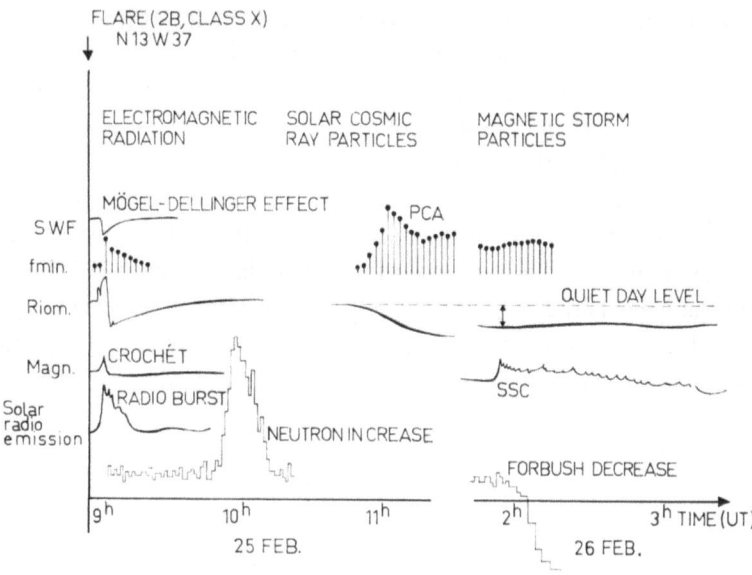

Fig. 8.   Ground effects of solar flare, February 25, 1969.

5.1. SIMULTANEOUS EFFECTS

Solar proton flares are usually accompanied by increased radiation in the 0–10 Å region. This radiation causes on the sunlight side of the earth, especially in the D-region of the ionosphere, the sudden ionospheric disturbances also called Mögel-Dellinger effects (SID). Following the flare of February 25 several of these effects were observed:

Sudden Enhancement of Signal Strength (SES)
(observed on VLF transmission)
Sudden Phase Anomaly (SPA)
(phase shifts of the sky wave on VLF recordings)
Sudden Enhancement of Atmospherics (SEA)
(of low frequency, about 27 kHz)
Sudden Cosmic Noise Absorption (SCNA)
(at about 18 MHz)
Short Wave Fadeout (SWF)
(field strength recordings of distant high frequency radio transmissions).

For the last type five examples of the recordings are given in Figure 9. The measurements were made by Fernmeldetechnisches Zentralamt Darmstadt, Germany. The absorption was reported to have been total. This is particularly clear from Figure 9 for the 9.177 MHz field strength recording.

The typical onset times for the ionospheric disturbances on February 25, 1969 (data from *Solar Geophysical Data*) were at about 9 h 02 min to 9 h 12 min with a maximum at 9 h 12 min to 9 h 18 min and recovering at 9 h 52 min to 10 h 55 min (UT).

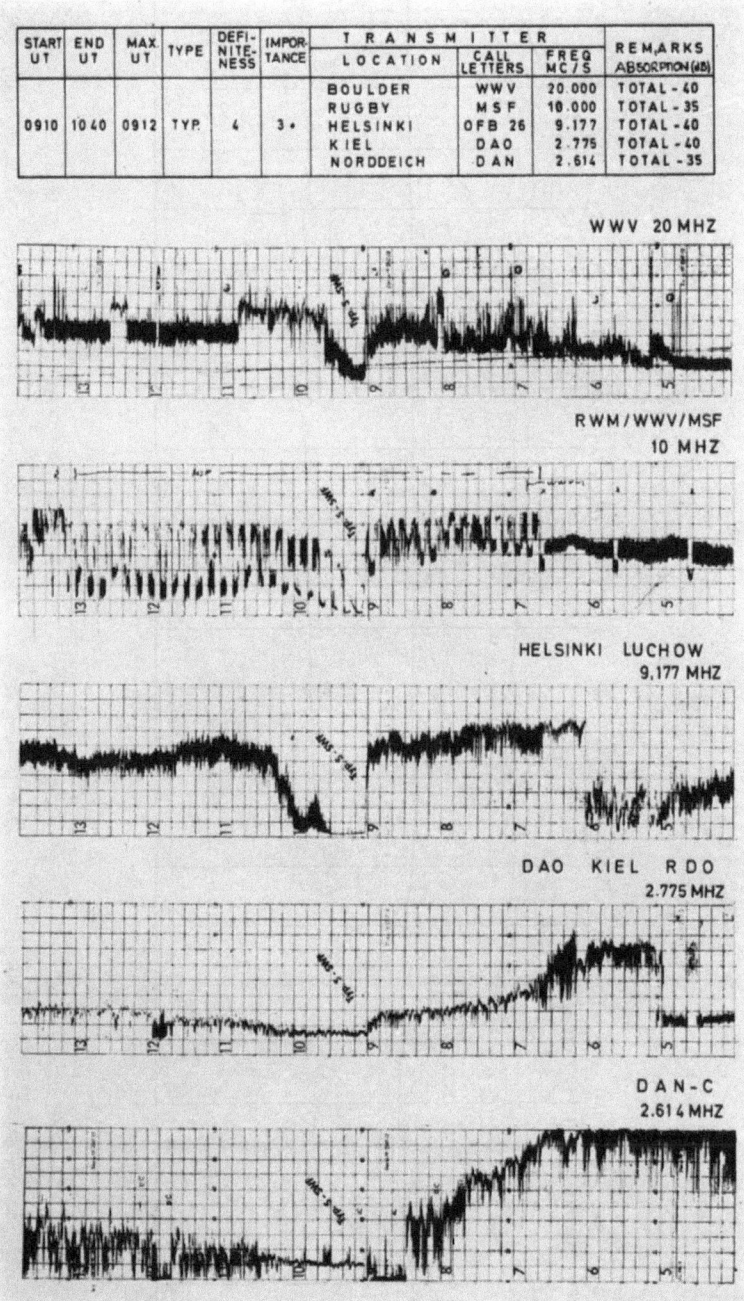

START UT	END UT	MAX. UT	TYPE	DEFI-NITE-NESS	IMPOR-TANCE	LOCATION	CALL LETTERS	FREQ MC/S	REMARKS ABSORPTION (dB)
0910	1040	0912	TYP.	4	3+	BOULDER	WWV	20.000	TOTAL - 40
						RUGBY	MSF	10.000	TOTAL - 35
						HELSINKI	OFB 26	9.177	TOTAL - 40
						KIEL	DAO	2.775	TOTAL - 40
						NORDDEICH	DAN	2.614	TOTAL - 35

WWV 20 MHZ

RWM/WWV/MSF
10 MHZ

HELSINKI LUCHOW
9,177 MHZ

DAO   KIEL   RDO
2.775 MHZ

DAN-C
2.614 MHZ

Fig. 9.   Short-wave fadeouts at February 25, 1969 (UT). Measurements made by
'Fernmeldetechnisches Zentralamt Darmstadt'.

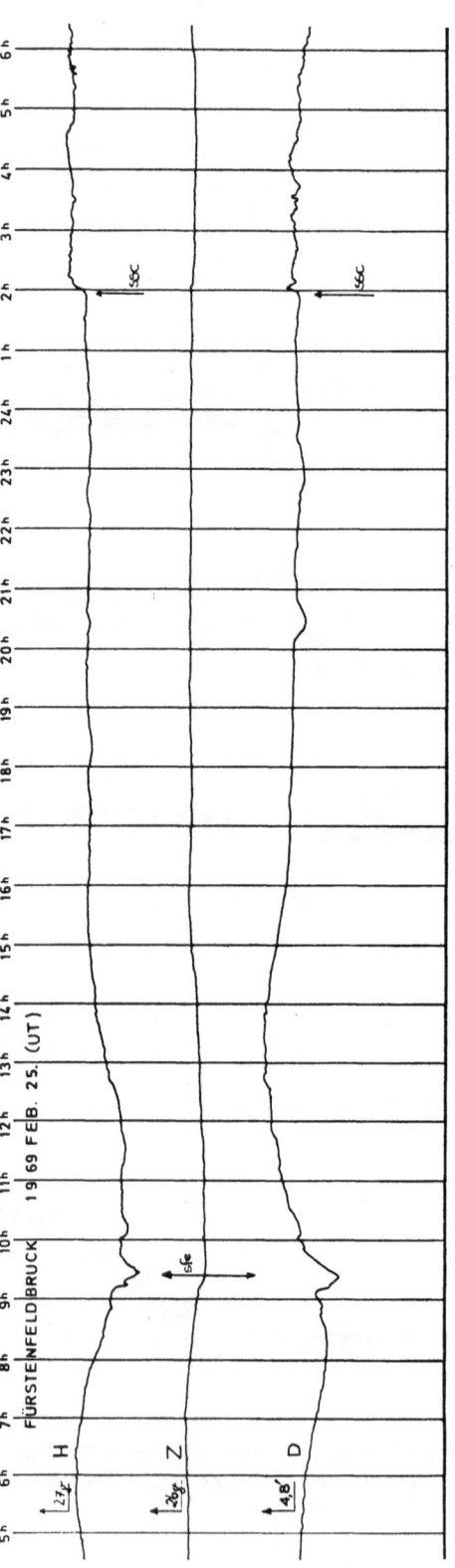

Fig. 10.

Also associated with the increase in ionisation caused by enhanced X-ray radiation is an effect recorded on geomagnetic variometers mainly also on the daylight side, the magnetic crochet, (see Figures 10, 12 and 20) often referred to as solar flare effect (SFE). A marked deviation from the usual daily variation is to detect, if at the time of the SID the geomagnetic field is relatively quiet. The three examples from Fürstenfeldbruck, Kiruna (ESRANGE-ESRO rocket launch range) and Göttingen show variations in the order of 20 γ to 40 γ in horizontal components.

Another method which is very often used to measure ionospheric disturbances is the recording of fmin, the minimum frequency reflected by the D-layer region of the ionosphere (see Figures 11, 16 and 17). On Figure 11 – fmin PLOT 1 – the fmin values (measured in MHz) are plotted against the universal and local time for 6 stations (the geographical latitude is printed after the station name), which were at 9 h UT on the daylight side of the earth. The increase in fmin towards the southern hemisphere and thus higher elevation of the sun is very pronounced. (The station Lindau is an exception since more frequent measurements were made at the time.) The other values are hourly measurements. The highest fmin values were observed in Tsumeb, Southwest Africa (for the event half hourly values were plotted). Increased absorption was recorded in Tsumeb for several hours. For the higher latitude station Nurmijärvi some indication of the PCA event is visible, demonstrated by increased absorption on February 25, during daylight time.

Figure 12 shows magnetometer and riometer recordings performed at ESRANGE, (Kiruna, Sweden). Three different frequencies are given: 20, 27.6, and 35 MHz. The abscissa is the time (UT), the ordinate the current, recorded in milliampere. The usual increase in absorption with decreasing frequencies is evident. At the 35 MHz riometer solar radio emission was recorded at about 9 h 10 min UT, characterised by the strong current increase followed by the sudden decrease due to the D-layer absorption. The X-magnetometer shows for the same time a pronounced crochet of ca. 40 γ maximum deviation. A slight absorption, indicating the beginning of a weak PCA event, s visible in the 27.6 MHz and 35 MHz riometer records starting at about 11 h UT.

5.2. Effects of Solar Cosmic Ray Particles

Figure 13 gives measurements of neutron monitors in higher latitudes for February 25 (the histograms are taken from *Solar Geophysical Data*). For the stations Deep River and Goose Bay the neutron counts (5 min interval) increase sharply reaching a maximum of ca. 16% increase at about 9 h 45 min, whereas the stations Inuvik and Alert have a much smaller maximum of ca. 4.5% increase at 10 h and about 10 h 30 min. For 3 other stations Churchill, Dallas and Victoria the neutron counts are plotted in Figure 14. Dallas has a maximum of ca. 4% increase at 9 h 30 min probably due to the most energetic particles, whereas Churchill and Victoria show a maximum at about 10 h UT of more than 10% increase.

The seven stations have the rigidity cutoff values (in BV) and geomagnetic latitudes as given in Table II. The fact that this ground level effect (GLE) was a pronounced one is demonstrated by Figure 15 (from *Solar Geophysical Data*), giving hourly

396 K. G. LENHART

f - MIN PLOT 1

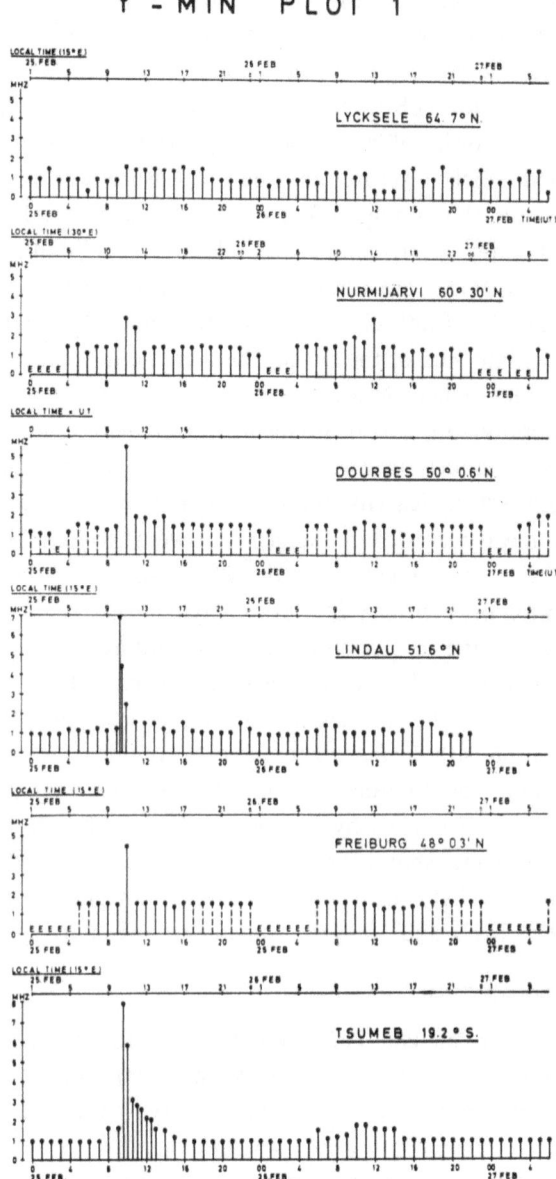

Fig. 11. —— unqualified value. ---- qualified 'ES'. -·-·- qualified 'S' only. 'B' numerical value not determined due to non-deviative absorption. 'C' measurement influenced by or impossible because of any non-ionospheric reason. 'E' measurement influenced by or impossible because of lower limit of frequency range. 'S' measurement influenced by or impossible because of interference or atmospherics.

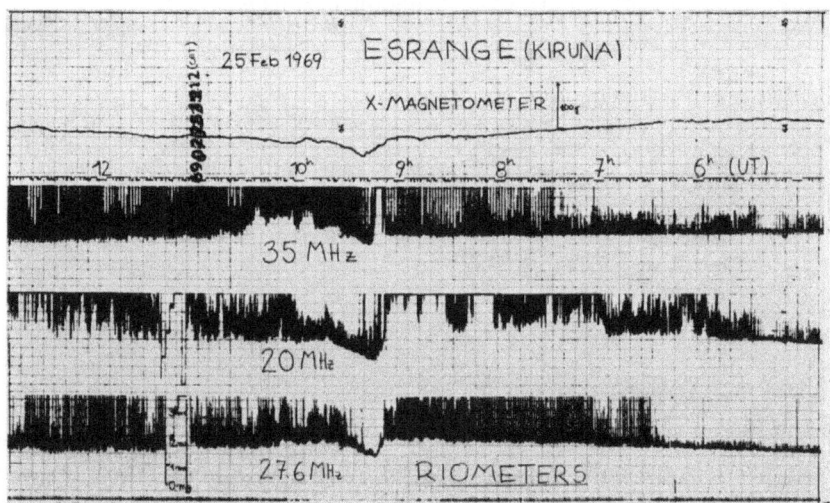

Fig. 12.

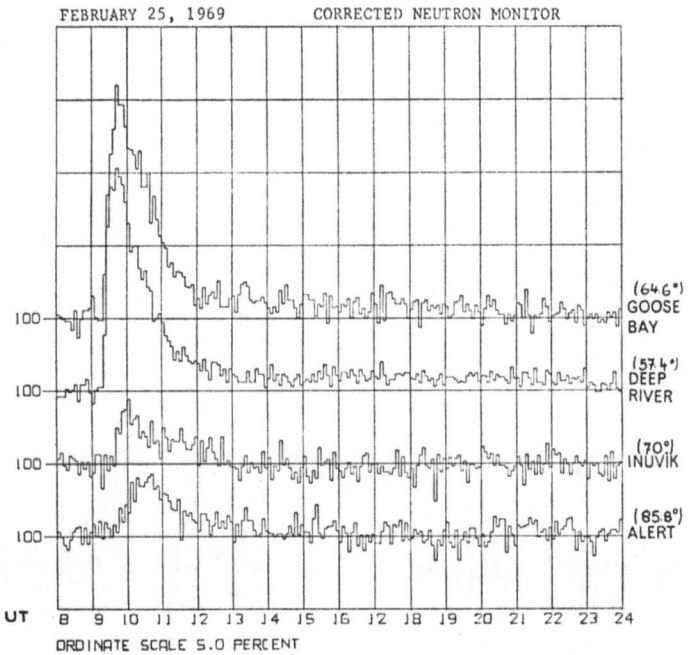

Fig. 13.   5-minute intervals of cosmic rays.

neutron monitor count data for Alert and Deep River for the months February and March, 1969. The event of February 25 is outstanding with its sharp increase and decay, followed by Forbush decrease during the next days.

TABLE II

	Cutoff	Geom. lat.
Alert	0.	86.8°
Inuvik	0.18	70.4°
Churchill	0.21	68.7°
Goose Bay	0.52	64.6°
Deep River	1.02	57.4°
Victoria	1.86	54.2°
Dallas	4.35	42.8°

5.3. Effects of Lower Energy Flare Particles

A proton flare is usually accompanied by a PCA event, where strong absorption of the ionosphere in the polar caps is recorded, normally within a few hours after the maximum phase of the flare. For the detection of the PCA ionospheric measurements are often used. Figure 16 – fmin PLOT 2 – shows for six stations in higher lati-

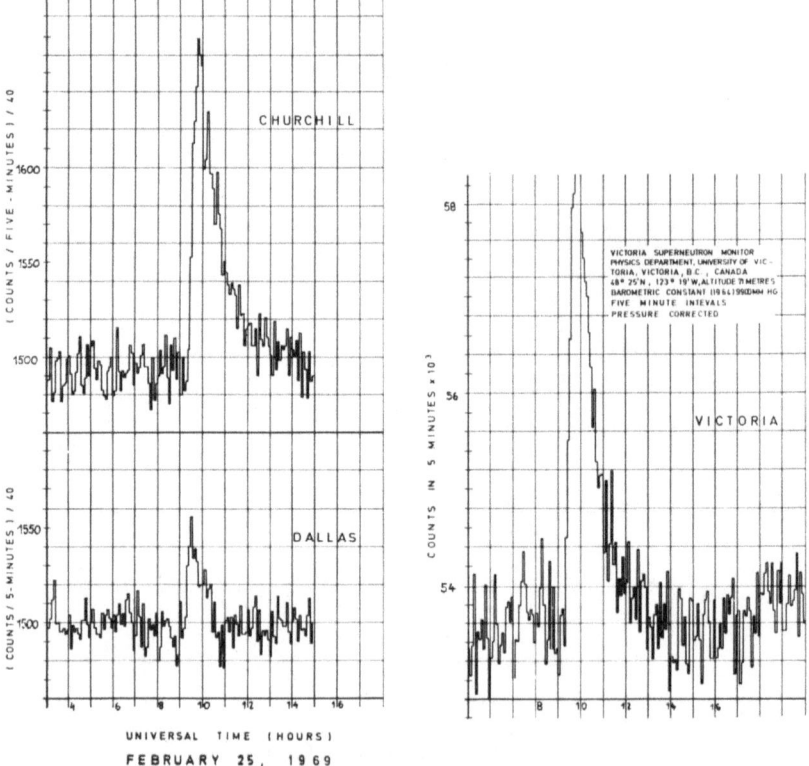

Fig. 14.

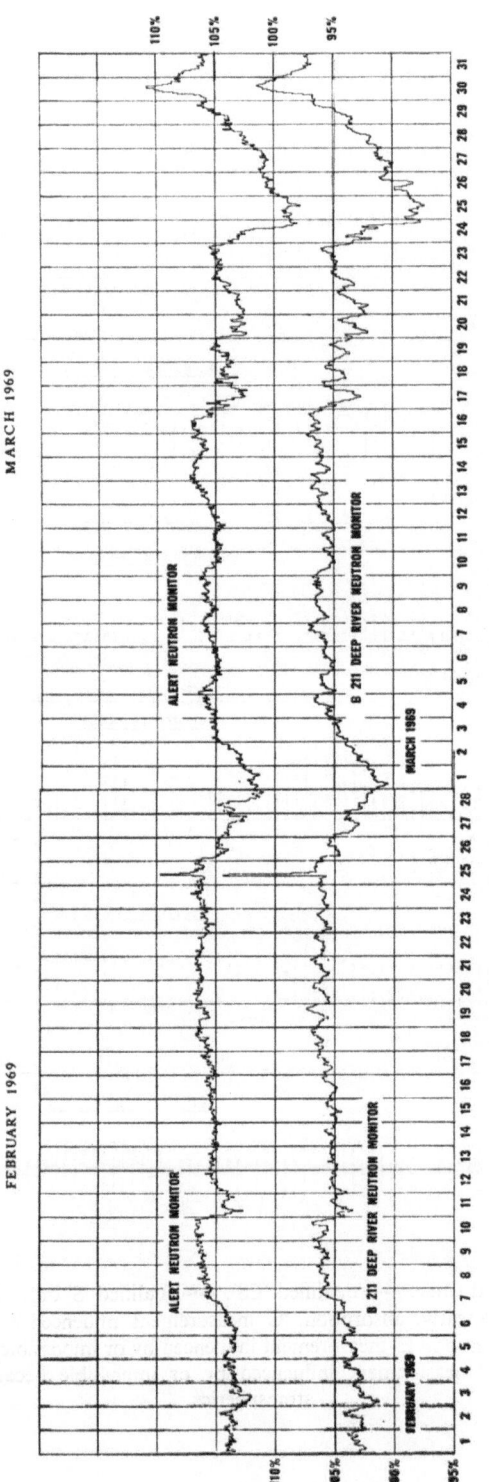

Fig. 15.   Cosmic ray indices. Pressure corrected hourly totals.

f - MIN   PLOT   2

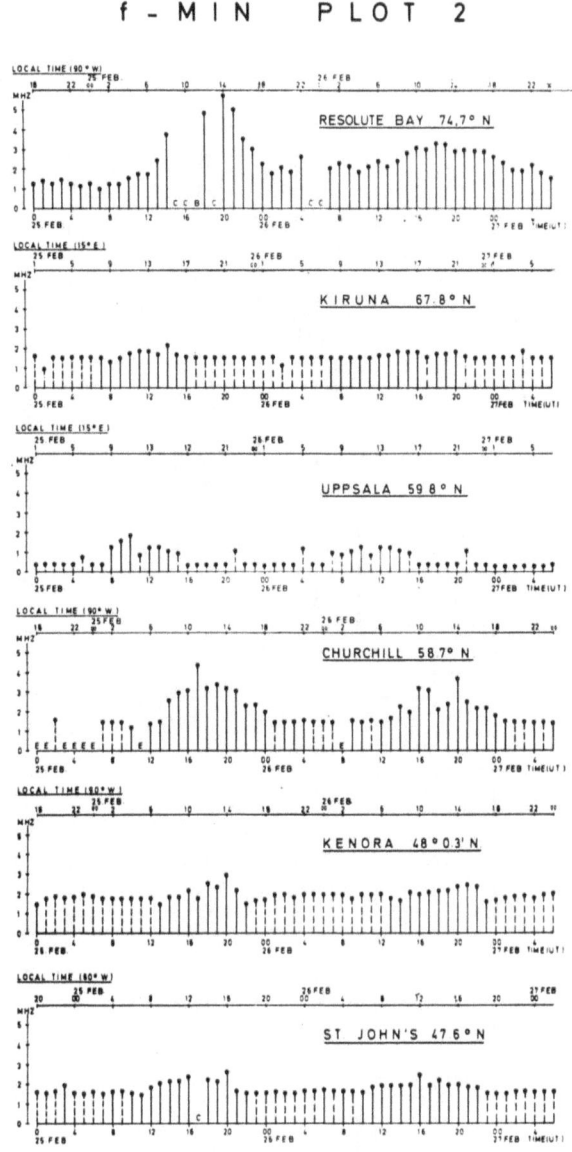

Fig. 16. —— unqualified value. ---- qualified 'ES'. -·--· qualified 'S' only. 'B' numerical value not determined due to non-deviative absorption. 'C' measurement influenced by or impossible because of any non-ionospheric reason. 'E' measurement influenced by or impossible because of lower limit of frequency range. 'S' measurement influenced by or impossible because of interference or atmospherics.

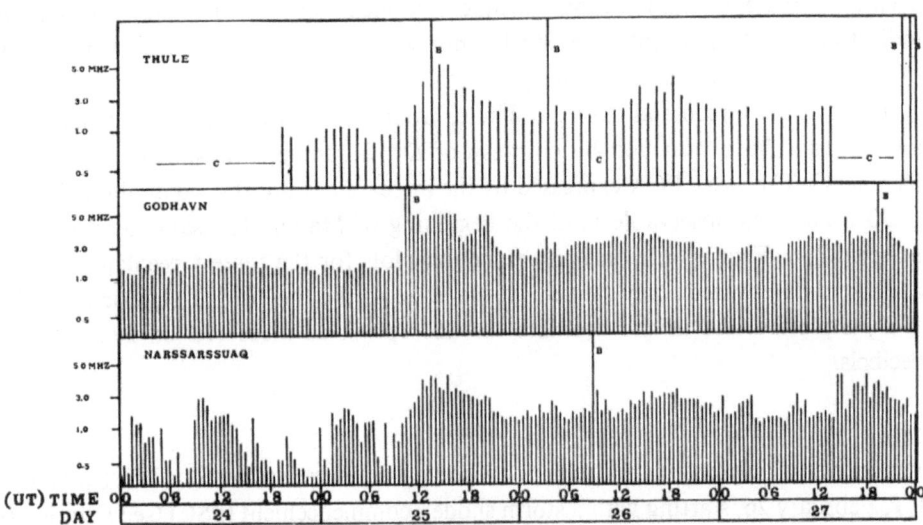

Fig. 17.   Recordings of f-min., February 24–27, 1969.

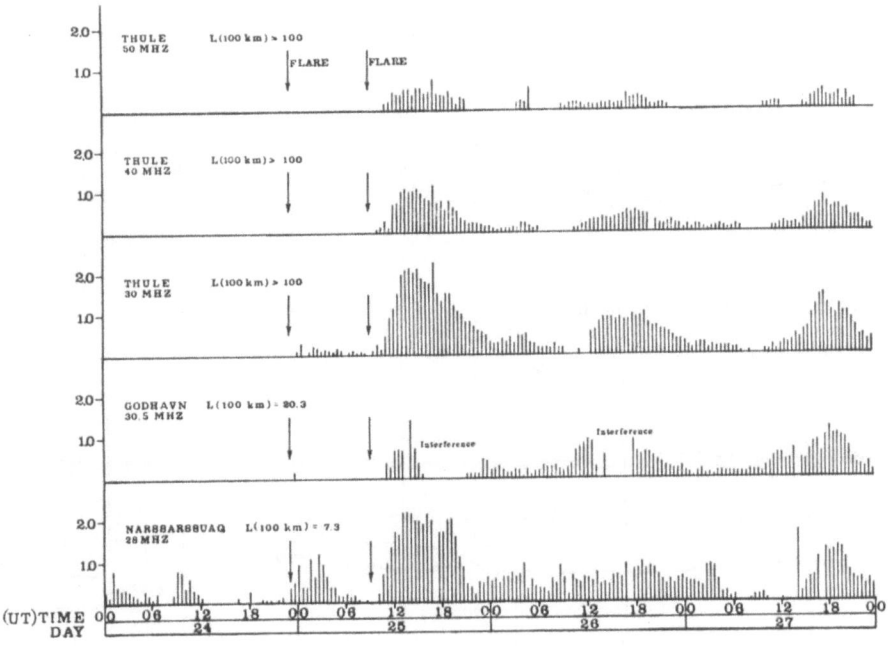

Fig. 18.   Cosmic noise absorption (dB), February 24–27, 1969.

tudes (the geographic latitude is printed after the station name) the hourly values
of fmin in MHz for February 25–27, 1969. At Resolute Bay, Churchill and Uppsala
a clear increase in fmin during daytime (see local time co-ordinate) hours is an
indication of a PCA event. The effect is not so pronounced in the other three plots
of Kiruna, Kenora and St. John's, but still discernible. Figure 17 shows fmin recordings at the Greenland stations: Thule, Godhaven and Narssarssuaq (Figures 17
and 18 have been provided by courtesy of Dr. Stauning, Ionosphere Laboratory,
Denmark). The weak PCA expressed by the increase in fmin started at about 10 h to
10.30 h and was observable until the beginning of March. The same tendency can
be seen from Figure 18, riometer absorption plots for the three Greenland stations
and five different frequency ranges. At the abscissa the time in UT (February 24–27,
1969) is plotted and at the ordinate the absorption as measured on riometers, in
decibels.

5.4. GEOMAGNETIC STORM EFFECTS

The arrival of low energy particles produced by the flare was recorded at 1 h 58 min
UT, February 26, starting with a storm sudden commencement (SSC) (see Figures 10,
19, 20). The effect was worldwide, but the strength of the storm very low. This
indicates that a small number of the low energy particles hit the earth.

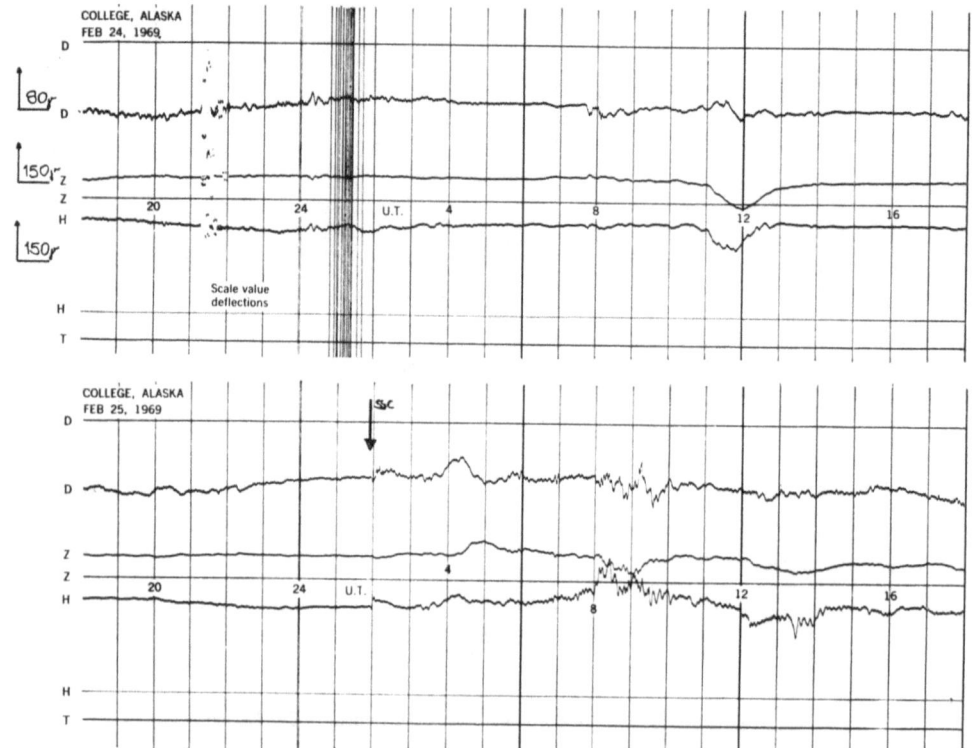

Fig. 19.

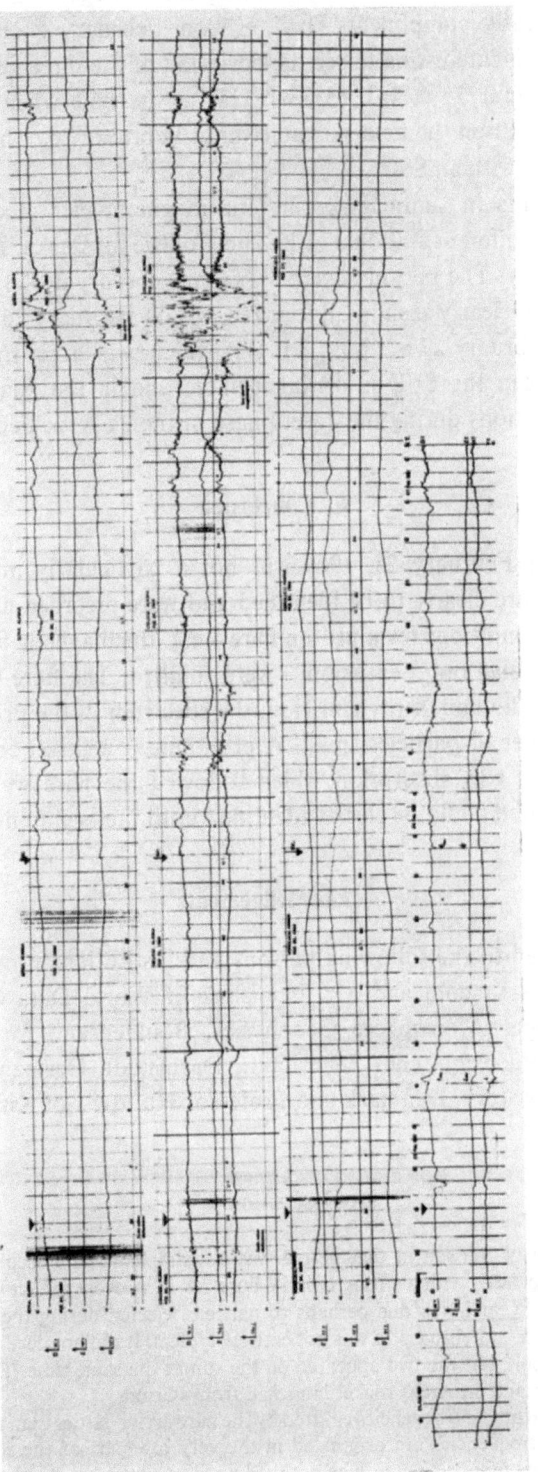

Fig. 20.

This is demonstrated in Figure 19, showing for College, Alaska, magnetometer recordings of the three components D, Z, H from February 24, 18 h UT to February 26, 18 h UT. It is interesting to see the deviation of the three field components on February 25 at about 11 h UT to 13 h UT, at the time when the PCA started. Apart from this deviation the field was quiet until February 26, 1 h 58 min, when the small storm started with an SSC.

Figure 20 presents in summary form four magnetometer recordings (D, Z, H components) from different stations: Sitka and College in Alaska, Honolulu, Hawaii, Göttingen, Germany. The period covered is from February 24–28, 1969. The simultaneous SSC can be clearly seen in all four records on February 26 and also the start of the storm on February 27 at about 3 h UT, which was larger than that of February 26. In Göttingen the SFE is distinguishable and in the Alaskan stations the magnetic field variations during the onset phase of the PCA on February 25, 1969.

6. Conclusion

The solar event of February 25, 1969 was not a particularly intense one but its geophysical effects are clearly to be identified and were recorded all over the world. Most of the usual conditions for a proton flare – e.g. the magnetic field configuration and structure, the solar radio emission – were fulfilled. The flare itself was accompanied by strong radio and X-ray emission, the spectrum of the protons ejected was hard, but the number of particles small. A pronounced neutron count increase was recorded. The polar cap absorption which followed the flare was not strong but lasted several days and could be clearly identified until the beginning of March.

Acknowledgements

I am deeply indebted to the following persons and institutions providing recordings for this event: Miss Lincoln and Mr. Paulishak of World Data Center A (of the subcenters for Upper Atmosphere Geophysics, Boulder and for Geomagnetism, Rockville); Fernmeldetechnisches Zentralamt, Darmstadt, Germany; Mr. Stauning of Ionosphere Laboratory, Denmark and Professor Hultqvist of Kiruna Geophysical Observatory.

Discussion

Švestka: The condition of the sun in that period was not simple. A flare occurred on February 24 at 23 00 h and was associated with a strong type IV burst. A PCA event in fact was recorded before the flare on February 25, and was due perhaps to particles ejected during the February 24 flare.

The flare in discussion on February 25 was not entirely typical. It did not develop into two ribbons from one common ribbon, but the two appeared on the sun at the same time. The proton flare was confirmed by a balloon and an Arcas rocket launched from Kiruna.

Question: Why do the higher frequencies emitted by the flare arrive earlier than the low frequencies?

De Jager: The high frequencies are originated in the very low part of the solar atmosphere, in the region where the flare first originates, whereas the low frequencies are mostly related to emission in the higher coronal part, during the latest expansion of the flare.

SUBSTORM INFLUENCES ON
VHF CONTINUOUS WAVE AURORAL BACKSCATTER

P. CZECHOWSKY and G. LANGE-HESSE

Max-Planck-Institut für Aeronomie, Lindau/Harz, Germany

1. Introduction

The Max-Planck-Institute for Aeronomy at Lindau/Harz, West-Germany operates since autumn 1967 a network of VHF beacon transmitters and receiving stations in Scandinavia and Northern Germany to study VHF continuous wave bistatic auroral backscatter communications in cooperation with the Royal Swedish Telecommuni-

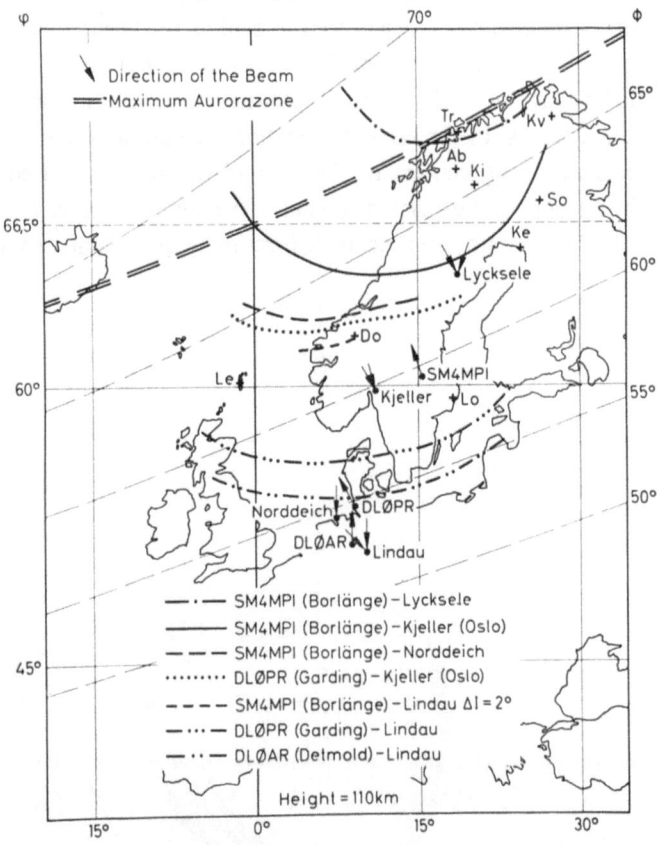

Fig. 1.   Position of the VHF-beacon transmitters (SM4MPI near Borlänge, Sweden, frequency: 145.960 MHz; DLØPR, Garding, Schleswig-Holstein, frequency: 145.971 MHz; DLØAR, Bielstein, Teutoburger Wald, frequency: 29.0 MHz) and of the receiving stations (Lycksele, Sweden; Kjeller, Norway; Norddeich and Lindau, West-Germany) with the appertaining backscatter curves for the pairs of stations mentioned in the figure. The arrow at the stations indicates the direction of the antenna beam (half-power width ± 30°).

V. Manno and D. E. Page (eds.), Intercorrelated Satellite Observations Related to Solar Events, 405–412. *All Rights Reserved*
Copyright © 1970 by D. Reidel Publishing Company, Dordrecht-Holland

cation Board, the Norwegian Defence Research Establishment and the German Federal Post. The network is shown in Figure 1.

According to the theory the backscattering of VHF radio waves by aurora not only is controlled by the visual aurora but obviously on a larger scale by the polar electrojet (PEJ). Some years ago Buneman (1963) and Farley (1963) pointed out independently that in an ionospheric current system like e.g. the equatorial electrojet (EEJ) plasma instabilities of a 'two-stream' type can occur so that acoustic plasma waves are generated. The plasma instability appears when the relative drift velocity between the ions and electrons in the electrojet exceds a certain critical speed which is close to the thermal velocity of the ions. It is said that the electrojet at this phase has exceeded the threshold. Acoustic plasma waves in shape of longitudinal density waves then are generated, which propagate along the electrojet transverse to the geomagnetic lines of force. The periodical density oscillations caused by the plasma waves (which represent a special fine structure of the electrojet) are the field aligned centers which give rise to the backscattering of VHF radio waves. Bowles *et al.* (1963) and Cohen and Bowles (1963) have shown that the VHF waves backscattering centers in the EEJ are caused by acoustic plasma waves. In the same publications the authors pointed out that the VHF waves backscattering centers in the PEJ obviously are caused by the same mechanism. The purpose of the observation network shown in Figure 1 is among others to check whether this assumption is correct. First reduction results from observations

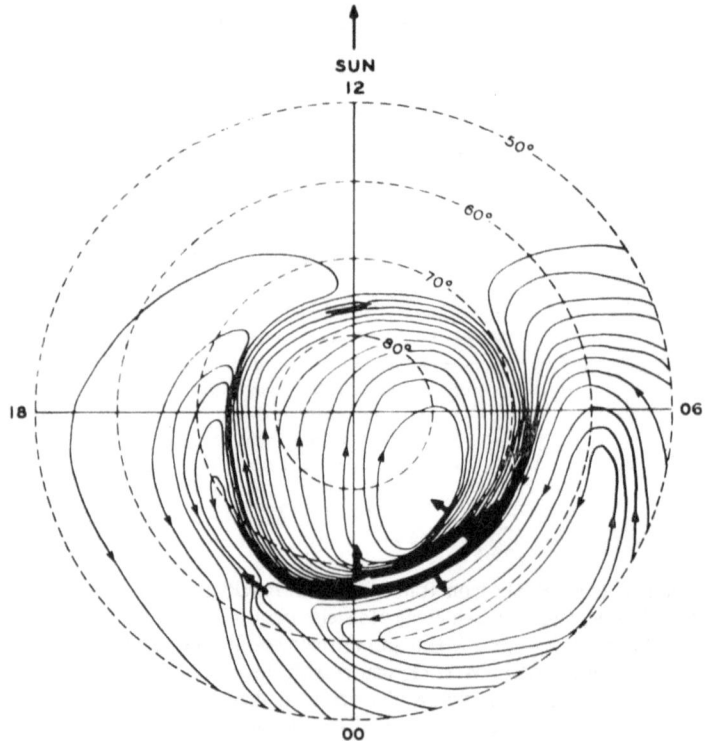

Fig. 2.   The polar electrojet. After Akasofu *et al.* (1965).

carried out with this network has been published by Czechowsky (1969). The observations shown in this paper are additional results in this direction.

The maximum of the current density of the PEJ fits together in a first approximation with the visual aurora. In former years therefore it was assumed that the normal ionization in the region of the visual aurora directly causes the backscattering of VHF radio waves. On the assumption that the theory of Buneman and Farley is also valid for auroral regions the fine structure of the PEJ, however, would be the controlling element of the VHF auroral backscatter phenomenon, and if therefore the PEJ in the phase beyond the threshold intersects a backscatter curve in Figure 1 then VHF auroral backscatter communication ought to occur between the pairs of stations belonging to the backscatter curve. For a pair of stations the appertaining backscatter curve represents the center of a region where the most favorable geometrical conditions (also called 'ideal backscatter conditions') are fulfilled for VHF auroral backscatter communications. These curves are calculated for a height of 110 km. Introductions and more details about VHF auroral backscatter see e.g. Czechowsky (1966, 1969), Lange-Hesse and Czechowsky (1966), Lange-Hesse (1967, 1968, 1969).

2. The Event from Feb. 27, 1969

On Feb. 26, 1969 at 0258 MEZ* (=0158 UT) a SSC was observed as a consequence of the events recorded on Feb. 25, 1969 (e.g. PCA, electromagnetic radiation, solar cosmic rays). The subsequent geomagnetic storm reached its peak disturbance

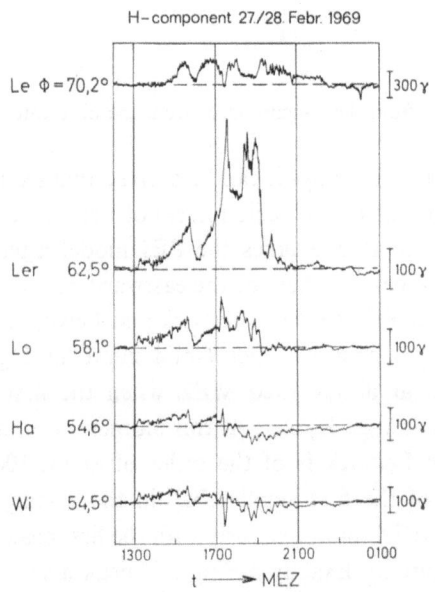

Fig. 3. Magnetograms from the *H*-component from the geomagnetic observatories Leirvogur (near Reykjavik/Iceland), Lerwick and Lovö (see Figure 1), Hartland/England and Witteveen/The Netherlands. MEZ = time 15°E.

* MEZ = Central European Time.

degrees up to $K_p = 6+$ from about 1300 to 2200 MEZ on Feb. 27. During this time VHF auroral backscatter echoes were recorded with the network in Northern Germany and Scandinavia shown in Figure 1. These observations are compared in this contribution with simultaneous geomagnetic records that means with the location, direction and intensity variation of the PEJ. The results of this comparison are discussed on the basis of the ion acoustic wave theory according to Farley (1963). In the magnetograms (Figures 3 and 4) bays occurred in the afternoon and early evening of

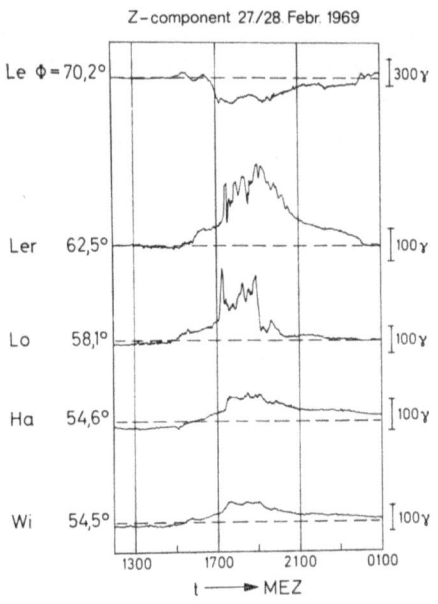

Fig. 4.   Magnetograms from the Z-component from the observatories mentioned in Figure 3.

Feb. 27. From the sign of the bays it can be derived that from about 1500–2000 MEZ (MEZ = time 15°E) the current is located north of Lerwick and south of Leivogur and flows from west to east. If one takes the PEJ model after Akasofu *et al.* (1965) (Figure 2) the observed current must be the eastward flowing backcurrent of the PEJ-system. This part of the PEJ between Lerwick and Leivogur intersects the region of the backscatter curve Borlänge-Kjeller (solid curve in Figure 1). This current is beyond the threshold at about 1520 MEZ when the first backscatter signals are recorded on the line Borlänge-Kjeller. At this moment the amplitude of the bay in the *H*-magnetogram from Lerwick is of the order of about 100 γ, which indicates the threshold. The peak of the first smaller bay in the *H*-magnetogram from Lerwick occurs at about 1545 MEZ simultaneously with the first peak in the backscatter signal Borlänge-Kjeller (Figure 5). Exactly during this peak a short backscatter signal with small amplitude appears on the more southern line Borlänge-Norddeich indicating that the PEJ extends to the south so that it could just intersect the region of the backscatter curve Borlänge-Norddeich (Figure 1). No backscatter signal, however, occurs during this time on the most southern line in Figure 5 (Borlänge-Lindau).

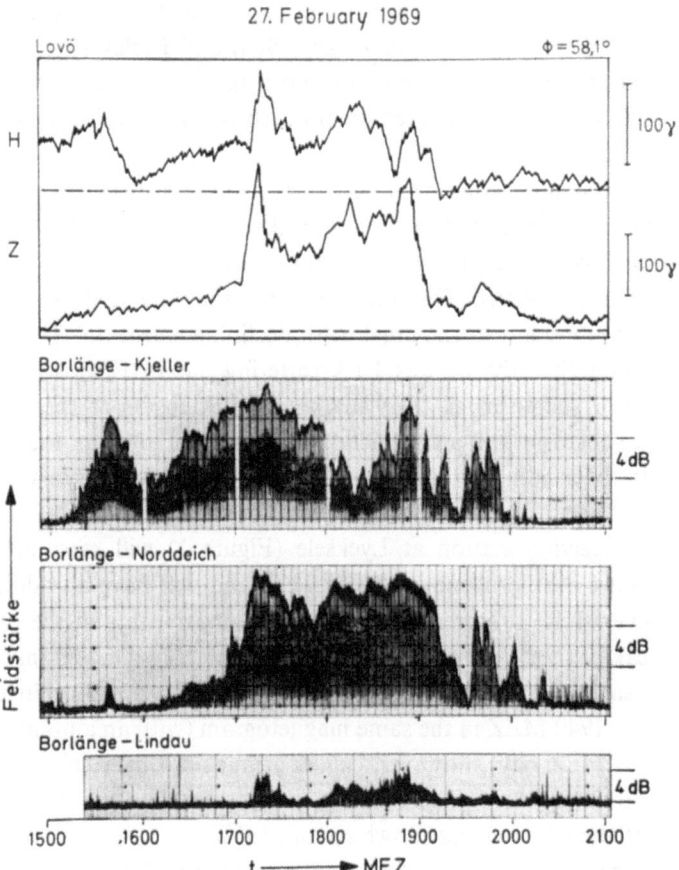

Fig. 5.   VHF auroral backscatter recordings from the beacon station Borlänge, SM4MPI, 145.960 MHz at three different stations and the magnetograms from the *H*- and *Z*-component from the Lovö observatory. See Figure 1. MEZ = time 15°E. Feldstärke = field strength.

The strongest bay with more than 300 γ amplitude appears in the *H*-magnetogram from Lerwick at about 1730 MEZ. Simultaneously peaks in the backscatter signals now occur on all lines. The amplitude of the peaks on the two northern lines now are of the same order, however, the amplitude on the most southern line Borlänge-Lindau is much smaller. This indicates that compared to the time 1545 MEZ the current further has extended to the south being able to intersect now the region of all three appertaining backscatter curves (Figure 1).

A second stronger bay with an amplitude of about the same order as the first appears from about 1820 to 1915 MEZ. This bay is bifurcated and shows two peaks. Simultaneously with the magnetic peaks, peaks in the backscatter signal occur on the lines from Borlänge to Norddeich and Lindau. The magnetic bifurcation is best reproduced in the backscatter signal trace Borlänge-Lindau. Strong backscatter signals are also recorded during the bifurcated bay on the most northern line Borlänge-Kjeller. If the VHF waves backscattering centers are produced in the PEJ by acoustic plasma waves according to the theory of Buneman (1963) and Farley (1963) a broad eastward flowing current beyond the threshold must have intersected on Feb. 27, 1969, from about 1820 to 1915 MEZ the region of the three appertaining backscatter curves (Figure 1) simultaneously.

The last bay of small amplitude occurs in the Lerwick H-magnetogram at about 1940 MEZ. About simultaneously a peak in the backscatter signal appears on the two lines Borlänge-Kjeller and Norddeich but no signal on the most southern line in Figure 5 (Borlänge-Lindau). This indicates that the PEJ and the backscattering centers show a contraction to the north with declining substorm. The results shown in Figure 5 exhibit that the centers backscattering the VHF waves show a similar behavior during a substorm as the visual aurora: extension to the south just after the start of the substorm and a contraction to the north towards the end. Whether there is also an extension to the north of the backscattering centers like the visual aurora just after the start of a substorm the authors hope to check after August 1969 when the new receiving station at Lycksele (Figure 1) will start operation. This station is the most northern one in the network and will have an additional equipment for Doppler shift measurements for to study movements of the backscattering centers. The peak of the bay at about 1545 MEZ in the H-magnetogram from Lerwick causes a short backscatter signal of low amplitude on the line Borlänge-Norddeich. The bay at about 1940 MEZ in the same magnetogram (with an amplitude lower than that of the 1545 MEZ bay), however, causes a much stronger signal than the 1545 MEZ bay in the same record. This indicates a hysteresis-like behavior which already was observed at the EEJ (see e.g. Bowles *et al.*, 1963).

In the following section the closer relation between VHF auroral backscatter echoes and the geomagnetic disturbance are investigated in detail. The variation of the backscattered echo signals (in scale units) as a function of the horizontal perturbation vector H_d (in γ) is shown in Figure 6. The H_d vector is computed from the values ΔH, ΔD and the base line value H_0. It can be seen from the upper diagram of this figure (Borlänge-Kjeller), that a threshold value of the order of 100 γ must be exceeded before the VHF auroral backscatter process starts at about 1520 MEZ (see Figure 5, diagram Borlänge-Kjeller). In the two lower diagrams of Figure 6 no threshold value of this order can be recognized. This means that in these latitudes no new irregularities in form of plasma acoustic waves are generated. Waves of this kind excited in the region of the backscatter curve Borlänge-Kjeller, must therefore have extended to the south during the development of the geomagnetic storm. A hysteresis-like behavior can also be seen in Figure 6 that means that a stronger PEJ current (higher H_d value) is necessary for the generation of the waves than for the maintenance of the plasma waves which give rise to the backscattering process. Hysteresis-like behavior of the PEJ was observed e.g. by Czechowsky (1969) and Lange-Hesse (1968). Another detailed description of some more events is published by Czechowsky *et al.* (1970).

This short contribution about a single event clearly demonstrates a close correlation between the PEJ and the backscattering of VHF radio waves. The geomagnetic deviations coincide in nearly all dctails with the amplitude variations of the backscattered signals. In a first approximation an extension to the south of the backscattering centers can be observed. The observations are in accordance with the results of the theory of Farley.

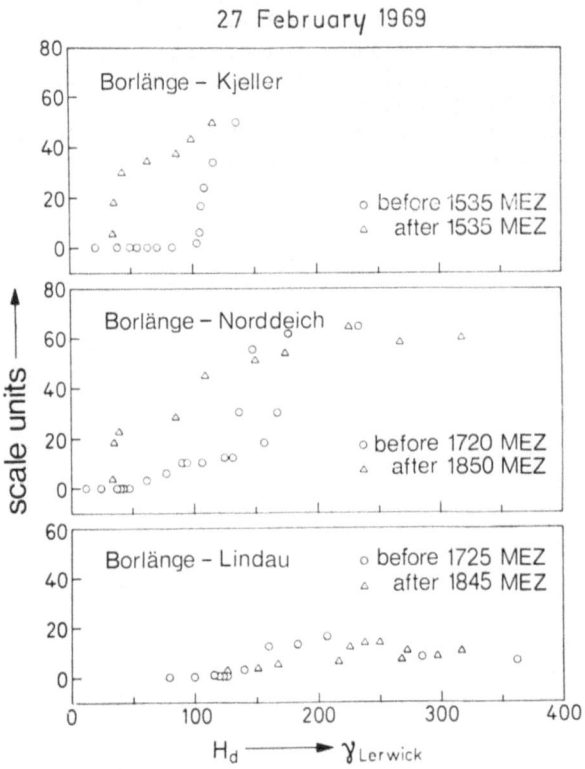

Fig. 6.   Relation between the amplitude of aurora-backscatter echoes (in scale units) and the variation of the horizontal disturbance vector H_d (in γ).

References

Akasofu, S.-I., Chapman, S., and Meng, C.-I.: 1965, 'The Polar Electrojet, *J. Atmos. Terr. Phys.* **27**, 1275–1303.

Bowles, K. L., Balsley, B. B., and Cohen, R.: 1963, 'Field-aligned E-Region Irregularities Identified with Acoustic Plasma Waves', *J. Geophys. Res.* **68**, 2485–2501.

Buneman, O.: 1963, 'Excitation of Field-Aligned Sound Waves by Electron Streams', *Phys. Rev. Letters* **10**, 285–287.

Cohen, R. and Bowles, K. L.: 1963, 'The Association of Plane-Wave Electron-Density Irregularities with the Equatorial Electrojet', *J. Geophys. Res.* **68**, 2503–2525.

Czechowsky, P., 1966, 'Analyse von Rückstreubeobachtungen ultrakurzer Wellen an Polarlichtern', Diplom-Arbeit (Master's Thesis) University of Göttingen, Germany.

Czechowsky, P.: 1969, 'Statistische Auswertung von Polarlicht-Rückstreubeobachtungen und Vergleich mit der Theorie der Plasma-Instabilität', *Kleinheubacher Berichte*, Band 13, published by Fernmeldetechnisches Zentralamt, Darmstadt, in press.

Czechowsky, P., Kochan, H., Lange-Hesse, G., Lauche, H., and Möller, H. G.: 1970, 'Simultane

Beobachtung verschiedener ionosphärischer Phänomene während des erdmagnetischen Sturmes vom 31. Okt. bis 2. Nov. 1968', *Z. f. Geophys.* **36** (in print).

Farley, D. T., Jr.: 1963, 'A Plasma Instability Resulting in Field-Aligned Irregularities in the Ionosphere', *J. Geophys. Res.* **68**, 6083–6097.

Lange-Hesse, G.: 1967, 'Radio Aurora, I. Observations, II. Comparison of the Observations with a Theoretical Model', in *Aurora and Airglow* (ed. by B. M. McCormac), Reinhold Publishing Corporation, New York-Amsterdam-London, pp. 519–562.

Lange-Hesse, G.: 1968, 'VHF Bistatic Auroral Backscatter Communication', in *Ionospheric Radio Communication* (ed. by K. Folkestad), Plenum Press, New York, pp. 174–205.

Lange-Hesse, G.: 1969, 'Radio Observations of the Aurora by Continuous Wave Transmissions', in *Atmospheric Emissions* (ed. by B. M. McCormac and A. Omholt), Van Nostrand Reinhold Company, New York-London, pp. 201–212.

Lange-Hesse, G. and Czechowsky, P.: 1966, 'VHF Bistatic Auroral Backscatter Communications. Comparison of the Observations with the Theory', *Archiv der elektr. Übertragung (AEÜ)* **20**, 365–375.

X-RAY OBSERVATIONS BETWEEN 1 AND 55 Å
OF THE FEBRUARY 25TH FLARE

A. C. BRINKMAN and W. DE GRAAFF

Space Research Laboratory, Utrecht, The Netherlands

and

M. L. SHAW

Space Science Department, (ESLAB), Noordwijk, The Netherlands

Abstract. The data from satellite observations of the February 25th solar flare at soft X-ray frequencies are presented. Temporal variations of the intensity at 4 different wavelength bands are shown and compared. The variation of the spectrum during the flare is considered.

1. Introduction

Ground-based solar observations indicate that the most significant solar flare of the period under consideration by the Symposium was the class X flare of importance 2B which occurred at 0902 UT on February 25th, 1969. Real-time data were acquired from the ESRO II satellite between 0907 and 0917 UT and the spacecraft was in sunlight for most of this time. The two X-ray experiments on board the spacecraft, which between them observe between 1 and 18 Å and 44 to 55 Å, were able to acquire data showing the increase phase of the flare.

2. The Instruments

The instruments essentially consist of seven proportional counters which have different transmission characteristics determined by the window materials and filling gases. Three detectors which nominally measure in the 1–3, 3–9, and 6–18 Å bands were in use at the time of this event in the S 36 experiment (Mullard Space Science Laboratory, University College London). The S 37 experiment (University of Utrecht) has two nearly identical detectors which measure in the 44–55 Å region. The sampling rate for S 36 during this flare was one 1–18 Å spectrum every 11 or 12 sec. S 37 takes measurements every 16 sec. The S 36 experiment incorporates an 8-channel pulse height analyser which is used to provide more detailed spectral information within each of the three wavelength bands observed. With continued use of the S 36 experiment since its launch in May 1968 some deterioration of the high voltage power supply has occurred and this implies that small gain changes may have occurred to the detectors. We are therefore loath to specify the exact limits of the wavebands under observation by this experiment but prefer to talk still of the nominal wavelength bands. The two S 37 detectors each incorporate a reference proportional counter which enables one to calculate small gain changes. One of the detectors has

V. Manno and D. E. Page (eds.), Intercorrelated Satellite Observations Related to Solar Events, 413–418. All Rights Reserved
Copyright © 1970 by D. Reidel Publishing Company, Dordrecht-Holland

deteriorated to such an extent that the data are not used here; the other detector shows a small increase in gain.

We are confident that none of these slight gain changes misrepresent the situation in any significant fashion.

3. Observational Results

The ESRO II data for late February, and also that from SOLRAD 9 (reported in *Solar Geophysical Data* – April 1969) indicates that the level of solar X-ray activity was quite low for several days preceding February 22nd. From that time onwards the level of activity increased until by late on the 24th considerable variability was evident. This high level of activity then continued for some days, its peak being on the 25th.

Figure 1 shows the variation of count rate with respect to time in the 1–3, 3–9, 6–18 and 44–55 Å bands and the 9500 MHz microwave flux for the period of the flare.

At the start of the period of data acquisition, the satellite had just emerged from the earth's shadow but was still behind a considerable absorbing path length of the atmosphere. It is necessary to assess the effect that this additional absorption has on the count rates before one is able to comment on the changes in count rates due to the flare.

This problem has been tackled in two ways; firstly, we have looked at the effect on count rates whilst the spacecraft is emerging from eclipse at a time of fairly constant

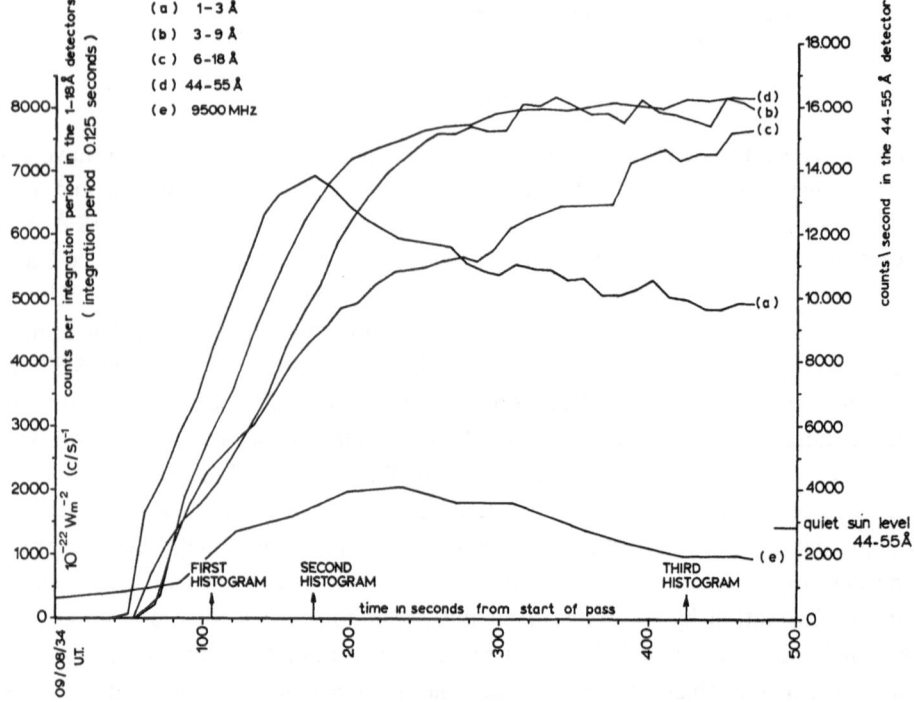

Fig. 1.   Variation of X-ray flux with time during the February 25 flare. Radio data by courtesy of the PTT Radio Laboratory and the observatory 'Sonnenborgh', Utrecht.

solar activity, and secondly, we have calculated the minimum height of the satellite-sun line above the earth in order to be able to assess when it is sufficiently high to be clear of atmospheric absorption.

Figure 2 shows the variation of the count rates during the period of constant activity. This is actually data taken from the spacecraft by the ESRO Fairbanks tracking station (which also took the data containing the flare data) exactly one orbit later, from 1046 UT onwards. This means that the height, latitude and local solar

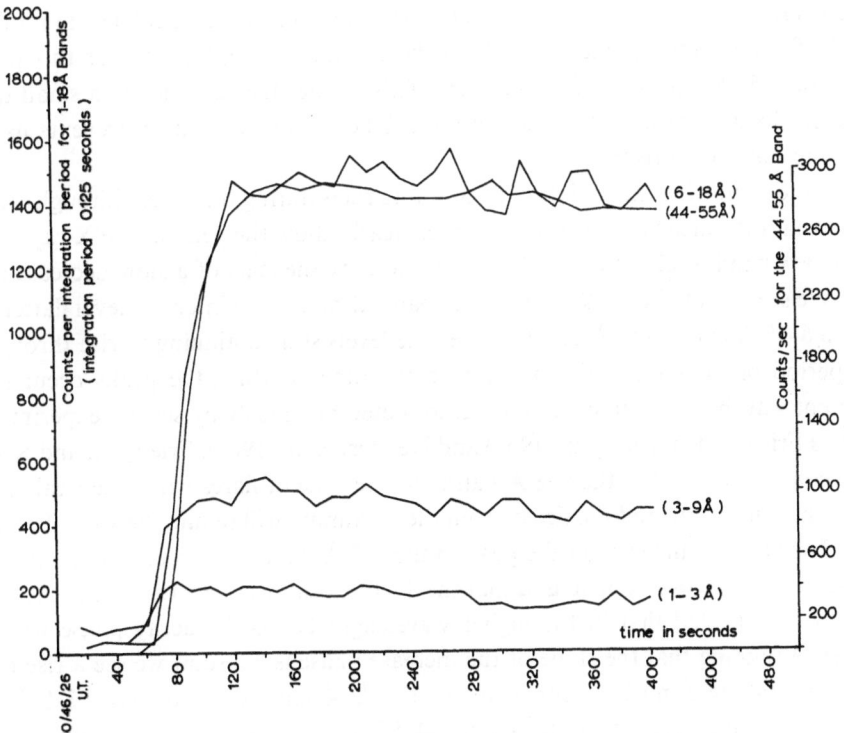

Fig. 2.   Variation of X-ray flux one orbit after the flare.

time of the spacecraft are about the same at both times. Data from the orbit immediately preceding the flare might have been more suitable but unfortunately were not available. As far as we are aware no new additional activity was in progress during this second period. The data show that for the 1–3 Å band the period between the emergence from the optical eclipse and the emergence from the X-ray eclipse is less than 70 sec. For 3–9 Å this period is less than 80 sec, for 6–18 Å less than 110 sec and for 44–55 Å less than 120 sec.

It is interesting whilst looking at this second period of data to note that not only has the X-ray flux decreased significantly, but also, as one would expect, the particle background which is near zero during the flare has increased substantially. This will doubtless be substantiated by the results from the Imperial College geiger counters on board the same spacecraft.

The minimum height between the satellite-sun line and the earth has been calculated for a few points at the onset of the X-ray increase. About 80 sec after the end of the optical eclipse (0908.46 UT) the minimum height is 127 km. Ten sec later this value has already increased to 156 km, where absorption is negligible. In principle, one could use the eclipse data curve to compensate for the absorption effect in the flare data. In order to do so the minimum height calculation has been repeated for a few points during the X-ray eclipse of the next orbit (Figure 2).

Here, we find that the measured X-ray intensity has reached about $\frac{2}{3}$ of its unattenuated value at a minimum height of 113 km. However, if we calculate the minimum height for the time of emergence from the optical eclipse, we find for this point a value of -57 km instead of about 0 km. This implies the existence of a small timing error in the data or an inaccuracy in the orbital parameters which prevents us from carrying out the correction.

The plots showing the variation of counting rates during the flare, although initially somewhat influenced by the X-ray eclipse, clearly show the peak of the X-ray flare in the 1–3 Å band at about 0911.29 UT, followed by the start of a slow decrease in the flux. We see no definite peak in the 3–9 Å band although the intensity level flattens off. In the 6–18 Å and 44–55 Å bands we see the levels still continuing to rise throughout the period of the data. This variation in the time at which the peaks occur in the different wavebands is an expected feature noted previously by several experimenters such as Friedman and Kreplin (1968) and Neupert *et al.* (1969). Friedman and Kreplin have commented that in the 0–8 Å band the rise time of flares will in general require between 2 and 5 min and the decay from the maximum will require between 10 and 30 min. They also comment that the peak in the 0–3 Å band will occur $\frac{1}{2}$ to 2 min ahead of the peak in the 0–8 Å band and that the 8–20 Å peak will be 3 to 10 min later. Also it is to be expected that in the higher wavelength bands the decay phase will take longer. Assuming that the slope of the increase phase is constant we see a rise to the peak flux taking almost $2\frac{1}{2}$ min between 1–3 Å, 4 min 50 sec between 3–9 Å, and more than 7 min between 6–18 Å and 44–55 Å. We also note that 3–9 Å peak is reached about 250 sec after the 1–3 Å peak.

It appears that the peak of the 1–3 Å radiation occurred just before the peak of the centimetric microwave burst which occurred at 0912 UT.

We have used the S 36 pulse height analyser to give some idea of the variation of the spectrum during this flare. In Figure 3 are shown count histograms from the three S 36 detectors. It should be stressed that the relation between a count histogram and the spectrum that produced it is not an easy thing to establish. The problem is considerably complicated by the discontinuous efficiency profiles of the detectors caused by the absorption edges and by the finite and energy dependent resolution of the detectors. However, the count histograms alone are sufficient to establish some additional details.

Each of the 8 channels in each of the 3 detectors is of equal energy width. As indicated above, the channel limits indicated on the plots have not been corrected for possible gain changes.

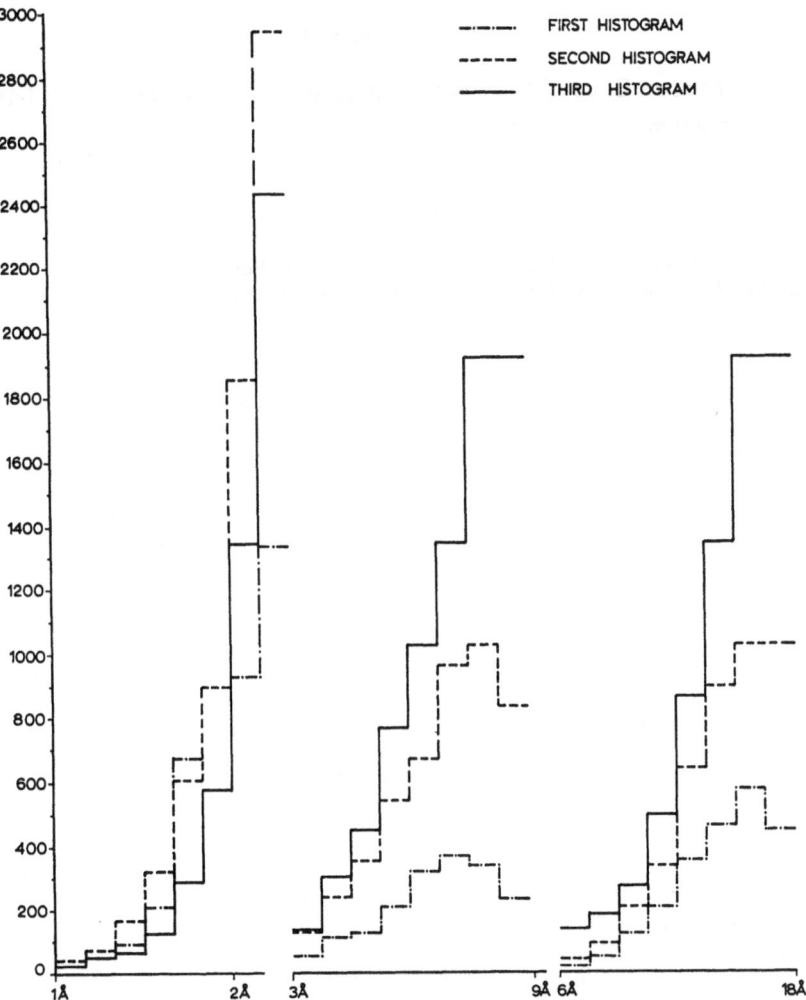

Fig. 3.   Variation of spectrum with time.

In each case the time at which the histogram was obtained is indicated on the plots of count rate against time.

In the 1–3 Å band, the three histograms are all very similar with the exception that the histogram taken during the rising part of the flux indicates considerable enhancement around 2 Å. This is doubtless due to the concentration of iron lines in that region which thus appear to be considerably enhanced before the peak of the flare, but are much less so during the decrease phase (see Neupert *et al.* 1969). In the other two wavebands it can be seen that as the flare progresses there is a softening of the spectrum i.e. the peaks in the detectors move towards the low energy end.

Approximate peak flux values for the three detectors measuring below 18 Å are 6.2×10^{-3} erg sec^{-1} cm^{-2} between 1 and 3 Å, 4.6×10^{-2} erg sec^{-1} cm^{-2} between 3 and 9 Å, and 7.1×10^{-2} erg sec^{-1} cm^{-2} between 6 and 18 Å.

Acknowledgements

We thank Professor Boyd of University College and Professor De Jager of Utrecht for their advice and encouragement.

References

Friedman, H. and Kreplin, R. W.: 1968, *Annals of the IQSY*, Vol. 3.
Neupert, W. M., White, W. A., Gates, W. J., and Swartz, M.: 1969, *Solar Phys.* 6, 183.

THE SOLAR PARTICLE EVENT OF FEBRUARY 25, 1969

HEOS-1 Interplanetary Magnetic Field Measurements

P. C. HEDGECOCK

*Physics Department, Imperial College of Science & Technology,
Prince Consort Road, London, England*

Abstract. HEOS-1 measurements of the interplanetary magnetic field during the particle event of February 25, 1969, are reported. The measurements provide virtually continuous coverage from 0130 Z on February 25 until 0700 Z on February 27, when HEOS-1 re-entered the magnetosphere. Preliminary values for the power spectra of the field fluctuations are quoted.

1. Instrumentation

The HEOS-1 satellite was launched on December 5, 1968 into an eccentric orbit with apogee occurring at a geocentric distance of 35 R_e, 52°, above the ecliptic plane. At the time of the particle event closely following the solar flare at 0914 Z on February 25, the satellite had just passed apogee at a solar ecliptic longitude of 304°. The subsequent magnetopause crossing occurred at approximately 0700 Z. on February 27 at solar ecliptic latitude +32°, longitude 340°.

The magnetic field measurements are made with a three axis fluxgate sensor mounted on a boom extending along the spin axis of the satellite. The sensor is located 2 m from the centre of gravity of the satellite. Exhaustive pre-flight tests indicated that the three components of the magnetic field of the satellite at the position of the triple sensor were each not larger than 0.1 γ.

The axial boom geometry, though avoiding deployment problems, restricted the weight of the sensor package to such an extent that it was not thought feasible to include a mechanical device to interchange the directions of the fluxgate sensors for calibrating the zero level of the axially pointing sensor. To overcome this problem and to provide data on the axial component of the magnetic field of the satellite, a second single-component sensor is mounted approximately half way along the boom. This sensor forms a single component gradiometer in conjunction with the axial sensor of the triple fluxgate package. Changes in the zero level of either of the axial sensors or changes in the axial component of the field of the satellite, caused, e.g., by the launch environment, appear as a variation in the gradiometer output from that measured prior to launch.

The in-flight data indicate no change in the remanent field of the satellite from the measured pre-launch values. Using the technique of power spectral analysis, the zero levels of the radial fluxgates can be established at least to an accuracy of $\pm 0.1 \gamma$; however, the gradiometer data and electrical bias measurements permit an estimation of the axial component zero level to the somewhat poorer accuracy of approximately $\pm 0.25 \gamma$. During the period of the measurements reported here the positive satellite spin axis was directed towards solar ecliptic latitude 0° longitude 270° thus provid-

V. Manno and D. E. Page (eds.), Intercorrelated Satellite Observations Related to Solar Events, 419–426. All Rights Reserved
Copyright © 1970 by D. Reidel Publishing Company, Dordrecht-Holland

ing unambiguous measurements of the N-S component of the interplanetary field.

The analogue magnetometer outputs are digitized in 0.5 γ steps, all three components of the interplanetary field being sampled within 2 msec corresponding to an angular rotation of 0.1° about the satellite spin axis. The pre-sampling analogue bandwidth is 5 Hz representing a suitable compromise between sensor noise level and excessive phase-shift due to the spin of the satellite. At this bandwith the measured noise level of the worst sensor was 0.06 γ r.m.s.

The digital data output is shared between two channels. One provides a single vector measurement every 50 sec, the other, operating in conjunction with a core memory, provides a 17 min record at one vector measurement every 1.5 sec repeated after each 5 hour replay period. An alternative faster field sampling rate is available for single component measurements.

The memory programmer provides a choice of several different measurement programs. For example, the recordings may either occur at random positions in the orbit as determined by the readout sequence or they may be started at times determined by ground command. Alternatively, the recordings may be replayed to the ground only if the field fluctuations during the recording sequence exceed a selected threshold amplitude.

2. Field Measurements during the February 25 Event

The 12 hour sequences of field data for the period 0130 Z on February 25 to 0700 Z on February 27 are plotted in Figures 1–5. In each figure the field magnitude and direction in solar ecliptic polar co-ordinates are plotted against Universal Time. The field values are individual vector measurements made at 50 sec intervals. Uncertainties in the field magnitude and direction are due mainly to the effects of digitisation, the error in field magnitude being $\pm 0.5\,\gamma$, while the errors in the angular co-ordinates are in the range $\pm 5 - \pm 45°$ depending on the magnitudes of the three field components registered by the magnetometers. These digitisation errors are indicated in each figure.

While the data available for presentation here cover too small an interval of time to permit any general statement concerning the macroscopic features of the interplanetary field at the time of this event it is perhaps worthwhile to point out the following features as being of some importance for the detailed analysis of this period.

An inspection of Figures 1 and 2 indicates that the field during February 25 is generally weaker than the typical 5 γ value, that the field is directed outward along the spiral arm (PHI = 135°) with relatively large fluctuations about this direction and that during the latter part of the day the field is inclined Southward at a large angle to the ecliptic plane. It will be noticed that this Southward component persists for the whole period covered by the data. These field disturbances almost certainly are related to weak solar activity during the preceding days.

The solar proton increase occurs shortly after 0900 Z in Figure 1. The correlation between these magnetic field measurements and the proton flux measurements recorded by experiments S24B and S24C are reported by Engel (1970) and Balogh (1970).

On February 26 at 0158 Z a weak geomagnetic storm sudden commencement associated with a relatively minor magnetic storm showing a Main Phase depression of 100 γ at low latitudes was recorded at equatorial ground stations. This was evidently caused by the arrival of the shock seen in Figure 3 at approximately 0150 Z. This shock is particularly interesting since the field signature closely follows the theoretical pattern described by Hundhausen (1970). Formisano (1970) has reported the HEOS-1 plasma data on this shock. The abrupt change in field amplitude and direction at 1130 Z appears not to be related to similar shock features in the plasma data.

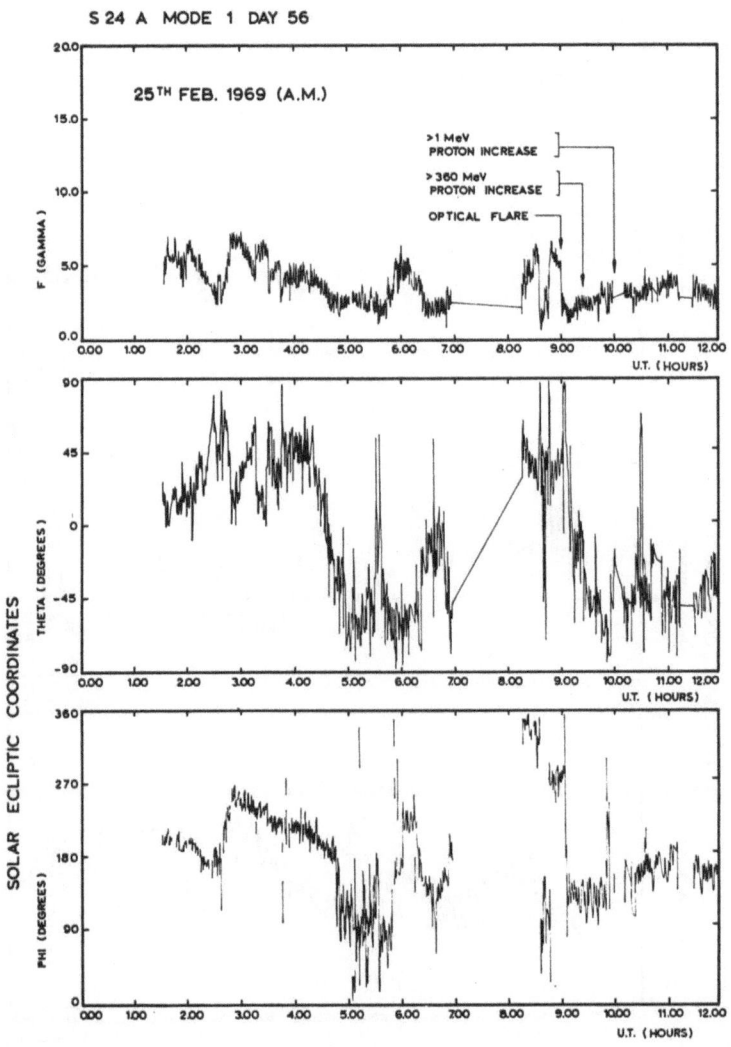

Fig. 1.

As confirmed by the S58/73 plasma data, HEOS-1 crossed through the Earth's Bow shock at 0225 Z on February 27 at a geocentric distance of approximately 14.5 R_e, solar ecliptic latitude $+39°$ and longitude 322°. The magnetic field signature on this occasion is particularly striking, as shown in Figure 5. At 0310 Z just after this bow shock crossing a weak storm sudden commencement occurred. This was followed an hour or so later by a minor geomagnetic storm. The enhanced plasma pressure associated with this event evidently displaced the bow shock inwards past the satellite at 0350 Z to produce a further multiple shock crossing at 0500 Z. Comparison with the HEOS-1 plasma data indicates that the magnetopause was crossed close to the end of the data sequence at 0700 Z on February 27. Though no core memory data are

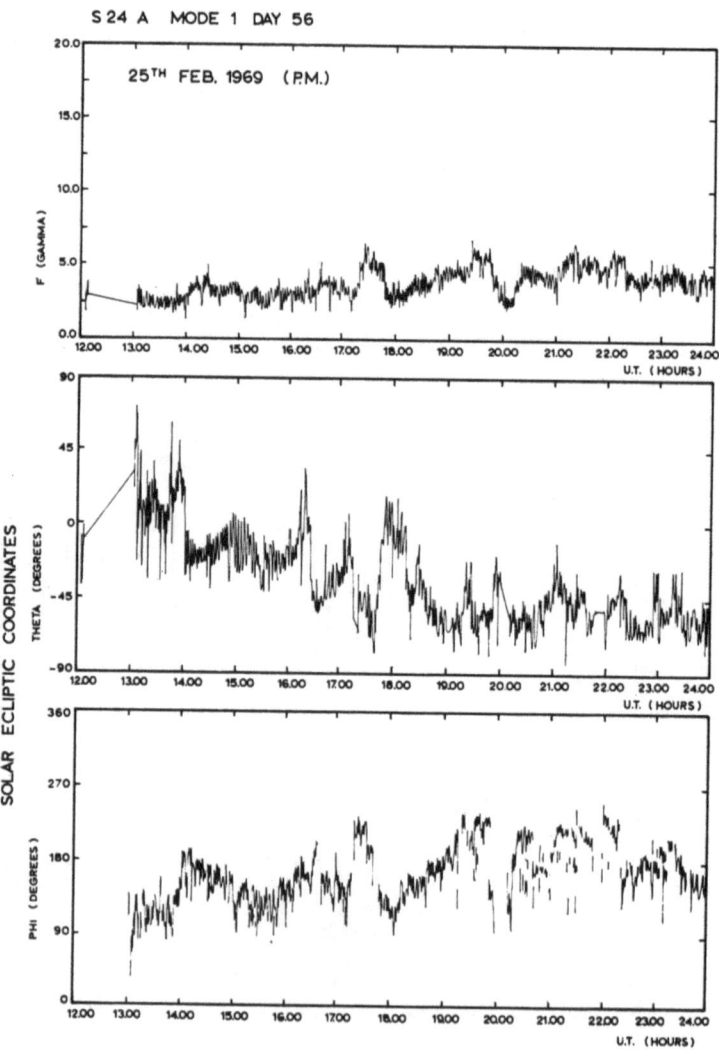

Fig. 2.

available for this period, the Vela measurements reported by Coleman (1970) confirm the variability of the shock position during this magnetic storm.

The power spectra of the field fluctation have been computed separately for the 24 hour periods of February 25 and 26, following the method of Blackman and Tukey (1958). Though this work is incomplete it is evident that on February 25, the power spectra of the three Cartesian solar ecliptic field components can be represented by a $1/f^{1.8}$ power law between frequencies of 6×10^{-5} and 2.5×10^{-3} Hz passing through a power spectral density level of 100 (γ^2)/Hz at 10^{-3} Hz. A similar power law passing through 450 $(\gamma)^2$/Hz at 10^{-3} Hz is appropriate for February 26. It is hoped to report details of this work at a future date.

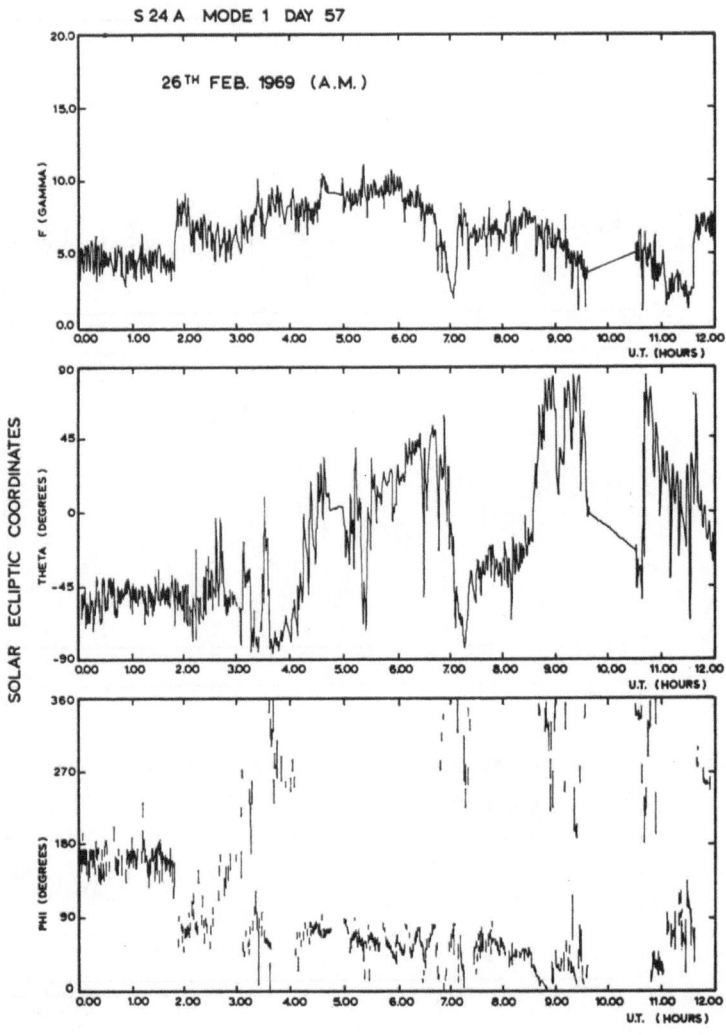

Fig. 3.

Acknowledgements

I would like to thank the ESRO project team led by Dr J. A. Vandenkerckhove, the ESLAB Project Scientist Dr B. G. Taylor and the Prime Contractor Junkers Flugzeug und Motorenwerke GmbH for their splendid execution of the HEOS-1 project.

I would like to thank Dr. O. G. Feil, Dr. E. Wunderer and Mr. A. Vuye for their careful attention to the magnetic cleanliness of the satellite.

I would also like to thank the Schonstedt Instrument Company for producing the magnetometer, McMichael Ltd. for producing the analogue to digital converter and

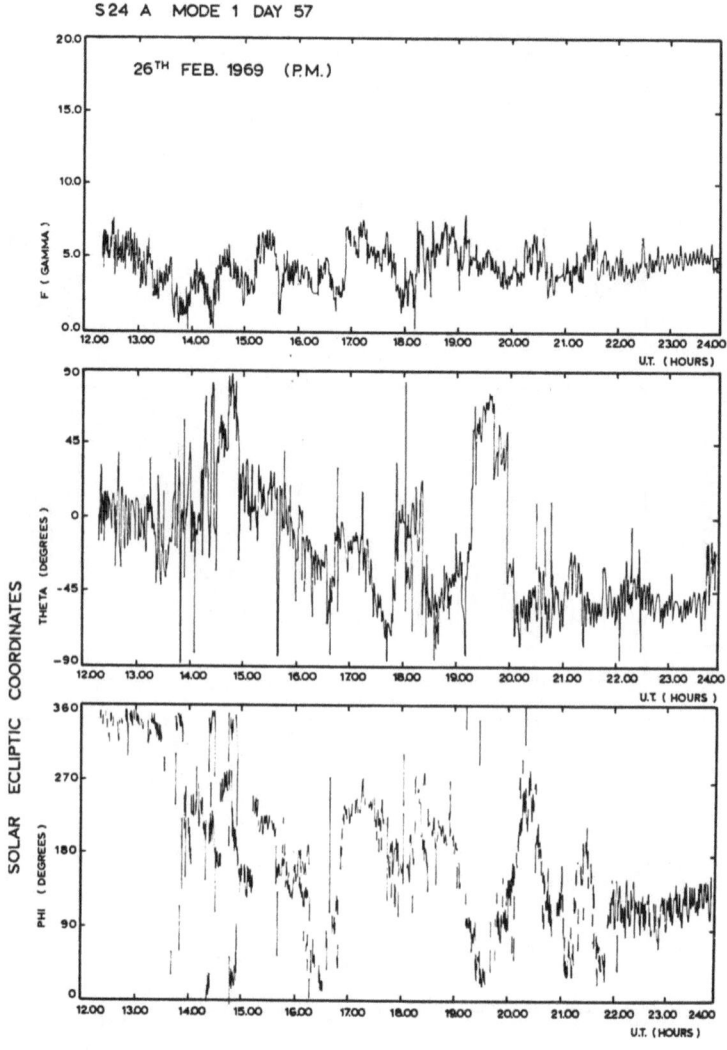

Fig. 4.

measurement programmer, and GEC-AEI Applied Electronics Laboratory for producing the Core Memory.

The computer processing of the data would not have been accomplished but for
the hard work and assistance of my colleagues at Imperial College, John Sear, Anne
Evans and Bob Hynds, also Jim Anscombe at ESOC Darmstadt.

The rewarding comparison of these results with the HEOS-1 plasma measurements
of Professor Bonnetti, Professor Pizzella, Dr Mariani, Dr Egidi and Dr Formisano,
also my discussions during the course of this conference with Dr Coleman, Dr Wilcox
and Dr Hundhausen are gratefully acknowledged.

Finally, I would like to thank Professor H. Elliot and Dr J. J. Quenby for their

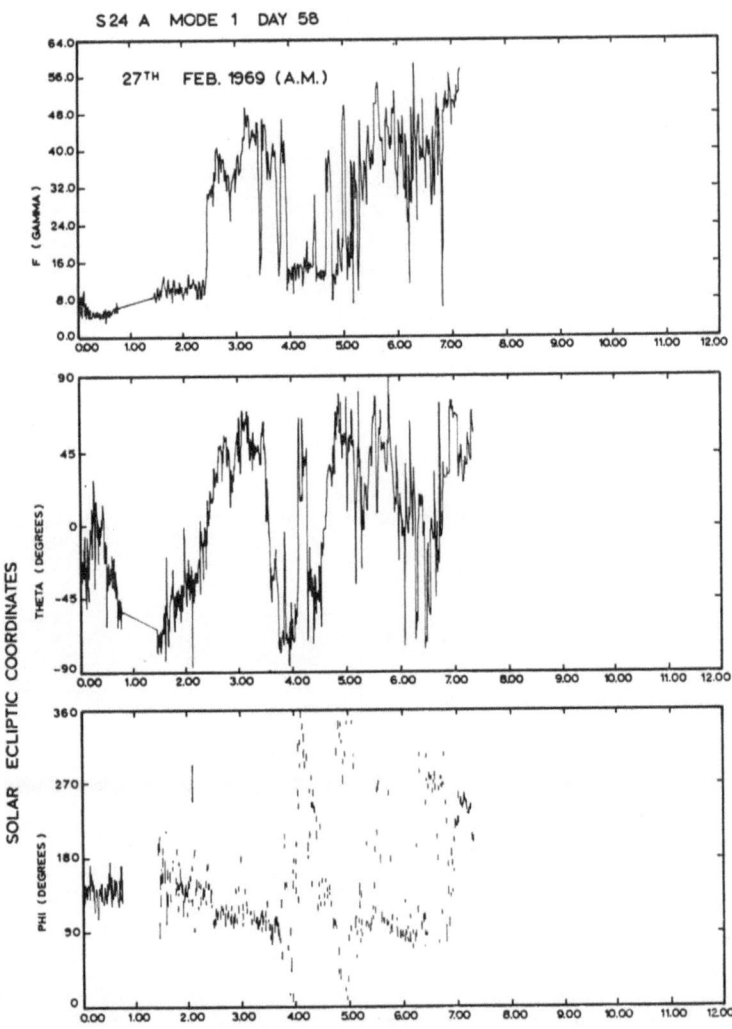

Fig. 5.

enthusiasm, guidance and encouragement during the course of the HEOS-1 project and Dr. N. F. Ness for bringing me into contact with many useful ideas and techniques during my infrequent visits to Goddard Space Flight Centre.

References

Balogh, A.: 1970, the present volume, p. 471.
Blackman, R. B. and Tukey, J. W.: 1958, *The Measurement of Power Spectra from the Point of View of Communications Engineering*, Dover Publications Inc., New York.
Coleman, P. J.: 1970, the present volume, p. 251.
Engel, A. R.: 1970, the present volume p. 478.
Formisano, V.: 1970, the present volume, p. 436.
Hundhausen, A. J.: 1970, the present volume, p. 111.

INTERPLANETARY MAGNETIC FIELD MEASURED BY
PIONEER 8 DURING THE 25 FEBRUARY 1969 EVENT

F. MARIANI and B. BAVASSANO

Istituto di Fisica, Universita Roma, Laboratorio per il Plasma nello Spazio del CNR, Roma, Italy

and

N. F. NESS

NASA-Goddard Space Flight Center, Greenbelt, Md., U.S.A.

1. Introduction

In this paper we present preliminary results from the magnetic field experiment on the space probe Pioneer 8 in interplanetary space during the 25 February 1969 event.

The spacecraft was injected into a solar orbit on Dec. 13, 1967. During the time interval in which we are interested, the sun to Pioneer line was pointing 23° east of the sun-earth line. The heliocentric distance of the probe was near to 1.52×10^8 km, i.e. slightly larger than the corresponding sun-earth distance of 1.48×10^8 km. The Pioneer 8 – earth distance was about 6×10^7 km.

Magnetic measurements were made by a single sensor mounted at an angle of 54° 45′ with respect to the spin axis of the probe. Vector components were then obtained by three consecutive measurements at 120° intervals from each other during the rotation of the spacecraft, which is 1.0 sec. Thus one complete vector measurement is obtained in 0.67 sec.

The quantization uncertainties were $\pm 0.125\ \gamma$ and $\pm 0.375\ \gamma$, respectively in the two possible ranges of operation, $\pm 32\ \gamma$ and $\pm 96\ \gamma$. Due to the low total bit rate of 16 bits/sec from the probe, one field vector is telemetered on the average each 7 sec. Each vector component is the average of four consecutive measurements which are obtained by the instrument and computed on-board the spacecraft by a special computer, the Time-Average-Unit (Scearce *et al.*, 1968). Because the reception of telemetry data was intermittent, unfortunately long gaps are present in the data. However, an approximate picture of the time evolution of the event in the distant interplanetary space can be given.

Results from the simultaneous plasma measurements on the same probe and comparison with those obtained by other near earth satellites will improve the interpretation of these discontinuous data.

2. Experimental Results

A summary of the results is shown in Figure 1 where averages over 10 min intervals of the usual magnetic field parameters are plotted: ϕ is the azimuth on the ecliptic plane ($\phi = 0$ when field points to the sun); ϑ is the elevation relative to the ecliptic

V. Manno and D. E. Page (eds.), Intercorrelated Satellite Observations Related to Solar Events, 427–435. All Rights Reserved
Copyright © 1970 by D. Reidel Publishing Company, Dordrecht-Holland

plane (positive when the field has a component pointing toward the north pole); F is the average field strength as computed by the averages of individual components; $\bar{F}$ is the average of the individual field strengths. Short period (i.e. $<10^m$) variations of the field are present when

$$\bar{F} - F > 0$$

The main features seen in an inspection of Figure 1 are the following:

(a) A rather steady field is observed in the early hours of Feb. 25 with an average intensity of 4 to 6 γ and an azimuth of $120°$–$130°$.

(b) A rapid increase of the field occurs slightly after 2000 UT: the variation from the pre-increase value to the compressed field near 10 γ occurs in a few minutes, between 2000 and 2022 UT. After that the field slowly increases to a maximum value of about 14 γ; sporadic coherent fluctuations are present when individual values are compared.

(c) Significant long period variations are seen between 0200 UT and 0500 Feb. 26 for each of the three magnetic parameters and also short period variations as the difference $\bar{F} - F$ occasionally becomes very large.

(d) A very quiet field of about 6 γ is observed between 2000 UT Feb. 26 and 0500 the next day. Also, short term variations are absent.

(e) At 2149 UT Feb. 27, when data are again telemetered the field is already at a high level of more than 10 γ. These high values are maintained for several hours. Between 2310 and 2320 a rapid, large azimuth variation occurs. Several polarity variations of the field occur as shown by the $180°$ variation of the azimuth.

(f) In the last time interval, between 2000 UT Feb. 28 and 0500 Feb. 29 the average field strength is 5.5 ± 1.5 γ, an approximately average interplanetary value. Quasi periodic variations with period ≈ 1.5 hours are found in the field strength and the inclination ϑ. Also several polarity changes are present.

A higher time-resolution plot of the field increase after 1957 UT Feb. 25 is shown in Figure 2, where 10-sec averages of the field elements $\bar{F}$, ϑ, ϕ are given. The structure of the field variation appears quite complicated: the field azimuth ϕ and the elevation ϑ change abruptly at 2008; however, a gradual increase of ϑ was already going on several minutes earlier. The field intensity is slowly varying during the time interval preceding a 5 min gap in the data coverage; it is steady but at a higher level at 2022, after the data gap. This last field increase is associated with a small variation in direction.

3. Discussion

A number of correlated events occurred in the period considered. A summary is given in Figure 3 where the timing sequence of the different events is illustrated. Several long lasting strong flares of importance 2 were observed on the sun, one cosmic ray proton event and three sudden commencement geomagnetic storms were observed on earth and sudden ionospheric disturbances were detected. A very active region, McMath No. 9946, was present in the northern hemisphere of the sun (Central

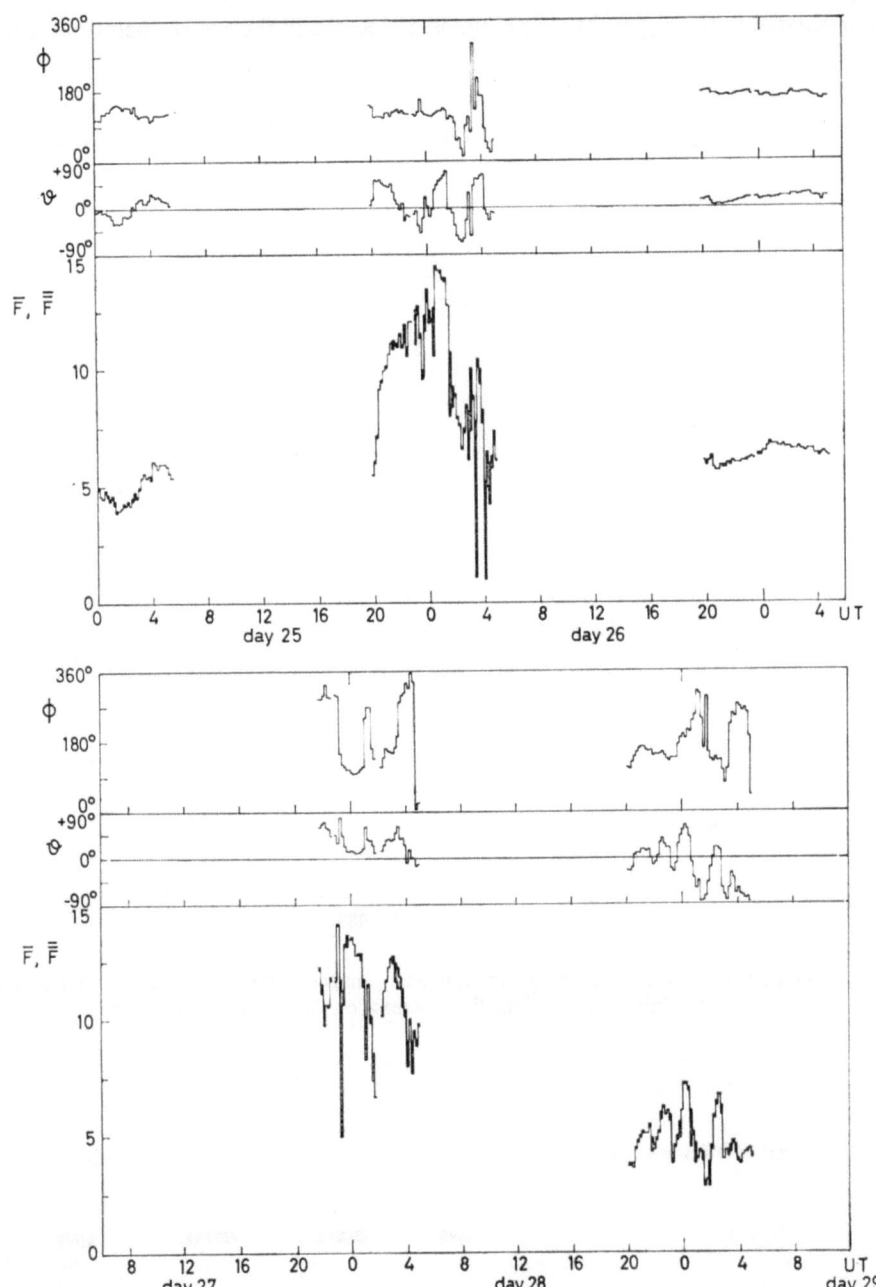

Fig. 1.   Pioneer 8 10-min averages of the interplanetary magnetic field elements F, $\bar{\bar{F}}$, ϑ and ϕ for the period Feb. 25 to 29. See text.

Meridian Passage ≈ 1200 UT February 23). The flares indicated by solid arrows all originated in the above perturbed region. The flare indicated by a dotted arrow (Figure 3) originated in the region McMath No. 9957 located in the eastern hemisphere of the sun.

Although neither of the last two flares is the source of the increased field observed on Pioneer 8, they can help to build the right time sequence of the event. It seems

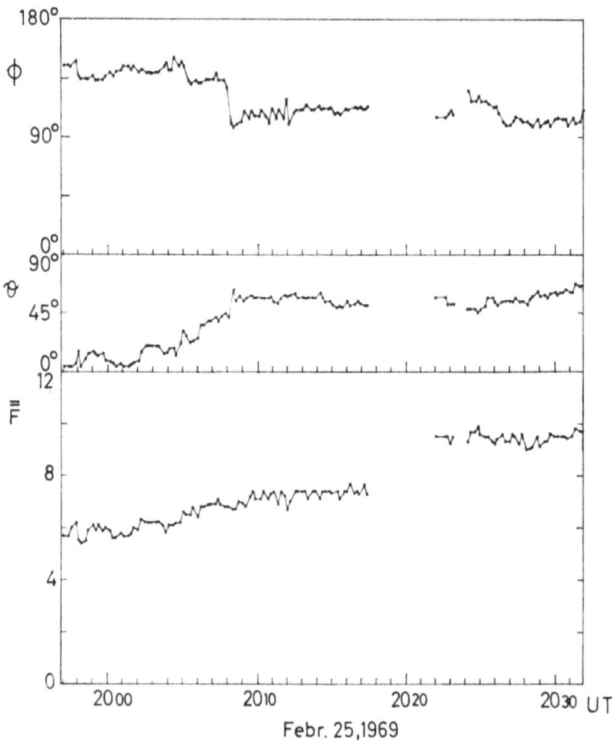

Fig. 2. Pioneer 8 10-sec averages of the interplanetary magnetic field elements $\bar{\bar{F}}$, ϑ and ϕ, during the initial phase of the field increase on 25 February 1969.

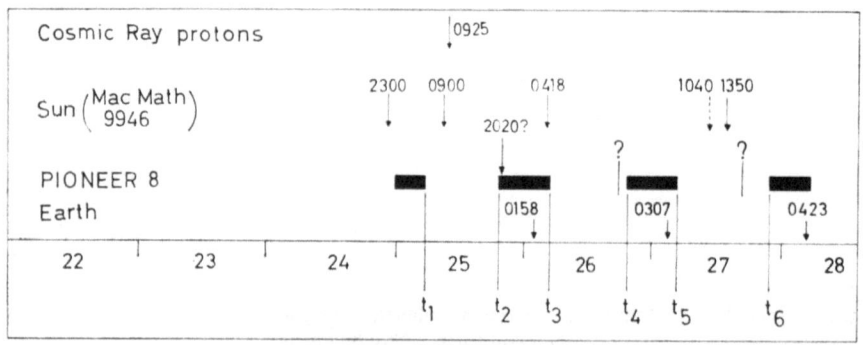

Fig. 3. Time sequence of the events on the sun, Pioneer 8 and on the earth. The data gaps are defined by the times t_1 to t_2, t_3 to t_4 and t_5 to t_6.

reasonable to associate the flares of February 24, 25 and 26 with the three s.s.c. on the ground, 25 hours apart from each other. This interpretation is confirmed by the time variations of cosmic rays on the earth and in cis-lunar space. A solar proton event on the ground was indeed detected by several radiation monitors, as well as by several satellites and space probes; its first appearance is put around 0911 UT Feb. 25; the Deep River neutron monitor indicated a strong intensity increase starting at 0925 UT. The following Forbush decrease on the ground started during the first few hours of February 27 in association with the s.s.c. observed at 0307 UT. The average velocities of the perturbation responsible for the s.s.c., as estimated by the sun-earth transit time are given in Table I, which includes the approximate origin time of the parent flare assumed to be responsible for the s.s.c.

TABLE I

Flare at			s.s.c. at	Sun-earth transit time	Estimated average velocity
Time	Heliographic				
	long.	lat.			
2300 Feb. 24	30°W	11°N	0158 Feb. 26	27 hours	1540 km/sec
0900 Feb. 25	37°W	13°N	0307 Feb. 27	42 hours	1000 km/sec
0418 Feb. 26	46°W	13°N	0423 Feb. 28	48 hours	860 km/sec
1040 Feb. 27	27°E	15°S			
1350 Feb. 27	65°W	13°N			

It seems reasonable to interpret the field strength increase observed by Pioneer 8 as having begun at approximately 2003 UT Feb. 25 and due to the effect of the flare of 2300 Feb. 24 which occurred about 21 hours before. If the scaling factors derived from the velocities of Table I are used, the next magnetic field increases would have been expected on Pioneer 8 about 33 and 37 hours after the corresponding flares, i.e. around 1800 Feb. 26 and 1700 Feb. 27 respectively, during the data gap period. Measurements taken between 2000 UT Feb. 26 and 0500 UT Feb. 27 do not show any significant perturbation: in the next data period, starting at 2200 UT Feb. 27, the field is already at a steady, high value which indicates that the perturbation was already present at Pioneer 8 location.

The directional change at 2008 UT Feb. 25 appears to be an MHD discontinuity with the field vector rotating around a direction approximately perpendicular to the field lines. The increase in the intensity a few minutes later, although it cannot be studied in detail due to the data gap, might be due to the effect of a fast shock discontinuity.

Actually two worldwide geomagnetic storms were observed on Feb. 26 and 27. Presence of shocks in the interplanetary medium can be expected to be associated with them (Ogilvie and Burlaga, 1969). Unfortunately the data gaps prevent unique identification of the field variation associated with the shock. The results presented in Figure 2, with the additional feature of a very steady field in the second half of the time interval 2000 to 2100 UT, when the fastest field variation occurs, lead us to suggest that a shock wave passed by Pioneer 8 around 2020 UT.

Previous evidence of interplanetary shocks is due to several authors.* Evidence for a shock wave in deep interplanetary space on board Mariner 2 was presented by Sonett *et al.* (1964) who found an impulsive field change from 6 to 16 γ in a time interval less than 3.7 min. The average shock velocity between the spacecraft location and the earth was estimated at 510 km/sec assuming a spherical wave front from the sun. The resulting velocity referred to the travelling plasma was about 130 km/sec. Van Allen and Ness (1967) detected an interplanetary shock on board Explorer 33 at a geocentric radial distance of about 70 R_E on July 8, 1966. The field abruptly changed from 12 to 20 γ with a rise time of 5 to 10 sec. The estimated near earth velocity of the shock was 890 ± 40 km/sec as compared with an average sun-satellite velocity $\bar{v} = 950$ km/sec and the shock normal was pointing in the direction $\phi = 182° \pm 5°$, $\vartheta = -27° \pm 5°$ Plasma data are not yet available to determine the shock wave velocity properly.

More recently, from the Vela 3 data, Gosling *et al.* (1968) have identified two shock-like structures in the deep interplanetary medium, on October 5, 1965 and January 20, 1966. The average velocity from the sun to the spacecraft in the two events have been respectively 2500 and 1670 km/sec; corresponding velocities past the spacecraft were 410 and 420 km/sec, which means 70 and 90 km/sec with respect to a reference system moving with the plasma. These results also indicate clearly a slowing down of the shock as it propagates away from the sun.

Ness and Taylor (1969) also studied this same event using in addition the magnetic field data from Pioneer 6, deep in interplanetary space. Unfortunately data gaps also precluded unique identification of the event on Pioneer 6.

Taylor (1969) using data from Explorer 28 (IMP-3) was able to study the geometry of many shocks in the near earth interplanetary space and associated terrestrial disturbances which occurred between June 1965 and January 1967. Unambiguous identification of the parent flare for each event was not possible so that a statistical study of shock velocity was not conducted.

Ogilvie and Burlaga (1969) using magnetic and plasma data from Explorer 34 identified 7 interplanetary shocks between May 1967 and January 1968. They found that the Rankine-Hugoniot conditions are satisfied at the shock front and that all the events were the cause of s.s.c. geomagnetic storms. Shock velocities with respect to the moving plasma were of the order of tens of km/sec.

In our case, the observed perturbation has also been interpreted as a shock, although until plasma data will be available not all the features of the shock can be uniquely found. The high average velocity of the shock can well be an effect of the long duration of the generating flare. An important question arises from the association of solar and magnetic events outlined above, and it is: why the field increase is detected on the spacecraft 6 hours earlier than on the ground? The time sequence of

---

* At the time this paper was prepared, the relevant review presented by Hundhausen (this volume, p. 111) was not available.

magnetic field variations on Pioneer and on the earth does not favour a spherically propagating shock. The Pioneer-sun distance was larger than the sun-earth distance: even a very high velocity, 1000 km/sec, of the shock wave would imply about one hour delay between earth and Pioneer field perturbations. One possible interpretation is that the disturbance emitted from the sun was travelling with different velocities at different azimuthal orientations. Evidence for non-spherical propagation was already given by Hirshberg (1968), Taylor (1969) and others. In this interpretation, the average velocity would be significantly higher toward the Pioneer 8 position, i.e. the isotime

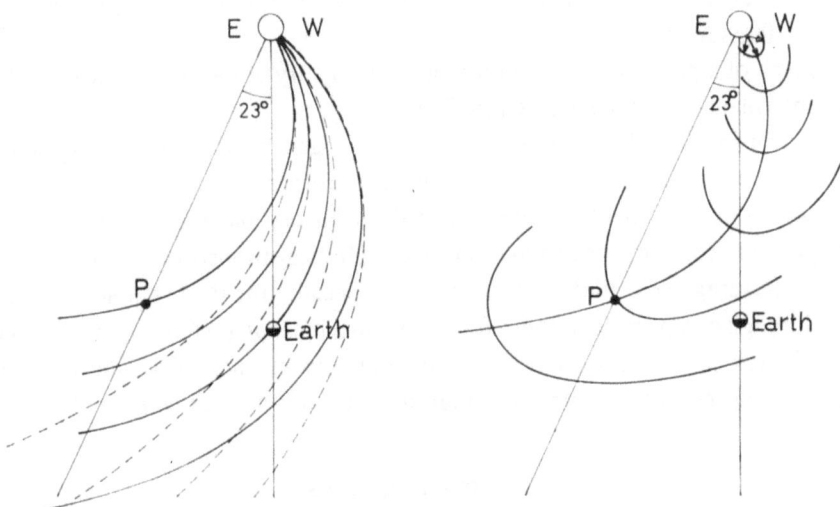

Fig. 4.   A sketch of the magnetic field lines (left) and of the shock front propagation into the inter-planetary medium (right); P is the position of Pioneer 8. Solid and dashed lines on the left hand represent the field configuration immediately before the time t_2 and between t_4 to t_5, as defined in Figure 3.

curves, which are those points in space reached by the perturbation at a given time, should not be spherical around the point where the flare originated. The azimuthal distribution of the velocity should exhibit a maximum in a direction east of the sun earth-line. If the shock actually occurred in the data gap between 2017 and 2022 Feb. 25, the normal to its surface can be estimated by considering of the **B** vector variation ahead and behind the shock. Using 5-min averages of the data on each side of the gap (SE field components ahead and behind the shock: $B_x = -1.6\ \gamma$, $B_y = 3.8\ \gamma$, $B_z = 6.0\ \gamma$ and $B_x = -2.3\ \gamma$, $B_y = 5.1\ \gamma$, $B_z = 7.5\ \gamma$ respectively) the corresponding field intensities and orientations are $B = 7.3\ \gamma$, $\phi = 113°$, $\vartheta = 55°$ and $B = 9.4\ \gamma$, $\phi = 115°$, $\vartheta = 53°$. Determination of the possible shock front based on use of continuity of the normal component of the field and coplanarity theorem may contain a large error due to the very small difference in direction of the field vectors ahead of and behind the shock. The normal to the shock front as estimated by the above field values corresponds to $\phi \approx 155°$ and $\vartheta \approx -35°$. This direction has a large component aligned with the average field, which suggests that a high velocity shock wave was propagating toward Pioneer 8 approximately along the interplanetary magnetic field lines. The angular relative position of the solar region where the flares took place and Pioneer 8, as well as the spiral geometry of the interplanetary magnetic field support this hypothesis. With the available data we cannot state how long the perturbation persisted, nor are we able to say whether or not an effect of the flare at 0900 UT Feb. 25 was detected by the spacecraft. However, a similar azimuthal distribution of the velocity can be inferred from the 27 Feb. field increase which also occurred on the spacecraft at least 6 hours before the 0423 Feb. 28 s.s.c. on the ground. In this case the required propagation velocity in the direction of Pioneer 8 is definitely lower, between 1300 to 1000 km/sec.

A sketch of a possible interpretation of the geometry of the events initiated on Feb. 25 in interplanetary space is given in Figure 4.

In conclusion, the observations on Pioneer 8 suggest but do not uniquely determine the passage of two fast shocks; one on Feb. 25, the other sometime on Feb. 27 between 0500 and 2200. The lack of information on the solar plasma on board the same spacecraft prevents complete determination of the physical properties of the shocks. The estimated average velocities of these shocks, especially for the first one, appear quite high although comparable to those found by Gosling et al. (1968). As the plasma data from Pioneer 8 become available, and simultaneous magnetic field and plasma data from other spacecraft, an improved analysis of this event shall be possible.

Acknowledgements

We are pleased to acknowledge the important contributions of our colleagues S. Cantarano of Rome and C. S. Scearce of GSFC in the development of the flight instrumentation, and L. F. Burlaga of GSFC for comments on an earlier draft of the manuscript.

References

Gosling, J. T., Asbridge, J. R., Bame, S. J., Hundhausen, A. J., and Strong, I. B.: 1968, 'Satellite Observations of Interplanetary Shock Waves', *J. Geophys. Res.* **73**, 43–50.
Hirshberg, J.: 1968, 'The Transport of Flare Plasma from the Sun to the Earth', *Planetary Space Sci.* **16**, 309–319.
Ness, N. F. and Taylor, H. E.: 1969, 'Observations of the Interplanetary Magnetic Field July 4–12, 1966', *Annals of IQSY* **3**, 366–374.
Ogilvie, K. W. and Burlaga, L. F.: 1969, 'Hydromagnetic Shocks in the Solar Wind', *Solar Phys.* **8**, 422–434.
Scearce, C. S., Ehrmann, C. H., Cantarano, S. C., and Ness, N. F.: 1968, Magnetic Field Experiment Pioneers 6, 7 and 8, NASA-GSFC Report X-616-68-370.
Sonett, C. P., Colburn, D. S., Davis, Jr., L., Smith, E. J., and Coleman, P. J.: 1964, 'Evidence for a Collision-Free Magnetohydrodynamic Shock in Interplanetary Space', *Phys. Rev. Letters* **13**, 153–155.
Taylor, H. E.: 1969, 'Sudden Commencement Associated Discontinuities in the Interplanetary Magnetic Field Observed by IMP-3, *Solar Phys.* **6**, 320–334.
Van Allen, J. A. and Ness, N. F.: 1967, 'Observed Particle Effects of an Interplanetary Shock Wave on July 8, 1966', *J. Geophys. Res.* **72**, 935–942.

Discussion

Wibberenz: Was there any sector boundary between the location of Pioneer 8 and the earth, since this would explain the difference in the arrival times of the shock?

Hedgecock: On February 25, 26, 27, the HEOS satellite was in interplanetary space and the field was pointing opposite to the sun; on the 28th, when HEOS came out of the magnetosphere, the field was pointing towards the sun very steadily. There could have been a sector transition during the day HEOS was in the magnetosphere. However, the data of February 28 just cover a period of 6 hours, and this may not be statistically significant.

OBSERVATION OF SOLAR WIND DISCONTINUITIES
FROM FEBRUARY 24 TO FEBRUARY 28, 1969

A. BONETTI and G. MORENO

Laboratorio per lo Studio delle Radiazioni Extraterrestri – CNR, Università di Firenze, Firenze, Italy

and

M. CANDIDI, A. EGIDI, V. FORMISANO, and G. PIZZELLA

Laboratorio Plasma Spaziale – CNR, Università di Roma, Roma, Italy

Abstract. Interplanetary-plasma data at 1 AU obtained with experiment S-73 on board of HEOS-A during the last week of February 1969, have been analyzed.

A well defined shock wave, travelling with a velocity of 600–700 km/sec, has been observed, most likely related with a large solar flare.

A second shock wave occurring a few days later is also recognizable.

A preliminary analysis of data within the magnetosheath is presented; the possible interaction between a solar wind discontinuity and the magnetosheath is reported.

1. Experimental Apparatus

The experimental apparatus and its operation have already been described in detail [1], [2]; here we will only briefly recall the main features.

The sensor (Figure 1) consists of a particle detector (a hemispheric electrostatic deflector and a Faraday cup), a fast electrometer-converter and a variable high voltage supply.

The field of view of the instrument is fan-shaped with an angular aperture of about

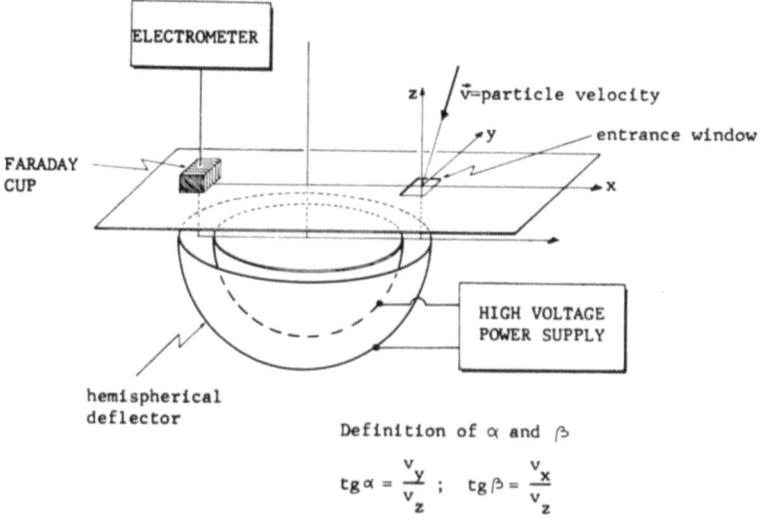

Definition of α and β

$$\mathrm{tg}\,\alpha = \frac{v_y}{v_z} \; ; \quad \mathrm{tg}\,\beta = \frac{v_x}{v_z}$$

Fig. 1.   Sketch of the electrostatic deflector mounted on board of HEOS-A. In the figure α and β define the direction of the velocity of the incoming particle.

V. Manno and D. E. Page (eds.), Intercorrelated Satellite Observations Related to Solar Events, 436–447. All Rights Reserved
Copyright © 1970 by D. Reidel Publishing Company, Dordrecht-Holland

$\pm 50°$ parallel to the spin axis (centered about the equatorial plane of the spacecraft) and $\pm 6°$ perpendicular to it.

The operation of the instrument is the following:

(1) for a given high voltage U applied across the electrodes, positively charged particles with a specified energy per unit charge E are detected. The width of the energy window $\Delta E/E$ is about 15% of the center value E.

The ratio E/U is about 2.28.

(2) The electrometer-converter measures the charge collected by the Faraday cup, delivering at its output a number of pulses proportional to the collected charge. The calibration of this circuit gives for the counting rate dC/dt due to an electrical current I the relationship $dC/dt = 112 + 40\, I$ (I in units of 10^{-13} A), which is linear with a very good accuracy over the entire range.

(3) The high voltage applied to the deflector plates is varied in steps in order to detect particles with different energies. There are 28 voltage steps covering approximately the energy range 200 eV to 16000 eV per unit charge (Figure 2).

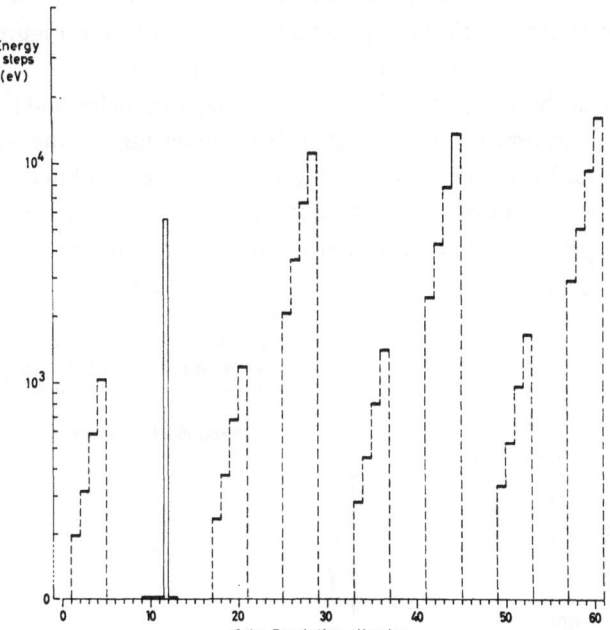

Fig. 2. Time sequence of energy steps: the energy E_i is related to the high voltage V_i applied to the hemispherical deflector, by the equation: $E_i = 2.28\, V_i$.

(4) During one complete spin revolution two measurements are made. One measurement consists in integrating the electrometer response during the half revolution when the detector looks towards the sun; the other one is the integral in the opposite half revolution. The high voltage is held constant during the complete spin revolution, then changed to a different value to examine a different portion of the energy spectrum.

A complete energy distribution is obtained every 6 min and 24 sec; however, the particular sequence chosen for the voltage steps applied to the deflector allows a rough survey of the whole energy spectrum about every 1.5 min (see Figure 2).

2. Interplanetary Region

2.1. In Figure 3 a typical example of energy spectrum in the interplanetary region is shown. In this figure the charge measured by the electrometer in units of counts C, corrected for the background counting rate as described in [2], is plotted vs. the value of the high voltage applied across the deflector.

Bars represent the experimental data.

The values of the parameters bulk velocity V, most probable thermal speed W and number density N are computed from the data following a method described in [2]. This method consists essentially in assuming a Maxwellian distribution for the particle speed and computing for given sets of V, W, N the expected response of the instrument. It is thus possible to determine a set of parameters for which the computed values best agree with the experimental data. Crosses in Figure 3 represent the values of counts C, as evaluated from the above method.

The method has been applied to both protons and α-particles, and the corresponding sets of V, W, N parameters are indicated. It is interesting to note that bulk velocity and thermal speed are approximately the same for the two kinds of particles. The density of α-particles in this particular case is 5% that of protons. Although a detailed analysis of the errors has not been carried out yet, estimates based on the method of

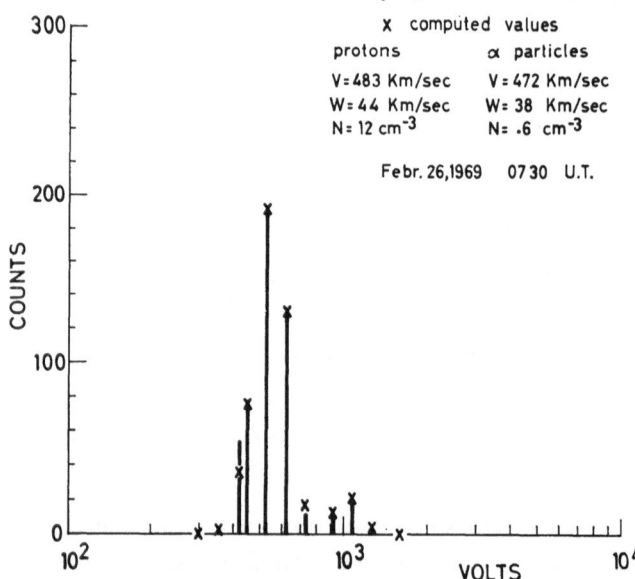

Fig. 3. A typical example of energy spectrum in the interplanetary region. The counts of the electrometer-converter are plotted vs. the high voltage applied to the deflector. Bars represent the experimental data and crosses the values computed assuming a Maxwellian distribution characterized by the parameters given in the upper right hand corner.

deriving the parameters and on the results obtained from 'quiet plasma' conditions show that the error on V and W is about ± 10 km/sec for protons and ± 20 km/sec for α-particles, while the error on density is about $\pm 20\%$ for both species.

2.2. A survey of the general properties of the proton component of the solar wind for the period Feb. 24 to Feb. 28 is shown in Figure 4.

The main features to be noted are the following:

(a) A somewhat disturbed situation exists during Feb. 24 and is followed during Feb. 25 by a gradual decrease of V, W, N down to relatively constant values.

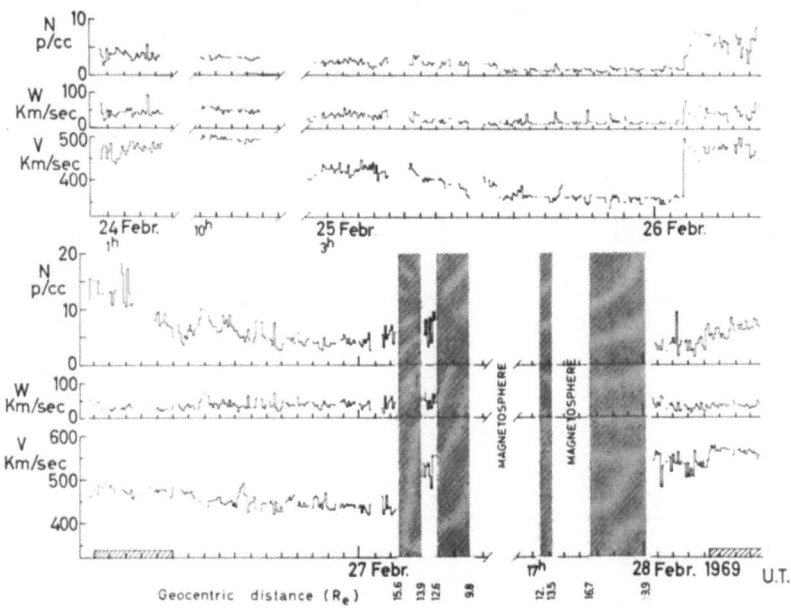

Fig. 4. Time history of the number density N, thermal speed W, bulk velocity V for the period of interest. The time scale is broken when no data are available. The horizontal shaded areas indicate the periods when α-particles were observable. The vertical shaded areas indicate the periods when the satellite was in the magnetosheath.

(b) At the beginning of Feb. 26 (0153 UT $\pm 1.5^m$) a sudden increase of all three parameters occurs: the bulk velocity increases from 365 to 500 km/sec, the thermal speed from 15 to 60 km/sec and the density from 1 to 3.5 p/cc.

(c) During the following 5 hours velocity and thermal speed remain at about the same high value, while the density goes up to 12–20 p/cc. Afterwards the three parameters undergo a slow irregular decrease down to values comparable to the disturbed situation observed at the beginning of Feb. 24.

(d) At the beginning of Feb. 27 the satellite overcomes the bow shock traversing the transition region (shaded area) during about 1.5 hours. Then apparently, the shock moves towards the earth faster than the satellite and again signals typical of the interplanetary region are detected during 1 hour showing a bulk velocity higher than before. This being the general pattern, the observations suggest the existence in the meanwhile of other fast oscillations of the shock front. Then the spacecraft enters ultimately the magnetosheath and 2 hours afterwards the magnetosphere.

(e) After perigee, on its way out, the satellite apparently crosses two times the magnetosheath during about 1 hour and 3.5 hours respectively before enetering the interplanetary region. A minor sudden discontinuity in the solar wind parameters is observed at 0418 UT $\pm 3^m$ of February 28.

Figure 5 shows the projection of the orbit on the *XY*- and *XZ*-plane in a solar ecliptic reference system. The blackened segments indicate the regions where the transition region is traversed (see (d) and (e) above).

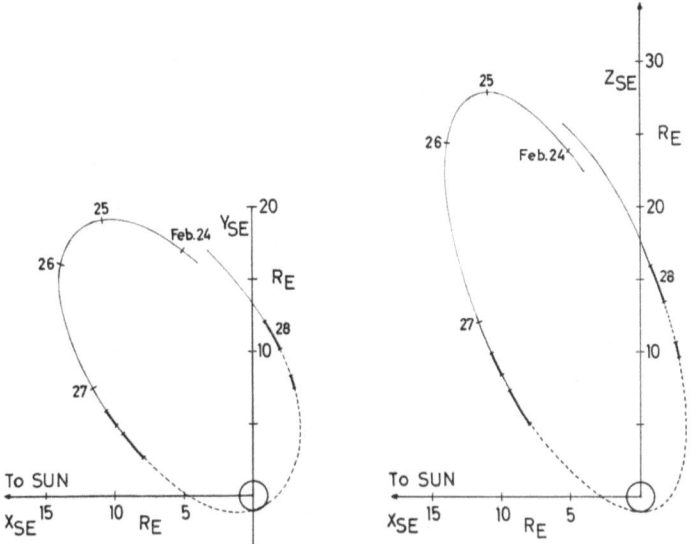

Fig. 5.   Orbit projections on the *xy*, *xz* plane of a solar-ecliptic frame of reference. Solid thin lines
indicate the interplanetary region; solid thick lines the magnetosheath and
dashed lines the magnetosphere.

2.3. At least two discontinuities in the solar wind parameters are clearly detected in the interplanetary region during the period under consideration.

Figure 6 shows in more detail the behaviour of proton parameters across the discontinuity of Feb. 26 at 0153 UT and the correlation with magnetic measurements on the ground. The event is detected by the ground stations with an average delay of about 3 min. A rough computation of the shock velocity, based on the proton parameters before and after the discontinuity gives a value of about 600 km/sec. The delay time computed from this velocity is consistent with the observed value of 3 min.

Figure 7 shows the sudden increase of the proton bulk velocity observed at 0418 UT February 28, and correlated with ground measurements. This discontinuity is much weaker than the previous one and appears to be preceded by an increase in particle density.

A third one is likely to have occurred between 0225 and 0400 UT, Feb. 27, while

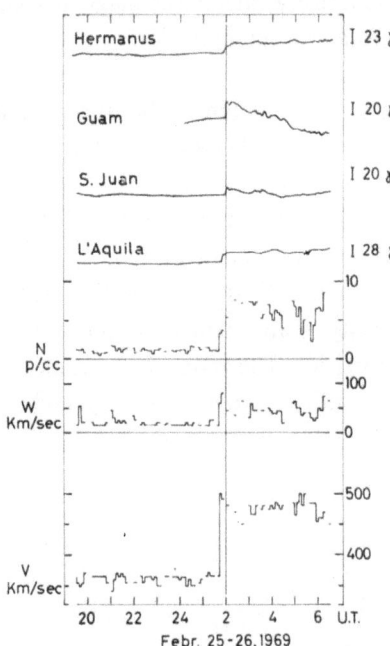

Fig. 6. Correlation between proton parameters observed by HEOS-A and ground magnetograms for the event of 0153 UT, February 26, 1969.

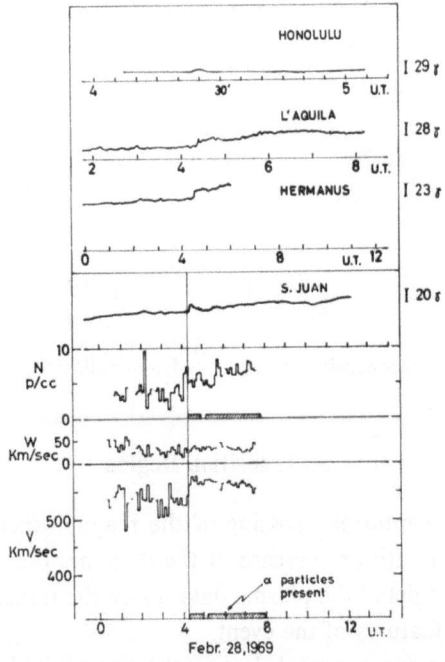

Fig. 7. Correlation between proton parameters observed by HEOS-A and ground magnetograms for the event of 0418 UT, February 28, 1969. Note the different time scale for Honolulu, L'Aquila, Hermanus magnetograms. The period is indicated when α-particles were observed.

the spacecraft, which had already entered the magnetosheath, was being overcome by the shock front subject to a swift compression.

2.4. A flux of α-particles high enough to be analyzed using the method outlined above is detected in two periods, between 0700 and 1200 UT, February 26, and between 0415 and 0730 UT, February 28.

In the first period α-particles are observed when N attains the largest values. In the second period α-particles appear immediately after the second discontinuity. Figure 8 shows the computed parameters N, W, V for both α-particles and protons. It can be noted that the bulk velocity is approximately the same, while the thermal speed is somewhat larger for α-particles. The ratio of α-particle to proton density has values ranging from 0.02 to 0.12, with an average value of 0.08.

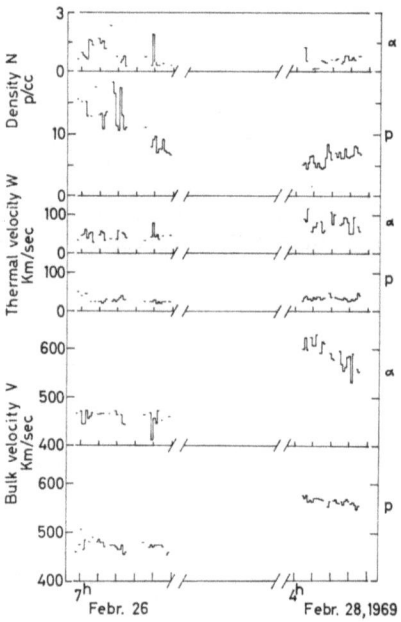

Fig. 8.   Comparison of number density, thermal speed, and bulk velocity of α-particles and protons.

3. Transition Region

3.1. The existence of a multiple crossing of the magnetosheath at the beginning of Feb. 27, in connection with an increase of the dynamic pressure of the solar wind, has led us to analyze in detail the plasma data inside the transition region. We report here only the general features of the event.

The character of the experimental data in this region is quite different from that observed in the interplanetary region. The time scale of the variation in this region is usually shorter than the 6 min required to perform a complete energy spectrum. On the other hand the energy spectrum is much wider than in the interplanetary region, covering as much as 20 energy windows.

This makes it possible to obtain meaningful information on the spectrum every 1.5 min, making use, as already mentioned, of the particular time sequence of the voltage steps.

Figure 9 shows a typical sequence of 4 successive 1.5 min spectra obtained within

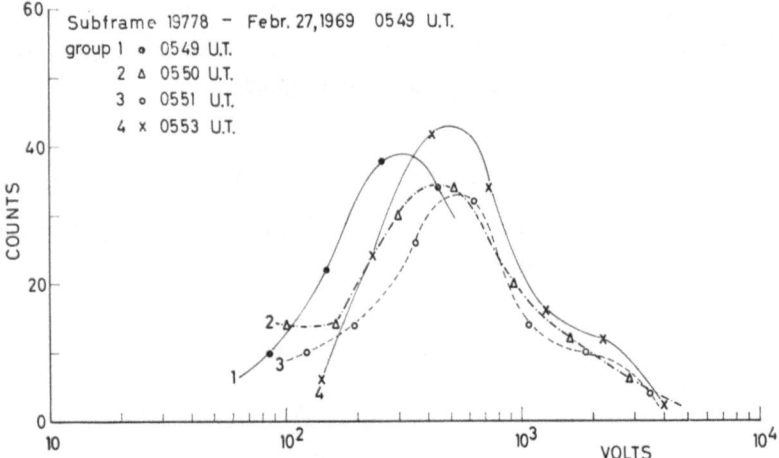

Fig. 9. Sample of energy spectrum in the magnetosheath. The four 1.5 min spectra are separately shown.

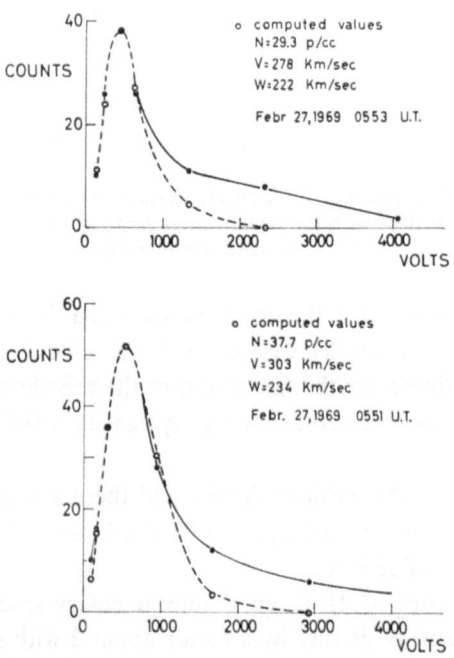

Fig. 10. Samples of 1.5 min spectra in the magnetosheath fitted with a Maxwellian distribution. The points represent the experimental data and the circles the computed values.

one telemetry subframe. It is evident from the figure that, while no meaning can be attributed to the global distribution of all points, it is possible to give a definite meaning to each of the four partial spectra.

The analysis of the partial spectra has been performed fitting the experimental points with a Maxwellian distribution. Figure 10 shows that, when a good agreement is obtained around the maximum value, the Maxwellian distribution (dotted line) does not agree with the experimental points on the high energy side, even allowing for a contribution from α-particles. Following the method previously adopted in the analysis of the electronic component in the magnetosheath [3], a quasi-Maxwellian K-distribution of the following form has been adopted:

$$f_K(V) = \frac{K!}{K^{3/2}(K-3/2)!\,\pi^{3/2}\,W^3}\,\frac{1}{(1+V^2/KW^2)^{K+1}}.$$

A fairly good agreement is obtained for $K=3$ (Figure 11): the contribution from α-particles has been neglected.

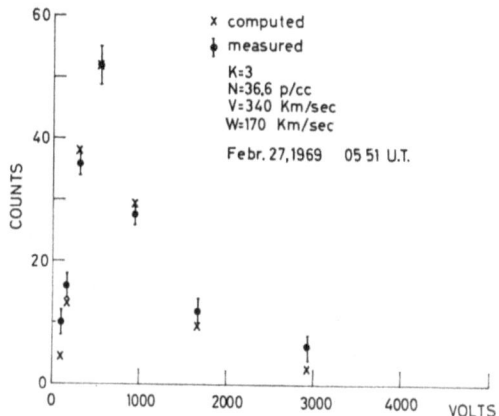

Fig. 11.   Sample of 1.5 min spectrum in the magnetosheath fitted with the K-distribution. The bars represent the experimental data with the quantization error, crosses are computed counts for the given set of parameters.

3.2. At approximately 0225 UT the satellite crosses the shock at a distance of 15.6 R_e as indicated from the sudden decrease of the bulk velocity and increase of thermal speed illustrated in Figure 12: the first graph at the top shows the ratio of the total proton energy density normalized to the energy density before entering the transition region.

A gradual increase of the particles density and thermal speed produces an evident increase of the total energy, starting from 0300 UT until the shock takes over the satellite, at a distance of 13.9 R_e.

The successive measurements, taken in interplanetary space, show that there has been an increase of energy density by a factor about 2 with respect to the measurements taken in interplanetary space before the first crossing.

It appears evident that a discontinuity in the interplanetary plasma has reached the

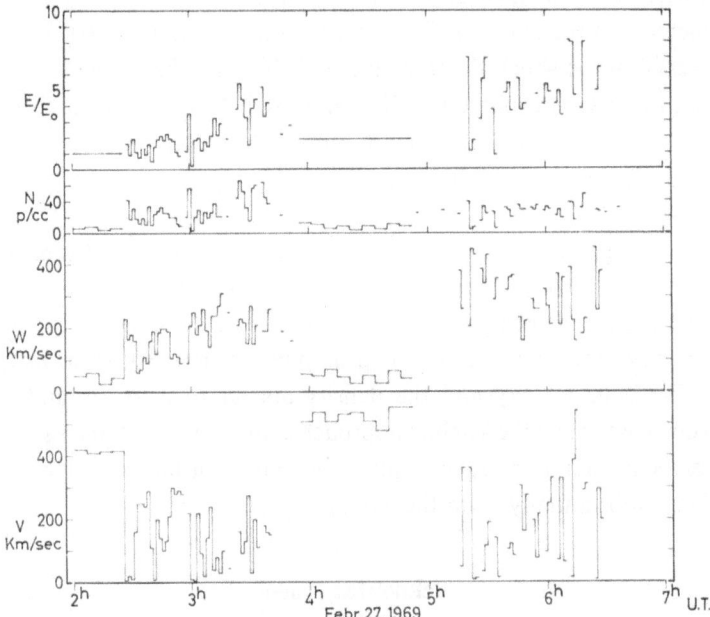

Fig. 12.   Time history of the proton parameters during the inbound magnetosheath traversal. The top graph shows the ratio of the total proton energy density normalized to the energy density before entering the transition region.

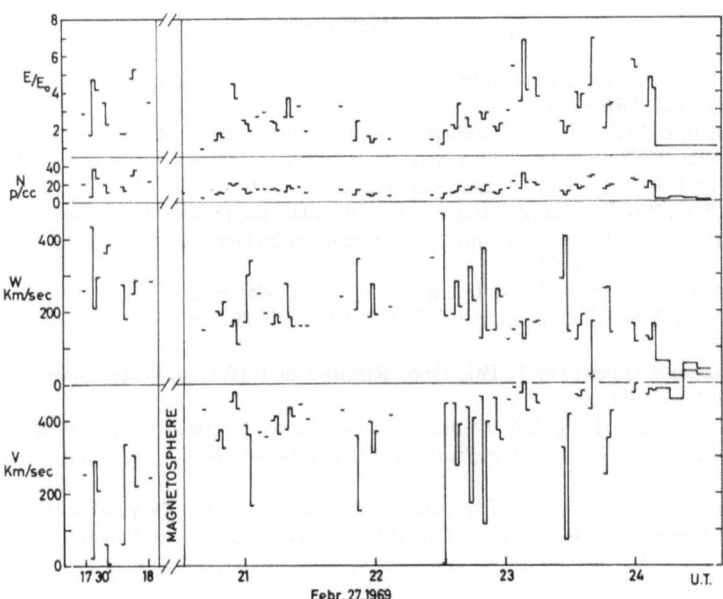

Fig. 13.   Time history of the proton parameters during the outbound magnetosheath traversal. The top graph shows the ratio of the total proton energy density normalized to the energy density found just outside the transition region.

bow shock between 0225 and 0355 UT; most probably at 0300 UT, in agreement with a sudden increase of the magnetic field detected on the ground at 0307 UT. After the second traversal of the shock the total energy is found to have kept its high value, indicating that the magnetosphere is still compressed. Probably the arrival of the discontinuity is also an explanation for the remarkably high variability of the plasma parameters inside the transition region.

3.3. After the perigee, the satellite crosses the magnetopause a first time at a distance of 12 R_e, travelling through the transition region for less than 1 hour. The satellite emerges from the magnetosphere for the last time at a distance of 16.7 R_e.

Figure 13 shows the behaviour of the proton parameters during the magnetosheath traversals. The bulk velocity and the density appear to increase, and the thermal velocity to decrease, when the satellite approaches the bow shock (crossed at 19.9 R_e). The general behaviour is relatively quiet: in only 9 instances out of 65 the bulk velocity differs substantially from the average value.

Acknowledgements

This research has been supported by the Consiglio Nazionale delle Ricerche.

The ground magnetograms were kindly supplied by the World Data Center A – Geomagnetic Data Center of Rockville, Md., U.S.A., and by Dr. Molina for L'Aquila.

We are pleased to acknowledge Dr. L. Palmiotto for her useful help in the data analysis.

References

[1] D. E. Page (ed.): 1967, Description of Scientific Experiments of HEOS-A Satellite Experiment S58/S73 – ESLAB/ESTEC.
A. Bonetti, B. Melchiorri, F. Melchiorri, S. Cantarano, A. Egidi, R. Marconero, F. Palutan, and G. Pizzella: 1967, 'Esperimento per la misura del vento solare sul satellite europeo HEOS-A', Atti ufficiali del VII Convegno Tecnico Scientifico sullo Spazio.
A. Egidi, G. Pizzella, and M. Terenzi: 1967, 'Risultati preliminari dello studio di un deflettore elettrostatico emisferico per la misura del plasma interplanetario a bordo del satellite europeo HEOS-A', *Ric. Sci.* **37**, 393.
[2] A. Bonetti, G. Moreno, S. Cantarano, A. Egidi, R. Marconero, F. Palutan, and G. Pizzella: 1969, 'Solar Wind Observations with ESRO Satellite HEOS-A in December 1968', *Nuovo Cimento* **64B**, 307.
[3] G. Moreno, S. Olbert, and L. Pai: 1966, 'Risultati di IMP-1 sul vento solare', *Quaderni della Ricerca Scientifica*, No. 45.
V. M. Vasyliunas: 1966, 'Observations of Low Energy Electrons with the OGO-A Satellite', Ph.D. Thesis, M.I.T.; 1966, 'Observations of 50 to 2000 eV Electrons with OGO-A', *Trans. A.G.U.* **47**, 142.
S. Olbert: 1968, 'Summary of Experimental Results from M.I.T. Detector on IMP-1', in *Physics of the Magnetosphere* (ed. by R. L. Carovillano), Reidel, Dordrecht, Holland.

Discussion

Hundhausen: During the February 26 event the results you presented show the solar wind momentum flux to jump at the shock and to keep on rising during 20–30 min after. Yet the low latitude magnetometers show the S.C. but no further compressions hereafter, even though the momentum flux of the solar wind is going up. This shows that the magnetosphere is not responding simply to the momentum flux of the solar wind, but something to do with internal currents is already taking place at this time.

Willcox: Based on your observations what is your opinion on the shock driver gas rich of Helium?

Answer: We have seen events where the increase in helium content after the discontinuity was between 2% and 20%.

OBSERVATIONS OF THE SPECTRA AND TIME HISTORY
OF PROTONS IN INTERPLANETARY SPACE,
FEBRUARY 25–28, 1969

E. BAROUCH*, J. ENGELMANN, M. GROS, L. KOCH, and P. MASSE†

Service d'Electronique Physique, CEN-Saclay, France

1. Introduction

During the flare sequence February 24–28, 1969, our laboratory was able to observe the time history of the particle flux close to earth in some detail. We wish to report on results obtained from the Saclay detectors aboard the HEOS A1 and IRIS satellites and aboard two Centaure Rockets launched from Andenes at 1633 and 2350 UT on February 25.

HEOS was launched on December 5, 1968 from Cape Kennedy and the S.E.P. experiment has performed very well since launch time. Data coverage by the ESRO network of space tracking stations in conjunction with NASA is of the order of 95% during the period of interest; however, the experiment was turned off several times for routine checks during this interval. IRIS was launched on May 17 from Cape Kennedy and the S.E.P. experiment has functioned satisfactorily since then. At the present time the on-board tape recorder is no longer in operation and data are only obtained while the satellite is in view of a tracking station.

The S.E.P. experiment R 75 was launched aboard two Centaure rockets in the course of an ESRO campaign investigating P.C.A. events. Both rockets functioned satisfactorily reaching an apogee of 140 km approximately. 222 sec of data above 65 km were obtained in both rocket shots.

2. Experimental Arrangement

Figure 1 shows the arrangement of the detectors in each module.

Each detector consists of an array of two or more silicon diode detectors in coincidence or single mode. In IRIS and R 75 counts from the first detector are sampled and sent to a 128 channel pulse height analyser, and the appropriate energy is then deduced from the energy loss versus incident energy curves. (All the detectors were calibrated with protons generated by various high energy accelerators in Saclay and Orsay.)

In HEOS a single channel pulse height analyser is used in order to define the energy loss: the analysis channel is cyclically taken over the entire range of possible energy losses from α-particles and protons in the first diode. Thus in IRIS and R 75 we obtain the total count number in the coincidence mode, and samples of the energy

---

* Centre National de la Recherche Scientifique.

† Institut National d'Astronomie et de Géophysique.

C. Manno and D. E. Page (eds.), Intercorrelated Satellite Observations Related to Solar Events, 448–459. All Rights Reserved
Vopyright © 1970 by D. Reidel Publishing Company, Dordrecht-Holland

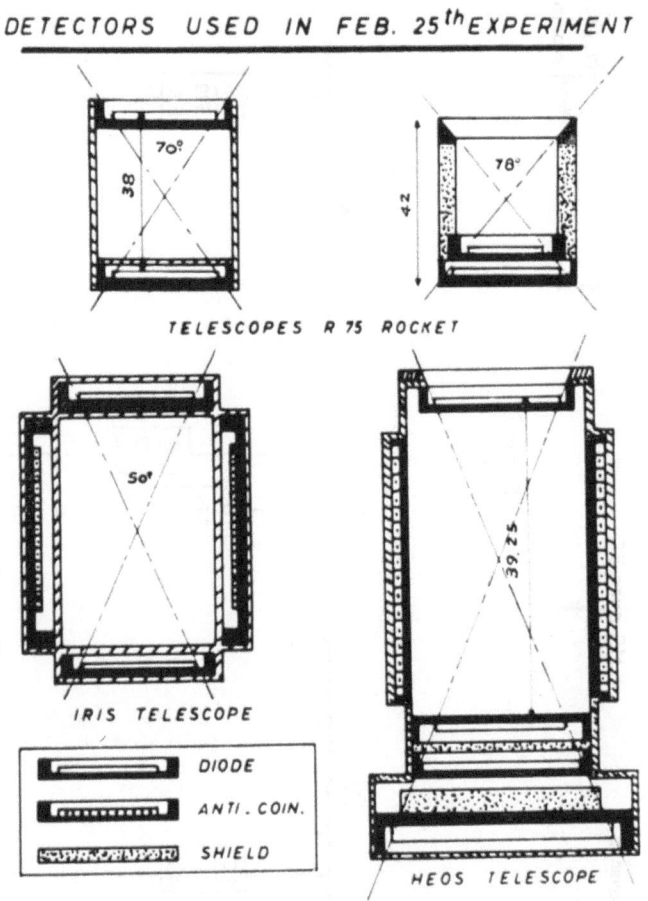

Fig. 1. Arrangement of the Saclay detectors aboard IRIS, HEOS A1 and the R 75 rockets. Each rocket bore two separate telescopes as shown.

spectrum. In R 75 we obtain in addition the total singles count in the first diode, and total singles and coincidence from a second detector, the level of which is biased to permit only particles whose energy loss is above 3 MeV to be detected in the first diode. In HEOS we obtain total counts in each of five energy loss channels together with an indicator of the diodes traversed. Because of the gold screening this arrangement helps discrimination between protons and α-particles. Each of these detectors is described in detail elsewhere.

3. Time Sequence

Table I shows the chronology of the visual and radio observations. The class 2B flare responsible of the proton increase, occurred on February 25, 1969 at N 14 W 36. The optical flare began at 0900 UT, the maximum phase was at 0915 UT. According to the 5-min plots of the neutron monitors of Goose Bay and Deep River, high energy

TABEL I

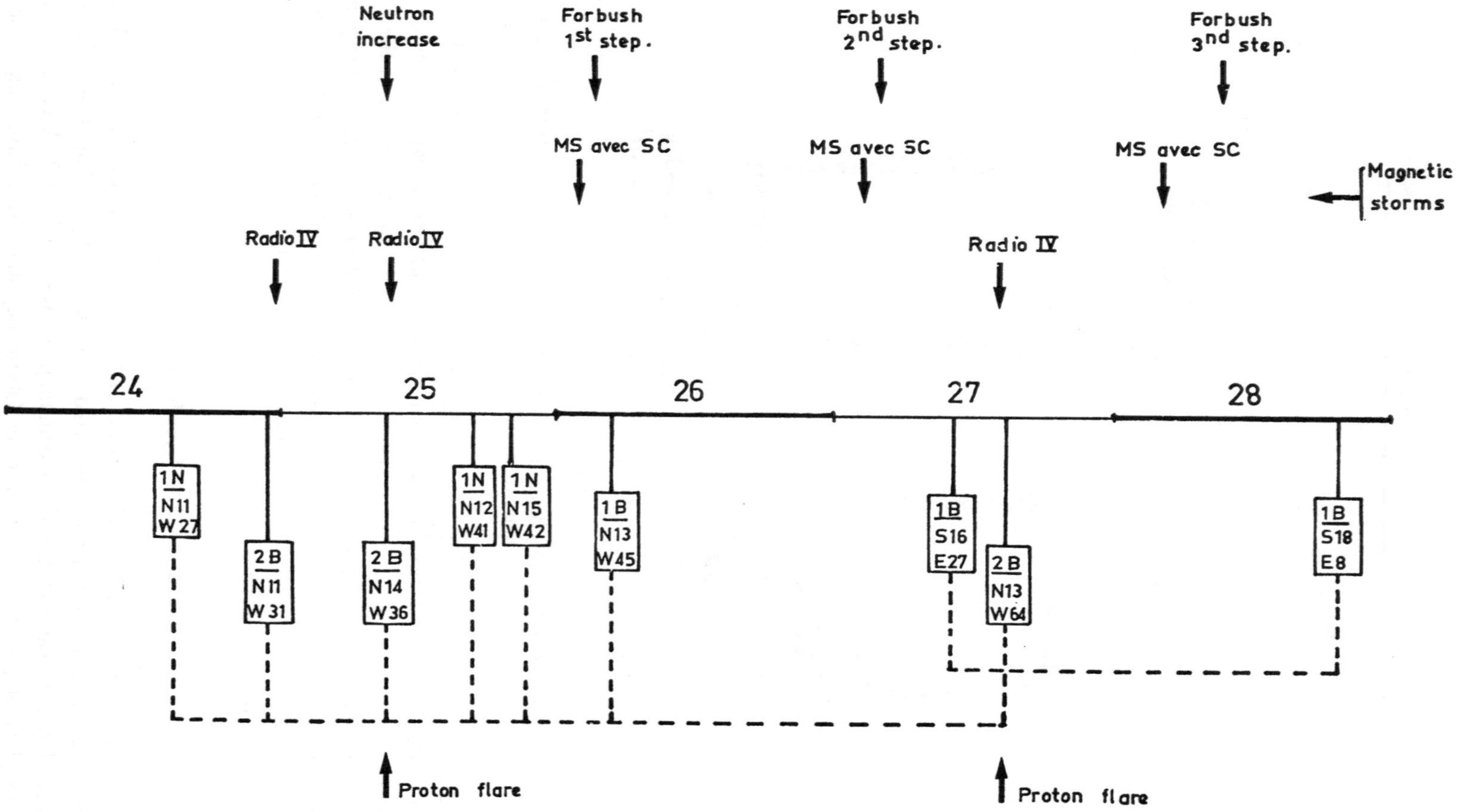

PROTON EVENT FEBRUARY 25 - 27

particles began to arrive at the earth around 0915–0920 UT (*Solar Geophysical Data,*
No. 296) [1]. From this time, we suppose the acceleration took place at the sun at
0906 UT, which corresponds to the maximum phase of the optical flare, taking into
account the travel time of the light to earth.

The history of the event is best depicted in the data from HEOS A. Figure 2 shows
the particle flux as a function of time in three energy windows (analysis could be
performed in 16 windows).

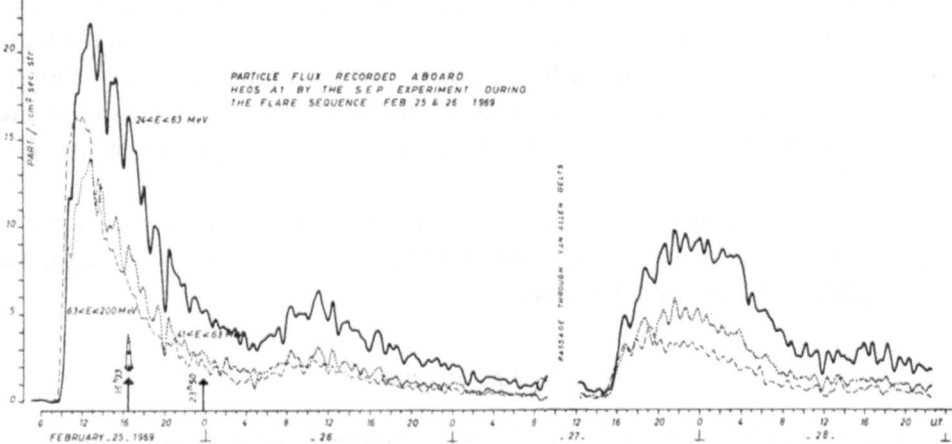

Fig. 2.   Measurements made aboard HEOS of the particle flux in the energy ranges 24–63 MeV,
41–63 MeV and 63–200 MeV vs time. The curves are plotted after smoothing by the computer.

The gross features of the time profile show a large bump after the first peak corresponding to normal diffusion of the particles in interplanetary space. According to
the classification of Lin *et al.* [2] and Kahler [3], we may try to interpret this bump as
being due either to ESP or to halo protons; in fact, as shown by Kahler these two
types of phenomena seem to have the same origin and the differences of structure are
mainly due to the position of the observer with respect to the flare.

We can attempt to picture a model of conditions in interplanetary space fitting our
observations. Table II shows the relationship between the flare sequence and the
magnetic storm sequence that we propose. All the flares included in this picture
arose in the same active region.

TABLE II

Flare No	Importance	Position		Time of flare	SSC	Plasma transit time
1	2B	N 11	W 31	Feb. 24, 2315 UT	Feb. 26, 0158 UT	27 h
2	2B (proton)	N 14	W 36	Feb. 25, 0901 UT	Feb. 27, 0308 UT	42 h
3	1B	N 13	W 45	Feb. 26, 0428 UT	Feb. 28, 0424 UT	48 h
4	2B (proton)	N 13	W 64	Feb. 27, 1415 UT		~ 50 h

The 25th February event was located W 36, and as we are led to believe, was responsible for the SSC on February 27 at 0308. At that time the active region was located W 59, slightly east of the line of force connecting to the earth at quiet time. But this proton event was preceded by a flare 2B on the same active region on February 24 2315 UT and we believe this flare to be responsible for the SSC of February 26 0158 UT. Therefore, the lines of force of the interplanetary field were stretched by this high velocity plasma and the active region of interest is probably connected to the earth when this first SSC occurs.

We believe that the second SSC on February 0308 UT is due to arrival of the shock front of the proton flare of February 25. And we see on the time profile that the bump of low energy particles is located between these two SSC. Therefore we may interpret this bump as being due to the satellite passing through magnetic field lines which are connected close to the flare region on the sun and which are populated by the 'halo protons'.

Studies are in progress in order to determine whether the data are sufficient to yield strong support to either of these hypotheses, 'ESP' or 'halo' as an adequate description of the events in this case.

4. Fine Structure

The time profile shown on Figure 2 displays noticeable fluctuations in time of the particle flux. These fluctuations are quasi periodic and we would like to understand their nature. One can think of several possible explanations a priori:

(1) Instrumental origin, due to a variation in sensitivity of the apparatus.

(2) Instrumental origin due to stroboscopic coupling between the satellite spin rates angle of view of the instrument with respect to the anisotropy of the incoming radiation and the sampling rate.

(3) Instrumental origin due to the data handling process.

(4) Physical origin in fluctuations at the source.

(5) Interaction with the bow shock wave of the earth.

(6) Magnetic field strength modulation by hydromagnetic waves in interplanetary space causing particle trapping between successive crests.

(7) Resonance interaction with hydromagnetic waves causing acceleration of certain groups of particles.

(8) Quasi-periodic filamentary structure of the interplanetary magnetic field.

We have rejected possibilities (1), (2) and (4) from quite simple arguments. Possibility (3) is most convincingly rejected by looking at the raw data. Figure 3 shows the raw count number per 128 sec frame (recording time 23.3 sec per frame) in the 24–63 MeV window.* A smooth curve has been drawn by hand through the data points, and parallel curves at $\pm\sqrt{N}$ and $\pm 2\sqrt{N}$ have been drawn. Figure 4 shows the same data, with the points averaged by pairs. The smoothing technique used to obtain figure 2

---

* This is a 'typical' window, in the sense that it is not the window in which fluctuations show up best or worst.

has been described elsewhere [8]. We think that the fluctuations in the counting rate are convincingly shown up by these graphs.

Possibility (5) is not very likely, due to the position of the satellite with respect to the bow shock. At the time of observation the satellite is several earth radii outside the bow shock.

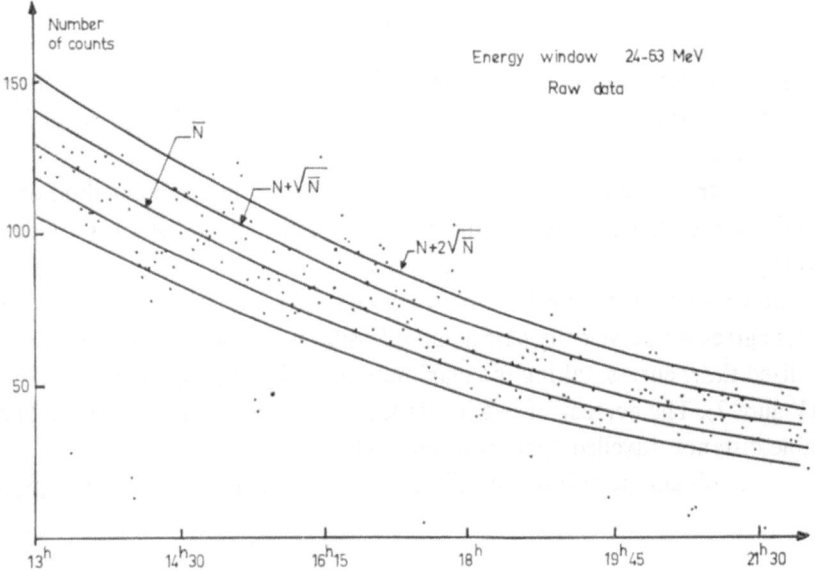

Fig. 3.   Raw data for the 24–63 MeV window. The curves N, to $N \pm \sqrt{N}$ and to $N \pm 2\sqrt{N}$ are fitted by eye to the trend.

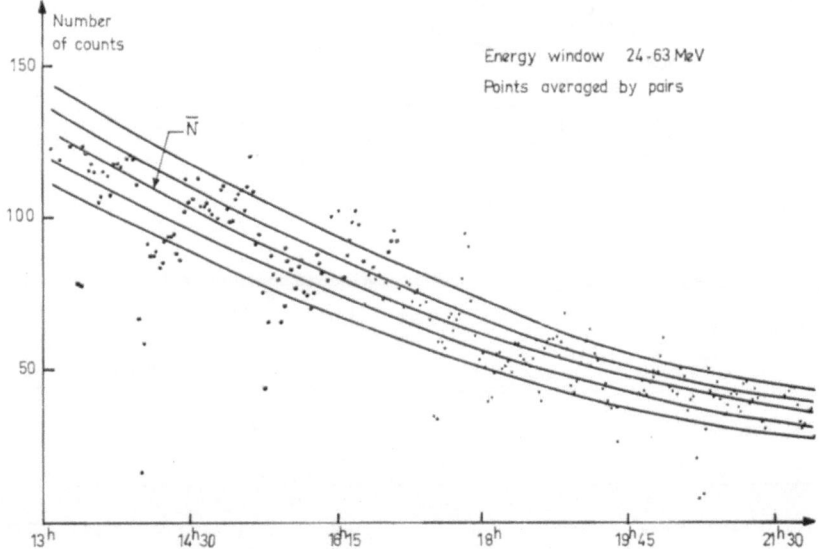

Fig. 4.   As in Figure 3, each point corresponding to the mean value of two adjacent points in the preceding figure. The curves correspond to the mean value N, to $N \pm \sqrt{N}/2$ and to $N \pm 2\sqrt{N}/2$.

We are trying to think out the consequences of possibilities (6), (7) and (8) in order to determine which fits our data best. We like possibility (8) best as it seems to us to explain the energy dependence of the magnitude of the fluctuations and the rather loose correlation with magnetometer data adequately: however we still wish to look at the data in more detail before making a definite statement.

5. Diffusion Phase

From the IRIS data we could build the time profile of 6 energy windows (Figure 5). Each point corresponds to the mean flux value measured during a passage above the North polar cap.

From that graph we deduced the distance profile, following the method of Bryant *et al.* [5] (Figure 6). The injection time was supposed to be at 0906 UT, as discussed previously.

It is apparent that during the first 15 hours after the event the propagation of particles agrees with a velocity dependent diffusion model. For confirmation, we have normalized these curves, taking the maximum intensity of each energy window equal to 1 (Figure 7). The agreement is good at least between 30 and 160 MeV. The most probable distance travelled appears to be 8 AU.

In order to obtain the diffusion coefficient, we tried to apply the Krimigis [9] model,

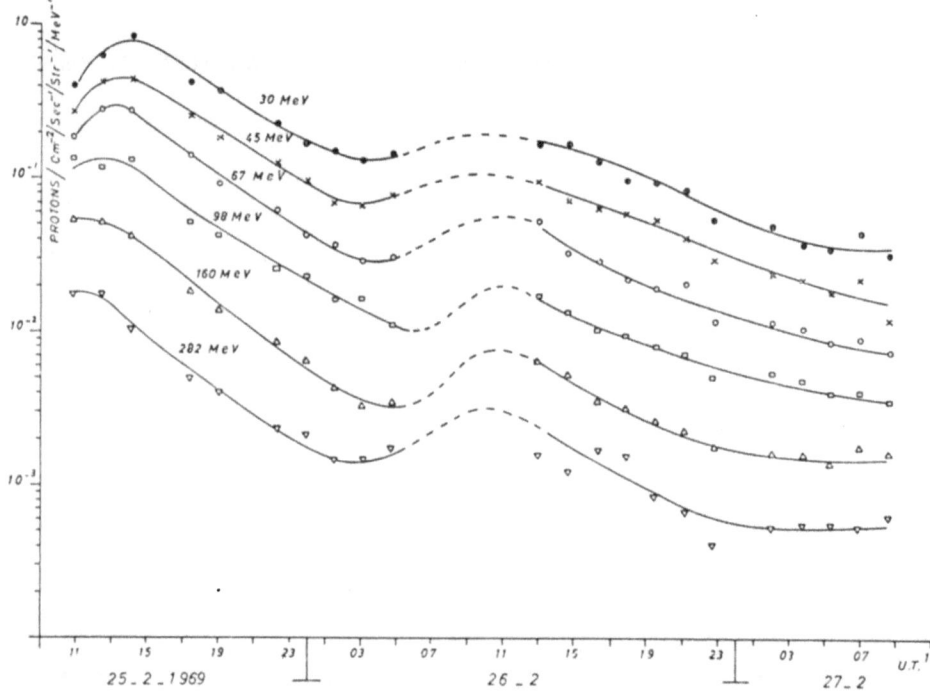

Fig. 5. Time profile of the differential flux measured aboard IRIS for 6 energy values. Each point corresponds to the mean flux value above the North polar cap.

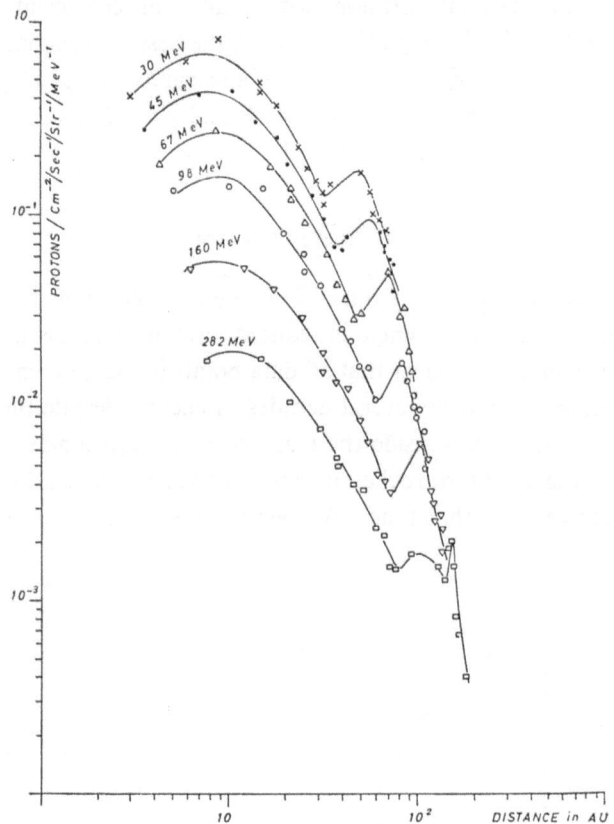

Fig. 6. Distance profile for 6 energy values, drawn from the IRIS data.

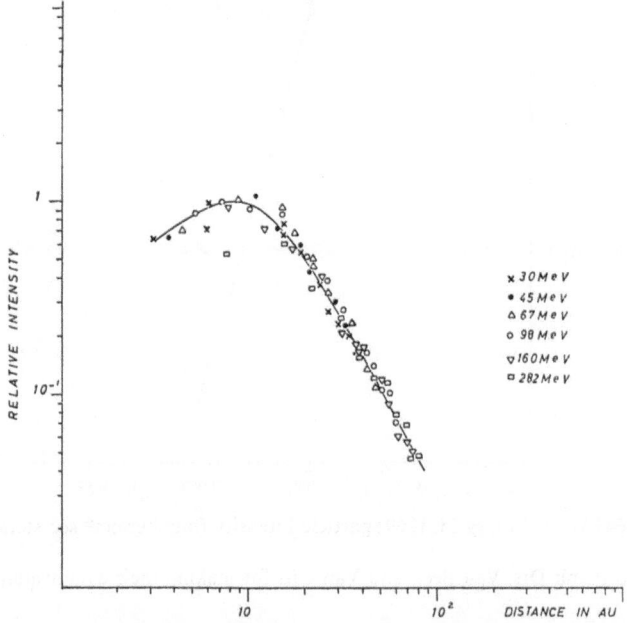

Fig. 7. Distance profile after normalization.

supposing three-dimensional diffusion, with a diffusion coefficient D varying with distance as r^β. When plotting $\log It^{3/2-\beta}$ vs $1/t$, the best approximation to a straight line was found for $\beta = \frac{2}{3}$. From the slope of the curve we could get the diffusion coefficient D. As an example for 30 MeV, we found $D = 0.07\ (AU)^2/h$, and the mean free path at the orbit of the earth for particles between 30 and 100 MeV was found to be 0.12 AU.

6. Energy Spectra

Data from all these detectors and from a similar detector flown by the Utrecht Group*
aboard the same rocket are presented in reduced form in Figures 8 and 9 for the times of the two rocket shots. It is clear that all data points in the differential spectrum fall on a fairly smooth curve over several decades in energy, despite the fact that some of the points are measured outside the magnetosphere and others very well within. (The apogee of the Centaure rocket is only 140 km whereas HEOS A1 was over 160000 km from earth at that time.) We believe this to indicate that for this event

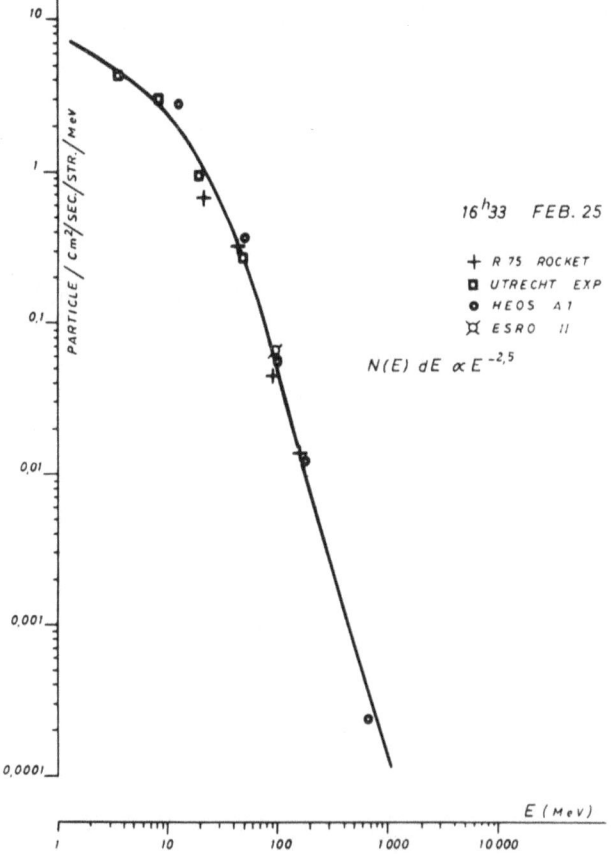

Fig. 8.   At 1643 UT February 25, 1969: particle intensity (number/cm² sec sterad. MeV) vs energy.

* We wish to thank Drs. Van Beek and Van Gils for making their data available to us.

solar cosmic rays could penetrate the magnetosphere quite freely, in agreement with Krimigis *et al.* [11].

The experimental data do not fit a power law in energy or momentum or an exponential law in momentum. If we had chosen data from a single detector this would not have been apparent, and in fact the spectra taken in R 75 over a limited range appear to be a power law with good accuracy (Figure 10). This shows the importance of observing solar flare phenomena with a battery of detectors permitting a broad spectral range to be examined simultaneously.

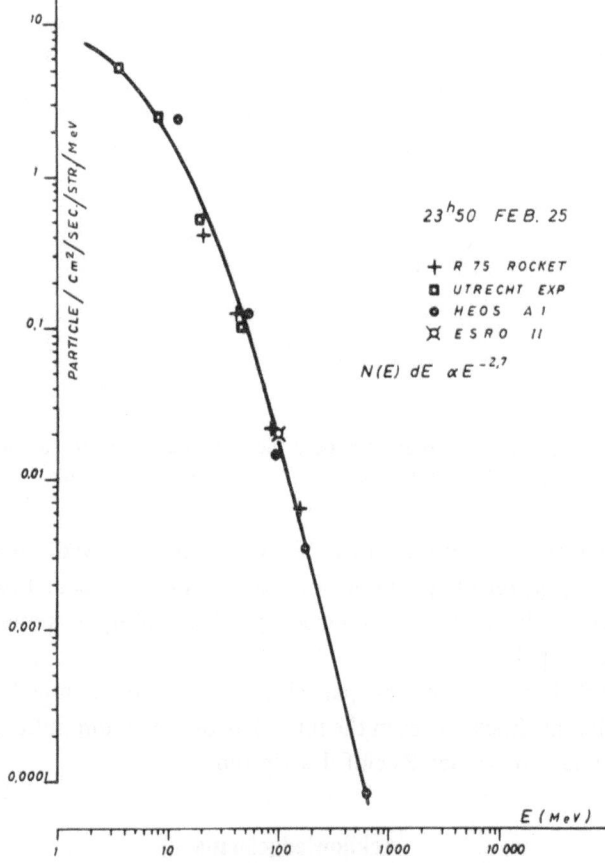

Fig. 9.   At 2350 UT February 25, 1969: particle intensity (number/cm² sec sterad. MeV) vs energy.

7. Conclusion

To summarize, we believe the following conclusions can be drawn from our data:

(1) Passage of solar cosmic rays into the magnetosphere was unhindered down to 2 MeV particles in this event, over the polar cap.

(2) Although falling on a smooth curve, the intensity spectrum does not follow a simple power or exponential law.

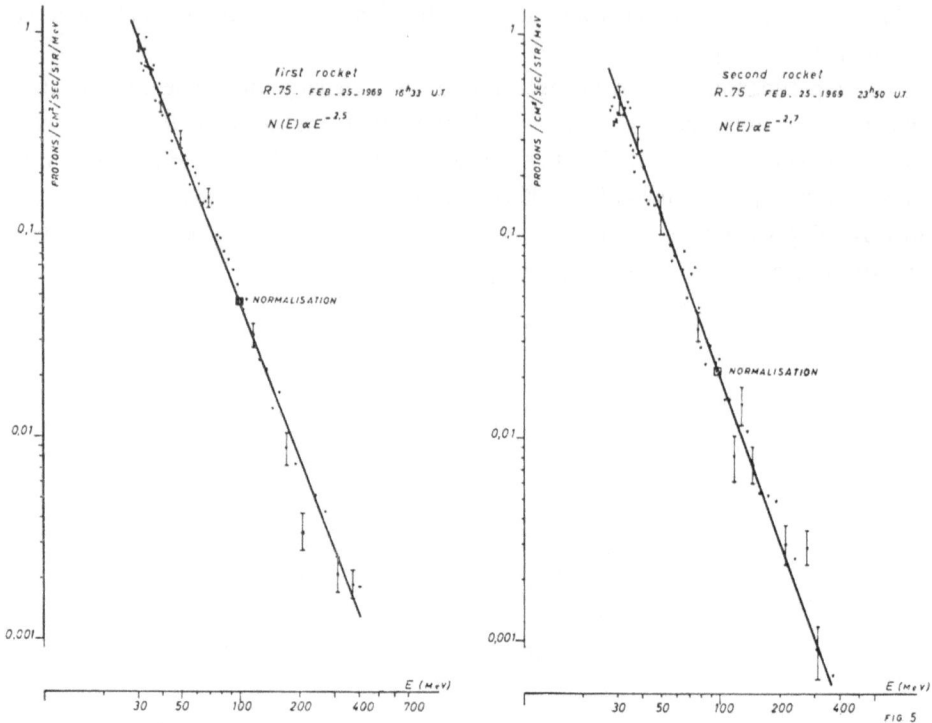

Fig. 10. Differential spectra observed aboard the R 75 rockets. These sampled spectra are normalized in this figure to provide an intensity scale.

(3) Fluctuations in intensity from 24–200 MeV were observed during the decrease of the event. To interpret the phenomenon, further work is needed to select between several hypotheses, the most probable being a filamentary structure of the interplanetary magnetic field.

(4) The overall behaviour of the particles flux can be explained by the detailed study of the different shock waves in the interplanetary medium but is not inconsistent with the Lin-Anderson-Kahler-Roelof description.

Acknowledgements

It is a pleasure to acknowledge contributions to this work made by Jacques Labeyrie, head of the Service d'Electronique Physique, Paul Keirle and Yves Amram, project engineer for HEOS and IRIS respectively, Celestine Jehanno and Claude Julliot, who fabricated and tested the silicon detectors, Georges Labade who built the mechanical structure.

We express our gratitude to the ESRO teams of IRIS and HEOS satellites and of C 49 rockets.

This work was supported in part by the Centre National d'Etudes Spatiales under contract 68/CNES/288.

References

[1] *Solar Geophysical Data*, 1969, Nos. 295, 296.
[2] R. P. Lin, S. W. Kahler, and E. C. Roelof: 1968, *Solar Phys.* **4**, 338–360.
[3] S. W. Kahler: 1969, *Solar Phys.* **8**, 166–185.
[4] D. J. Williams: 1969, International Seminar on Physics of Interplanetary Space, Leningrad.
[5] D. A. Bryant, T. L. Cline, U. D. Desai, and F. B. McDonald: 1965, *Astrophys. J.* **141**, 478.
[6] C. Y. Fan, J. E. Lamport, J. A. Simpson, and D. R. Smith: 1966, *J. Geophys. Res.* **71**, 3289.
[7] W. C. Bartley, R. P. Bukata, K. G. McCracken, and U. R. Rao: 1966, *J. Geophys. Res.* **71**, 3297.
[8] E. Barouch: 1965, *J. Phys.* **26**, 241A.
[9] S. M. Krimigis: 1965, *J. Geophys. Res.* **70**, 2943.
[10] H. F. Van Beek and J. W. Van Gils: 1969, Cospar Symposium, Prague.
[11] S. M. Krimigis, J. A. Van Allen, and T. P. Amstrong: 1967, *Phys. Rev. Letters* **18**, 1204.

Discussion

Lanzerotti: You talked about the model of Lin and Kähler, for the tail of protons. In general, the model would predict that at lower energies the peak should get larger and larger, if the analogy to energetic storm particles is correct. Is that what you have seen? It is in fact unusual to find a peak at such high energies, because the data of Kähler and Lin generally don't show that.

Engelmann: There are two or three peaks. These may be due to the tail of protons or to the third flare. We cannot make a decision because of the few data so far analysed.

It would certainly be interesting to explain this large bump at high energies.

Wibberenz: Another explanation is possible. A shock front would be very effective in keeping stored particles which have been accelerated in the same flare of the ones which populate the main peak.

Roederer: I would like to draw your attention on some papers published about ten years ago and referring to some important solar flares, like the one of July 1961, where neutron monitors data show the same features as the ones you describe.

SOLAR COSMIC RAY INCREASE ON
25 FEBRUARY – 5 MARCH, 1969 ACCORDING TO
MEASUREMENTS FROM VENUS-5 AND VENUS-6 SPACE PROBES

S. N. VERNOV, N. N. KONTOR, G. P. LYUBIMOV,
N. V. PERESLEGINA, and E. A. CHUCHKOV

Moscow State University, Moscow, U.S.S.R.

Abstract. During the period from 25 Feb. to 5 March, 1969 the Venus-5 and Venus-6 space probes detected an increase in solar cosmic ray protons with $E_p > 1$ MeV and electrons with $E_e > 0.05$ MeV. The increase had a complex form and was due to a series of solar flares occurring in a single active region. The main flares occurred on 24 and 27 Feb. 1969. A scheme for the development of a disturbance in the interplanetary medium caused by the above mentioned flares is considered on the basis of the analysis of the form of increase for various proton energies, anisotropy of protons with $E_p = 1 \div 4$ MeV, and the increase in electrons with $E_e > 0.05$ MeV together with solar and geophysical events.

1. Introduction

The method of investigation used below is based upon consideration of low-energy cosmic ray interaction with quasistationary and moving magnetic fields in the interplanetary medium. This method is a generalization of the results of the experimental study of solar cosmic ray bursts which were observed by the Zond-3, Venus-2, 3 and Venus-4 space probes which flew in 1965–67. The results are set forth in detail in [1–8].

Low-energy solar cosmic ray propagation characterizes the state of the interplanetary medium within the region of propagation. Since the magnetic field lines are, on the average, in the form of an Archimedes spiral form, the particles will move along these lines at velocities determined by their energies and pitch-angles in the absence of magnetic inhomogeneities the size of which are smaller than or comparable to Larmour radius. In this case the region of propagation of such particles will be determined by their initial distribution over the magnetic field lines near the sun and the specific form of these spiral lines in interplanetary space. In cases of such propagation occurring we have not been able to observe particles which were produced in the flares which were not connected with the observation point by a field line. However, increases in the 1–5 MeV proton intensity relating to flares occurring on the eastern part of solar disk were experimentally observed.

Probably, each flare generates protons which propagate in two ways: (1) a portion of the particles having a high velocity and strong anisotropy move along field lines which are not yet disturbed by the shock wave and have an Archimedes spiral form; (2) quasistationary particles, the flux of which is almost anisotropic, propagate with a low velocity behind the shock wave.

Observation of one or the other type of particle propagation depends on the location of the flare on the sun with respect to the detector, i.e. on the longitude of the flare with respect to the visible central meridian of the sun [4–6].

Consider first the second mode of particle propagation which is the more difficult to understand, i.e. the transfer with the shock wave velocity. In the case of this mode of propagation, solar cosmic rays are observed (at a distance of about 1 AU) about 1–3 days after the flare on the sun. In such cases the anisotropy of the proton flux with $E_p = 1 \div 5$ MeV does not usually exceed some 20%. Simultaneously with the appearance of the solar cosmic rays, a Forbush decrease in galactic cosmic rays, increases in strength and changes to the direction of magnetic field [9], and an increase in the solar plasma stream [10] were observed. These events corresponded to the arrival of the shock wave and plasma from the flare.

In accordance with Parker's model for shock wave propagation from solar flares [11], consider three space regions: '1' – in front of the shock wave front, '2' – between the wave front and plasma front; '3' – between the plasma front and sun.

In region '1', which corresponds to the undisturbed medium, the galactic or solar cosmic ray particles accelerated at the shock front which remain from previous flares or which constitute quiet quasistationary fluxes, or the particles which have passed through the front may be observed. These particles may form a peak before the beginning of the Forbush decrease.

In region '2' which corresponds to the fall phase of the Forbush decrease of the galactic cosmic rays, i.e. which in other words, comprises a magnetic field of increased strength which is compressed and bent between the shock front and the plasma front, localized solar cosmic rays generated by the flare but retarded in this space region are observed. These particles propagate together with layer '2'. When propagating and moving along a bend in the magnetic field line the solar cosmic rays may partially pass through the bend, i.e. stream from the localization layer and replenish regions '1' and '3'. Thus, the solar proton anisotropy may be fairly high along the field direction and low perpendicular to the field, i.e. along the direction of the shock front propagation. The particles which leave layer '2', or have been injected by the next flare or derive from steady streaming from active regions, reside in region '3' which corresponds to the recovery phase of the galactic cosmic ray intensity. The first two sources, i.e. the particles which leave the layer or which are injected by the next flare should be characterized by a two-direction anisotropy due to particle movement between the layer and sun. The characteristic feature of solar cosmic rays streaming for a long time from active region is a positive anisotropy due to the action of the source.

We turn now to the description of the direct particle movement along the magnetic field lines. In this case we observe solar cosmic rays several hours after the flare on the sun. The particles arrive mainly from the sun, anisotropy is close to 100%. In this case the level of galactic cosmic rays does not undergo a very considerable decrease. In some cases a delayed particle flux returning to the sun with respect to the direct flux is observed at the initial time. This fact is indicative of the presence of some peculiar 'mirror' at a distance of about 1.5 AU from the sun. It should be noted that in other cases a small return flux is observed simultaneously with the direct one which is indicative of the reflection and scattering of particles from a number of distributed and semitransparent 'mirrors', i.e. of the presence of a number of inhomogeneities on magnetic field lines near the detector. The maximum value of anisotropy for the case of direct arrival of solar protons occurs at the time of the intensity increase. However, the anisotropy remains high (higher than 60%) for 16–20 hours. According to calculations which use Monte-Carlo methods [7, 8, 12] such a high value of anisotropy is possible only at the beginning of a flare. Therefore, from the point of view of the authors of [12] the $1 \div 5$ MeV proton injection is not instantaneous. The author of [6] believes that prolonged observation of particles may be caused by replenishment of the flux by a secondary source of particles which is retarded behind the shock front, but it is difficult too to exclude proton generation by the shock wave [8, 13]. During a flare on the sun the particles populate quite a large beam of magnetic field lines. The corotation of the beam of magnetic field lines with the sun with respect to the observation point corresponds to the passage of field lines with various particle populations through the detector. Thus, if particle leakage with time (and that due either to prolonged injection from the sun, or due to the second source behind the shock front) is neglected the intensity profile will be determined by the particle distribution over the spiral flux cross-section.

When a flare is located in the western hemisphere of the sun, approximately between the central meridian and 60°, one can observe the arrival of solar cosmic rays at a given point of space due to both the first and second modes of propagation, i.e. two $1 \div 5$ MeV proton peaks may be distinguished. The presence of the first peak corresponds to the arrival at the detector (due to corotation with the sun) of the central field line with maximum population, and the second peak corresponds to the arrival of solar cosmic rays localized behind the shock front and it is observed together with the intensity decay phase of the Forbush decrease in the galactic cosmic rays.

Evaluations of proton energy spectra in the $1 \div 30$ MeV range for the two cases of direct arrival of the particles and of particle transfer behind the shock front reveal substantial differences [5, 6, 8, 14]. The spectra for protons localized behind the shock fronts are softer than those for direct particle arrival since the spectrum becomes softer and is stabilized after formation of the region of solar cosmic ray localization.

For the case of direct particle arrival during the intensity increase period the exponent of the integral spectrum is also increased, i.e. the spectrum becomes softer. This is connected with the fact that protons of lower energies arrive later at the observation point, i.e. the particles are divided in space in accordance with the differences of their natural velocities determined by kinetic energy.

2. Increase in Solar Cosmic Rays from 25 February to 5 March, 1969

During the Venus-5 and Venus-6 flight in January–May, 1969, solar and galactic cosmic rays were measured using equipment similar to that installed on the Zond-3 and the Venus-2, 3 and 4 space probes. Junction detectors measured the $1 \div 4$ MeV proton fluxes streaming from the sun, to the sun and at an angle of 90° to the space probe – sun line and $4 \div 12$ MeV proton fluxes streaming to and from the sun. Window gas-discharge counters were used to detect electrons with $E_e > 0.05$ MeV and protons with $E_p > 1$ MeV between the directions of 90° to the space probe – sun line. Gas-discharge counters installed inside the space probes measured protons with $E_p > 30$ MeV and electrons with $E_e > 2$ MeV.

Considered below in detail is the increase in solar cosmic rays from 25 February to 5 March, 1969 which is shown in Figure 1. Shown in the left part of Figure 1 as a function of time (UT) are: '1' – the intensity of protons with $E_p > 30$ MeV (log-linear scale); '2' – the intensity of electrons with $E_e > 0.05$ MeV; '3' – the intensity of protons with $E_p = 4 \div 12$ MeV; '4' – the intensity of protons with $E_p = 1 \div 4$ MeV; above which is shown the anisotropy $(A\%)$ of the proton flux for $E_p = 1 \div 4$ MeV. Below is shown hourly data from the neutron monitor at Sulphur, two groups of chromospheric proton eruptions (vertical shadowing) and three-hourly values of the K_p-index of geomagnetic disturbance with marks to show sudden pulses. The characteristics of cosmic rays obtained from Venus-5, 6 (curves '1', '2', '3', '4', and $A\%$) are represented in the form of 4-hour average values; greater averaging intervals are marked (limited) by vertical shadowing. The anisotropy $A\%$ was calculated from the relation $A\% = n_p^+ - n_p^+ / n_p^+ + n_p^- \cdot 100\%$ where n_p^+ and n_p^- are the proton fluxes with $E_p = 1 \div 4$ MeV streaming from and to the sun respectively. The intensity of electrons with $E_e > 0.05$ MeV was determined using the difference method; by subtracting the values measured with the junction proton detectors and the gas-discharge counter installed inside the space probes from the readings of the window gas-discharge counter taking into account the corresponding geometrical factors. It should be noted that, due to a comparatively large inaccuracy in the difference method for determining the electron flux this characteristic is only approximate.

Shown in the centre and to the right of Figure 1 looking in the projection of the ecliptic plane from its northern pole are schemes of the mutual positions of the sun, earth, Venus-5, and Venus-6; and the directions of the solar flares (arrows with indication of coordinates) which are labelled with the numbers 1a and 2 in the left part of Figure 1 and in Table I, and the Archimedes spirals for a 350 km/sec solar wind velocity drawn from the active region on the sun in which the flares occurred in the night from 24 to 25 February and on 27 February, 1969. On these days Venus-5 and 6 were at about 140×10^6 km mean distance from the sun at an angle of about 1 °E to

TABLE I

Solar flare						Radioemission		Maximum
No. in Figure 1	Date in Feb., 1969	Beginning, h	m	Coordinates: latitude, longitude	Class	Maximum flux[a]	Spectrum, type	X-ray flux[b]
Ia	24	23	00	11N 31W	2B	580	II, IV	1400
Ib	25	09	00	14N 38W	2B		IV	
Ic	25	16	47	13N 41W	− N	144	IV	
Id	25	19	38	15N 42W	IN	158	IV	
Ie	26	04	18	13N 45W	2B		II, cont.	5000
2	27	13	48	13N 65W	2B	810	II, IV	5000

[a] $\times 10^{-22}$ W m^{-2} cps^{-1} at a 2800 Mcps.
[b] $\times 10^{-4}$ erg cm^{-2} sec^{-1} at a $1 \div 8$ Å wavelength.

the sun-earth line. The space probes were about 2×10^6 km apart in the radial direction from the sun.

Shown with dots in the right lower part of Figure 1 is the time dependence of the intensity of protons with $E_p > 30$ MeV (in the decrease section and after subtracting the background level) plotted on a log-log scale. The Venus-6 data beginning with the peaks of two intensity increases of protons with $E_p > 30$ MeV are used in this plot. Time

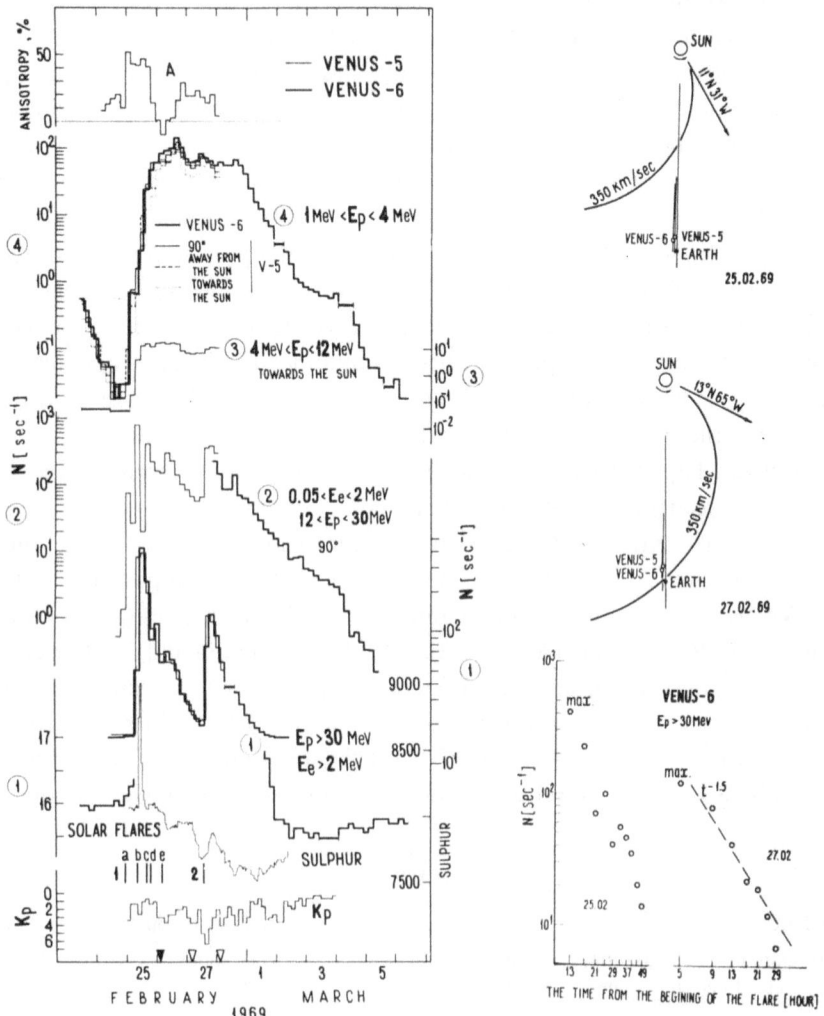

Fig. 1.   Increase in solar cosmic rays on 25 February – 5 March, 1969 (at the left) according to Venus-5 and 6 data together with solar and geophysical data: '1' – $E_p > 30$ MeV, '2' – $E_e > 0.05$ MeV, '3' – $E_p = 4 \div 12$ MeV, '4' – $E_p = 1 \div 4$ MeV, $A\%$ – anisotropy of protons with $E_p = 1 \div 4$ MeV; neutron monitor data (Sulphur); moments of solar flares (1a, b, c, d, e, 2), K_p-index, S.C. schemes of location (top at the right and in the center) of sun, earth, Venus-5, and Venus-6; directions of flares and magnetic force line from the region of flare for 25 and 27 February, 1969. Intensity of protons with $E_p > 30$ MeV (bottom at the right) according to Venus-6 data as a function of time elapsed from the moment of flare on the sun (log-log scale).

is counted from the instants of commencement of the chromospheric eruptions 1a and 2 (Table I) respectively. For comparison the dependence $t^{-1.5}$ is plotted with a dashed line.

Beside the Venus-5 and 6 data the solar and geophysical data published in [15, 16] are used in Figure 1, in Table I, and in the text.

The increase on 25 February, 1969 – 5 March, 1969 is of complex form and has a long duration. These features are connected, firstly with the fact that the increase was cuased by a series of 6 powerful proton flares occurring from 24 to 27 February, 1969 in a single active region (No. 9946) during the passage of this region across the solar disk; secondly, with the superposition of an additional solar cosmic ray flux on the decrease in the intensity of protons with $E_p = 1 \div 4$ MeV and electrons with $E_e > 0.05$ MeV with its peak on about the 3rd March 1969. However, the protons with $E_p > 30$ MeV (curve '1') show two very well separated narrow peaks; two broader increases in protons with $E_p = 4 \div 12$ MeV can be clearly seen (curve '3'); and two groups of electrons with $E_e > 0.05$ MeV consisting of several peaks may be isolated on curve '2'. These two main increases on about the 25th and toward the end of the 27th February, 1969 are separated on curves '1', '2', '3', with a very distinct broad intensity decrease toward the end of the 26th and the beginning of the 27th February, 1969. Such separation corresponds well to two groups of solar flares which are also separated from each other with an interval of about 1.5 day (Figure 1, Table I). It is very difficult to see a separation for the protons with $E_p = 1 \div 4$ MeV (curve '4'). However, the fact that the low energy protons can be easily transferred behind the shock waves helps to separate a single broad increase (about 3 days) and to match individual small peaks to solar flares in accordance with marks indicating sudden commencements (SC). Figure 1 shows three SC: on 26 Feb. 1969 at 01^h58^m, 27 Feb. 1969 at 03^h08^m and 28 Feb. 1969 on 04^h24^m. In Table I the flares Nos. 1a, e, 2 were accompanied by radioemission of type II the presence of which is usually connected with plasma ejection during a flare and with shock wave formation. If one considers that the above mentioned SC correspond to these flares the mean velocities of the shock waves will be 1550, 1810, and 2870 km/sec respectively. Taking into account the high values of the calculated velocities, the distance between the earth and Venus-5 and 6 (about 10^7 km) and the time resolution (4 hours) on the curves examined, one may consider the indications of SC on the earth as the approximate times of passage of the shock wave through the orbits of Venus-5 and 6. This conclusion is confirmed by a considerable particle increase and by the corresponding structure of the stream of the solar interplanetary plasma which is outlined by the measurements made on Venus-5 and 6 which are given by Gringauz et al. (this volume, p. 130). Then the highest peak on the 26th Feb., 1969 in graph '4' will correspond to solar cosmic ray particles from flares Nos. 1a, b, c and 1d retarded behind the front of the first shock wave; the next, smaller, peak on the 27th Feb., 1969 will correspond to protons from flare No. 1e behind the front of its shock wave; and the even smaller peak on the 28th Feb. 1969 will correspond to particles retarded behind the shock wave from flare No. 2. A narrow minimum of 4 hours duration corresponding to the separation of particles from flares of group No. 1 and flare No. 2 is observed between the pair of the first two peaks and the third one from the 27th to the 28th Feb., 1969. The direct arrival of protons with $E_p = 1 \div 4$ MeV can be distinguished only for the first flare (No. 1a) and for the last flare (No. 2). These times are marked with a small tooth at the front edge of the intensity increase (dotted curve '4') at about 18^h00^m on the 25th Feb. 1969 and a small bump on the flat part of curve '4' (thick line) at about 06^h00^m on the 28th Feb. 1969 (after the minimum).

Now it is possible to describe the whole increase in the proton flux with $E_p = 1 \div 4$ MeV (curve '4') in connection with the flares (Table I).

Within 3 ± 2 hours after the beginning of flare No. 1a an exponential increase in the intensity of protons arriving from the sun having a time constant $\tau_R = 2.8$ hours began and lasted for 16 ± 2 hours. An abrupt increase in anisotropy ($A\%$) with a peak of more than 50% was seen to correspond to the beginning of detection of these particles. After this the anisotropy slowly decreased.

This phase of the development of the low-energy solar cosmic ray burst is determined by the following factors: the particle population within the magnetic field line beam within an angle of about $\pm 30°$ with respect to the direction of the burst; the direct propagation of these particles with a velocity corresponding to their kinetic energy (and pitch-angle) along field lines; the corotation of this spiral flux with the sun; and the approach of the central channel to the detector in the presence of a long acting source of solar cosmic rays (scheme for 25 Feb. 1969 in Figure 1). After this the intensity increase stops (for about 4 hours) and then continues again (for about 20 ± 2 hours), following approximately an exponential law with $\tau_R = 12.6$ hours and reaches a peak value of about 10^3 particle $cm^{-2} sec^{-1} ster^{-1}$ for protons with $E_p = 1 \div 4$ MeV arriving from the sun. The instant of the decrease in the steepness of the intensity increase corresponds to an abrupt fall in the anisotropy, an increase in the plasma stream (see the report by Gringauz et al., this volume, p. 130) and the sudden commencement of the magnetic storm on the earth on the 26th Feb. 1969. During this period of increase and the subsequent very slow decrease in the intensity (with the secondary peak described above) down to the minimum at 01 ± 2^h on the 28th Feb. 1969 the value of the anisotropy fluctuates between approximately 0 and 20%.

This phase of the development of the solar cosmic ray burst is indicated by the arrival of shock waves from the first group of flares at the detector and by the presence of solar cosmic rays retarded behind and between the fronts.

The last section of the curve contains the two small peaks at 06 ± 2^h and 18 ± 2^h on the 28th Feb. 1969 mentioned above; the first of them being connected with the direct arrival of solar cosmic rays from flare No. 2 (more accurately, probably, these particles are further accelerated at the front edge of the shock wave from this flare) and the second one being connected with the protons localized behind the front of the shock wave from this flare. In addition it shows the exponential intensity decay for the whole increase which was violated on the 3–4th March, 1969 by an additional solar cosmic ray flux which is not analyzed in the present work. The time constant of the decrease for the $1 \div 4$ MeV protons for directions perpendicular to the sun-detector line is $\tau_D = 11.7$ hours over about 48 hours.

We shall not linger on the description of curve '3' for the $4 \div 12$ MeV proton intensity since only a limited section of this curve is available for particles streaming toward the sun and the protons of these energies have propagation characteristics intermediate between those of movement along force lines and those of the diffusion mechanism. Note only that the first increase in protons of these energies caused by the first group of flares is not so prolonged as that for the 1–4 MeV protons and its peak at 06 ± 2^h on the 26th Feb. 1969 corresponds, probably, to the particles behind the front of the first shock wave and the maximum flux is about 2×10^2 particle cm^{-2} sec^{-1}ster^{-1}.

Examine curve '1' in Figure 1, i.e. the plot of intensity of protons with $E_p > 30$ MeV. As was noted above two distinct peaks of these particles can be easily identified in accordance with the occurrence of the two groups of solar flares. The long duration of the first peak is probably connected with the number and total duration of all the flares in the first group. Each increase consists of a peak of the diffusion wave type and an additional peak due to particles transferred behind the shock fronts. The presence of particles transferred behind the shock fronts results in not only an increase in the total duration of the intensity decrease but also in a slower character of this decrease. These features are clearly seen in plot '1' of Figure 1 from the times of sudden commencements and Forbush decreases (on the curve for the Sulphur neutron monitor) and to the right part of Figure 1 where the disagreement between the intensity decrease law $t^{-1.5}$ and the experimental points can be seen.

The first peak is observed at about 11^h on 25 Feb. 1969 and is delayed by about 12 hours with respect to flare No. 1a. The second peak is observed at about 18^h on 27 Feb. 1969 and is delayed by about 5 hours with respect to flare No. 2. The moments of the peaks have been selected taking account of the Venus-5 and 6 data. On the basis of the delays found, the velocities and diffusion coefficients are, respectively: $V_{dif}(1) = 3200$ km/sec, $Q(1) = 0.8 \times 10^{21}$ cm^2 sec^{-1} and $V_{dif}(2) = 7800$ km/sec, $Q(2) = 1.8 \times 10^{21}$ cm^2 sec^{-1} for protons with $E_p > 30$ MeV. The results presented show that the diffusion wave of protons with $E_p > 30$ MeV propagates faster from flare No. 2 due to the fact that the diffusion for this flare is mainly along the magnetic field (see schemes for 25 and 27 Feb. 1969 in Figure 1). The maximum proton fluxes with $E_p > 30$ MeV for the first and second increases are ~ 30 particle cm^{-2} sec^{-1} ster^{-1} and 10 part cm^{-2} sec^{-1} ster^{-1} respectively. Ground neutron monitors detected the solar cosmic ray burst on 25 Feb. 1969 with only a 15% maximum increase in cosmic rays over the background level [15]. The velocity and diffusion coefficients calculated on the basis of a 10.5 hour delay of the peak of the cosmic ray burst detected by the neutron monitors are $V_{dif}(1) = 4000$ km/sec, $Q(1) = 10^{21}$ cm^2 sec^{-1}. To complete the description of the increase of 25 Feb. – 5 March, 1969 we shall linger on the features of the electron flux with $E_e > 0.05$ MeV (curve '2' of Figure 1).

The maximum electron flux was observed at 10 ± 2^h on 25 Feb. 1969 and was about 6×10^3 particles cm^{-2} sec^{-1} ster^{-1} from the direction perpendicular to the sun-detector line. As was noted above, the whole increase is divided into two individual increases, however, in contrast to curve '4' for protons with $E_p = 1 \div 4$ MeV, each of them is strongly jagged. Thus, four peaks are pronouncedly distinguishable in the first increase from 25 to 26 Feb. 1962 and two peaks can be seen in the second increase. The times of solar flares correlate well with the first five peaks and in the majority of cases they are within a 4-hour duration of the peak (for flares Nos. 1b, c, d); flare No. 1 begins 1 hour earlier, flare No. 2 is 1.5 hour and flare No. 1e is 4 hours earlier. The rapid arrival of particles after the flare is probably connected with the high velocity of electrons and the very small Larmour radius. These features ensure a high

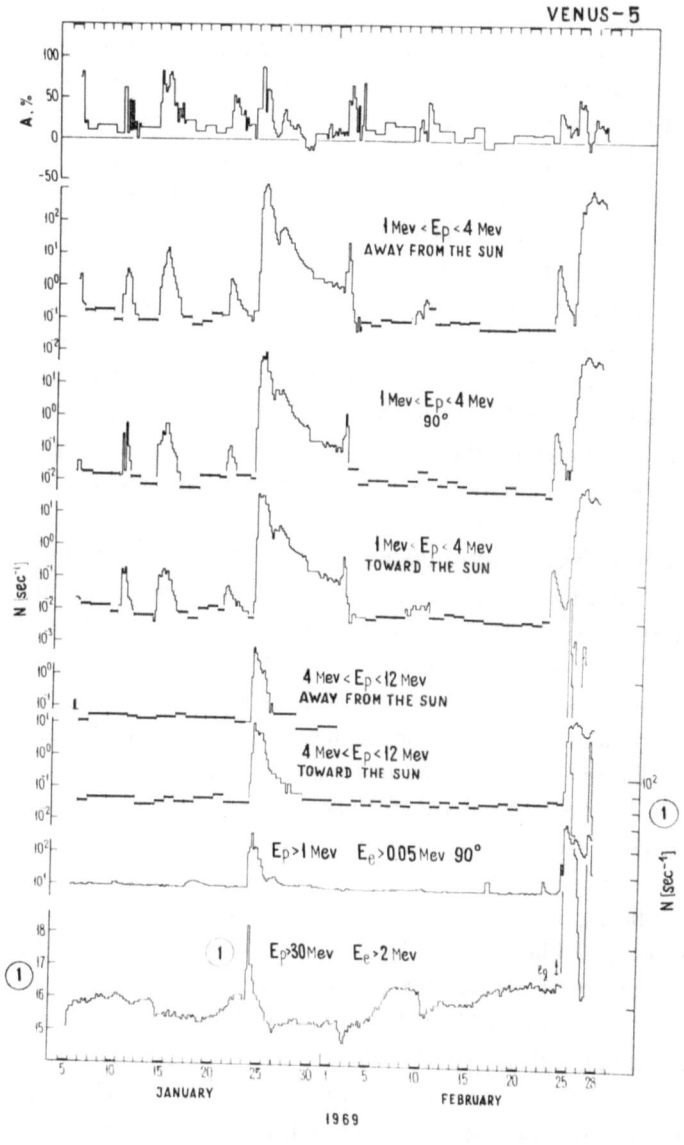

Fig. 2.   Solar cosmic ray intensity in the period from 5 January to 28 February, 1969 measured on Venus-5 and anisotropy (A %) of the proton flux with $E_p = 1 \div 4$ MeV (upper curve).

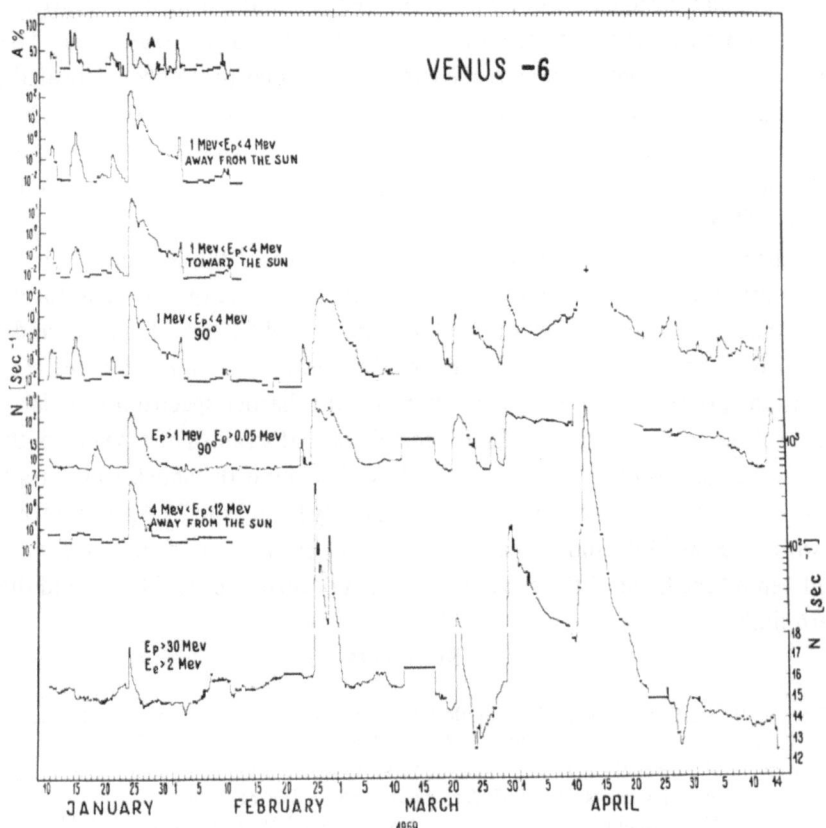

Fig. 3.   Solar cosmic ray intensity in the period from 10 January to 14 May, 1969 measured from Venus-6 and anisotropy ($A\%$) of proton flux with $E_p = 1 \div 4$ MeV (upper curve).

mobility of the electrons, i.e. the capability of fast propagation from the injection points with passage through bends in magnetic field line and, in particular, through shock fronts.

Increasing duration and extension of the rear front may be noted for the entire sequence of peaks. This effect may be explained by the accumulation of electrons from the sequence of bursts. However, it is necessary to assume, taking account of their high mobility, either the closeness of the magnetic field lines in the observation region or a very large and prolonged electron flux from the sun. The electron intensity decrease from 1 to 2 March, 1969 is exponential, with $\tau_D = 18.6$ hours, i.e. the time constant for electrons with $E_e > 0.05$ MeV is 1.6 time as great as that for the protons with $E_p = 1 \div 4$ MeV which may be also indicative of the formation of a large magnetic trap as a result of the action of all of the flares.

3. Conclusion

The analysis in the previous part of this report has shown that the increase in the low-energy solar cosmic rays observed from Venus-5 and 6 from 25 Feb. to 5 March, 1969 is a complex burst of solar cosmic rays within a broad energy range. The complexity of its form is determined by the effects of six solar proton flares located on the Western part of the solar disk and by the combination of the direct arrival of solar cosmic rays and their transfer beyond the shock wave fronts. During this flare the electron flux with $E_e > 0.05$ MeV was about 10 times higher than the proton flux with $E_p = 1 \div 4$ MeV.

The character of the intensity change is indicative of the accumulation of solar cosmic rays in interplanetary space. It should also be noted that the analysis of other solar cosmic ray bursts detected during the Venus-5 and 6 flights (Figures 2 and 3) and set forth in [8] shows that on January to May, 1969 one could observe: (1) an increase in the power of solar cosmic ray bursts; (2) a harder spectrum of solar cosmic rays; (3) effective accumulation of solar cosmic rays at a radius of 1 AU from the sun; (4) considerable proton fluxes with $E_p > 30$ MeV behind the shock wave fronts.

During one of the solar cosmic ray bursts, namely from 11 to 22 April, 1969 a proton flux with $E_p > 30$ MeV and a peak of about 200 particle $cm^{-2} sec^{-1} ster^{-1}$ was observed behind the front of the shock wave which occurred, probably, behind the sun's eastern limb.

References

[1] Vernov, S. N., Chudakov, A. E., Vakulov, P. V., Logachev, Yu. I., Lyubimov, G. P., Nikolaev, A. G., and Pereslegina, N. V.: 1966, *Doklady Akad. Nauk SSSR* **171**, 847.

[2] Vernov, S. N., Chudakov, A. E., Vakulov, P. V., Logachev, Yu. I., Lyubimov, G. P., Nikolaev, A. G., and Pereslegina, N. V.: 1967, *Izv. Akad. Nauk SSSR* **31**, 1296.

[3] Vernov, S. N., Chudakov, A. E., Vakulov, P. V., Gorchakov, E. V., Kontor, N. N., Kuznetsov, S. N., Logachev, Yu. I., Lyubimov, G. P., Nikolaev, A. G., Pereslegina, N. V., and Tverskoy, B. A.: 1968, *Proc. V All-Union Winter School on Cosmophysics*, Apatity.

[4] Vernov, S. N., Kontor, N. N., Kuznetsov, S. N., Logachev, Yu. I., Lyubimov, G. P., and Pereslegina, N. V.: 1969, Report at International Symposium on Solar-Terrestial Physics, Crimean Astrophysical Observatory, 1968.

[5] Vernov, S. N., Lyubimov, G. P., and Pereslegina, N. V.: 1969, Report at VI Winter School of Cosmophysicists, Apatity.

[6] Lyubimov, G. P.: 1969, Thesis of dissertation, Institute of Nuclear Physics, Moscow, State University, 1969.

[7] Vernov, S. N., Vakulov, P. V., Gorchakov, E. V., Kontor, N. N., Logachev, Yu. I., Lyubimov, G. P., Pereslegina, N. V., Timofeev, G. A., and Chudakov, A. E.: 1969, Report at the Seminar of Nuclear Physics Branch of the Academy of Sciences of USSR, Leningrad, 1969.

[8] Vernov, S. N., Chudakov, A. E., Vakulov, P. V., Gorchakov, E. V., Kontor, N. N., Logachev, Yu. I., Lyubimov, G. P., Pereslegina, N. V., and Timofeev, G. A.: 1969, Report at Int. Conf. on Cosmic Ray Physics, Budapest, 1969.

[9] Dolginov, Sh. Sh., Eroshenko, E. G., and Zhuzgov, A. N.: 1968, *Space Research*, Vol. 4.

[10] Gringauz, K. I., and Solomatina, E. K.: 1968, *Space Research*, Vol. 4.

[11] Parker, E.: 1965, *Dynamical Processes in Interplanetary Medium* (Russian translation), Mir, Moscow.

[12] Vernov, S. N., Gorchakov, E. V., and Timofeev, G. A.: 1969, *Geomagnetism and Aeronomy*, No. 5.

[13] Tverskoy, B. A.: 1967, *Doklady Akad. Nauk SSSR* **53**, 1417.

[14] Charakhchyan, A. N., Tulinov, V. F., and Charakhchyan, T. N.: 1961, *Zh. Eksp. Teor. Fiz.* **41**, 735.

[15] 1969, Compilation of *Solar-Geophys. Data*, Boulder, Colo.

[16] Neutron monitor data in WDC-B2.

LOW ENERGY PROTON MEASUREMENTS IN INTERPLANETARY SPACE ON BOARD THE HEOS-1 SATELLITE

A. BALOGH and R. J. HYNDS

Physics Department, Imperial College, London, England

Abstract. Directional measurements of low energy (1 to 13 MeV) protons in interplanetary space are presented which were obtained during the 25th February 1969 solar proton event. The time history of the anisotropies is presented and discussed in connection with magnetic field measurements taken on the same spacecraft.

1. Introduction

This paper contains results obtained by experiment S-24C on board the HEOS-1 satellite. This is an ESRO satellite and was launched into a highly elliptic orbit on 5th December 1968. The characteristics of the spacecraft and its orbit have been described elsewhere (Page, 1970). S-24C was part of a package of three related experiments designed by the Imperial College group and flown on HEOS-1.

HEOS-1 was near its apogee at about 35 earth radii when a west limb (39°) solar flare of importance 2B occurred ot 09.00 UT on 25th February 1969. After the occurrence of the solar flare HEOS-1 stayed in interplanetary space for about two days. Data have become available for the 3-day period from 25/2 to 28/2. Some results from the analysis of these data will be presented.

Correlation between the direction of the interplanetary magnetic field and the propagation path of solar flare particles is well known to exist (McCracken and Ness, 1966). In order to study this correlation, particle experiments with varying degrees of directional resolution have been flown on a number of spacecraft (e.g. Bartley *et al.*, 1966). However, due to the complex 'fine' structure of the interplanetary magnetic field (Ness, 1969; Hedgecock, 1970), much more observational data are required to understand fully its role in the propagation of low energy solar protons.

The originality of experiment S-24C consists in the possibility to perform directional analysis in the plane perpendicular to the ecliptic plane. Experiment S-24B, also presented at this Symposium (Engel, 1970), also has the same facility. The possibility derives from the attitude control system of the HEOS-1 satellite.

2. Description of the Experiment

The experiment is designed to measure the intensities and anisotropies of solar protons. For this purpose two nominally identical particle telescopes, both consisting of four semi-conductor silicon detectors are mounted in the satellite. One of them, S-24C1, is mounted with its axis perpendicular to the spin axis. As the satellite spins this telescope scans in the satellite equatorial plane. The other telescope, S-24C2, is mounted parallel to the spin axis.

V. Manno and D. E. Page (eds.), Intercorrelated Satellite Observations Related to Solar Events, 471–477. All Rights Reserved
Copyright © 1970 by D. Reidel Publishing Company, Dordrecht-Holland

The satellite delivers eight equi-spaced sun signals during each revolution. These are used in the data handling section of the experiment in such a way that particle counts are recorded in one of eight data channels, depending on the angle between the telescope axis and the satellite-sun direction. At the time of the solar flare the spin axis of the spacecraft was in the ecliptic plane. The 45° sectors, corresponding to the eight data channels, together with the orientation of the spacecraft in solar ecliptic coordinates, are shown on Figure 1. The calibration of the sectors is performed once every 384 sec by experiment S-24B (Engel, 1970).

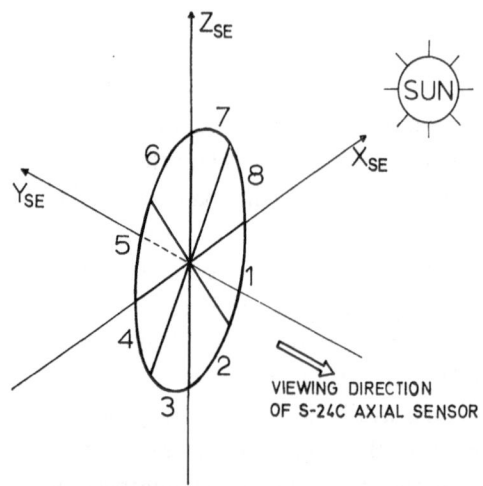

Fig. 1.   Experiment S-24C viewing directions in solar ecliptic coordinates at the time of the 25th February 1969 solar event.

Data are recorded in the eight directional channels during an integral number of satellite revolutions. The integral number is set by the telemetry sampling time (48 sec) and the spacecraft spin rate. At the present spin rate (11 r.p.m.) data accumulation occurs over either 8 or 9 spins and has been averaged over 16 or 17 spins. The satellite housekeeping provides data for resolving the ambiguity and for calculating the exact sampling times. All the sectors are sampled during identical time periods.

Experiments S-24C was designed to measure protons in four energy ranges from 1 to 25 MeV. Channel 1 measures protons of energy 1–13 MeV using only the shielded front detector of the telescope. The geometric factor for this channel is 0.21 cm² sterad. Channels 2, 3 and 4 (which require coincidences between 2 or more detectors) have much smaller geometric factors. This paper contains measurements from S-24C1 (radial sensor) Channel 1 only.

The relative response of Channel 1 sectors is shown on Figure 2. This is obtained from the static differential geometric factor $A(\theta)$ by calculating its convolution with the geometric sector of 45° width. It will be observed that neighbouring sectors overlap. The amount of overlap is indicated by the fact that the counts recorded in each sector contain, in an isotropic flux, 20% contamination from both the neighbouring sectors. The overlap results in some smoothing of the actual anisotropies. The cross-section sensitivity of the S-24C sector in the plane containing its axis and the satellite spin axis has a half opening angle of 45°.

Automatic switching (in time) is performed between the four energy channels. Each channel is sequentially sampled, on average for 96 sec, during each 6 min 24 sec interval. Therefore the Channel 1 data in which we are interested are obtained in this sampled form. It was convenient, when calculating averages, to take an integral number (e.g. 10) of these 6 min 24 sec intervals.

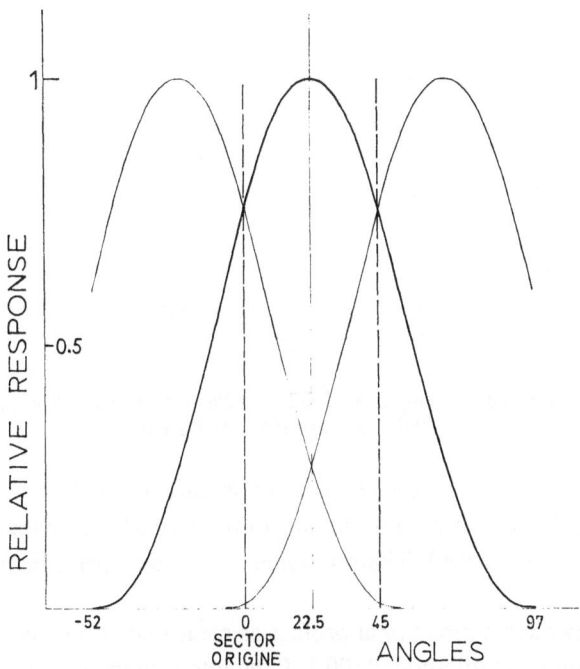

Fig. 2.   Directional sensitivity of S-24C Radial Sensor sectors in the $X_{SE}Z_{SE}$ plane.

3. The Solar Proton Event of 25th February, 1969

In the absence of solar activity the background measured by the experiment is essentially zero. The particles detected by the experiment before the beginning of the solar flare were therefore of solar origin. As we as yet have no data for the days before 25/2 we cannot discuss their origin.

The intensity increase which can be associated with the flare activity began at 10.00 UT on 25/2. Figure 3 shows the complete time history of the event. We plotted there the average counting rate in the radial sensor, that is the average in the XZ plane in solar ecliptic coordinates. There are noteworthy fluctuations of large amplitude between 1600 and 2100 on 25/2 and 0100 and 0700 on 26/2. Both these periods of intensity fluctuations can be associated with specific features in the microstructure of the magnetic field. The disturbed period on 26/2 will be discussed again later in this paper.

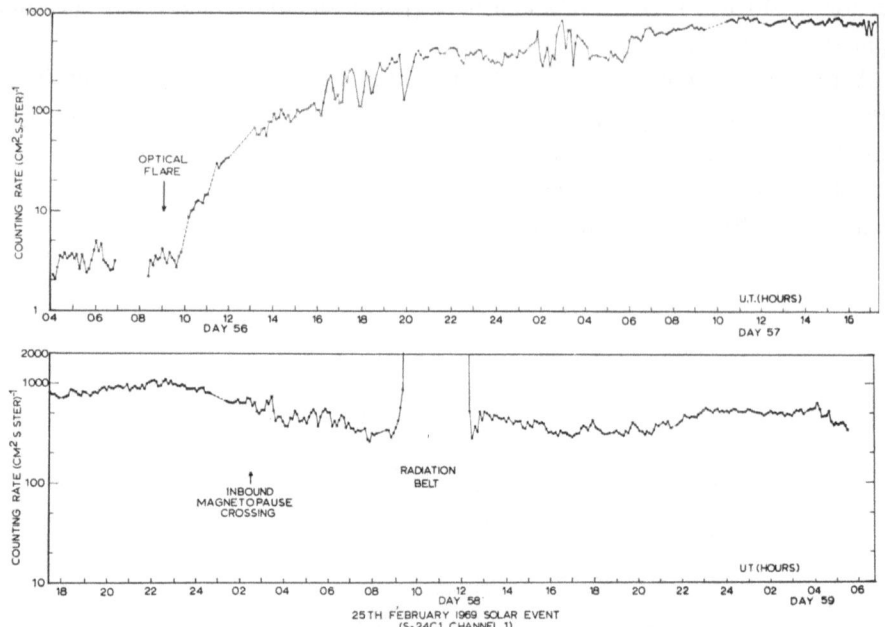

Fig. 3.   Spin-averaged counting rate of S-24C radial sensor in the $X_{SE}Z_{SE}$ plane for the
25th February 1969 solar event.

The maximum intensity is reached at 2230 on 26/2, its value is about 10^3 protons/ cm^2 sec sterad. After the in-bound crossing of the magnetopause the intensity declines and becomes more disturbed. There is some evidence of periodicity in the intensity fluctuations.

Figure 4a shows some directional profiles as measured at the beginning of the flare in the $X_{SE}Z_{SE}$ plane. Until about 1100 UT the intensity increase is quite isotropic. It then becomes anisotropic, when there is some evidence of backscatter. Later profiles show a quasi-sinusoidal distribution with the maximum intensity from the direction of the sun. The maximum intensity is about 3 to 4 times the minimum intensity.

The variations of the intensities from the sun and anti-sun directions are shown for the whole event on Figure 4b. The points represent data averaged over successive 64-min intervals. They were obtained by adding sectors 1 and 8 (sun direction) and sectors 4 and 5 (anti-sun direction). Two things are to be noted:

(a) On the whole, there is good correlation between the two curves, implying that on this time scale the backscatter is proportional to the direct flux.

(b) After the discontinuity observed in the magnetic field at 0155 on 26/2, for about three hours the backscattered particles overtook in average intensity the direct stream. This time period will be examined in more detail later.

Figure 5a represents the direction of the magnetic field vector projected onto the ecliptic and solar meridian planes together with the particle anisotropies similarly projected. It is important to observe that there are considerable differences between the field direction and anisotropy direction. We have in fact defined the direction of

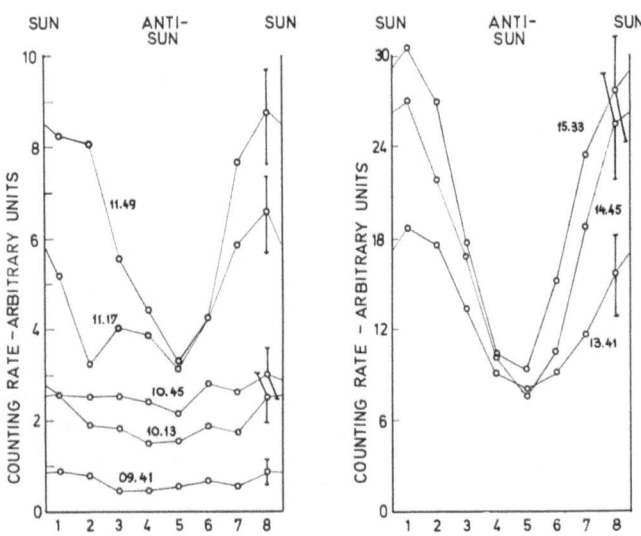

Fig. 4a.   Directional profiles at the beginning of the solar event in the $X_{SE}Z_{SE}$ plane.

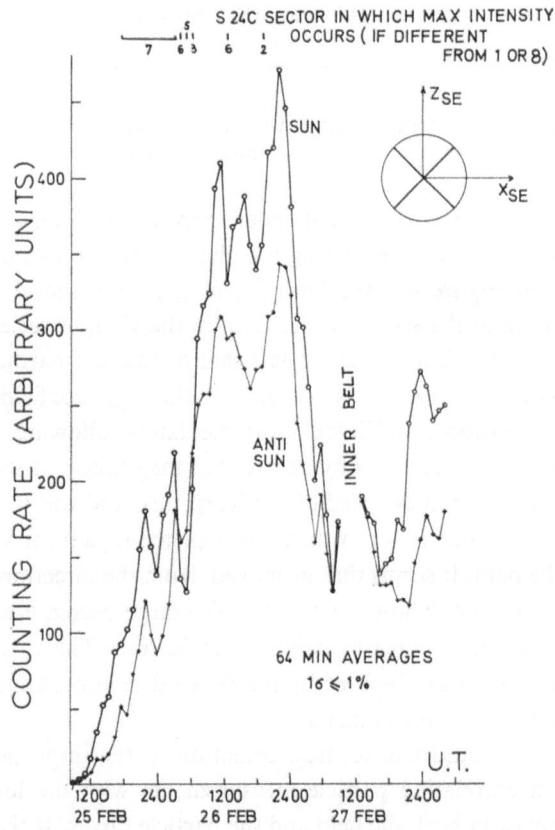

Fig. 4b.   Sun-anti sun anisotropy for low energy (1 to 13 MeV) protons for the 25th February 1969 solar event.

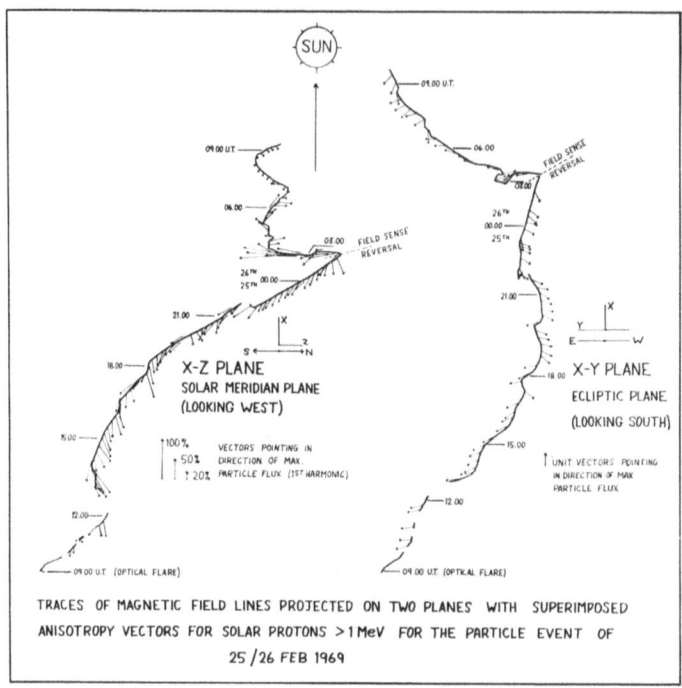

Fig. 5a.   Projections of the magnetic field and the low energy particle anisotropies in the solar meridian and solar ecliptic planes.

the anisotropy as the phase of the first-order term in a Fourier analysis of the eight sector data. This is equivalent to finding the phase of the best sinusoidal fit to curves similar to those on Figure 4a. We have found that the second harmonic, i.e. the amplitude and phase of the second order term in the Fourier series is often far from negligible; sometimes its importance equals that of the first harmonic.

The disturbances following the discontinuity in the magnetic field which occurred at 0155 UT on 26/2 are shown on Figure 5b. Immediately following the shock front the intensity in some sectors decreases by an order of magnitude. It seems that following the shock there was a region depleted of particles; this void was in fact situated below the ecliptic plane. The direction in which this depleted region (or void) moved can be estimated from the data. It seems that it 'moved' from the direction making about 45° with the satellite-sun line, below the ecliptic. The large decrease was preceded in the same sectors by a gradual increase lasting about 30 min. The variations observed at this time by S-24C were not observed by the S-24B detectors (Engel, 1970) indicating that they were low energy phenomena.

In conclusion, we would like to draw attention to the importance of micro- and meso-structures in correlating particle measurements with the local magnetic field. These structures exist in both the field and the particle fluxes. It therefore seems to us that rapid progress in the understanding of events such as that of 25th February require the simultaneous consideration of both magnetic field and directional particle measurements made on the same satellite. Thus the so-called 'Integrated Payload' approach seems to be the most promising for the future.

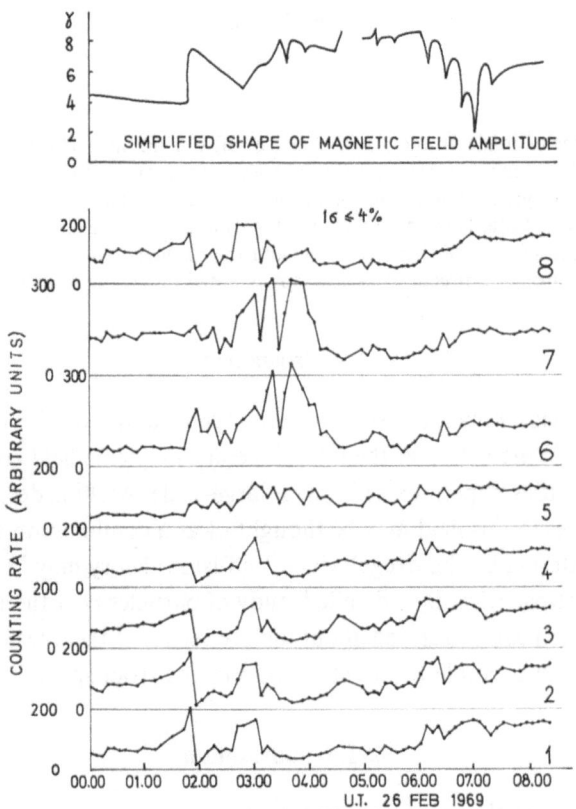

Fig. 5b.   Counting rates in the eight sectors of S-24C radial sensor in the solar meridian plane at the time of an interplanetary shock wave through the HEOS-1 spacecraft.

Acknowledgements

The HEOS-1 satellite was successfully launched by the ESRO project team. S-24C was built by McMichael Ltd., of Slough, England. We would like to thank Dr. Hedgecock for supplying the magnetic field data; Professor H. Elliot for his constant encouragement and advice and Dr. J. J. Quenby for a number of useful discussions.

References

Bartley, W. C., Bukata, R. P., McCracken, K. G., and Rao, U. K.: 1966, *J. Geophys. Res.* **71**, 3297.
Engel, A. R.: 1970, this volume, p. 478.
Hedgecock, P. C.: 1970, this volume, p. 419.
McCracken, K. G. and Ness, N. F.: 1966, *J. Geophys. Res.* **71**, 3315.
Ness, N. F.: 1969, 13th International Conference on Cosmic Rays, Budapest.
Page, D. E.: 1970, this volume, p. 367.

THE SOLAR PARTICLE EVENT OF 25 FEBRUARY, 1969

Directional Measurements of Solar Protons with Energies greater than

360 MeV on HEOS-1

A. R. ENGEL

Physics Department, Imperial College, London, England

Abstract. Measurements made on the HEOS-1 spacecraft during the particle event of 25 February, 1969 are reported. Directional measurements were made on protons > 360 MeV and omnidirectional measurements on protons > 30 MeV. Results indicate that diffusion of the particles has occurred in interplanetary space. Initially the particle flux is highly anisotropic. There is an evident discrepancy between the direction of maximum particle flow and the magnetic field direction measured on the same spacecraft. A strong bi-directional anisotropy is noted indicating some backscattering process beyond the earth.

1. Introduction

Magnetic field measurements (Ness *et al.*, 1964) and cosmic ray observations (McCracken, 1962) indicate that the interplanetary magnetic field consists essentially of small irregularities superimposed on a large-scale Archimedes spiral field. The motion of particles in this field may be thought of as a combination of guiding-centre motion with scattering by the irregularities. Diffusion theory may be used to describe the distribution of particles. By a detailed study of particles near the earth originating in solar flares, much can be learned about the properties of the interplanetary medium in regions so far inaccessible to direct measurements from spacecraft.

2. Instrumentation

It is the purpose of this paper to present measurements on particles > 360 MeV which can complement other energy ranges measured on the same or other spacecraft. The HEOS-1 spacecraft was launched on 5 December, 1968, into a highly eccentric orbit with apogee 222 kkm and period $4\frac{1}{2}$ days. At the time of the particle event on 25 February, 1969, it was near apogee at 35 earth radii and solar ecliptic latitude 50° North and longitude 56° West. The spacecraft was spin-stabilized with a period of 6 sec. and the spin axis pointing almost 90° East (see Figure 1).

On board were three complementary experiments, designated S-24A, S-24B and S-24C, supplied by Imperial College, London. S-24A is a fluxgate magnetometer and S-24C is a low energy particle telescope both of which are described in accompanying papers. This paper is concerned with the results of S-24B.

Figure 2 is a diagram of the S-24B sensor. It consists of an axial and radial mounted counter telescope the first element of each being a unidirectional perspex Cerenkov radiator and the second element a shared plastic scintillator. The geometric factor of each telescope is approximately 7 cm^2 ster. Cerenkov-scintillator coincidences correspond to > 360 MeV protons determined by the threshold for Cerenkov light.

V. Manno and D. E. Page (eds.), Intercorrelated Satellite Observations Related to Solar Events, 478–485. All Rights Reserved

As it rotates with the spacecraft, the radial telescope scans the spacecraft equatorial plane and the counts are divided into eight radial channels corresponding to eight viewing directions. The viewing cone of the axial telescope remains fixed in space and the counts are accumulated in a single axial channel. At the time of the particle event the radial telescope was scanning the X-Z plane (see Figure 1) while the axial telescope was pointing 90° West.

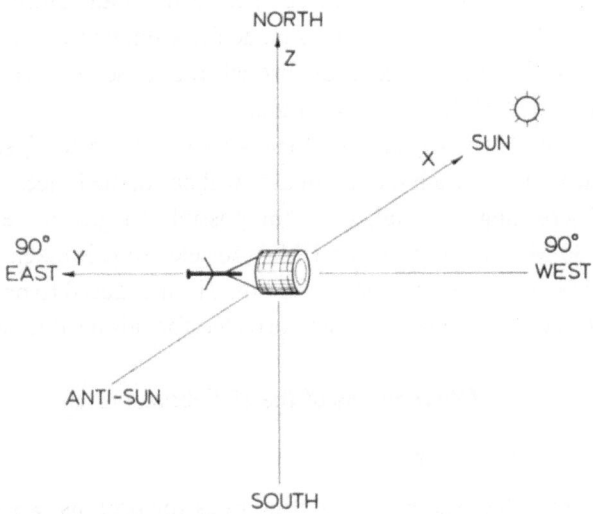

Fig. 1.   Definition of coordinate system used in the text.

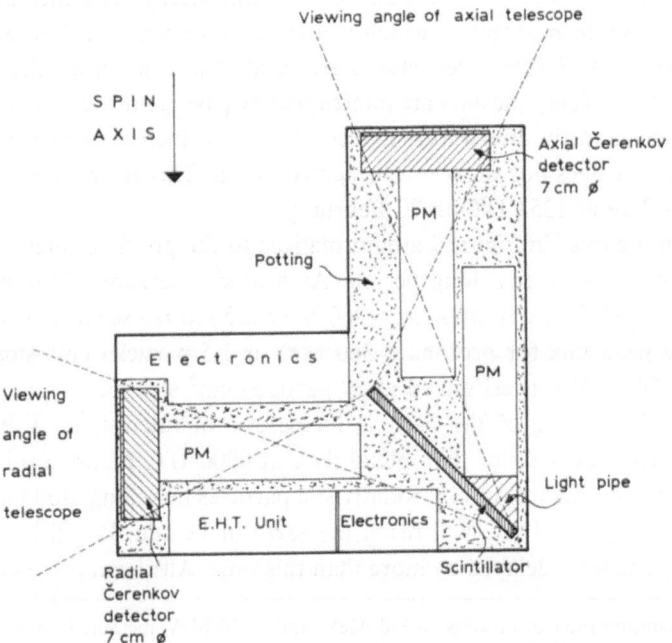

Fig. 2.   Diagram of S-24B proton telescope carried on HEOS-1.

The nine channels are each sampled every 48 sec but to improve statistics, the counting rates are usually averaged over 6.4 min. Two corrections are made to the raw data. The exact sampling period for each radial channel is determined by sun-sensors mounted round the spacecraft. These may not be quite evenly spaced resulting in a false anisotropy. A calibration mode allows a correction to be made with an accuracy of 2 parts in 10^4. The axial counting rate is also normalized to the mean radial rate using data from an undisturbed period when any anisotropy is known to be small ($<1\%$). A second correction is made for a small fraction of particles which are detected when they pass backwards through the telescope due to imperfect uni-directional properties of the Cerenkov radiator.

The counting rate of the scintillator alone is also sampled for $\frac{2}{3}$ sec every 6.4 min, and any particle that can reach the scintillator will be counted. Because of the complex disposition of screening material, it is not possible to give a well defined energy threshold, but 30 MeV for protons or 1 MeV for electrons is taken as a typical value for a geometric factor of 43 cm^2 ster. A dead time is introduced to prevent this channel overflowing; the counting rates are then corrected for this dead time.

3. Observations of the 25 February Event

A. TIME HISTORY OF THE EVENT

Figure 3 shows the time history of counting rates for protons >360 MeV and >30 MeV. A class 2B flare on the West side of the solar disk, beginning at 0900 UT, appears to be the origin of the particle increase which is characterized by a fast rise to a peak, followed by a slow, essentially exponential decay. The increase seen by the Deep River Neutron Monitor, which is a measure of rather higher energy particles, is also shown. A Forbush decrease, starting almost a day after the event, may be noted. On 27 February the data are interrupted by passage through the radiation belts near perigee. Shortly afterwards, a second particle increase occurred at 1500 UT, which however does not show in the neutron data. This is probably the result of a second 2B flare at 1358 UT on 27 February.

All counting rates in Figure 2 are normalized to the pre-flare rates which had been relatively constant over a long period. As a crude measure of the anisotropy, the mean radial and the axial rates for >360 MeV are plotted separately (dotted and full lines). The peak flux for protons >360 MeV is 1.5 particles/cm^2 ster sec while for protons >30 MeV it is estimated at 50 particles/cm^2 ster sec.

The detailed timing of the particle increase is of interest. In Table I times are measured from the start of the optical flare at 0900 UT. Errors are indicated when limited by the timing resolution. Relativistic particles travelling along a smooth spiral field line of length 1.3 AU will reach the earth in 11 min. It will be noted that the arrival of particles is delayed by more than this time. Although some of the difference

---

Fig. 3.   Counting rates of protons >360 MeV and >30 MeV measured by S-24B on HEOS-1 for the particle event 25–28 February, 1969. The Deep River neutron monitor rate is also shown. All rates are normalized to the pre-flare cosmic ray background.

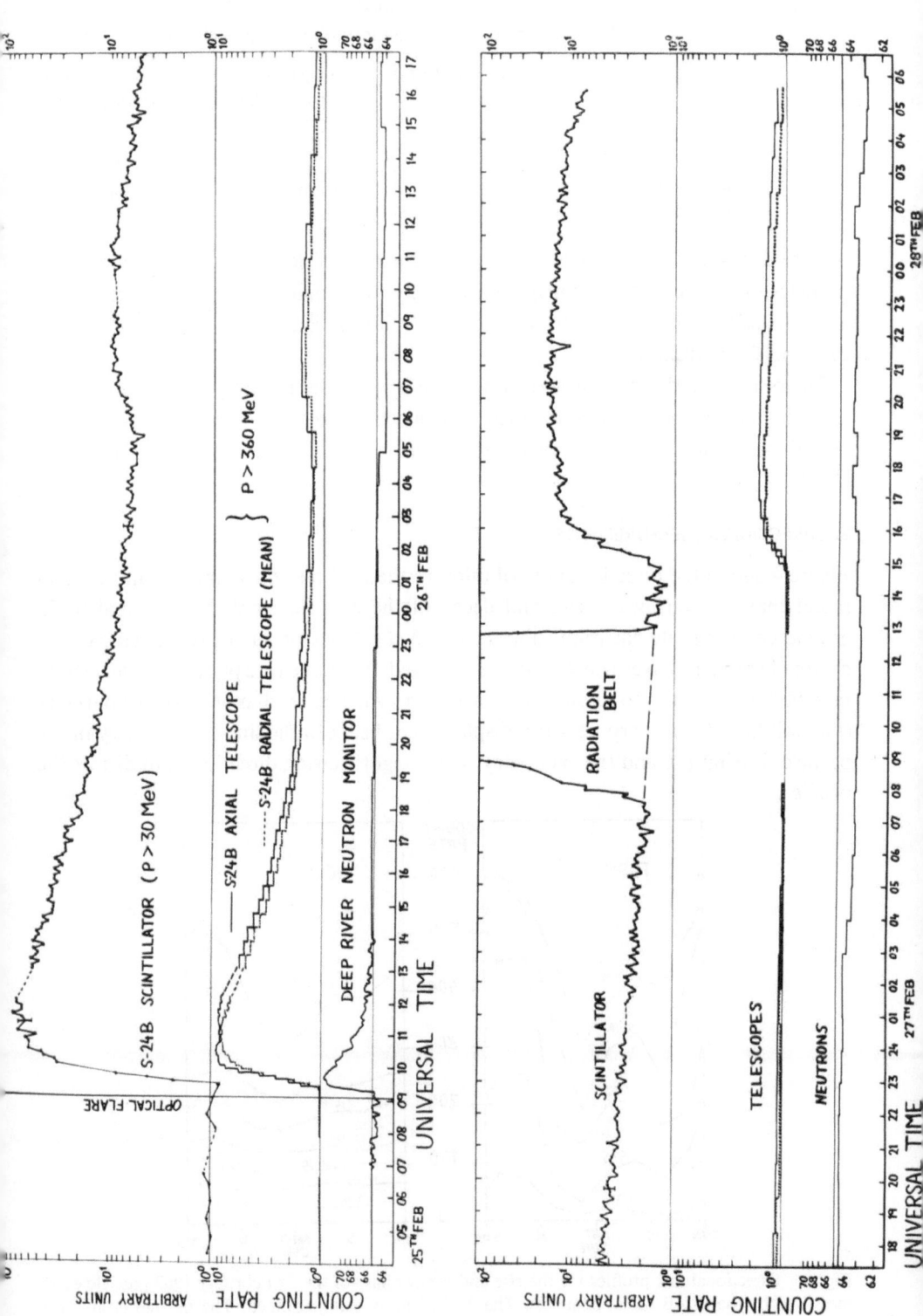

TABLE I

Time in minutes of the main features in the particle event

	Arrival of first particles	Peak of particle flux	Decay to $\frac{1}{2}$ rate at peak
Neutron data	23 ± 2	43 ± 2	90
$P > 360$ MeV	27 ± 2	120	270
$P > 30$ MeV	33 ± 6	180	360

may be due to a delay between the optical flare and acceleration and injection of the particles, diffusion during propagation is likely to account for most of it (Hofmann and Winckler, 1963). A much longer delay (2 hours) is associated with the second event on 27 February.

The results clearly show how the lower energy particles rise and decay more slowly than the higher ones. If it is assumed that injection occurred over a relatively short period this picture is consistent with an interplanetary diffusion process (Bryant *et al.*, 1965).

B. DIRECTIONAL MEASUREMENTS

Figure 4 shows the directional distribution in the X-Z plane obtained from the eight radial channels during the rise and decay of the particle event. The normal background counting rate has been subtracted and the times on each profile refer to the centre of the period over which data are averaged. Note, the true picture will have been modified by the finite solid angle in space from which each channel receives particles (this will broaden a sharp peak); a displacement between the line of symmetry in the particle distribution and the symmetry of the eight viewing directions can distort the profiles.

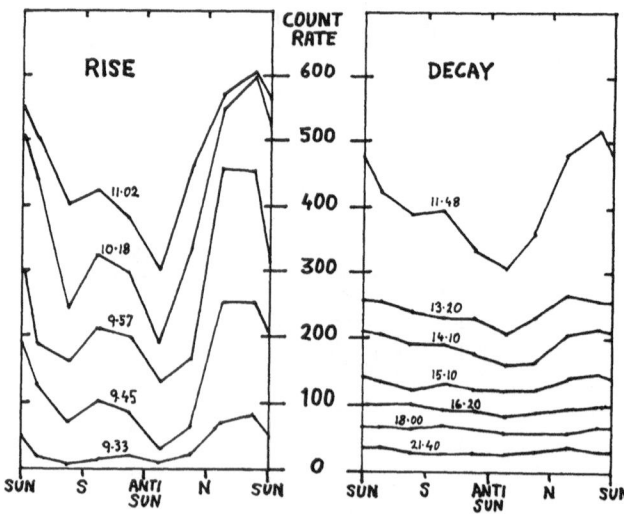

Fig. 4.   Directional flux profiles for the rise and decay phase of the 25 February 1969 particle event measured by the S-24B radial channels. The X-Z plane is being scanned. The normal cosmic ray background has been subtracted.

Initially there is a large anisotropy with most particles arriving from a direction about 30° North of the sun. Other directions fill in as the event progresses, but some anisotropy remains throughout. A second smaller peak will be noted, indicating a flow of particles in the reverse direction to the main flow. Similar bi-directional anisotropies for lower energy particles have been reported (Rao *et al.*, 1967) usually occurring later in an event and associated with a Forbush decrease. This peak is present right from the start and is a feature throughout the event.

The presence of this strong bi-directional anisotropy can be explained either by a magnetic field configuration which connects to the sun in both directions (Gold model tongue), or by the presence of strong backscattering from a field irregularity beyond the earth. Figure 5 has been drawn to investigate the latter possibility. Using data from S-24A, a magnetic field line projected on the *X-Y* plane has been traced out by the vector addition of 6.4 min averages of the field as it is swept past the spacecraft by the solar wind. This will be an approximate representation of the true field, provided the field magnitude does not change greatly. The Larmor radius of a 360 MeV proton is indicated on the diagram and is seen to be of comparable dimensions to the kink in the field. Wentzel (1964) has shown that such a kink mainly affects particles with small pitch angles and could reflect them. This, or a similar feature could therefore be responsible for the backscattered particles. It may also be noted that the earth's magnetosphere was in a favourable position relative to the spacecraft, to produce a similar result, though it would be hard to account for the large backward flux observed. It is not possible to decide on the responsible mechanism from these results alone, but any explanation must be able to account for the prompt appearance of the second peak which is present from the start within the resolution obtained (± 3 min).

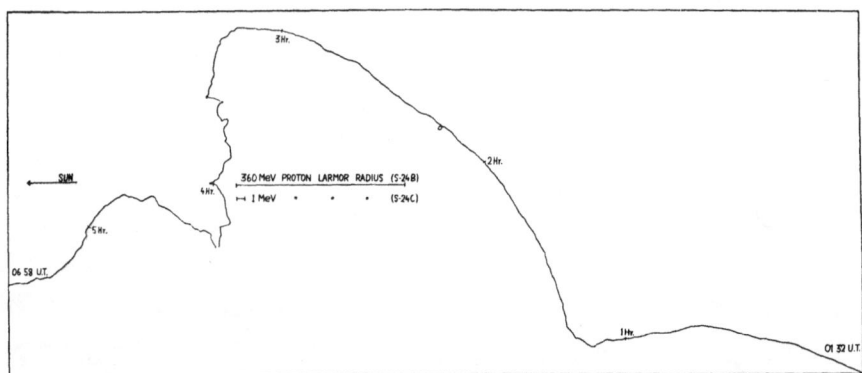

Fig. 5.   Magnetic field line trace projected on the *X-Y* plane for a period just preceding the 25 February, 1969, particle event. One hour intervals are marked on the line.

C. ANISOTROPY RELATED TO MAGNETIC FIELD

An attempt has been made to correlate the particle anisotropy with the magnetic field direction measured by S-24A. As a quantitative measure of the anisotropy, the amplitude and phase of the best fit sine wave to the eight radial channel rates has been calculated following the method used by Bartley *et al.* (1966). This corresponds to a first harmonic with the phase giving the direction of maximum particle flow and the ratio of amplitude to mean flux giving the magnitude of the anisotropy. In this analysis the reverse peak shows up as a second harmonic component. Thus the projection of the anisotropy in the X-Z plane can be measured. By assuming cylindrical symmetry around a direction of maximum flux and knowing the axial telescope rate it is also possible to deduce the angle of anisotropy projected in the X-Y plane. This has only been done at the beginning of the event, where the statistics are sufficiently good.

Data from S-24A have been used to obtain the dotted curves which represent magnetic field line traces projected in the X-Z and X-Y planes. Vector averages over 12.8 min were used. At positions where the field direction changes by the order of 180°, which is assumed to be associated with a sector type structure, the vector addition for the field trace is continued in the opposite sense so that the field is always streaming away from the sun. In order to obtain good statistics, the particle data have been averaged over longer periods, especially later in the event, and the corresponding mean field is shown by a solid line.

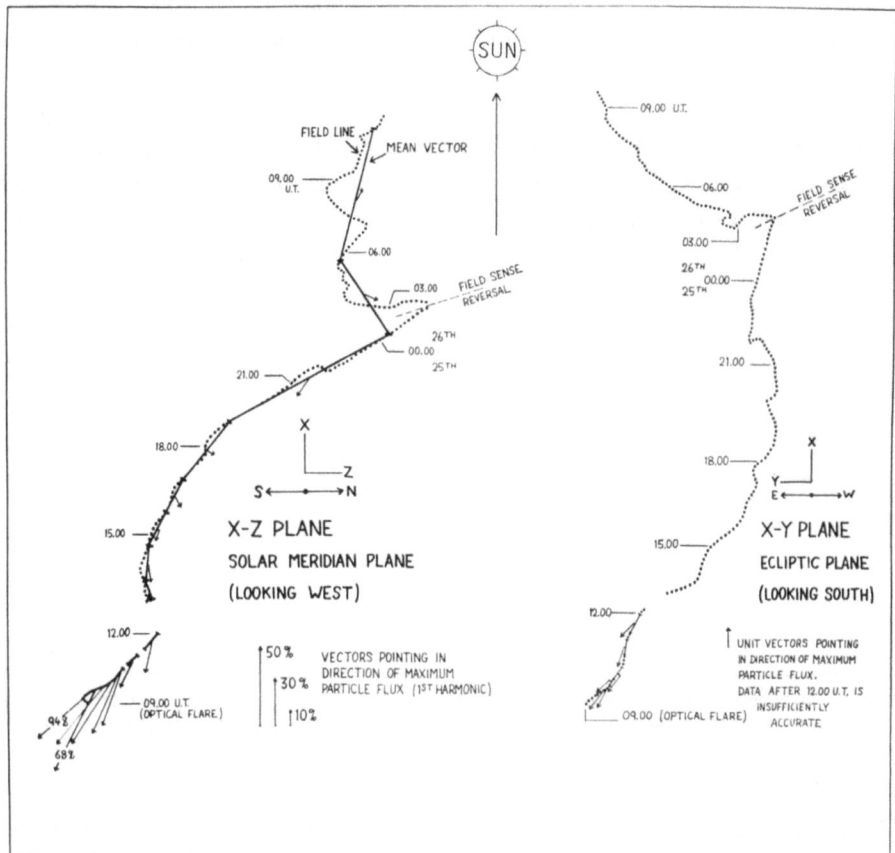

Fig. 6.   Magnetic field line trace together with particle anisotropy vectors projected on the X-Z plane and X-Y plane for the 25 February, 1969, particle event.

In the X-Y plane, the particle vectors line up well with the field. However, in the X-Z plane a divergence, predominantly in one direction, will be noted, even where the field is reasonably smooth for a long period. (The Larmor radius for a 360 MeV proton would correspond to somewhat less than one hour's data on this diagram.) It thus appears that particles of this energy do not follow the detail of the local field. Results for lower energy particles measured by S-24C show a similar trend. McCracken and Ness (1966) looking at lower energy particles have also noted persistent anisotropies, but tracking with the magnetic field direction was much better.

The initial large anisotropy for particles > 360 MeV smoothly decays at the beginning of the event from a figure of about 100% and then remains around 10% till the next flare. The initial decay is consistent with the prediction of Fisk and Axford (1968) for an anisotropy inversely proportional to time.

Acknowledgements

The author wishes to acknowledge the aid of the ESRO project team, managed by Dr. J. A. Vandenkerckhove, the project scientist Dr. B. Taylor and the prime contractor Junkers F. M. GmbH who all contributed to the successful launch of HEOS-1. The instrumentation described was built by McMichael Ltd. Help from Mrs. A. Evans, Dr. R. J. Hynds and Mr. G. Morfill in data reduction by computer is also gratefully acknowledged. Magnetic field data have been kindly supplied by Dr. P. C. Hedgecock. Finally I wish to thank Professor H. Elliot and Dr. J. J. Quenby for their guidance in preparing and interpreting the results.

References

Bartley, W. C., Bukata, R. P., McCracken, K. G., and Rao, U. K.: 1966, *J. Geophys. Res.* **71**, 3297.
Bryant, D. A., Cline, T. L., Desai, U. D., and McDonald, F. B.: 1965, *Astrophys. J.* **141**, 478.
Fisk, L. A. and Axford, W. I.: 1968, *IPAPS* 68/69, 268.
Hofmann, D. J. and Winckler, J. R.: 1963, *J. Geophys. Res.* **68**, 2067.
McCracken, K. G.: 1962, *J. Geophys. Res.* **67**, 447.
McCracken, K. G. and Ness, N. F.: 1966, *J. Geophys. Res.* **71**, 3315.
Ness, N. F., Scearce, C. S., and Seek, J. B.: 1964, *J. Geophys. Res.* **69**, 3531.
Rao, U. R., McCracken, K. G., and Bukata, R. P.: 1967, *J. Geophys. Res.* **72**, 4325.
Wentzel, D. G.: 1964, *Astrophys. J.* **140**, 1013.

Discussion

Wibberenz: What assumptions have been made regarding the symmetry of the pitch angle distribution? I don't see any reason why the particle flux should be symmetric about the field line since, if you have an azimuthal gradient in density, this would of course give rise to strong anisotropies in the same way as the expected density gradient perpendicular to the ecliptic plane gives rise to semi-diurnal variation. What has been assumed about the three-dimensional angular distribution?

Engel: We assumed cylindrical symmetry for obtaining projections in the ecliptic plane, but I agree that this may not be the case. In fact some of our directional profiles do not appear to be symmetrical and we only regard the projections in the ecliptic plane as approximate.

OBSERVATIONS RECORDED BY THE
LEEDS UNIVERSITY COSMIC RAY DETECTOR
ON BOARD THE ESRO II SPACECRAFT
DURING THE SOLAR FLARE OF FEBRUARY 25th, 1969

R. JAKEWAYS, P. L. MARSDEN and I. R. CALDER

The Physics Department, The University of Leeds, England

1. Introduction

The Leeds detector, S 29, records (a) electrons of energy >2.5 GeV, (b) protons of energy >250 MeV ('low energy protons') and (c) protons of energy >25 GeV ('high energy protons').

Results are presented showing the fluxes of electrons and low energy protons observed during the solar flare of February 25th, 1969.

2. The Detector

An outline of the detector is shown in Figure 1 and full details of its operation will appear elsewhere. Briefly, it consists of a gas Čerenkov counter with an energy threshold of ~ 25 rest masses followed by a simple lead-scintillator sandwich shower detector in which particles which have traversed the Čerenkov counter pass through 12.5 mm of lead, a scintillation counter, a further 37.5 mm of lead and finally another scintillation counter. Showers initiated by electrons (and a proportion of high energy protons) are measured by the scintillation counters and the extent of development of the shower gives information as to the nature of the initiating particle and, if it is an electron, its energy. The geometrical factor for the detector in this mode is 3.5 cm² sr.

The Čerenkov counter photomultiplier has a scintillator disc 0.12 mm thick attached to its front face which serves to identify particles which pass through the face and which might otherwise give a light flash of similar intensity to the Čerenkov light flash.

Such particles may also penetrate the full thickness of lead and, if they are protons, (as most of them will be) they require an energy of at least 250 MeV (0.7 GV) to accomplish this. Triple coincidences occurring in this way are recorded as low energy protons, and the geometrical factor of the detector in this mode is 3.2 cm² sr.

3. Results

Data from the detector have been analysed over a period of several days on either side of the flare event and it is relevant to note the direction of pointing of the detector during this period.

S 29 lies on the spin axis of the spacecraft, hence points, in the short term, in a

V. Manno and D. E. Page (eds.), Intercorrelated Satellite Observations Related to Solar Events, 486–491. All Rights Reserved
Copyright © 1970 by D. Reidel Publishing Company, Dordrecht-Holland

fixed direction in space. Prior to the flare the electron counter accepted particles from an open cone of mean semi-angle of about 15° centred on R.A. 251°, Dec + 33°. After the flare, attitude changing manoeuvres reduced the declination to nearly 0° in three steps, the R.A. remaining constant (except during the manoeuvre). The detector thus pointed in a generally westerly direction and at right angles to the solar vector.

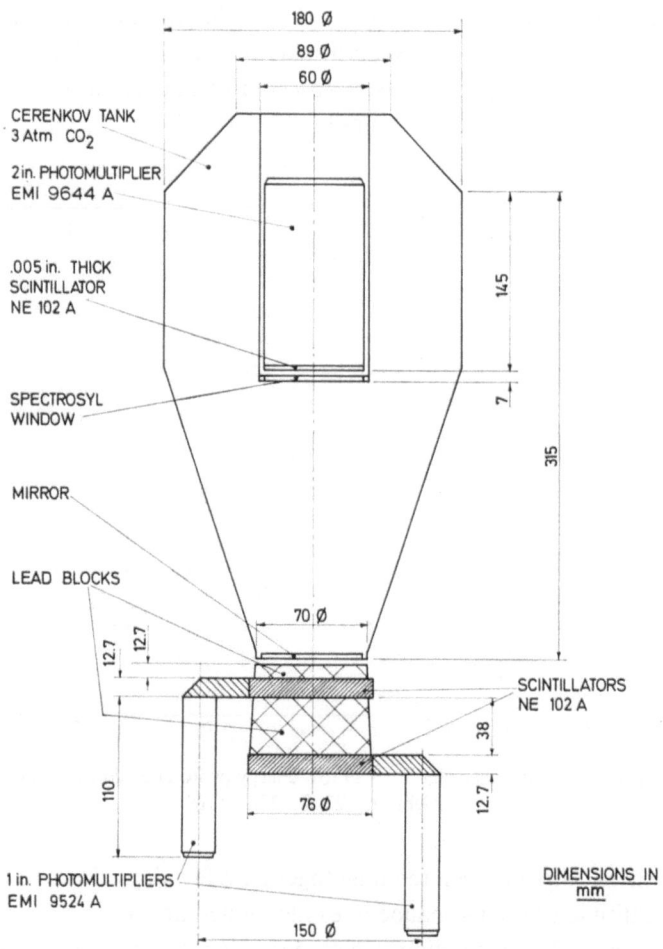

Fig. 1.   The University of Leeds Electron and Proton Detector.

The electron data are not shown diagrammatically since the counting statistics are too poor for an illustration to have any great significance. Each real time pass occupies about 10 min and, under quiet conditions, the detector records an average of about two electrons (with $E > 2.5$ GeV) during this time. No significant difference from this rate was observed throughout the flare, but in any case due to the limited statistics the increase in the electron counting rate would have had to exceed 100% to be detectabled

In Figure 2 the mean north polar counting rate for low energy protons – average.

over ~5 min intervals – is shown for 6 days around the flare period. The mean rates were calculated over those parts of the orbit when the spacecraft was moving in the direction indicated in the figure and when the calculated threshold rigidity was lower than the instrumental cut off.

The difference between the two sets of points is due to the fact that in the low energy proton mode of detection the counter is 'double ended', and at the beginning of the polar pass one end is 'looking' at the earth, whereas when the spacecraft is moving south at latitudes ~70°N both ends of the detector look into space. The counting rate of low energy protons is observed to double during the course of the pass under normal conditions (i.e. pre-flare).

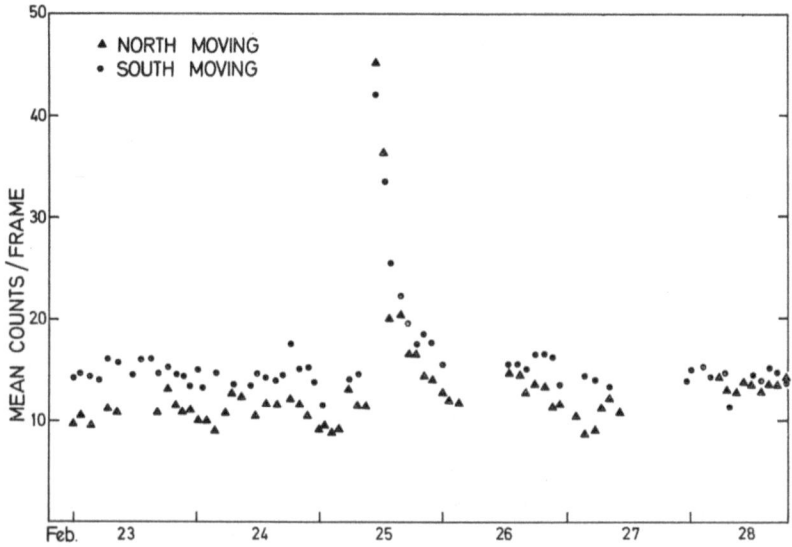

Fig. 2.   Mean counting rates recorded by S 29 low energy proton detector over north polar cap from February 23rd – 28th, 1969.

The two mean rates are seen to come together a few days after the flare and this is due to the attitude changing manoeuvres mentioned above.

At the observed flare maximum (which occurs about an hour later than the maximum as recorded by ground based neutron monitors) the mean counting rate was nearly a factor of 4 higher than before the flare. There is no significant difference between the north moving and south moving rates and this is consistent with the expected unidirectional nature of the flux of solar particles. Taking this into account the maximum observed rate of flare particles was $1.4 \text{ cm}^{-2} \text{ sr}^{-1} \text{ sec}^{-1}$.

Figure 3 shows the counting rate of low energy protons at the observed flare maximum for one pass over the north polar cap. The striking feature of this is the considerable variation in counting rate which appears in the form of peaks occurring on both the north moving and the south moving sections of the pass. (North moving corresponds to the night-time section of the orbit.)

The peak rates are approximately the same hence this enhanced flux has the same directional characteristics as the 'steady' flux.

There is some evidence from our data that the 'southbound' enhancement still exists in the next pass over the polar cap although the counting statistics are rather poor. There is little sign of the northbound peak. There is no evidence of any enhancement in the pass following this one, although flare particles are still being detected at this time.

An analysis of several pre-flare polar cap passes produced no evidence at all of similar peaks in counting rate, thus this is strictly a phenomenon associated with the flare.

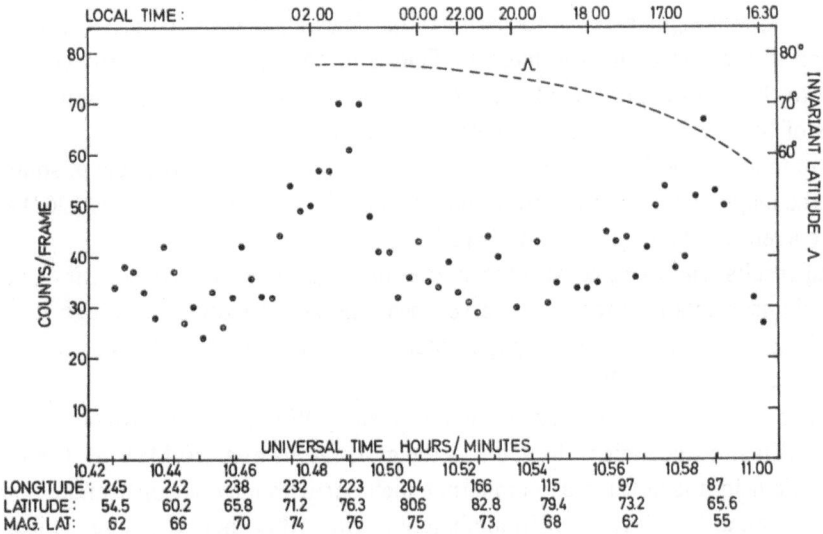

Fig. 3.   Variation of S 29 low energy proton counting rate during the north polar cap pass of orbit 4156, February 25th, 1969.

4. Discussion

The maximum mean counting rate which was observed corresponded to a flux of solar protons of 1.4 ± 0.2 particles $cm^{-2} sr^{-1} sec^{-1}$, but since the first observation from the spacecraft was made about one hour after the maximum enhancement in neutron monitor rates observed on the ground it is quite likely that the actual maximum flux was rather higher than this. The solar proton rate was observed to decay exponentially with a time constant of 225 ± 20 min and there is no evidence from our data of any further substantial fluctuations in the low energy proton counting rate during the next few days.

The most interesting feature of the data is the existence of the fluctuations in rate observed during the polar cap pass at maximum. Spatial variations of a similar nature have been noted by other observers of solar flare phenomena and, in particular, differences have been noted between north and south polar data (Reid and Sauer,

1967). Unfortunately no data were received from ESRO II covering a south polar cap pass during this particular event.

There is always the possibility that such fluctuations may be temporal rather than spatial but the evidence is against this. In particular, detectors on board HEOS A (Hynds *et al.*, 1969) recorded an almost constant intensity during the time that the enhancements were observed.

It is difficult to see how protons of such an energy as we recorded can be constrained to be deposited over such a restricted area of the polar region. The northbound enhancement has a duration of about 2 min, during which time the spacecraft travelled nearly 1000 km. Looked at in terms of geographic coordinates the enhancement region could form a ring having a width, in latitude, of the order of 4° at a latitude of $\sim 70°$. There is thus a high degree of focussing of particles since the gyro radius of a 250 MeV proton in the interplanetary magnetic field is of the order of 1.5×10^5 km, whereas close to the earth the radius of gyration is ~ 60 km, which is smaller than the dimensions of the regions showing enhanced counting rates.

Two models have been suggested to describe the entry of fast particle of solar origin into the magnetosphere: the connected line model (e.g. Reid and Sauer, 1967) and the ring current model (e.g. Williams and Bostrom, 1969).

Our results tend to agree with the second of these since, in particular, it can explain the enhancements in intensity observed near the auroral zones, although it must be noted that the discussion in the paper cited is concerned with protons of considerably lower energy (~ 10 MeV).

The radius of the ring current system is given by Williams and Bostrom as about 20 earth radii and since this is the same as the gyro-radius of a 250 MeV proton in a 20 γ magnetic field it is not unreasonable that such particles may be constrained inside the ring system, although it is not immediately obvious that they would be subject to the same diffusion processes that lead to a concentration of lower energy particles near the circumference of the ring and hence in the auroral zones.

Hynds *et al.* (1969) suggest that some random and fortuitous configuration of magnetic lines has produced the situation where lines of force at the region in question connect directly with interplanetary space, whereas lines from the surrounding region are confined to the magnetosphere. This is possible, providing that the enhancement region is confined in longitude as well as in latitude, but there are not sufficient data to enable this to be determined. Further investigation of this phenomenon is clearly needed.

It is interesting to follow the track of the spacecraft during the pass in question in geomagnetic co-ordinates and Figure 4 shows a sketch of this. The cross-hatched portions of the track (indicated by the arrowed line) show the regions where an enhanced intensity is observed together with approximate L values. The $L = 5$ region is close to the trapping boundary and it might be suggested that we are recording trapped or quasi-trapped, particles. This is rather unlikely, however, since this phenomenon is only seen during the solar flare and it is very difficult to see how our detector can respond to such particles. The $L > 20$ region is well away from both the trapping boundary and the auroral zone, hence this probably represents a different phenomenon from that which produces the lower latitude enhancement.

Finally, we have no information on the time of arrival of particles at the polar cap, although neutron monitor data indicated a considerable time delay (for high energy particles) between the flare and the first increase in particle flux that would be better explained by the ring current model.

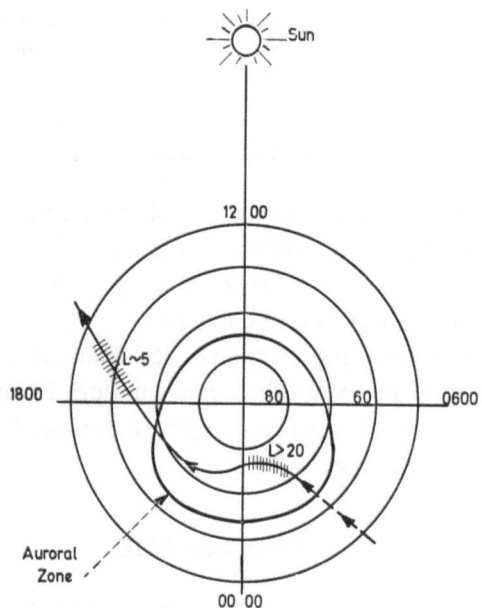

Fig. 4. A sketch of the track (shown arrowed) of ESRO II during the solar cap pass of Figure 3. The circles are lines of magnetic latitude and the auroral zone is indicated. Regions of enhanced intensity are shown cross-hatched.

Acknowledgements

This work has been supported by the Science Research Council and we gratefully acknowledge the assistance and cooperation of ESRO throughout all stages of the project.

References

Hynds, R. J., Balogh, A., Bewick, A., Durney, A. C., Engel, A. R., Elliot, H., Hedgecock, P. C., and Quenby, J. J.: 1969, *Proc. of 11th International Conference on Cosmic Rays*, Budapest, 1969.
Reid, G. C. and Sauer, H. H.: 1967, *J. Geophys. Res.* **72**, 4383.
Williams, D. J. and Bostrom, C. O.: 1969, *J. Geophys. Res.* **74**, 3019.

ESRO I SATELLITE OBSERVATIONS OF 1–30 MeV PROTONS DURING THE 25th FEBRUARY 1969 SOLAR EVENT

G. R. THOMAS, R. DALZIEL and W. DONALDSON

Science Research Council, Radio and Space Research Station, Ditton Park, Slough, Bucks, England

Abstract. The paper describes observations of solar protons over the northern polar cap during the 25th February 1969 solar event. The time-history of the proton distributions is described, and an asymmetry is found between the morning and evening sides. It is suggested that in the early stages of the event the protons had access to the polar cap both by diffusion into the geomagnetic tail, and by direct entry through the dayside neutral point on the high-latitude magnetopause.

1. Introduction

This paper describes results obtained from one of the five particle experiments flown on the ESRO I satellite. The experiment, designated S71E in ESRO terminology, measures the spectrum of protons in the energy range 1–30 MeV. An estimate of the integral intensity above 30 MeV is also obtained.

The results relate to the solar proton event of 25th February 1969, when the satellite orbit (inclination 94°) was in a dawn-dusk configuration. The data were recorded on passes over the northern polar cap, and complete coverage of the polar cap was obtained for about 60% of the passes.

2. Experiment S71E

The S71E experiment consists of a totally depleted silicon surface barrier detector and a caesium iodide (thallium activated) scintillator and photomultiplier, operated with coincidence logic to form a 'telescope'. Pulse height analysis of the photomultiplier output gives a three-point differential spectrum in the energy range 6–30 MeV. Anti-coincidence logic is used for counts in the solid-state detector and in the scintillator; these give the intensity at 1–6 MeV and the integral intensity above 30 MeV.

The viewing cone has a semi-angle of 27° for counts above 6 MeV, and the geometric factor is 0.28 cm² ster. For the 1–6 MeV counts the corresponding figures are 56° and 2.3 cm² ster. Eight different counts are made sequentially, the count period being 1.55 sec, and a complete spectrum is obtained every 12.8 sec.

The ESRO I satellite is magnetically stabilized, and the S71E detector is consequently aligned within about 10° of the geomagnetic field. Over the northern polar cap the detector measures 'precipitated' protons.

3. Results

Figure 1 shows the 1–6 MeV proton counts for a northern polar cap pass 22 hours after the solar flare which occurred at 09.00 UT on 25th February. The unsmoothed

V. Manno and D. E. Page (eds.), Intercorrelated Satellite Observations Related to Solar Events, 492–498. All Rights Reserved
Copyright © 1970 by D. Reidel Publishing Company, Dordrecht-Holland

points are plotted at 12.8 sec time intervals, and the corresponding geographic latitude and satellite height are also indicated. The evening and morning sides of the orbit are at the left and right of this figure, respectively.

A fairly uniform count rate is seen over the polar cap, decreasing abruptly on both sides to the background level of about 10 counts per measuring period. This count rate is due to an internal calibration source, and represents the lower limit of the dynamic range of the experiment. The upper limit, determined by the electronic counter capacity, is 8×10^3 counts per measuring period.

From a series of plots such as Figure 1 we have examined how the 'plateau' count level varies with time following the flare. Figure 2 illustrates this variation for the 1–6 MeV protons, during the period 25th February – 3rd March. Also indicated are three solar flares, all class X and importance 2B (*ESSA Solar Geophysical Data –* March 1969) which occurred in this period, and the subsequent sudden commencement magnetic storms (SSC).

The first three points on the 25th are near the internal calibration source level which corresponds to a differential intensity of about $1.25 \ \mathrm{cm}^{-2} \ \mathrm{sec}^{-1} \ \mathrm{ster}^{-1} \ \mathrm{MeV}^{-1}$. The next six passes, following the solar flare at 09.00 UT, indicate an increase of intensity by two orders of magnitude. It should be noted, however, that for these six passes the intensity was not uniform over the polar cap as in Figure 1. The actual distributions with latitude are discussed later, but the six points which are plotted in Figure 2 represent average intensities over the polar cap. The remainder of the points represent genuine plateau levels over the polar cap.

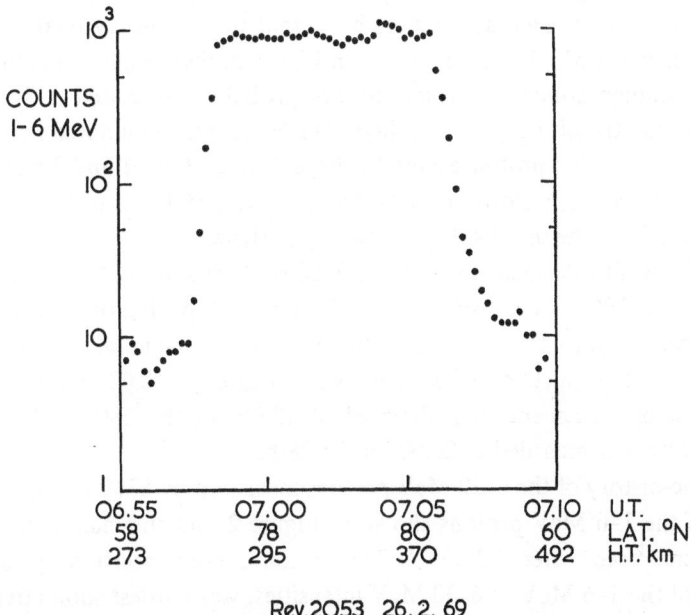

Fig. 1. Counts of 1–6 MeV protons per 1.55 sec observed over the northern polar cap on rev. 2053, 26th February 1969, plotted against universal time with corresponding geographic latitude and satellite height also indicated.

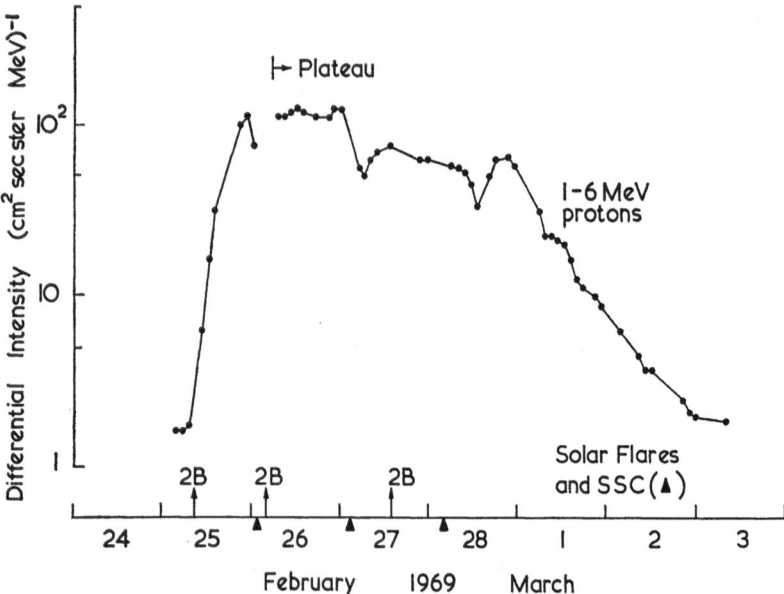

Fig. 2.   Northern polar 'plateau' values of differential intensity of 1–6 MeV protons observed in the period 25th February–3rd March 1969; solar flares and sudden commencement magnetic storms (SSC) are also indicated.

Following the initial peak in the proton intensity due to solar cosmic-ray particles from the first 2B flare, there is a broad maximum followed by two further maxima before the intensity decays over a period of days towards the background level. Although we cannot exclude the possibility that the secondary increases are due to other minor flares which are not shown in Figure 2, they begin soon after the second and third sudden commencements and are probably associated with solar plasma streams having transit times of 1–2 days. This effect has been noted in interplanetary observations of solar proton events by Bryant *et al.* (1965) and Lanzerotti (1969), and the term 'energetic storm protons' has been used (Obayashi, 1967) to distinguish the particles from the usual solar cosmic-ray particles.

Preliminary VELA satellite results for 3–20 MeV protons (*ESSA Solar Geophysical Data* – March 1969) are consistent with this 1–6 MeV proton time-history. The counting rate was enhanced by more than three orders of magnitude at 14.00 UT on the 25th, and a further injection of particles was indicated at 12.00 on the 27th. A fourth, and largest enhancement was observed at 12.00 on the 28th and the maximum counting rate was recorded at 20.00 on the 28th.

The time-history of the 6–30 MeV protons measured by S71E was generally similar to that of the 1–6 MeV protons shown in Figure 2, but the maximum intensity was only $2.5 \text{ cm}^{-2} \text{ sec}^{-1} \text{ ster}^{-1} \text{ MeV}^{-1}$. The proton energy spectrum, given roughly by the ratio of the 1–6 MeV to 6–30 MeV intensities, was hardest soon after the 2B flare on the 25th, and it also became harder after the 2B flare on the 27th. These spectral variations are similar to those observed by Thomas and Dalziel (1970) for the 18th November 1968 solar event.

For each polar plateau such as that shown in Figure 1 it is possible to define cut-off latitudes where the intensity begins to decrease towards the background level (Paulikas *et al.*, 1968). Figure 3 shows the variations of the cut-off latitudes of the 1–6 MeV protons on the evening and morning sides during the event. The protons are always observed at lower invariant latitudes on the evening side, the average asymmetry being 3–4°, similar to that given by Taylor's (1967) calculations for 1.2 MeV protons.

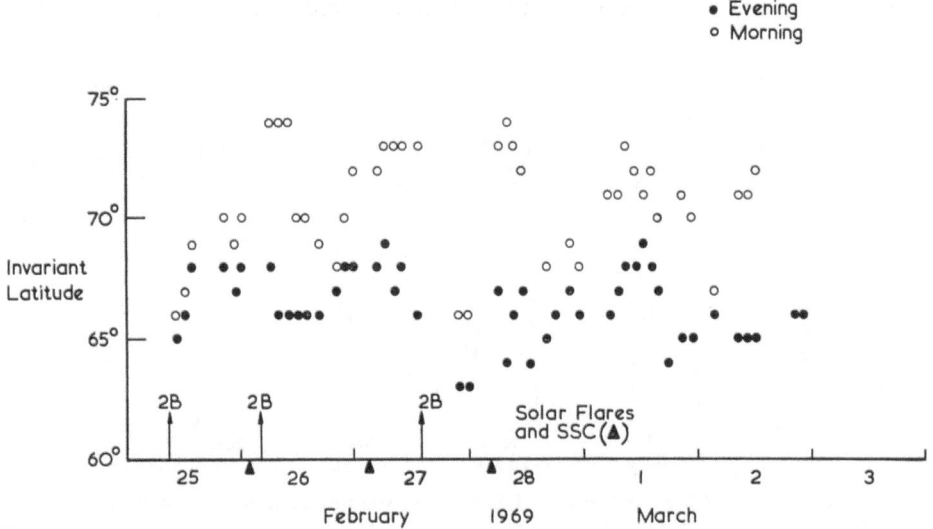

Fig. 3.   Cut-off latitudes of 1–6 MeV protons on evening and morning sides of northern polar cap for period 25th February–3rd March 1969; approximate local times are 18 hrs and 08 hrs, respectively; solar flares and sudden commencement magnetic storms are also indicated.

Further, they are observed at the lowest latitudes on both evening and morning sides after the 2B flares on the 25th and the 27th. Similar behaviour has been reported by Thomas and Dalziel (1970) for the 18th November 1968 event, with the protons extending to latitudes which are well within the closed field line region of magnetospheric models such as that of Fairfield (1968). However, the models are appropriate to quiet magnetic conditions whereas our observations generally refer to disturbed conditions when ring current effects are expected to lower the cut-off latitudes (Akasofu *et al.*, 1963).

We consider now the polar-cap distributions of 1–6 MeV protons seen at the start of the event. Figure 4 shows the counts measured on the first seven passes for which data were obtained following the 2B flare at 09.00 UT on the 25th. The counts are plotted against time, so that they are non-linear in latitude, and the approximate UT of each pass is indicated. Passes 2041–2049, with 2047 being repeated in the figure for ease of reference, correspond to the six points mentioned above in connection with Figure 2. Pass 2053, shown in detail in Figure 1, was the first which showed the uniform distribution over the polar cap.

It is likely that some of the variation observed for pass 2041 can be ascribed to uncollimated high energy particles which were present early in the event, but the remainder of the first six passes shown in Figure 4 exhibit a characteristic triple-peaked structure. In order to investigate the development of this structure with time, Figure 5 shows the peak values of the differential intensity observed over the central polar cap, and at the 'auroral zone' on the evening and morning sides as a function of time following the 2B flare on the 25th. The central polar cap intensities, which are consistently greater than the 'auroral zone' peak intensities, reach a maximum 14 hours after the flare. Coincident with this maximum, there is a small minimum in the

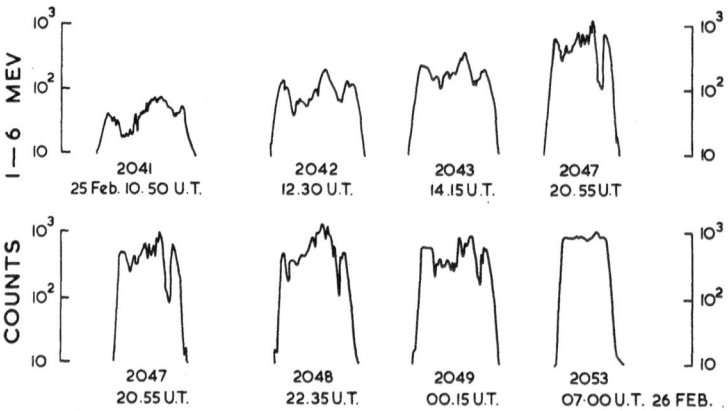

Fig. 4. Counts of 1–6 MeV protons per 1.55 sec observed for seven northern polar cap passes on 25th–26th February 1969; pass 2047 is repeated for ease of reference; the abscissa for each profile is universal time.

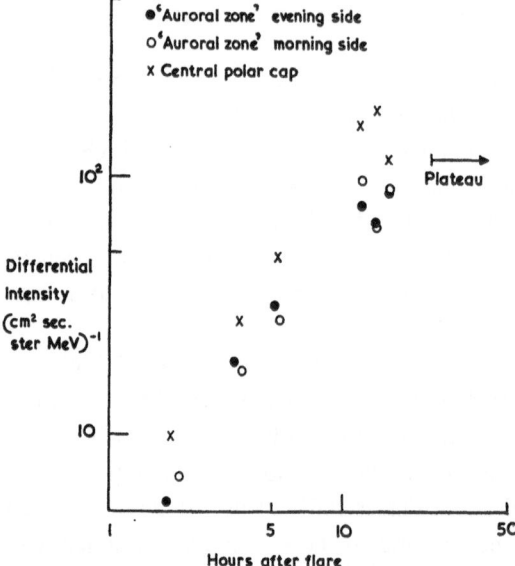

Fig. 5. Peak values of differential intensities of 1–6 MeV protons observed over northern central polar cap and on evening and morning sides of 'auroral zone', plotted against time following 09.00 UT on 25th February (logarithmic scale).

'auroral zone' intensities, and all three intensities then converge to the plateau value indicated (cf. Figure 2).

Figure 6 illustrates the corresponding variations in the invariant latitudes of the three peaks. The behaviour of the 'auroral zone' peaks is similar to that shown in Figure 3 for the cut-off latitudes – the peaks are seen at the lowest latitudes soon after the flare and then move to higher latitudes. However, they later move to lower latitudes, particularly on the evening side, at the same time as the minimum occurs in the intensity (Figure 5). The central polar cap peaks, except for the first point which corresponds to pass 2041 (Figure 4), occur at invariant latitudes of 80–84°, and at magnetic local times of 06–09 hours.

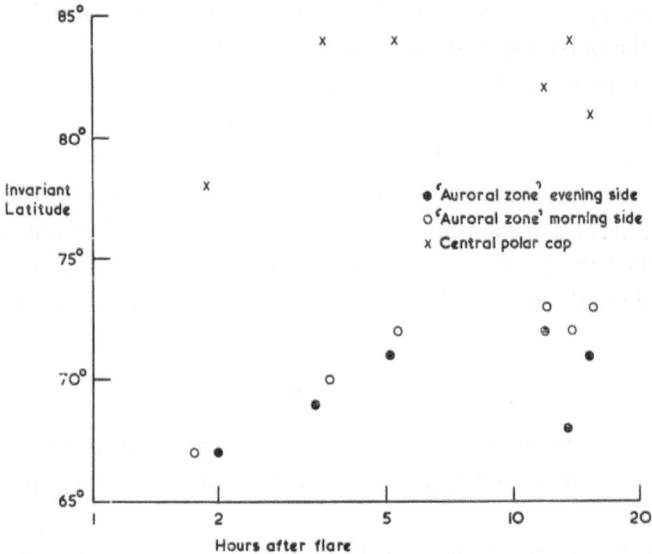

Fig. 6.   Invariant latitudes of the 1–6 MeV proton intensity peaks observed over northern central polar cap and on evening and morning sides of 'auroral zone', plotted against time following 09.00 UT on 25th February (logarithmic scale).

In view of the different temporal characteristics of the intensities observed over the central polar cap and at the lower latitudes, it would be profitable to examine the corresponding variations of proton intensity outside the magnetosphere. Such information would help to elucidate the mechanism of access of the protons to the polar regions.

The occurrence of the 'auroral zone' peaks and the delay of up to 22 hours before the observation of uniform intensity over the polar cap is qualitatively consistent with diffusion of the protons into the geomagnetic tail, as discussed by Michel and Dessler (1965). Similar behaviour was observed by Thomas and Dalziel (1970) for the 18th November 1968 event.

Enhancements of 1.2–2.2 MeV solar protons at invariant latitudes of about 80° have been interpreted by Williams and Bostrom (1967) as the effect of a high-latitude neutral line on the magnetopause, or as an indication of the latitude at which geomagnetic field lines first connect with the interplanetary field. The central polar cap peaks observed in the present results could also be interpreted in this way, but it would be necessary to examine simultaneous interplanetary magnetic field data to determine whether interconnection of the geomagnetic and interplanetary fields was possible. However, the latitudes at which the peaks are observed (Figure 6) are consistent with Mead's (1967) magnetospheric model in which the field lines from the neutral points in the noon magnetic meridian intersect the earth's surface at geomagnetic latitudes of about 80°–83°.

4. Conclusions

The distribution of 1–6 MeV protons over the northern polar cap was investigated for the 25th February 1969 solar event. Except in the early stages of the event the distribution over the polar cap was fairly uniform, and increases in intensity were associated with the arrival at the earth of streams of low-energy solar plasma. The protons extended to the lowest latitudes soon after the 2B flares on the 25th and the 27th, and the cut-offs then moved to higher latitudes. There was an asymmetry of 3–4° invariant latitude, the protons extending to lower latitudes on the evening side. In the early stages of the event a triple-peak structure was observed, and it was suggested that protons reached the polar cap both by diffusion into the geomagnetic tail, and by direct entry through the dayside neutral point on the high-latitude magnetopause.

Acknowledgements

The authors are grateful to Mr. R. J. Pratt for assistance with computer programming and to Dr. D. M. Willis for many helpful discussions concerning the interpretation of these results.

The initial processing of the ESRO I satellite data was performed at the ESRO Operations Centre (ESOC), Darmstadt. The subsequent analysis was carried out at the Radio and Space Research Station of the Science Research Council and the paper is published with the permission of the Director.

References

Akasofu, S.-I., Lin, W. C., and Van Allen, J. A.: 1963, *J. Geophys. Res.* **68**, 5327.
Bryant, D. A., Cline, T. L., Desai, U. D., and McDonald, F. B.: 1965, *Astrophys. J.* **141**, 478.
Fairfield, D. H.: 1968, *J. Geophys. Res.* **73**, 7329.
Lanzerotti, L. J.: 1969, *J. Geophys. Res.* **74**, 2851.
Mead, G. D.: 1967, *Space Sci. Rev.* **7**, 158.
Michel, F. C. and Dessler, A. J.: 1965, *J. Geophys. Res.* **70**, 4305.
Obayashi, T.: 1967, in *Solar-Terrestrial Physics* (ed. by J. W. King and W. S. Newman), Academic Press, New York, p. 107.
Paulikas, G. A., Blake, J. B., and Freden, S. C.: 1968, *J. Geophys. Res.* **73**, 87.
Taylor, H. E.: 1967, *J. Geophys. Res.* **72**, 4467.
Thomas, G. R. and Dalziel, R.: 1970, in *Space Research X*, North-Holland Publ. Co., Amsterdam, in press.
Williams, D. J. and Bostrom, C. O.: 1967, *J. Geophys. Res.* **72**, 4497.

ROCKET MEASUREMENTS OF PROTONS AND α-PARTICLES
DURING THE FEBRUARY 25, 1969 SOLAR EVENT

G. WIBBERENZ and M. WITTE

Institut für Reine und Angewandte Kernphysik, University of Kiel, Germany

Abstract. Two Centaure rockets were launched during an ESRO PCA campaign 7.3 and 14.6 hours after the February 25, 1969, solar event. Energy spectra of protons between 30 and 120 MeV and of α-particles between 24 and 75 MeV/nucleon are presented and compared with other particle measurements on board the same rockets. Proton-to-α ratios for the energy/nucleon range 24 to 75 MeV/N are 88 ± 15 and 190 ± 50 for the two flights. Implications of the time varying proton-to-α ratio are discussed in terms of a diffusion model of propagation.

1. Introduction

A large number of observations have been made after the February 25, 1969 solar event (see e.g. Barouch *et al.*, 1969a; Green *et al.*, 1969; Engel *et al.*, 1969; Van Beek and Van Gils, 1969; Datlowe *et al.*, 1969; Quenby *et al.*, 1969). Since many other papers during this symposium deal with the same event, we shall concentrate on a single point, namely the ratio of protons and α-particles measured at two time intervals during the event.

Two Centaure rockets were launched at 1633 and 2350 UT on February 25, 1969 as part of an ESRO PCA campaign from Andoya rocket range, Norway. The payloads contained charged particle experiments from the groups at Saclay, Utrecht and Kiel. Some details and preliminary results from our $E - dE/dx$-arrangement were described by Green *et al.* (1969). The geometry factor was 3.6 cm^2 sr for 22 MeV protons and 1.7 cm^2 sr for 133 MeV protons. The optical axis of the telescope was mounted under 50° with respect to the rocket axis.

2. Energy Spectra

Two-dimensional pulse height distributions were constructed from coincidences between the dE/dx- and E-counter for 230 sec of flight time above approximately 70 km. Protons and α-particles in the energy range 22 to 133 MeV/nucleon could be identified. The resulting energy spectra for protons and α-particles at about 7.3 and 14.6 hours after the flare are plotted on an energy/nucleon scale in Figure 1, with the α-particle fluxes multiplied by a factor 10. The total number of registered α-particles was 71 for F1 and 22 for F2, which leads to the large statistical uncertainties.

The intensities have been corrected

(1) for absorbing layers of 14 mg/cm^2 aluminium above the first scintillator and of 50 mg/cm^2 between the dE/dx- and E-scintillator,

(2) for deadtime effects of 40% during F1 and 21% during F2. The deadtime cor-

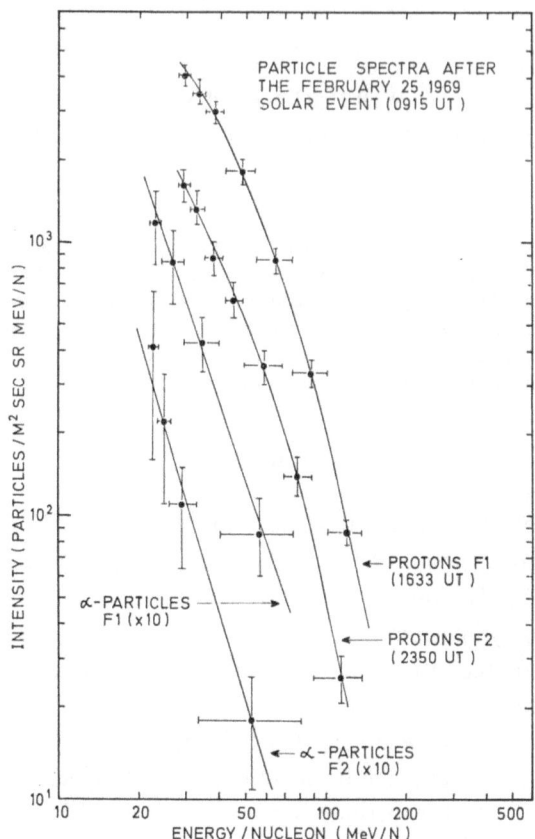

Fig. 1.   Energy/nucleon spectra of protons and α-particles obtained with an $E - dE/dx$-telescope 7.3 and 14.6 hours after the February 25, 1969 solar event. α-particle intensities are multiplied by a factor 10.

rected count rates were in agreement with the rates of two- and threefold coincidences transmitted through a separate channel.

The proton spectra above 30 MeV are gradually steepening as a function of energy. Conversion to an exponential in rigidity, $dN/dP = \text{const} \exp(-P/P_0)$ gives the following e-folding rigidities:

$$P_0 = 90 \pm 10 \text{ MV for F1, protons between 240 and 400 MV,}$$
$$P_0 = 76 \pm 10 \text{ MV for F2, protons between 240 and 400 MV,}$$
$$P_0 = 96 \pm 20 \text{ MV for F1, } \alpha\text{'s} \qquad \text{between 400 and 600 MV.}$$

Energy spectra of protons and α-particles as measured in the same rocket payload by Van Beek and Van Gils (1969) extend smoothly down to energies of about 2 MeV/nucleon. They are shown in Figure 2 together with our own points. Smooth curves are drawn through all data to show the interconnection of data obtained with different experiments. Statistical errors are indicated.

Proton spectra obtained during the same time intervals in the course of the event are presented by Barouch *et al.* (1969a), with part of the data being obtained also in the same rocket payload. The slope of the energy spectra is quite similar to the ones shown in Figure 2, the absolute intensities are systematically larger, in case of F2 by more than a factor 2. The reason for this discrepancy which lies outside the statistical uncertainties is presently not clear.

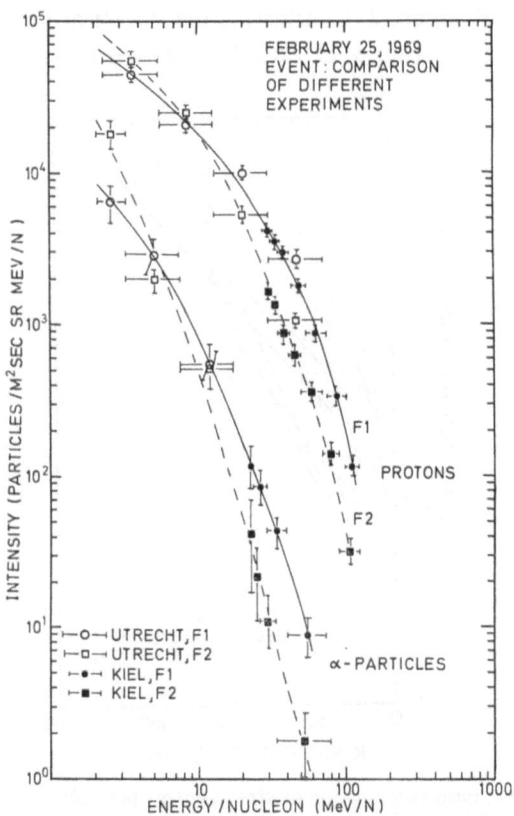

Fig. 2.   Comparison of proton and α-particle energy spectra obtained with two different experiments on board the same rocket (ESRO payload C49). Open points after Van Beek and Van Gils (1969), full points after Green *et al.* (1969), see Figure 1.

3. Proton-to-α Ratios

If we take the absolute number of α-particles observed over the energy range 24–75 MeV/nucleon and compare them with the number of protons in the same interval (see Figure 1), we obtain the following ratios of protons to α-particles:

$$88 \pm 15 \text{ for F1 (7.3 hours after the flare)}$$
$$190 \pm 50 \text{ for F2 (14.6 hours after the flare).}$$

Combination with the results from Van Beek and Van Gils (1969), see Figure 2, allows to derive this ratio as a function of energy between about 3 and 60 MeV/N. This ratio increases as a function of energy/nucleon in accordance with many earlier observations, as shown in Figure 3. Curve A through F and points G and H are taken from the summary article by Fichtel and McDonald (1967). The general trend of the ratio for the February 25, 1969, solar event is indicated by the dash-dotted line (F1, 7.3 hours after the flare) and the dotted line (F2, 14.6 hours after the flare); these results have been obtained by taking the ratios between the smooth curves drawn through the points of Figure 2. Considering the statistical uncertainties for a given energy/nucleon interval it cannot be excluded that the ratio is *independent* of time.

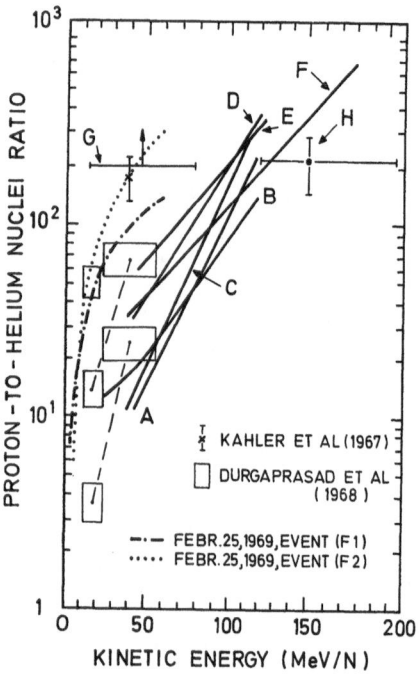

Fig. 3.   The proton-to-α ratio as a function of kinetic energy per nucleon for various events. Curves A through F and points G and H are taken from Fichtel and McDonald (1967). The result by Kahler *et al.* (1967) is an estimate for the March 24, 1966, event. Data by Durgaprasad *et al.* (1968) were obtained for times 8.7, 16.5, and 35.5 hours after the Sept. 2, 1966 event; points belonging to the same time interval are connected by dashed lines. The dotted and dot-dashed curves for the Febr. 25, 1969, event for times 7.3 and 14.6 hours after the flare are derived from the smooth curves in Figure 2.

However, the general trend of the curves as well as the ratios derived directly from the measured particle numbers (see above) show that for energies above 30 MeV/nucleon the ratio should increase as a function of time.

This increase has also been observed earlier, e.g. after the November 15, 1960, event (Biswas *et al.*, 1963) or after the September 2, 1966, event (Durgaprasad *et al.*, 1968). Data from the latter experiment, where 3 rockets were launched 8.7, 16.5 and 35.5 hours after the flare, are also added to Figure 3. The points were derived from the ratios of protons to medium nuclei given by Durgaprasad *et al.* (1968), assuming a constant ratio of helium-to-medium nuclei of 60. This ratio is shown to vary very little from one solar event to the other and during the event (see Biswas and Fichtel, 1965; Durgaprasad *et al.*, 1968). One more point in Figure 3 was obtained from an estimate of Kahler *et al.* (1967) for the March 24, 1966, solar event.

4. Discussion

The large variation of the proton-to-α ratio from one solar event to the other (see Biswas and Fichtel, 1965; Freier and Webber, 1963) is one of the interesting features of solar cosmic rays. Before one can attribute these variations quantitatively to the acceleration process, it should be known, how the variation of this ratio during one event can be explained by propagation effects and how the particles can be traced back to the sun in order to obtain this ratio during the acceleration and injection phase.

Let us consider a few consequences if the particle propagation is described by a diffusion model. For particles walking at random in a given magnetic field structure the diffusion tensor at a given point **x** in space can generally be written as

$$K_{ij}(\mathbf{x}, v, P) = v f_{ij}(\mathbf{x}, P) \tag{1}$$

for particles of velocity v and magnetic rigidity P. This result can be obtained from the general scattering formalism (Jokipii, 1966; Hasselmann and Wibberenz, 1968). f has the dimension of a length and can be identified with $\frac{1}{3}$ of the mean free path for a three-dimensional isotropic diffusion process; its value should be the same for particles of the same rigidity (gyroradius), since they scan the same spatial structures in the magnetic field (see also Parker, 1963).

Bryant *et al.* (1965) have successfully used the method of plotting the observed intensities as a function of travelled distance. Conversion of the diffusion equation to the variable $v \cdot t$ gives intensity-distance-profiles which are identical for particles of the same rigidity; this holds even for the general case of anisotropic diffusion.

In the following we shall neglect azimuthal effects; it is well known that in many cases the observed intensity-time-profiles can adequately be described by a simple model of three-dimensional isotropic diffusion. This means essentially that the observed intensity profile does not depend critically on where the diffusion actually occurs; a variety of assumptions about the distribution of scattering processes between source and observer leads to actually the same result – at least as long as the solar particles are measured at one point in space only.

The validity of the general assumption (1) can be checked by simultaneous measurements of protons and α-particles, since the rigidity of an α-particle is two times as large as that of a proton of the same velocity. At least two methods could be used:

(i) Intensity-distance-profiles for protons and α-particles of the same rigidity are compared. They should be identical if (1) is valid. Lanzerotti (1969) has published data for the May 28, 1967, solar event, which could possibly be used for this purpose. One difficulty lies in the fact that the proton-to-α ratio may undergo sudden changes following interplanetary disturbances (see Lanzerotti and Robbins, 1969).

(ii) The proton-to-α ratio for the same energy/nucleon (velocity) is studied as a function of time. A propagation which is only velocity dependent over a large range of proton rigidities has been observed for many events (see Bryant et al., 1965; Fichtel and McDonald, 1967). In terms of the diffusion picture this would lead to a mean free path which is independent of rigidity. In this case the proton-to-α ratio should be independent of time.

In case of the February 25, 1969, solar event it has been observed by Barouch et al. (1969b) that the propagation seems to be purely velocity dependent in the proton energy range 30 to 160 MeV, with the most probable travel distance around 8 AU. One would therefore not expect the proton-to-α ratio during this event to vary with time, in contradiction to the result of Section 3. Before drawing any conclusions from this observation, it should certainly be confirmed by data of better statistical accuracy and over a larger time interval.

The proton-to-α ratio has been observed to vary during one event in several cases; so far this has been interpreted qualitatively as a propagation effect indicating a rigidity dependent mean free path (see e.g. Biswas et al., 1963; Durgaprasad et al., 1968; Biswas and Fichtel, 1965). It seems worthwhile to study quantitatively the prediction for the proton-to-α ratio, once a diffusion model with a set of parameters has been successfully used.

In the case analyzed by Krimigis (1965) the mean free path depends on radial distance and rigidity as

$$\lambda(r, P) = \lambda_E(P)(r/r_E)^{s(P)}, \quad r_E = 1 \text{ AU}. \tag{2}$$

Let $I_p(0, v)$ and $I_\alpha(0, v)$ be the injection spectra of protons and α-particles of velocity v at $t = 0$. The proton-to-α ratio at time t can easily be found by the closed solutions of the diffusion equation given by Krimigis (1965). Consider for illustration the simple case $\partial s / \partial P = 0$ over the range of interest. Then one obtains

$$\frac{I_p(t)}{I_\alpha(t)}\bigg|_{v=\text{const}} = \frac{I_p(0, v)}{I_\alpha(0, v)} \left(\frac{\lambda_\alpha}{\lambda_p}\right)^{3/(2-s)} \exp\left\{-\frac{3}{2-s}\left(1 - \frac{\lambda_p}{\lambda_\alpha}\right)\frac{t_m^{(p)}}{t}\right\}. \tag{3}$$

Here the time is measured in units of the time of maximum intensity of the proton group under consideration, $t_m^{(p)}$. In general we have $\partial \lambda / \partial P > 0$, so that $\lambda_\alpha > \lambda_p$. As a consequence, the ratio approaches a constant finite value for large t. This asymptotic ratio is identical with the ratio at injection except for the factor $(\lambda_\alpha/\lambda_p)^{3/(2-s)}$, varying between about 1 and 8 in extreme cases.

The logarithm of the ratio (3) will vary proportional to $-t^{-1}$. Numerical results for some parameter values are plotted in Figure 4. The ratios are normalized to 1 for $t = t_m^{(p)}$. Instead of the ratios λ_α/λ_p we have allocated to the curves the rigidity dependence of the mean free path, $\lambda \sim P^k$. Therefore, $\lambda_\alpha/\lambda_p = 2^k$.

It can be seen that for given ratio λ_α/λ_p the time dependence is more pronounced for larger s-values, and that for a given s the curve becomes steeper for increasing ratios λ_α/λ_p.

Also shown in Figure 4 are the experimental results for the proton-to-α ratios in the range 9.5–24 MeV/N derived from Durgaprasad *et al.* (1968); the statistical uncertainties are slightly too large to decide whether a straight line fit is possible or not. A value $t_m^{(p)} = 28$ hours has been estimated from the temporal behaviour of protons around 10 MeV obtained during the three rocket flights.

If we choose $s = 1$, the best fit is obtained by the dashed line, corresponding to a rigidity dependence $\lambda \sim P^{0.65}$. A parameter $s = 1$ is shown to give a good fit of a diffusion model to the intensity-time-profile in several cases (see Krimigis, 1965). Of

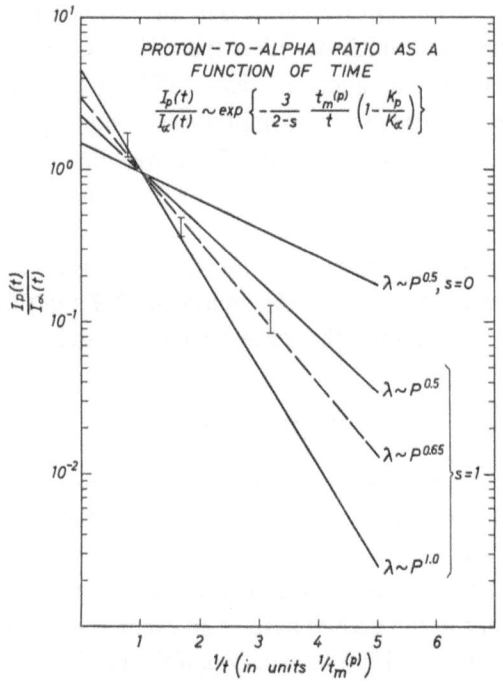

Fig. 4. Model calculations for the proton-to-α ratio as a function of time (Equation (3)) for different parameter values. The curve with $\lambda \sim P^{0.65}$, $s = 1$ is fitted to experimental values derived from Durgaprasad *et al.* (1958). Time is measured in units $t_m^{(p)}$; the ratio is normalized to 1 for $t = t_m^{(p)}$.

course, there is large arbitrariness in the choice of parameters; it would be necessary to determine the parameters beforehand by a good fit of the proton time profile to a diffusion model and to derive afterwards the prediction for the time dependence of the proton-to-α ratio.

It is interesting to note, that the result for the ratio at injection, $I_p(0)/I_\alpha(0)$ does *not* depend sensitively on the choice of parameters. The result on the basis of Equation (3) is $I_p(0)/I_\alpha(0) \approx 27$ in the energy/nucleon range 9.5–24 MeV/N for the September 2, 1966 event. The injection ratio turns out to be about 65 in the energy/nucleon range 24 to 54 MeV/N on the basis of the same model assumptions.

Under the simplifying assumption that the radial dependence of the mean free path is the same over the corresponding range of rigidities, we always obtain asymptotically a constant ratio p/α. The reason is that we obtain for both particle types a

t^{-k}-dependence (with constant k) late in the event. A similar result has been obtained earlier by Hakura (1967). He considers protons and α-particles of the same rigidity, which is relevant for the penetration of cosmic ray particles to a given location in the geomagnetic field. In this case the ratio of protons to α-particles of the same *rigidity* is decreasing as a function of time; Hakura has used this result to discuss the possible effect of solar α-particles in the late phase of a PCA event.

It has been shown that even for the restricted model assumptions contained in Equation (2) a large variety of solutions for $I_p(t)/I_\alpha(t)$ is possible. It should be noted that in constructing the curves of Figure 4 only the simplest possible assumption has been used: the value of s which describes the radial dependence of the mean free path is the same for protons and α-particles differing by a factor 2 in rigidity. A rigidity dependence of s leads to an asymptotic behaviour $I_p(t)/I_\alpha(t) \sim t^a$ for large t. For $s_p \lesssim s_\alpha$ we have $a \gtrless 0$; in principle, therefore, we could expect arbitrarily large as well as arbitrarily small proton-to-α ratios during the course of an event, simply from propagation effects.

The finite sphere model (no diffusion beyond a radial distance R_D) gives an exponential decay late in the event (Parker, 1963; Burlaga, 1967); in this case we would also expect $I_p/I_\alpha \to \infty$ for $\lambda_\alpha > \lambda_p$ since the decay time is inversely related to the diffusion coefficient.

5. Final Remarks

In the discussion we have neglected processes of convection by the solar wind, adiabatic deceleration and transport processes under the influence of time varying magnetic fields (as e.g. moving magnetic field structures with trapping properties). Fisk and Axford (1968) have shown that the adiabatic deceleration may cause appreciable deviations in the decay phase of an event. Parker (1963) has pointed out that one obtains a different diffusion equation, if the particle scattering is the result of displacements by the scattering medium. It should be noted that the general relation (1) will also no longer hold in this case.

Further studies of the proton-to-α ratios during solar events should result in

(i) determination of the injection ratios,

(ii) check of the model assumptions about the radial dependence of the mean free path,

(iii) indications for necessary refinement of the propagation model.

Point (ii) will allow a better determination of the local mean free path (λ_E), which is up to now not uniquely defined from a fit of the diffusion model to the observed intensity-time-profile. It is of particular interest to obtain the local mean free path, since it can also be derived theoretically from the interplanetary magnetic field spectral densities (Jokipii, 1966, 1967; Wibberenz et al., 1969).

Acknowledgements

We thank Professor Dr. E. Bagge for the continuous support of this work and his interest in it.

We thank our colleagues G. Green, R. Krieger, and E. Rode for their contributions to development, construction and testing of the instruments and for assistance during launch preparations.

We express our gratitude to the ESRO teams responsible for the successful C 49 rocket launchings.

We acknowledge useful discussions with Dr. Barouch, Dr. Engelmann and Dr. Van Beek concerning the particle measurements during this solar event.

This work was supported by the Bundesministerium für wissenschaftliche Forschung.

References

Barouch, E., Engelmann, J., Gros, M., Koch, L., and Masse, P.: 1969a, XIth Cosmic Ray Conference, Budapest, 1969 (in press).
Barouch, E., Engelmann, J., Gros, M., Koch, L., and Masse, P.: 1969b, this volume, p. 448.
Biswas, S. and Fichtel, C. E.: 1965, *Space Sci. Rev.* **4**, 7091.
Biswas, S., Fichtel, C. E., Guss, D. E., and Waddington, C. J.: 1963, *J. Geophys. Res.* **68**, 3109.
Burlaga, L. F.: 1967, *J. Geophys. Res.* **72**, 4449.
Bryant, D. A., Cline, T. L., Desai, U. D., and McDonald, F. B.: 1965, *Astrophys. J.* **141**, 478.
Datlowe, D., L'Heureux, J., and Meyer, P.: 1969, XIth Cosmic Ray Conference, Budapest, 1969 (in press).
Durgaprasad, N., Fichtel, C. E., Guss, D. E., and Reames, D. V.: 1968, *Astrophys. J.* **154**, 307.
Engel, A. R., Balogh, A., Elliot, H., Hynds, R. J., and Quenby, J. J.: 1969, XIth Cosmic Ray Conference, Budapest, 1969 (in press).
Fichtel, C. E. and McDonald, F. B.: 1967, *Ann. Rev. Astron. Astrophys.* **5**, 351.
Fisk, L. A. and Axford, W. I.: 1968, *J. Geophys. Res.* **73**, 4396.
Freier, P. S. and Webber, W. R.: 1963, *J. Geophys. Res.* **68**, 1605.
Green, G., Krieger, R., Wibberenz, G., and Witte, M.: 1969, XIth Cosmic Ray Conference, Budapest, 1969 (in press).
Hakura, Y.: 1967, *J. Geophys. Res.* **72**, 1461.
Hasselmann, H. and Wibberenz, G.: 1968, *Z. Geophys.* **34**, 353.
Jokipii, J. R.: 1966, *Astrophys. J.* **146**, 480.
Jokipii, J. R.: 1967, *Astrophys. J.* **149**, 405.
Kahler, S. W., Primbsch, J. H., and Anderson, K. A.: 1967, *Solar Phys.* **2**, 179.
Krimigis, S. M.: 1965, *J. Geophys. Res.* **70**, 2943.
Lanzerotti, L. J.: 1969, *J. Geophys. Res.* **74**, 2851.
Lanzerotti, L. J. and Robbins, M. F.: 1969, 'Solar Flare α-to-Proton Ratio Changes following Interplanetary Disturbances', Preprint.
Parker, E. N.: 1963, *Interplanetary Dynamical Processes*, Interscience, New York.
Quenby, J. J., Balogh, A., Engel, A. R., Elliot, H., Hedgecock, P. C., Hynds, R. J., and Sear, J. R.: 1969, XIth Cosmic Ray Conference, Budapest 1969 (in press).
Van Beek, H. F. and Van Gils, J. N.: 1969, COSPAR Symposium, Prague (to be published in *Space Research*, Vol. 9).
Wibberenz, G., Hasselmann, K., and Hasselmann, D.: 1969, XIth Cosmic Ray Conference, Budapest, 1969 (in press).

ROCKET OBSERVATIONS OF PROTONS AND α-PARTICLES
AT ANDØYA AFTER THE SOLAR FLARES OF
24TH–25TH FEBRUARY 1969

H. F. VAN BEEK and G. A. STEVENS

Space Research Laboratory, Utrecht, The Netherlands

Abstract. During the Polar Cap Absorption event occurring after the flares of 24th and 25th February 1969, three rockets have been launched successfully at Andenes (Norway).

Riometer and magnetometer recordings show that the P.C.A. event is a fairly weak one.

In this paper the energy spectra of protons and α-particles in the energy range from 2.3 to 70 MeV, obtained during these flights are given. The main results are given in Figure 5. The α/proton ratio found, is 0.013 in the range 4–20 MeV/nucleon at both flights. The mean free path of the protons is estimated to be 0.06 AU.

1. Introduction

The PCA campaign of the European Space Research Organization started in August 1968 at Kiruna and was moved later on to Andenes, Andøya (geographic latitude 69° 30′).

Two Centaure and seven Arcas rockets were available for this campaign. Apart

Fig. 1.   Photograph of the instruments.

V. Manno and D. E. Page (eds.), Intercorrelated Satellite Observations Related to Solar Events, 508–514. All Rights Reserved

from instruments of other groups, particle detectors of the Space Research Laboratory Utrecht have been launched with these rockets.

The Arcas instrument (experiment R 127) consists of one solid state detector telescope for the detection of protons. The Centaure instrument (experiment R 3) is equiped with two particle telescopes (see Figure 1). The proton telescope is similar to the one used in the Arcas instrument, the other is particularly designed to detect α-particles.

2. Description of the Instruments

The proton telescopes of both experiments consist of a stack of three solid state detectors with absorbers in between. In fact this configuration justifies the name telescope (see Figure 2). In this design fully depleted surface barrier detectors have been used with thicknesses of 200 and 400 microns. The mass and energy of a particle can be derived from the energy losses in two detectors of a telescope.

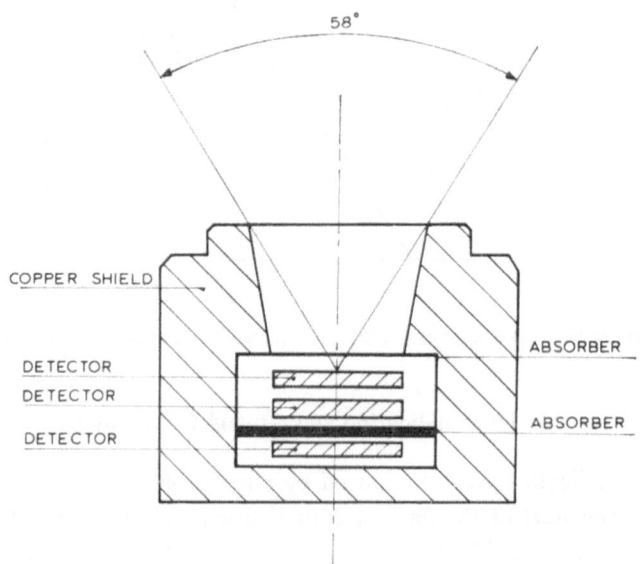

Fig. 2.   Schematic cross section of a particle telescope.

Protons can be analysed in four energy ranges viz.: 2.3 to 5.5, 5.5 to 13, 13 to 30 and 30 to 70 MeV.

The Centaure α-particle telescope differs slightly from a proton telescope because it was particularly designed to detect α-particles in the energy range from 5.5 to 70 MeV, viz.: 5.5 to 8.5, 8.5 to 13, 13 to 27 and 31 to 72 MeV.

Due to a suitable choice of detector and absorber thicknesses the rejection of interactions caused by other particles than those for which a telescope is designed, is better than 99%.

The copper shield around the telescope is an effective protection against low energy particles. At the same time this shield defines the opening angle of about 0.7 steradian.

Events caused by high energy particles, coming from directions outside of the cone of view and penetrating through the copper shield will be rejected by very strict impulse height and coincidence requirements. Therefore, these events will not be recorded in the registers. The sensitive area of the detector is 2 cm². The angle between detector axis and rocket spin axis is 30°, the angle between magnetic field vector and spin axis is 49° (see Figure 3). Due to this configuration an integration of the particle flux over all different pitch angles will be obtained.

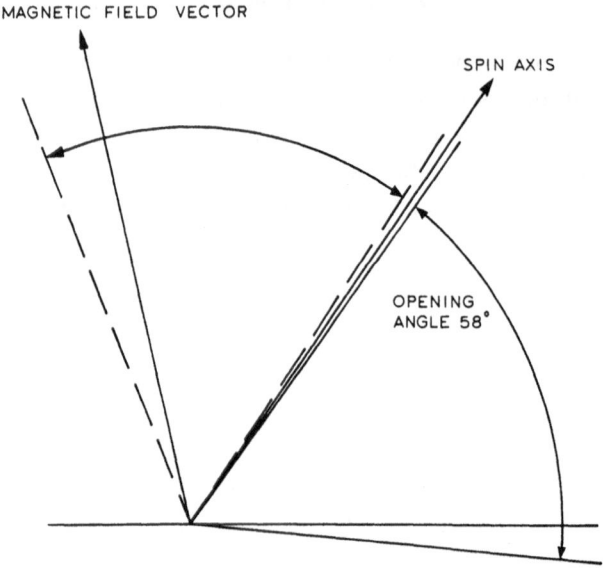

MAGNETIC FIELD VECTOR

SPIN AXIS

OPENING
ANGLE 58°

Fig. 3.   Position of the rocket spin axis and opening angle with regard to the magnetic field vector.

3. Flare Indication and Ground Observations

The first flare indication was announced by Kiruna observatory about a quarter of an hour after the start of the flare of 24th February 1969 which was at 23h08 UT. Afterwards this flare was identified as an importance 2B flare which arose in the N13 W33 group. A 10 cm radioburst which occurred simultaneously had a duration of 74 min and a maximum intensity of 410% above the quiet sun level. The first flare indication, however, was obtained by Very Low Frequency observations, which showed a Sudden Phase Anomaly effect.

Riometer absorption started at 01h30 UT and continued during about half an hour.

A second flare appearing at 09h07 UT of 25th February accompanied by a type IV radioburst of 600% with a duration of 50 min gave a riometer absorption during the day in the range 0.5 to 1.5 dB only (see Figure 4a). Effects caused by flares appearing in the afternoon of the same day might have interfered with those from the second flare. The riometer frequency was 27.6 MHz.

The magnetometer gave some indication of a magnetic storm on the 27th February, as can be seen in Figure 4b.

The 25th February at noon VLF measurements gave a clear indication of some proton precipitation; but the event could only be classified as a medium size PCA event.

Balloon observations at Kiruna at about 13h00 UT indicated a six times higher proton flux than during normal circumstances. After the confirmation of a Polar Cap Absorption event obtained both by VLF measurements and balloon observations, it was decided to launch the first Arcas rocket.

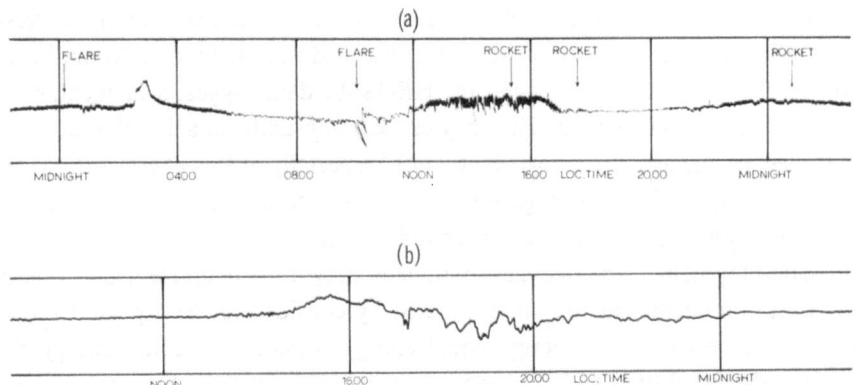

Fig. 4.   (a) Recording of riometer 25th February 1969. – (b) Recording of magnetometer 27th February 1969.

4. Launching Sequence

At 14h21 UT the Arcas rocket made a successful flight. Immediately after the flight the proton intensities obtained with this instrument lead to the decision to launch the first Centaure rocket. This launch took place at 16h33 UT. After that flight the particle flux appeared to be decreased already, as compared with the flux measured by the Arcas instrument. Therefore it was decided to launch the second Centaure at midnight instead of the next morning as was planned in the original time schedule. A second Arcas launch at 21h37 UT was a failure.

After the second Centaure launch the particle flux and riometer absorption were considered to be too low to justify further Arcas firings.

5. Results

Least square fittings of the counting rates vs. energy to an exponential or power law function of either energy or rigidity showed a preference for the power spectrum.

The measurements together with a fitting according to $dn(E)/dE = CE^{-\gamma}$ is shown in Figure 5. The boundaries of the energy channels in which the particles have been observed are marked. Along the horizontal axis the energy of the particles is given from 1 to 100 MeV. The number of particles counted is given along the other axis per MeV cm^2 sec. steradian from 10^{-2} to 10^2. Also the statistical error in the measurements is indicated. In Figure 6 the intensity-time profiles are given, one for each channel.

6. Discussion

Figure 5 shows clearly a steepening of the spectra with increasing time, suggesting a maximum in the proton intensity close to the time of the first flight (i.e. 5.1 hours after the flare of 25th February). The absolute intensity, however, is not a simple decreasing function of time, especially at the lower energy channels as shown in Figure 6. The increase of the intensity in these channels with time after the second flight stress the impossibility to fit the intensity-time data into a diffusion model. As indicated by Barouch *et al.* (this volume, p. 448), the HEOS A1 data suggest a filamentary structure of the interplanetary field. This may cause a non-uniform azimuthal distribution of particles, visible as an intensity fluctuation, especially at low energies. So one may conclude that apart from propagation effects, also spatial distributions play a part. One may still get an idea about the magnitude of the mean free path λ, by considering the time development of the steepness parameter only. First of all the general character of the diffusion coefficient may be estimated by considering the α/proton ratio as a function of time for equal energy, equal energy/nucleon and equal energy/charge. Supposing the fitted lines to be the right ones it turns out that this ratio is the same within 10% for equal energy/nucleon (equal velocity) in the range 4–20 MeV/nucleon, the average ratio being 0.013. In the simple isotropic diffusion model, this means a rigidity independent mean free path λ. Supposing the source spectrum to be $\sim E^{-s}$

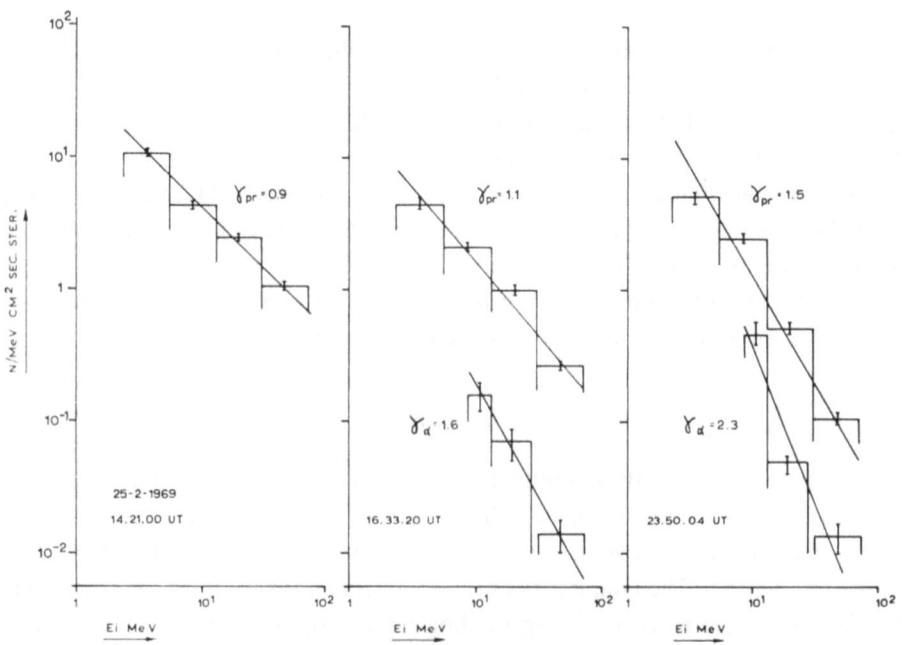

Fig. 5. Particle energy spectra obtained during three flights.

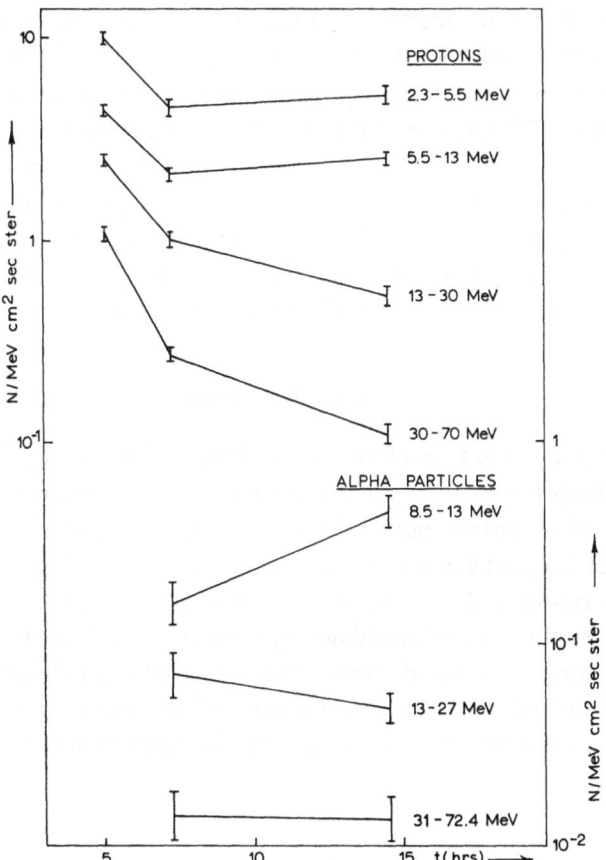

Fig. 6. Particle intensity as a function of time.

and putting $D = \frac{1}{3}\lambda v$, the intensity as a function of time and energy in this model reads:

$$I = nv = CE^{-s-1/4}(1/\lambda t)^{3/2}\exp(-r^2/4\,\lambda vt).\qquad(1)$$

Azimuthal distributions can be included in C. One can see from (1) that for low energies and small λt the exponential may play a part in the energy spectrum, expressing itself in a bending over of the spectrum in a $\ln I - \ln E$ plot at the lower energy side, becoming less important at later times. This seems to be confirmed in Figure 2 given by Wibberenz and Witte (this volume, p. 499), which shows a combined picture of their and our results.

A fitting of the spectra to such a function, however, requires data at high as well as at low energies. So a different procedure is followed to estimate the value of λ. From (1) the steepness time evolution is derived to be:

$$d\ln I/d\ln E = -(s+\tfrac{1}{4}) + 3/8\lambda vt.\qquad(2)$$

By the differentiation the spatial distribution effects have disappeared. The fitting parameters γ in our measurements can be looked upon as averages of (2) over the energy ranges concerned. Interpreting it that way, one finds from the last two flights $\lambda_{pr} = 0.037$ AU and $\lambda_{\alpha} = 0.033$ AU, the average λ_{pr} of the three flights being 0.06 ± 0.02 AU. The values confirm the rigidity independency of the mean free path after the second flight (i.e. 7.3 hours after the flare). The average value for the protons indicate a maximum for 30 MeV protons at about 5 hours after the flare. Although these values are subjected to great uncertainties, the general picture is one of a very fast propagation of particles from the source to the earth, in agreement with neutron monitor data. This might be due to the flare on the preceding day (23h08 UT) at the same active region, stretching the field lines in the interplanetary medium.

Acknowledgements

Grateful thanks are due to Professor C. de Jager and Dr. L. D. de Feiter for their stimulating discussions and the encouragement during the course of the work. The electronic part of the instruments was built under supervision of Ir. J. N. van Gils. The instrument was designed and built by L. van den Brink, B. F. Drenth, G. J. van Dijen, H. P. C. van Heusden, J. E. Kikken, T. Koppen, J. Mulder, J. A. P. Weber and W. Zandee. Assistance at the launching range was given by H. B. Rouws and A. Vermeer. Thanks are due to Mr. P. Polak (Institute for Nuclear Research, Amsterdam) who provided the radioactive calibration sources. The riometer and magnetometer recordings given in Figure 4 were provided by the Norwegian Defense and Research Establishment.

Reference

Van Gils, J. N., Van Beek, H. F., De Feiter, L. D., and Hendrickx, R. V.: 1969, *Planetary Space Sci.* **17**, 255.

ELECTRON DENSITY OBSERVATIONS
DURING THE PCA EVENT OF 25 FEBRUARY 1969

M. JESPERSEN

Danish Space Research Institute, Lyngby, Denmark

and

J. TRØIM and B. LANDMARK

Norwegian Defence Research Establishment, Kjeller, Norway

Abstract. The paper discusses results from rocket observations of electron density during the PCA event of 25 February 1969. Results were obtained in two Sidewinder-Arcas rockets. The electron densities are related to results from high energy particle measurements. The effective loss rates derived are interpreted in terms of a two-ion model of recombination. It is shown that the interpretation is consistent with similar results obtained during quiet and auroral absorption conditions.

1. Introduction

During the PCA event after the solar flare of 25 February 1969, a series of Centaure and Sidewinder-Arcas rockets were launched by ESRO. These rockets carried a number of experiments to measure protons and α-particles.

In two of the rockets measurements were also made of the *D*-region electron density, by means of the Faraday rotation technique. It is the main purpose of the present paper to present the results from these observations, but some preliminary results obtained from a comparison of these data with those obtained from simultaneous particle measurements will also be included.

In Section 2 the results from the electron density observations will be presented and compared with similar results obtained during quiet and auroral absorption conditions. Results were obtained in two Sidewinder-Arcas rockets. One of these was only partially successful in that a protective belt did not release so that the particle measurements could be made. This rocket only reached a peak altitude of 68 km.

In Section 3 the results will be related to those of particle measurements in order to derive effective loss rates in the *D*-region, and the interpretation of these results will be discussed in Section 4.

2. Electron Density Observations

Linearly polarized radio waves were transmitted from the ground and received on a linearly polarized antenna in the spinning rocket. Measurements were made at the three frequencies 3.8, 7.8, and 15 MHz. From the observed amplitude patterns the electron density can be derived.

In Figure 1 the electron density profiles observed in the two Sidewinder-Arcas rockets are shown. The riometer at the range, which is operated at 27.6 MHz, showed

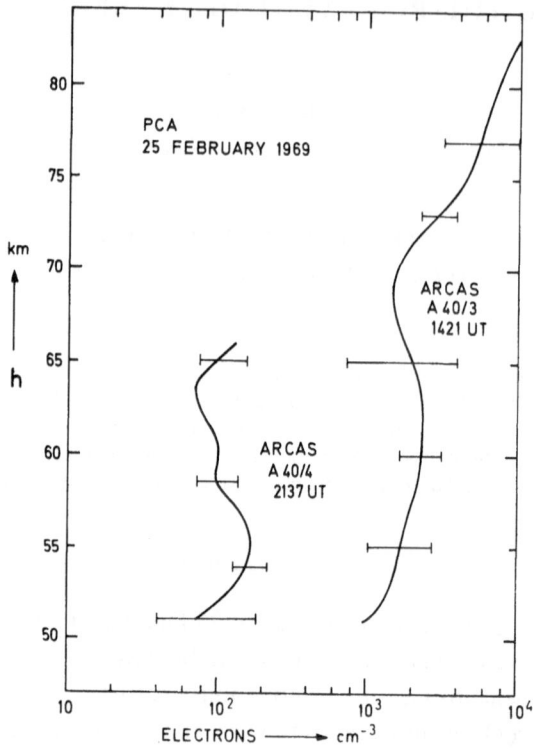

Fig. 1.   Electron density profiles obtained during the PCA event.

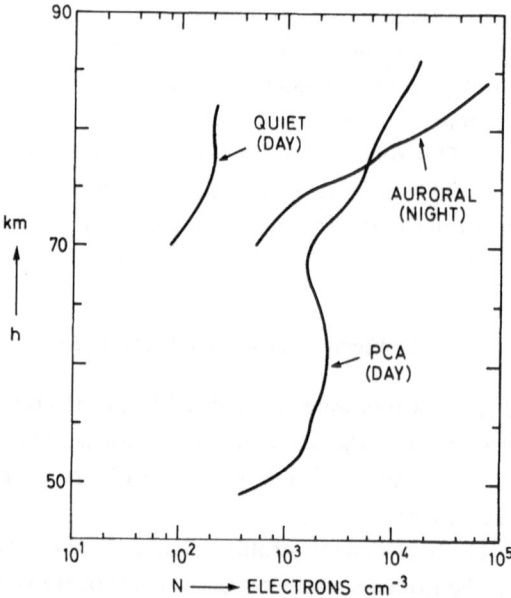

Fig. 2.   Electron density profiles obtained during different conditions.

an absorption of about 2 decibels at 1421 UT when the rocket A 40/3 was launched, and 0.5 decibels at 2137 UT when A 40/4 was launched.

It is seen that for the day time firing a nearly constant electron density of about 10^3 electrons per cm^3 is observed over the heights range from 50 to 70 km. At greater height the density is increasing with height. For the night time firing, the electron density is reduced by about a factor of 10.

In Figure 2 the electron density profile obtained in A 40/3 is compared with a profile obtained during auroral absorption conditions discussed by Folkestad and Armstrong [1] and a quiet day electron density profile discussed by Reid [2]. This profile was obtained by Mechtly and Smith [3] in a rocket flight over Wallops Island. The auroral data were obtained at night, whereas the mid-latitude quiet day curve refers to measurements at a solar zenith angle of 60° during the month of September 1965.

3. Deduced Loss Rates

From the results of particle measurements reported by Van Beek [4], it is possible to make an estimate of the ion production rates at the time of the rocket A 40/3. Measurements were made of both protons and α-particles, but the results show that only the protons were of importance for the ion production process.

Measurements were made in 4 energy intervals, and a proton energy-spectrum was obtained from these results. The protons were then divided into 1 MeV wide intervals, and the production rate for the particles in each interval was obtained.

The total production rate as a function of height was obtained by adding the contributions from the particles within the different intervals. In this treatment it was

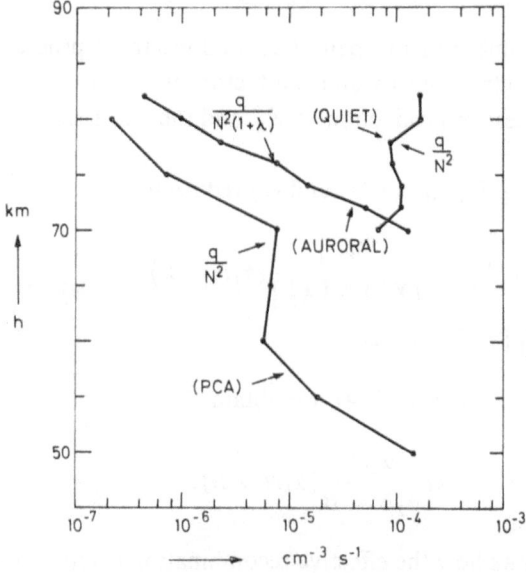

Fig. 3. Effective recombination coefficients for different conditions.

assumed that all particles are coming in vertically. For the qualitative arguments of this paper this assumption is satisfactory.

Figure 3 shows the effective recombination coefficient α_{eff}, defined as q/N^2, as a function of height for the three cases considered. In the auroral rocket [1], λ (ratio of negative ions to electrons) was also measured, and in this case $\alpha_{eff}/(1+\lambda)$ is plotted. Negative ions are probably not important during day time conditions for heights above 70 km. In deriving the effective recombination coefficient for quiet conditions, Reid [2] assumed that ionization of NO by Lyman-α radiation dominates, and made use of reasonable Lyman-α intensities and a modified Barth distribution (Barth [5]) for the NO density.

4. Discussion

A noticeable feature of the results presented in Figure 3 is the large difference between the results obtained during disturbed conditions (both PCA and Auroral) and quiet conditions, particularly at the higher altitudes.

Folkestad and Armstrong [1] have shown that the results obtained during auroral conditions are consistent with those obtained during quiet conditions in terms of a two-ion recombination model suggested by Haug and Landmark [6].

In their treatment, Haug and Landmark considered the two ions X_1^+ and X_2^+. The one ion, X_1^+, has an electron-ion recombination coefficient that is much lower than that of the second ion, X_2^+. They further assumed that the production rate of the ion X_2^+ is proportional to the density of the ion X_1^+. In that case one has, neglecting ion-ion recombinations

$$d[X_1^+]/dt = q - B[X_1^+] - \alpha_1[X_1^+] \cdot N \tag{1}$$

$$d[X_2^+]/dt = B[X_1^+] - \alpha_2[X_2^+] \cdot N \tag{2}$$

where q is the number of ion pairs (X_1^+ and electron) produced per unit time and volume, N the electron density and B a factor given as the product of the rate coefficient for the process of production of X_2^+ and the densities of the species involved in this process.

For steady state, Equations (1) and (2) reduce to

$$q = N(1 + \lambda) \frac{[X_1^+]}{[X_1^+] + [X_2^+]} (\alpha_1 N + B) \tag{3}$$

$$[X_2^+]/[X_1^+] = B/\alpha_2 N. \tag{4}$$

Combining Equations (3) and (4), we obtain

$$q = N(1 + \lambda) \frac{\alpha_2 N}{\alpha_2 N + B} (\alpha_1 N + B). \tag{5}$$

Equation (5) shows how the effective recombination coefficient α_{eff} depends upon the number density of free electrons.

For values of B in the range $\alpha_1 N \ll B \ll \alpha_2 N$, we obtain from Equation (5)

$$q = (1 + \lambda) BN. \qquad (6)$$

In this case, therefore, N is proportional to q rather than to $q^{1/2}$ as normally assumed.

In Figure 4 the height dependance of q/N ($q/N (1+\lambda)$ for the auroral case) is shown. The illustration brings out the point already shown by Folkestad and Armstrong [1] that the results obtained during auroral absorption conditions are consistent with those obtained during quiet conditions in terms of the proposed model of recombination.

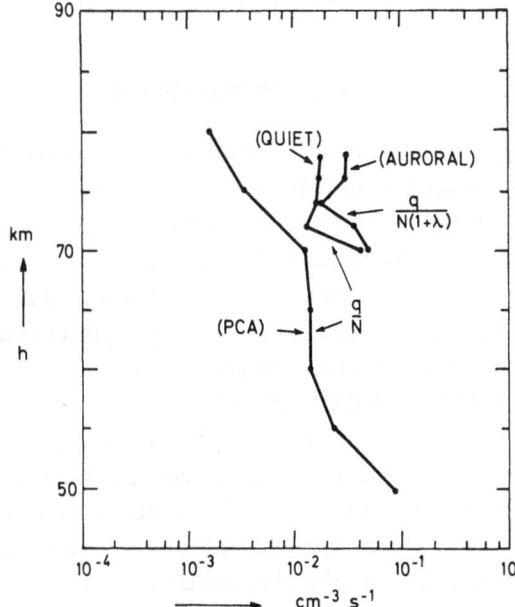

Fig. 4.   The loss rate B for different conditions.

The results obtained during the PCA event show smaller values of the loss rate whether we consider a q/N^2 or a q/N process. The best agreement, however, is obtained with the last type of law.

In order to explain the rapid recombination observed in the normal D-region, Reid [2] suggested that hydrated cluster-ions may be important. If this is so, the quantity B may be related to the water vapor content in the D-region. It is pointed out that a stratospheric warming occurred in the period 21–25 February 1969 and that this might result in a reduction of the water vapor content at D-region levels and thus in the quantity B.

References

[1] K. Folkestad and R. J. Armstrong: 1970, *J. Atmos. Terr. Phys.* (in press).
[2] G. Reid: *J. Geophys. Res.* (in press).
[3] E. A. Mechtly and L. G. Smith: 1968, *J. Atmos. Terr. Phys.* **30**, 1555.
[4] H. F. van Beek: this volume, p. 508.
[5] C. A. Barth: 1966, *Ann. Geophys.* **22**, 198.
[6] A. Haug and B. Landmark: *J. Atmos. Terr. Phys.* (in press).

ROCKET-BORNE D-REGION PROBE MEASUREMENTS

J. MURDIN and T. S. BOWLING

Mullard Space Science Laboratory, Dorking, Surrey, England

1. Introduction

Two Arcas rockets, which were part of the ESRO PCA campaign, were launched from Andøya on 25th February 1969 after the onset of a solar proton event. A positive ion probe experiment formed part of the payload instrumentation, enabling the relative positive ion density profiles from 60 to 70 km for the two firings to be determined.

2. Experimental Method

The experiment utilised a cylindrical electrostatic probe which consisted of a 10 cm long, rhodium plated steel wire of diameter 0.05 cm biased at -2.7 volt with respect to the rocket body. The probe current was measured by a d.c. transistor amplifier, which typically had a dynamic range of 0 to 5.0×10^{-9} amps. For the initial portion of the flight the probe was protected by a belt which was released at approximately 55 km altitude, then the probe hinged into its deployed position, approximately 70 cm behind the tip of the nose cone and perpendicular to the longitudinal rocket axis. A similar system was used by Bowling *et al.* (1967).

The amplifier zero level was determined after probe deployment. For a daytime firing only measurements made when the probe was completely shadowed by the rocket were considered. These times were determined by reference to a solar sensor which was part of the payload instrumentation. The shadowed probe current was averaged over approximately 30° of rocket rotation. For a night-time firing the probe current was averaged over approximately 30° of rocket rotation about its maximum value in the rotation.

Since, in the *D*-region, the particle mean free path is usually less than the debye length, for example below 75 km there are at least ten molecular mean free paths within a debye length if the positive ion density is less than 10^3 cm^{-3}, collisions play an important role in charged particle collection by electrostatic probes. Hoult (1965) has shown that for a blunt subsonic electrostatic probe ion collection is mobility controlled. The Arcas rocket travels through the *D*-region with a supersonic speed but Sonin (1967) has shown that in the limit of a strongly attracting field the ion collection of a blunt probe is independent of flow effects and the expression for the probe current is the same as that for a subsonic or static probe (Hoult, 1965).

$$i = eN_+ KEA$$

where i is the probe current, e is the ion charge, N_+ is the ion density, K is the ion mobility, A is the probe collecting area, and E is the magnitude of the electric field at the probe.

V. Manno and D. E. Page (eds.), Intercorrelated Satellite Observations Related to Solar Events, 520–523. All Rights Reserved

For a cylindrical probe this reduces to

$$i = eN_+ 2\pi LKV_p / \ln(S/r) \tag{1}$$

where L is the probe length, r is the probe radius, S is the sheath radius, and V_p is the magnitude of the probe to plasma voltage.

Two August 1968 firings of an Arcas borne probe experiment consisting of three different diameter collectors swept in voltage showed that below 80 km the probe current was proportional to the probe voltage for attracting potentials greater than approximately 1 volt. Also the relative probe currents for the three probes were in good agreement with these predicted by Equation (1).

At a fixed altitude the relative positive ion density between two firings, if subscript 1 and 2 represent the two firings, is given by

$$R = \frac{N_{+1}}{N_{+2}} = \frac{i_1 K_2 V_{p2}}{i_2 K_1 V_{p1}}.$$

Now K is inversely proportional to p, the gas pressure, and a weak function of temperature (McDaniel, 1964) and since p, over the altitude range of interest and at a

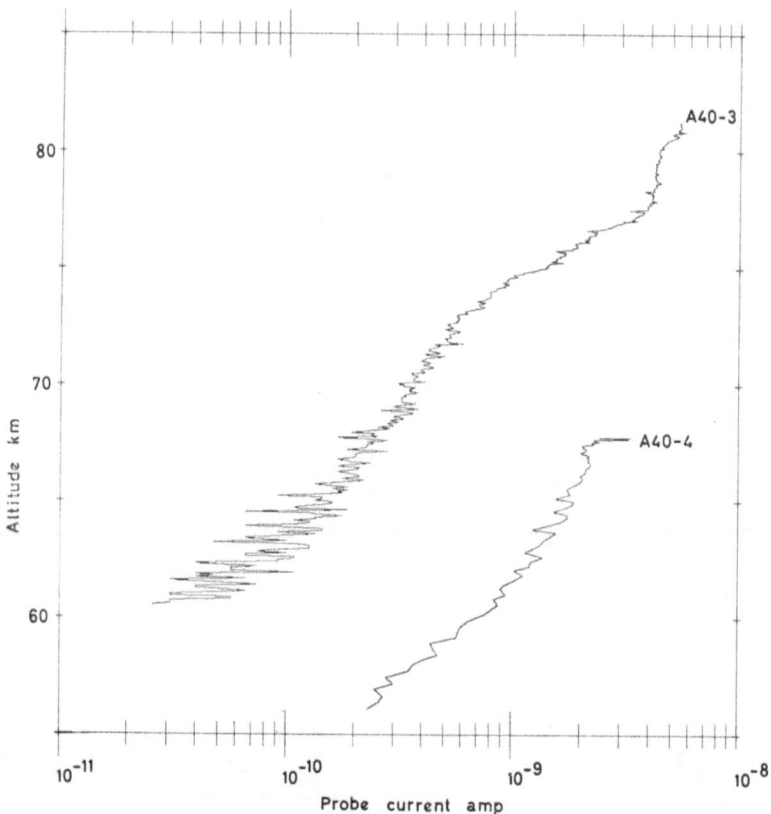

Fig. 1. Probe current for A40-3 and A40-4 as a function of altitude.

fixed altitude, does not have a diurnal variation greater than 10%, (Cole and Kantor, 1965) the greatest uncertainty in R is due to the uncertainty in V_p. The two August 1968 Arcas firings, previously mentioned, indicated a daytime rocket potential as high as $+1$ volt below 80 km, and presumably the rocket would take a negative potential for a night-time firing. The unknown contact potential between the probe and rocket introduces further uncertainty in the value of V_p.

3. Results

Two Arcas rockets, with the positive ion probe experiment as part of their instrumentation, were launched from Andøya on the 25th February 1969. One, designated A40-3, was launched at 1421 UT, had an apogee of 86 km and a spin rate approximately 5 rps. The other, designated A40-4, was launched at 2137 UT, with an apogee of 68 km and a low spin rate. The probe current, for the ascents of A40-3 and A40-4, is shown as a function of altitude in Figure 1. In the altitude range 61 to 68 km at any particular altitude the A40-4 probe current is approximately an order of magnitude higher than the A40-3 probe current. Taking the rocket potential as (1 ± 1) volt for the

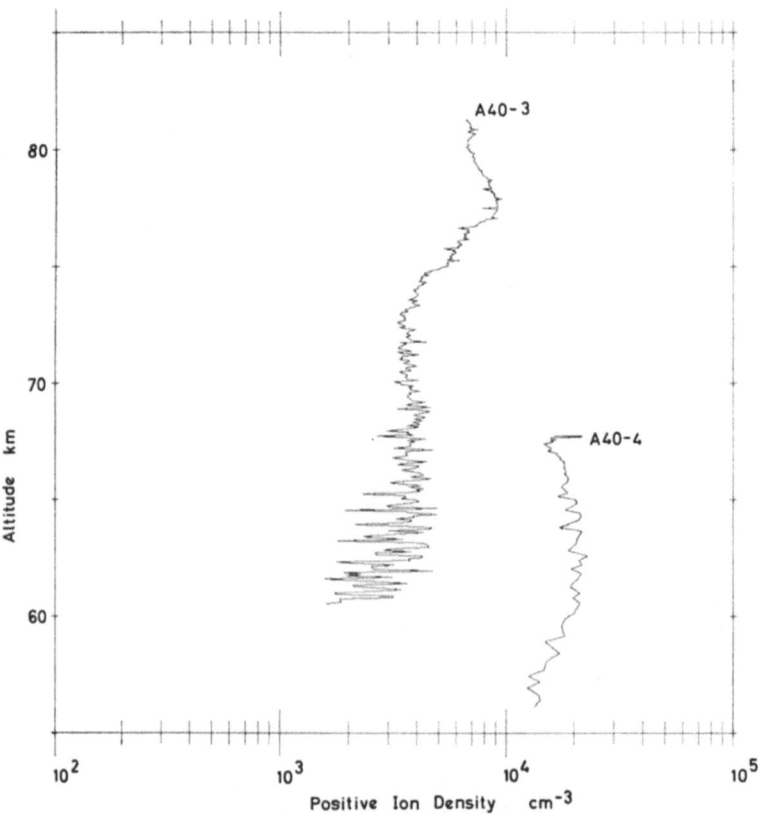

Fig. 2.   Positive ion density for A40-3 and A40-4 as a function of altitude.

day firing (A40-3) and (-1 ± 1) volt for the night firing (A40-4) and ignoring variations in gas density at a fixed height between the two firings we obtain at a fixed altitude in the range 61 to 68 km.

$$R = \frac{N_+(\text{A40-4})}{N_+(\text{A40-3})} = 4.5$$

with lower and upper limits of 1.5 and 10. Whatever expression for probe current is used, R would lie within these limits as long as at a fixed height the probe current is proportional to the positive ion density and no stronger a function of probe voltage than linear (Bowling *et al.*, 1967). The positive ion density was derived from the probe current employing Equation (1) and using a reduced mobility of 1.8 cm^2 volt^{-1} sec^{-1} and probe to plasma voltage of -3.7 volt for A40-4 and -1.7 volt for A40-3. The density profiles are shown in Figure 2, and may be in error by up to a factor 3, due to the uncertainties in V_p and K.

References

Bowling, T. S., Norman, K., and Willmore, A. P.: 1967, *Planetary Space Sci.* **15**, 1035.
Cole, A. E. and Kantor, A. J.: 1965, *Handbook of Geophysics and Space Environments*, AFCRL.
Hoult, D. P.: 1965, *J. Geophys. Res.* **70**, 3183.
McDaniel, E. W.: 1964, *Collision Phenomena in Ionised Gases*, Wiley.
Sonin, A. A.: 1967, *J. Geophys. Res.* **72**, 4547.

Discussion

Reid: Did you compare your measurements with Dr. Landmark's electron density measurements?

Murdin: For the daytime firing the positive ion densities are about the same as the electron densities measured by Dr. Landmark.

Landmark: The electron densities measured during night-time at a height of about 55 km are much smaller than the positive ion densities. It is difficult to explain this high positive ion density content by any mechanism. For the daytime measurements there is no dependence on Λ, which is difficult to understand.

ELECTRON DENSITY MEASUREMENTS IN THE
THERMAL PLASMA OF THE MAGNETOSPHERE USING
A LANGMUIR PROBE

R. FREEMAN, K. NORMAN, and A. P. WILLMORE

University College London,
Mullard Space Science Laboratory

1. Introduction

An experiment designed to measure the ambient electron density and temperature in the magnetosphere, was provided by University College London for inclusion in the payload of the NASA Orbiting Geophysical Observatory satellite, OGO-V. The satellite was launched into an eccentric, near-equatorial orbit on March 4, 1968, and the experiment has continued to operate from that time until well after the solar flare event of February 25, 1969. Results taken at the time of the February 1969 event are presented in this paper and compared with earlier satellite passes during periods of different geomagnetic activity. The data consist of unrefined electron density profiles plotted directly from the processed data tapes. A pass early in the life of the satellite is used to identify the dominant features of the magnetosphere because it has been possible to compare the electron density data with those obtained simultaneously by other experiments on the same satellite.

2. The Experiment

The configuration of the OGO-V spacecraft is shown in Figure 1. A spherical Langmuir probe, 6 cm diameter, was mounted on a 50 cm stem projecting from the end of the EP-1 boom, which was approximately 2.5 m long. The associated electronics module was mounted within the main body section of the spacecraft.

The probe was swept repetitively over a 10 V range at 1 V per second and the extremes of the sweep range relative to the spacecraft were variable on command from −20 V to +20 V, to allow for variation of this order in the satellite potential. An A.C. modulation technique was used to analyse the current-voltage relationship for the probe, from which the local electron temperature and density were deduced.

The current to a Langmuir probe in the electron retarding region just negative of local plasma potential is given by the expression in Equation (1)

$$ i = i_0 \exp\left(\frac{-eV}{kT}\right). \tag{1} $$

Equation (1) is the form of the electron current to the probe but in the region considered, the positive ion contribution is negligible and 'i' becomes the total probe

V. Manno and D. E. Page (eds.), Intercorrelated Satellite Observations Related to Solar Events, 524–534. All Rights Reserved
Copyright © 1970 by D. Reidel Publishing Company, Dordrecht-Holland

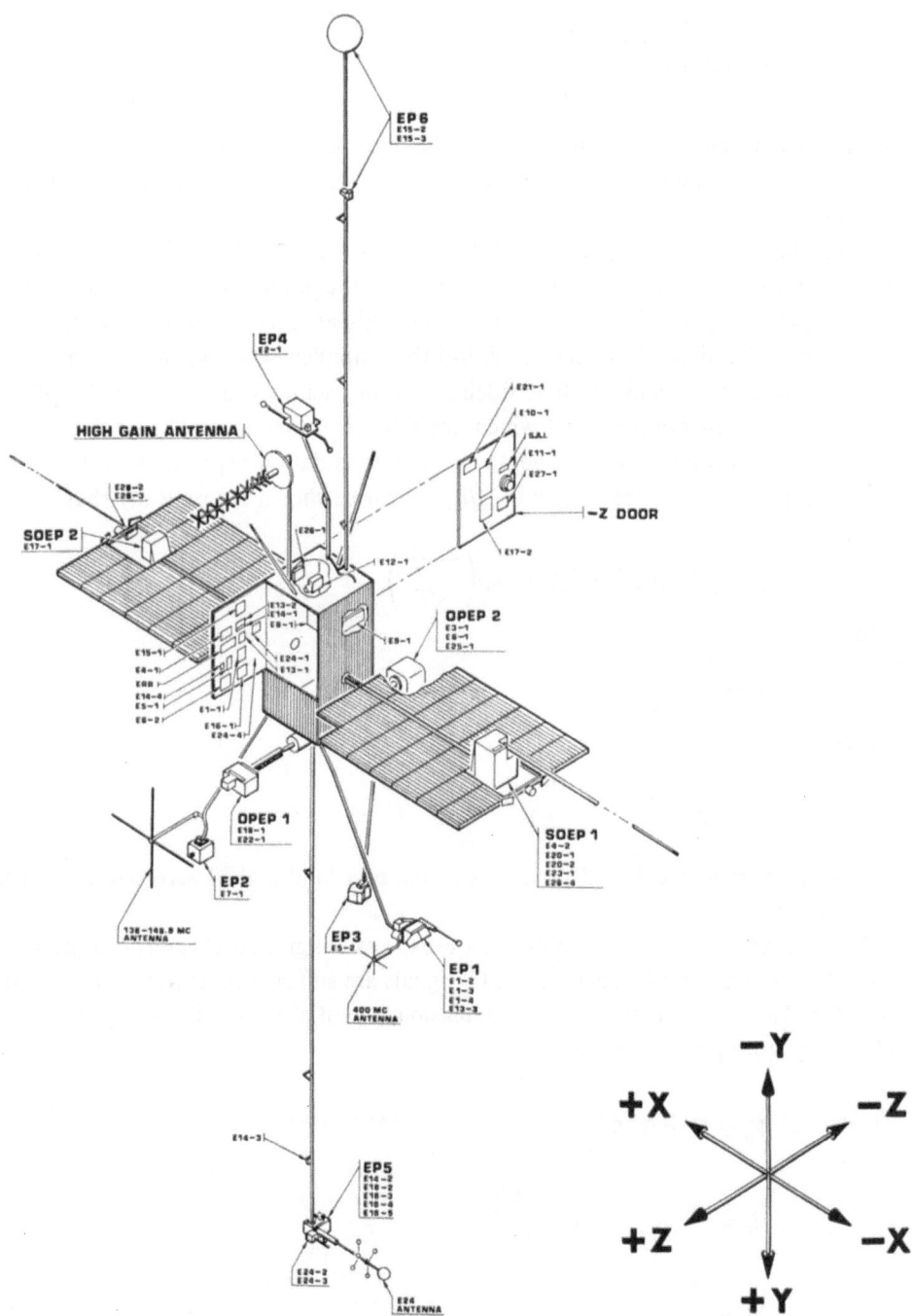

Fig. 1. Drawing of the OGO-5 satellite showing the experiment locations.

current. Negative ions are neglected. The probe current at plasma potential 'i_0' is given by Equation (2).

$$i_0 = nAe \left(\frac{kT}{2\pi m} \right)^{1/2},$$ (2)

where m=the electronic mass, e=the electronic charge, n=the electron density, T=the electron temperature, k=Boltzmann's constant, and A=the surface area of the probe.

Two sinusoidal voltage signals of frequencies 'f_a' and 'f_b' and amplitudes 'V_a' and 'V_b' respectively are simultaneously applied to the probe and the A. C. component of the current is coupled into an amplifier with an acceptance bandwidth $f_a \pm f_b$ (where f_a is the higher frequency). Within the amplifier network, the current at the centre frequency, i_a, and the first sideband component, i_{ab}, are separated and two quasi-D.C. signals are produced which are calibrated functions of the r.m.s. carrier current, I_a, and the modulation depth, m. The modulation depth is the ratio of the first sideband to the carrier current (i_{ab}/i_a). It can be shown theoretically that

$$I_a = \sqrt{2} \cdot i_0 \cdot I_0(\beta) \cdot I_1(\alpha) \cdot \exp\left(\frac{-eV}{kT} \right)$$ (3)

and

$$m = \frac{2 \cdot I_1(\beta)}{I_0(\beta)},$$ (4)

where

$$\alpha = \frac{eV_a}{kT} \quad \text{and} \quad \beta = \frac{eV_b}{kT}.$$

I_0 and I_1 are modified Bessel functions of the first kind and of zero and first order respectively.

Equations (3) and (4) are simplified to the expressions given in (6) and (7) respectively when the voltage amplitudes of the A.C. signals are sufficiently small to satisfy Equation (5). The r.m.s. carrier current at plasma potential ($V=0$) is then given by (8) after substituting for 'i_0' in (6)

$$V_a \simeq V_b < 0.25 \frac{kT}{e}$$ (5)

$$I_a = \frac{i_0}{\sqrt{2}} \frac{eV_a}{kT} \cdot \exp\left(\frac{-eV}{kT} \right)$$ (6)

$$m = \frac{eV_b}{kT}$$ (7)

$$I_{a0} = \frac{AV_a e^2 n}{2(\pi mkT)^{1/2}}.$$ (8)

Equation (5) may be satisfied by using A.C. voltages which are sufficiently small at the lower end of the electron temperature range, but this would seriously limit the sensitivity of the experiment at the upper end of the range. Therefore, the oscillator voltages are switched high at the beginning of each probe sweep and are reduced proportionately by a feedback signal when $m = 0.2$ to prevent the modulation depth from exceeding this value. The maximum sensitivity is then maintained over a wide temperature range.

The electron temperature is derived directly from Equation (7) or from the slope of a semi-logarithmic graph of 'I_a' versus 'V' (Equation (6)). The electron density is obtained by substitution of 'T' in Equation (8).

The designed operating range for the experiment was approximately:

$$700 \text{ K} \leqslant T \leqslant 10^5 \text{ K}$$

and

$$10 \text{ cm}^{-3} \leqslant n \leqslant 5 \times 10^4 \text{ cm}^{-3}.$$

The lower limit of the electron density range is determined by the point at which the Debye length becomes comparable with the length of the satellite boom.

3. The Satellite

A diagram of the satellite is shown in Figure 1 but only those details relevant to the results are described. The co-ordinate axes refer to the main body section of the satellite and the attitude is controlled around the orbit to keep the $+Z$-axis pointing towards the centre of the Earth. Additionally, the solar panels are maintained orthogonal to the satellite–Sun line and rotate through an angle of $180°$ about the X-axis. Thus, in spacecraft co-ordinates, the Sun always lies in the Y-Z-plane and in the $-Y$-direction. The Langmuir probe on the EP-1 boom is part of the time in sunlight and part in the shadow of the solar panel. The probe lies on the $+Z$-side of the spacecraft body due to the complex angle of the EP-1 boom.

Soon after launch, the OGO-V orbit extended to a geocentric distance of 24 R_e at apogee, and a height above the Earth's surface of only 300 km at perigee. The projection of the line of apsides on to the ecliptic plane was inclined at approximately $40°$ to the solar direction and increasing at about $1°$ per day. Plasma measurements could therefore be made from just above the F-region maximum out into the solar wind beyond the bow shock. Figure 2 shows the position of the satellite orbit projected onto the ecliptic plane when viewed from the north. The satellite travels anti-clockwise around the orbit with a period of 62.5 hours and the orbit precesses clockwise at $1°$ per day. The orbits are drawn for March and June 1968 and February 1969. During this time, the height of perigee increased to 1500 km in June 1968 and to 6900 km in February 1969. There was a similar decrease in the altitude of apogee. The bow shock, magnetopause and plasmapause lines represent only typical positions for these boundaries.

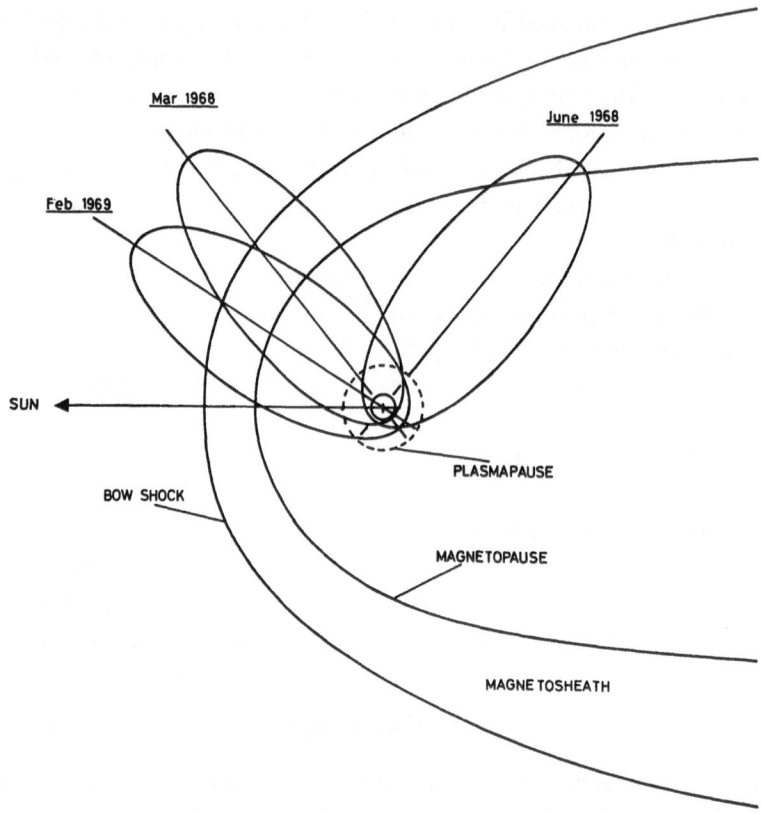

Fig. 2.   Projection of the OGO-5 orbit onto the ecliptic plane for three periods
in the life of the satellite.

4. Experimental Results

The electron density profiles shown in Figures 3–7 are computer drawn graphs of
unrefined satellite data. Measurements made at 10 sec intervals are plotted against
a linear time scale and each figure compares two passes through perigee using a
common geocentric distance scale. Perigee is marked by the letter 'P' on each profile
and the satellite is inbound on the left and outbound on the right. Gaps in the profile
indicate no real time data available or that the probe sweep range was incorrect. Data
drop-out points, indicated by vertical lines down to the abscissa or horizontal straight
lines, should be ignored.

The electron temperature of the local plasma could be determined only when the
probe was in the shadow of the solar panel or near perigee, where the plasma current
was much larger than the photo-emission current. However, it was possible to determine in orbit, the magnitude of the photo-current as a function of probe voltage by
analysing data outside the magnetosheath where the ambient plasma density was
extremely low. This photo-current contribution was then subtracted from the total
current when the probe was in sunlight, to yield the plasma contribution. The sub-

traction was not sufficiently accurate near the low current, or negative end, of the probe characteristic to give the electron temperature discretely, but it was possible to calculate the current at plasma potential as given by Equation (8). Thus the ratio $nT^{-\frac{1}{2}}$ was obtained.

Measurements of electron temperature, when the probe was in shadow varied from 3×10^3 K near perigee up to a maximum of 3.5×10^4 K just inside the magnetopause. Therefore, the shape of the electron density profile was little changed by assuming a constant temperature of 10^4 K and it allowed data for the probe in sunlight and shadow to be plotted continuously. The maximum error introduced was less than a factor 2 and comparison with the more accurately calculated density profiles when the probe was in shadow showed that the approximation error was usually less than 50%.

The profiles for two successive passes of the satellite through perigee on March 12 and 14, 1968, are shown in Figure 3. Comparison with magnetometer, plasma wave and low energy particle experiments on the same satellite confirmed the positions of the bow shock and magnetopause boundaries, as indicated by the letters 'B' and 'M' respectively. The plasma pause is in the region of very steep density gradient at a distance of 3–5 R_e, and within this region the electron density measurements were in reasonably good agreement with comparable experiments on the satellite.

Outside the plasmapause, the distance of the probe from the spacecraft was in-sufficiently large compared to the Debye length in the ambient plasma, and secondary electrons from the spacecraft were able to reach the probe. They are identified by their relatively low energy of approximately 1 eV which remains constant from just outside the plasmapause into the solar wind. The number of secondary electrons reaching the probe depends on the position and attitude of the satellite and they appear to result from both sunlight and higher energy particles striking one or more surfaces on the side of the main structure close to the probe. It will be seen from Figure 3 that the measured plasma density outside the bow shock is much higher on the inbound side than on the outbound side of both passes and the same is true for the region between the plasmapause and magnetopause. The increase from less than 10 cm^{-3} at the end of the first profile to about 50 cm^{-3} at the beginning of the second occurred smoothly near apogee and its cyclic variation over successive orbits is consistent with the change in attitude of the spacecraft and therefore is most probably due to photo-electrons. The photo-electron contribution is very small on the outbound side and a number of other profiles at about this time showed an ambient electron density in the outer magnetosphere similar to that of March 12; less than 10 cm^{-3}. However, the profile of March 14, with densities as high as 20–30 cm^3 indicates considerable variation in this region.

The sharp change in the electron density profile as the magnetosheath was en-countered, was caused by a change in the satellite environment, but the observed electron densities of 200–300 cm^{-3} were not consistent with other measurements. It seems likely that a large number of the electrons observed in this region are the result of secondary emission from the spacecraft due to bombardment by higher energy

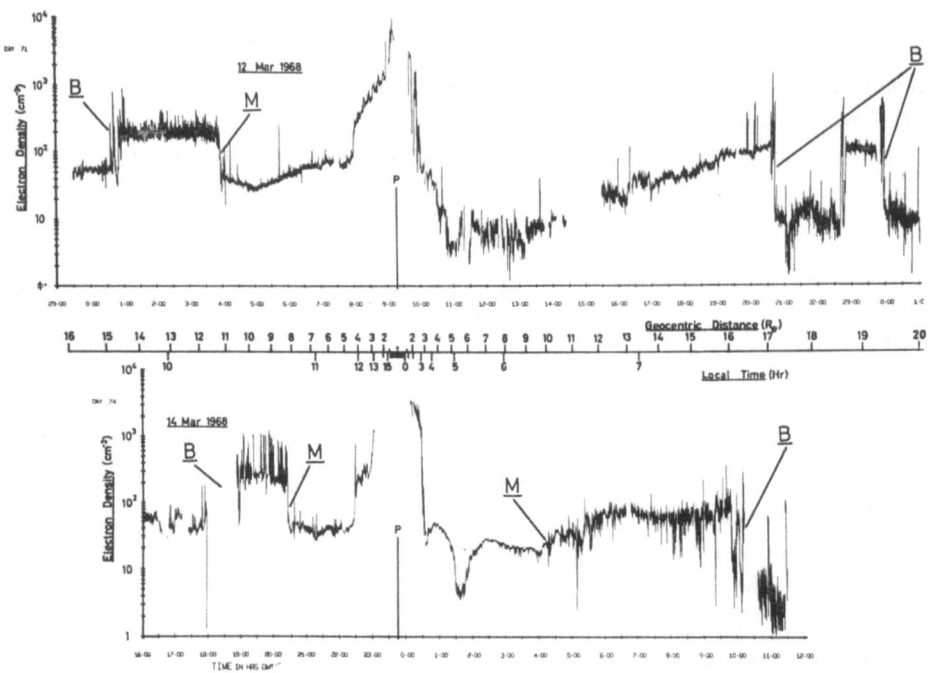

Fig. 3. Electron density profiles obtained in the magnetosphere on March 12 and 14, 1968.

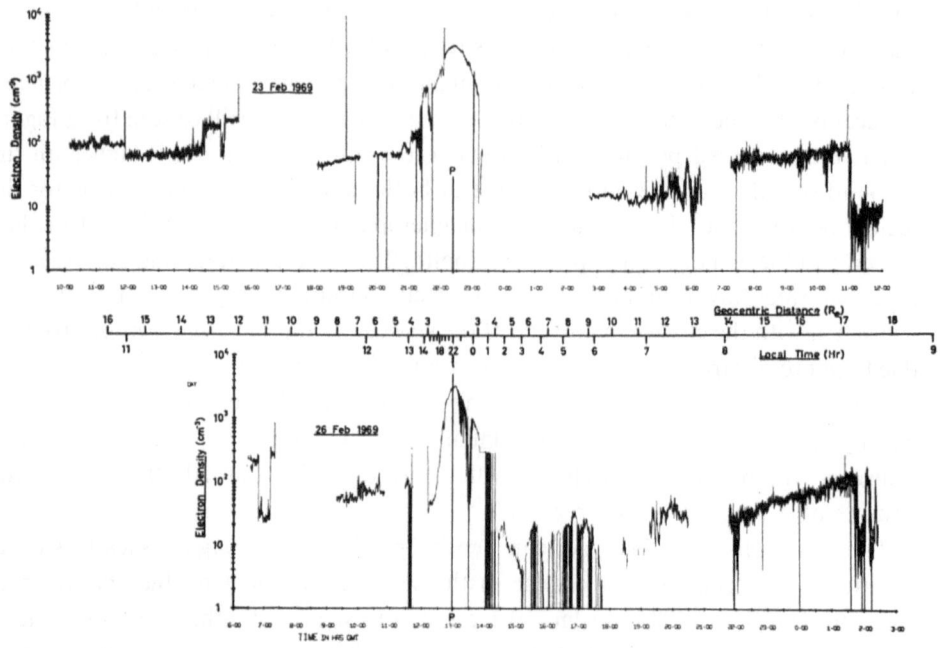

Fig. 4. Electron density profiles obtained in the magnetosphere on February 23 and 26, 1969.

particles. The superimposed fluctuations could be caused by variations either in the flux or direction of the incident particles. Only a small attitude effect was observed which is consistent with a fairly isotropic particle flux distribution.

The foregoing discussion of the Figure 3 profiles illustrates the uses and limitations of the results in their present form. The electron density scale is accurate only within the plasmasphere but a considerable amount of information on the behaviour of the magnetosphere may be deduced by continuing the profiles to beyond the bow shock. The position of the magnetopause and bow shock is seen to change considerably between successive passes and during the outbound pass of March 12, 1968, the bow shock region appeared to expand past the satellite from 17 to 19.3 R_e, causing the satellite to cross it twice.

Figure 4 shows two successive passes of the satellite through perigee on February 23 at 2220 UT and February 26 at 1300 UT 1969. The sudden commencement as a result of the solar flare occurred at 0150 UT on February 26, thus the profiles were obtained before and after the event. The bow shock positions are similar for both orbits on the outbound pass at approximately 17 R_e. Due to gaps in the data only one magnetopause crossing is actually observed at approximately 11.2 R_e outbound on the earlier orbit. The general size of the magnetosphere cavity, however, is the same for both passes and outside the plasmasphere no large differences can be

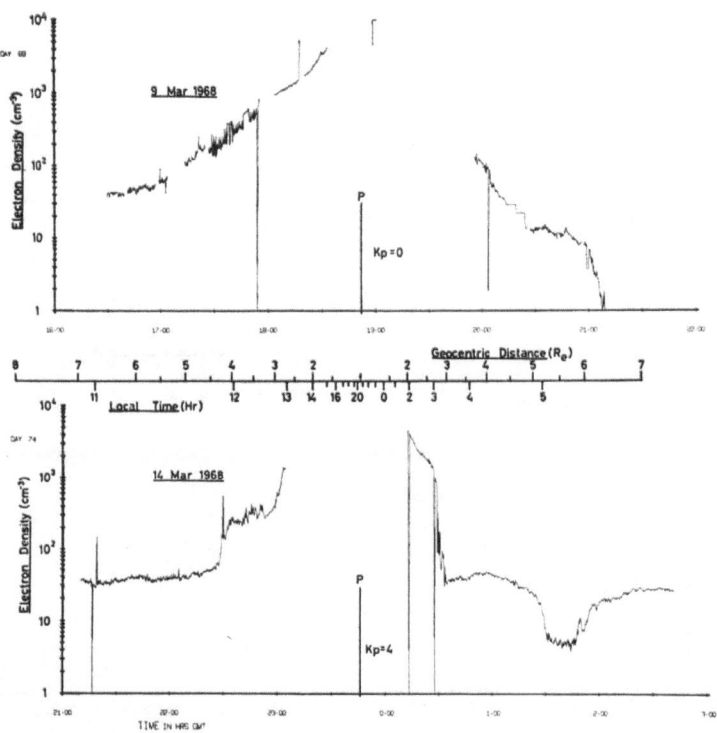

Fig. 5.   Electron density profiles obtained near perigee on March 9 and 14, 1968.

observed. The plasmapause boundary however is closer to the Earth on the second pass and the geomagnetic K_p index is correspondingly higher.

Figure 5, 6 and 7 show the structure of the plasmapause boundary for six satellite passes at various levels of geomagnetic activity. The data for February 1969 are on an expanded time scale in Figure 7. On March 9 and May 26, 1968, the K_p index was zero and the electron density decreased smoothly from perigee out to approximately 6 R_e when the limit of resolution of the experiment was reached. At higher values of K_p, the plasmapause became more definite but often appeared to have a double structure, as on March 14 and June 11, 1968. It was the inner, and much more pronounced, density gradient at 2–3 R_e which corresponded to the position of the plasmapause as predicted by the empirical formula of Binsack (1967), but there was a secondary steep gradient at 4–6 R_e. Some other authors (Carpenter, 1963, 1966; Carpenter and Smith, 1964) have described the dependence of the height of the plasmapause on the geomagnetic activity index, K_p, and a summary of the results is given in a recent review paper by Gringauz (1969).

5. Conclusions

The experimental observations of the plasmapause boundary position and its relation

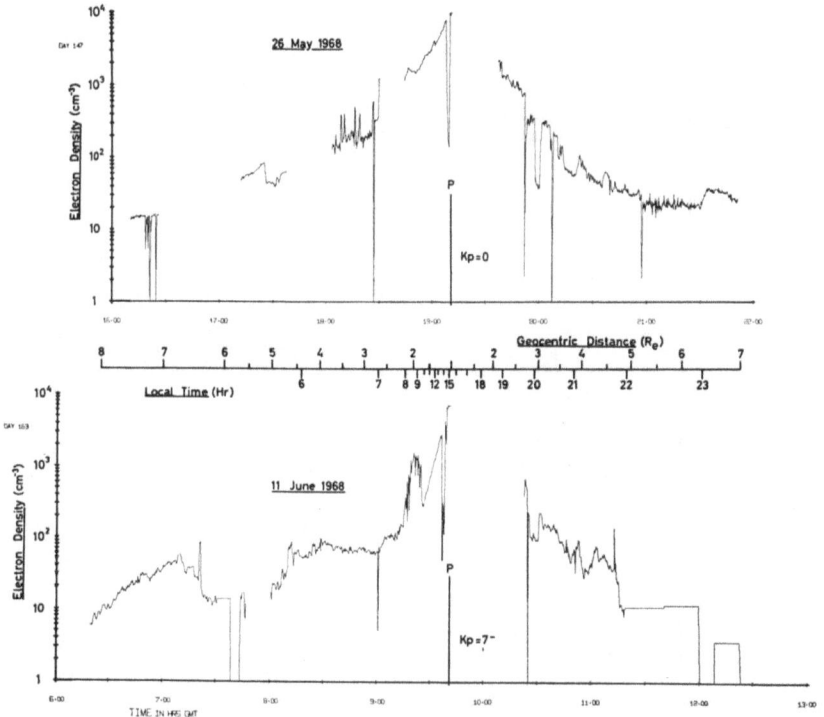

Fig. 6.   Electron density profiles obtained near perigee on May 26 and June 11, 1968.

to the K_p index are consistent with the results published in previous papers but a detailed, quantitative survey of a larger number of plasmapause crossings will be made. The sudden commencement of February 26, 1969, appeared to cause no major disturbances in the thermal plasma in the magnetosphere and no significant changes in the positions of the bow shock and magnetopause were observed on the two satellite passes investigated. The position of the plasmapause boundary was consistent with the level of magnetic activity.

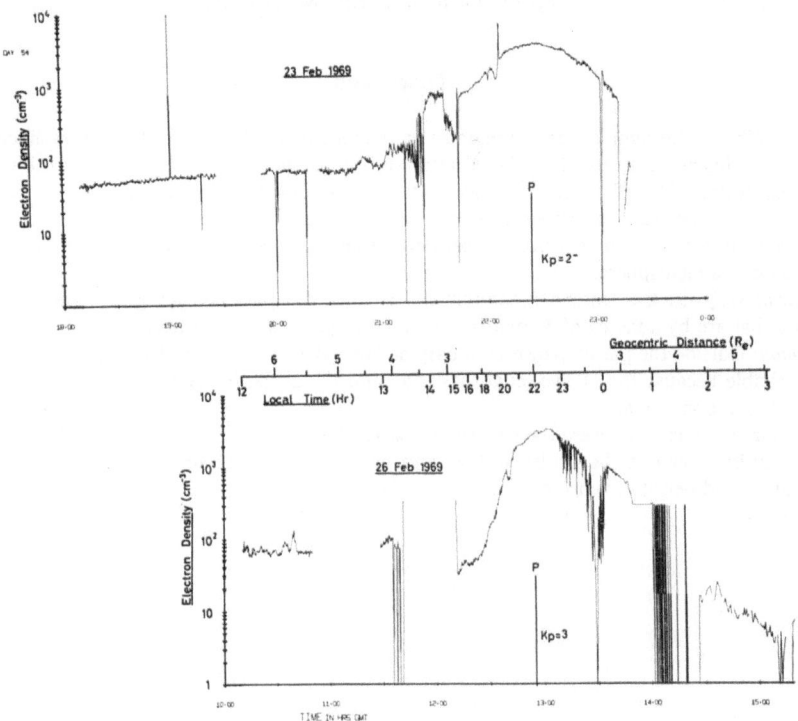

Fig. 7.   Electron density profiles obtained near perigee on February 23 and 26, 1969.

Acknowledgements

We wish to thank Professor R. L. F. Boyd for his encouragement and interest in the project; Mr. J. Blades and Mr. R. Wood of Pye Ltd., for the design and manufacture of the experiment electronics module; Mr. P. Sheather for the mechanical design and the NASA for the opportunity to fly the experiment. The interpretation of the results has also been assisted by comparison of data with a number of experimenters on the same satellite, notably: Scarf, Fredricks and Green at TRW; Coleman at UCLA; Smith and Holtzer at JPL; Sharp and Harris at Lockheed; Heppner and Ogilvie at GSFC.

References

Binsack, J. H.: 1967, 'Plasmapause Observations with the MIT Experiment on IMP 2', *J. Geophys. Res.* **72**, 5231–5237.

Carpenter, D. L.: 1963, 'Whistler Evidence of a 'Knee' in the Magnetospheric Ionisation Density Profile', *J. Geophys. Res.* **68**, 1975–1982.

Carpenter, D. L.: 1966, 'Whistler Studies of the Plasmapause in the Magnetosphere. I: Temporal Variations in the Position of the Knee and some Evidence of Plasma Motions near the Knee', *J. Geophys. Res.* **71**, 693–709.

Carpenter, D. L. and Smith, R. L.: 1964, 'Whistler Measurements of Electron Density in the Magneto-sphere', *Rev. Geophys.* **2**, 415–441.

Gringauz, K. I.: 1969, 'Low Energy Plasma in the Earth's Magnetosphere', *Rev. Geophys.* **7**, 339–378.

Discussion

Tunaley: What is the significance of the noise on the measurements just inside the bowshock?

Norman: This noise is caused by the variation of the secondary electron density in the vicinity of the probe. It probably reflects the variation in the intensity of the higher energy particle flux.

Gringauz: Which electron temperatures did you measure?

Norman: The electron temperatures increase from about 2000–3000 K at perigee to about $3-4 \times 10^4$ K in the outer magnetosphere.

Gendrin: Why do Carpenter's measurements beyond the plasmapause give values of electron densities that are by a factor of 5 smaller compared to your measurements?

Norman: Outside the plasmapause boundary at low values of electron density, the measurements are unreliable because the Debye length is large, and the probe is within the space charge sheath surrounding the spacecraft.

Gringauz: In which way were the electron densities derived?

Norman: By sweeping the probe voltage and using the standard techniques to derive electron temperature and density from the probe characteristics.

IONOSPHERIC MEASUREMENTS BY THE ESRO-I SATELLITE DURING THE FEBRUARY 25th 1969 SOLAR PROTON EVENT

W. J. RAITT

University College London,
Mullard Space Laboratory, Dorking, Surrey, England

Abstract. Measurements on local electron density and temperature variations for complete near earth satellite orbits are presented for periods before and after the solar proton event of 25th February, 1969.

No large scale effects on the ionospheric structure can be detected from this relatively small sample of data.

1. Introduction

The orbit, attitude control and general configuration of the ESRO-I satellite have been described elsewhere (Page, 1970).

As explained the spacecraft has two ionospheric experiments designated S44 and S45, making direct measurements on electrons and positive ions respectively in the vicinity of the vehicle. The measurements on the plasma are made by using forms of Langmuir probe as sensors, the general configuration of the sensors being shown in Figure 1.

Quite early in the life of ESRO-I a problem affecting the operation of the probes became apparent. This was the drift of the spacecraft potential to negative values greater than -6 volts when the attitude was within certain limits. This seriously affected the operation of the ion probe (S45) and also affected one of the electron probes

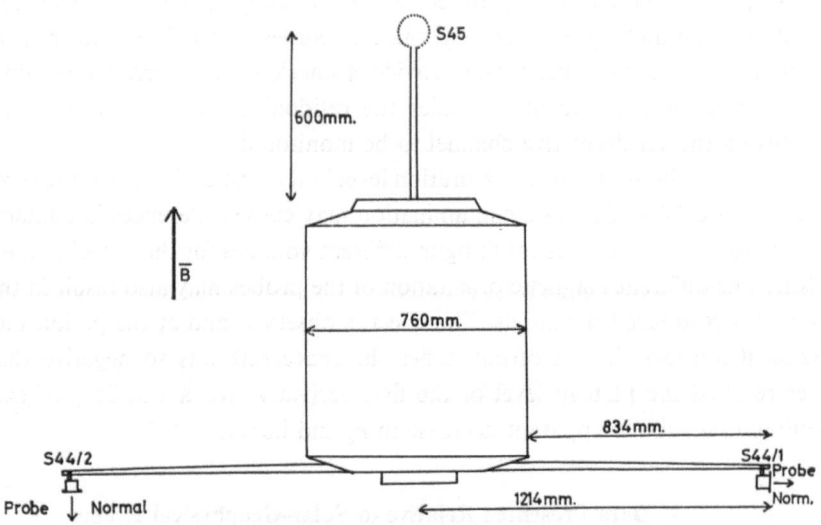

Fig. 1. Positions of ionospheric probes on ESRO-I.

V. Manno and D. E. Page (eds.), Intercorrelated Satellite Observations Related to Solar Events, 535–548. All Rights Reserved
Copyright © 1970 by D. Reidel Publishing Company, Dordrecht-Holland

(S44/1) over part of the orbit at the time of the February 25th event. For these reasons only the electron measuring experiment will be considered in this paper, and of the two probes only results from that designated S44/2 will be presented.

2. Experimental Technique

The S44 probes consist of circular planar Langmuir probes 2 cm in diameter surrounded by a guard electrode 4 cm in diameter. The current voltage characteristic of the probe is partly analysed electronically by applying a.c. signals to the probe as it is swept from positive to negative values. The details of the technique have been fully described elsewhere (Boyd, 1961; Wrenn, 1969).

The output of the equipment consists of two signals:

$$\alpha \log\left(\frac{\mathrm{d}i}{\mathrm{d}V_p}\right) \tag{1}$$

$$\alpha \, \mathrm{d}^2 i / \mathrm{d}V_p^2 / \mathrm{d}i / \mathrm{d}V_p \tag{2}$$

where i = probe current, V_p = probe potential relative to the plasma.

The maximum value of the first signal is related to electron density n_e and temperature T_e by

$$(\mathrm{d}i/\mathrm{d}V_p)_{V_p>0} \, \alpha \, \frac{n_e}{T_e^{1/2}}$$

while the second signal at its maximum is related to electron temperature only

$$[(\mathrm{d}^2 i / \mathrm{d}V_p^2)/(\mathrm{d}i/\mathrm{d}V_p)]_{V_p=0} \, \alpha \, 1/T_e.$$

In ESRO-I the peak values are recorded once per sweep on the internal tape recorder, that is approximately once per 10 sec, while the complete characteristics are transmitted over the high speed telemetry link and curves of the form shown in Figure 2 are obtained. The high speed data provide a check on the correct operation of the peak storage circuits and also enables the residual current at the negative probe voltages on the 1st derivative channel to be monitored.

Both curves should show the saturation level in the first derivative at the same probe voltage, but effects of surface contamination may cause differences in contact potential between the two probes resulting in different voltages for the 'knee' in the characteristic. The different magnetic orientation of the probes may also result in this effect due to $V \times B$ induced potentials. The effect is observed and at the period under discussion it appears that at certain times the spacecraft was so negative that S44/1 never reached the plateau level of the first derivative for about 25% of each orbit resulting in a sudden apparent decrease in n_e and increase in T_e.

3. Data Presented Relative to Solar-Geophysical Events

Figure 3 shows three groups of orbits which were available before and during the solar proton event of 25th February 1969. Orbits 2023–2036 are included as a reference for undisturbed data, the K_p index is fairly low and the time covered precedes the occurrence of both class 2B solar flares at about 2300 on February 24th and at 09.00 on February 25th. The second group of orbits 2047–2049 are taken after the detection of energetic particles by both satellites and ground stations, but before the magnetic disturbance indicated by a sudden commencement at 0158 on February 26th. The final group of orbits 2053–2057 are taken after the sudden commencement and are at similar universal times to those on February 24th and consequently cover a similar geographic position to orbits 2023–2026 as is illustrated later.

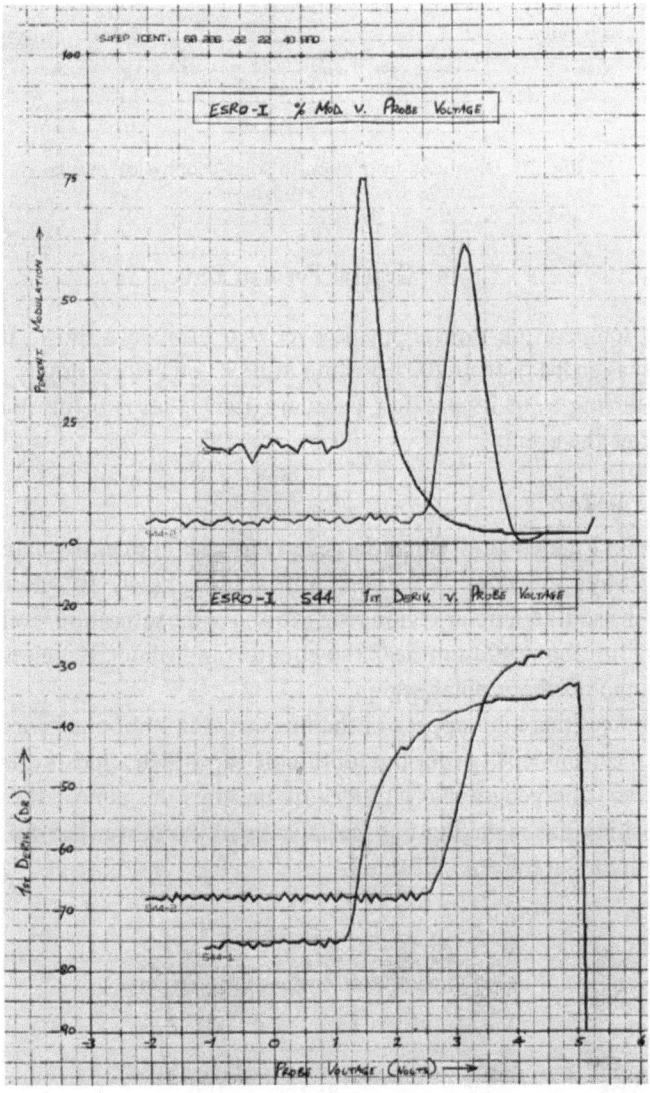

Fig. 2.   Typical electron density probe characteristics.

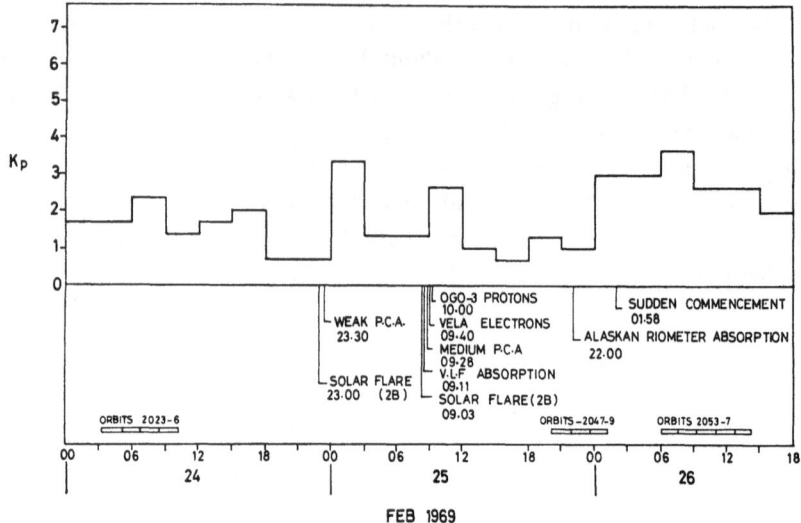

Fig. 3. Data used in relation to solar-geophysical events.

4. General Form of Data

Density and temperature measurements over two orbits are shown in Figures 4–7. The data are taken from orbit 2025 starting at 08.43 on 24th February 1969 and from orbit 2055 starting at 09.28 on 26th February 1969, that is before the solar proton event and after the event.

4.1. DENSITY DATA

It can be seen in Figures 4 and 5 that the general form of the density data is very similar for the two orbits showing a variation from approximately 10^4 electrons/cm^3 near apogee over the southern polar region rising to above the saturation level of about 2.10^5 electrons/cm^3 at about 400 km near the equator and about 10^5 electrons/cm^3 near perigee over the northern polar region.

Any differences there are appear in the fine structure, but one pronounced feature is the dip in electron density over the south pole cap which appears almost identically in both curves. The geographic latitudes of the dips are almost the same, but the magnetic position has moved to a lower latitude in the post-event dip.

The coordinates are given in Table I.

TABLE I

Orbit	Position of electron density minimum				
	Long	Lat	Mag Long	Mag Lat	L
2025	340	−81.2	21.0	−71.5	8.3
2055	303	−80.0	6.0	−68.5	6.7

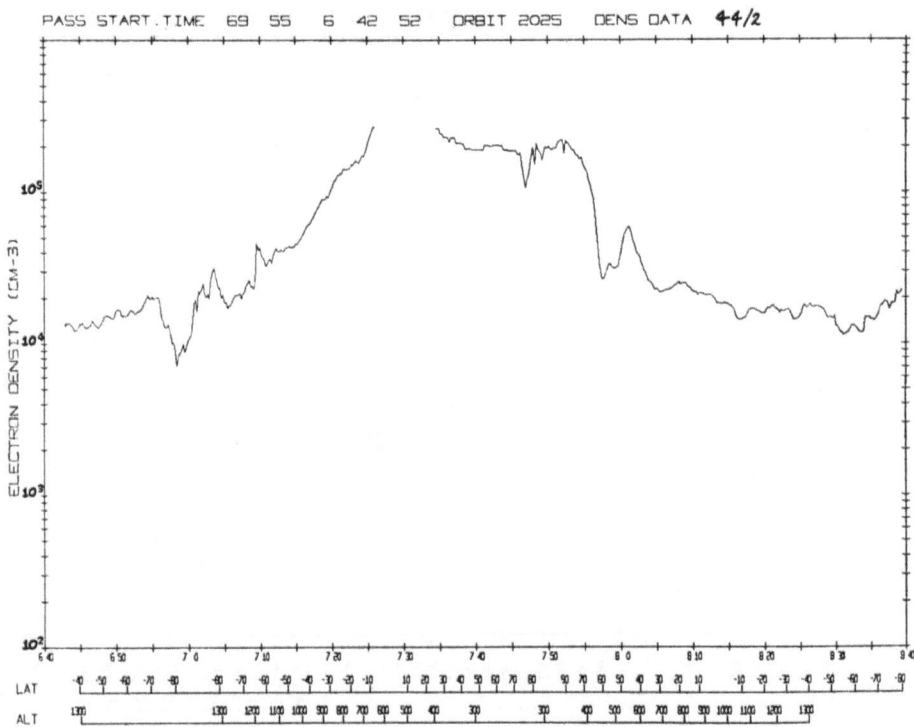

Fig. 4.   Orbit 2025 electron density profile.

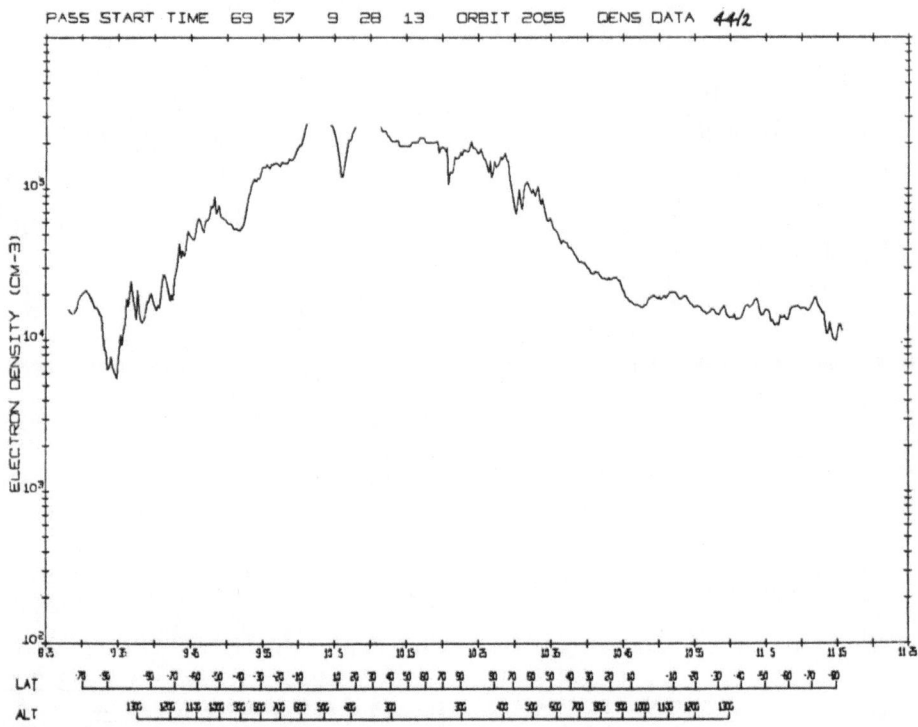

Fig. 5.   Orbit 2055 electron density profile.

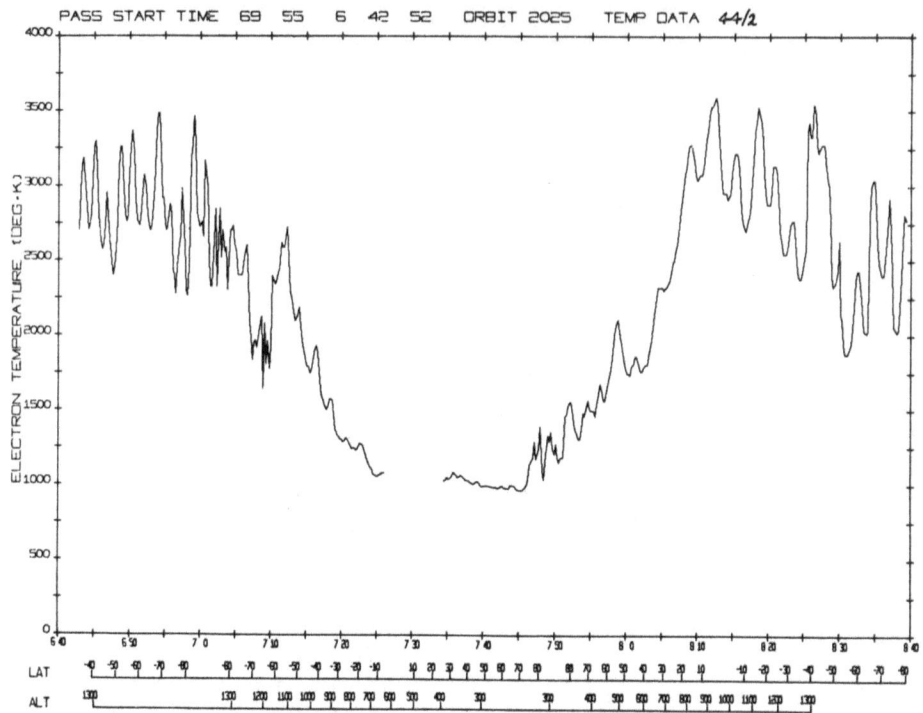

Fig. 6.   Orbit 2025 electron temperature profile.

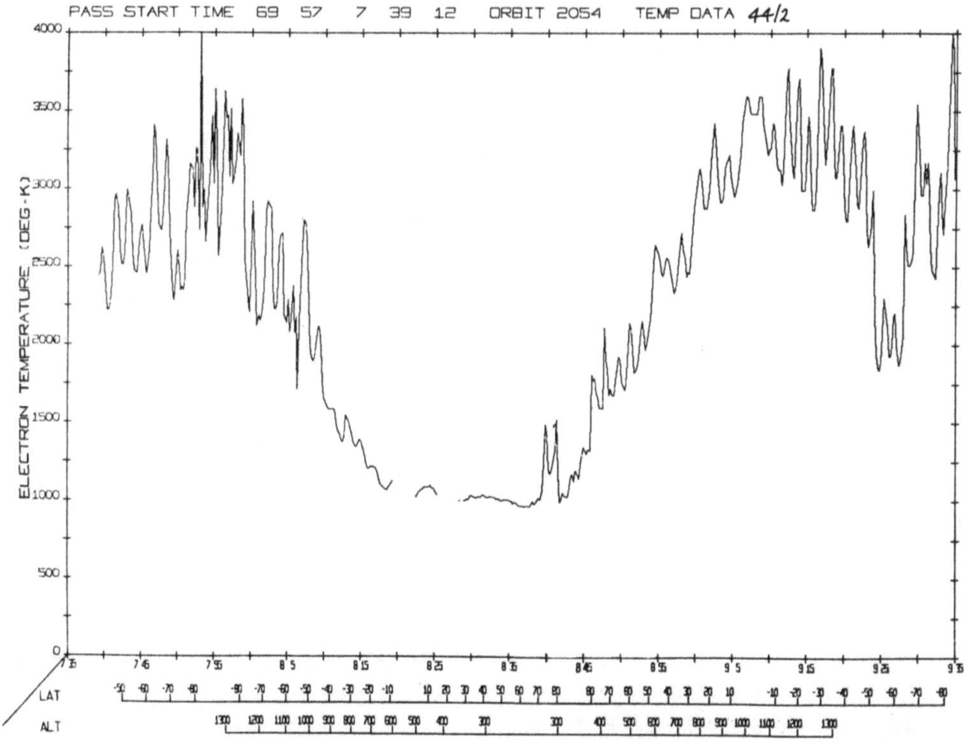

Fig. 7.   Orbit 2054 electron temperature profile.

The most obvious difference between the curves is the very rapid decline in electron density observed during the transit of the satellite from the North pole towards the equator on orbit 2025, but not on orbit 2055. The position of this ledge of ionisation is defined by the parameters given in Table II.

TABLE II

	Long	Lat	Mag. Long	Mag. Lat	L
Northern edge	359	64	93	66	5.2
Southern edge	357	59	86	62	3.8

These features are not unique to all orbits taken near the two used as examples and so generalisations cannot be drawn from the examples given.

4.2. TEMPERATURE DATA

The variation of electron temperature around the orbit is shown in Figures 6 and 7. In dealing with similarities first it can be seen that at mid and low latitudes on the northbound leg of the orbit there is excellent agreement while on the southbound leg there is general agreement but much greater fluctuations in temperature.

The difference here being local time (morning on the southbound equator crossing and evening on the northbound crossing), and altitude, although the latter is probably

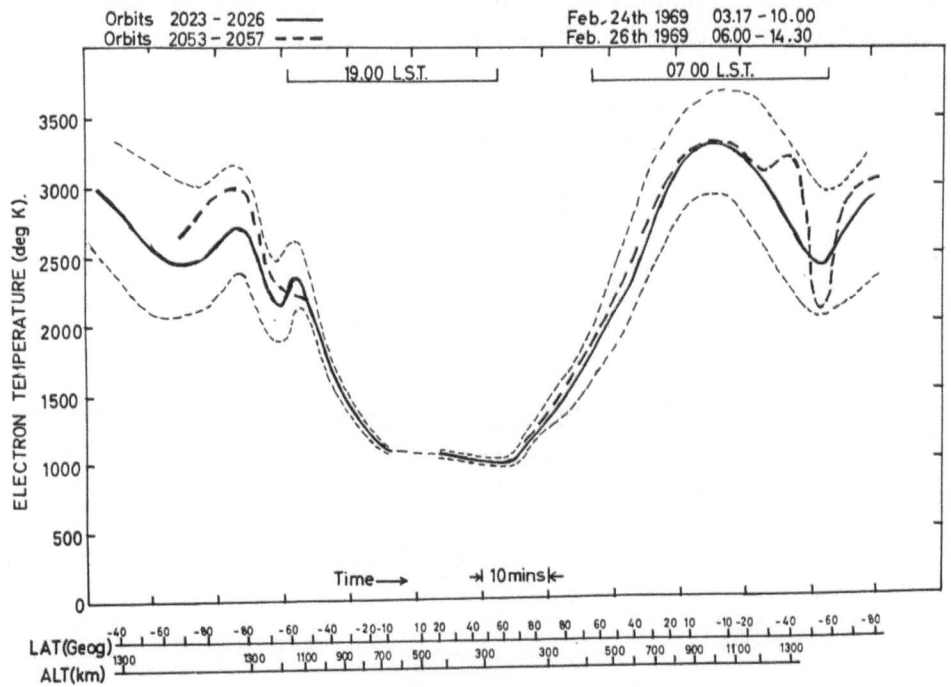

Fig. 8.   Orbits 2023–6 and 2053–7 median electron temperature profiles.

less important since part of the mid latitude crossings cover the same altitude range.

The fluctuations observed on morning mid latitude temperatures are also in evidence over the polar regions where variations in temperature of the order of 500 K in times of the order of 2 to 3 min are seen. These fluctuations tend to mask any trends in electron temperature before and after the solar event, so an attempt was made to ignore the fluctuations and draw a median temperature curve for the group of orbits 2023–6 and another for the group of orbits 2053–7. These median temperature curves are shown in Figure 8.

It can be seen that away from the south polar region there is good agreement between the temperature profiles. Near the south pole there is a dip in electron temperature which is sharper in the later orbits than in those taken before the solar event and also asymmetry between morning and evening sides. It is difficult however to place too much reliance on this result because of the large scatter indicated by the lightly drawn broken line. It would be desirable to observe the effects of more solar proton events to produce better statistical evidence.

5. Investigation of Polar Data

Before discussing more detailed plots of data over the polar regions reference to Figures 9 and 10 shows the orbit configurations of the three groups relative to the sun and the geomagnetic pole. The orbits over a period of two days change very little in local time, so the position of the dawn/dusk lines at ground level and the orbital track remain fixed as shown. However, for each orbit the position of the magnetic pole differs. The locus of the pole position is a circle centered on the geographic pole and

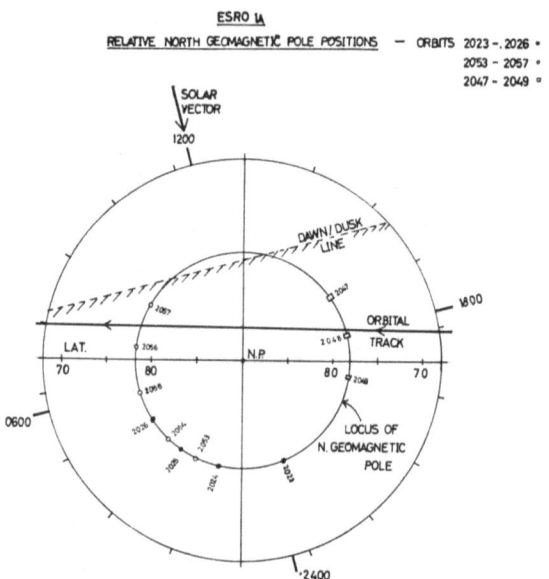

Fig. 9.   Passes over north polar cap relative to sun and magnetic pole.

the positions for the traverses of the various orbits over the pole are shown on the diagram. It can be seen that relative to the orbit the pole is in similar positions for the groups of orbits 2023-6 and 2053-7, being in the 24.00–06.00 quadrant for North polar passes and 12.00–18.00 for South pole passes. While for the group of orbits 2047-9 the pole is around 18.00 for North polar passes and around 06.00 for the South polar passes.

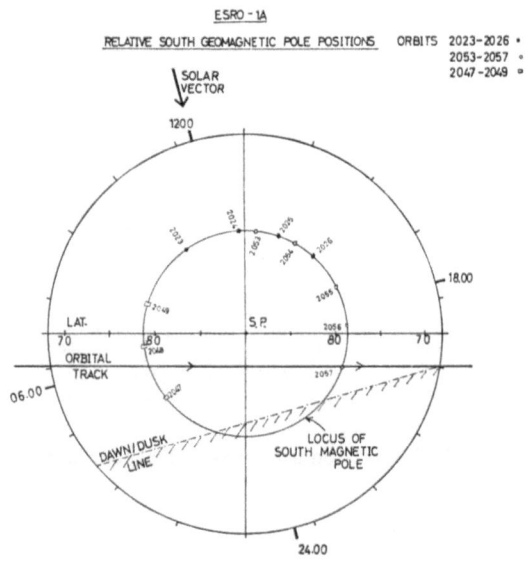

Fig. 10.   Passes over south polar cap relative to sun and magnetic pole.

The sequence of Figures 11–16 show variations of electron density over the South pole cap (Figures 11–13) and electron temperature over the North pole cap (Figures 14–16) for the three groups of orbits referred to previously. Although at first sight the data from orbits 2047-9 differ from the others this can be explained in terms of a different altitude range covered by the particular range of geomagnetic latitudes. Variations in altitude are probably the cause of much of the scatter between successive orbits in the groups selected since plotting against geomagnetic latitude does not give a constant altitude scale. No attempt has been made to take out the altitude variations because of lack of knowledge of the mean ionic mass over the polar regions.

A feature which is apparent in the temperature curves is the sudden transition between the very steady and repeatable temperatures on the equator side of the North polar cap on northbound passes to a higher and fluctuating temperature at latitudes around 60° to 70°. This does show a tendency to start at lower latitudes after the sudden commencement, but as with other observations there are considerable variabilities within the groups of orbits selected. There is also some asymmetry between the latitude and magnitude of the effect best seen in Figure 15, but rather masked by the effects of altitude and the regular fluctuations referred to earlier. There is also some

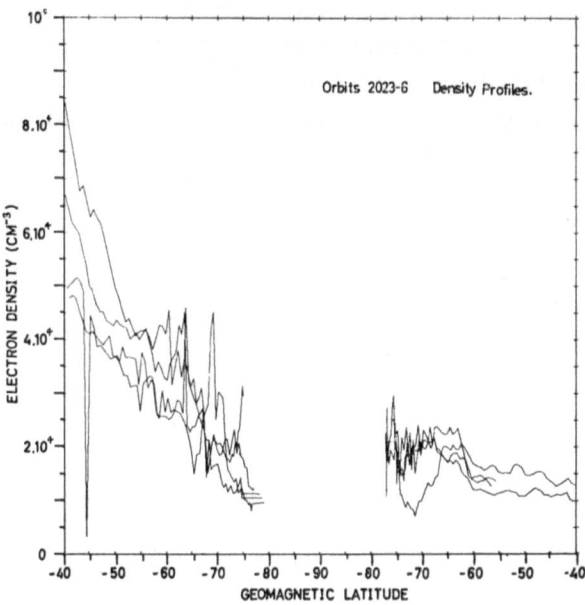

Fig. 11.   Orbits 2023–6 electron density vs geomagnetic latitude (south pole).

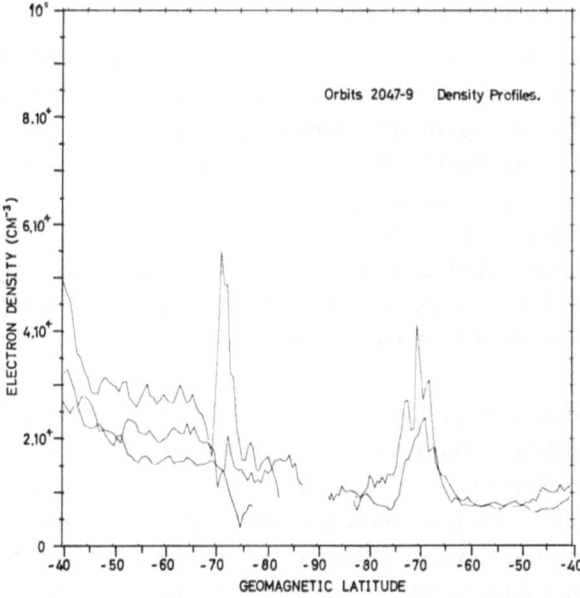

Fig. 12.   Orbits 2047–9 electron density vs geomagnetic latitude (south pole).

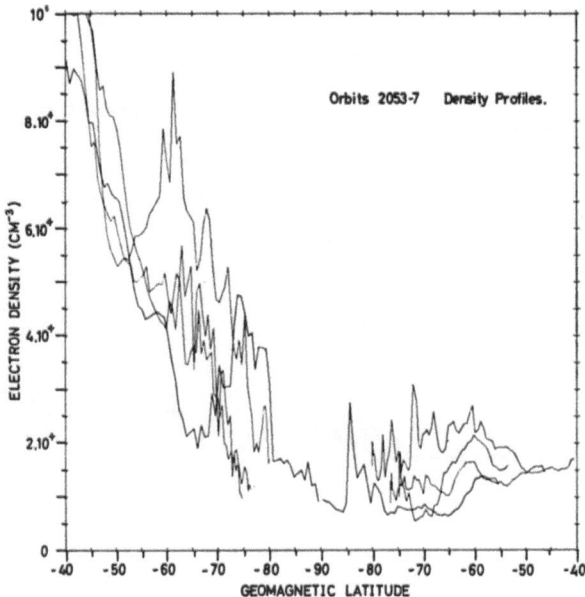

Fig. 13.   Orbits 2053–7 electron density vs geomagnetic latitude (south pole).

evidence for the whole polar cap showing more disturbed and elevated temperatures after the sudden commencement seen by comparing Figures 15 and 16 unfortunately the geomagnetic latitude in the earlier group of orbits never reached a high enough value to give further verification of this phenomenon.

The density profiles show greater variability both within groups of orbits and from group to group due largely to altitude variations. It is therefore very difficult to draw any conclusions on the effect of the solar proton event on electron densities over the Southern polar cap.

6. Conclusions

A study of isolated groups of ionospheric data measured in the altitude range 300–1300 km over all geomagnetic latitudes has shown no obvious, large scale effects of the type of solar proton event which occurred on February 25th 1969.

It might be expected that energetic particle effects at these altitudes would be masked by solar ionisation since during the period under discussion the orbit of ESRO-I was fully sunlit. However, it is difficult to explain the rapid fluctuations in temperatures and densities in the polar region without resorting to a particle source of ionisation.

The data analysis is as yet in an early stage, but it would seem that a better understanding of the interactions of the disturbed particle flux over the polar regions and the high altitude ionosphere will have to wait for either a statistical study of a number of solar proton events or more collaboration between the individual ESRO-I experimenters.

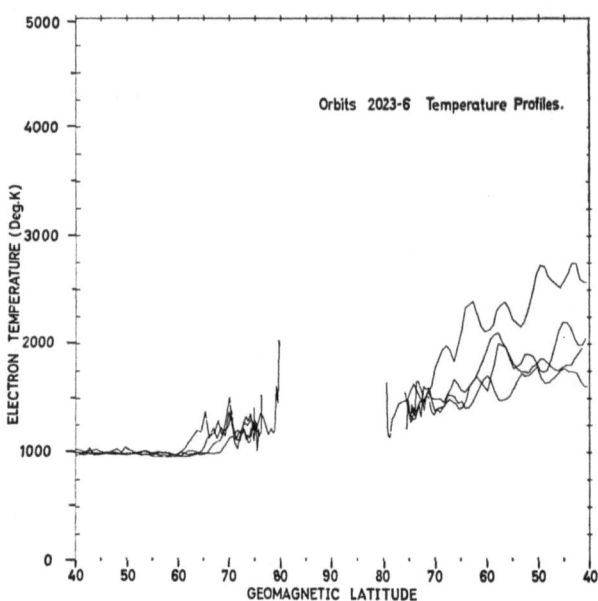

Fig. 14.   Orbits 2023–6 electron temperature vs geomagnetic latitude (north pole).

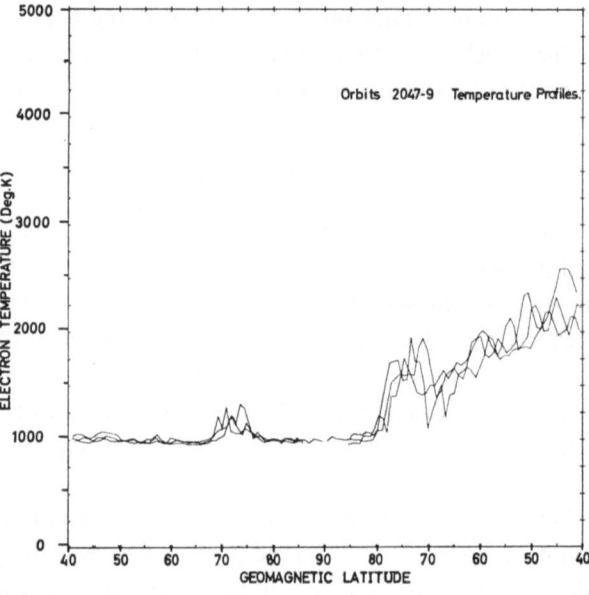

Fig. 15.   Orbits 2047–9 electron temperature vs geomagnetic latitude (north pole).

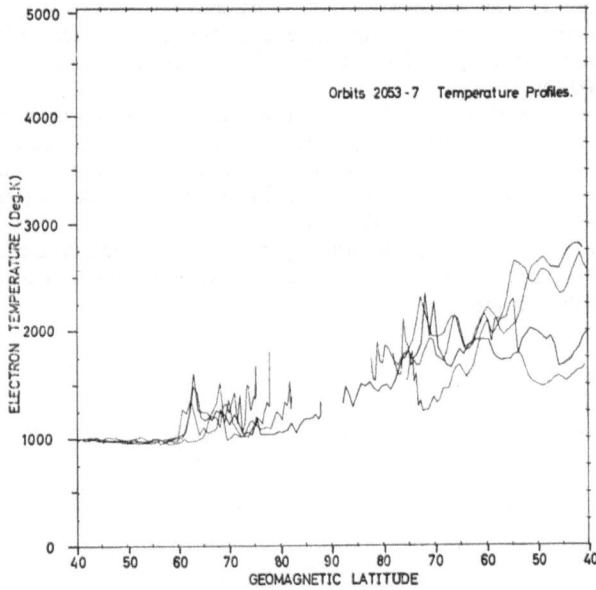

Fig. 16.   Orbits 2053–7 electron temperature vs geomagnetic latitude (north pole).

Acknowledgements

The author would like to acknowledge the interest and advice received from Professors R. L. F. Boyd and A. P. Willmore and also the experiment integration work and many discussions with Mr. T. S. Bowling.

References

Boyd, R. L. F.: 1961, *J. Brit. I.R.E.* **22**, 405.
Page, D. E.: 1970, this volume, p. 367.
Wrenn, G. L.: 1969, *Proc. I.E.E.E.*, **57**, 1072.

Discussion

Lanzerotti: Why was there more variation of temperature on the morning side; was the temperature raised because the altitude was higher?

Raitt: Yes, the altitude was increasing. The plots involving geomagnetic latitude could be improved by taking into account variations in altitude.

Lanzerotti: Is the altitude more constant on the dusk side?

Raitt: Yes, because perigee occurs just before the pole on the morning side. Corrections in the temperature might be easy but for the density there is the problem of knowing what is the mean ionic mass. This we do not yet know, particularly in the polar regions.

Gringauz: How did you separate the effects of variations in altitude and latitude?

Raitt: No corrections have been made for altitude; the difficulties as in the previous question.

Gringauz: Did you compare the density data for altitudes below F_{max} with ground-based observation?

Raitt: No, but this should be possible; the perigee of ESRO I at this time was about 270 km so that it should have been at or just below the peak of the F-region.

Question: Do you know the spatial scale of the variations?

Raitt: The smooth fluctuations in temperature take place in a time scale of 2–3 min for all the orbits I have examined in this period, where the satellite velocity is 7 km/sec. This indicates a spatial scale of about 1000 km.

Reid: The sudden transition in the electron temperature profile was presumably the plasma-pause crossing. Why do you see a sharp plasma-pause on one side but not on the other?

Raitt: Although altitude variations cause the curves to differ, a sharp transition should show up on each one. It is probably just a question of the morning side being different from the evening.

Reid: Is the satellite always in sunlight?

Raitt: Yes, but it is a dawn-dusk orbit with a fairly high inclination. You change local time very rapidly as you cross the pole; even up to 70° geographic latitude the local time has changed only one hour from that at the equator.

ESRO I SATELLITE OBSERVATIONS OF 45–450 keV ELECTRONS
DURING THE PERIOD 24TH FEBRUARY – 3RD MARCH 1969

G. R. THOMAS, P. A. SMITH, and R. DALZIEL

S.R.C., Radio and Space Research Station, Ditton Park, Slough, Bucks, England

Abstract. Results are presented for trapped and precipitated electrons observed at northern high latitudes during morning- and evening-side passes of the ESRO I satellite in the period 24th February – 3rd March 1969. Variations of the integral intensities > 50 keV, and of the latitudes at which the peak intensities occur, are examined in relation to magnetic activity and local time. The energy spectra of the trapped and precipitated electrons are compared, and variations of the spectra are related to ocal t ime and latitude.

1. Introduction

The paper describes results obtained from the ESRO I satellite experiment S71A. This experiment uses scintillators and photomultipliers to measure the spectrum of both trapped and precipitated electrons in the energy range 45–450 keV.

The results are derived from tape-recorded data for northern high latitudes during the period 24th February – 3rd March 1969. At this time the satellite orbit, with inclination 94°, was approximately along the dawn-dusk meridian.

2. Description of the Experiment

Two pairs of detectors are used, each detector consisting of a collimator, plastic scintillator and photomultiplier. Pulse height analysis of the photomultiplier output is used to derive a six-point integral spectrum for electron energies between 45 keV and 450 keV. In order to estimate the background radiation, one detector of each pair has a magnet across the collimator to remove electrons with energies below about 1 MeV.

The satellite is magnetically stabilised and one pair of detectors, aligned at 90° to the satellite axis, measures 'trapped' electrons over the complete orbit. The other pair of detectors is aligned at 10° to the axis and measures 'precipitated' electrons at northern high latitudes. Each collimator has a rectangular cross-section which defines a field of view of about $18° \times 9°$, and the geometric factor is 5.6×10^{-4} cm^2 ster.

The experiment operates both in a low-speed telemetry mode, with recorded and real-time coverage, and in a high-speed real-time mode. In this paper we have used only low-speed data, in which the six-point spectrum is derived from simultaneous measurements lasting 750 msec. The spectra of the trapped and precipitated electrons and the corresponding background spectra are measured sequentially, the complete sequence taking 12.8 sec.

3. Results

3.1. INTEGRAL INTENSITIES > 50 keV

In this section we consider the measurements of electrons above the lowest energy

V. Manno and D. E. Page (eds.), Intercorrelated Satellite Observations Related to Solar Events, 549–556. All Rights Reserved

threshold (about 50 keV). Figure 1 shows the counts of trapped and precipitated electrons for a typical morning-side pass at northern high latitudes. The satellite at this time was moving southward, with its altitude increasing from about 600 km to 700 km. The peak counts of the trapped and precipitated electrons are approximately equal and occur at about the same latitude. Several of the results below relate to the variations of intensity and position of such peaks. At higher latitudes the counts decrease towards the poleward boundary of the outer radiation zone, and at lower latitudes the precipitated electron counts decrease more rapidly than those of the trapped electrons. The background level of 20–30 counts for the precipitated electrons is due to the internal calibration source of the detector. These observations are consistent with those of O'Brien (1964) and Fritz (1968), who have reported enhanced electron precipitation and approximately isotropic fluxes in the region of the high-latitude outer-zone boundary.

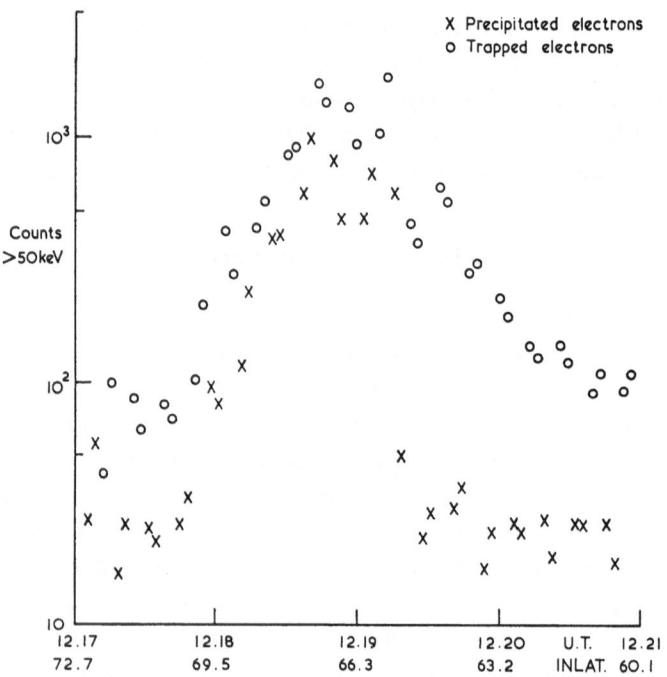

Fig. 1.   Counts of trapped and precipitated electrons > 50 keV as a function of universal time and invariant latitude; rev. 2113, 2nd March 1969, morning side.

Figure 2 shows the peak intensities observed on the morning and evening sides during the period 24th February – 3rd March 1969, the minimum measurable intensity being about 5×10^4 cm^{-2} sec^{-1} ster^{-1}. The local times were approximately 08 and 18 hours. Precipitated electrons at these energies were rarely observed on the evening side, and this is qualitatively consistent with the observed morning maximum in auroral absorption (Hartz *et al.*, 1963). The trapped electron intensity was also never very high on the evening side, and was usually less than that on the morning side. The trapped and precipitated intensities on the morning side showed large but similar variations from pass to pass, and the trapped intensity was almost always greater than the precipitated intensity. There are no obvious effects which can be directly associated with any of the major solar flares which occurred in this period. We have examined the variability of the peak electron intensities in relation to

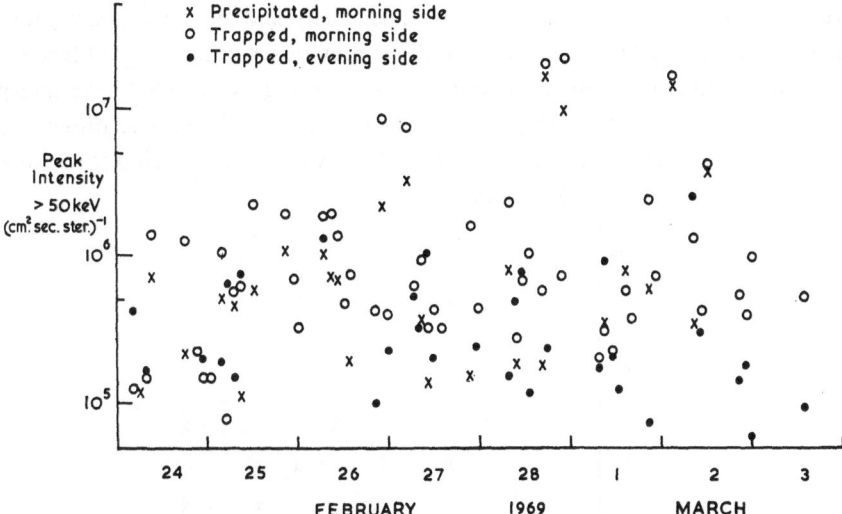

Fig. 2. Peak intensities > 50 keV (cm² sec ster)⁻¹ of precipitated electrons on the morning side and trapped electrons on the morning and evening sides, during the period 24th February – 3rd March.

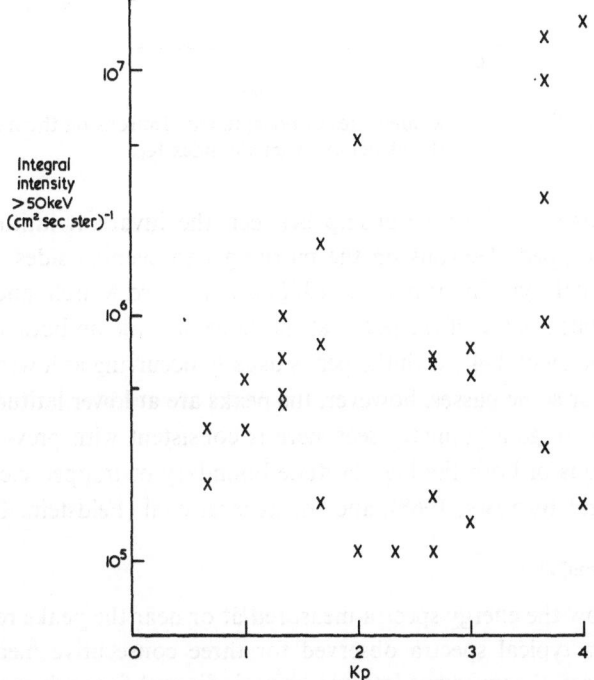

Fig. 3. Peak intensities > 50 keV (cm² sec ster)⁻¹ of precipitated electrons on the morning side against the planetary magnetic index K_p.

magnetic activity, and Figure 3 is a plot of the peak intensities of precipitated electrons on the morning side against the K_p index. The scatter of the points increases with K_p, and the greatest intensities occur for the highest values of K_p. Figure 4 shows the corresponding variations in the invariant latitude at which the peaks occur. Again there is a large scatter in the points, but the precipitation occurs at the lowest latitudes when K_p is highest. An increase in intensity and an equatorward movement with increase in K_p have been observed by previous workers (e.g. O'Brien, 1964; Maehlum and O'Brien, 1963), while other workers (e.g. Rao, 1969) have interpreted such variations in terms of magnetospheric substorm activity. The present results, which are limited to $K_p \leqslant 4$, do not indicate a very high correlation between the electron intensities and the K_p index.

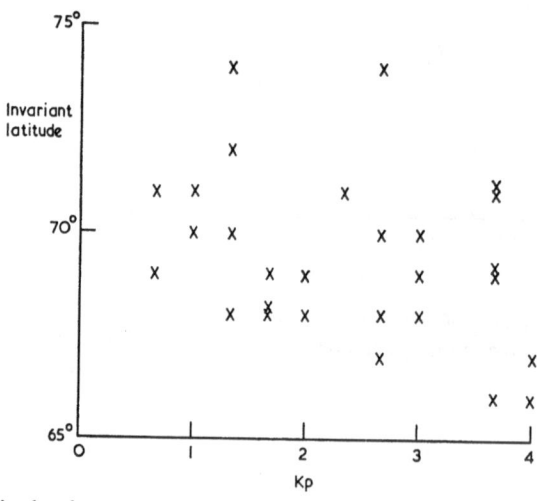

Fig. 4.   Invariant latitude of peak intensities of precipitated electrons on the morning side, against the planetary magnetic index K_p.

Figure 5 illustrates the relationship between the invariant latitudes of the peak intensities of trapped electrons on the morning and evening sides. Each point corresponds to a single pass in the period 24th February – 3rd March, and the broken line represents the occurrence of the peaks at the same latitude on both sides. There is an average asymmetry of 3–4°, with the peaks usually occurring at lower latitudes on the evening side. For some passes, however, the peaks are at lower latitudes on the morning side. The average asymmetry seen here is consistent with previous work on the diurnal variations of both the high latitude boundary of trapped electrons > 40 keV (McDiarmid and Burrows, 1968), and the auroral oval (Feldstein, 1966).

3.2. ENERGY SPECTRA

We consider now the energy spectra measured at or near the peaks referred to above. Figure 6 shows typical spectra observed for three consecutive measurements on a morning-side pass, the invariant latitude being indicated for each measurement. Only spectral points below 200 keV are shown; in the case of the precipitated electrons there is no measurable intensity for the two remaining spectral points at higher energy. It is clear that the precipitated electrons have a softer spectrum than the trapped electrons. Fitting exponential spectra over the energy range 50–150 keV, the e-folding energies E_0 are near 50 keV for the trapped electrons and 26 keV for the precipitated electrons.

This difference in hardness is illustrated more generally by plotting for all morning-

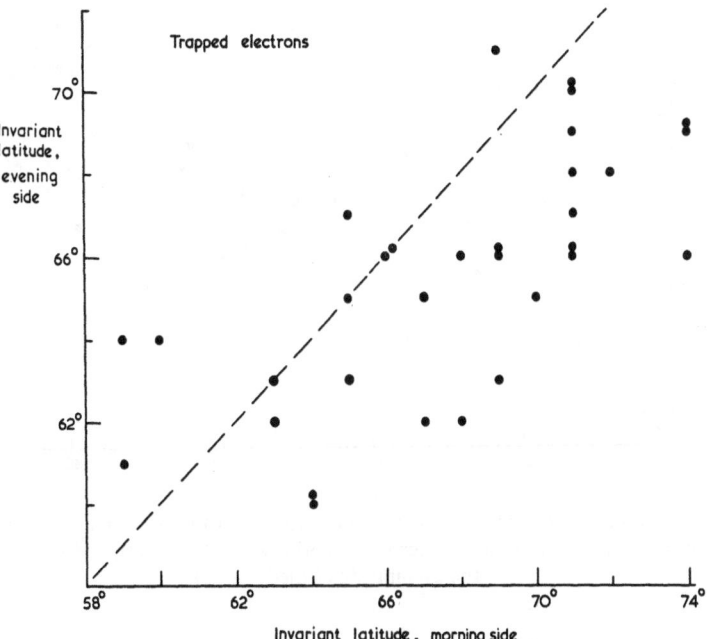

Fig. 5. Invariant latitude of the peak intensities of trapped electrons on the evening side, against the invariant latitude of the peak intensities of trapped electrons on the morning side.

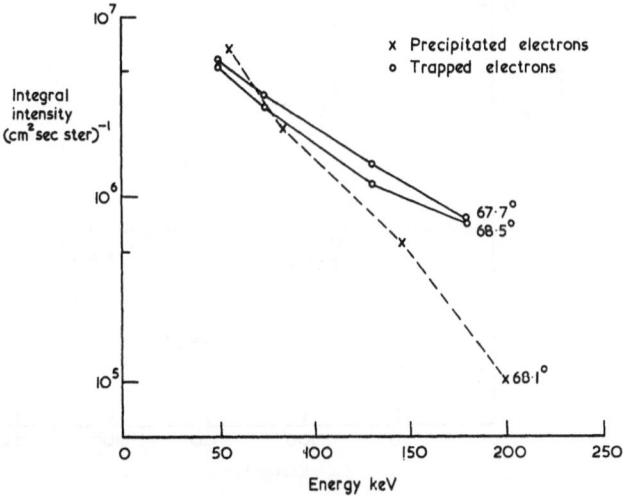

Fig. 6. Energy spectra measured consecutively for trapped and precipitated electrons, with invariant latitudes indicated; rev. 2100, 1st March 1969, 14.24 UT, morning side.

side passes the E_0 values of the trapped electrons against the E_0 values of the precipitated electrons. These are represented by the crosses in Figure 7, each point corresponding to a single pass, and the broken line represents the same spectrum for the trapped and precipitated electrons. With few exceptions, the spectrum is hardest for the trapped electrons.

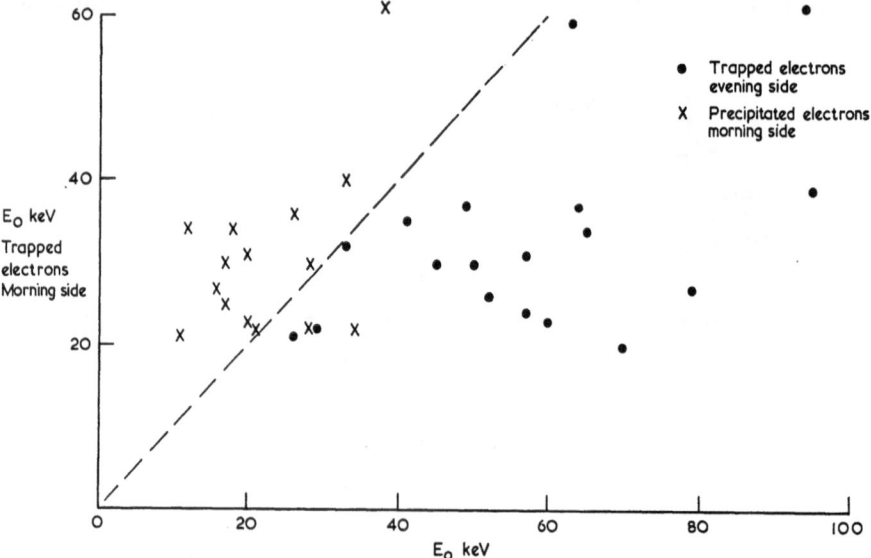

Fig. 7. e-folding energy E_0 for trapped electrons on the morning side, against E_0 for precipitated electrons on the morning side (crosses), and against E_0 for trapped electrons on the evening side (filled circles).

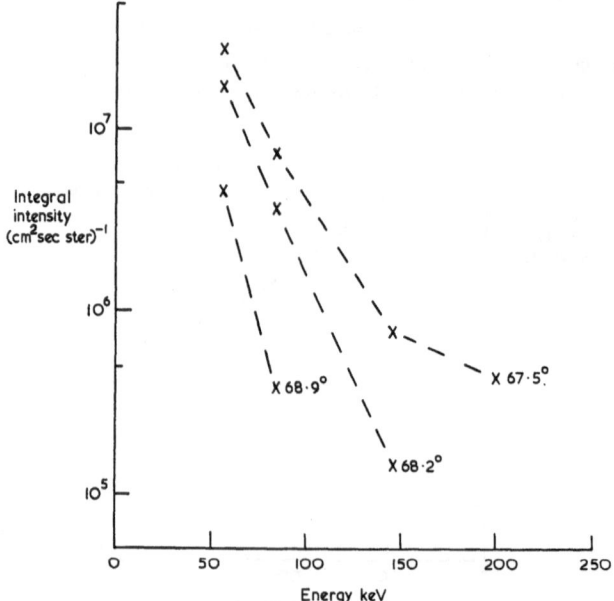

Fig. 8. Energy spectra measured consecutively for precipitated electrons, with invariant latitudes indicated; rev. 2113, 2nd March 1969, 12.18 UT, morning side.

The filled circles in Figure 7 are plots of the values of E_0 for the trapped electrons on the morning and evening sides. Softer spectra are observed on the morning side, in qualitative agreement with the diurnal variation of auroral electron precipitation inferred from X-ray measurements (Bewersdorff et al., 1966). However, in comparing the spectra on the morning and evening sides, we have not taken into account the variation of satellite altitude. During the period of these observations, the altitude was 270–320 km for the evening-side passes and 400–700 km for the morning-side passes, and at least part of the variation which we observe may be due to this difference in altitude.

Previous satellite measurements (e.g. Johnson et al., 1967) have indicated that the spectrum of precipitated electrons softens with increasing latitude. This effect is illustrated in Figure 8, in which three spectra are plotted for the morning-side pass shown in Figure 1. In the range of energies up to 150 keV the deduced values of E_0 are 20 keV, 17 keV and 11 keV at invariant latitudes of 67.5, 68.2 and 68.9, respectively, indicating the typical softening of the spectrum with increasing latitude.

4. Summary of Results

Electrons with energies > 50 keV were observed at northern high latitudes during the period 24th February – 3rd March. The peak intensities of the trapped and precipitated electrons on the morning side of the orbit were approximately equal, and occurred at about the same latitude. Precipitated electrons were rarely observed on the evening side, and on the morning side there were large variations in the trapped and precipitated peak intensities from pass to pass. The greatest intensities of precipitated electrons on the morning side occurred for the highest values of K_p, which also corresponded to the occurrence of the precipitation at the lowest invariant latitudes. There was an average asymmetry of 3–4° in the latitudes at which the peak intensities of the trapped electrons occurred on the morning and evening sides, the latitudes being lower on the evening side.

A softer energy spectrum was observed for the precipitated electrons than for the trapped electrons on the morning side. For the trapped electrons, harder spectra were found on the evening side than on the morning side. The spectrum of the precipitated electrons became softer with increasing latitude.

Acknowledgements

The authors wish to acknowledge computer-programming assistance from Mr. R. J. Pratt of the Radio and Space Research Station.

The initial processing of the ESRO I satellite data was performed at the ESRO Operations Centre (ESOC), Darmstadt. The subsequent analysis was carried out at the Radio and Space Research Station of the Science Research Council and the paper is published with the permission of the Director.

References

Bewersdorff, A., Dion, J., Kremser, G., Keppler, E., Legrand, J. P., and Riedler, W.: 1966, *Ann. Geophys.* **22**, 23.

Feldstein, Y. I.: 1966, *Planetary Space Sci.* **14**, 121.

Fritz, T. A.: 1968, *J. Geophys. Res.* **73**, 7245.

Hartz, T. R., Montbriand, L. E., and Vogan, E. L.: 1963, *Can. J. Phys.* **41**, 581.

Johnson, R. G., Meyerott, R. E., and Evans, J. E.: 1967, in *Aurora and Airglow* (ed. by B. M. McCormac), Reinhold, N.Y., p. 169.

McDiarmid, I. B. and Burrows, J. R.: 1968, *Can. J. Phys.* **46**, 49.

Maehlum, B. N. and O'Brien, B. J.: 1963, *J. Geophys. Res.* **68**, 997.

O'Brien, B. J.: 1964, *J. Geophys. Res.* **69**, 13.

Rao, C. S. R.: 1969, *J. Geophys. Res.* **74**, 794.

ESRO I MEASUREMENTS OF LOW-ENERGY AURORAL
PARTICLES FROM FEBRUARY 23 TO MARCH 2, 1969

W. RIEDLER

Kiruna Geophysical Observatory, Sweden*

Abstract. Results of satellite electron and proton measurements in the energy range 1 to 13 keV are reported. Three different electron spectra can be distinguished: a 'hard' type, a high-intensity 'soft' type and a low-intensity 'soft' type. These appear in distinct regions: two hard zones, one probably coinciding with the auroral oval, one with the auroral zone, a high-intensity soft belt and a low-intensity soft polar cap region. Proton precipitation was highest in the dusk and noon sectors. Anisotropies and changes of zone boundaries during the observed period February 23 to March 2, 1969, are reported.

1. Introduction

The experiment S71-B included in the ESRO I (Aurorae) scientific payload is capable of measuring electron and proton fluxes in the energy range 1 to 13 keV and in two directions ($10°$ and $80°$) with the satellite axis. Assuming stabilization of the spacecraft along the magnetic field vector, this means corresponding pitch angles of the measured particles.

Figure 1 gives an idea of the experimental technique used in the instrument. Channel multipliers together with electrostatic analyzers serve as sensor elements. They were arranged in two groups of six each, at $10°$ and $80°$, respectively. In one of these six sensor elements in each box, the analyzer was replaced by a Ni-63 radioactive source, providing a reference background. This enables one to follow a possible deterioration effect of the channel multipliers. The experiment has been switched on in orbit for the first time on October 14, 1968, and this effect has been small hitherto (September 1969).

The geomagnetic factors are in the order of 10^{-4} to 10^{-5}, but in this paper count rates rather than absolute values will be discussed. A detailed description of the instrument will be published elsewhere (Riedler *et al.*, 1969).

The experiment makes use of the high-speed real-time (HSRT) telemetry system of ESRO I (Aurorae), with read-out stations at Redu (Belgium), Tromsø and Spitsbergen (Norway) and Fairbanks (Alaska). It is switched on and off on ground command and data are obtained during suitable satellite passes. The duration of the data reception periods varies between 2 and 12 min, depending on the geographical location of the pass relative to the read-out station and the satellite altitude. For a long-lasting event like the one considered in this Symposium, the experiment thus provides 'snapshots', which sample the event at more or less regular time intervals (see Section 2).

The time resolution which can be achieved by the instrument is 41.8 msec, if the flux is sufficiently high (more than 64 counts during 41.8 msec). In this paper, no reference is made to this high time resolution (corresponding to approximately 400 m

* New address: Dept. of Communications and Wave Propagation, Technical University, Graz, Austria.

V. Manno and D. E. Page (eds.), Intercorrelated Satellite Observations Related to Solar Events, 557–566. All Rights Reserved

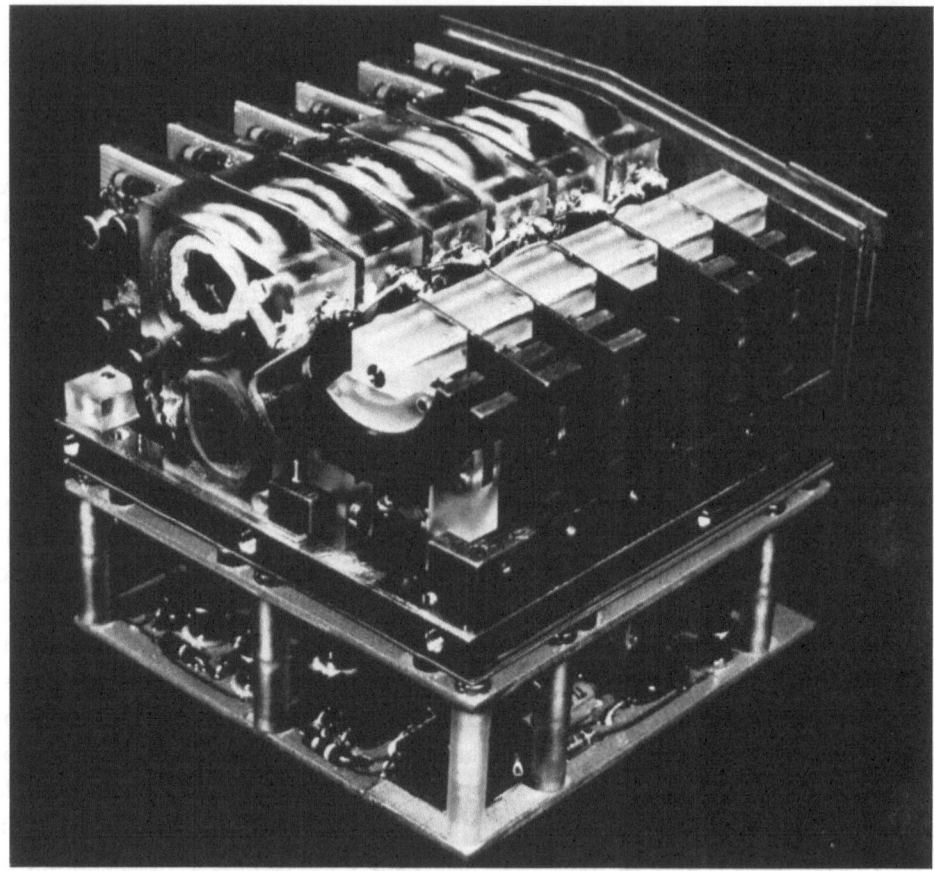

Fig. 1. Sensor box of experiment S71-B. Upper deck: electrostatic analyzers, channel multipliers and preamplifiers. Lower deck: high-voltage supply unit. In front of one channel multiplier is mounted a Ni-63 radioactive source for reference purposes.

spatial resolution). Instead, maximum count rate values, observed during 10 successive sub-frames (8 sec) are used.

The instrument is, of course, not suited to measure solar particles directly, as they are present during a PCA event. However, if there is a sufficiently high flux of protons above ~ 50 MeV, these can penetrate the satellite material and cause counts of the detectors. An example for this effect can be seen in Figure 2, where measurements during the Nov. 18, 1968 PCA event are shown.

The constant count rates of all detectors, clearly to be seen after 19.37.30 UT are caused by these penetrating protons. In the auroral zone, these count rates are superimposed by the regular counts, from 1 to 13 keV particles. No such effect could be observed during the February 25, 1969, event which agrees well with the fact, that it was a weak event of much less intensity than the one of November 18, 1968. According to other measurements on ESRO I (Søraas, private communication), fluxes of protons > 50 MeV were more than a factor of 20 less than on November 18, 1968.

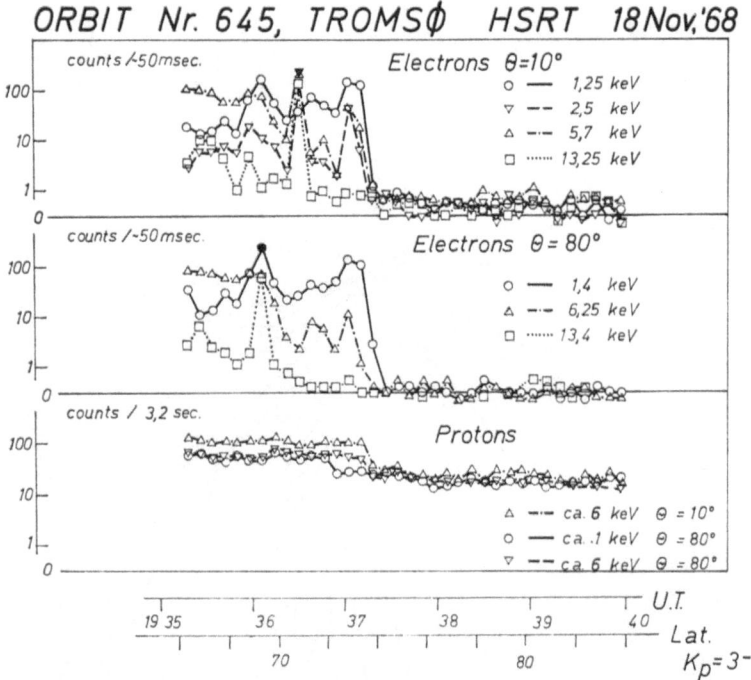

Fig. 2. Count rates observed during the November 18, 1968 PCA event. UT and geographical latitude are given.

2. Measurements Between February 23 and March 2, 1969

The passes of ESRO I (Aurorae) over a suitable read-out station, where S71-B has been switched on, are summarized in Figure 3 for the period February 23 to March 2, 1969. The (very few) Fairbanks passes are not included on the map, however. Southward Tromsø passes occurred during the morning (~ 6 UT), northward Tromsø passes during evening (~ 18 UT). Spitsbergen passes were typically at noon. No midnight passes are available. Unfortunately, satellite perigee was close to the read-out stations in question, which caused the short duration and therefore the small latitudinal coverage of the individual data reception periods. This is also the main reason, why some of the successive orbit numbers are missing, i.e. S71-B was not switched on during these very short passes.

The passes were deliberately grouped into three time classes: February 23 and 24, February 25 to 27 and February 28 to March 2, 1969. These groups correspond roughly to the development of the PCA event in question. This can be seen more clearly from Figure 4, where the riometer absorption curve at 35 MHz and the X-component of the geomagnetic field, as observed in Kiruna, have been plotted together with the times of available satellite passes with S71-B switched on. (The standard riometer recordings of 27.6 MHz showed too much interference noise to be included here.)

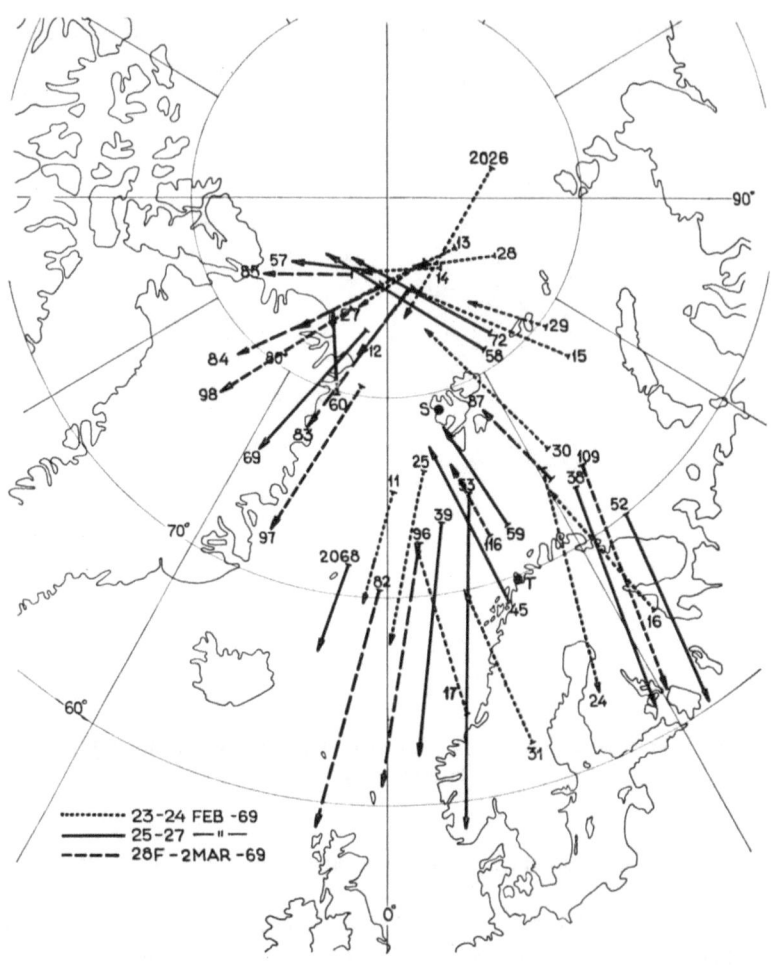

Fig. 3.   Summary of ESRO I (Aurorae) passes over Scandinavian HSRT read-out stations with
S71-B switched on in the period February 23 to March 2, 1969 (orbit numbers 2011 to 2116).
T = Tromsø, S = Spitsbergen.

Following the flare on February 23 at 0444 UT, solar particles started to arrive at
the earth around 0500 UT, as could be seen from VLF phase measurements at Kiruna
(Westerlund, private communication), and from the riometer recordings. This first,
very weak event (February 23 and 24) was enhanced, when a new flare on February 24
at 2308 UT caused higher particle fluxes, giving stronger riometer absorption from
February 25, 0100 UT about. Still more absorption and VLF disturbances could be
observed from 0928 UT, following another flare at 0903 UT. This situation prevailed
for some time (February 25 to 27) until the storm effects of the plasma cloud were
fully developed, giving rise to magnetic disturbances and auroral type absorption
from February 27, 1500 UT. The third group of data (February 28 to March 2) thus
refers to a situation, where the decaying solar particle effects and substorm type
disturbances were superimposed.

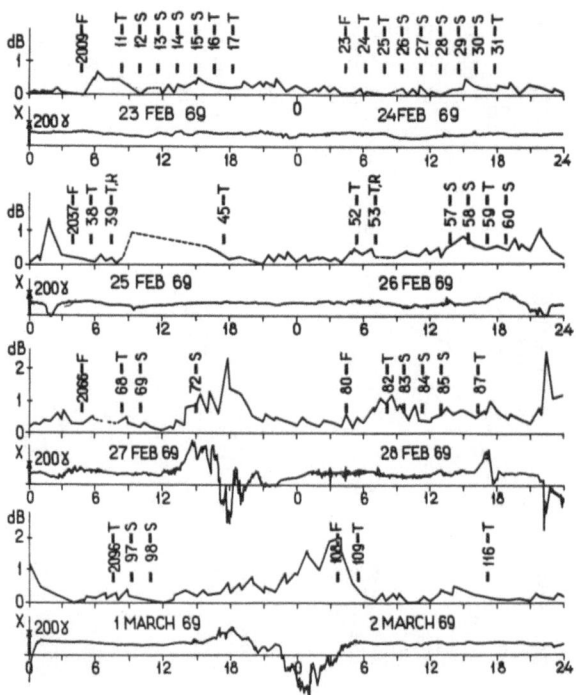

Fig. 4.   Riometer absorption (35 MHz) and *X*-component of the geomagnetic field at Kiruna, together with available satellite passes. T = Tromsø, S = Spitsbergen, F = Fairbanks, R = Redu.

In Figure 5 two distinct types of electron count rates are shown, which occurred in three successive passes (orbits 2026, 2027, 2028 on February 24, 1969, all recorded at Spitsbergen in HSRT). They correspond to relatively hard and soft spectra, respectively. The difference is apparent from the curves in Figure 5, which are based on 8-sec maximum values. For the soft types, practically no counts have been obtained for energies other than ~ 1 keV. The two spectral types can also be distinguished by their temporal (latitudinal) behaviour. The hard spectral type (pass S2026) is quite smooth, whereas the soft type exhibits pronounced peaks, if the intensities are high, as can be seen on passes S2027 and S2028. Whether these peaks are temporal or spatial in character, can not be decided, however. The soft type can occur at lower *L*-values than the hard type. There is no pronounced pitch angle anisotropy in either case. On pass S2028 (Figure 5) a decrease of intensity can be seen for *L*-values greater than 25. This is not a coincidence only, but it seems to be a systematic effect. The soft spectral type thus has to be subdivided into two groups. One with high intensities and pronounced peaks and one with much lower intensities, which usually is smoother in appearance.

In order to obtain a schematic picture of the 'hard' and 'soft' zones, the measurements from the period February 23 to March 2, 1969 have been summarized in Figure 6, using a local magnetic time (eccentric dipole time, EDT) and invariant latitude $[\Lambda = \cos^{-1}\sqrt{(1/L)}]$ as coordinate system. The 'hard' electron spectral type is

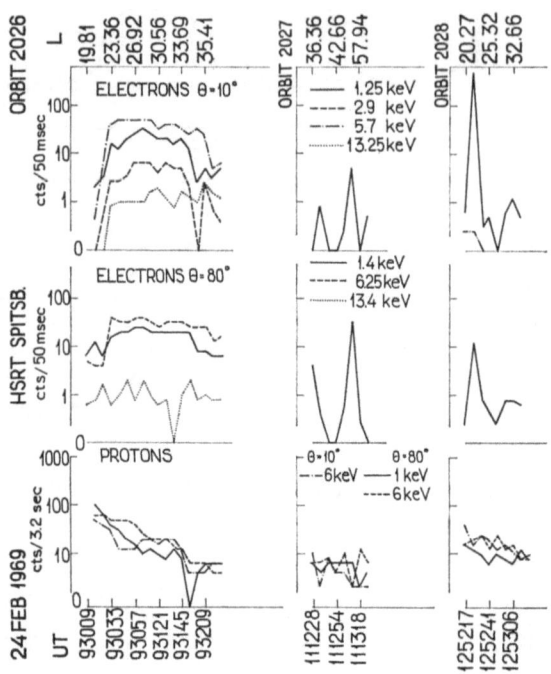

Fig. 5.   Hard and soft electron spectral types, observed on three successive passes
on February 24, 1969.

indicated by heavy bars, the 'soft' type by dots. The hardest spectra are always observed in the dawn-noon sector at relatively low L-values (see examples Figure 7).
The boundaries are not sharply defined, contrary to the noon-dusk sector, where
quite steep boundaries are observed. In the morning region, a pronounced softening
of the spectrum is usually seen near the southern flux boundary. The highest electron
fluxes for the hard spectral type are observed in the dawn-noon and dusk-midnight
regions (at 1 keV $\sim 10^8$ el/cm^2 sec ster keV). The soft-type peak fluxes can be higher,
however. Observed values at 1 keV are $> 10^9$ el/cm^2 sec ster keV. There is a pronounced difference of these high, spiky, fluxes, which are observed in a narrow
($\sim 2°$) zone north of the 'hard' electron zone and the low-intensity soft region. The
peak fluxes can be three orders of magnitude less, with quite sharply defined boundaries between the two (see examples in Figures 5 and 9). All electron fluxes on February
23 and 24 seem to be quite isotropic, with no systematic exceptions. This situation was
changed on February 25, 1969, as illustrated in Figure 7. In the morning region, the
80°-electron counts extend more to the south than the 10° counts. In this region the
spectrum is softer than where the flux is isotropic.

    The observed regions of proton fluxes are also indicated in Figure 6. Proton flux
intensities are usually highest in the evening sector for $\Lambda < 75°$ (at 1 keV typically
5×10^5 pr/cm^2 sec ster keV), whereas very few proton counts were obtained in the
low-intensity soft electron region, located on the polar cap. As can be seen from

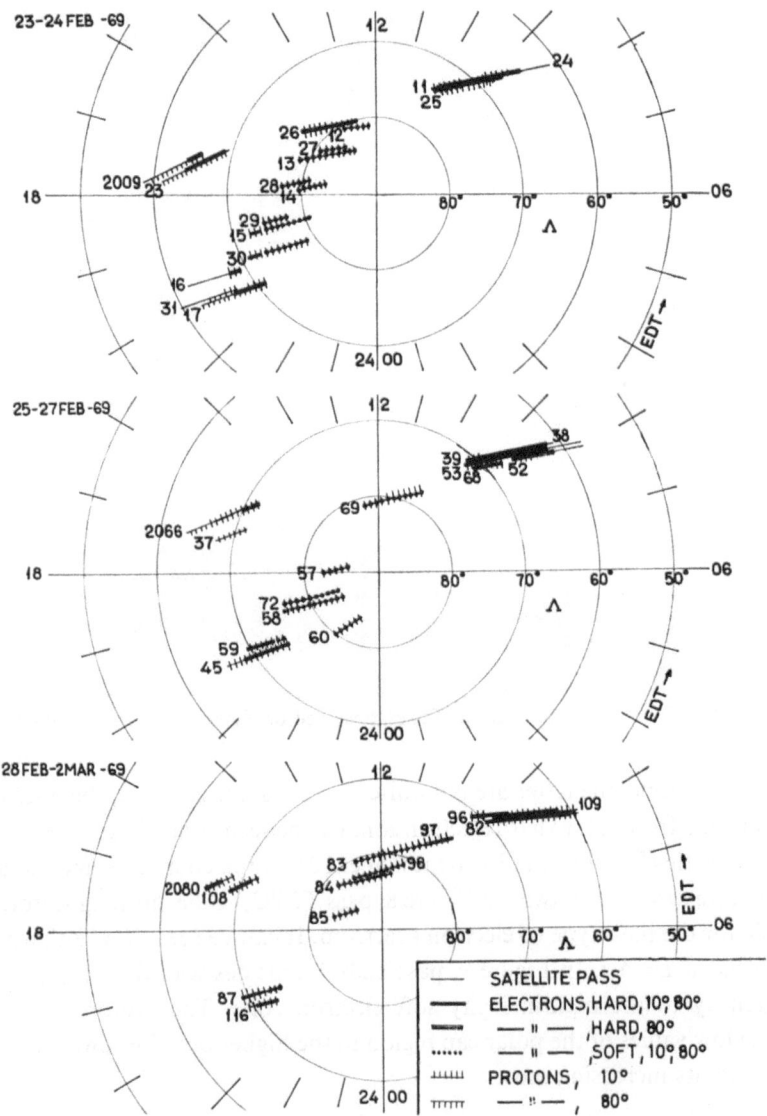

Fig. 6. S71-B HSRT passes from February 23 to March 2, 1969, using eccentric dipole time (EDT) and invariant latitude $A = \cos^{-1}\sqrt{(1/L)}$ as coordinate system.

Figure 6, 80°-proton counts are often observed much to the south of a more or less isotropic region in the evening sector, whereas in the morning the (few) proton counts for 10° and 80° decline together towards the south (cf. Figure 7).

A pronounced tendency towards proton anisotropy with 10° count rates much higher than the 80° count rates can be observed at noon on passes S2069, S2083 and S2097. Figure 8 shows this anisotropy, which coincides with high-intensity soft electron spectra and the transition to the 'hard' spectra. Because S2069 is close to the

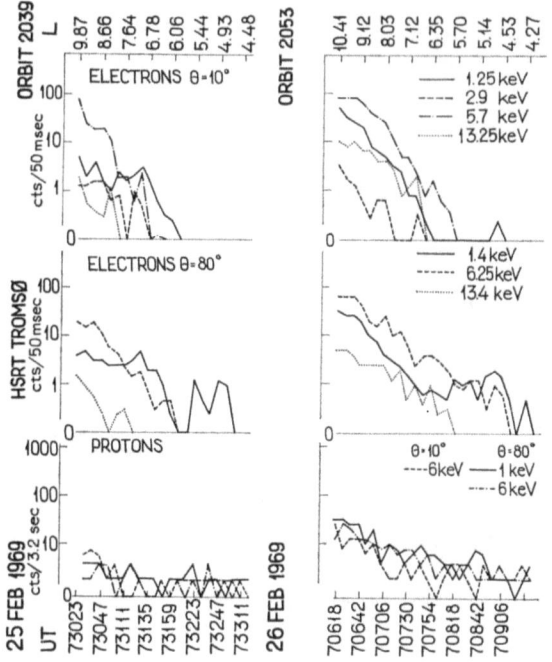

Fig. 7.   Anisotropic electron fluxes, observed on February 25 and 26, 1969.

hard electron zone, the latter are not quite as soft as, for example, the ones on S2028.
The transition from soft to hard electron zone can be seen clearly in Figure 9, where pass
S2097 shows this transition and also the higher 10° proton count rate, mentioned above.

Two more cases are shown. In Figure 8, pass T2082, in the morning sector, is a good
example for the hard type of electron spectrum. It can be seen, how a sharp softening
takes place at $L = 5.5$. In Figure 9, pass S2098 coincides with the transition from the
low-intensity to the high-intensity soft electron zone, The proton counts increase
from the low values of the polar cap region to the higher ones for lower latitudes, with
the 10° counts increasing most.

3. Summary and Discussion

For the local magnetic times available in the analyzed period, the hard and soft types
of electron spectra seem to belong to well defined zones. If we consider Figure 6,
February 23 and 24, 1969, there is a 'hard' zone, consisting of passes 2026, parts of
2015, 2030, 2016, and 2017. This zone is asymmetric in the noon-midnight direction
and coincides roughly with the auroral oval (Feldstein, 1966). There is a second, hard
zone, however, consisting in this case of parts of passes 2009, 2023, 2011, 2024 and
2025. This zone is quite narrow in the evening (2023), broader in the morning and
seems to coincide with the auroral zone. Within the oval 'hard' zone, the high-intensity
soft spectral type has been observed in a region, approximately 2° to 3° wide (passes

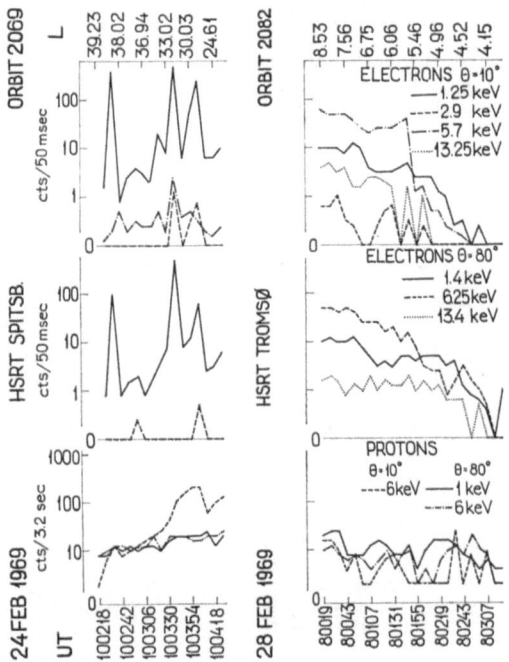

Fig. 8.   Examples of soft and hard electron spectral types observed on February 27 and 28, 1969 respectively. Proton counts on pass S2069 show increasing 10°-flux.

2012, 2027 and 2029, and parts of passes 2013, 2028, 2015 and 2030). Inside this high-intensity soft belt, a region of low-intensity could be observed. For example, pass 2014 lies entirely inside this region. Unfortunately, for the period analyzed here, no passes in the midnight-morning sector are available and it is not possible to make any statements about a down-dusk symmetry or asymmetry.

It seems, that the three different zones, discussed above, are the ones reported by Burch (1968) and Hoffman (1969) for other local times. The low-intensity soft zone could be identical with the polar cap region, mentioned by Burch.

On February 25, 1969, changes took place, as to the positions of the zones. In the dusk region, the southern boundary moved from $L=5$ ($\Lambda=63.5$) to $L=8.5$ ($\Lambda=70°$). From the data, nothing can be said about the width of this zone after the change. Even around 2000 EDT, a northward shift of the hard electron zone of $\sim 2°$ could be noticed. The width remained approximately the same. In the dawn region, the relative increase of 80° electron counts at the southern boundary could be observed, which was discussed in Section 2. From February 28 to March 2, 1969, this anisotropy in count rates disappeared again. On pass 2109, both the 10° and 80° count rates decrease together.

Concerning the proton zones, the boundary for isotropic fluxes in the dusk region was shifted southwards by at least 4° on February 25, 1969, as can be seen from

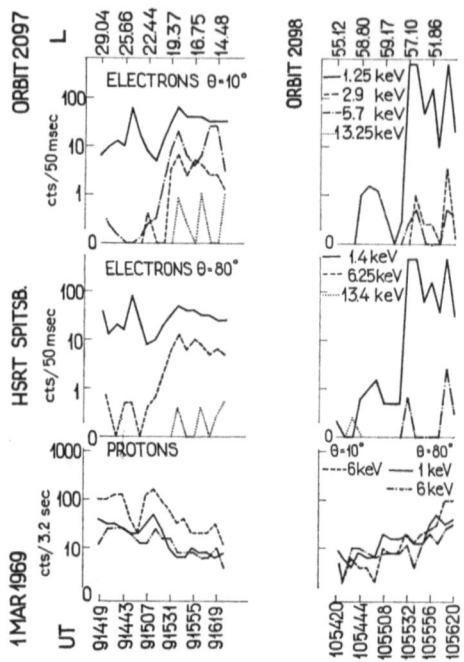

Fig. 9.   Transitions from soft and hard electron zones and from high to low intensity in the soft
zone, observed on March 1, 1969.

Figure 6. Whether or not the pronounced anisotropy around noon was present already on February 24, 1969, can not be inferred from the available data.

On the whole, it is not clear at present, how much, if any, of the observed effects is connected with the particular event of February 25, 1969. A data analysis is under way, which will reveal the 'normal' behaviour of the low-energy electron and proton fluxes. This analysis will also cover more local times and will thus be suited to give information on the dawn-dusk symmetry situation.

References

Burch, J. L.: 1968, 'Low-Energy Electron Fluxes at Latitudes above the Auroral Zone', *J. Geophys. Res.* **73**, 3585.

Feldstein, Y. I.: 1966, 'Peculiarities in the Auroral Distribution and Magnetic Disturbance Distribution in High Latitudes Caused by the Asymmetrical Form of the Magnetosphere', *Planetary Space Sci.* **14**, 121.

Hoffman, R. A.: 1969, 'Low-Energy Electron Precipitation at High Latitudes', *J. Geophys. Res.* **74**, 2425.

Riedler, W., Hultqvist, B., and Olsen, S.: 1969, 'A Satellite Instrument for Particle Measurements Using Channel Multipliers', to be published.

AN OBSERVATION OF THE FEBRUARY 26, 1969
INTERPLANETARY SHOCK WAVE

A. J. HUNDHAUSEN, S. J. BAME, and M. D. MONTGOMERY

University of California, Los Alamos Scientific Laboratory, Los Alamos, N.M., U.S.A.

1. Introduction

Only a sparse amount of solar wind data was obtained by the Vela satellites during the February 25 solar event and the subsequent interplanetary disturbance, so that a detailed time history of solar wind properties cannot be given. However, data are available from 0004 to 0240 UT on February 26, when the Vela 4B spacecraft was near the earth's bow shock at 19.2 earth radii, 15° solar ecliptic longitude, and −19° solar ecliptic latitude. An interplanetary shock wave, which caused the sudden commencement reported at ground stations shortly before 0200 UT, was observed at 0152 UT. An observational description of the shock and a brief discussion of its relationship to the solar events of February 25 will be given here.

2. The Shock Observation

The Vela 4B electrostatic analyzer system and the technique employed in interpreting data are described in Montgomery *et al.* (1968, 1969). During the period of observation under discussion here the spin axis of the earth-oriented satellite is at an angle of ~25° to the satellite-sun line. The angular acceptance windows of the analyzer extend over ±55° in spacecraft latitude; thus the highly directional beam of solar wind positive ions, coming from near the direction of the sun, falls near the edge of these windows, and only a fraction of the total flux is observed. This precludes any determination of the proton density. However, the solar wind speed (related to the position of the flux peak in an energy-per-charge spectrum) can be determined with good accuracy, while the proton temperature (related to the energy and angular widths of the flux peak) can be roughly determined. The spin axis orientation introduces no problem into the observation of the nearly isotropic flux of solar wind electrons, and both the electron density and temperature can be determined.

Figure 1 shows the solar wind speed, electron density, proton temperature, and electron temperature measurements made early on February 26. As solar wind particle distribution functions are often anisotropic, two temperatures are shown for both the solar wind protons and electrons. The observed flow speed and temperature values indicate that Vela 4B was in the magnetosheath until 0052 UT when it crossed the bow shock into the solar wind. After 50 min in the solar wind, the satellite re-entered the magnetosheath, where the measured plasma properties were close to those from the period before 0052. Between 0152 and 0153 UT, an abrupt change to conditions characteristic of the solar wind occurred, but with a higher flow speed, electron

V. Manno and D. E. Page (eds.), Intercorrelated Satellite Observations Related to Solar Events, 567–570. All Rights Reserved
Copyright © 1970 by D. Reidel Publishing Company, Dordrecht-Holland

density, proton temperature, and electron temperature than were measured during the earlier excursion beyond the bow shock. We interpret this change in solar wind properties as evidence for the passage of an interplanetary shock at 0152 UT. The increased momentum flux in the post-shock solar wind has pushed the bow shock closer to the earth, leaving the satellite, which had moved into the magnetosheath only ten minutes earlier, in interplanetary space.

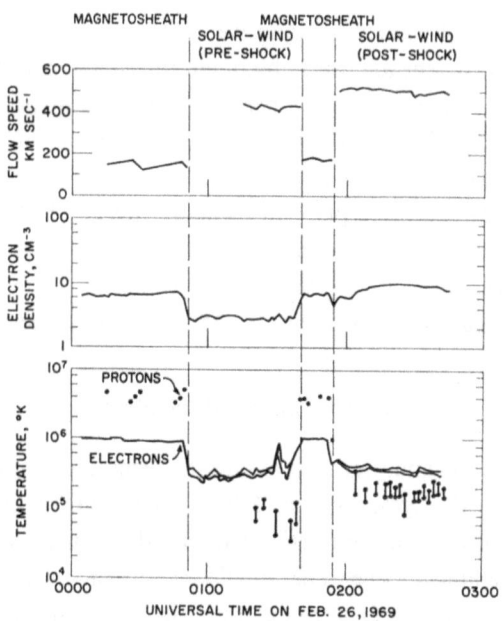

Fig. 1.   Vela 4B measurements of the solar wind speed, electron density, proton temperature, and electron temperature on February 26, 1969. Maximum and minimum temperature values are given for both electrons and protons in the solar wind, as the distribution functions of both particle species are often anisotropic.

3. Properties of the February 26 Shock

The pre-shock solar wind properties are the values obtained between 0052 and 0142 UT; post-shock solar wind properties are those obtained after 0152 UT. Table I gives

TABLE I

Solar wind properties measured before and after the February 26, 1967, interplanetary shock

	Pre-Shock	Post-Shock
Density	2.9 cm^{-3}	7.5 cm^{-3}
Flow speed	421 km sec^{-1}	512 km sec^{-1}
Electron temperature	3.6×10^5 K	3.9×10^5 K
Proton temperature	$\sim 6 \times 10^4$ K	$\sim 2 \times 10^5$ K

Change in electron characteristics occurs between 0152 and 0153.
Satellite is at 15.3° SE longitude, −19.1° SE latitude.

the values of solar wind speed, electron density, and electron temperature averaged over twenty minutes of pre- and post-shock measurements. Proton temperature values are given based on the measurements judged to be most reliable.

Table II gives some properties of the shock derived from the plasma parameters given in Table I. The shock speed has been computed from the mass continuity equation under the assumption that the shock normal was in the radial direction. An upper limit to shock speed is also given, based on the requirement of subsonic flow in the post-shock gas (again assuming a radial shock normal). In the frame of reference of the shock, the ambient plasma flows in at 149 km sec^{-1}, or at twice the sound speed, while the post-shock plasma flows out at 48 km sec^{-1}, or at half the sound speed.

TABLE II

Properties of the February 26, 1967 interplanetary shock, derived from the measured solar wind parameters of Table I

| Shock speed (assuming radial propagation) | 570 km sec^{-1} | |
| Upper limit on shock speed | 603 km sec^{-1} | |
	Pre-shock	Post-shock
Sound speed	76 km sec^{-1}	91 km sec^{-1}
Flow speed in shock frame	$+$ 149 km sec^{-1}	$-$ 48 km sec^{-1}
Mach number	2.0	0.5

4. Discussion

The February 26, 1969 interplanetary shock appears to be entirely typical of the shocks observed during a decade of interplanetary measurements (see Table II of Hundhausen, 1970). In particular, the propagation speed of 570 km sec^{-1} is only slightly higher than the representative value of 500 km sec^{-1}.

As the pre-shock solar wind speed is close to 400 km sec^{-1}, a predicted transit time from the sun can be obtained from Figure 9 of Hundhausen (1970). If the shock had propagated from the sun at the observed speed, the transit time would have been 73 hours. If the shock were decelerated in interplanetary space, a shorter transit time would result. In the blast wave limit, a propagation speed at 1 AU of 570 km sec^{-1} implies a transit time of 53 hours. These values are upper and lower limits for propagation through an ambient solar wind with no irregularities in the density.

It is tempting to associate the February 26 shock with one of the flares observed on February 24–25. However, the range of possible transit times inferred above raises considerable difficulties for any such association. Even the earliest of the 2B flares, that of ~ 2300 UT on February 24, gives a transit time of 27 hours, or a mean transit speed of 1540 km sec^{-1}. Several possibilities exist for salvaging such a flare association:

(1) The shock normal may have been tilted away from the radial direction, so that the shock propagation speed has not been correctly derived. However, an extremely large tilt ($\sim 70°$) would be required to raise the speed as high as 1500 km sec^{-1}.

(2) The shock may have been decelerated near the earth by a thin region of dense interplanetary plasma. That is, it propagated to nearly 1 AU at a speed in excess of 1500 km sec^{-1} and was then rapidly slowed to the observed speed. However, densities large enough to produce this effect in a short distance (e.g., deceleration within a region 0.1 AU thick implies a density more than an order of magnitude higher than usual) have rarely been observed in 2-$\frac{1}{2}$ years of detailed Vela 3 observations. Although neither of these possibilities can be ruled out, both appear to be unlikely.

(3) The disturbance may have been produced by a small flare, such as the 1B flare observed at 0440 UT on February 23.

The possibility that the February 26 disturbance was not flare-associated, but was part of a long-lived interplanetary stream, should be given serious consideration. Reconstruction of a more complete time history of solar wind properties by combining data from several satellites would be useful in finding an interpretation for this event.

Acknowledgements

This research was done as part of the Vela nuclear test detection satellite program, jointly administered by the Advanced Research Projects Agency of the Department of Defense and the U.S. Atomic Energy Commission, and managed by the U.S. Air Force.

References

Hundhausen, A. J.: 1970, this volume, p. 111.
Montgomery, M. D., Bame, S. J., and Hundhausen, A. J.: 1968, *J. Geophys. Res.* **73**, 4999.
Montgomery, M. D., Asbridge, J. R., and Bame, S. J.: 1969, submitted to *J. Geophys. Res.*

Discussion

Lanzerotti: In the February 24, 09.00 flare we see the high energy particles substantially before the flare. It seems clear that there are other things going on before the 09.00 flare.

Hundhausen: I agree. In a complex series of events these associations are not at all unambiguous. When you have the information how fast things are going to the earth it points in one direction but it is not always the one we have taken (topped to set).

Question: Can you say anything about the direction in which the shock is propagating?

Hundhausen: We cannot determine it from the set of variables we have. If it would be tilted over at about 45°, it would not make a large correction to the propagation speed yet.

THE ASSOCIATION OF ENERGETIC STORM PARTICLES
WITH INTERPLANETARY SHOCK WAVES

SIDNEY SINGER

University of California, Los Alamos Scientific Laboratory, Los Alamos, N.M., U.S.A.

1. Introduction

At the earth, the major features of a classic solar particle event are observed to be a velocity dependent onset time (measured from the time of the flare presumed to accelerate the particles), a velocity dependent rise time of some several hours, and a nearly exponential decay time of many hours. The delayed onset and rise is usually explained in terms of a complex particle path whose great length arises because of scattering or diffusion produced by irregularities and fluctuations in the curved interplanetary field; the long decay is interpreted as a slow leakage of particles out of the inner solar system, perhaps through a diffusing barrier far beyond the earth where the solar wind undergoes a subsonic transition (Parker, 1958).

As the structure of the interplanetary field becomes better understood, modifications to the simple diffusion model have been required. Explicit allowances for an anisotropic diffusion must be made, and one must consider the possibility that the diffusion may take place predominantly near the sun instead of throughout the inner solar system. The presence of large scale persistent structures (i.e., sectors and filaments) in the interplanetary field is known to produce deviations from the predictions of the simple diffusion model.

One important deviation consists of the large enhancement of 1–10 MeV protons that sometimes occurs several days after the onset. The time histories of these 'Energetic Storm Particles' (ESP) (Bryant *et al.*, 1965) are not consistent with predictions of a conventional diffusion model. ESP events are usually seen in connection with SC magnetic storms and Forbush decreases, and only when solar energetic particles from an earlier flare are present in the solar system. Most investigators have concluded that ESP events are in some way connected with the arrival of the interplanetary shock wave that produces the SC. Van Allen and Ness (1967) have presented evidence that an ESP event arises from a Fermi acceleration of particles at the shock front. Rao *et al.* (1967, 1968), on the other hand, discount that possibility, as well as the one wherein particles are swept up and possibly trapped in the shock; they propose instead that the ESP's result from acceleration in the shock itself.

In this paper, observations of several ESP events associated with the February 25, 1969, May 1967, and January 9, 1968, solar energetic particle events are reported. The following conclusions have been drawn from these observations:

(a) ESP events are associated with shock waves and are not seen when other kinds of solar wind discontinuities (e.g., tangential discontinuities) are detected.

(b) ESP events are associated with protons of energy less than several MeV; the

V. Manno and D. E. Page (eds.), Intercorrelated Satellite Observations Related to Solar Events, 571–582. All Rights Reserved
Copyright © 1970 by D. Reidel Publishing Company, Dordrecht-Holland

intensity and energy spectrum of protons >20 MeV are not usually affected by the event.

(c) The ESP event is characterized by a slow increase in flux intensity beginning several hours before shock arrival and, frequently, by an abrupt, stepped decay, as reported by Axford and Reid (1963), and Rao *et al.* (1967, 1968); however, the *peak* of the event (the 'shock spike') occurs *essentially simultaneously* with the plasma discontinuity which defines the shock passage.

(d) The degree of enhancement of the particle flux is variable, and perhaps depends on the strength of the shock.

(e) Comparisons of time history of ESP events seen by Vela satellites *inside* and *outside* the magnetopause show that ESP's do not have free and immediate access to the magnetotail.

2. Instrumentation

The observations to be described here are taken from data from the Vela 4 Energetic Particle experiment and the Electron-Proton Spectrometer. Detailed descriptions of these instruments have been given elsewhere (Singer *et al.*, 1969; Montgomery *et al.*, 1968), and a brief description is contained in another part of this volume (Singer, 1970).

3. Observations

A. THE NATURE OF THE ESP EVENT

Following the February 25, 1969, proton event, three SC's were detected at earth: 0157 UT on February 26, ~0308 UT on February 27, and 0424 UT on February 28. Each of these is probably due to an interplanetary shock produced in a solar flare ~48–72 hours earlier (Hundhausen, 1970a, b). The SC of February 28 was probably produced by the same flare which gave rise to the solar energetic particle event of February 25.

The time history of the event associated with the February 26 shock is shown in Figure 1. The absolute differential fluxes for four energy channels covering the range 0.65 to 14.2 MeV are plotted for the period 0000–0800 UT on February 26. At 0051 UT, the spacecraft, which had been in the magnetosheath, crossed the bow shock and entered the interplanetary medium. At 0113 UT, the spacecraft briefly re-entered the magnetosheath, returned to interplanetary space, and finally returned to the magneto-sheath at 0140 UT. The interplanetary shock arrived at 0155 UT and pushed the magnetosheath back, so that the spacecraft found itself in the disturbed interplanetary medium. An examination of the energetic particle flux shows a spike that occurs within a few minutes of the shock arrival. The slow buildup in flux intensity that usually begins some hours before shock arrival is not detectable here; but the drop to 80% of the preshock flux (0.65–0.91 MeV) intensity is evident. It is also apparent that the ESP event is confined to the low energy end of the spectrum; that an ESP event had occurred would scarcely be discernible if only >4.5 MeV data were available.

In Figure 2, the proton flux for five channels from 0.55–13.6 MeV are plotted for the ESP event of February 28. Also shown is the counting rate of the Vela 4 neutron monitor, whose deviations from background are produced by 20–40 MeV protons. At the moment of shock arrival (0420 UT at the spacecraft), the spacecraft was outside the magnetopause and probably in the interplanetary medium. Beginning at about 0300 UT, a slow increase in the flux intensity can be detected, and the spike associated with the ESP event occurs only four minutes before detection of the SC at earth. At the spacecraft, the time difference was probably less. Within an hour after shock arrival, the 0.65–0.91 MeV intensity dropped to ~40% of the preshock level; the step-like decrease resembles the Forbush decrease that sometimes accompanies ESP events (Rao, 1968). The ESP's are again seen to be relatively soft; the proton fluxes above 4.5 MeV appeared to be unaffected by the passage of the shock.

B. RELATIONSHIP OF THE ESP EVENT TO THE INTERPLANETARY SHOCK

In most events, the intensity spike that characterizes the ESP event occurs essentially simultaneously with the arrival of the interplanetary shock at the spacecraft; the shocks of May 24, 1967, and May 30, 1967, will be used to illustrate this point.

The solar wind positive ion temperature, density, and bulk velocity, and the 0.68–0.95 MeV proton flux are plotted for the time interval 1200–2000 UT on May 24,

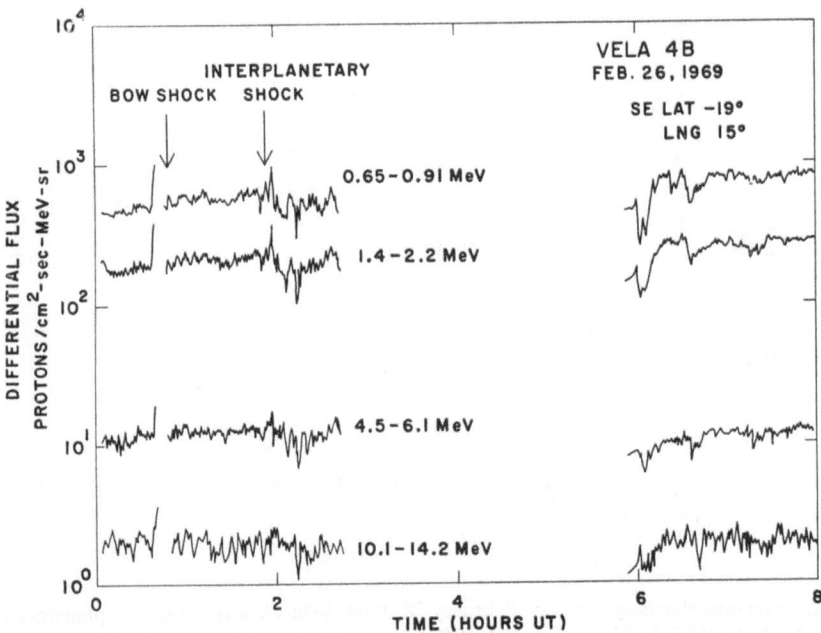

Fig. 1. Interplanetary shock of February 6, 1969. Vela 4B was in the magnetosheath at 0156 UT on February 26, 1969 when the interplanetary shock wave pushed the bow shock back and left the spacecraft in the interplanetary medium. Four different energies of energetic protons are plotted to define the ESP event. The shock spike is detected essentially simultaneously with the shock arrival, and the postshock decrease in the energetic particle flux can be seen. The ESP's cannot be detected at energies above 10 MeV.

1967, in Figure 3. The proton bulk velocity, density, and temperature of ~ 200 km/sec, 20/cm³, and 2×10^6 K, respectively, indicate (Bame, 1968) that 4A was in the magneto-sheath. At 1732 UT, the bulk velocity and density increased to 300 km/sec and $\sim 70/\text{cm}^3$, and the temperature decreased to $\sim 1.3 \times 10^6$ K. The interplanetary disturb-ance which caused these changes was a shock wave (Bame, 1969) that produced an SC at the earth at 1726 UT. The reason for the temperature *decrease* is not clearly understood. The energetic particle flux showed a slow rise beginning at ~ 1545 UT, but the ESP spike began at precisely the satellite shock arrival time. The rise time of the spike was about 3 min, and its width was ~ 10 min.

A similar set of data is presented for the interplanetary shock of May 30, 1967, in Figure 4, where the energetic particle flux and plasma parameters are plotted from 1030–2000 UT. During the entire period, Vela 4A was in the interplanetary medium. The shock arrival at 1426 UT was signified by simultaneous changes in the velocity, density, and temperature of from 410 to 580 km/sec, 7 to 14 protons/cm³, and 4×10^4 to 2×10^5 K, respectively. Beginning at ~ 1300 UT the energetic particle flux showed the slow rise common to many ESP events; and the spike, whose peak rises more than an order of magnitude above the prestorm level, occurred within one or two minutes

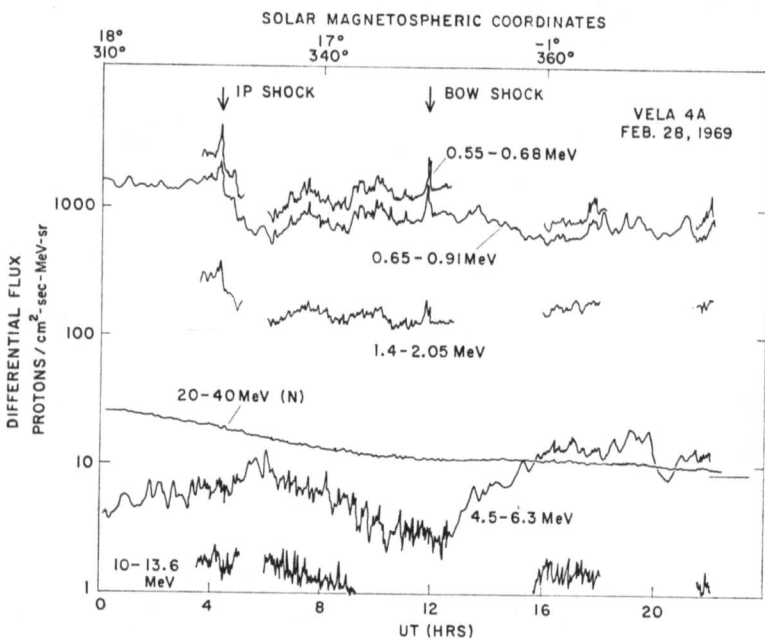

Fig. 2.   The interplanetary shock of February 28, 1969. Vela 4A was in the interplanetary medium when the shock of 0424 UT, February 28, 1969, arrived. The time history of the ESP event is defined by the five energetic proton channels and the 20–40 MeV signal in counts/sec from the Vela neutron monitor (whose cosmic ray background level is shown at the right edge of the graph at ~ 9 counts/sec). The apparent gaps in some of the channels are due to the limited data storage available when the spacecraft is operated in the memory mode. The slow buildup prior to shock arrival, the shock spike, and the sharp decreases following the shock are clearly evident. The ESP's are seen only below proton energies of 4.5 MeV, as are the modulations produced in the *bow shock* crossing near 1200 UT.

of the shock arrival time at the spacecraft. The spike width of ~ 5 min is not very different from the May 24 shock spike shown in Figure 3. The first of two step decreases in intensity can be seen at ~ 1700 UT.

Shock waves are only one sort of disturbance in the solar plasma that can produce geomagnetic disturbances such as SC's and SI's. Colburn and Sonnett (1966) have pointed out that the solar plasma behind tangential discontinuities can exert a different pressure on the magnetosphere than the upstream plasma, and Gosling et al. (1967) gave several examples of SC and SI events that were probably due to tangential discontinuities. In order to determine whether ESP events can also be produced by such discontinuities, all SC or SI disturbances for the period May 24, 1967 – January 31, 1968, and February 1, 1969 – February 28, 1969 were correlated with the solar energetic particle data, and the results are shown in Table I. A total of 47 events were found: 15 are known to be shocks (Ogilvie and Burlaga, 1969), 31 are probably tangential discontinuities, and 1 may be a shock. Of the 15 shocks, 9 had ESP's associated with them, 2 did not, 1 showed only a step decrease in energetic particle flux, 1 occurred at a time for which no data are available, and 2 occurred when no solar energetic particles pre-existed (and none were detected at the shock front). Of the 31 nonshock events, 19 showed no ESP's, 3 occurred when no data were available, and 2 occurred when no energetic particles pre-existed (and none could be detected at the

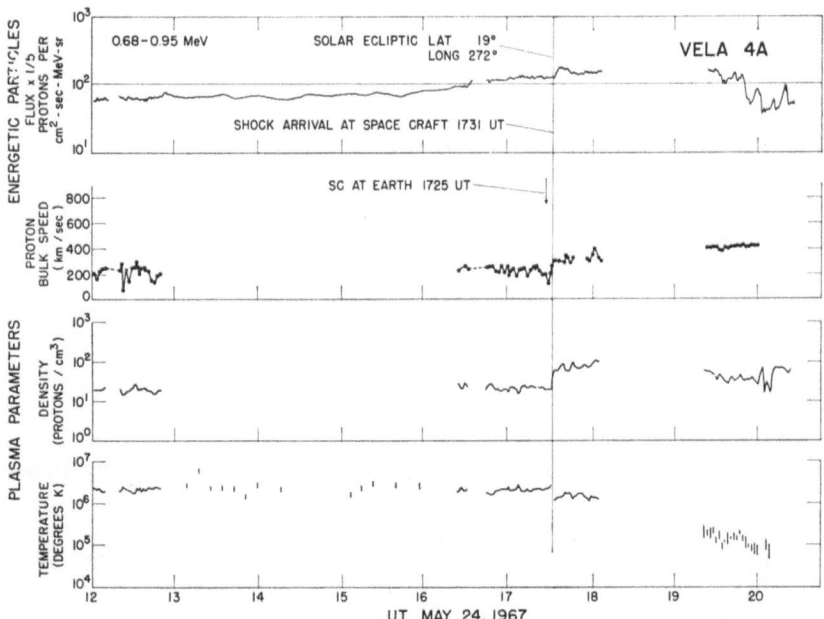

Fig. 3. Relation of shock spike to shock arrival. On May 24, 1967, the arrival of an interplanetary shock at 1731 pushed the bow shock back beyond Vela 4A, leaving it in the interplanetary medium. The temperature, bulk velocity, and density of the solar plasma are compared with the flux of energetic particles before and after shock arrival to show the simultaneity of the shock spike with the jump in plasma parameters. The vertical strokes used for some of the plasma temperature data indicate the magnitude of directional anisotropy in the temperature.

discontinuity). In 6 events, transient disturbances in the energetic particle flux were approximately associated with the discontinuity, and in one event, a step decrease in flux intensity was clearly associated with an SC. Each of the 6 transient fluctuations were detected at a time when other similar fluctuations were occurring concurrently, and none of the six had the characteristics of ESP's. No particle spike could be detected for the step-decrease event, and it is not clear whether this remaining possibility was an ESP event. We therefore conclude that ESP's are almost always associated with interplanetary shocks and are almost never associated with other types of discontinuities.

C. PENETRATION OF ENERGETIC STORM PARTICLES INTO THE MAGNETOTAIL

By comparing data from both Vela 4 spacecraft, it is possible to determine whether the ESP's have direct and immediate access to the magnetptail In Figure 5 (Montgomery and Singer, 1969) the 0.65–0.9 MeV proton fluxes from Vela 4A and 4B are plotted for the May 30, 1967, shock. Spacecraft 4A was in the interplanetary medium for the entire day, whereas 4B was in the high latitude magnetotail. Prior to the start of the event at ~ 1000 UT, the proton fluxes from both spacecraft were in good agreement; afterward, little resemblance existed between the two. Even as late as 2.8 hours after the start of the event, no traces of penetrated particles could be detected in the magnetotail. During the decaying phase of the event, however, the magnetotail signal was

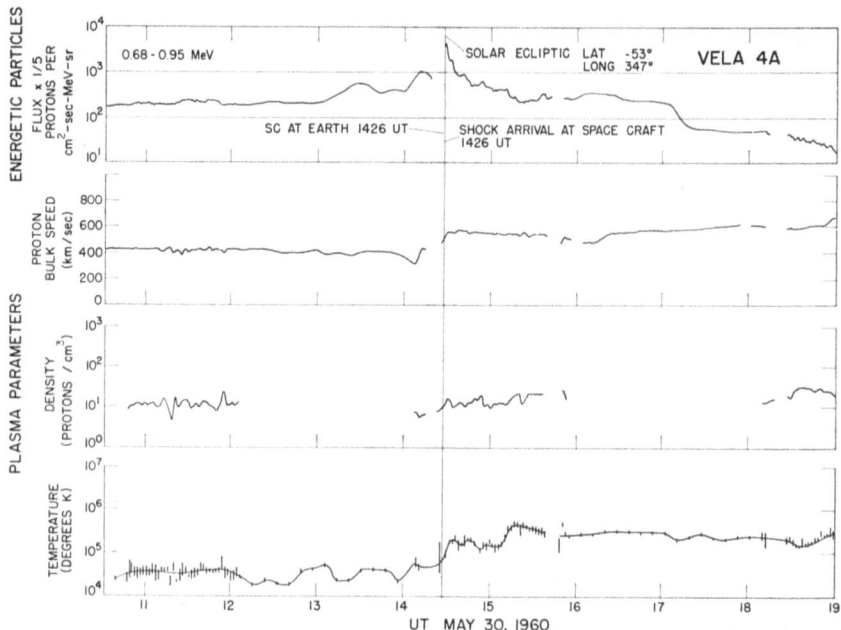

Fig. 4.   Relation of shock spike to shock arrival. The energetic particle flux and the solar wind temperature, bulk velocity, and density are compared for the May 30, 1967 shock to show that the shock spike occurs at the same time as the arrival of the interplanetary shock. The vertical strokes used to display solar wind temperature indicate the magnitude of directional anisotropy in the solar wind protons.

TABLE I

Comparison of ESP events with SC and SI events[a]

Date	UT	Shock	ESP event
May 24, 1967	1726	yes	yes
25	1235	yes	yes
28	0311	no	no
	0548	no	no
	1303	no	no data
30	1730	yes	yes
June 5	1914	yes	yes
25	0221	yes	no data
	0929	no	no
26	1458	yes	yes
July 10	2337	no	no
Aug 4	0700	no	yes (step decrease only)
11	0556	yes	step decrease
29	1738	yes	no protons
31	0825	no	no protons
Sept 13	0345	yes	no
19	1958	yes	no
20	0210	no	no
	1737	no	possible
28	0535	no	no
Oct 28	1637	no	no
	1728	no	no
30	0225	no	no
	0426	no	possible
Nov 3	0914	no	no
	1439	no	no data
	1626	no	no data
10	1205	no	no
29	0512	yes	yes
Dec 5	0628	no	no
15	0501	no	no
18	0538	no	possible
29	2225	no	no
30	0228	no	no
	1513	no	possible
Jan 2, 1968	0222	no	no
	0525	no	no
11	1251	yes	yes
12	0843	no	possible
16	1942	no	no
26	0815	no	no
	1440	no	no
Feb 2, 1969	1500	yes[b]	yes (step decrease)
10	2024	no	no protons
26	0157	yes	yes
27	0308	yes	no data
28	0424	yes	yes

[a] Compiled in part from *Solar Geophysical Data*, ESSA Research Laboratories, U.S. Department of Commerce.
[b] Tentative identification.

delayed by only 15–30 min. Occurrences where the leading edge of a transient disturbance in the interplanetary flux intensity is delayed for a much longer time than the decaying edge are not uncommon (Montgomery and Singer, 1969).

In Figure 6, data from the January 11, 1968, interplanetary shock are used to demonstrate the delayed penetration of ESP's into the magnetotail. During the period 1100–1500 UT, 4A was in the interplanetary medium, and 4B was in the midlatitude magnetotail not far from the midnight meridian. The plot shows the time history of the 0.68–0.95 MeV and 4.5–6.3 MeV proton fluxes from both spacecraft. The onset of the event is clearly seen at ∼1130 UT, and the leading edge of the large spike occurs within 1–2 min of the SC produced at the earth by the shock. The decay was more gradual than that of the May 30 event, but step-like decreases did occur after 1500 UT. Tracking coverage for Vela 4B was lacking during the early part of the event, but the data available in the one hour period after shock arrival show a peak at 1325 UT that must be comprised of particles from either the interplanetary

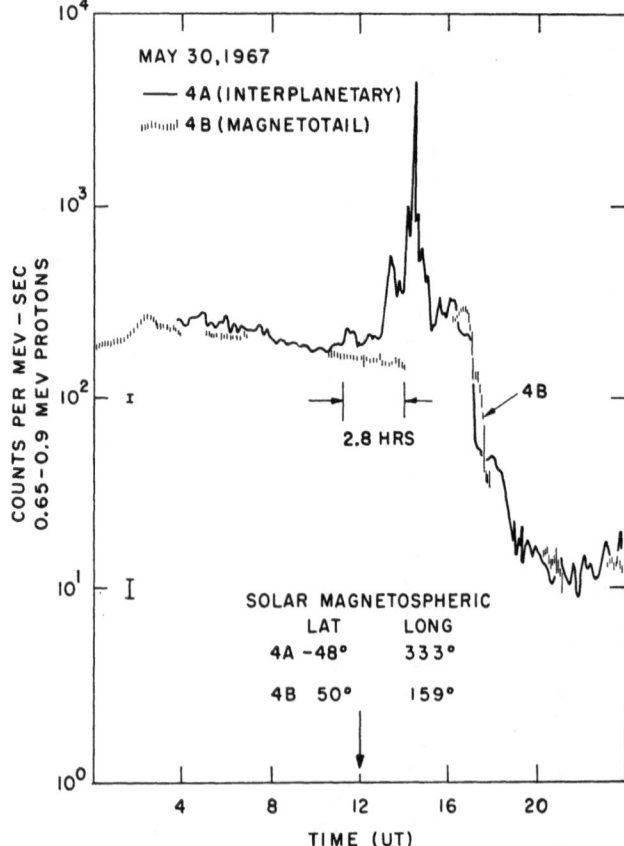

Fig. 5.   Penetration of ESP's into the magnetotail. When the May 30, 1967, interplanetary shock arrived at earth, Vela 4A was in the interplanetary space whereas 4B was in the high latitude magnetotail. A comparison of the 0.65–0.9 MeV proton flux intensity from each spacecraft shows that ESP's were not detected in the magnetotail until at least 2.8 hours after the onset of the increases that signify an ESP event.

peak at 1227 UT or the one at 1254 UT. Comparison of the height, width, and small scale fluctuations associated with these peaks led to the conclusion that the 1254 UT (shock) spike was the precursor, and that the magnetotail signal was delayed by ~30 min and did not undergo serious shape distortion. (Had the 1227 UT peak been chosen as precursor, the delay would have been ~1 hr, and the penetration mechanism would somehow have had to produce an *increase* in peak amplitude and a *narrowing* in time of the penetrated particle peak.)

4. Discussion and Conclusions

Three explanations for ESP's have been proposed: (a) Fermi acceleration of pre-existing energetic particles by scattering between the approaching shock and the magneto-sphere (Axford and Reid, 1963); (b) trapping of solar accelerated particles in a magnetic tongue (Gold, 1959); and (c) acceleration of the pre-existing energetic particles to greater energies in the shock front itself (Parker, 1965; Rao *et al.*, 1967).

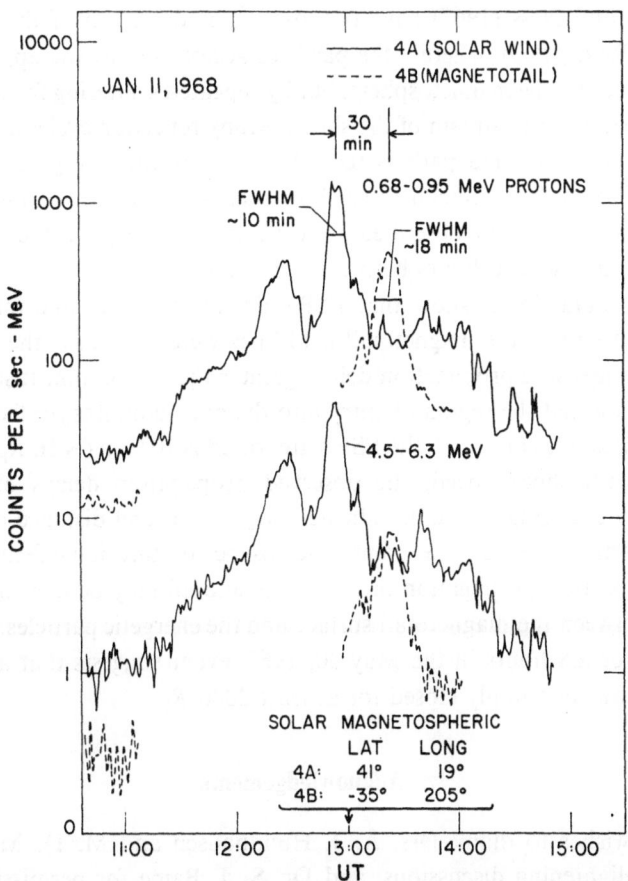

Fig. 6.   Penetration of ESP's into the magnetotail. 0.68–0.95 MeV and 4.5–6.3 MeV proton data from Vela 4A (interplanetary medium) and 4B (magnetotail are compared to show that the penetration of ESP's from the January 11, 1968, interplanetary shock wave was delayed by at least 30 min.

The Vela 4 data indicate that the ESP fluxes are normally unidirectional until at least after the peak of the event (see also Rao *et al.*, 1967); hypotheses (a) and (b), however, require that an ESP event be characterized by essentially bi-directional fluxes. The fact that the ESP spike, the most characteristic feature of the event, occurs at precisely shock arrival time is evidence for acceleration of the particles in the shock front itself.

The thickness of the region in which acceleration may occur can be estimated from the shocked plasma characteristics or from the width of the shock spike. The plasma data indicate that the jump from preshock to postshock conditions occurs in less than ~ 2–3 min, and that a typical shock velocity near earth is ~ 500 km/sec (Montgomery, 1969). The shock front thickness would then be 0.6–1×10^5 km. The shock spike is 5–10 min wide, and if it propagates at shock speed, the spike would have a spatial extent of ~ 1.5–3×10^5 km. This is somewhat larger than the estimate obtained from the plasma data; a somewhat broader shock spike can be expected because the turbulence associated with the acceleration mechanism (Parker, 1965) is likely to enhance the diffusion of energetic particles perpendicular to the field.

Van Allen and Ness (1967) have proposed a modification of the explanation of Axford and Reid (1963), wherein the particles accelerated by the approaching shock are reflected not by the magnetosphere, but by repeated scattering from interplanetary field irregularities far upstream of the shock. Many reflection cycles are required, and if the diffusion mean free path is 0.05 AU, the overall energization time would probably be hours. It is difficult to see how such a mechanism can explain the ESP spike or the stepped decay described above; but it may explain the slow buildup of ESP's that begins several hours before shock arrival.

The Vela 4 data clearly show that ESP's do not have free and immediate access to the magnetotail. If the magnetotail field lines were open *near* the earth, it would be difficult to measure propagation delays greater than ~ 10 min. Instead, the observations indicate that the region of entry into the magnetotail is *far* from the earth: if the band of field lines on which the ESP's are found is 10^5 km (~ 16 R_E) wide, and if it propagates with shock speed, the observed propagation delays of 30–>60 min (Montgomery and Singer, 1969), indicate entry at a radial distance of 500–1000 R_E behind the earth. For these ESP events, the absence of diffusion effects noted in other examples of particle propagation into the magnetotail may be due to the small area of contact between the magnetotail surface and the energetic particles. The absence of penetration for 2.8 hours in the May 30, 1967, event suggests that at that time the magnetotail was essentially closed for at least 2000 R_E.

Acknowledgements

The author wishes to thank Drs. A. J. Hundhausen and M. D. Montgomery for numerous enlightening discussions, and Dr. S. J. Bame for permission to use the Vela 4 Electron-Proton Spectrometer data. The efforts of the Vela Data Reduction Group, under the direction of Earl Tech, made possible the analysis of the Vela data.

This research was done as a part of the Vela Nuclear Test Detection Satellite Program, jointly administered by the Advanced Research Projects Agency of the Department of Defense and the U.S. Atomic Energy Commission, and managed by the U.S. Air Force.

References

Axford, W. I. and Reid, G. C.: 1963, *J. Geophys. Res.* **68**, 1793.
Bame, S. J.: 1968, in *Earth Particles and Fields* (ed. by B. M. McCormac), Reinhold Book Corporation, New York, p. 373.
Bame, S. J.: 1969, Private communication.
Bryant, D. A., Cline, T. L., Desai, U. D., and McDonald, F. B.: 1965, *Astrophys. J.* **141**, 478.
Colburn, D. S. and Sonnett, C. P.: 1966, *Space Sci. Rev.* **5**, 439.
Gold, T.: 1959, *J. Geophys. Res.* **64**, 1665.
Gosling, J. T., Asbridge, J. R., Bame, S. J., Hundhausen, A. H., and Strong, I. B.: 1967, *J. Geophys. Res.* **72**, 3357.
Hundhausen, A. J.: 1970a, this volume, p. 111.
Hundhausen, A. J.: 1970b, this volume, p. 155.
Montgomery, M. D.: 1969, Private communication.
Montgomery, M. D. and Singer, S.: 1969, *J. Geophys. Res.* **74**, 2869.
Montgomery, M. D., Bame, S. J., and Hundhausen, A. J.: 1968, *J. Geophys. Res.* **73**, 4999.
Ogilvie, K. W. and Burlaga, L. F.: 1969, 'Hydromagnetic Shocks in the Solar Wind', to be published.
Parker, E. N.: 1958, *Phys. Rev.* **109**, 1874.
Parker, E. N.: 1965, *Phys. Rev. Letters* **14**, 55.
Rao, U. R., McCracken, K. G., and Bukata, R. P.: 1967, *J. Geophys. Res.* **72**, 4325.
Rao, U. R., McCracken, K. G., and Bukata, R. P.: 1968, *Canadian J. Phys.* **46**, 5844.
Singer, S., Aiello, W. P., Conner, J. P., and Klebesadel, R. W.: 1969, *IEEE Trans. on Nuclear Science*, **NS-16**, 336.
Singer, S.: 1970, this volume, p. 170.
Van Allen, J. A. and Ness, N. F.: 1967, *J. Geophys. Res.* **72**, 935.

Discussion

Question: Could it be that the change in the time delay before and after the shock is related in some way to the change in the magnetosphere that probably took place due to the arrival of the shock?

Singer: That's possible, in fact quite likely. But it is also possible to see the same phenomenon occur for disturbances not caused by shocks, i.e. one sees fluctuations in the interplanetary proton counting rate which are probably due to the filamentary structure in the interplanetary medium, and not associated with shocks. One sees the same sort of delays on the leading edge but frequently much shorter delays on the following edge of these fluctuations. I still think that it is likely that there is some change in the magnetosphere configuration which permits the latter end to propagate in more rapidly.

Balogh: I feel that the large increase one sees in low energy protons was just before the shock front and the drop in intensity occurred just at the crossing. We see both the increase and the drop to be extremely directional.

Singer: That's right. I should add that at the time of the large increase and also during the time of the decrease one frequently sees very large increases in the anisotropy ratio of the protons. This is not simply a change in the streaming direction which in a single instrument appears to be a change in the anisotropy ratio. This is literally a change in the ratio of streaming particles to background particles. I think that future results will show that there is association with the presence of the shock front accelerated particles. I don't believe that these are particles that are trapped behind the shock, i.e. particles of an energy spectrum comparable to the existing protons and which are visible to the spacecraft instrumentation after the shock has gone by. If that were the case one clearly should also see this effect in the high energy particles. This is not seen. This is strictly a lower energy proton phenomenon.

Schindler: I wonder whether it would be possible to estimate the electric field of the shock that would be necessary to accelerate the particles. Do you know the energy and width?

Singer: I give you some numbers and you may make the estimate. If one draws an energy spectrum before and after the shock and makes no attempt to normalize the two, one observes that above about 10–20 MeV there is no detectable difference between the energy spectra. The effect is at the lower energies and it is as if a 0.5 MeV particle had been increased in energy by possibly 100 or 200 keV. This agrees with previous estimates of van Allen on observations in July 1965. As far as the width is concerned I would estimate 10 minutes at 500 km/sec.

GM OBSERVATIONS OF AURORAL PARTICLES AND GROUND OBSERVATIONS DURING THE FEBRUARY 25TH SOLAR EVENT

PETER STAUNING

Ionosphere Laboratory, Lyngby, Denmark

GUNNAR SKOVLI

Norwegian Defence Research Establishment, Kjeller, Norway

and

AXEL BAHNSEN

Danish Space Research Institute, Lyngby, Denmark

1. Introduction

The S71D experiment on board the polar orbiting ESRO-I satellite is an array of Geiger counters. The experiment is primarily devoted to the study of fast variations and fine structure in the intensities and in the angular spectra of energetic (>40 keV) auroral electrons. The experiment is briefly described below, and some preliminary results from measurements made during the February 25th solar event are presented.

2. Experiment

The experiment comprises three Geiger counters of the type Lionel 222 and one (at 160°) of the type EON 6222.

The detectors are oriented in directions of 0°, 45°, 90° and 160° with respect to the satellite axis, which – to some approximation – is antiparallel to the local magnetic field direction. The counters have end-windows to detect electrons above 40 keV and protons above 0.5 MeV. The geometrical factor is 8.32×10^{-3} cm^2 ster and the total opening angle is 16°6 for the 0°, 45° and 90° detectors. The 160° detector has a geometrical factor of 3.13×10^{-2} cm^2 ster and a total opening angle of 32°1.

The experiment operates only in a high speed, real time (HSRT)-mode. One complete sample comprising the counts from all the 4 detectors is obtained each 25 msec. The geographical coverage is restricted to passes over telemetry stations among which the Tromsö station in Northern Norway has far the best location in connection with auroral studies.

Besides the incomplete geographical and temporal coverage, the Geiger counter results are – of course – also limited by the lack of energy resolution and by the problem of proper particle identification.

3. Measurements

Most of the passes considered below are low altitude (300–500 km) passes taken from

V. Manno and D. E. Page (eds.), Intercorrelated Satellite Observations Related to Solar Events, 583–591. All Rights Reserved
Copyright © 1970 by D. Reidel Publishing Company, Dordrecht-Holland

Tromsö or Spitsbergen during the February 25th event. At auroral latitudes the local time at the satellite position was around 08 in the morning sector.

The counting rates for all detectors are very low (detected flux $< 10^3$ cm^{-2} sec^{-1} ster^{-1}), when the satellite is close to the magnetic pole (inv. lat. $> 80°$). At such low fluxes and well inside the polar cap the high energy (> 0.5 MeV) solar protons could make a significant contribution to the fluxes observed during the solar flare event. However, during auroral zone passes, the relative contribution from high energy solar protons or radiation belt protons is very small compared to the contribution from the electrons and might be neglected for most purposes. This has been confirmed by comparisons with the S71C data on energetic proton fluxes.

4. Energetic Electron Precipitation Pattern

The amount of data now available from the S71D experiment is still too limited for an analysis of the general morphology of energetic particle precipitation at high latitudes. This subject has, however, been considered extensively in the literature (see e.g. Hultqvist, 1968, and the references quoted herein). The summary of results shown in Figure 1 from the INJUN III Geiger counter experiment (O'Brien, 1966) indicates the general features of energetic electron (> 40 keV) precipitation at auroral latitudes. The diagram will be considered a reference for the S71D data presented below in order to classify events on the basis of a statistically small number of samples.

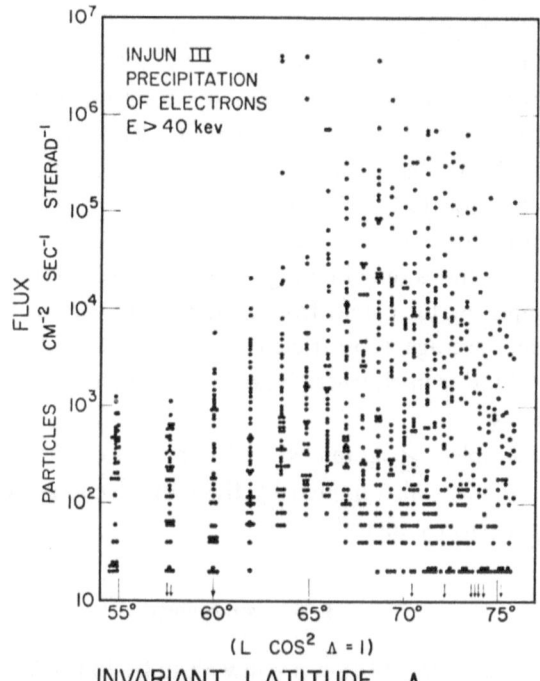

Fig. 1. Variation of precipitation with invariant latitude. (From O'Brien, 1966.)

From Figure 1 one notices that some precipitation is always present at invariant latitudes around 66° slightly equatorward of the center of the auroral zone (Feldstein, 1960). The most intense events and the greatest variability is seen at about 70° invariant latitude.

The S71D data are generally in good agreement with these earlier results and we may therefore concentrate on the more detailed description of the dynamical behaviour of the energetic electron population as observed during selected passes. At the times, when these passes were taken, the satellite was located at approximately constant local solar time, at about the same local magnetic time and in a rather narrow range of heights. The observed fluxes can therefore be directly compared in a study of their dynamical features.

The pass from orbit 1967 in Figure 2 (a) represents a 'typical' morning pass (cf. also Figure 1). At that time K_p was $2-$. The observed fluxes of precipitated (0° and 45°), locally trapped (90°) and backscattered (160°) electrons are rather low. The flux of trapped electrons varies smoothly with latitude, while bursts of precipitated electrons are observed within a zone located at invariant latitudes from 65° to 68°. This zone coincides roughly with the Feldstein (1960) auroral zone.

In what exists of the pass from orbit 2052, shown in Figure 2 (b), the precipitation appears to have roughly the same structure and location as observed during orbit 1967. The K_p-index was 3 during orbit 2052. The fluxes of both trapped and precipitated electrons observed during the pass of orbit 2052 are somewhat larger than those measured during the pass of orbit 1967.

The pass from orbit 2052 extends only to an invariant latitude of 67°5 but in the succeeding pass of orbit 2053, a zone of intense precipitation has appeared poleward of the normal precipitation zone by more than 3° of invariant latitude. In the energetic electron fluxes shown in Figure 2 (c) the angular spectrum of the downward flux closely approaches isotropy during periods of maximum intensity. At this pass one notices also the shock-like appearance of the equatorward boundary of intense precipitation. The increase in the precipitated fluxes at 0° and at 45° is rather abrupt and the fluxes just poleward of the front of intense precipitation are characterized by large-amplitude oscillations. Figure 2 (d) gives an expanded profile of this transition.

One might consider the enhanced fluxes poleward of the normal zone of precipitation, the isotropic spectra for the downward flux and the variability in the trapped electron intensities as indicative of a fresh injection of energetic particles. As time progresses after such a injection one would expect that the fluxes at the highest latitudes would decay most rapidly, that the particles with the smallest pitch angles would first be lost into the atmosphere and that large variations in space or time of the trapped fluxes would be smeared out by particle drift and diffusion. The low altitude, high latitude electron fluxes would then approach a profile like the one shown in Figure 2 (a). The precipitation event observed during orbit 2053 is not a very intense event, but still large enough, however, to require an active source of energetic electrons. This question will be briefly considered in a later section.

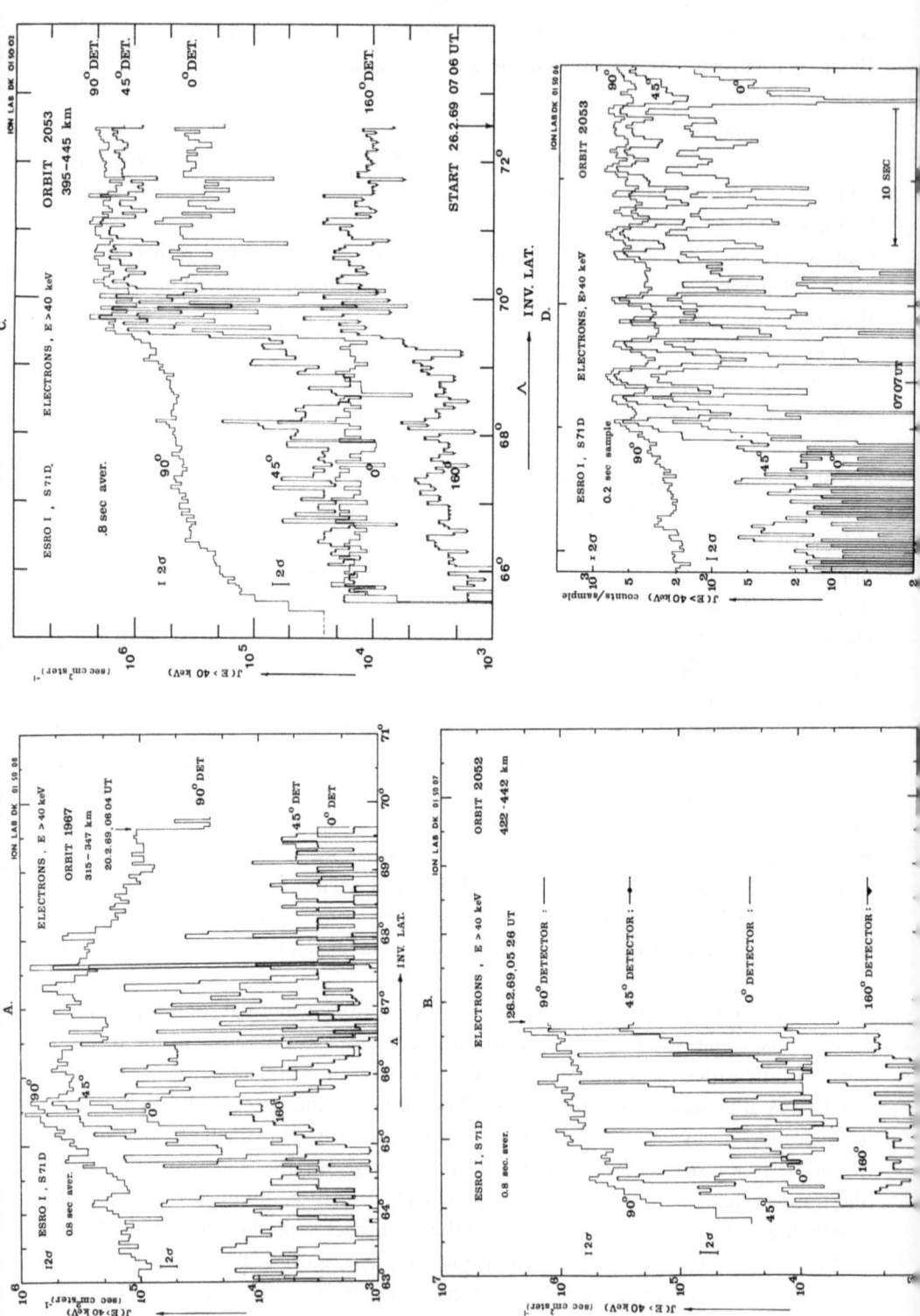

5. Angular Distribution

The angular distribution of energetic (> 40 keV) electrons has been analysed for some ESRO I passes at auroral latitudes during the February 25th event. During all the passes considered the satellite was at a low altitude (300–500 km) in the morning sector of local time. As a first impression, the angular spectra appear to have the same variability as the intensities. This is evident in Figure 2 by comparing the fluxes measured at different angles.

For a more detailed investigation of the angular spectra, the ratio between the fluxes detected by the different Geiger counters has been plotted as function of the intensities in Figure 3.

The angles used for the flux directions in the diagrams of Figure 3 refer to the satellite axis. These angles, however, represent pitch angles to within $\pm 10°$ as the misalignment angle between the satellite axis and the magnetic field direction was less than $10°$ during the passes selected.

The half angle of the loss cone was in the range from $65°$ to $75°$ during all the passes. The fluxes $J(0°)$ and $J(45°)$, therefore, are directional fluxes of precipitated electrons; $J(90°)$ is a flux of locally trapped electrons and $J(160°)$ is a flux of backscattered electrons.

The upper and lower diagrams to the left in Figure 3 show the ratio of precipitated flux ($0°$ and $45°$) to trapped flux ($90°$) vs. the flux of trapped electrons. It is seen in both diagrams, that the angular distribution for the downward flux approaches isotropy as the total flux increases. This result is in agreement with many earlier observations (see e.g. O'Brien, 1964).

The backscattered flux is considered in the three diagrams to the right in Figure 3. The upper diagram shows the ratio of backscattered flux at $160°$ to locally trapped flux at $90°$. The two lower diagrams show the ratio of backscattered flux to precipitated flux at $0°$ and at $45°$ respectively.

It appears from the two lower diagrams that the directional backscattered flux is approximately 10% of the directional precipitated flux, when the fluxes are low. It is, however, evident in both diagrams, that the ratio of backscattered to precipitated flux decreases very markedly as the precipitated flux increases.

This effect has been discussed earlier (see e.g. Cummings et al., 1966; Maehlum, 1968). The observed decrease in the ratio of backscattered to forward flux could be expected if the upward flow of backscattered particles was suppressed by the precipitation mechanism or by secondary effects from the precipitated electrons.

However, the precipitation of energetic electrons is not very well correlated to spectacular changes in the upper atmosphere which could be indicative of processes that might affect substantially the upward flow of backscattered particles. Such changes (auroras, magnetic disturbances, ionospheric heating etc.) are probably more

Fig. 2.   Intensity profiles for the fluxes at various angles of energetic (> 40 keV) electrons during high latitude ESRO I passes.

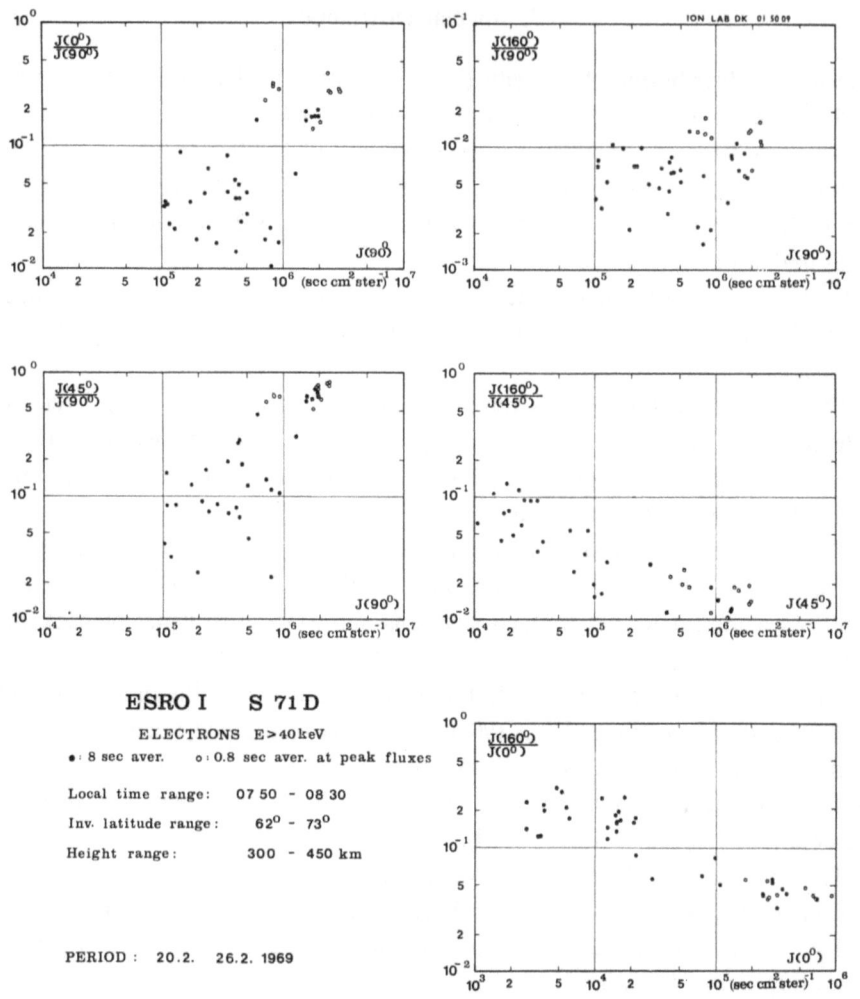

Fig. 3.   Ratio of fluxes detected by different detectors plotted vs. flux intensities.

directly related to the low energy (<40 keV) particle fluxes. One would therefore expect that the backscatter processes for the energetic electrons should be independent of flux intensity.

Referring now to the upper right diagram of Figure 3 it appears – although the scatter is rather large – that the ratio of backscattered flux at $160°$ to trapped flux at $90°$ tends to increase slightly, when the flux of trapped particles increases.

It is tempting to suggest from interpolation, that the ratio of backscattered flux at $160°$ to downward flux at some angle between $45°$ and $90°$ – nearest to $90°$ – is independent of flux intensity (but still a function of the particle energy spectrum and of the parameters of the atmosphere).

Taking into account that the loss cone half angle is about $70°$ for the selected passes, the results indicate that most of the backscattered energetic electrons are particles whose pitch angles were close to the edge of the loss cone before they entered the atmosphere.

Such particles will experience an almost glancing incidence upon the atmosphere. After possible pitch angle changes through collisions, they could escape the atmosphere much more easily than the more deeply penetrating particles of smaller initial pitch angles. This effect has been demonstrated by Stadsnes and Maehlum (1965) in their numerical calculations of electron backscattering.

With the reservations imposed by the limited amount of data considered, the result suggests, that the dominant part of the backscattered flux of energetic electrons is related to the directional flux at approximately the loss cone half angle and not directly related to the fluxes at pitch angles well inside the loss cone, that constitute the bulk of precipitated radiation.

6. Source Mechanisms

Rough comparisons of the available S71D data with geomagnetic recordings from observatories at auroral latitudes indicate generally a good correlation between bay-like geomagnetic activity related to substorm activity and enhanced energetic particle precipitation at high latitudes in agreement with earlier observations (see e.g. O'Brien, 1964; Parks and Winckler, 1968).

The processes related to substorm activity appear to be an important source for the precipitated fluxes of energetic auroral electrons, but in some cases a more direct relation to solar wind conditions may exist.

For the period of February 26th comprising the ESRO I morning passes previously discussed, the magnetograms from a number of high latitude observatories covering a wide range of local hours have been inspected. Recordings of the horizontal component (for some stations the X-component) of the geomagnetic field during February 26th are shown in Figure 4 for some auroral and a few equatorial stations. From about 05 UT to 08 UT no indication was found of substorm activity that could be related to the enhanced precipitation of energetic electrons at very high latitudes observed during the pass of orbit 2053 starting at 0706 UT.

During this period, however, the interplanetary conditions at the magnetospheric boundary changed markedly. The solar wind parameters were monitored at that time by the HEOS A satellite from a location of approximately 25 R_E upstream in the solar wind. Measurements of the solar wind density and velocity have been reported by Bonetti et al. (this volume, p. 436), while measurements of the interplanetary field have been reported by Hedgecock (this volume, p. 419).

During a period of about 1 hour preceding the pass starting at 0526 of the ESRO I orbit 2052, the solar wind conditions at the satellite did remain at an enhanced but rather constant post-shock level. The density was approximately 5 part/cm³, the velocity 470 km/sec., the magnetic field intensity 8 γ and the field inclination to the solar ecliptic plane about 0°. From 06 UT in the period between the pass of ESRO I orbit 2052 and the pass starting at 0706 UT of orbit 2053, the solar wind density increased by a factor of 3 to reach a maximum of 16 part/cm^3 shortly before 07 UT. The solar wind velocity remained at a rather constant level. The interplanetary magnetic field decreased during the same period from about 9 γ reaching a minimum of 2 γ at 0702. At the onset of the intensity decrease the magnetic field had a northward component. From about 0625 to 0645 large amplitude oscillations in the field direction were observed and at 0650 UT the inclination started to change rather abruptly from about 55 °N reaching almost 90 °S at 0706.

Including a delay of about 6 min for propagation of the solar wind from the HEOS A position to the earth, the changes in the solar wind density and magnetic field

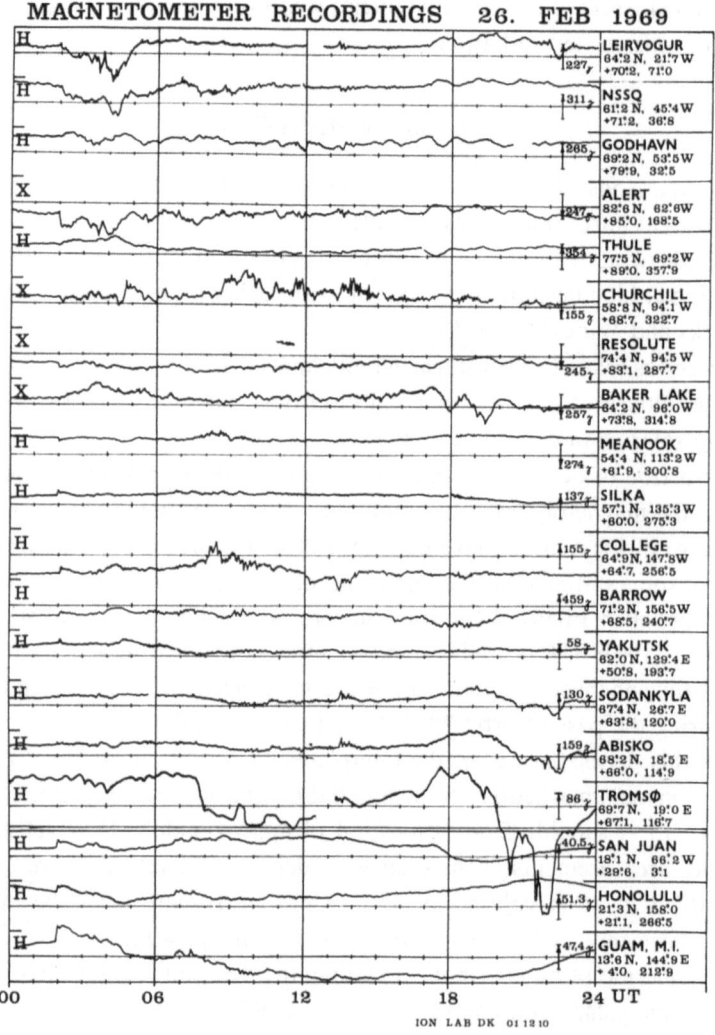

Fig. 4.   Magnetograms for February 26th, 1969, from some high latitude stations and a few equatorial stations. The high latitude stations are arranged according to their geographical longitude.

coincide with the very high latitude energetic electron precipitation event shown in Figure 2 (c) and (d) which were observed during the ESRO I orbit 2053.

One might add to the above observations, that approximately half an hour later the onset of a bay-like geomagnetic disturbance can be seen in the magnetograms from several high latitude observatories. The above energetic electron event might be a precursor of high latitude disturbances which possibly were directly related to the variations in the solar wind.

7. Concluding Remarks

The amount of available data from the S71D experiment as well as their present evaluation is still too sparse to permit the drawing of definitive conclusions. The purpose of the above presentation of preliminary results is primarily to demonstrate, that the instrument and telemetry link has the ability to resolve to a large degree the fast variations and the fine structure in the intensities and spectra of energetic auroral particles. We expect that the progressing analysis of data from the experiment – in particular when combined with results from the other experiments on board the ESRO I satellite as well as relevant observations from other satellites and from ground observatories – will provide the basis for interesting excursions in the field of auroral physics.

Acknowledgements

The large efforts invested in the experiment by colleagues at our institutes and by other individuals are gratefully appreciated. We are in particular indebted to P. Jonasson for his contribution to the experiment and its integration into the satellite and to A. Furu for programming the major part of the data processing.

References

Cummings, D., LaQuey, R. D., O'Brien, B. J., and Walt, M.: 1966, 'Rocket-Borne Measurements of Auroral Zone Particles', *J. Geophys. Res.* **71**, 1399.

Feldstein, Y. I.: 1960, 'Investigation of the Aurorae', *Academy of Sciences of the USSR* **4**, 61.

Hultqvist, B.: 1968, 'Auroral Particles', in *The Birkeland Symposium on Aurora and Magnetic Storms* (ed. by A. Egeland and J. Holtet), C.N.E.T.

Maehlum, B. N.: 1968, 'Satellite and Rocket Particle Observation in the Auroral Zone', in *The Birkeland Symposium on Aurora and Magnetic Storms* (ed. by A. Egeland and J. Holtet), C.N.E.T.

O'Brien, B. J.: 1964, 'High Latitude Geophysical Studies with Satellite INJUN III. 3. Precipitation of Electrons into the Atmosphere', *J. Geophys. Res.* **69**, 13.

O'Brien, B. J.: 1966, 'Precipitation of Electrons and Protons into the Atmosphere', in *Radiation Trapped in the Earth's Magnetic Field* (ed. by B. M. McCormac), Reidel, Dordrecht, The Netherlands.

Parks, G. K. and Winckler, J. R.: 1968, 'Acceleration of Energetic Electrons Observed at the Synchronous Altitude During Magnetospheric Substorms', *J. Geophys. Res.* **73**, 5786.

Stadsnes, J. and Maehlum, B.: 1965, Scattering and Absorption of Fast Electrons in the Upper Atmosphere, NDRE Report E-53.

OBSERVATION OF LOW-ENERGY SOLAR PROTONS
DURING THE FEBRUARY 25, 1969 EVENT

F. SØRAAS, K. AARSNES, and H. R. LINDALEN

Dept. of Physics, University of Bergen, Norway

1. Introduction

This paper reports on initial results from the S71C experiment (University of Bergen, Norway and Danish Space Research Institute, Denmark) pertinent to solar protons. The experiment is flown on board the ESRO I (Aurorae) satellite launched on October 3, 1968 into a near-polar orbit with inclination 94°, apogee 1540 km and perigee 260 km. On February 25, 1969 perigee was over the northern polar region. The satellite is magnetically stabilized. The S71C experiment is designed to study trapped and dumped protons in the energy range 100 keV to 6 MeV. The instrument will measure protons belonging to the radiation belt and transient phenomena such as solar and auroral protons. There is an enormous variation both in intensities and spectral hardness of the protons observed during these events. The intensity of solar protons above 100 keV is very small compared with what is normally present above this energy during a proton aurora. Thus, data obtained over a larger part of the polar cap were averaged in order to improve the counting statistics about the solar proton fluxes observed during the February 25 event.

The S71C experiment employs three totally depleted surface barrier detectors, two unshielded detectors D_1 and D_2, and one background detector D_3, shielded by 0.3 mm of aluminium. The detectors have the following orientation: D_1 along, D_2 normal and D_3 45° with the magnetic field. The orientation of the detectors in the northern hemisphere is such that D_1 detects particles that will be absorbed in the atmosphere, whereas D_2 responds to particles that mirror at the satellite altitude. A collimator restricts the solid angle of the detector to 0.074 sr and the directional geometric factor is about 0.037 cm² sr. Magnets in front of the apertures sweep away electrons with energy less than 500 keV. Each detector is followed by a charge-sensitive amplifier and a main amplifier. The signals from the amplifiers are fed to a six-channel differential pulse-height analyzer. In the first five channels both protons and α-particles will be recorded, while channel six will respond to α-particles only.

The different energy channels are counted sequentially taking two adjacent channels each time. The counting period being 0.345 sec. A complete sample of all three detectors is obtained every 3.6 sec. A full description of the instrument is given by Søraas *et al.* (1969).

2. Results and Discussion

Figure 1 shows the energy response of one of the unshielded detectors and the background detector. The background detector will respond to protons with energies

V. Manno and D. E. Page (eds.), Intercorrelated Satellite Observations Related to Solar Events, 592–598. All Rights Reserved

above 5.7 MeV only. By using data from all five energy channels on each detector, information about the proton energy spectrum over a large energy range can be obtained.

A major flare, importance 2B occurred in the McMath plage region 9946, at a solar position of 38° west and 14° north on 0900 UT on February 25. The flare location was east of the foot of the theoretical garden hose interplanetary field line connected to the earth. Other papers at the symposium have in detail dealt with the solar conditions, and this will therefore be omitted here.

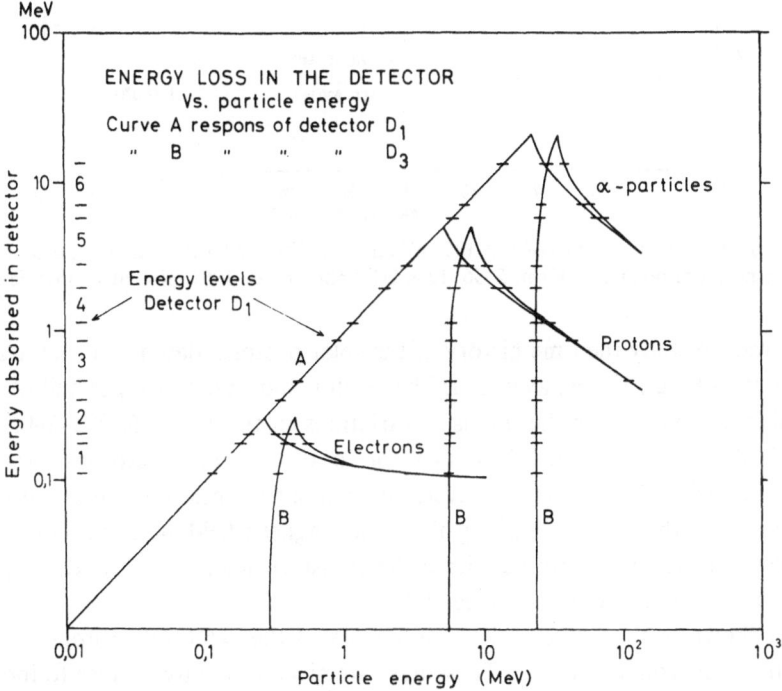

Fig. 1.   Energy loss in detector D_1 and in the background detector D_3 as a function of particle energy.

Figure 2 illustrates the invariant latitude variation of locally mirroring protons in the energy ranges 120–185 keV and 500–920 keV obtained over the north polar region on orbit 2056 (1206 UT on February 26). The intensity variations for both energy ranges are fairly similar and uniform over the polar cap and decreases at lower latitude. The 120–185 keV protons have though a pronounced peak at an invariant latitude of 73° at the dawn side. At a local time of 0700 hours the geomagnetic cuf-off is 71° for the 120–185 keV protons and about 70° invariant latitude for the 500–920 keV protons. Below 68° invariant latitude the intensity of the 120–185 keV protons starts to increase associated with their normal presence in the outer radiation belt. At 1900 local time the cut-off latitude for the 500–920 kV protons can be found at 67°. It is not possible to define a cut-off latitude for the 120–185 keV protons at this local time as the solar protons merge into the radiation belt protons. The K_p value was 3⁻.

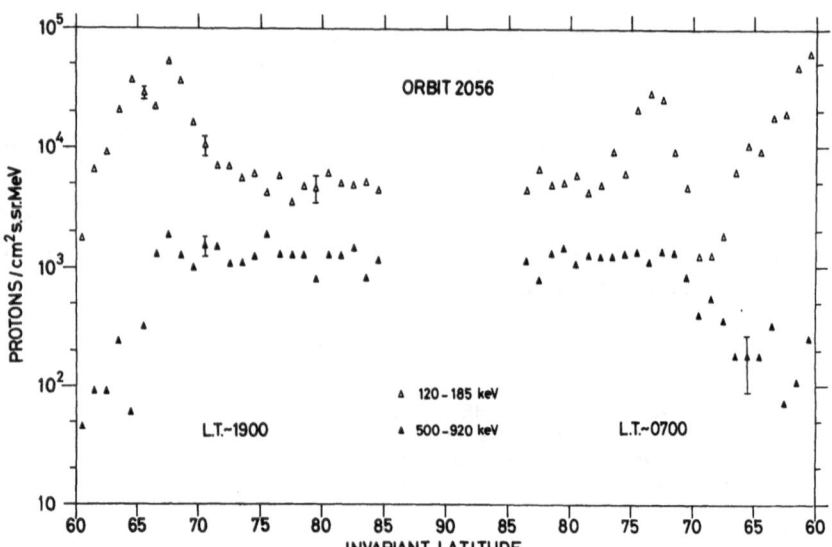

Fig. 2. Differential intensities of 120–185 keV and 500–920 keV locally mirroring protons observed over the northern polar cap. Orbit 2056. 1206 UT February 26, 1969, plotted vs. invariant latitude.

In order to study the time history of the solar protons, data from each pass above invariant latitude 75° were averaged. The result is shown in the upper half of Figure 3. The different curves give the intensity variations in the 120–185, 225–360, 500–920, 1170–2140 and 3260–7000 keV ranges for locally mirroring protons for a $4\frac{1}{2}$-day period following the February 25 event. By comparing these intensities with the ones obtained from the detector looking along the magnetic field lines, it is confirmed that the proton fluxes are isotropic within the statistical uncertainty of the experiment. The statistical uncertainty is typical 10%.

Also shown in the figure is the time history of the >6 MeV protons as obtained from the background detector. These data are given as relative values to indicate the behaviour of the more energetic proton component during the event.

In the bottom half of the figure the motion of the flare region on the solar disc is shown. The occurrence of flares of importance 1B and 2B are indicated. Solar wind velocity data from Vela 3 and 4 (*Solar-Geophysical Data*, 1969) are shown. Knowing the solar wind velocity the distance between sun and earth and the solar angular velocity, the solar longitude of the spiral field line from the sun to the earth has been calculated. The variation in time of this longitude is shown. The onset of the flare causing the February 25 solar proton event is approximately 0900 UT. The count rates of all detector channels have started to rise by 1030 UT. The >6 MeV protons had by that time reached their maximum and started to decay smoothly. The count rate in the 3260–7000 keV channel was increasing. The count rates of the lowest energy channels had reached a temporary maximum. This maximum can not be due to solar protons, as the transit time from the sun to the earth is too short for protons of these energies 100–400 keV.

The maxima is most likely due to prompt electrons. These electrons must have had an energy > 500 keV and will thus not be excluded from the detector by the broom magnet. From Figure 1 it is evident that electrons can only be recorded in the two lowest energy channels. If the high energy electron passes sidewise through the detector it can be recorded in a higher channel. The efficiency for detecting such electrons is small, however, due to the smaller geometric factor for particles travelling in this direction.

That this is a likely interpretation is supported by data from the Vela satellite, which at 0940 UT records electrons in the 0.45–2.5 MeV range (*Solar-Geophysical Data*, 1969). Figure 3 indicates that there was a significant flux of low energy protons even before the February 25 event. They could be due to flare activity in the days before the 25th. No higher energy protons were, however, recorded during this time period.

By 1424 UT the two lowest energy channels reach a minimum. The count rates in

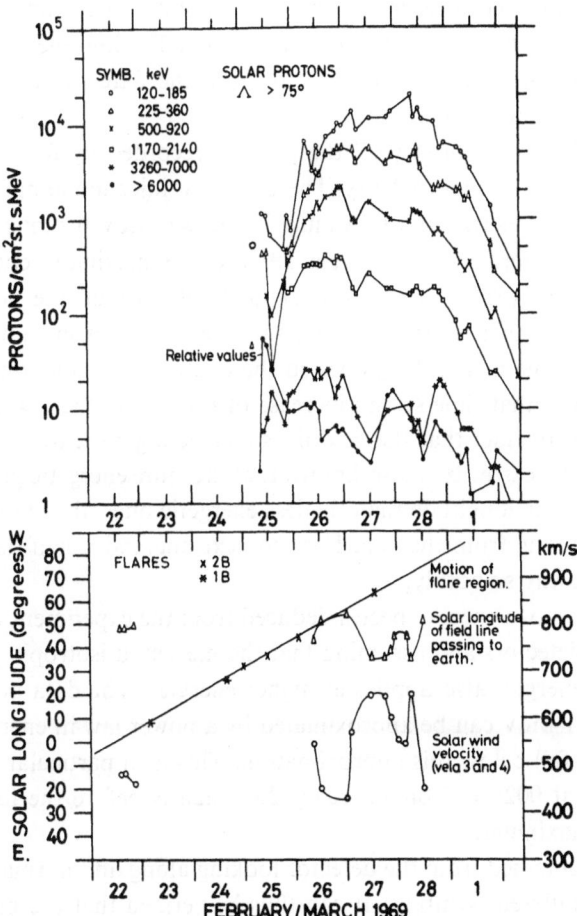

Fig. 3. Average values of differential intensities of 120–185 keV, 225–360 keV, 500–920 keV, 1170–2140 keV and 3260–7000 keV protons above invariant latitude 75° plotted vs. time. Relative values for the > 6 MeV protons are also given. The lower half of the figure gives as a function of time the solar longitude of the McMath plage region 9946 together with the solar longitude of the field lines connecting sun–earth.

the five lower energy channels increased rapidly between 2100 and 2236 UT. This corresponds to an elapse time about 12 hours after the flare and agrees well with what one should expect for protons in this energy range travelling a distance somewhat larger than 1 AU. While the intensities of the lower energy protons are rising the protons with energies >6 MeV are still decaying.

The sudden commencement on February 26 0158 UT occurs after the above 100 keV protons had been increasing for many hours. It is thus highly unlikely that these protons have been trapped behind the storm plasma. At about 0706 UT the low energy protons seem to have reached a plateau value.

The reason for the increase and levelling off of the low energy protons is probably due to the fact that the flare site is positioned near the foot of the interplanetary field lines which connect to the earth, as can be seen from the lower half of Figure 3 where the solar longitude of the fieldlines passing to the earth is shown. What we observe is probably the core or and the halo structure of the low energy particles, a concept which was introduced by Lin *et al.* (1968). The low-energy protons observed are thus prompt protons arriving directly from the flare site along the spiral field connecting the sun with the earth.

From 1324 UT on February 27 the protons recorded by four lowermost energy channels show a fairly smooth decay. There is a new injection of particles on February 27 (noon) which gives rise to a maximum in the >6 MeV protons around 2100 UT. At 2000 UT on February 28 the Vela satellite sees maximum counting rate in the 3–20 MeV channel (*Solar-Geophysical Data*, 1969). About one hour later the 3 to 7 MeV proton channel on the ESRO I satellite reaches a maximum. The time resolution at this time is about 3 hours so there is no contradiction between the two results. We can at that time see no increase of protons in the lower energies. This is probably due to the fact that the field lines connecting back to the flare region have corotated past the earth. It is also known that the more energetic particles are spreading further in solar longitude than the less energetic ones, thus filling a larger sector of field lines coming from the sun. For the next one and a half day the low energy protons decay away smoothly.

Figure 4 shows two energy spectra deduced from the experiment using information from all three detectors and assuming that the measured isotropic pitch-angle distribution at low energies also applies at higher energies. The data points obtained for energies above 1 MeV can be approximated by a power law in energy. Below 1 MeV the data points fall below this approximation. This is in particular true for the spectrum obtained at 0024 UT on February 26 which is before the lower energies had reached their maximum.

Data points obtained from the detector looking along and normal to the magnetic-field are given different symbols. It can thus be verified that the proton flux had an almost isotropic pitch angle distribution.

For the time period considered, the differential energy spectrum above 0.5 MeV have been fitted by a spectrum of the form $dn/dE = N_0 E^{-\gamma}$. The exponent γ is plotted in Figure 5 as a function of time. The exponent is increasing up to the beginning of

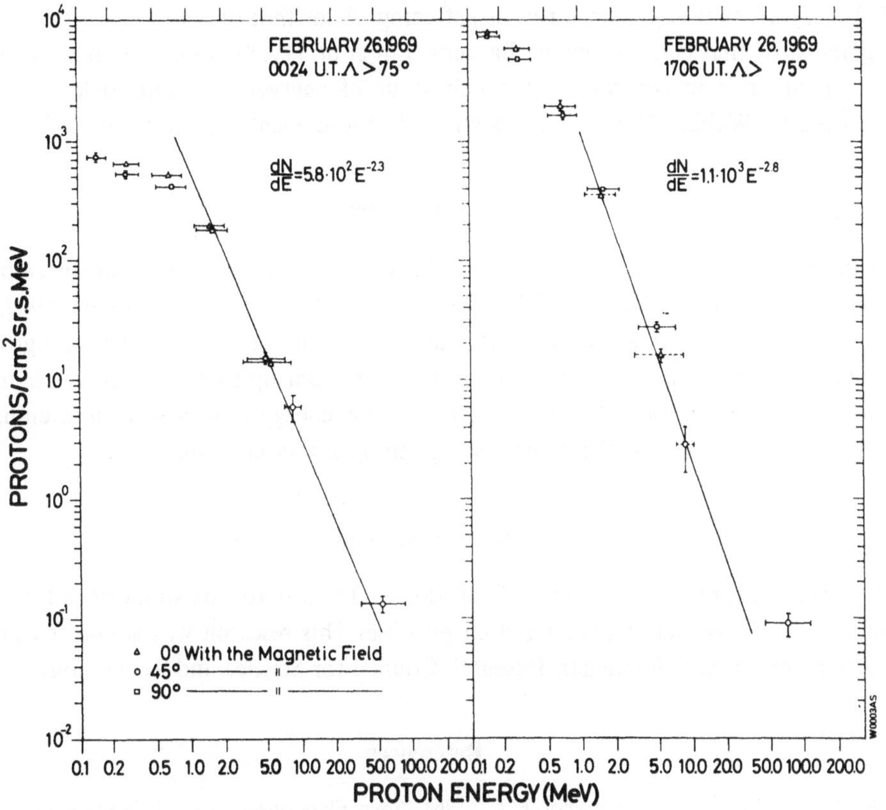

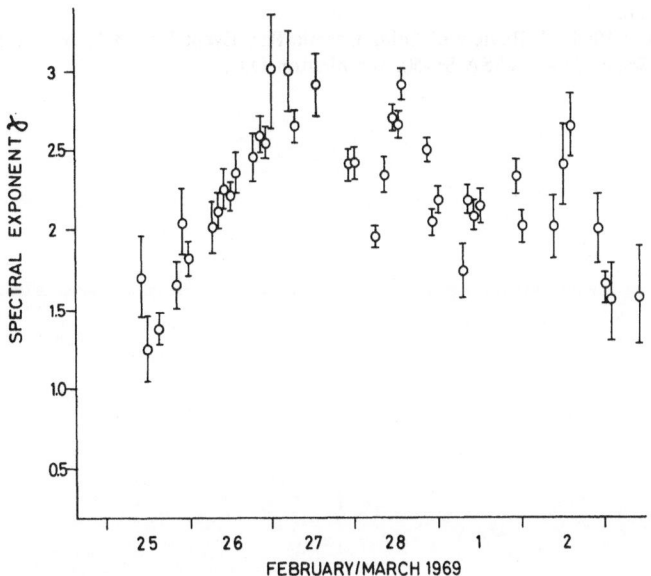

Fig. 4.   Differential energy spectra of solar protons obtained over the northern polar cap on February 26, 0024 UT and 1706 UT.

Fig. 5.   The spectral index γ ($dN/dE = N_0 E^{-\gamma}$) plotted vs. time.

February 27, that is, the spectrum is softening. After that time however, γ decreases significantly, that is the spectrum hardens. Lin *et al.* (1968) point out that this kind of behaviour is in contrast to the behaviour of energetic stormparticle events as reported by Webber (1964). The spectrum in those events continues to soften with time.

3. Conclusion

Protons in the energy range 100 keV to 6 MeV were observed over the northern polar cap for a period of five days following the solar flare of 25th February 1969. By combining the information from the unshielded detectors with the background detector information about the solar proton spectrum up to an energy of about 50 MeV could be obtained. The increase in the low energy protons on the evening of February 25 seems to fit the description given by Lin *et al.* (1968).

Acknowledgements

We wish to express our thanks to Professor B. Trumpy for his support and interest and to ESRO for making the research possible. This research was supported financially by the Royal Norwegian Research Council for Science and Technology.

References

Lin, R. P., Kahler, S. W., and Roelof, E. C.: 1968, 'Solar Flare Injection and Propagation for Low-Energy Protons and Electrons in the Event of 7–9 July, 1969', *Solar Phys.* **4**, 338–360.
Solar-Geophysical Data, March 1969, ESSA Research Laboratories.
Søraas, F., Aarsnes, K., Lindalen, H. R., and Møhl Madsen, M.: 1969, 'A Satellite Instrument for Measuring Protons in the Energy Range 0.1 MeV to 6 MeV', University of Bergen, Dept. of Physics, Report No. 8.
Webber, W. R.: 1964, 'A Review of Solar Cosmic Ray Events', in *AAS-NASA Symposium on the Physics of Solar Flares*, NASA SP-50, Washington D.C.

HIGH ENERGY SOLAR ELECTRONS

C. J. BLAND*, G. DEGLI ANTONI, C. DILWORTH,

D. MACCAGNI, and E. G. TANZI

Istituto di Scienze Fisiche dell'Università e Gruppo GIFCO del CNR, Milano, Italy

and

Y. KOECHLIN, A. RAVIART, and L. TREGUER

Service Électronique Physique, C.E.R.N., Saclay, France

1. Introduction

This contribution deals with the electron events originated by solar activity which have been observed by the Milan-Saclay electron detector on board of HEOS-A1 since the launch of the satellite (5th December 1968) to the end of June 1969. Solar flare electrons in the keV energy range have been first observed by Van Allen and Krimigis in 1965 and many of these events have been studied by Lin and Anderson (1967). Lin (1968) has shown the existence of a correlation between these electron events and solar radio and X-ray burst emission. Contemporary observations (Anderson and Lin, 1966) made in different locations in space showed the events to be largely anisotropic. Our observations follow those of Cline and McDonald (1968) and extend their energy range.

2. The Detector

The detector (experiment S 79, Bland *et al.*, 1969b) consisting of a directional ($\pm 20°$) gas Cerenkov counter, a solid-state counter and a shower detector phoswich combination of a lead glass Cerenkov (3 radiation lengths deep) and a CaF_2 scintillation counter (Figure 1), was designed to operate in the electron energy range from 20 MeV to 500 MeV. After launch, the high tension supply to the photomultipliers increased, with a consequent rise in gain of a factor 2, thus displacing the energy range registered to 8 to 300 MeV. The geometrical factor of the instrument is (1.1 ± 0.2) cm^2 sterad and the maximum detection efficiency for electrons is 50% at 50 MeV.

All events registered require a coincidence between the gas Cerenkov counter and a fast pulse from the lead-glass-CaF_2 phoswich counter, together with a pulse from the solid-state counter corresponding to between 0.5 and 1.5 minimum ionization equivalent.

A slow pulse from the CaF_2 scintillator above threshold (set at one eight minimum ionization) indicates that the event has penetrated the 3 radiation lengths of the lead glass (Mode B). Under normal operating conditions these events comprise essentially high energy electrons and protons of energy greater than the gas Cerenkov

---

* Now at the University of Calgary, Calgary, Canada.

V. Manno and D. E. Page (eds.), Intercorrelated Satellite Observations Related to Solar Events, 599–609. *All Rights Reserved*

threshold (10 GeV). The increase in the high tension in flight however caused a rise in noise level of the gas Cerenkov counter photomultiplier by a factor of about 20, and spurious coincidences between this noise and protons of energy greater than 250 MeV crossing the solid-state counter and the lead-glass-CaF$_2$ complex became dominant. About 80% of the penetration events can be ascribed to these relatively low energy protons.

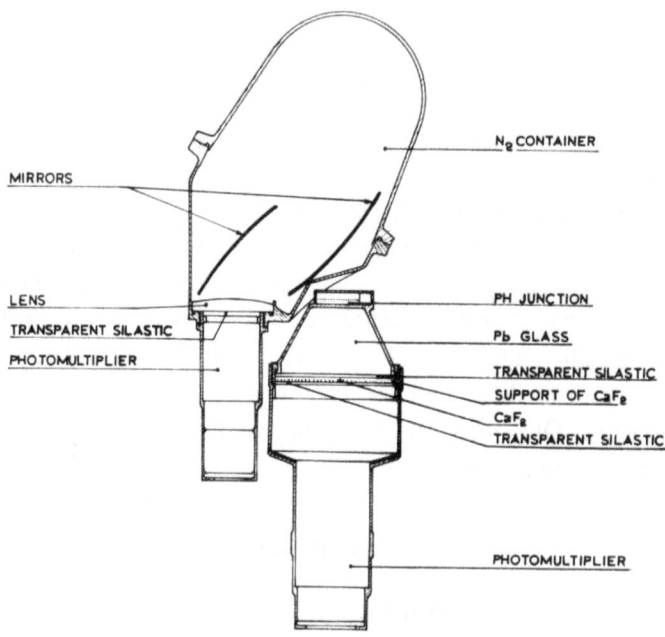

Fig. 1.   Sketch of the sensors of the experiment S 79.

The absence of a slow pulse from the CaF$_2$ scintillator indicates the total absorption of the event within the leadglass (Mode A). In normal operating conditions these events are low-energy electrons with a contamination of the order of 0.5% by protons. With the increased high tension, this contamination can be expected to rise to about 10%.

A second flight model, calibrated before launch with respect to the first, has been operated at sea level with the gain of its photomultipliers increased to the value observed in flight. The analysis of the cosmic ray sea level spectrum under the old and new operating conditions, has led to an estimate of the change in the energy vs. pulse-height relation, and in the efficiency and absorption factors.

3. Experimental Results

During the period of observation S-79 has registered four enhancements of the electron flux, namely on February 25th, February 27th, March 30th and April 12th,

1969. The February 27th and April 12th enhancements are small compared with the two main events. The data coverage is complete for the first two events (Feb. 25th and 27th) while we have not yet received complete data for the others.

Two principle areas of the solar disc exhibited activity prior to the event of the 25th February, the McMath plage regions 9946 and 9957. An importance 2B optical event was observed at solar longitude 36 W and solar latitude 14 N, with a maximum in intensity at 0925 UT. This was accompanied by a 10 cm radio outburst at 0910 UT which lasted 50 minutes.

At the time of the flare HEOS-A1 was at an altitude of 218 000 km and its attitude was such that our experiment was pointing westward with its axis parallel to the ecliptic plane.

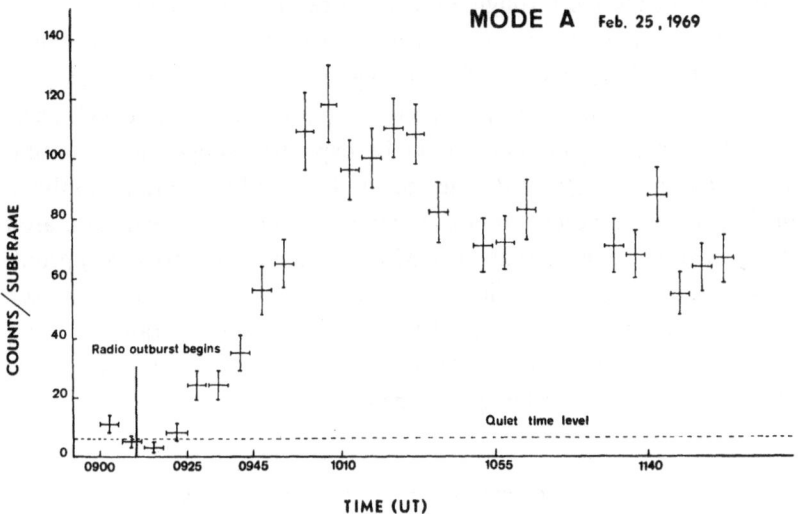

Fig. 2. Time profile of the electron (~ 12 MeV) event of February 25, 1969.

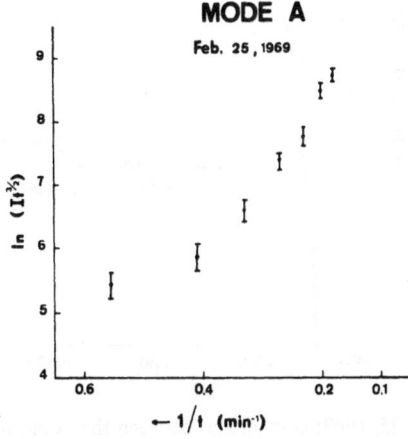

Fig. 3. Event of February 25, 1969; fit to a simple diffusion model of the onset phase.

Figure 2 shows the time profile of this event registered in Mode A. S-79 starts recording the flare electrons at 0920 UT and the maximum intensity is reached at 1004 UT. Then we have at least two oscillations before a marked decrease phase begins. The onset phase cannot be fitted either to a simple diffusion model (Figure 3) or to the ADB model (Figure 4) and an exponential law for the decay phase (Figure 5) does not represent a good fit for many experimental points between 1000 and 1200 UT. The maximum flux in the direction of observation is 1 el/cm² sec sterad. A comparison of the neutron monitors records of different stations (e.g. Calgary and Uppsala) shows the event to be anisotropic and since the fit to the ADB model proves to be difficult also for neutron monitor data it is possible that the degree of anisotropy has changed during the flare. The anisotropy and the time characteristics of this flare event have led us to suppose that the electrons we observed were coming directly from the sun. If we are directly connected through a line of force of the interplanetary magnetic field to the flare location on the sun, our instrument would be able to detect those particles coming spiralling along this line of force (the radius of curvature for electrons in this energy range is about 30 000 km) with pitch angle between 38° and 78°. Then the time profile of the onset of this flare could be explained supposing a continuous and constant emission of electrons during the whole period of the radio outburst (Figure 6), thus implying that acceleration to relativistic energies and emission are two contemporary phenomena at the sun. The Mode B (penetrating particles) profile (Figure 7) of this same event is very similar to a spike with its maximum intensity some 10–15 min after the electron maximum. The increase in counting rate is mainly due to protons of energy greater than 250 MeV; electrons whose showers have triggered the CaF_2 scintillator are 10% of the quiet time counting rate.

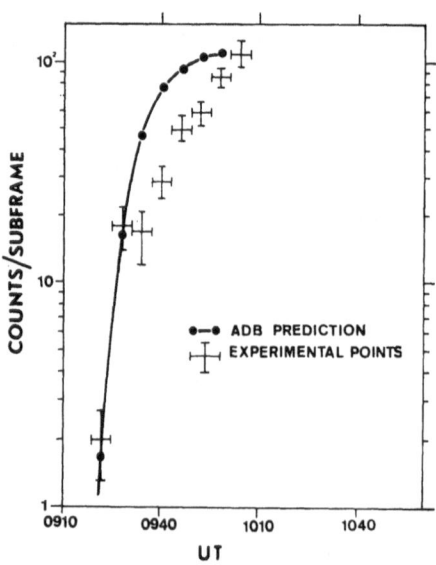

Fig. 4.   Event of February 25, 1969; comparison between the experimental points and the onset as predicted by the ADB model.

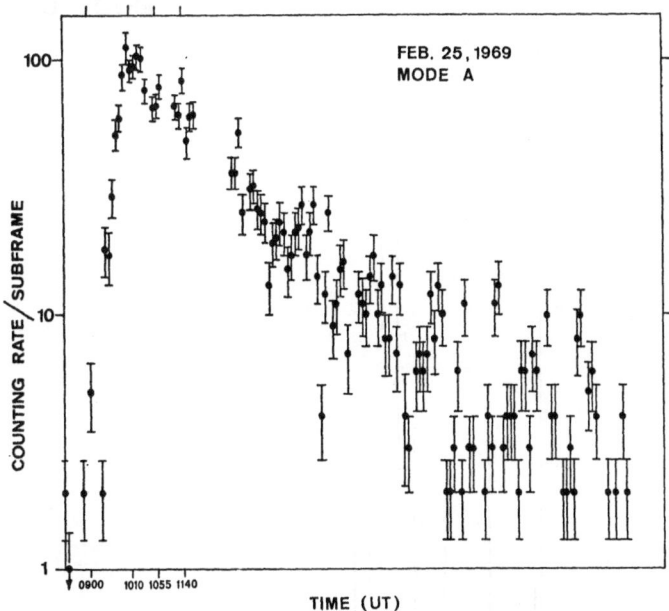

Fig. 5. Event of February 25, 1969; log-linear plot of excess counting rate above quiet time average.

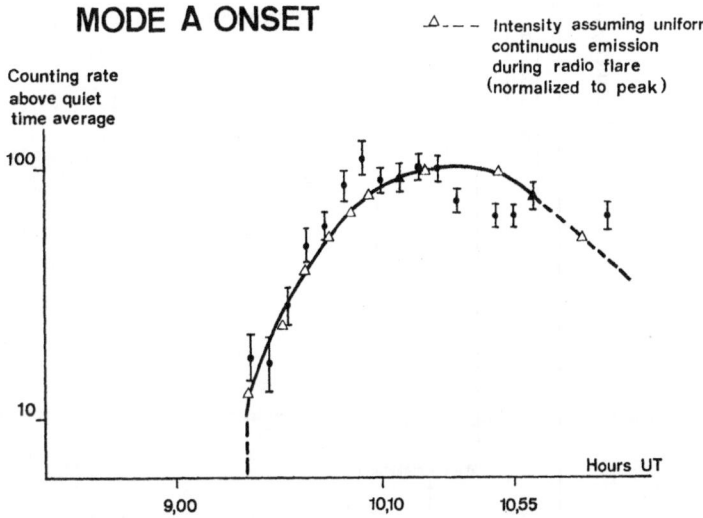

Fig. 6. Event of February 25, 1969; time profile of the onset phase compared to that expected assuming continuous constant emission of electrons during the radio burst and their direct spiralling along magnetic field lines to the detector.

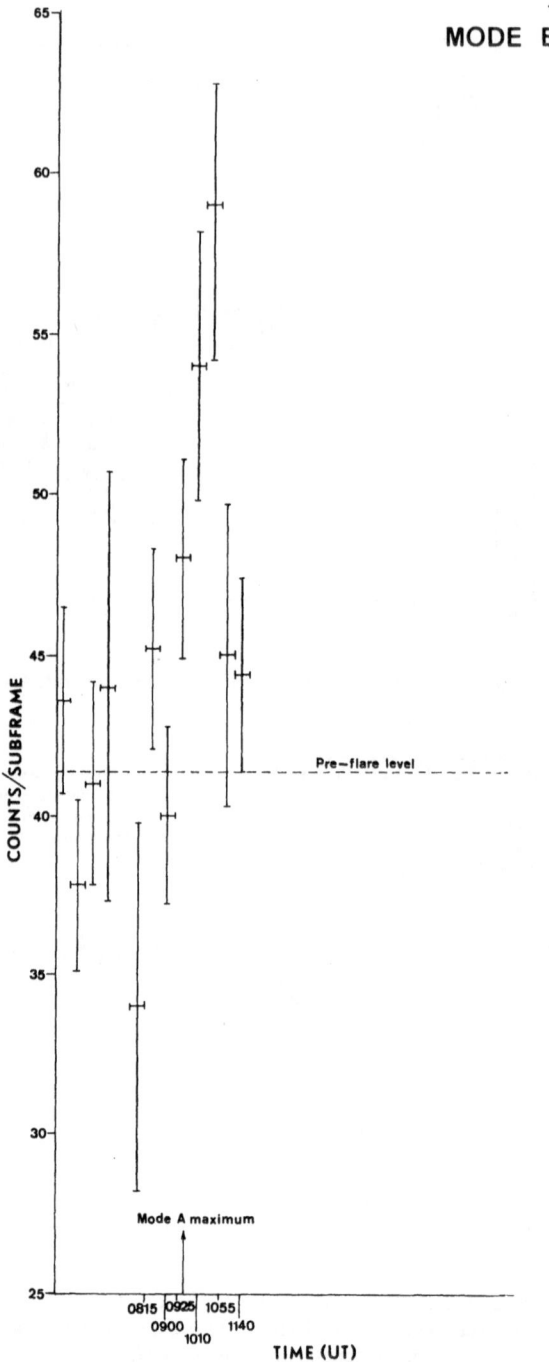

Fig. 7.   Time profile of the protons (> 250 MeV) event of February 25, 1969.

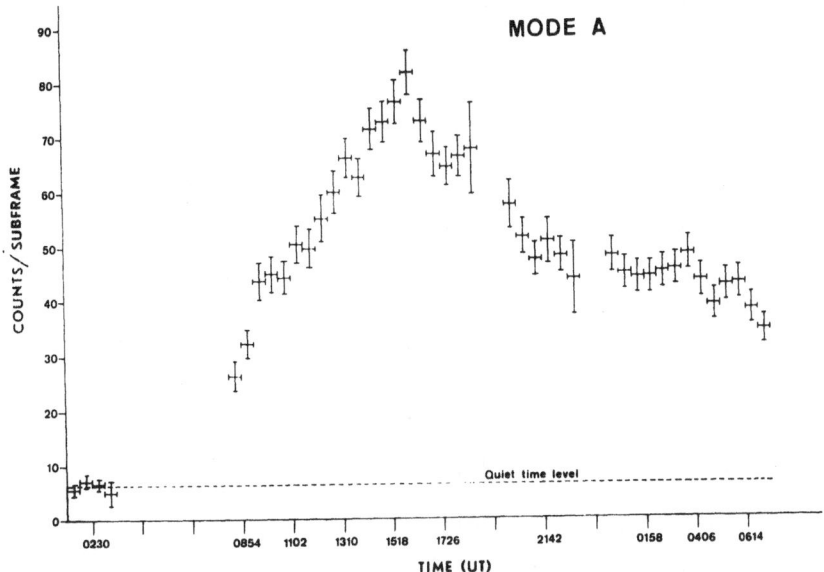

Fig. 8.   Time profile of the electron ($\sim$ 12 MeV) event of March 30–31, 1969.

Figures 8 and 9 show the time profile of the electron and proton event respectively of the 30th–31st March 1969. The optical flare was located on the sun's west limb and a radio outburst was recorded at Manila starting at 0247 UT. The attitude of HEOS-A1, which was at an altitude of 170 000 km on the 30th March, was such that the axis of our instrument was pointing north. Therefore we could only observe the flare particles by diffusion, however plots of ln $(It^{3/2})$ vs. $1/t$ (I, Mode A intensity) with different emission times would rather indicate that diffusion has taken place with two different diffusion coefficients and also the plot of ln $(It^{5/2})$ vs. $1/t$ (Figure 10, ADB model) cannot be assimilated to a straight line. This time the Mode B time profile shows a maximum before the maximum intensity is reached by the electrons. It should be noticed that the interplanetary magnetic field was extremely perturbed at the time of this flare as a consequence of the Forbush decrease of the 24th March and this can possibly be the cause of the composite time profile of the event.

By the same experimental procedure described elsewhere (Bland et al., 1969a), we have obtained the energy spectrum of the electrons for the two flares (Figure 11). The slope is 3.5 ± 0.2 and 3.2 ± 0.2 for the 25th February and 30th–31st March events respectively, very close to the slope of the quiet time low energy portion of the spectrum (Bland et al., 1969a). The average energy of the electrons registered, taking into account the effective efficiency of the instrument, is 12 MeV.

Let us consider the two electron events of February 27th and April 12th (Figures 12 and 13). Then comparing them with the other two events we observed we find some common characteristics: (a) their small size; (b) no correlation with large radio bursts; (c) no correlation with neutron monitor enhancements; (d) they both occur a few days after what is a big event for us.

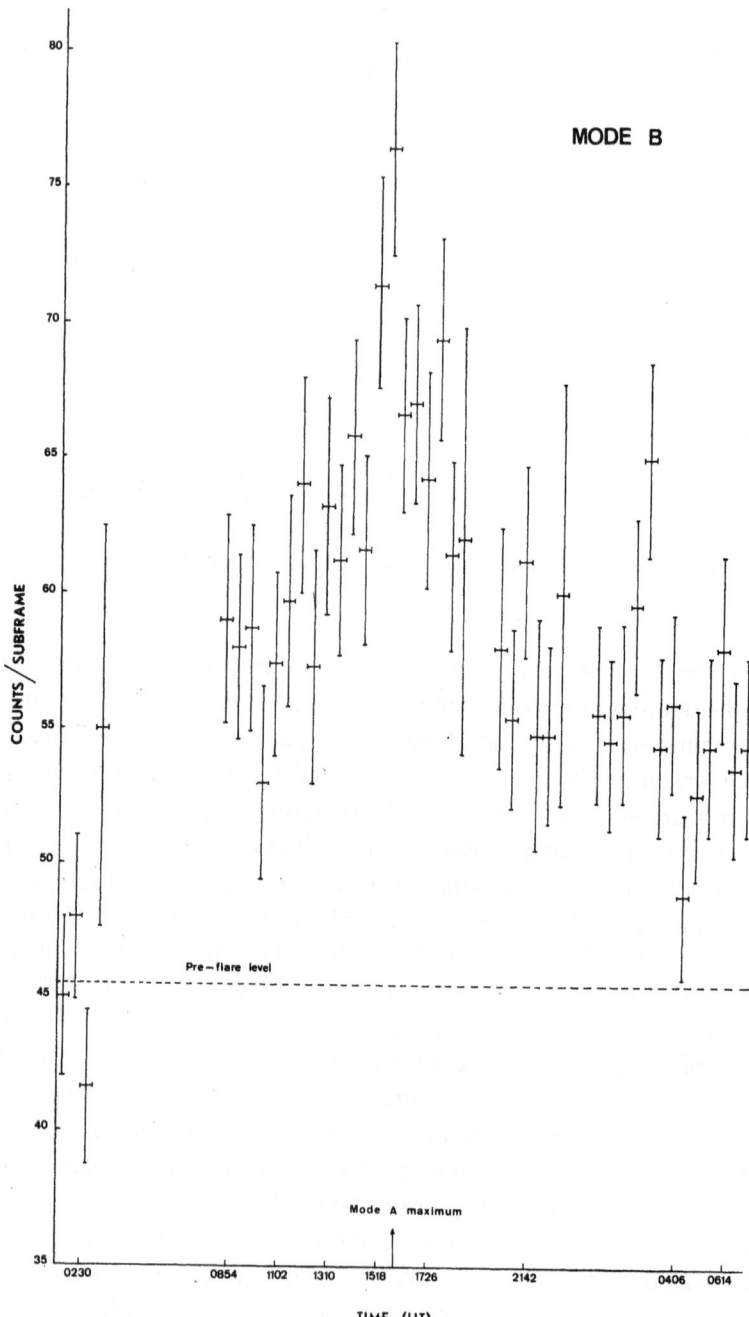

Fig. 9.   Time profile of the proton (> 250 MeV) event of March 30–31, 1969

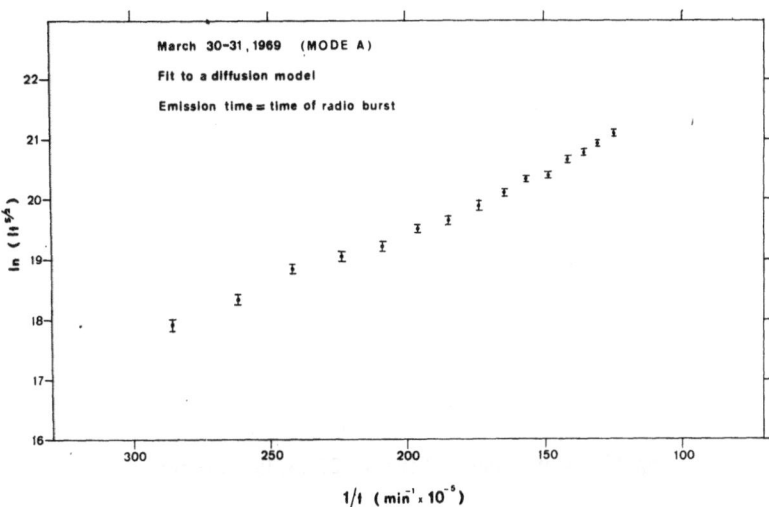

Fig. 10. Check of the ADB model for the onset phase of the electron event of March 30–31, 1969 assuming the emission has taken place at 0247 UT.

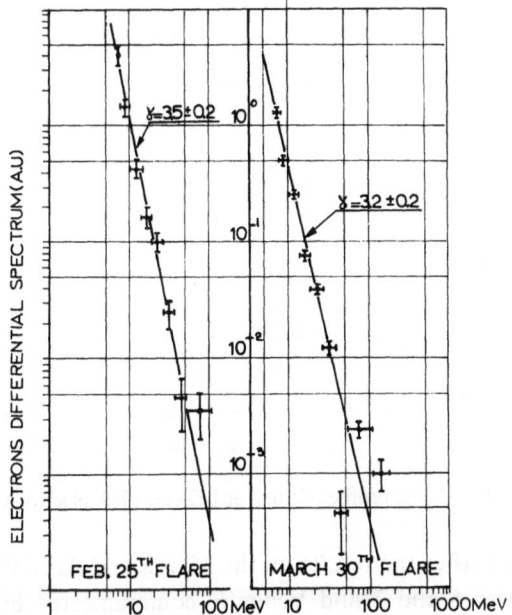

Fig. 11. Energy spectrum of electrons emitted during the events of February 25 and March 30–31, 1969.

Lacking information about the measurements carried out by other experiments and since a direct correlation with solar activity is not self-evident for these events, we have been tempted to suppose the existence of a semi-transparent barrier at some 10–20 AU from the sun which would reflect back part of the electrons produced during the large flares. Such a hypothesis of course needs to be supported by more evidence than we can produce and other explanations can be found for our two small

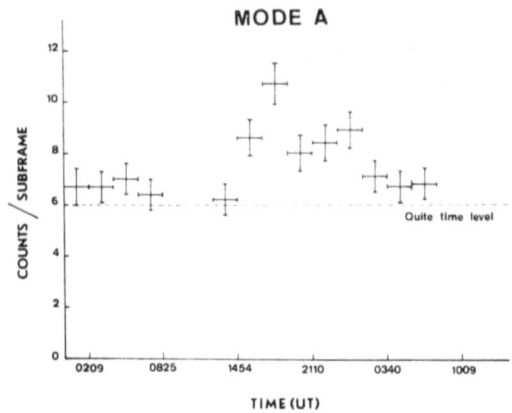

Fig. 12.   Time profile of February 27–28, 1969 electron event.

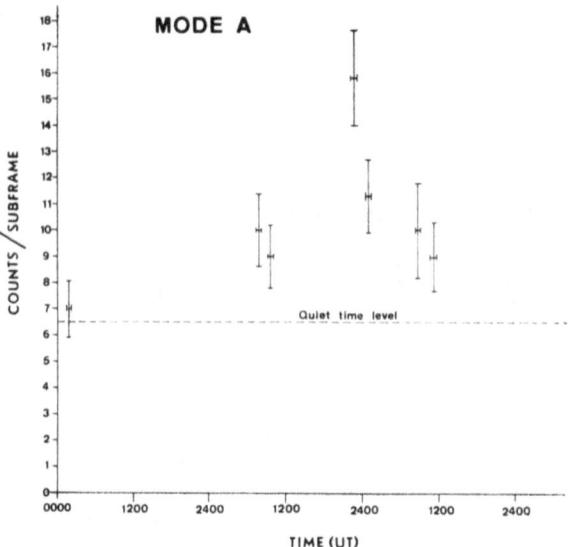

Fig. 13.   Time profile of the April 12–13, 1969, electron event.

events. Furthermore, after the results on the March and April flares obtained by the space probes Pioneer 6 and 7 and Venus 6, communicated by Prof. Vernov (this volume, p. 460), we can consider electrons we detected to be due to invisible solar activity, perhaps from the same active areas on the sun as the large flares which precede them. It is odd that while for Venus 6 April flare is the biggest one in 1969, much larger than the one of March 30th, for us it is just the opposite.

References

Anderson, K. A. and Lin, R. P.: 1966, *Phys. Rev. Letters* **16**, 1121.
Bland, C. J., Degli Antoni, G., Dilworth, C., Maccagni, D., Tanzi, E. G., Koechlin, Y., Raviart, A., Treguer, L.: 1969a, in *Proceedings of the 11th Int. Conf. on Cosmic Rays*, Budapest.

Bland, C. J., Degli Antoni, G., Dilworth, C., Maccagni, D., Tanzi, E. G., Koechlin, Y., Raviart, A., Treguer, L.: 1969b, 'A 20 to 500 MeV Satellite Electron Detector', to be published.
Cline, T. L. and McDonald, F. B.: 1968, *Can. J. Phys.* **46**, S761.
Cline, T. L. and McDonald, F. B.: 1968, *Solar Phys.* **5**, 507.
Lin, R. P. and Anderson, K. A.: 1967, *Solar Phys.* **1**, 446.
Lin, R. P.: 1968, *Can. J. Phys.* **46**, S757.

Discussion

Švestka: The small event which you showed for February 27 could very well be related to a flare on that same day which also caused a peak in the high energy proton count. This flare and the one on February 26 were typical flares which one, according to the very strong type-4 bursts, would expect to have been sources of high energy particles.

OGO-5 MEASUREMENTS OF ELECTRONS ABOVE 500 MeV

B. N. SWANENBURG

Cosmic Ray Working Group, Kamerlingh Onnes Laboratorium, Leiden, The Netherlands

Abstract. A cosmic ray electron detector is flown on board the OGO-5 satellite. The long term variation of the electron intensity above 500 MeV is clearly observed. Short term decreases related to solar events show a similar energy dependence.

A cosmic-ray electron detector was launched on board the OGO-5 satellite on March 4, 1968. Since then the instrument has been almost continuously monitoring the cosmic-ray electron spectrum above 500 MeV as well as the integral intensity of protons and α-particles above approximately 400 MeV/nucleon.

Results regarding both the quiet time spectrum and the energy dependence of the solar modulation were presented at the XIth International Conference on Cosmic

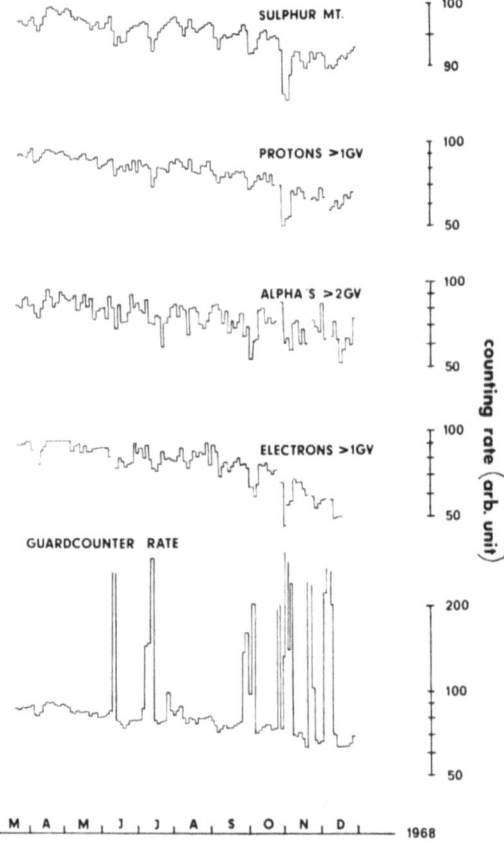

Fig. 1.   Orbital averages of the counting rates on OGO-5 from March to December 1968, uncorrected for a gradual change in the detector. The Sulphur Mountain counting rate is indicated for comparison.

V. Manno and D. E. Page (eds.), Intercorrelated Satellite Observations Related to Solar Events, 610–613. All Rights Reserved
Copyright © 1970 by D. Reidel Publishing Company, Dordrecht-Holland

Rays (Bleeker *et al.*, 1969). For the purpose of this symposium we summarize the most relevant data.

A chronological account of the observations during 1968 is given in Figure 1. Data are averaged over one orbital period (2.5 days). The three most relevant curves give: (a) protons above 1 GV rigidity, (b) α-particles above 2 GV rigidity, and (c) electrons above 1 GV rigidity, which have typical statistical errors of 1%, 5% and 4%, respectively. No corrections were applied for a gradual change in detector sensitivity. For comparison the Sulphur Mountain neutron monitor rates and the rate of the guard

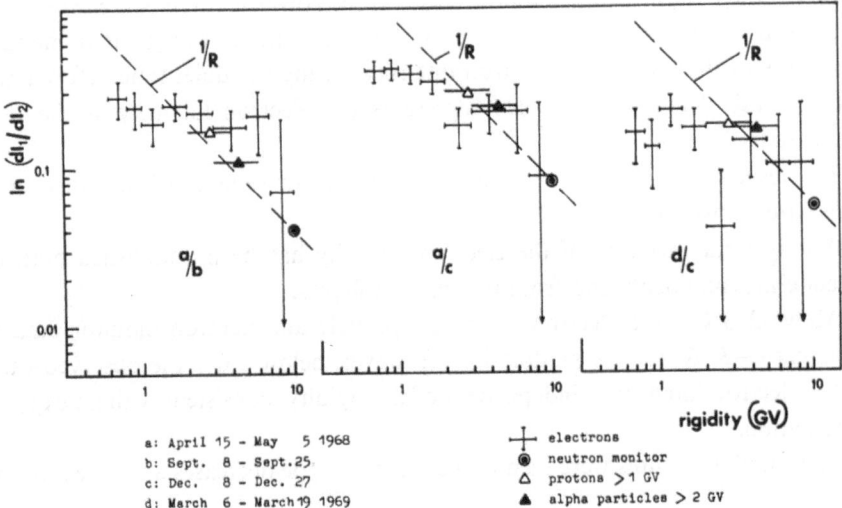

Fig. 2.   The natural logarithm of the relative modulation of electrons, $\ln(dI_1/dI_2)$, versus rigidity, observed over periods of several months as specified. Changes in proton and α-particle intensities and neutron monitor counting rate are shown for comparison.

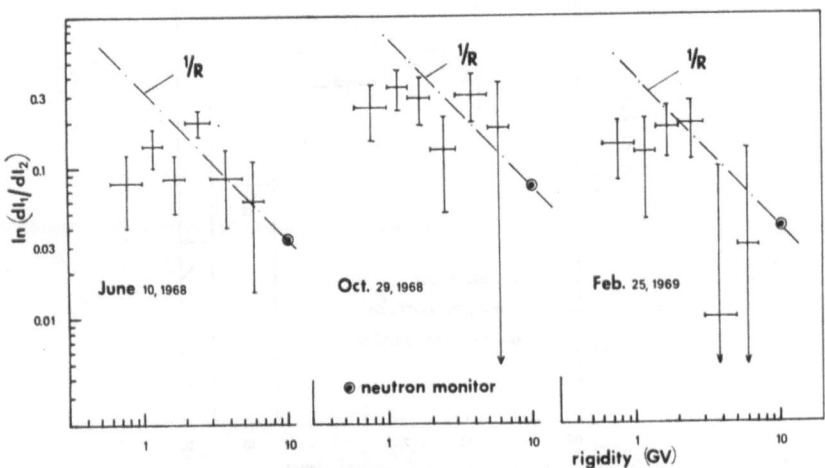

Fig. 3.   The natural logarithm of the relative modulation of electrons, $\ln(dI_1/dI_2)$, vs rigidity, observed during the Forbush decreases of: (a) June 10, 1968, (b) Oct. 29, 1968 and (c) Feb. 25, 1969. The Sulphur Mountain neutron monitor rate changes are also indicated.

counter (sensitive to particles of lower energy) are also presented. A clearly positive correlation is seen between the neutron monitor, proton and α-rate and the integral electron rate. This indeed is expected if the electrons have a non solar origin.

In-flight calibration allows us to correct for the gain drift in the counters. After applying these corrections, the rigidity dependence of the long term variation near solar maximum is shown in Figure 2. The similarity between the decreasing phase between April and December 1968 and the increasing phase between December 1968 and March 1969 is striking.

The observations also permitted a study of changes of the electron spectrum during periods when the neutron monitor recorded a Forbush decrease. Results of three such events are presented in Figure 3. In the presentation of the relative modulation, dI_1 refers to the differential electron intensity during the quiet time before and after the Forbush decrease, while dI_2 represents the electron intensity at the Forbush decrease itself.

The results of balloon observations carried out in 1966 and 1968 (Bleeker *et al.*, 1969) are shown in Figure 4.

A long term variation of the electron intensity has been established both by the satellite measurements and from the balloon flights.

Above 2–3 GV the electron, proton, α-particle and neutron monitor data fit the form $\exp(-K/R)$ for the modulation. However, below 2 GV the observed variation of the electron intensity is independent of the rigidity, consistent with an $\exp(-K/R)$ modulation.

Very similar features were seen in the Forbush type decreases and it seems unlikely

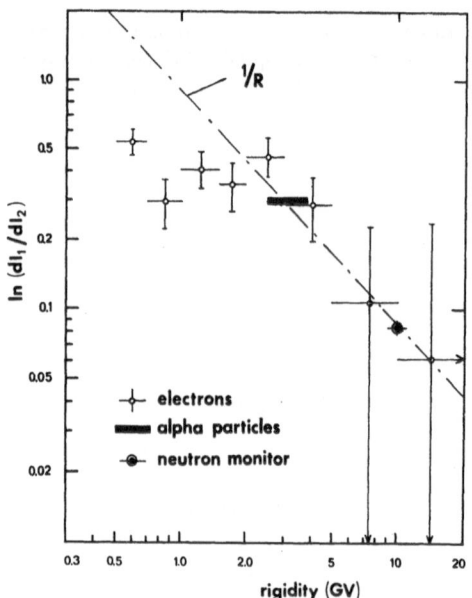

Fig. 4.   The natural logarithm of the observed relative modulation of electrons, $\ln(dI_1/dI_2)$, vs rigidity; $t_1 = 1966$, $t_2 = 1968$. The Sulphur Mountain neutron monitor counting rate change is also indicated.

that this agreement is fortuitous. Again, the observed points strongly deviate from the $\exp(-K/R)$ dependence extrapolated from the neutron monitor point. They are consistent with absence of rigidity dependence below a few GV, but the statistical uncertainties prohibit an accurate determination of the functional form.

The present data support the diffusion-convection theory of the solar modulation mechanism. The intriguing similarity between the long-term variation and the Forbush decreases should be indicative for more detailed models. Correlation of the present data with observations on protons and α-particles in the overlapping rigidity range will serve as a sensitive test for such models.

References

Bleeker, J. A. M., Burger, J. J., Deerenberg, A. J. M., Van de Hulst, H. C., Scheepmaker, A., Swanenburg, B. N., and Tanaka, Y.: 1969, Papers presented at the XIth Int. Conf. on Cosmic Rays, Budapest, 1969. (Paper OG34/35 and OG36.)

Discussion

Occhialini-Dilworth: Some data from Dr. Maccagni's experiment show examples of Forbush decreases in the low energy range (10–20 MeV) and a good correlation with neutron monitor data. These decreases do occur in what we believe to be solar electrons, because after the period of flares these electrons remain with an enhanced intensity for a period of months. Since low energy solar electrons apparently feature the same type of modulation as high energy electrons during a Forbush decrease, one cannot say that particles are galactic, because they suffer from Forbush decreases.

Swanenburg: Since our experiment only covers the energy range above 500 MeV, there probably is an extension of the energy range where we more safely can assume that particles are indeed galactic.

McDonald: As yet we regard it an open question whether these particles are solar or galactic. We expect, however, that they are galactic.

8. LECTURE IN CONCLUSION

SUGGESTIONS FOR FUTURE OBSERVATIONS

W. I. AXFORD

*Dept. of Physics and Dept. of Applied Physics and Information Science,
University of California, San Diego, La Jolla, Calif., U.S.A.*

In proposing possible new measurements to be carried out in space in connection with solar events, it seems reasonable to take it for granted that present experimental techniques will continue to improve. In fact, the detail which can now be achieved in measurements of electron and proton distribution functions is remarkable, and indeed is beginning to strain our ability to absorb and comprehend the data. There are however at least two areas in which further improvements should be sought; these involve increased coverage (especially with multiple space-craft), and particle composition measurements.

It is evident that if we wish to understand the structure of solar disturbances in the interplanetary medium, it will be necessary to place spacecraft in regions of space other than the very narrow region near the ecliptic between the orbits of Venus and Mars which has so far been explored. In particular, data obtained from spacecraft at high heliographic latitudes, and also in the ecliptic within ~0.5 AU of the Sun, would be extremely valuable. At present our only sources of information concerning these regions of space involve comet tail observations and radio star scintillation observations, both of which are indirect and limited in content. Missions which could provide the necessary data are being considered (e.g. a Venus-Mercury fly-by, and a Jupiter fly-by out of the ecliptic to high latitudes), however, there appears to be a good case for making a more intensive effort to have spacecraft put in these regions. The possibility of having a spacecraft in a 'solar-synchronous' orbit (at about 0.2 AU heliocentric distance) is a rather interesting one in this respect.

Increased coverage, in terms of multiple satellites, is also required in the case of magnetospheric studies. The need is most pressing in the case of auroral zone observations where disentangling spatial and temporal variations is of prime concern. The problem of the aurora, and of the substorm phenomenon in general, is the outstanding problem in magnetospheric physics. No matter how well it is instrumented a single geo-stationary satellite, for example, cannot provide the information necessary to solve the problem. It is necessary to have at least one other satellite, with similar instrumentation, in an orbit which can provide as many interesting correlations as possible in local time and L-value separation over a reasonable period of time. This might be achieved by combining a geostationary satellite with a similar satellite orbiting near the equatorial plane between say $L=4$ and $L=9$. An ideal supplement to this pair would be a low-altitude polar orbiting satellite, again with similar instrumentation.

Composition observations are capable of providing a great deal of new information concerning the origin and mode of transport of both high energy particles and low

V. Manno and D. E. Page (eds.), Intercorrelated Satellite Observations Related to Solar Events, 617–619. All Rights Reserved
Copyright © 1970 by D. Reidel Publishing Company, Dordrecht-Holland

energy plasma. Thus if the low energy end of the cosmic ray spectrum is of solar origin as has been suggested, this could be confirmed by observations showing that the particles concerned have a composition similar to that found during solar energetic particle events. In particular it is necessary to demonstrate that secondary nuclei (e.g. He^3, Li, Be, B) have abundances appropriate to passage of the primary beam through the corona rather than the interstellar medium. Relatively few observations have been made of the behavior of nuclei other than protons during solar energetic particle events. It is quite apparent however that concurrent observations of the behavior of α-particles should yield a great deal of additional information concerning the mode of propagation of the particles from their region of origin.

It is very important that detailed observations of the composition and state of ionization of the solar wind plasma be carried out on a continuing basis. The information obtained can provide us with evidence concerning the temperature of the solar corona and the presence of neutral gas in the interplanetary region. In addition the appearance of flare ejecta is often made evident by a sudden enhancement of the relative concentration of α-particles; thus studies of the structure of interplanetary disturbances can be greatly aided by such observations. Unfortunately most solar wind experiments have not been designed specifically to provide this type of information, and ions other protons and α-particles have been detected on only a few special occasions. In order to distinguish between various ions it is normally necessary to measure two distinct quantities (e.g. the energy per unit charge and the particle momentum), but in the one experiment in which this has been tried, only protons and α-particles were selected. The aluminum foil experiment first carried out during the Apollo 11 mission to the Moon can give the relative concentration of a number of interesting species in the solar wind (especially the more abundant of the inert gases); the usefulness of the experiment as it is carried out at present is limited however due to the need to recover the foil.

There are a number of interesting composition measurements that can be carried out within the magnetosphere. The presence of relatively large fluxes of energetic particles makes it possible to obtain useful results with counter techniques, and a few observations of α-particles have been made in the aurora and in the radiation belts. The observations should be persued much further however, since it has yet to be demonstrated that the auroral primary ions and the energetic ions in the magnetosphere have originated in the solar wind. If it is possible to distinguish between He^+ and He^{++} this question could be decided, since it is difficult to argue on the one hand that He^{++} ions originate in the ionosphere, and on the other that He^+ ions originate in the solar wind in significant quantities. It should also be possible to observe the 'aging' of a freshly-injected He^{++} population as it is changed to He^+ and ultimately neutral helium due to charge-exchange with the neutral hydrogen geo-corona.

An alternative method of determining the origin of energetic magnetospheric and auroral primary ions involves the detection of He^3. The relative concentration of He^3 to He^4 is about 10^3 times as great in the solar wind as it is in the earth's atmosphere. Consequently it is expected that the flux of energetic He^3 ions in the magnetosphere and aurora should be about 10^{-3} times that of He^4 ions if the particles originate in the solar wind, and essentially otherwise. Measurements of these quantities could be carried out both in the outer magnetosphere and in the aurora at low altitudes. It is believed that the terrestrial economy of He^3 is largely controlled by auroral precipitation of ions captured from the solar wind and thermal escape of neutrals. Thus the measurements described are of interest from the point of view of geo-chemistry as well as of magnetospheric physics.

INDEX OF NAMES

When a name appears in the references at the end of a paper, the page number is italicized

ASTROPHYSICS AND SPACE SCIENCE LIBRARY

Edited by

J. E. Blamont, R. L. F. Boyd, L. Goldberg, C. de Jager, Z. Kopal, G. H. Ludwig, R. Lüst,
B. M. McCormac, H. E. Newell, L. I. Sedov, Z. Švestka, and W. de Graaff

1. C. de Jager (ed.), *The Solar Spectrum. Proceedings of the Symposium held at the University of Utrecht, 26–31 August, 1963.* 1965, XIV + 417 pp. Dfl. 50.—

2. J. Ortner and H. Maseland (eds.), *Introduction to Solar Terrestrial Relations. Proceedings of the Summer School in Space Physics held in Alpbach, Austria, July 15–August 10, 1963 and Organized by the European Preparatory Commission for Space Research.* 1965, IX + 506 pp. Dfl. 65.—

3. C. C. Chang and S. S. Huang (eds.), *Proceedings of the Plasma Space Science Symposium, Held at the Catholic University of America, Washington, D.C., June 11–14, 1963.* 1965, IX + 377 pp. Dfl. 68.—

4. Zdeněk Kopal, *An Introduction to the Study of the Moon.* 1966, XII + 464 pp. Dfl. 72.—

5. Billy M. McCormac (ed.), *Radiation Trapped in the Earth's Magnetic Field. Proceedings of the Advanced Study Institute, Held at the Chr. Michelsen Institute, Bergen, Norway, August 16–September 3, 1965.* 1966, XII + 901 pp. Dfl. 130.—

6. A. B. Underhill, *The Early Type Stars.* 1966, XIII + 282 pp. Dfl. 56.—

7. Jean Kovalevsky, *Introduction to Celestial Mechanics.* 1967, VIII + 427 pp. Dfl. 30.—

8. Zdeněk Kopal and Constantine L. Goudas (eds.), *Measure of the Moon. Proceedings of the Second International Conference on Selenodesy and Lunar Topography held in the University of Manchester, England, May 30–June 4, 1966.* 1967, XVIII + 479 pp. Dfl. 90.—

9. J. G. Emming (ed.), *Electromagnetic Radiation in Space. Proceedings of the Third ESRO Summer School in Space Physics, held in Alpbach, Austria, from 19 July to 13 August, 1965.* 1968, VIII + 307 pp. Dfl. 58.—

10. R. L. Carovillano, John F. McClay, and Henry R. Radoski (eds.), *Physics of the Magnetosphere. Based upon the Proceedings of the Conference held at Boston College, June 19–28, 1967.* 1968, X + 686 pp. Dfl. 130.—

11. Syun-Ichi Akasofu, *Polar and Magnetospheric Substorms.* 1968, XVIII + 280 pp. Dfl. 55.—

12. Peter M. Millman (ed.), *Meteorite Research. Proceedings of a Symposium on Meteorite Research held in Vienna, Austria, 7–13 August, 1968.* 1969, XV + 941 pp. Dfl. 160.—

13. Margherita Hack (ed.), *Mass Loss from Stars. Proceedings of the Second Trieste Colloquium on Astrophysics, 12–17 September, 1968.* 1969, XII + 345 pp. Dfl. 65.—

14. N. D'Angelo (ed.), *Low-Frequency Waves and Irregularities in the Ionosphere. Proceedings of the 2nd ESRIN-ESLAB Symposium, held in Frascati, Italy, 23–27 September, 1968.* 1969, VII + 218 pp. Dfl. 43.—

15. G. A. Partel (ed.), *Space Engineering. Proceedings of the Second International Conference on Space Engineering, held at the Fondazione Giorgio Cini, Isola di San Giorgio, Venice, Italy, May 7–10, 1969.* 1970, XI + 728 pp. Dfl. 140.—

p.t.o.

16. S. Fred Singer (ed.), *Manned Laboratories in Space. Second International Orbital Laboratory Symposium.* 1969, XIII + 133 pp. Dfl. 30.—

17. B. M. McCormac (ed.), *Particles and Fields in the Magnetosphere. Symposium Organized by the Summer Advanced Study Institute, held at the University of California, Santa Barbara, Calif., August 4–15, 1969.* 1970, XI + 450 pp. Dfl. 85.—

18. Jean-Claude Pecker, *Experimental Astronomy.* 1970, approx. 114 pp. approx, Dfl. 22.—